The Structural Econometric Time Series Analysis Approach

Bringing together a collection of previously published work, this book provides a timely discussion of major considerations relating to the construction of econometric models that work well to explain economic phenomena, predict future outcomes, and be useful for policy-making. Analytical relations between dynamic econometric structural models and empirical time series MVARMA, VAR, transfer function, and univariate ARIMA models are established with important application for model-checking and model construction. The theory and applications of these procedures to a variety of econometric modeling and forecasting problems as well as Bayesian and non-Bayesian testing, shrinkage estimation, and forecasting procedures are also presented and applied. Finally, attention is focused on the effects of disaggregation on forecasting precision and the new Marshallian macroeconomic model (MMM) that features demand, supply, and entry equations for major sectors of economies is analyzed and described. This volume will prove invaluable to professionals, academics and students alike.

ARNOLD ZELLNER is H. G. B. Alexander Distinguished Service Professor Emeritus of Economics and Statistics, Graduate School of Business, University of Chicago and Adjunct Professor, University of California at Berkeley. He has published books and many articles on the theory and application of econometrics and statistics to a wide range of problems.

FRANZ C. PALM is Professor of Econometrics, Faculty of Economics and Business Administration, Maastricht University. He has published many articles on the theory and application of econometrics and statistics to a wide range of problems.

The Structural Econometric Time Series Analysis Approach

Edited by

Arnold Zellner and Franz C. Palm

CAMBRIDGE UNIVERSITY PRESS
Cambridge, New York, Melbourne, Madrid, Cape Town, Singapore,
São Paulo, Delhi, Dubai, Tokyo, Mexico City

Cambridge University Press
The Edinburgh Building, Cambridge CB2 8RU, UK

Published in the United States of America by Cambridge University Press, New York

www.cambridge.org
Information on this title: www.cambridge.org/9780521187435

First published 2004
First paperback edition 2010

A catalogue record for this publication is available from the British Library

Library of Congress Cataloguing in Publication data
The structural econometric time series analysis approach / edited by Arnold Zellner and
Franz C. Palm.
p. cm.
Includes bibliographical references and index.
ISBN 0 521 81407 3 (cloth)
1. Econometric models. I. Zellner, Arnold. II. Palm, Franz C., 1948–
HB141.S87 2004
330′.01′519232 – dc 21 2003055141

ISBN 978-0-521-81407-2 Hardback
ISBN 978-0-521-18743-5 Paperback

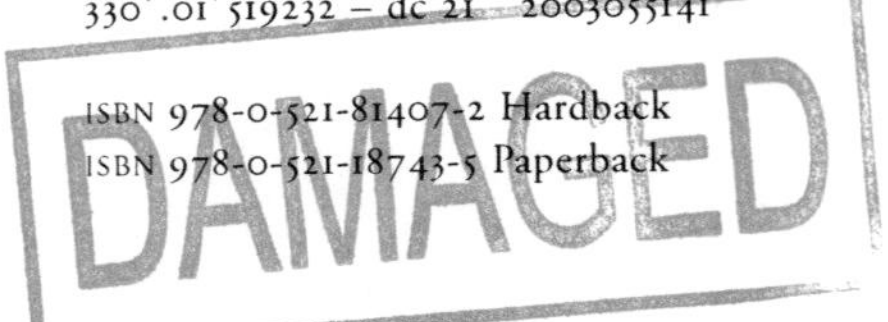

Contents

Part IV Disaggregation, forecasting, and modeling

Contributors

AHKING, FRANCIS. W., Department of Economics, University of Connecticut, Storrs, CT

BELSLEY, DAVID A., Professor of Economics, Department of Economics, Boston College, Boston, MA

CHEN, BIN, Chicago Partners, LLC, Chicago, IL

CHOW, GREGORY C., Professor of Economics, Department of Economics, Princeton University, Princeton, NJ

CHRIST, CARL F., Professor Emeritus of Economics, Department of Economics, Johns Hopkins University, Baltimore, MD

EVANS, PAUL, Professor of Economics, Ohio State University, Columbus, OH

GARCIA-FERRER, ANTONIO, Professor of Economics, Departamento de Analsis Economico: Economia Cuantitava, Universidad Autonoma de Madrid

GULATI, GAURANG M., Georgetown University, Law Center, Washington, DC

HIGHFIELD, RICHARD A., Dean, School of Business Administration 364, School of Business, New York University at Albany, NY

HONG, CHANSIK, Department of Economics, Sookmyung Women's University, Seoul

HOOGSTRATE, ANDRÉ J., Ministry of Justice, Netherlands Forensic Institute, Rijswijk

KUH, EDWIN, Professor of Economics, Sloan School of Management, MIT, Cambridge, MA

LESAGE, JAMES P., Professor of Economics, Department of Economics, University of Toledo, Toledo, OH

LOMBRA, RAYMOND E., Professor of Economics, Department of Economics, Pennsylvania State University, University Park, PA

MAGURA, MICHAEL, Department of Economics, University of Toledo, Toledo, OH

MARAVALL, AUGUSTÍN, Chief Economist, SIMC, Banco de España, Madrid

MATHIS, ALEXANDRE, SIMC, Banco de España, Madrid

MILLER, STEPHEN M., Department of Economics, College of Nevada Las Vegas, Las Vegas, NV

MIN, CHUNG-KI, Department of Economics, Hankuk University of Foreign Studies, Seoul

PALM, FRANZ C., Professor of Econometrics, Faculty of Economics and Business Administration, Universiteit Maastricht

PFANN, GERARD A., Professor of Econometrics, Faculty of Economics and Business Administration, Universiteit Maastricht

PLOSSER, CHARLES I., Professor of Economics, Simon Graduate School of Business Administration, University of Rochester, Rochester, NY

ROBINSON, PETER M., Professor of Econometrics, Department of Economics, London School of Economics and Political Science, London

ROTHENBERG, THOMAS J., Professor of Economics, Department of Economics, University of California, Berkeley, CA

SIMS, CHRISTOPHER A., Professor of Economics, Department of Economics, Princeton University, Princeton, NJ

TOBIAS, JUSTIN, Department of Economics University of California, Irvine, CA

TRIVEDI, PRAVIN K., Professor of Economics, Department of Economics, Indiana University, Bloomington, IN

WEBB, ROGER I., McIntire School of Commerce, University of Virginia, Charlottesville, VA

ZELLNER, ARNOLD, H. G. B. Alexander Distinguished Service Professor Emeritus of Economics and Statistics, Graduate School of Business, University of Chicago, Chicago, IL

Acknowledgments

The following chapters are reprinted with the kind permission of the publishers listed below. Chapter 1, from the *Journal of Econometrics*, and chapters 1, 5, 10, 12, 16, 17, 19, and 20, from the *Journal of Econometrics*, by North-Holland Publishing Company; chapter 2, from the *Journal of the American Statistical Association*, and chapters 13, 18, and 20, from the *Journal of Business and Economic Statistics*, by the American Statistical Association; chapters 3 and 9, from the Bureau of the Census, US Department of Commerce, by the US Government Printing Office; chapters 4 and 22, from the *Journal of Forecasting*, by John Wiley & Sons; chapter 6, from *Sankhya: The Indian Journal of Statistics*, by the Indian Statistical Institute; chapters 7 and 8, from the *International Economic Review*, by the Wharton School of Finance and Commerce, University of Pennsylvania and the Osaka University Institute of Economic Research Association; chapter 11, from the *Review of Economics and Statistics*, by MIT Press; chapter 21, from the *Journal of Regional Science*, by the Regional Science Research Institute, in cooperation with the Wharton School of Finance and Commerce, University of Pennsylvania, Blackwell Publishing Ltd; chapter 15, by North-Holland Publishing Company; chapter 23, by Kluwer Academic Publishers; and chapter 24 by Cambridge University Press.

Introduction

In the early 1970s we were concerned about the relationships between multivariate and univariate time series models, such as those brilliantly analyzed by Quenouille (1957) and Box and Jenkins (1970) and multivariate dynamic structural econometric models that had been and are widely employed in explanation, prediction and policy-making. Fortunately, we discovered the relationships and reported them in our paper, Zellner and Palm (1974) that is included in part I of this volume (chapter 1). See also the other general chapters in part I discussing general features of our approach, the reactions of leading researchers, and many useful references to the literature.

Having discovered the algebraic relations connecting statistical time series and structural econometric models, we next considered how this discovery might be used to produce improved models. In this connection, we thought it important not only to emphasize a philosophical preference for sophisticatedly simple models that is discussed in several chapters in part I and Zellner, Keuzenkamp, and McAleer (2001), but also operational techniques that would help researchers actually produce improved models. As illustrated in the chapters included in this volume, our approach involves (1) deducing algebraically the implied marginal processes and transfer functions for individual variables in a multi-equation model, e.g. a vector autoregression (VAR) or a structural econometric model (SEM), and (2) comparing these derived equations' forms and properties with those derived from the data by use of empirical model identification and testing techniques. See Palm and Zellner (1980), included in part I (chapter 5) for some early estimation and testing procedures that have been improved over the years. If the information in the data is compatible with the empirically determined, simple time series models and not with those implied by a VAR or SEM, then we conclude that the VAR or SEM needs reformulation and improvement. See, for example our (1975) paper in part II (chapter 6) analyzing monetary models of the US economy and other papers for applications of this approach to many other problems including Trivedi (1975) on modeling inventory behavior (chapter 7), Evans (1978) on the German hyperinflation (chapter 8), Plosser (1976) on seasonality (Chapter 9), Webb (1985) on behavior of speculative prices (chapter 10), Ahking and Miller (1987) on exchange rate models (chapter 11), and Maravall and Mathis (1994) on diagnosis of VAR models using French macroeconomic data (chapter 12). These studies demonstrate well the usefulness of our SEMTSA approach in analyzing, comparing, and improving models.

Since there is often no satisfactory model available, in part III we illustrate how relatively simple forecasting equations have been developed, studied, and tested in point and turning point forecasting experiments using modern estimation and forecasting techniques. Here the objective is to get forecasting equations that work well in point and turning point forecasting and have reasonable dynamic properties. Then the objective is to produce reasonable economic models to rationalize the good empirical performance of these empirical forecasting equations. Thus we do not in the present instance go from theory to the data but reverse the process by going from what works well empirically to theory that explains this unusual empirical finding. As mentioned in several chapters in part III, the empirical forecasting equations for countries' annual GDP growth rates have been rationalized by Hong (1989) in terms of a Hicksian [IS-LM macroeconomic model, by Min (1992) in terms of a generalized real business cycle model that he formulated and by Zellner and Anton (1986) in terms of an aggregate demand and supply model. Thus the empirical relations studied intensively in the chapters included in part III have some theoretical as well as empirical support. Note, too, that many methodological tools were developed and tested in the chapters on empirical forecasting work in part III – namely, Bayesian shrinkage estimation and prediction, optimal turning point forecasting techniques, optimal Bayesian model-combining or pooling methods, etc. Also, comparisons of forecasting root mean-squared errors (RMSEs) and mean absolute errors (MAEs) indicate that various simple forecasting equations' performance is competitive with that of certain large-scale macroeconometric models for many economies. See, for example some comparisons reported in Garcia-Ferrer *et al.* (1987) and Hoogstrate, Palm, and Pfann (2000) (chapter 13 and 18) for some improved results that utilize various pooling techniques in analysis and forecasting of panel data for eighteen countries.

While the studies in part III provide useful, improved macroeconomic results, it is the case that aggregation of output and other kinds of data, say over sectors of an economy, can involve a loss of valuable information, as has been discussed many times in the past. Thus part IV presents chapters dealing with disaggregation, forecasting, and modeling. A simple experiment, reported in Zellner and Tobias (2000) (chapter 22) shows empirically how disaggregation can result in improved forecasting precision in connection with forecasting the annual medians of eighteen countries' growth rates. In chapters 20 and 21 by LeSage and Magura (1990) and LeSage (1990), it is shown how shrinkage point and turning point forecasting procedures perform using regional data. Then in Zellner (2000) and in Zellner and Chen (2000) (chapters 23 and 24), Marshallian sector models of industrial sectors are formulated, building on the earlier work of Veloce and Zellner (1985), and tested in forecasting experiments using annual data for eleven sectors of the US economy. The annual output forecasts of the sectors are added to get a forecast of total GDP and its growth rate year by year. Such forecasts are compared with forecasts derived from models implemented with aggregate data. In this instance, it was found that it pays to disaggregate. Further work to improve and expand the Marshallian sector model in line with the SEMTSA approach is described in these chapters.

In summary, pursuing the SEMTSA approach over the years has been an exciting experience that has led to new empirical findings, improved and novel methodological tools, and improved models. We thank all those who have contributed to these positive developments and hope that future developments will be even better. Also, thanks to the US National Science Foundation and the H. G. B. Alexander Endowment Fund, University of Chicago, for financial support. Ashwin Rattan at Cambridge University Press, provided much help in arranging for the publication of our book, for which we are most grateful.

EDITORS' NOTE

References cited in this introduction appear in chapters in the text except for Zellner, H. Keuzenkamp, and McAleer (2001).

Minor editorial intervention has been made to update the text of some chapters.

BIBLIOGRAPHY

Zellner, A., H. Keuzenkamp, and M. McAleer, 2001, *Simplicity, Inference and Econometric Modeling* (Cambridge, Cambridge University Press)

Part I

The SEMTSA approach

1 Time series analysis and simultaneous equation econometric models (1974)

Arnold Zellner and Franz C. Palm

1 Introduction

In this chapter we take up the analysis of dynamic simultaneous equation models (SEMs) within the context of general linear multiple time series processes such as studied by Quenouille (1957). As noted by Quenouille, if a set of variables is generated by a multiple time series process, it is often possible to solve for the processes generating individual variables, namely the "final equations" of Tinbergen (1940), and these are in the autoregressive-moving average (ARMA) form. ARMA processes have been studied intensively by Box and Jenkins (1970). Further, if a general multiple time series process is appropriately specialized, we obtain a usual dynamic SEM in structural form. By algebraic manipulations, the associated reduced form and transfer function equation systems can be derived. In what follows, these equation systems are presented and their properties and uses are indicated.

It will be shown that assumptions about variables being exogenous, about lags in structural equations of SEMs, and about serial correlation properties of structural disturbance terms have strong implications for the properties of transfer functions and final equations that can be tested. Further, we show how large sample posterior odds and likelihood ratios can be used to appraise alternative hypotheses. In agreement with Pierce and Mason (1971), we believe that testing the implications of structural assumptions for transfer functions and, we add, final equations is an important element in the process of iterating in on a model that is reasonably in accord with the information in a sample of data. To illustrate these general points and to provide applications of the above methods,

Research financed in part by NSF Grant GS-2347 and by income from the H.G.B. Alexander Endowment Fund, Graduate School of Business, University of Chicago. Some of the ideas in this chapter were presented in econometrics lectures and at a session of the Econometric Society's meeting in 1971 by one of the authors. The second author received financial support from the Belgian National Science Foundation.

Originally published in the *Journal of Econometrics* 2 (1974), 17–54.

a dynamic version of a SEM due to Haavelmo (1947) is analyzed using US post-Second World War quarterly data.

The plan of the chapter is as follows. In section 2, a general multiple time series model is specified, its final equations are obtained, and their properties set forth. Then the implications of assumptions needed to specialize the multiple time series model to become a dynamic SEM for transfer functions and final equations are presented. In section 3, the algebraic analysis is applied to a small dynamic SEM. Quarterly US data are employed in sections 4 and 5 to analyze the final and transfer equations of the dynamic SEM. Section 6 provides a discussion of the empirical results, their implications for the specification and estimation of the structural equations of the model, and some concluding remarks.

2 General formulation and analysis of a system of dynamic equations

As indicated by Quenouille (1957), a linear multiple time series process can be represented as follows:[1]

$$\underset{p\times p}{H(L)}\ \underset{p\times 1}{\boldsymbol{z}_t} = \underset{p\times p}{F(L)}\ \underset{p\times 1}{\boldsymbol{e}_t}, \quad t = 1, 2, \ldots, T, \tag{2.1}$$

where $\boldsymbol{z}_t' = (z_{1t}, z_{2t}, \ldots, z_{pt})$ is a vector of random variables, $\boldsymbol{e}_t' = (e_{1t}, e_{2t}, \ldots, e_{pt})$ is a vector of random errors, and $H(L)$ and $F(L)$ are each $p \times p$ matrices, assumed of full rank, whose elements are finite polynomials in the lag operator L, defined as $L^n z_t = z_{t-n}$. Typical elements of $H(L)$ and $F(L)$ are given by $h_{ij} = \sum_{l=0}^{r_{ij}} h_{ijl} L^l$ and $f_{ij} = \sum_{l=0}^{q_{ij}} f_{ijl} L^l$. Further, we assume that the error process has a zero mean, an identity covariance matrix and no serial correlation, that is:

$$E\boldsymbol{e}_t = 0, \tag{2.2}$$

for all t and t',

$$E\boldsymbol{e}_t\boldsymbol{e}_{t'}' = \delta_{tt'} I, \tag{2.3}$$

where I is a unit matrix and $\delta_{tt'}$ is the Kronecker delta. The assumption in (2.3) does not involve a loss of generality since correlation of errors can be introduced through the matrix $F(L)$.

The model in (2.1) is a multivariate autoregressive-moving average (ARMA) process. If $H(L) = H_0$, a matrix of degree zero in L, (2.1) is a

[1] In (2.1), $\boldsymbol{z}_t$ is assumed to be mean-corrected, that is $\boldsymbol{z}_t$ is a deviation from a population mean vector. Below, we relax this assumption.

moving average (MA) process; if $F(L) = F_0$, a matrix of degree zero in L, it is an autoregressive (AR) process. In general, (2.1) can be expressed as:

$$\sum_{l=0}^{r} H_l L^l \boldsymbol{z}_t = \sum_{l=0}^{q} F_l L^l \boldsymbol{e}_t, \tag{2.4}$$

where H_l and F_l are matrices with all elements not depending on L, $r = \max_{i,j} r_{ij}$ and $q = \max_{i,j} q_{ij}$.

Since $H(L)$ in (2.1) is assumed to have full rank, (2.1) can be solved for $\boldsymbol{z}_t$ as follows:

$$\boldsymbol{z}_t = H^{-1}(L)F(L)\boldsymbol{e}_t, \tag{2.5a}$$

or

$$\boldsymbol{z}_t = [H^*(L)/|H(L)|]F(L)\boldsymbol{e}_t, \tag{2.5b}$$

where $H^*(L)$ is the adjoint matrix associated with $H(L)$ and $|H(L)|$ is the determinant which is a scalar, finite polynomial in L. If the process is to be invertible, the roots of $|H(L)| = 0$ have to lie outside the unit circle. Then (2.5) expresses $\boldsymbol{z}_t$ as an infinite MA process that can be equivalently expressed as the following system of finite order ARMA equations:

$$|H(L)|\boldsymbol{z}_t = H^*(L)F(L)\boldsymbol{e}_t. \tag{2.6}$$

The ith equation of (2.6) is given by:

$$|H(L)|z_{it} = \boldsymbol{\alpha}_i'\boldsymbol{e}_t, \quad i = 1, 2, \ldots, p, \tag{2.7}$$

where $\boldsymbol{\alpha}_i'$ is the ith row of $H^*(L)F(L)$.

The following points regarding the set of final equations in (2.7) are of interest:

(i) Each equation is in ARMA form, as pointed out by Quenouille (1957, p. 20). Thus the ARMA processes for individual variables are compatible with some, perhaps unknown, joint process for a set of random variables and are thus not necessarily "naive," "ad hoc" alternative models.

(ii) The order and parameters of the autoregressive part of each equation, $|H(L)|\, z_{it}, i = 1, 2, \ldots, p$, will usually be the same.[2]

(iii) Statistical methods can be employed to investigate the form and properties of the ARMA equations in (2.7). Given that their forms, that is the degree of $|H(L)|$ and the order of the moving average

[2] In some cases in which $|H(L)|$ contains factors in common with those appearing in all elements of the vectors α_i', e.g. when H is triangular, diagonal or block diagonal, some cancelling will take place. In such cases the statement in (ii) has to be qualified.

errors, have been determined, they can be estimated and used for prediction.

(iv) The equations of (2.7) are in the form of a restricted "seemingly unrelated" autoregressive model with correlated moving average error processes.[3]

The general multiple time series model in (2.1) can be specialized to a usual dynamic simultaneous equation model (SEM) if some prior information about H and F is available. That is, prior information may indicate that it is appropriate to regard some of the variables in $\boldsymbol{z}_t$ as being endogenous and the remaining variables as being exogenous, that is, generated by an independent process. To represent this situation, we partition (2.1) as follows:

$$\begin{pmatrix} H_{11} & H_{12} \\ H_{21} & H_{22} \end{pmatrix} \begin{pmatrix} \boldsymbol{y}_t \\ \boldsymbol{x}_t \end{pmatrix} = \begin{pmatrix} F_{11} & F_{12} \\ F_{21} & F_{22} \end{pmatrix} \begin{pmatrix} \boldsymbol{e}_{1t} \\ \boldsymbol{e}_{2t} \end{pmatrix}. \tag{2.8}$$

If the $p_1 \times 1$ vector $\boldsymbol{y}_t$ is endogenous and the $p_2 \times 1$ vector $\boldsymbol{x}_t$ is exogenous, this implies the following restrictions on the submatrices of H and F:

$$H_{21} \equiv 0, \quad F_{21} \equiv 0, \quad \text{and} \quad F_{12} \equiv 0. \tag{2.9}$$

With the assumptions in (2.9), the elements of $\boldsymbol{e}_{1t}$ do not affect the elements of $\boldsymbol{x}_t$ and the elements of $\boldsymbol{e}_{2t}$ affect the elements of $\boldsymbol{y}_t$ only through the elements of $\boldsymbol{x}_t$. Under the hypotheses in (2.9), (2.8) is in the form of a dynamic SEM with endogenous variable vector $\boldsymbol{y}_t$ and exogenous variable vector $\boldsymbol{x}_t$ generated by an ARMA process. The usual structural equations, from (2.8) subject to (2.9), are:[4]

$$\underset{p_1 \times p_1}{H_{11}(L)} \underset{p_1 \times 1}{\boldsymbol{y}_t} + \underset{p_1 \times p_2}{H_{12}(L)} \underset{p_2 \times 1}{\boldsymbol{x}_t} = \underset{p_1 \times p_1}{F_{11}(L)} \underset{p_1 \times 1}{\boldsymbol{e}_{1t}}, \tag{2.10}$$

while the process generating the exogenous variables is:

$$\underset{p_2 \times p_2}{H_{22}(L)} \underset{p_2 \times 1}{\boldsymbol{x}_t} = \underset{p_2 \times p_2}{F_{22}(L)} \underset{p_2 \times 1}{\boldsymbol{e}_{2t}}, \tag{2.11}$$

with $p_1 + p_2 = p$.

Analogous to (2.4), the system (2.10) can be expressed as:

$$\sum_{l=0}^{r} H_{11l} L^l \boldsymbol{y}_t + \sum_{l=0}^{r} H_{12l} L^l \boldsymbol{x}_t = \sum_{l=0}^{q} F_{11l} L^l \boldsymbol{e}_{1t}, \tag{2.12}$$

where H_{11l}, H_{12l} and F_{11l} are matrices the elements of which are coefficients of L^l. Under the assumption that H_{110} is of full rank, the reduced

[3] See Nelson (1970) and Akaike (1973) for estimation results for systems similar to (2.7).

[4] Hannan (1969, 1971) has analysed the identification problem for systems in the form of (2.10).

form equations, which express the current values of endogenous variables as functions of the lagged endogenous and current and lagged exogenous variables, are:

$$\boldsymbol{y}_t = -\sum_{l=1}^{r} H_{110}^{-1} H_{11l} L^l \boldsymbol{y}_t - \sum_{l=0}^{r} H_{110}^{-1} H_{12l} L^l \boldsymbol{x}_t + \sum_{l=0}^{q} H_{110}^{-1} F_{11l} L^l \boldsymbol{e}_{1t}. \tag{2.13}$$

The reduced form system in (2.13) is a system of p_1 stochastic difference equations of maximal order r.

The "final form" of (2.13), Theil and Boot (1962), or "set of fundamental dynamic equations" associated with (2.13), Kmenta (1971), which expresses the current values of endogenous variables as functions of only the exogenous variables, is given by:

$$\boldsymbol{y}_t = -H_{11}^{-1}(L) H_{12}(L) \boldsymbol{x}_t + H_{11}^{-1}(L) F_{11}(L) \boldsymbol{e}_{1t}. \tag{2.14}$$

If the process is invertible, i.e. if the roots of $|H_{11}(L)| = 0$ lie outside the unit circle, (2.14) is an infinite MA process in $\boldsymbol{x}_t$ and $\boldsymbol{e}_{1t}$. Note that (2.14) is a set of "rational distributed lag" equations, Jorgenson (1966), or a system of "transfer function" equations, Box and Jenkins (1970). Also, the system in (2.14) can be brought into the following form:

$$|H_{11}(L)| \boldsymbol{y}_t = -H_{11}^{*}(L) H_{12}(L) \boldsymbol{x}_t + H_{11}^{*}(L) F_{11}(L) \boldsymbol{e}_{1t}, \tag{2.15}$$

where $H_{11}^{*}(L)$ is the adjoint matrix associated with $H_{11}(L)$ and $|H_{11}(L)|$ is the determinant of $H_{11}(L)$. The equation system in (2.15), where each endogenous variable depends only on its own lagged values and on the exogenous variables, with or without lags, has been called the "separated form," Marschak (1950), "autoregressive final form," Dhrymes (1970), "transfer function form," Box and Jenkins (1970), or "fundamental dynamic equations," Pierce and Mason (1971).[5] As in (2.7), the p_1 endogenous variables in $\boldsymbol{y}_t$ have autoregressive parts with identical order and parameters, a point emphasized by Pierce and Mason (1971).

Having presented several equation systems above, it is useful to consider their possible uses and some requirements that must be met for these uses. As noted above, the final equations in (2.7) can be used to predict the future values of some or all variables in $\boldsymbol{z}_t$, given that the forms of the ARMA processes for these variables have been determined and that

[5] If some of the variables in $\boldsymbol{x}_t$ are non-stochastic, say time trends, they will appear the final equations of the system.

parameters have been estimated. However, these final equations cannot be used for control and structural analysis. On the other hand, the reduced form equations (2.13) and transfer equations (2.15) can be employed for both prediction and control but not generally for structural analysis except when structural equations are in reduced form ($H_{110} \equiv I$ in (2.12)) or in final form [$H_{11} \equiv I$ in (2.10)]. Note that use of reduced form and transfer function equations implies that we have enough prior information to distinguish endogenous and exogenous variables. Further, if data on some of the endogenous variables are unavailable, it may be impossible to use the reduced form equations whereas it will be possible to use the transfer equations relating to those endogenous variables for which data are available. When the structural equation system in (2.10) is available, it can be employed for structural analysis and the associated "restricted" reduced form or transfer equations can be employed for prediction and control. Use of the structural system (2.10) implies not only that endogenous and exogenous variables have been distinguished, but also that prior information is available to identify structural parameters and that the dynamic properties of the structural equations have been determined. Also, structural analysis of the complete system in (2.10) will usually require that data be available on all variables.[6] For the reader's convenience, some of these considerations are summarized in table 1.1.

Aside from the differing data requirements for use of the various equation systems considered in table 1.1, it should be appreciated that before each of the equation systems can be employed, the form of its equations must be ascertained. For example, in the case of the structural equation system (2.10), not only must endogenous and exogenous variables be distinguished, but also lag distributions, serial correlation properties of error terms, and identifying restrictions must be specified. Since these are often difficult requirements, it may be that some of the simpler equation systems will often be used although their uses are more limited than those of structural equation systems. Furthermore, even when the objective of an analysis is to obtain a structural equation system, the other equation systems, particularly the final equations and transfer equations, will be found useful. That is, structural assumptions regarding lag structures, etc. have implications for the forms and properties of final and transfer equations that can be checked with data. Such checks on structural assumptions can reveal weaknesses in them and possibly suggest alternative structural assumptions more in accord with the information in the data. In the following sections we illustrate these points in the analysis of a small dynamic structural equation system.

[6] This requirement will not be as stringent for partial analyses and for fully recursive models.

Table 1.1 *Uses and requirements for various equation systems*

Equation system	Uses of equation systems: Prediction	Control	Structural analysis	Requirements for use of equation systems
1. Final equations[a] (2.7)	yes	no	no	Forms of ARMA processes and parameter estimates
2. Reduced form equations (2.13)	yes	yes	no	Endogenous–exogenous classification of variables, forms of equations, and parameter estimates
3. Transfer equations[b] (2.15)	yes	yes	no	Endogenous–exogenous classification of variables, forms of equations, and parameter estimates
4. Final form equations[c] (2.14)	yes	yes	no	Endogenous–exogenous classification of variables, forms of equations, and parameter estimates
5. Structural equations (2.10)	yes	yes	yes	Endogenous–exogenous variable classification, identifying information,[d] forms of equations, and parameter estimates

Notes:
[a] This is Tinbergen's (1940) term.
[b] These equations are also referred to as "separated form" or "autoregressive final form" equations.
[c] As noted in the text, these equations are also referred to as "transfer function," "fundamental dynamic," and "rational distributed lag" equations.
[d] That is, information in the form of restrictions to identify structural parameters.

3 Algebraic analysis of a dynamic version of Haavelmo's model

Haavelmo (1947) formulated and analyzed the following static model with annual data for the United States, 1929–41:

$$c_t = \alpha y_t + \beta + u_t, \tag{3.1a}$$

$$r_t = \mu(c_t + x_t) + v + w_t \tag{3.1b}$$

$$y_t = c_t + x_t - r_t \tag{3.1c}$$

where c_t, y_t and r_t are endogenous variables, x_t is exogenous, u_t and w_t are disturbance terms, and α, β, μ and v are scalar parameters. The

definitions of the variables, all on a price-deflated, *per capita* basis, are:
c_t = personal consumption expenditures,
y_t = personal disposable income,
r_t = gross business saving, and
x_t = gross investment.[7]

Equation (3.1a) is a consumption relation, (3.1b) a gross business saving equation, and (3.1c) an accounting identity.

In Chetty's (1966, 1968) analyzes of the system (3.1) employing Haavelmo's annual data, he found the disturbance terms highly autocorrelated, perhaps indicating that the static nature of the model is not appropriate. In view of this possibility, (3.1) is made dynamic in the following way:

$$c_t = \alpha(L)y_t + \beta + u_t, \tag{3.2a}$$

$$r_t = \mu(L)(c_t + x_t) + v + w_t \tag{3.2b}$$

$$y_t = c_t + x_t - r_t \tag{3.2c}$$

In (3.2a), $\alpha(L)$ is a polynomial lag operator that serves to make c_t a function of current and lagged values of income. Similarly, $\mu(L)$ in (3.2b) is a polynomial lag operator that makes r_t depend on current and lagged values of $c_t + x_t$, a variable that Haavelmo refers to as "gross disposable income." On substituting for r_t in (3.2b) from (3.2c), the equations for c_t and y_t are:

$$c_t = \alpha(L)y_t + \beta + u_t, \tag{3.3a}$$

$$y_t = [1 - \mu(L)](c_t + x_t) - v - w_t. \tag{3.3b}$$

With respect to the disturbance terms in (3.3), we assume:

$$\begin{pmatrix} u_t \\ -w_t \end{pmatrix} = \begin{pmatrix} f_{11}(L) & f_{12}(L) \\ f_{21}(L) & f_{22}(L) \end{pmatrix} \begin{pmatrix} e_{1t} \\ e_{2t} \end{pmatrix}, \tag{3.4}$$

where the $f_{ij}(L)$ are polynomials in L, e_{1t} and e_{2t} have zero means, unit variances, and are contemporaneously and serially uncorrelated.

Letting $\boldsymbol{z}_t' = (c_t, y_t, x_t)$, the general multiple time series model for $\boldsymbol{z}_t$, in the matrix form (2.1), is:

$$\underset{3\times3}{H(L)}\ \underset{3\times1}{\boldsymbol{z}_t} = \underset{3\times1}{\boldsymbol{\theta}} + \underset{3\times3}{F(L)}\ \underset{3\times1}{\boldsymbol{e}_t}, \tag{3.5}$$

[7] In Haavelmo's paper, gross investment, x_t, is defined equal to "government expenditures + transfers − all taxes + gross private capital formation," while gross business saving, r_t, is defined equal to "depreciation and depletion charges + capital outlay charged to current expense + income credited to other business reserves − revaluation of business inventories + corporate savings".

where $\boldsymbol{e}_t' = (e_{1t}, e_{2t}, e_{3t})$ satisfies (2.2)–(2.3) and $\boldsymbol{\theta}' = (\theta_1, \theta_2, \theta_3)$ is a vector of constants. In explicit form, (3.5) is:

$$\begin{pmatrix} h_{11}(L) & h_{12}(L) & h_{13}(L) \\ h_{21}(L) & h_{22}(L) & h_{23}(L) \\ h_{31}(L) & h_{32}(L) & h_{33}(L) \end{pmatrix} \begin{pmatrix} c_t \\ y_t \\ x_t \end{pmatrix} = \begin{pmatrix} \theta_1 \\ \theta_2 \\ \theta_3 \end{pmatrix} + \begin{pmatrix} f_{11}(L) & f_{12}(L) & f_{13}(L) \\ f_{21}(L) & f_{22}(L) & f_{23}(L) \\ f_{31}(L) & f_{32}(L) & f_{33}(L) \end{pmatrix} \begin{pmatrix} e_{1t} \\ e_{2t} \\ e_{3t} \end{pmatrix}. \quad (3.6)$$

To specialize (3.6) to represent the dynamic version of Haavelmo's model in (3.3) with x_t exogenous, we must have $\theta_1 = \beta, \theta_2 = v$,

$$\begin{array}{lll} h_{11}(L) \equiv 1 & h_{12}(L) \equiv -\alpha(L) & h_{13}(L) \equiv 0 \\ h_{21}(L) \equiv -[1-\mu(L)] & h_{22}(L) \equiv 1 & h_{23}(L) \equiv -[1-\mu(L)] \\ h_{31}(L) \equiv 0 & h_{32}(L) \equiv 0 & h_{33}(L) \end{array} \quad (3.7a)$$

and

$$f_{13}(L) \equiv f_{23}(L) \equiv f_{31}(L) \equiv f_{32}(L) \equiv 0. \quad (3.7b)$$

Utilizing the conditions in (3.7), (3.6) becomes:

$$\begin{bmatrix} 1 & h_{12}(L) & 0 \\ h_{21}(L) & 1 & h_{23}(L) \\ 0 & 0 & h_{33}(L) \end{bmatrix} \begin{bmatrix} c_t \\ y_t \\ x_t \end{bmatrix} = \begin{bmatrix} \theta_1 \\ \theta_2 \\ \theta_3 \end{bmatrix} + \begin{bmatrix} f_{11}(L) & f_{12}(L) & 0 \\ f_{21}(L) & f_{22}(L) & 0 \\ 0 & 0 & f_{33}(L) \end{bmatrix} \begin{bmatrix} e_{1t} \\ e_{2t} \\ e_{3t} \end{bmatrix}. \quad (3.8)$$

Note that the process on the exogenous variable is $h_{33}(L)x_t = f_{33}(L)e_{3t} + \theta_3$ and the fact that x_t is assumed exogenous requires that $h_{31}(L) \equiv h_{32}(L) \equiv 0$ and that $F(L)$ be block diagonal as shown in (3.8).

In what follows, we shall denote the degree of $h_{ij}(L)$ by r_{ij} and the degree of $f_{ij}(L)$ by q_{ij}.

From (3.8), the final equations for c_t and y_t are given by:

$$\begin{aligned}(1 - h_{12}h_{21})h_{33}c_t = \theta_1' &+ (f_{11} - f_{21}h_{12})h_{33}e_{1t} \\ &+ (f_{12} - f_{22}h_{12})h_{33}e_{2t} + f_{33}h_{12}h_{23}e_{3t}\end{aligned} \quad (3.9)$$

and

$$\begin{aligned}(1 - h_{12}h_{21})h_{33}y_t = \theta_2' &+ (f_{21} - f_{11}h_{21})h_{33}e_{1t} \\ &+ (f_{22} - f_{12}h_{21})h_{33}e_{2t} - f_{33}h_{23}e_{3t},\end{aligned} \quad (3.10)$$

Table 1.2 *Degrees of lag polynomials in final equations*

Final equation	Degrees of AR polynomials[a]	Degrees of MA polynomials for errors[b] e_{1t}	e_{2t}	e_{3t}
(3.9): c_t	Maximum of r_{33} and $r_{12}+r_{21}+r_{33}$	Maximum of $r_{33}+q_{11}$ and $r_{33}+r_{12}+q_{21}$	Maximum of $r_{33}+q_{12}$ and $r_{33}+r_{12}+q_{22}$	$r_{12}+r_{23}+q_{33}$
(3.10): y_t	r_{33} and $r_{12}+r_{21}+r_{33}$	$r_{33}+q_{21}$ and $r_{33}+r_{21}+q_{11}$	$r_{33}+q_{22}$ and $r_{33}+r_{21}+q_{12}$	$r_{23}+q_{33}$
(3.11): x_t	r_{33}	–	–	q_{33}

Notes:
[a] r_{ij} is the degree of h_{ij}. Note from (3.7a), $h_{21} \equiv h_{23} \equiv -[1-\mu(L)]$, and thus $r_{21} = r_{23}$.
[b] q_{ij} is the degree of f_{ij}.

with $h_{21} \equiv h_{23}$ and θ_1' and θ_2' being new constants. Note that the AR parts of (3.9) and (3.10) have the same order and parameters. The degrees of the lag polynomials in (3.9) and (3.10) and in the process for x_t,

$$h_{33}x_t = f_{33}e_{3t} + \theta_3, \tag{3.11}$$

are indicated in table 1.2.

As mentioned above, the AR polynomials in the final equations for c_t and y_t are identical and of maximal degree equal to $r_{12}+r_{21}+r_{33}$, as shown in table 1.2, where r_{12} = degree of $\alpha(L)$ in the consumption equation, r_{21} is the degree of $\mu(L)$ in the business saving equation, and r_{33} is the degree of h_{33}, the AR polynomial in the process for x_t. Also, if the disturbance terms u_t and w_t are serially uncorrelated and if all the q_{ij} in table 1.2 are zero, the following results hold:

(i) In the final equation for c_t, the degree of the AR part is larger than or equal to the order of the MA process for the disturbance term; that is $r_{12}+r_{21}+r_{33} \geqq max(r_{12}+r_{23}, r_{33}+r_{12})$, with equality holding if $r_{33}=0$, since $r_{21}=r_{23}$, or if $r_{21}=r_{23}=0$.

(ii) In the final equation for y_t, the degree of the AR polynomial is larger than or equal to the order of the MA process for the disturbance term; i.e., $r_{12}+r_{21}+r_{33} \geqq r_{33}+r_{21}$ with equality holding if $r_{12}=0$. Thus if the process for x_t is purely AR and the structural disturbance terms u_t and w_t are not serially correlated, (i) and (ii) provide useful implications for properties of the final equations that can be checked with data as explained below.

Further, under the assumption that the structural disturbance terms u_t and w_t are serially uncorrelated, all q_{ij} other than q_{33} in table 1.2 will be equal to zero. If the process for x_t is analyzed to determine the degree of h_{33}, r_{33}, and of f_{33}, q_{33}, this information can be used in conjunction with the following:

(iii) In the final equation for c_t, the degree of the AR polynomial will be smaller than or equal to the order of the MA disturbance if $q_{33} \geqq r_{33}$. (Note $r_{21} = r_{23}$.) If $q_{33} < r_{33}$, the degree of the AR polynomial will be greater than the order of the MA disturbance term.

(iv) In the final equation for y_t, the degree of the AR polynomial will be greater than the order of the MA disturbance term given that $r_{12} + r_{33} > q_{33}$ and $r_{12} > 0$. They will be equal if $r_{12} = 0$ and $r_{33} \geqq q_{33}$ or if $r_{12} + r_{33} = q_{33}$. The latter will be greater if $r_{12} + r_{33} < q_{33}$.

In what follows, post-Second World War quarterly data for the United States, 1947–72, are employed to analyze the final equations for c_t, y_t and x_t and to check some of the implications mentioned above.

From (3.8), the dynamic structural equations of the dynamized Haavelmo model are:

$$\begin{pmatrix} 1 & h_{12} & 0 \\ h_{21} & 1 & h_{23} \end{pmatrix} \begin{pmatrix} c_t \\ y_t \\ x_t \end{pmatrix} = \begin{pmatrix} \theta_1 \\ \theta_2 \end{pmatrix} + \begin{pmatrix} f_{11} & f_{12} \\ f_{21} & f_{22} \end{pmatrix} \begin{pmatrix} e_{1t} \\ e_{2t} \end{pmatrix}, \tag{3.12a}$$

or

$$\begin{pmatrix} 1 & h_{12} \\ h_{21} & 1 \end{pmatrix} \begin{pmatrix} c_t \\ y_t \end{pmatrix} = \begin{pmatrix} \theta_1 \\ \theta_2 \end{pmatrix} + \begin{pmatrix} 0 \\ -h_{23} \end{pmatrix} x_t + \begin{pmatrix} f_{11} & f_{12} \\ f_{21} & f_{22} \end{pmatrix} \begin{pmatrix} e_{1t} \\ e_{2t} \end{pmatrix}. \tag{3.12b}$$

From (3.12b), the transfer equations, the analogue of (2.15) are:

$$\begin{vmatrix} 1 & h_{12} \\ h_{21} & 1 \end{vmatrix} \begin{pmatrix} c_t \\ y_t \end{pmatrix} = \begin{pmatrix} \theta_1'' \\ \theta_2'' \end{pmatrix} + \begin{pmatrix} 1 & -h_{12} \\ -h_{21} & 1 \end{pmatrix} \begin{pmatrix} 0 \\ -h_{23} \end{pmatrix} x_t + \begin{pmatrix} 1 & -h_{12} \\ -h_{21} & 1 \end{pmatrix} \begin{pmatrix} f_{11} & f_{12} \\ f_{21} & f_{22} \end{pmatrix} \begin{pmatrix} e_{1t} \\ e_{2t} \end{pmatrix}, \tag{3.13}$$

or

$$(1 - h_{12}h_{21})c_t = \theta_1'' + h_{12}h_{23}x_t + (f_{11} - f_{21}h_{12})e_{1t} + (f_{12} - f_{22}h_{12})e_{2t} \tag{3.14}$$

and

$$(1 - h_{12}h_{21})y_t = \theta_2'' - h_{23}x_t + (f_{21} - f_{11}h_{21})e_{1t} + (f_{22} - f_{12}h_{21})e_{2t}, \tag{3.15}$$

where θ_1'' and θ_2'' are constant parameters that are linear functions of θ_1 and θ_2.

The following properties of the transfer equations, (3.14) and (3.15) are of interest:

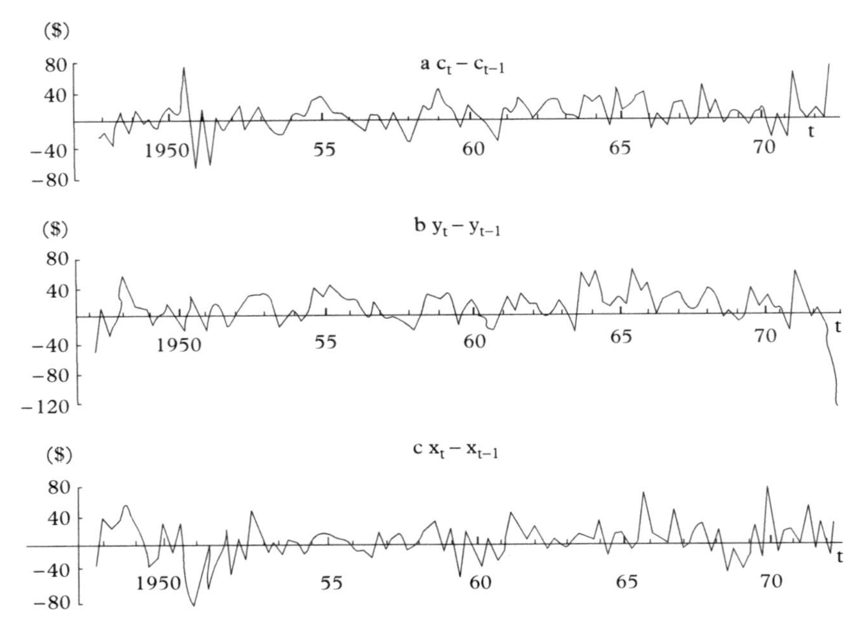

Figure 1.2 Plots of data for $c_t - c_{t-1}$, $y_t - y_{t-1}$, and $x_t - x_{t-1}$, 1950–1970

of the variables may induce stationarity. For the reader's benefit, plots of the first differences of the variables are presented in figure 1.2. It is clear from the plots of the first differences that they are less subject to trend than are the levels of the variables. However, a slight trend in the magnitudes of the first differences may be present if the levels are subject to a relatively constant proportionate rate of growth. For this reason, we also performed analyzes based on second differences.

In figure 1.3, we present the estimated autocorrelation function for the series $c_t - c_{t-1}$, the first difference of consumption.[11] Also indicated in figure 1.3, is a $\pm 2\hat{\sigma}$ confidence band for the autocorrelations where $\hat{\sigma}$ is a large sample standard error for the sample autocorrelations.[12] It is seen that all estimated autocorrelations lie within the band except for that of lag 2. This suggests that the underlying process is not purely AR. If the autocorrelation estimate for lag 2 is regarded as a cut-off, the results

[11] The computer program employed was developed by C. R. Nelson and S. Beveridge, Graduate School of Business, University of Chicago.

[12] $\hat{\sigma}^2$ is an estimate of the following approximate variance of r_k, the kth sample serial correlation, given in Bartlett (1946). With $\rho_v = 0$ for $v > q$, var $(r_k) \doteq (1 + 2\sum_{v=1}^{q} \rho_v^2)/T$, for $k > q$. The $\pm 2\hat{\sigma}$ bounds for r_k, $k = 1, 2, \ldots, 12$, are calculated under the assumption, $\rho_v = 0, v > 0$. For $k > 12$, they are calculated assuming $\rho_v = 0, v > 12$.

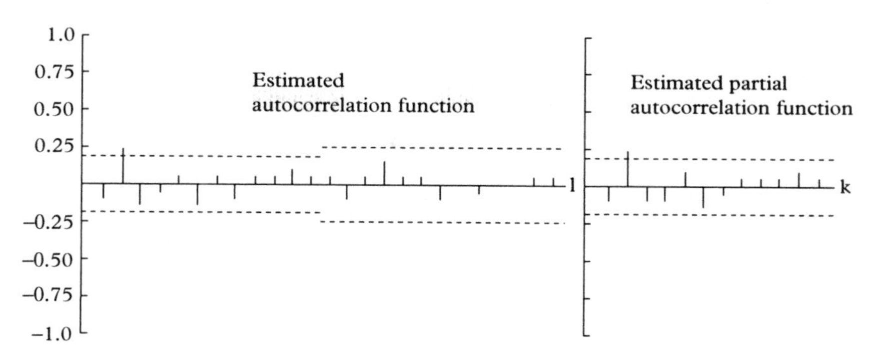

Figure 1.3 Estimated autocorrelation function and estimated partial autocorrelation function for $c_t - c_{t-1}$

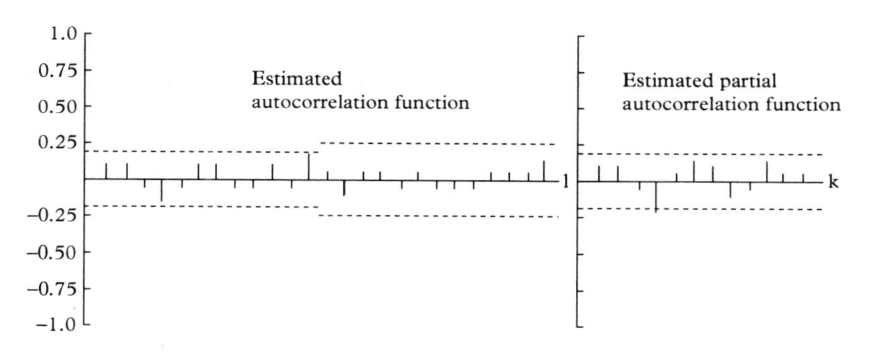

Figure 1.4 Estimated autocorrelation function and estimated partial autocorrelation function for $y_t - y_{t-1}$

suggest that a second order MA process may be generating the first differences of c_t. The estimated partial autocorrelation function, also shown in figure 1.3, does not appear to contradict this possibility. Estimation of a second order MA model for the first differences of consumption, led to the following results using the BJ non-linear algorithm:

$$c_t - c_{t-1} = e_t + \underset{(0.101)}{0.0211}e_{t-1} + \underset{(0.101)}{0.278}e_{t-2} + \underset{(2.96)}{10.73} \quad s^2 = 530, \tag{4.1}$$

where s^2 is the residual sum of squares (RSS) divided by the number of degrees of freedom and the figures in parentheses are large sample standard errors.

For income, y_t, a plot of the first differences is given in figure 1.2. From the plot of the estimated autocorrelations for the first differences in figure

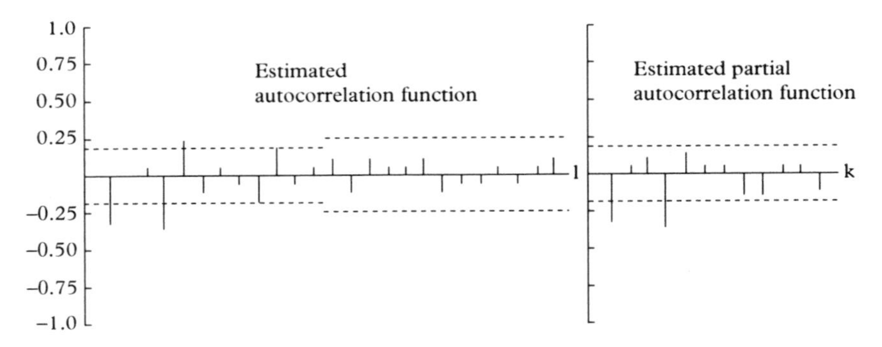

Figure 1.5 Vertical areas of figures 1.3–1.5 for $x_t - x_{t-1}$: on the right: $\hat{\phi}_{\hat{k}\hat{k}}$, on the left: r_l

1.4, it appears that none of the autocorrelations is significantly different from zero, a finding that leads to the presumption that the underlying model is not AR. Estimates of the partial autocorrelations for lag 4 and lag 10 lie close to the limits of the $\pm 2\hat{\sigma}$ band – see figure 1.4. Other partial autocorrelations appear not to differ significantly from zero. If all autocorrelations and partial autocorrelations are deemed not significantly different from zero, then the conclusion would be that the first differences of income are generated by a random walk model which was estimated with the following results:

$$y_t - y_{t-1} = e_t + \underset{(8.336)}{10.03} \quad s^2 = 842. \tag{4.2}$$

For the first differences of investment, $x_t - x_{t-1}$ – see the plot in figure 1.2 – the estimated autocorrelation and partial autocorrelation functions are given in figure 1.5. The autocorrelations alternate in sign and show some significant values for lags less than or equal to 5 which suggests an AR model. The partial autocorrelation function has a cut-off at lag 4, supporting the presumption that the model is AR and indicating a fourth order AR scheme. Also, the partial autocorrelation function for the second differences has a cut-off at lag 3 while the autocorrelations alternate in sign for lags less than 11, findings which support those derived from analysis of first differences. In view of these findings, a fourth order AR model has been fitted with the data:

$$(1 + \underset{(0.0942)}{0.263}L - \underset{(0.0976)}{0.0456}L^2 + \underset{(0.0970)}{0.0148}L^3 + \underset{(0.0933)}{0.376}L^4)(x_t - x_{t-1})$$
$$= e_t + \underset{(3.265)}{7.738} \quad s^2 = 939. \tag{4.3}$$

In contrast to the processes for the first differences of c_t and y_t in (4.1) and (4.2), that for the first differences of investment, x_t, in (4.3) has an AR part. Thus the requirement of the structural form that all endogenous variables have identical AR parts of order equal to or greater than that for x_t – see (3.9)–(3.10) above – is not satisfied given the results in (4.1)–(4.3). Using the notation of table 1.2 with h_{ij} of degree r_{ij} regarded as an element of $H(L)/(1 - L)$, the degree of the AR polynomial in (4.31) is $r_{33} = 4$ while that of the error process is $q_{33} = 0$. In the case where no cancelling occurs in (4.1)–(4.2), it is clear that the conditions (3.9) and (3.10) of table 1.2 can not be met. Even if h_{23} in (3.8) satisfies $h_{23} \equiv 0$ so that c_t and y_t are generated independently of x_t, the conditions on the final equations are not met by the results for the final equations in (4.1)–(4.3).[13] Thus while (4.1)–(4.3) appear to be consistent with the information in the data, they are not compatible with the dynamized Haavelmo model specified in section 3, (3.2a)–(3.2c).

At this point, the following are considerations that deserve attention:

(1) Although the fits of the models in (4.1)–(4.3) are fairly good, it may be that schemes somewhat more complicated than (4.1)–(4.3) are equally well or better supported by the information in the data and are compatible with the implications of the Haavelmo model. This possibility is explored below.

(2) To compare and test alternative final equations for each variable, it would be desirable to have inference methods that are less "judgmental" and more systematically formal than are the BJ methods. In the next subsection, we indicate how likelihood ratios and posterior odds ratios can be used for discriminating among alternative final equation models.

(3) It must be recognized that there are some limitations on the class of AR models that can be transformed to a stationary process through differencing. That is, only those AR models whose roots lie on the boundary or inside the unit circle can be transformed to stationary models by differencing. Other transformations, say logarithmic, have to be used for models with roots outside the unit circle.

(4) Differencing series may amplify the effects of measurement errors present in the original data and seriously affect estimates of the autocorrelation and partial autocorrelation functions. Of course, this problem arises not only in the BJ approach but also in any analysis of ARMA processes, particularly those of high order.

[13] If $h_{23} \equiv 0$, then (4.1)–(4.2) imply $r_{12} + r_{21} = 1; r_{12} + q_{21}, q_{11}, q_{12}, r_{12} + q_{22} \leqq 2$ (with at least one equality); and $r_{21} + q_{11}, q_{21}, r_{21} + q_{12}, q_{22} \leqq 0$. These conditions imply $q_{11} = q_{12} = q_{21} = q_{22} = r_{21} = 0, r_{21} = 1$, and $r_{21} = 2$ which cannot hold simultaneously.

4.2 *Analyzes of final equations utilizing likelihood ratios and posterior odds*

The purpose of this section is to provide additional procedures for identifying or determining the forms of final equations. These procedures involve use of likelihood ratios and Bayesian posterior odds. After showing how to obtain likelihood ratios and posterior odds, some of the results are applied in the analysis of Haavelmo's model.

Consider the following ARMA model for a single random variable z_t,

$$\phi(L)z_t = \theta(L)\varepsilon_t, \quad t = 1, 2, \ldots, T, \tag{4.4}$$

where $\phi(L)$ and $\theta(L)$ are polynomials in L of degree p and q, respectively. Assume that the ε_t's are normally and independently distributed, each with zero mean and common variance, σ^2. Let $u_t \equiv \theta(L)\varepsilon_t$. Then given the "starting values" for ε_t and z_t, ε_0 and $\boldsymbol{z}_0$, the vector $\boldsymbol{u}' = (u_1, u_2{}' \ldots,$ $u_T)$ has a T-dimensional multivariate normal distribution with zero mean vector and covariance matrix Σ, that is:

$$p(\boldsymbol{u}|\boldsymbol{\phi}, \boldsymbol{\theta}, \boldsymbol{\sigma}^2, \boldsymbol{z}_0, \varepsilon_0) = (2\Pi)^{-T/2} \exp|\Sigma|^{-\frac{1}{2}}\left\{-\tfrac{1}{2}\boldsymbol{u}'\Sigma^{-1}\boldsymbol{u}\right\}, \tag{4.5}$$

where $\phi' = (\phi_1, \phi_2, \ldots, \phi_p)$ and $\boldsymbol{\theta}' = (\theta_1, \theta_2, \ldots, \theta_q)$. The matrix Σ is a $T \times T$ positive definite symmetric matrix with elements given by:

$$\begin{aligned} \sigma_{t,t-k} &= \sigma^2\left(1 + \sum_{i=1}^{q}\theta_i^2\right), && \text{for } k = 0,\\ \sigma_{t,t-k} &= \sigma^2\left(-\theta_k + \sum_{i=k+1}^{q}\theta_{i-k}\theta_i\right), && \text{for } 0 < k \leqq q,\\ \sigma_{t,t-k} &= 0, && \text{for } k > q. \end{aligned} \tag{4.6}$$

Also, the joint probability density function (pdf) for the ε_ts, given by

$$\varepsilon_t = z_t - \phi_1 z_{t-1} - \cdots - \phi_p z_{t-p} + \theta_1\varepsilon_{t-1} + \cdots + \theta_q\varepsilon_{t-q}, \quad t = 1, 2, \ldots, T, \tag{4.7}$$

is:

$$p(\varepsilon|\boldsymbol{\phi}, \boldsymbol{\theta}, \boldsymbol{\sigma}^2, \varepsilon_0, \boldsymbol{z}_0) = (2\Pi\sigma^2)^{-T/2}\exp\left\{-\frac{1}{2\sigma^2}\sum_{t=1}^{T}\varepsilon_t^2\right\}. \tag{4.8}$$

Since the Jacobian of the transformation from the ε_ts to the z_ts is equal to one, the joint pdf for the z_ts, the likelihood function, is:

$$p(\boldsymbol{z}|\boldsymbol{\phi}, \boldsymbol{\theta}, \boldsymbol{\sigma}^2, \varepsilon_0, \boldsymbol{z}_0) = (2\Pi\sigma^2)^{-T/2} \exp\left\{-\frac{1}{2\sigma^2}\sum_{t=1}^{T}(z_t - \phi_1 z_{t-1} - \cdots - \phi_p z_{t-p} + \theta_1\varepsilon_{t-1} + \theta_2\varepsilon_{t-2} + \cdots + \theta_q\varepsilon_{t-q})^2\right\}. \quad (4.9)$$

In this context, (4.9) is convenient since Marquardt's non-linear computational algorithm can be applied to obtain maximum likelihood (ML) estimates.

If we have an alternative ARMA model,

$$\phi_a(L)z_t = \theta_a(L)\varepsilon_{at}, \quad t = 1, 2, \ldots, T, \quad (4.10)$$

where $\phi_a(L)$ is of degree p_a, $\theta_a(L)$ of degree q_a and the error process ε_{at} is $NID(0, \sigma_a^2)$, then the likelihood ratio, λ, for (4.4) and (4.10) is

$$\lambda = (\max_{\phi,\theta,\sigma} l(\boldsymbol{\phi}, \boldsymbol{\theta}, \boldsymbol{\sigma}|\boldsymbol{z}))/(\max_{\phi_a,\theta_a,\sigma_a} l(\boldsymbol{\phi}_a, \boldsymbol{\theta}_a, \boldsymbol{\sigma}_a|\boldsymbol{z})), \quad (4.11)$$

where $l(\boldsymbol{\phi}, \boldsymbol{\theta}, \sigma|\boldsymbol{z})$ denotes (4.9) viewed as a function of its parameters and similarly for $l(\boldsymbol{\phi}_a, \boldsymbol{\theta}_a, \boldsymbol{\sigma}_a|\boldsymbol{z})$. The ratio of maximized likelihood functions in (4.11) reduces to:

$$\lambda = \left(\hat{\sigma}_a^2/\hat{\sigma}^2\right)^{T/2}, \quad (4.12)$$

where

$$\hat{\sigma}^2 = \frac{1}{T}\sum_{t=1}^{T}(z_t - \hat{\phi}_1 z_{t-1} - \cdots - \hat{\phi}_p z_{t-p} + \hat{\theta}_1\hat{\varepsilon}_{t-1} + \cdots + \hat{\theta}_q\hat{\varepsilon}_{t-q})^2 \quad (4.13a)$$

and

$$\hat{\sigma}_a^2 = \frac{1}{T}\sum_{t=1}^{T}(z_t - \hat{\phi}_{1a} z_{t-1} - \cdots - \hat{\phi}_{p_a a} z_{t-p_a} + \hat{\theta}_{1a}\hat{\varepsilon}_{a,t-1} + \cdots + \hat{\theta}_{q_a a}\hat{\varepsilon}_{a,t-q_a})^2 \quad (4.13b)$$

are the ML estimates for σ^2 and σ_a^2.

If model (4.10) is nested in model (4.4), i.e. $p_a \leqq p$ and/or $q_a \leqq q$, with at least one strict inequality, and under the assumption that (4.10) is the true model, $2\ln\lambda$ is approximately distributed as χ_r^2 with r being the number of restrictions imposed on (4.4) to obtain (4.10); that is, $r = p + q - (p_a + q_a)$ – see Silvey (1970, pp. 112–13). In choosing a significance level for this test, it is very important, as usual, to consider errors of the first and second kind. Rejecting the nested model when it is "true" appears to us to be a less serious error than failing to reject it when the broader model is "true". That is, using the restricted model when the restrictions are not "true" may lead to serious errors. Use of the broader model, when the restricted model is "true," involves carrying along some extra parameters which may not be as serious a problem as giving these parameters incorrect values. This argues against using extremely low significance levels, e.g. $\alpha = 0.01$ or $\alpha = 0.001$. Also, these considerations rationalize somewhat the usual practice of some degree of over-fitting when the model form is somewhat uncertain. More systematic analysis and study of this problem would be desirable.

In order to compare (4.4) and (4.10) in a Bayesian context, we have to specify a prior distribution on the parameter space. In the problem of comparing *nested* models, this prior distribution has a mixed form with weights whose ratio is the prior odds on alternative models – see, e.g., Jeffreys (1961, p. 250), Zellner (1971, pp. 297ff.), and Palm (1972). Formally, the posterior odds ratio relating to (4.4) and (4.10) is given by:

$$K_{1a} = \frac{\Pi}{\Pi_a} \frac{\int p(\boldsymbol{\phi}, \boldsymbol{\theta}, \boldsymbol{\sigma}) l(\boldsymbol{\phi}, \boldsymbol{\theta}, \boldsymbol{\sigma} | \boldsymbol{z}) \, \mathrm{d}\boldsymbol{\phi} \, \mathrm{d}\boldsymbol{\theta} \, \mathrm{d}\boldsymbol{\sigma}}{\int p(\boldsymbol{\phi}_a, \boldsymbol{\theta}_a, \boldsymbol{\sigma}_a) l(\boldsymbol{\phi}_a, \boldsymbol{\theta}_a, \boldsymbol{\sigma}_a | \boldsymbol{z}) \, \mathrm{d}\boldsymbol{\phi}_a \, \mathrm{d}\boldsymbol{\theta}_a \, \mathrm{d}\boldsymbol{\sigma}_a}, \qquad (4.14)$$

where K_{1a} is the posterior odds ratio, Π/Π_a is the prior odds ratio, and $p(\phi, \theta, \sigma)$ and $p(\phi_a, \theta_a, \sigma_a)$ are the prior pdfs for the parameters. Before (4.14) can be made operational, it is necessary to formulate the prior pdfs and to evaluate the integrals, either exactly or approximately.[14]

We now compute likelihood ratios to compare alternative formulations of the final equations of Haavelmo's model. The information in table 1.2 and empirical results in the literature on quarterly consumption relations suggest higher order AR and MA schemes than those fitted in section 4.1. For example, a fourth order AR model for the second differences of

[14] Note however, as pointed out by Lindley (1961), the likelihood functions in the numerator and denominator of (4.14) can be expanded about ML estimates. If just the first terms of these expansions are retained, namely $l(\hat{\boldsymbol{\phi}}, \hat{\boldsymbol{\theta}}, \hat{\boldsymbol{\sigma}} | \boldsymbol{z})$ and $l(\hat{\boldsymbol{\phi}}_a, \hat{\boldsymbol{\theta}}_a, \hat{\boldsymbol{\sigma}}_a | \boldsymbol{z})$, and if the prior pdfs are proper, (4.14) is approximated by:

$$K_{1a} \doteq [\Pi/\Pi_a [l(\hat{\boldsymbol{\phi}}, \hat{\boldsymbol{\theta}}, \hat{\boldsymbol{\sigma}} | z)/l(\hat{\boldsymbol{\phi}}_a, \hat{\boldsymbol{\theta}}_a, \hat{\boldsymbol{\sigma}}_a | \boldsymbol{z})],$$

i.e. a prior odds ratio, Π/Π_a, times the usual likelihood ratio. As Lindley points out, additional terms in the expansions can be retained and the resulting expression will involve some prior moments of parameters. Thus on assigning a value to Π/Π_a, the prior odds ratio, the usual likelihood ratio is transformed into an approximate posterior odds ratio for whatever non-dogmatic, proper prior pdfs employed.

consumption with third order MA error terms is a scheme tentatively suggested by considerations presented in section 4.1. This scheme has been fitted with both the consumption and income data with results shown for c_t and y_t in tables 1.3 and 1.4 and those for x_t in table 1.5 with figures in parentheses being large sample standard errors. Also shown in table 1.3 are the results for the simple schemes of section 4.1 and results for several other specifications. It should be noted that use of the broader schemes for c_t and y_t results in decreases in the value of the residual sum of squares divided by degrees of freedom of about 8–12 percent. However, it must be noted that the large sample standard errors associated with the point estimates are rather large in a number of instances.

To put the comparison of alternative schemes on a more formal basis, likelihood ratios have been computed and are reported in table 1.6. Using these ratios as a basis for large sample χ^2 tests, it is found that it is possible to reject the simpler versions at reasonable significance levels. The results of the tests indicate that it is reasonable to retain the model (5, 1, 4) for consumption and income and (4, 1, 0) for investment. Given that these models are tentatively accepted, it is the case that the AR and MA polynomials for the consumption and income processes have identical degrees. However, the point estimates of the AR parameters of consumption and income processes are not very similar, a finding that must be tempered by the fact that standard errors associated with coefficient estimates are rather large, particularly for the AR parameters of the income process. It would be very desirable to develop joint estimation techniques for the two equations in order to increase the precision of estimation and joint test procedures for testing the hypothesis that the AR parameters are the same for the two processes.

What are the implications of retaining (5, 1, 4) models for c_t and y_t and a (4, 1, 0) model for x_t? As noted above, the empirical finding that first differencing appears adequate to induce stationarity for all three variables suggests that the model can be expressed in first difference form. That is, we rewrite (3.5) as follows:

$$\overline{H}(L)(1 - L)\boldsymbol{z}_t = \boldsymbol{\theta} + F(L)\boldsymbol{e}_t, \tag{4.15}$$

where $\overline{H}(L)$ has elements that are the elements of $H(L)$ divided by $1 - L$. With the polynomials $h_{ij}(L)$ of degree r_{ij} considered elements of $\overline{H}(L)$ rather than $H(L)$ and if no cancelling occurs in (4.15), then under the restrictions imposed on the Haavelmo model in the preceding section (see table 1.2), we have $r_{33} = 4; q_{33} = 0; r_{12} + r_{21} + r_{33} = 5;$

$$\begin{aligned} &r_{33} + q_{11}, r_{33} + r_{12} + q_{21}, r_{33} + q_{12}, r_{33} + r_{12} + q_{22}, \\ &\quad r_{12} + r_{23} + q_{33} \leqq 4 \end{aligned}$$

Table 1.3 *Estimated final equations for consumption*

Model	RSS, residual			Estimates of the AR part					Estimates of the MA part				
(p, d, q)[a]	sum of squares	DF	RSS/DF	AR1	AR2	AR3	AR4	AR5	MA1	MA2	MA3	MA4	Constant
(1) (0, 1, 2)	52,012	98	530						−0.021 (0.101)	−0.278 (0.101)			10.727 (2.96)
(2) (4, 2, 3)	45,807	92	497	−0.727 (0.365)	−0.199 (0.255)	−0.077 (0.178)	−0.123 (0.115)		0.354 (0.344)	0.258 (0.302)	0.452 (0.210)		0.283 (0.256)
(3) (5, 2, 2)	48,411	92	525	−0.651 (0.458)	0.226 (0.142)	−0.0567 (0.160)	−0.230 (0.155)	0.0534 (0.162)	0.380 (0.452)	0.581 (0.409)			0.247 (0.164)
(4) (5, 1, 4)	44,634	91	490	0.587 (0.101)	0.514 (0.199)	−0.166 (0.246)	−0.655 (0.0849)	0.075 (0.140)	0.678 (0.0958)	0.215 (0.198)	0.178 (0.246)	−0.684 (0.134)	6.751 (1.851)

Note:

[a] (p, d, q) denotes an ARMA model for the dth differences of a variable that has AR polynomial of degree p and MA polynomial of degree q.

Table 1.4 *Estimated final equations for income*

Model	RSS, residual			Estimates of the AR part					Estimates of the MA part				
$(p, d, q)^a$	sum of squares	DF	RSS/DF	AR1	AR2	AR3	AR4	AR5	MA1	MA2	MA3	MA4	Constant
(1) (0, 1, 0)	85,042	101	842										10.03 (8.336)
(2) (0, 1, 4)	79,362	99	797									0.302 (0.107)	10.90 (2.012)
+ restrictions													
(3) (4, 1, 1)	78,546	95	823	0.0803 (0.353)	0.0861 (0.118)	−0.0156 (0.112)	−0.307 (0.110)		−0.016 (0.372)				12.182 (4.788)
(4) (4, 2, 3)	78,827	92	821	−0.429 (0.418)	−0.154 (0.532)	0.222 (0.153)	−0.063 (0.120)		0.440 (0.384)	0.227 (0.423)	0.384 (0.580)		0.128 (0.395)
(5) (4, 1, 4)	65,428	92	703	−0.225 (0.255)	−0.390 (0.244)	−0.437 (0.168)	−0.258 (0.224)		−0.295 (0.249)	−0.733 (0.215)	0.668 (0.210)	−0.057 (0.276)	22.545 (8.950)
(6) (5, 1, 4)	65,351	91	705	0.0678 (1.207)	−0.552 (0.380)	−0.164 (0.771)	−0.211 (0.266)	−0.0048 (0.357)	0.0210 (1.214)	−0.833 (0.349)	−0.290 (1.073)	0.0417 (0.373)	18.834 (23.14)

Note:
[a] (p, d, q) denotes an ARMA model for the dth differences of a variable that has AR polynomial of degree p and MA polynomial of degree q.

Table 1.5 *Estimated final equations for investment*

Model	RSS, residual			Estimates of the AR part					Estimates of the MA part				
(p, d, q)[a]	sum of squares	DF	RSS/DF	AR1	AR2	AR3	AR4	AR5	MA1	MA2	MA3	MA4	Constant
(1) (4, 1, 0)	90,519	96	939	− 0.263 (0.0942)	0.0456 (0.0976)	− 0.0148 (0.0970)	− 0.376 (0.0933)						7.738 (3.265)
(2) (4, 2, 3)	91,124	92	978	− 0.653 (0.284)	− 0.183 (0.294)	− 0.0334 (0.132)	− 0.325 (0.127)		0.587 (0.282)	0.301 (0.246)	0.115 (0.274)		− 0.127 (0.232)
(3) (5, 2, 1)	87,428	93	929	− 0.247 (0.0924)	0.0395 (0.0991)	− 0.0245 (0.0980)	− 0.337 (0.0991)	0.106 (0.096)	1.014 (0.012)				− 0.109 (0.102)
(4) (5, 2, 2)	90,391	92	970	− 0.659 (0.631)	− 0.076 (0.192)	− 0.0076 (0.127)	− 0.364 (0.118)	− 0.065 (0.276)	0.552 (0.609)	0.425 (0.630)			− 1.04 (0.187)

Note:

[a] (p, d, q) denotes an ARMA model for the dth differences of a variable that has AR polynomial of degree p and MA polynomial of degree q.

Table 1.6 *Results of large-sample likelihood ratio tests applied to final equations of Haavelmo's model*

Models compared[a]	$\lambda = \frac{\mathcal{L}114(X\mid H_1)}{\mathcal{L}114(X\mid H_0)}$	2 ln λ	r	Critical points for χ_r^2 $\alpha = 0.05$	$\alpha = 0.10$	$\alpha = 0.20$
1. *Consumption* c_t						
$H_0 : (0, 1, 2)$ vs. $H_1 : (4, 2, 3)$[b]	573.547	–	–	–	–	–
$H_0 : (0, 1, 2)$ vs. $H_1 : (5, 1, 4)$	2098.29	15.230	7	14.07	12.02	9.80
$H_0 : (5, 2, 2)$ vs. $H_1 : (4, 2, 3)$[b]	15.871	–	–	–	–	–
$H_0 : (4, 2, 3)$ vs. $H_1 : (5, 1, 4)$	3.654	2.592	2	5.99	4.61	3.22
2. *Income* y_t						
$H_0 : (0, 1, 0)$ vs. $H_1 : (0, 1, 4)$[c]	31.697	6.912	1	3.84	2.71	1.64
$H_0 : (0, 1, 0)$ vs. $H_1 : (4, 1, 4)$	4937.0×10^2	26.219	8	15.51	13.36	11.03
$H_0 : (0, 1, 0)$ vs. $H_1 : (5, 1, 4)$	5236.0×10^2	26.337	9	16.92	14.68	12.24
$H_0 : (0, 1, 0)$ vs. $H_1 : (4, 2, 3)$[b]	44.453	–	–	–	–	–
$H_0 : (4, 2, 3)$ vs. $H_1 : (5, 1, 4)$	117.8×10^2	18.748	2	5.99	4.61	3.22
3. *Investment* x_t						
$H_0 : (4, 1, 0)$ vs. $H_1 : (5, 2, 1)$[b]	5.678	–	–	–	–	–

Notes:
[a] $H : (p, d, q)$ denotes an ARMA model for the dth difference of a variable that has AR polynomial of degree p. and MA error polynomial of degree q.
[b] These are non-nested hypotheses.
[c] Here there are three restrictions on the parameters of the MA error process.

and

$$r_{33} + q_{21}, r_{33} + r_{21} + q_{11}, r_{33} + q_{22}, r_{33} + r_{21} + q_{12}, \\ r_{23} + q_{33} \leqq 4,$$

with at least one equality holding in both cases. These restrictions imply $r_{12} + r_{21} = 1$, all $q_{ij} = 0$ *and* $r_{12} = r_{21} = 0$, conditions that cannot hold simultaneously. Also, if we retain a (5, 2, 1) model for investment, we end up with a contradiction.

If we make the assumption that the joint process for Δc_t and Δy_t is independent of Δx_t, i.e. $h_{23} \equiv 0$ in (3.8), an assumption that may appeal to some Quantity of Money theorists but not to most Keynesians, the degrees of the polynomials reported in table 1.2 are reduced by r_{33} and we have the following restrictions on the degrees of the AR polynomials in the processes for Δc_t and $\Delta y_t : r_{12} + r_{21} = 5; q_{11}, q_{12}, r_{12} + q_{21}, r_{12} + q_{22} \leqq 4$; and $q_{21}, q_{22}, r_{21} + q_{11}, r_{21} + q_{12} \leqq 4$, with at least one equality holding in both cases. With further assumptions, e.g. $r_{12} = 2$ and $r_{21} = 3$, it is possible to determine compatible values for the degrees of the structural equations' lag polynomials and error term polynomials. However, this compatibility is attained only with the controversial assumption that the joint process for Δc_t and Δy_t is independent of the process for Δx_t, Haavelmo's investment variable.[15] A major implication of this last assumption is that the analysis of the transfer functions should reveal no dependence of either Δc_t or Δy_t on Δx_t, a point that is checked in the next section where we analyze the transfer equations (3.14)–(3.15).

An alternative way to achieve compatibility of the results of the final equation analyzes with structural assumptions is to assume that $h_{23}(L) \equiv h_{33}(L)$. This assumption implies that the investment variable influences Δc_t and Δy_t only through its disturbance term. With this assumption, h_{33} cancels in (3.9) and (3.10) and the empirical findings, combined with the results in table 1.2, imply that $r_{33} = 4, r_{23} = r_{33} = 4$, by assumption, $q_{33} = 0, r_{12} + r_{21} = 5$,

$$q_{11}, q_{12}, r_{12} + q_{21}, r_{12} + q_{22} \leqq 4$$

and

$$q_{21}, q_{22}, r_{21} + q_{11}, r_{21} + q_{12} \leqq 4,$$

with at least one equality in each case. Further, the autoregressive parts of the final equations for c_t and y_t are identical to the autoregressive parts in their transfer equations. These implications of the assumption, $h_{23}(L) \equiv h_{33}(L)$, and of final equation findings for the forms of the transfer equations will be checked in the next section.

5 Empirical analyses of transfer equations (3.14)–(3.15)

We now turn to an analysis of the transfer functions, shown in (3.14)–(3.15), associated with the dynamized Haavelmo model. These equations express c_t and y_t as functions of their own lagged values, of current and lagged values of x_t, and of current and lagged error terms. The first step in the analysis of the transfer functions is the determination or identification of the degrees of the lag polynomials. In general, a transfer function can be written as an infinite moving average process in exogenous variables plus an error term, u_t, with zero mean, that is,

$$y_t = v(L)x_t + u_t, \tag{5.1}$$

where $v(L) = \sum_{i=0}^{\infty} v_i L^i$. Often this infinite process can be well approximated by a finite distributed lag model of order k, that is $v(L) =$

[15] In terms of (3.8), this assumption implies that $h_{23}(L) \equiv 0$. With this assumption, the structural equations are, from (3.3): $\Delta c_t = \alpha(L)\Delta y_t + \beta + u_t$ and $\Delta y_t = [1 - \mu(L)]\Delta c_t - v - w_t$. That is, current and lagged values of y_t affect consumption and current and lagged consumption affect income.

$\sum_{i=0}^{k} v_i L^i$. Solving the Yule–Walker equations for a kth order approximation, we obtain

$$Ey_t x_{t-\tau} = v_0 Ex_t x_{t-\tau} + v_1 Ex_{t-1} x_{t-\tau} + \cdots + v_k Ex_{t-k} x_{t-\tau}, \quad \tau = 0, 1, 2, \ldots, k. \tag{5.2}$$

Rough estimates of the v_is can be obtained by replacing the expectations in (5.2) by corresponding sample moments and solving for the v_is. This is equivalent to regressing y_t on current and lagged values of x_t. Note, too, that $v(L)$ can be written as the ratio of two lag polynomials of degrees s and r, $\omega_s(L)$ and $\theta_r(L)$, as follows:

$$v(L) = [\omega_s(L)/\theta_r(L)]L^b, \tag{5.3}$$

with b some non-negative integer. Introduction of $b \neq 0$ allows for some "dead time" in the response pattern of y_t to x_t. Using (5.3) and the preliminary estimates of the v_is, obtained as described above, preliminary estimates of the parameters $\omega_s(L)$, ω_js, $j = 0, 1, 2, \ldots, s$, and of $\theta_r(L)$, θ_is, $i = 1, 2, \ldots, r$, can be found. As Box and Jenkins (1970, p. 378) point out, the v_js, the impulse response weights, consist of:

(1) b zero values, $v_0, v_1, \ldots, v_{b-1}$,
(2) a further $s - r + 1$ values, $v_b, \ldots, v_{b+s-r}$, following no fixed pattern (if $s < r$, no such values occur), and
(3) values v_j, with $j \geqq b + s - r + 1$, following a pattern given by an rth order difference equation with starting values $v_{b+s} \ldots v_{b+s-r+1}$.

Properties (1)–(3) can help to determine the values of b, s, and r from the preliminary estimates of the vs. Then the residuals $\hat{u}_t = y_t - \hat{v}(L)x_t$ are analyzed to determine the degrees of the AR and MA parts of the error process using estimated autocorrelation and partial autocorrelation functions. Final estimation of the transfer function so determined can be accomplished in the BJ approach by use of Marquardt's non-linear algorithm.

It is important to observe that the results of final equation analyses can be employed to obtain some information about the degrees of transfer functions' lag polynomials. In fact, if assumptions regarding structural equations' forms are in accord with information in the data, there should be compatibility between the final equations' and transfer equations' forms that we determine from the data.[16] That is, final equation analysis led us to (5, 1, 4) processes for c_t and y_t and to a (4, 1, 0) process for x_t, namely, $\phi_{(4)}(L)\Delta x_t = e_t$, or $\Delta x_t = \phi_{(4)}^{-1}(L)e_t$. If we difference the transfer functions in (3.14) and (3.15) and then substitute

[16] Here we abstract from the possibility that $h_{23}(L) \equiv 0$ since in this case transfer functions show no dependence on x_t, a point that is checked below.

$\Delta x_t = \phi_{(4)}^{-1}(L)e_t$, we obtain the final equations for c_t and y_t. To obtain compatibility with the empirically determined (5, 1, 4) final equations for c_t and y_t, the transfer functions must have polynomials hitting Δx_t with degree $s \leqq 3$ in the numerator and degree $r = 0$ in the denominator. In addition, the ratio of lag polynomials operating on the transfer functions' error terms should have a numerator of degree zero and denominator of degree one.

Under the assumption $h_{23}(L) \equiv h_{33}(L)$, introduced tentatively in the previous section, the final equation analyzes yield the following implications for the transfer functions' lag structures in (3.14)–(3.15):

(1) The AR parts of both transfer functions are identical with the AR parts of the final equations and are of degree $r_{12} + r_{21} = 5$.

(2) The polynomial, h_{23}, hitting Δx_t in the transfer function for income, is identical with the AR part of the final equation for x_t and has degree $r_{23} = r_{33} = 4$.

(3) The polynomial operating on Δx_t in the consumption transfer function has degree of at least 4.

(4) The order of the moving average error process in each transfer equation is equal to 4.

We shall check points (1)–(4) in the empirical analyzes that follow. In this connection, it is the case that there is no assurance that the information in the data will be in accord with compatible findings for the final equations and transfer functions of the Haavelmo model since model specification errors, measurement errors, imperfect seasonal adjustment, etc. can affect analyzes to produce incompatible results.

To determine the degrees of the lag polynomials in (3.14)–(3.15) and to get starting values for the vs, different values for k were employed in connection with (5.1)–(5.2) that provided preliminary estimates of the vs. For $k = 8$ and the first difference of c_t, the v_is with $i = 0$, 1 and 6 appear to be significantly different from zero and the behavior of the estimated v_is is very irregular. The fact that v_0 is significantly different from zero implies that $b = 0$. With respect to determining values for r and s, the degrees of the polynomials in (5.3), the results are not very precise. The values indicated by our final equation analysis are used and starting values for the ω_is are based on the estimates of the v_is for alternative values of s and r. Further, the analysis of the residuals from an eighth order distributed lag model for the first difference of c_t suggests a mixed first order AR and second order MA error process. However, it is thought that this determination of the transfer function's properties is very tentative and thus it was thought worthwhile to proceed to estimate transfer functions in forms suggested by our final equation analyzes. Some estimation results for these forms are shown in (5.4)–(5.5): with $s = 3$,

$$\Delta c_t = (\underset{(0.0689)}{-0.129} + \underset{(0.0704)}{0.188}L + \underset{(0.070)}{0.0875}L^2 - \underset{(0.068)}{0.037}L^3)\Delta x_t + e_t/(1 + \underset{(0.105)}{0.0208}L) + \underset{(2.297)}{10.41}, \tag{5.4}$$

with residual sum of squares (RSS) equal to 41,617, and with $s = 2$,

$$\Delta c_t = (\underset{(0.0685)}{-0.149} + \underset{(0.0709)}{0.172}L + \underset{(0.0677)}{0.0830}L^2)\Delta x_t + e_t/(1 - \underset{(0.105)}{0.0047}L) + \underset{(2.34)}{10.03}, \tag{5.5}$$

with RSS = 43,226.

Under the assumption $h_{23}(L) \equiv h_{33}(L)$, an estimate of the transfer function form suggested by the final equation analysis is:

$$\Delta c_t = -\frac{\underset{(0.068)}{0.0349} + \underset{(0.106)}{0.171}L + \underset{(0.102)}{0.264}L^2}{1 + \underset{(0.289)}{0.575}L - \underset{(0.480)}{0.187}L^2}\Delta x_t + \frac{1 - \underset{(0.388)}{0.550}L - \underset{(0.465)}{0.559}L^2}{1 - \underset{(0.327)}{0.484}L - \underset{(0.431)}{0.695}L^2 + \underset{(0.200)}{0.288}L^3} e_t + \underset{(8.573)}{10.55}, \tag{5.6}$$

with RSS = 30,945.

With respect to the first differences of y_t, with $k = 8$, implementation of (5.1)–(5.2) resulted in just v_0 being significantly different from zero, suggesting that $b = 0$. The estimated v_is appear to follow a damped wave-like pattern. The difference between the values of r and s is thus thought to be small but this inference is very uncertain. In view of this, the values of s and r implied by the final equation analysis, $s \leqq 3$ and $r = 0$ have been employed along with a ratio of polynomials for the error process with numerator of degree 0 and denominator of degree 1. Some estimates reflecting these considerations follow.

For $s = 3$,

$$\Delta y_t = (\underset{(0.066)}{0.385} + \underset{(0.0675)}{0.139}L + \underset{(0.067)}{0.0938}L^2 - \underset{(0.0658)}{0.124}L^3)\Delta x_t + \frac{e_t}{(1 + \underset{(0.105)}{0.055}L)} + \underset{(2.15)}{10.83}, \tag{5.7}$$

with RSS = 38,908, and for $s = 2$,

$$\Delta y_t = (\underset{(0.067)}{0.355} + \underset{(0.0696)}{0.121}L + \underset{(0.066)}{0.117}L^2)\Delta x_t + e_t/(1 + \underset{(0.103)}{0.0077}L) + \underset{(2.301)}{10.01}, \tag{5.8}$$

with RSS = 41,547.

Under the assumption that $h_{23}(L) \equiv h_{33}(L)$, the transfer function for income suggested by the final equation analyzes has been estimated with the following results:

$$\Delta y_t = \frac{\underset{(0.075)}{0.417} + \underset{(0.165)}{0.063}L + \underset{(0.163)}{0.068}L^2}{1 - \underset{(0.316)}{0.0628}L - \underset{(0.325)}{0.111}L^2 + \underset{(0.172)}{0.389}L^3}\Delta x_t + \frac{1 + \underset{(0.315)}{0.517}L}{1 + \underset{(0.303)}{0.551}L - \underset{(0.114)}{0.147}L^2} e_t + \underset{(2.396)}{11.02}, \tag{5.9}$$

with RSS = 36,462.

The estimates reported in (5.4)–(5.9) are in accord with the implications of final equation analyzes for the forms of the transfer functions. Further, we see that for (5.4)–(5.5) and (5.7)–(5.8), the AR polynomials for Δc_t and Δy_t are almost identical, a requirement that the transfer functions must satisfy given that the variables are generated by a joint process with x_t exogenous. Further, from (3.14), (a) the AR part of the transfer function for Δc_t should be identical, up to degree 1, to that operating on Δx_t in the same equation if the Haavelmo model is adequate, and (b) the lag operator acting on Δx_t in the equation for Δc_t should be a multiple of that for Δx_t in the income equation. The first of these requirements is not satisfied by the results in (5.4)–(5.5) since the polynomials acting on Δc_t and Δx_t have differing degrees. However, the requirements (a) and (b) are satisfied, as far as the degrees are concerned, for (5.6) and (5.9).[17]

Last, as mentioned above, one way to have the empirically determined final equations compatible with the dynamized Haavelmo model in (3.8) is to assume $h_{23} \equiv 0$, i.e. that the process for Δx_t is independent of the joint process for Δc_t and Δy_t. This implies no dependence of Δc_t and of Δy_t on Δx_t in the transfer equations. The dependence that has been found

[17] This suggests that the restriction $h_{21} \equiv h_{23}$, originally imposed, is probably not in accord with the information in the data.

above might be interpreted as due to specification errors (e.g. x_t might not be exogenous) or to other complicating factors (e.g. measurement errors, poor seasonal adjustment, etc.). On the other hand, it may be that the alternative assumption $h_{23} \equiv h_{33}$ is more in accord with the information in the data. Note the substantial reduction in RSS associated with (5.6) and (5.9) relative to the RSS for other models.

To explore this last point more systematically, some alternative transfer function models, formulated without taking into account results of final equation analyzes, have been estimated, with results shown in tables 1.7 and 1.8. For comparison, results with models implied by the final equation analyzes are also presented. A quick look at the residual sum of squares (RSS) indicates that for consumption, alternative model M_5 yields about a 12 percent reduction in RSS relative to M_1 while M_7 yields about a 20 percent reduction. For income, M_5 yields about a 3 percent reduction in RSS relative to M_1 while M_7 provides the lowest RSS, about 6 percent lower than for M_1.

For nested models, a large sample likelihood ratio test procedure has been employed to compare alternative formulations, with results reported in table 1.9. For consumption, M_2 is preferred to M_1 at the 5 percent level. However, pairwise comparisons of M_2 against M_4, M_5 and M_6 favor the latter relative to M_2. However, in comparisons with M_7, both M_2 and M_4 are rejected. Thus the results of the likelihood ratio tests favor M_7, a model that is compatible with the results of the final equation analysis under the assumption $h_{23}(L) \equiv h_{33}(L)$. The results for income transfer functions, shown in the bottom of table 1.9, indicate that M_2 is rejected in favor of M_1 while M_4 performs better than M_3 or M_6. Compared with M_7, models M_2, M_4 and M_6 are rejected at the 5 percent significance level, it appears that M_7 is not significantly different from M_5. The results of these comparisons suggest that it is reasonable to accept tentatively, models M_1, M_5 or M_7 as being in accord with the information in the data. If we retain models M_7 for consumption and M_7 for income,[18]

[18] Other possibilities, e.g. M_7 for Δc_t and M_1 for Δy_t or M_7 for Δc_t and M_5 for Δy_t, lead to incompatibilities with the requirements that the final and transfer equations must satisfy. In the first case, M_7 for Δc_t and M_1 for Δy_t, the AR parts of the transfer functions are not identical as required in (3.14)–(3.15) and even possible cancelling will not be sufficient to satisfy the condition on the polynomials hitting Δx_t in (3.14)–(3.15). If we retain M_7 for Δc_t and M_5 for Δy_t, their autoregressive parts have the same order, $r_{12} + r_{21} = 5$, and the degrees of the polynomials for Δx_t are, respectively, $r_{12} + r_{23} = 5$ and $r_{23} = 3$, implying $r_{21} = 2$. However, the assumption $h_{21}(L) \equiv h_{23}(L)$, implying $r_{21} = r_{23}$, is no longer satisfied. In addition, there is incompatibility with the analysis of the final equations requiring $r_{23} = 4$.

Table 1.7 *Estimated transfer functions for consumption*

Model	RSS	DF	RSS/DF	Estimates of the AR and MA parts of Δx_t	Estimates of the AR and MA parts of the error process	Constant
M_1	41,617	89	467.6	$-\underset{(0.0689)}{0.129}+\underset{(0.0704)}{0.188}L+\underset{(0.070)}{0.0875}L^2-\underset{(0.068)}{0.037}L^3$	$1/(1+\underset{(0.105)}{0.0208}L)$	$\underset{(2.297)}{10.41}$
M_2	43,286	91	475.0	$-\underset{(0.0685)}{0.149}+\underset{(0.0709)}{0.172}L+\underset{(0.0677)}{0.0830}L^2$	$1/(1-\underset{(0.105)}{0.0047}L)$	$\underset{(2.34)}{10.03}$
M_3	40,212	90	446.8	$\underset{(0.067)}{0.370}+\underset{(0.068)}{0.126}L+\underset{(0.0677)}{0.0829}L^2-\underset{(0.066)}{0.114}L^3$	$1-\underset{(0.106)}{0.0376}L$	$\underset{(2.202)}{10.62}$
M_4	37,941	84	451.7	$\dfrac{-\underset{(0.074)}{0.080}+\underset{(0.108)}{0.085}L+\underset{(0.0897)}{0.219}L^2}{1+\underset{(0.351)}{0.717}L-\underset{(0.349)}{0.143}L^2}$	$\dfrac{1+\underset{(0.530)}{495}L}{1+\underset{(0.537)}{0.445}L-\underset{(0.117)}{0.250}L^2-\underset{(0.192)}{0.003}L^3}$	$\underset{(2.83)}{10.16}$
M_5	36,504	82	445.2	$\dfrac{-\underset{(0.0699)}{0.111}+\underset{(0.0767)}{0.183}L+\underset{(0.083)}{0.0526}L^2+\underset{(0.074)}{0.123}L^3}{(1+\underset{(0.140)}{0.111}L)+\underset{(0.114)}{0.794}L^2}$	$\dfrac{1-\underset{(0.334)}{0.386}L}{1-\underset{(0.320)}{0.389}L-\underset{(0.115)}{0.236}L^2+\underset{(0.122)}{0.252}L^3}$	$\underset{(0.221)}{10.74}$
M_6	37,645	85	442.3	$\dfrac{-\underset{(0.072)}{0.129}+\underset{(0.103)}{0.050}L+\underset{(0.076)}{0.187}L^2}{1+\underset{(0.246)}{0.838}L+\underset{(0.331)}{0.275}L^2+\underset{(0.203)}{0.392}L^3}$	$\dfrac{1+\underset{(0.219)}{0.530}L}{1+\underset{(0.212)}{0.487}L-\underset{(0.112)}{0.204}L^2}$	$\underset{(2.601)}{10.75}$
M_7	33,453	83	403.1	$\dfrac{-\underset{(0.068)}{0.0349}+\underset{(0.106)}{0.171}L+\underset{(0.102)}{0.264}L^2}{1-\underset{(0.289)}{0.575}L-\underset{(0.480)}{0.187}L^2}$	$\dfrac{1-\underset{(0.388)}{0.550}L-\underset{(0.465)}{0.559}L^2}{1-\underset{(0.327)}{0.484}L-\underset{(0.431)}{0.695}L^2-\underset{(0.200)}{0.228}L^3}$	$\underset{(8.573)}{10.5}$

Table 1.8 *Estimated transfer functions for income*

Model	RSS	DF	RSS/DF	Estimates of the AR and MA parts of Δx_t	Estimates of the AR and MA parts of the error process	Constant
M_1	38,908	89	437.2	$\underset{(0.066)}{0.385}+\underset{(0.0675)}{0.139}L+\underset{(0.067)}{0.0938}L^2-\underset{(0.0658)}{0.124}L^3$	$1/(1+\underset{(0.105)}{0.055}L)$	$\underset{(2.15)}{10.83}$
M_2	41,547	91	456.6	$\underset{(0.067)}{0.355}+\underset{(0.0696)}{0.121}L+\underset{(0.066)}{0.117}L^2$	$1/(1-\underset{(0.103)}{0.0077}L)$	$\underset{(2.301)}{10.01}$
M_3	43,149	90	479.4	$-\underset{(0.0692)}{0.145}+\underset{(0.0710)}{0.173}L+\underset{(0.071)}{0.075}L^2-\underset{(0.069)}{0.0265}L^3$	$1-\underset{(0.106)}{0.0127}L$	$\underset{(2.34)}{10.17}$
M_4	43,751	92	475.6	$-\underset{(0.0683)}{0.156}+\underset{(0.0706)}{0.164}L+\underset{(0.0676)}{0.088}L^2$	$1+\underset{(0.105)}{0.0046}L$	$\underset{(2.33)}{9.84}$
M_5	37,775	87	434.2	$\dfrac{\underset{(0.070)}{0.408}+\underset{(0.165)}{0.064}L}{1-\underset{(0.325)}{0.129}L-\underset{(0.164)}{0.168}L^2+\underset{(0.134)}{0.389}L^3}$	$\dfrac{1-\underset{(0.344)}{0.228}L}{1-\underset{(0.326)}{0.181}L-\underset{(0.107)}{0.172}L^2}$	$\underset{(2.63)}{11.11}$
M_6	42,819	89	481.1	$\dfrac{\underset{(0.075)}{0.390}-\underset{(0.292)}{0.110}L-\underset{(0.209)}{0.180}L^2}{1-\underset{(0.816)}{0.462}L-\underset{(0.620)}{0637}L^2+\underset{(0.214)}{0.103}L^3}$	$1-\underset{(0.110)}{0.0364}L$	$-\underset{(13.84)}{13.41}$
M_7	36,462	85	429.0	$\dfrac{\underset{(0.075)}{0.417}+\underset{(0.165)}{0.063}L+\underset{(0.163)}{0.068}L^2}{1-\underset{(0.316)}{0.0628}L-\underset{(0.325)}{0.111}L^2+\underset{(0.172)}{0.389}L^3}$	$\dfrac{1+\underset{(0.315)}{0.517}L}{1+\underset{(0.303)}{0.551}L-\underset{(0.114)}{0.147}L^2}$	$\underset{(2.396)}{11.02}$

Table 1.9 *Results of large-sample likelihood ratio tests applied to transfer functions of Haavelmo's model*

Models compared		$\lambda = \frac{L(y\|H_1)}{L(y\|H_0)}$	$2 \ln \lambda$	r	Critical points for χ_r^2 $\alpha = 0.05$	$\alpha = 0.10$	$\alpha = 0.20$
1. *Consumption*							
(1) H_0 : M_2	H_1 : M_1	6.66	3.79	1	3.84	2.71	1.64
(2) H_0 : M_1	H_1 : M_5	702.27	13.11	5	11.07	9.24	7.29
(3) H_0 : M_2	H_1 : M_4	678.89	13.04	5	11.07	9.24	7.29
(4) H_0 : M_2	H_1 : M_5	4679.79	16.90	6	12.59	10.64	8.56
(5) H_0 : M_2	H_1 : M_6	992.41	13.80	5	11.07	9.24	7.29
(6) H_0 : M_3	H_1 : M_5	126.11	9.67	5	11.07	9.24	7.29
(7) H_0 : M_4	H_1 : M_5	6.893	3.86	5	11.07	9.24	7.29
(8) H_0 : M_2	H_1 : M_7	3.5×10^5	25.56	5	11.07	9.24	7.29
(9) H_0 : M_4	H_1 : M_7	5.02×10^2	12.44	1	3.84	2.71	1.64
2. *Income*							
(1) H_0 : M_2	H_1 : M_1	26.609	6.563	1	3.84	2.71	1.64
(2) H_0 : M_4	H_1 : M_3	1.999	1.385	1	3.84	2.71	1.64
(3) H_0 : M_4	H_1 : M_6	2.935	2.153	3	7.81	6.25	4.64
(4) H_0 : M_2	H_1 : M_7	683.83	13.06	5	11.07	9.24	7.29
(5) H_0 : M_4	H_1 : M_7	9,065	18.22	5	11.07	9.24	7.29
(6) H_0 : M_6	H_1 : M_7	3,088	16.07	2	5.99	4.61	3.22
(7) H_0 : M_5	H_1 : M_7	5.86	3.54	1	3.84	2.71	1.64

we have $r_{12} + r_{21} = 5$ for the order of the AR polynomials acting on Δc_t and Δy_t, the degrees of the polynomials operating on Δx_t are of degrees, 5 and 4, respectively, and the error processes are each of order 4. Under the assumption that $h_{23}(L) \equiv h_{33}(L)$, these results are in accord with the requirements that the final equations must satisfy (see table 1.2).

6 Summary of results and implications for structural equations

In table 1.10, we present the preferred final equation and transfer function models for the dynamized Haavelmo model. From the information provided in table 1.10, the following are the implied restrictions on the lag structures appearing in the structural equations of the model where the r_{ij}s refer to the degrees of elements of $\overline{H}(L)$, the matrix $H(L)$ divided by $(1 - L)$:

1. $r_{12} = 1; r_{33} = r_{23} = r_{21} = 4; q_{11} = q_{12} = 0$; and $q_{21}, q_{22} \leqq 3$, with at least one equality holding.

Table 1.10 *Final equation and transfer function models for dynamized Haavelmo model*[a]

Systems of equations	Model	Order of AR part	Degree of lag polynomial for Δx_t	Order of MA error process
1. *Final equations*				
Δc_t	$(5, 1, 4)^b$	$r_{12} + r_{21} = 5$	–	$q_{11}, q_{12}, r_{12} + q_{21}, r_{12} + q_{22} \leqq 4$ (at least one equality holding)
Δy_t	$(5, 1, 4)^b$	$r_{12} + r_{21} = 5$	–	$q_{21}, q_{22}, r_{21} + q_{11}, r_{21} + q_{12} \leqq 4$ (at least one equality holding)
Δx_t	$(4, 1, 0)^b$	$r_{33} = 4$	–	$q_{33} = 0$
2. *Transfer functions*				
Δc_t	$M_7{}^c$	$r_{12} + r_{21} = 5$	$r_{12} + r_{23} = 5$	$q_{11}, q_{12}, r_{12} + q_{21}, r_{12} + q_{22} \leqq 4$ (at least one equality)
Δy_t	$M_7{}^c$	$r_{12} + r_{21} = 5$	$r_{23} = 4$	$q_{21}, q_{22}, q_{11} + r_{21}, q_{12} + r_{21} \leqq 4$ (at least one equality)

Notes:

[a] It is assumed that $h_{23} \equiv h_{33}$.

[b] See tables 1.3–1.5 where estimated models are presented.

[c] See table 1.8 where estimated models are presented.

2. The transfer functions show a dependence of Δc_t and of Δy_t on Δx_t. Under the assumption that $h_{23} \equiv h_{33}$, the final equations and transfer functions selected by the likelihood ratio tests are compatible insofar as the degrees of the relevant lag polynomials are considered.
3. Explicitly, a structural representation compatible with the results of the final equation and transfer function analyses is:

$$\begin{bmatrix} 1 & -\alpha^{(1)} & 0 \\ \mu_1^{(4)} - 1 & 1 & \mu_2^{(4)} - 1 \\ 0 & 0 & \mu_2^{(4)} - 1 \end{bmatrix} (1 - L) \begin{bmatrix} c_t \\ y_t \\ x_t \end{bmatrix} \tag{6.1}$$

$$= \begin{bmatrix} \theta_1 \\ \theta_2 \\ \theta_3 \end{bmatrix} + \begin{bmatrix} f_{11}^{(0)} & f_{12}^{(0)} & 0 \\ f_{21}^{(\leq 3)} & f_{22}^{(\leq 3)} & 0 \\ 0 & 0 & f_{33}^{(0)} \end{bmatrix} \begin{bmatrix} e_{1t} \\ e_{2t} \\ e_{3t} \end{bmatrix},$$

where the superscripts in parentheses denote the degrees of lag polynomials that were determined from the final equation and transfer function analyzes. Note that these polynomials are equal to the polynomials of the matrix $H(L)$ in (3.8) divided by a common factor $1 - L$. The factor $1 - L$ hitting the variables c_t, y_t and x_t puts them in first difference form, a transformation that appears adequate to induce stationarity in all three variables, a condition required for the correlogram analysis of the variables. That the same differencing transformation induces stationarity in all variables is not necessary for all models but is an empirical finding in the present case. Also, to achieve compatibility, it is necessary that $h_{11}(L) \equiv h_{22}(L) \equiv 1$, a special case of what was assumed in (3.7a). Last, it should be noted that $\mu_1^{(4)}(L)$ and $\mu_2^{(4)}(L)$ are not necessarily identical.

The system in (6.1) can alternatively be expressed in the form of (3.2a)–(3.2c) as follows:

$$\begin{bmatrix} 1 & 0 & -\alpha^{(1)} & 0 \\ -\mu_1^{(4)} & 1 & 0 & -\mu_2^{(4)} \\ -1 & 1 & 1 & -1 \end{bmatrix} (1 - L) \begin{bmatrix} c_t \\ r_t \\ y_t \\ x_t \end{bmatrix} = \begin{bmatrix} \beta \\ \nu \\ 0 \end{bmatrix} + \begin{bmatrix} u_t \\ w_t \\ 0 \end{bmatrix}, \tag{6.2}$$

with u_t a serially uncorrelated disturbance term and w_t following a third order moving average process. Further, u_t and w_t will generally be correlated.

Using the identity, $\Delta y_t = \Delta c_t + \Delta x_t - \Delta r_t$, we can eliminate Δr_t from (6.2) to obtain:

$$\Delta c_t = \alpha_0 \Delta y_t + \alpha_1 \Delta y_{t-1} + \beta + u_t, \tag{6.3a}$$

and

$$\begin{aligned} \Delta y_t &= \left(1 - \mu_1^{(4)}\right)\Delta c_t + \left(1 - \mu_2^{(4)}\right)\Delta x_t - \nu - w_t, \\ &= \sum_{i=0}^{4} \gamma_i \Delta c_{t-i} - \nu + w_t', \end{aligned} \tag{6.3b}$$

where $\alpha^{(1)} \equiv \alpha_0 + \alpha_1 L$, $1 - \mu_1^{(4)} \equiv \sum_{i=0}^{4} \gamma_i L^i$, and $-(1 - \mu_2^{(4)})\Delta x_t = f_{33}^{(0)} e_{3t}$ have been used and $w_t' \equiv -(w_t + f_{33}^{(0)} e_{3t})$. The two-equation system in (6.3) is a simultaneous equation model with dynamic lags and contemporaneously correlated disturbance terms, u_t and w_t', the former non-autocorrelated and the latter following a third order MA process. We can estimate the parameters of (6.3) employing "single equation" or "joint" estimation techniques as explained briefly below.[19]

[19] These estimation procedures will be treated more fully in future work. A . . . paper by Byron (1973) treats some of these problems from the likelihood point of view. Also, it will be noted that non-unique estimates for certain parameters are available from the final equation and transfer function analyzes. In certain instances, these latter estimates

For single-equation estimation of (6.3a), we consider it in conjunction with the final equation[20] for y_t, namely a (5, 1, 4) ARMA process that we write as:

$$\Delta y_t = \sum_{i=1}^{5} \delta_i \Delta y_{t-i} + \phi_1 + \sum_{i=0}^{4} \lambda_{1i} a_{1t-i}, \tag{6.4}$$

where a_{1t} is a non-autocorrelated error with zero mean and constant finite variance. The parameters of (6.4) have already been estimated above. We now substitute for Δy_t in (6.3a) from (6.4) to obtain

$$\Delta c_t = \alpha_0 \widetilde{\Delta y}_t + \alpha_1 \Delta y_{t-1} + \beta' + v_{1t}, \tag{6.5}$$

where $\widetilde{\Delta y}_t \equiv \sum_{i=1}^{5} \delta_i \Delta y_{t-i}$, $\beta' \equiv \phi_1(\alpha_0 + \alpha_1)$ and $v_{1t} \equiv w'_t + \alpha_0 \sum_{i=0}^{4} \lambda_{1i} a_{1t-i}$ a fourth order MA process. Given consistent estimates of the δ_is in $\widetilde{\Delta y}_t$, we can calculate consistent estimates of α_0, α_1, β and parameters of the MA process for v_{1t}. The results of this approach are presented and discussed below.

With respect to single-equation estimation of (6.3b), we consider it in conjunction with the (5, 1, 4) ARMA final equation for c_t that was estimated above and is expressed as:

$$\Delta c_t = \sum_{i=1}^{5} \eta_i \Delta c_{t-i} + \phi_2 + \sum_{i=0}^{4} \lambda_{2i} a_{2t-i}, \tag{6.6}$$

where a_{2t} is a non-autocorrelated error with zero mean and constant finite variance. Then on substituting for Δc_t in the second line of (6.3b) from (6.6), we have:

$$\Delta y_t = \gamma_0 \widetilde{\Delta c}_t + \sum_{i=1}^{4} \gamma_i \Delta c_{t-i} + \nu' + v_{2t}, \tag{6.7}$$

where $\widetilde{\Delta c}_t \equiv \sum_{i=1}^{5} \eta_i \Delta c_{t-i}$, $\nu' \equiv \nu + \gamma_0 \phi_2$, and $v_{2t} \equiv w'_t + \gamma_0 \sum_{i=0}^{4} \lambda_{2i} a_{2t-i}$, a fourth order MA process. Since consistent estimates of the η_i are available from the analysis of (6.6), they can be used in conjunction with (6.7) to obtain consistent estimates of the γs, ν, and the parameters of the process for v_{2t}.

As regards joint estimation of (6.5) and (6.7), single-equation analysis yields residuals that can be used to estimate the covariance matrix for the disturbances, the v_{1t}s and v_{2t}s. For a two-equation system, this matrix will be generally a $2T \times 2T$ matrix with four submatrices in the form of

are obtained from estimates of ratios of lag polynomials and thus are probably not very reliable.

[20] Alternatively, the transfer function for Δy_t could be employed. However, it is not clear that use of the transfer function is to be preferred.

Table 1.11 *Single-equation estimates of parameters of consumption equation (6.5)*

1. *Using $\tilde{\Delta}y_t$ from final equation for Δy_t*

$$\Delta c_t = -\underset{(0.180)}{0.333}\,\tilde{\Delta}y_t + \underset{(0.111)}{0.251}\,\Delta y_{t-1} + \underset{(2.297)}{10.45} + (1 - \underset{(0.127)}{0.180}L + \underset{(0.128)}{0.022}L^2 - \underset{(0.118)}{0.375}L^3 - \underset{(0.119)}{0.079}L^4)a_{1t}$$

RSS = 40,701 DF = 86

2. *Using $\tilde{\Delta}y_t$ from transfer function for Δy_t*

$$\Delta c_t = -\underset{(0.092)}{0.034}\tilde{\Delta}y_t + \underset{(0.116)}{0.300}\Delta y_{t-1} + \underset{(2.240)}{7.216} + (1 - \underset{(0.129)}{0.258}L + \underset{(0.116)}{0.193}L^2 - \underset{(0.119)}{0.217}L^3 + \underset{(0.115)}{0.003}L^4)a'_{1t}$$

RSS = 40,646 DF = 86

band matrices characteristic of MA processes. Let this matrix be denoted Ω and an estimate of it, $\hat{\Omega}$. Then with $\boldsymbol{v}' = (\boldsymbol{v}'_1\boldsymbol{v}'_2)$, where the vector $\boldsymbol{v}_1$ has elements v_{1t} and $\boldsymbol{v}_2$ elements v_{2t}, minimization of $\boldsymbol{v}'\hat{\Omega}^{-1}\boldsymbol{\nu}$ can be done to provide joint estimates of the parameters.[21]

In table 1.11, we present various single-equation consistent estimates of the parameters of the consumption equation in (6.3a). In the first line of the table, the final equation for Δy_t was employed to substitute for $\tilde{\Delta}y_t$ in the consumption function while in the second line the transfer function for Δy_t was employed.[22] It is seen that in both cases the point estimate for α_0 is negative. However, the standard errors are large so that a confidence interval at a reasonable level would include positive values. The estimates of α_1, the coefficient of Δy_{t-1} in (6.3a), are in the vicinity of 0.3 with a standard error of about 0.1.[23] That α_0 and α_1 are not very precisely estimated is probably due to collinearity of $\tilde{\Delta}y_t$ and Δy_{t-1}. Use of an informative prior distribution for α_0 and α_1 in a Bayesian analysis could help to improve the precision of inferences. To specify a prior distribution for α_0 and α_1 and also to interpret the results in table 1.11, it may be useful to regard Δc_t^p, the planned change in expenditures, including durables, to be linked to permanent income change, Δy_t^p, and transitory income change, Δy_t^t, as follows: $\Delta c_t^p = k\Delta y_t^p + \alpha_0\Delta y_t^t + \alpha_1\Delta y_{t-1}^t$. In planning consumption expenditures for the tth period, note that Δy_t^t is as yet *unrealized* transitory income whereas Δy_{t-1}^t is realized transitory income for period $t-1$. We believe that consumer reactions to realized transitory income will be much greater than those to as yet unrealized transitory income, i.e. $\alpha_1 > \alpha_0$ with α_0 small. Using $\Delta c_t = \Delta c_{t_1}^p + u_t$

[21] The new residuals can be employed to re-estimate Ω and thus iteration of the process on Ω (and also on the parameters in $\tilde{\Delta}c_t$ and $\tilde{\Delta}y_t$) is possible.

[22] Note that the estimation of the consumption equation using the final equation expression for Δy_t is not linked to the assumption that Δx_t is exogenous whereas use of the transfer function expression for Δy_t is.

[23] As explained below, α_1 can be viewed as the coefficient of realized transitory income change and thus an estimate of α_1 in the vicinity of 0.3 seems reasonable.

and $\Delta y_t = \Delta y_t^p + \Delta y_t^t$ in connection with the relation for Δc_t^p above, we have $\Delta c_t = \alpha_0 \Delta y_t + \alpha_1 \Delta y_{t-1} + \beta + u_t$ with $\beta = (k - \alpha_0)\Delta y_t^p - \alpha_1 \Delta y_{t-1}^p$, assumed constant.[24] Within this framework, given the hypothesis that reaction to *unrealized* transitory income change, Δy_t^t is rather small, if not zero, while reaction to realized transitory income change Δy_{t-1}^t is positive, probably an α_1 between zero and one, the results in table 1.11 appear plausible.

In conclusion, we believe that the techniques presented above can be very helpful in checking the specifying assumptions of many existing linear or linearized models and in "iterating in" on models that are suitable approximations to the information in our data and that may predict well. Some topics that will receive attention in future work include further development of estimation techniques for different equation systems, joint testing procedures for nested and non-nested hypotheses, analyses of the comparative predictive performance of final equation, transfer function and structural equation systems, Bayesian procedures utilizing informative prior distributions, and applications. Finally, we cannot resist remarking that the present work lends support to the notion that so-called "naive" ARMA time series models are not all that naive after all.

APPENDIX DATA SOURCES

Personal consumption expenditures, disposable personal income, gross investment data

Series 1946–65:

United States Department of Commerce/Office of Business Economics, 1966, The National Income and Product Accounts of the United States, 1929–65, Statistical Tables (Washington, DC)

Series 1966–72:

United States Department of Commerce/Office of Business Economics, Survey of Current Business (Washington, DC)

Consumer price index:

United States Department of Commerce/Office of Business Economics, Survey of Current Business (Washington, DC)

Population data:

US Bureau of the Census, Current Population Reports: Population estimates, Series P-25 (Washington, DC)

[24] Alternatively, we could assume $(k - \alpha_0)\Delta y_t^p - \alpha_1 \Delta y_{t-1}^p = \beta + \varepsilon_t$, where ε_t is a non-autocorrelated random error with zero mean and constant variance.

BIBLIOGRAPHY

Akaike, H., 1973, "Maximum likelihood identification of Gaussian autoregressive-moving average models," *Biometrika* 60, 255–65

Bartlett, M. S., 1946, "On the theoretical specification of the sampling properties of autocorrelated time series," *Journal of the Royal Statistical Society* B 8, 27–41

Box, G. E. P. and G. M. Jenkins, 1970, *Time Series Analysis, Forecasting and Control* (San Francisco, Holden-Day).

Byron, R. P., 1973, "The computation of maximum likelihood estimates for linear simultaneous systems with moving average disturbances," Department of Economics, Australian National University, manuscript

Chetty, V. K., 1966, "Bayesian analysis of some simultaneous equation models and specification errors," Doctoral dissertation, University of Wisconsin, Madison, unpublished

1968, "Bayesian analysis of Haavelmo's models," *Econometrica* 36, 582–602

Dhrymes, P. J., 1970, *Econometrics, Statistical Foundations and Applications* (New York, Harper & Row)

Haavelmo, T., 1947, "Methods of measuring the marginal propensity to consume," *Journal of the American Statistical Society* 42, 105–22; reprinted in W. Hood and TC. Koopmans (eds.), *Studies in Econometric Methods* (New York, John Wiley, 1953)

Hannan, E .J., 1969, "The identification of vector mixed auto-regressive-moving average systems," *Biometrika* 57, 233–5

1971, "The identification problem for multiple equation systems with moving average errors," *Econometrica* 39, 715–65

Jeffreys, H., 1961, *Theory of Probability* (Oxford, Clarendon Press)

Jorgenson, D. W., 1966, "Rational distributed lag functions," *Econometrica* 34, 135–49

Kmenta, J., 1971, *Elements of Econometrics* (New York, Macmillan)

Lindley, D. V., 1961, "The use of prior probability distributions in statistical inference and decision," in J. Neyman (ed.), *Proceedings of the Fourth Berkeley Symposium on Mathematical Statistics and Probability*, I, 453–68

Marschak, J., 1950, "Statistical inference in economics, an introduction," in T. C. Koopmans (ed.), *Statistical Inference in Dynamic Economic Models* (New York, John Wiley)

Nelson, C. R., 1970, "Joint estimation of parameters of correlated time series," Graduate School of Business, University of Chicago, manuscript

Palm, F. C., 1972, "On mixed prior distributions and their application in distributed lag models," CORE Discussion Paper 7222, University of Louvain

Pierce, D. A. and J. M. Mason, 1971, "On estimating the fundamental dynamic equations of structural econometric models," Paper presented at the Winter Meeting of the Econometric Society, New Orleans

Quenouille, M. H., 1957, *The Analysis of Multiple Time series* (London, C. Griffin and Co.)

Silvey, S. D., 1970, *Statistical Inference* (Baltimore, Penguin)

Theil, H. and J. C. D. Boot, 1962, "The final form of econometric equation systems," *Review of the International Statistical Institute* 30, 136–52; reprinted in A. Zellner (ed.), *Readings in Economic Statistics and Econometrics* (Boston, Little Brown, 1968)

Tinbergen, J., 1940, "Econometric business cycle research," *Review of Economic Studies* 7, 73–90

Wold, H., 1953, *Demand Analysis: A Study in Econometrics* (New York, John Wiley)

Zellner, A., 1959, "Review of 'The analysis of multiple time-series' by M. H. Quenouille," *Journal of Farm Economics* 41, 682–4

1971, *An introduction to Bayesian Inference in Econometrics* (New York, John Wiley).

After completing this paper, the following Ph.D. Thesis, dealing with related topics, was brought to our attention by Dennis Aigner:

Haugh, L. D., 1972, "The identification of time series interrelationships with special reference to dynamic regression models," Department of Statistics, University of Wisconsin, Madison

2 Statistical analysis of econometric models (1979)

Arnold Zellner

1 Introduction

Substantial progress has been made in developing data, concepts, and techniques for the construction and statistical analysis of econometric models. Comprehensive data systems, including national income and product accounts, price, wage and interest rate data, monetary data, and many other measures, have been developed for almost all countries. In many cases, annual measurements have been augmented by quarterly and monthly measurements of a broad array of economic variables. In recent years, scientifically designed sample surveys have been employed to expand the data bases of a number of countries. While research continues to improve data bases, we must recognize that the work that produced our current, extensive data bases is a major accomplishment in the field of scientific measurement and enables economic analysts to avoid the charge of "theory without measurement."

In reviewing the development of concepts for the statistical analysis of econometric models, it is very easy to forget that in the opening decades of [the twentieth] century a major issue was whether a statistical approach was appropriate for the analysis of economic phenomena. Fortunately, the recognition of the scientific value of sophisticated statistical methods in economics and business has buried this issue. To use statistics in a sophisticated way required much research on basic concepts of econometric modeling that we take for granted today. It was necessary to develop fundamental concepts such as complete model, identification, autonomous structural relationships, exogeneity, dynamic multipliers, and stochastic

Research . . . financed by NSF Grant SOC 7305547 and by income from the H. G. B. Alexander Endowment Fund, University of Chicago. The author is grateful to David C. Hoaglin, Stephen E. Fienberg, and two anonymous referees for helpful comments.
An earlier version of this chapter was presented to the American Statistical Association's meeting in Chicago, August 1977.
Originally published in the *Journal of the American Statistical Association* 74 (1979), 628–51.

equilibrium, to name a few, that play an important role in linking statistical analyzes and economic theory.

Many statistical estimation, testing, and prediction techniques have been developed for use in connection with many different kinds of econometric models, including linear and non-linear interdependent structural models, models involving qualitative and quantitative variables, models with time series complications, models for combined time series and cross-section data, and models with random parameters. This research on statistical techniques, and computer programs implementing them, a joint product of statisticians and econometricians, has been extremely important in the development of modern econometric modeling techniques.

Given this past record of solid achievement in the areas of measurement, concepts, and statistical techniques, it is relevant to ask how current statistical analyzes of econometric models can be improved so as to yield models with better forecasting and policy-analysis performance. To answer this question, I shall first try, in section 2, to summarize the main features of current or traditional econometric modeling techniques. Traditional econometric analyzes, like many statistical analyzes, tend to concentrate attention mainly on given models and not on procedures for discovering and repairing defects of proposed models. Section 3 describes an approach that emphasizes the latter aspect of econometric model construction and is a blend of traditional econometric techniques and modern time series techniques. While this approach, called structural econometric modeling time series analysis (SEMTSA), is not a panacea for all problems, it probably will be helpful in improving the quality of econometric models. A concluding section 4 considers prospects for the future.

2 The traditional econometric modeling approach

In this section, I shall attempt to characterize traditional econometric modeling techniques, to provide a summary of statistical procedures used in econometric modeling, and to describe some of the statistical needs of traditional econometric model builders.

2.1 Overview of the traditional approach

The schematic diagram in figure 2.1 represents, in broad outline, the activities of many econometric modelers. Whatever the problem, there is usually a statement of objectives, although, at times, the statement may not be so clear-cut and specific as could be desired. Sometimes, objectives are so ambitious that, given our present knowledge, data, and techniques,

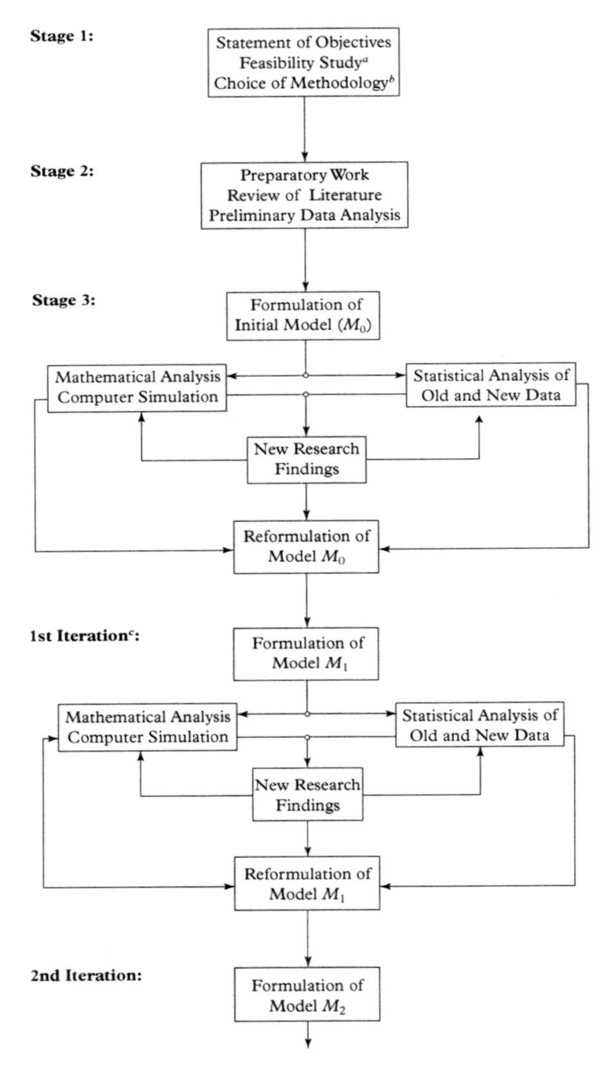

Figure 2.1 Econometric modeling

Note: For further discussion of this approach to modeling, see Hamilton *et al.* (1969) and Zellner (1970).

[a] It is assumed that this study shows the project to be feasible.

[b] It is assumed that a modeling approach is selected.

[c] The iterative procedure may disclose problems in the original formulation of goals, feasibility, and methodology so that refining and reformulation of the effort may not be confined solely to the model itself. Also, it is possible that other feedback loops, not shown in the figure, may be important in the process of converging on a satisfactory variant of a model.

they may be practically unattainable. The next steps in traditional econometric modeling involve a review of the theoretical and empirical literature bearing on the objectives of a modeling project, preparation of a data base, and preliminary data analysis. The objective of these activities is the formulation of an initial variant of an econometric model. Unfortunately, most econometrics and statistics texts are woefully silent on the basic methodology of how to formulate an initial variant of a model. General prescriptions, such as "use relevant economic theory" and "formulate as simple a model as possible," are valuable guidelines. Often the relevant economic theory does not yield precise information regarding functional forms of relationships, lag structures, and other elements involved in a stochastic specification of a model. Further, model simplicity has yet to be defined in a completely satisfactory manner. Still, it is worthwhile to emphasize the importance of using elements of economic theory, other outside information, and simplicity in formulating an initial variant of a model. For example, models that imply unexploited profit opportunities probably will be unsatisfactory because exploitation of such profit opportunities will generally upset properties of the proposed model that contains them.

Once an initial variant of a model, denoted by M_0, has been formulated, it is traditionally subjected to a number of mathematical, statistical, computer simulation, and judgmental checks. These include simple mathematical checks on the number of equations and number of endogenous variables, consistency of variables' units of measurement, conditions for parameter identification, and compatibility with results from mathematical economic theory. Computer simulation experiments are often employed to gain information about local and global dynamic and other properties of M_0. Statistical checks involve formal hypothesis testing procedures, forecasting tests, residual analysis, data evaluation, and other diagnostic checks. In evaluating the adequacy of M_0, a good deal of judgment or prior information is employed, usually informally. For example, the algebraic signs and magnitudes of parameter estimates are reviewed to ascertain whether they are compatible with results provided by economic theory, by previous studies, and by judgmental information.

If M_0 is found to be inadequate in certain respects, work is undertaken to reformulate M_0 and to produce a new variant of the model, M_1. Then M_1 is subjected to the battery of checks mentioned previously. This process of checks and reformulation continues, using as much new data as possible, until a satisfactory version of the model is obtained, satisfactory in the senses that it passes diagnostic checks satisfactorily and accomplishes the objectives of the model-building project.

In connection with realizing the objectives of a model-building project, it is useful to have formulated as simple a model as possible. If the objectives require the model builder to capture much detail, the model probably will be large, but with care in model building it can still be sophisticatedly simple. Large and simple models seem preferable to large and complicated models. In fact, a very disturbing feature of some large, complicated models in the literature is that it is not known whether they have a unique solution or many solutions.

In the past, model builders have used some or sometimes all the elements of the approach described above, but generally have not been vigorous enough in applying the various checks. Mathematical analyzes have often been superficial and incomplete. Simulation experiments have not been very extensive or well designed in general. Statistical checks on the quality of data and on specifying assumptions have not been pursued vigorously enough. The relationship of models' properties to relevant economic theory has not been examined thoroughly in a number of instances. Finally, many econometric model builders have not stressed simplicity enough. Some currently operating econometric models are highly complex systems of a hundred or more non-linear stochastic difference equations with hundreds of parameters that have to be estimated from highly aggregated time series data. Failure to take account of Ockham's Razor, the Jeffreys–Wrinch Simplicity Postulate, and the Principle of Parsimony in formulating econometric models has had very serious consequences in much traditional econometric model building. See Jeffreys (1957, 1967) for evidence of the importance of simplicity in science.

These criticisms of traditional econometric models have to be tempered, however, because many methodological techniques needed in a sensible model-building process are not yet available. Good formal sequential testing procedures for model construction remain to be developed. Even for a given structural econometric model, exact finite-sample tests and optimal finite-sample estimates and predictors have not been available. Good or optimal designs for simulation experiments remain to be derived. The problems of missing and imperfect data have not been completely solved. Tried and tested economic theory dealing with stochastic markets, dynamic reactions, and a number of other important issues has not been available. Thus econometric model building has been a mixture of economic and statistical theory and empirical practice. It is probable that such interaction between theory and practice will produce improvements in both.

To illustrate elements of recent statistical practice in traditional econometric model building, I next review some estimation, testing, and prediction techniques and provide some indications of current developments and open problems.

2.2 *Statistical estimation problems*

Learning the values of parameters appearing in structural econometric models (SEMs) is important in checking the implications of alternative economic theories and in using SEMs for prediction, forecasting, and policy-making. Thus, econometric research has placed a heavy emphasis on statistical estimation problems.

2.2.1 Asymptotically justified estimation procedures Research since the 1940s has resulted in a greatly enhanced understanding of estimation problems associated with SEMs and a relatively large number of operational procedures for obtaining consistent, asymptotically normally distributed, and efficient parameter estimates for some or all parameters of linear and non-linear, static, and dynamic SEMs with serially uncorrelated or serially correlated errors. These procedures, which are discussed at length in econometric textbooks and the econometric literature, include maximum likelihood, two- and three-stage least squares, $\mathcal{K}$-class, double $\mathcal{K}$-class, instrumental-variable, non-linear maximum likelihood, non-linear two- and three-stage least squares, and other procedures. Further, many of the parameter estimates produced by such procedures approximate Bayesian posterior means of parameters in large samples. A most important result of this research, aside from providing asymptotically justified estimation procedures, has been to rule out a number of proposed inconsistent and/or asymptotically inefficient estimation procedures. For example, it is well known by now that misapplication of the classical least squares (CLS)[1] estimation procedure to estimate structural parameters produces inconsistent estimates except in the very special case of a fully recursive SEM.

Choice among alternative asymptotically justified estimates has often been made on the basis of ease of computation. For example, with systems linear in the parameters, calculation of two- and three-stage least squares estimates involves just simple algebraic operations, whereas computation of maximum likelihood estimates involves more complex numerical procedures. Some current computer packages compute a number of asymptotically justified estimates and leave the difficult choice among them to the user. Of course, in truly large samples, asymptotically equivalent estimates should not be very far different. If in practice such estimates, based on a given large sample of data, are radically different, this may be interpreted as indicating that the asymptotic properties of different estimates

[1] Some use the term *ordinary least squares* (OLS); I prefer *classical least squares* (CLS), since the least squares principle is not ordinary.

take hold at different sample sizes or, more likely, that specification errors are present and affect alternative estimates differently. Unfortunately, not much analysis is available on the sensitivity of alternate asymptotically justified estimates to various kinds of specification errors; one . . . paper in this area is Hale, Mariano, and Ramage (1978). More systematic analysis of this range of problems and production of asymptotically justified estimates that are relatively robust to specification errors would be welcome and would serve as a useful additional guide to users in selecting estimates when the sample size is truly large. On the other hand, if the sample size is not truly large, even if a SEM is correctly specified, various asymptotically justified estimates of the same parameter can assume quite different values.

Students and others invariably ask for a definition of what constitutes a truly large sample. An easy answer to this question is hard to give. The sample size alone is not usually all that is relevant. Values of the parameters and features of input or exogenous variables also must be considered. Because parameter values usually are unknown and the object of estimation, prior information about them is needed before one can say with any confidence what is a truly large sample in connection with the estimation of a specific SEM. Needless to say, if the sample size is not truly large, the asymptotic justifications for estimation and other large-sample inference procedures become dubious. In a Bayesian context, one can compute the posterior distribution for a parameter and check to see that it is approximately normal with posterior mean equal to the maximum likelihood estimate and posterior variance equal to the relevant element of the inverse of the estimated Fisher information matrix. If so, large-sample conditions have been encountered. These considerations do not give a justification for using the large-sample normal approximation to the posterior distribution without computing the exact finite-sample posterior distribution.

2.2.2 Finite-sample problems and procedures Recognition that large-sample justifications for estimation procedures do not contain explicit information on how large a sample must be for them to hold and that practical workers often must deal with limited data has prompted considerable research on the finite-sample properties of estimation procedures. A good deal of research has been concentrated on obtaining analytically the exact finite-sample distributions of certain asymptotically justified estimators, for example, maximum likelihood (ML), two-stage least squares (2SLS), and other estimators for parameters in relatively simple models. This ingenious and difficult distributional work unfortunately has shown that the finite-sample distributions of estimators, derived in the main from an underlying non-central Wishart distribution, are rather complicated and involve a number of parameters with unknown values. The latter fact makes the application of these distributional results to concrete problems difficult. This research has shown that asymptotically equivalent estimators have very different finite-sample properties. For example, the (limited-information) ML estimator does not possess finite moments of any order, and in certain frequently encountered cases the 2SLS estimator does not possess a mean or higher moments. Also, certain asymptotically unbiased estimators can have serious finite-sample biases. Further, and perhaps surprising, under some conditions, the inconsistent CLS or OLS estimator has a smaller mean squared error (MSE) than consistent estimators that possess a finite-sample second moment. Of course, if an estimator fails to possess a second moment, it has infinite MSE and is clearly inadmissible. This is not to say that MSE is the only criterion for judging estimators, but it has received considerable attention in this area of research. As stated before, these results have been surprising to many, particularly those who narrowly emphasize unbiasedness, or minimum MSE, or minimum-variance unbiasedness as criteria for judging estimators or who uncritically accept asymptotic justifications. To illustrate that these criteria are inadequate even for the simple case in which a structural parameter θ is equal to the reciprocal of a reduced form regression coefficient, π, that is, $\theta = 1/\pi$, the ML and almost all other asymptotically justified estimation procedures would recommend estimating θ by $\hat{\theta} = 1/\hat{\pi}$, where $\hat{\pi}$ is the least squares estimator of the regression coefficient π. Because $\hat{\pi}$ is normally distributed, $\hat{\theta}$ is the reciprocal of a normally distributed variable and hence does not possess finite moments of any order. Thus $\hat{\theta}$ has infinite risk and is inadmissible relative to quadratic and many other loss functions.

In addition to exact distributional work on the finite-sample properties of asymptotically justified estimators, research has provided approximations to the moments of these estimators, surprisingly even sometimes when moments do not exist. As Anderson (1977) has pointed out, these moment expressions approximate moments of truncated Taylor or other series approximations to the estimators and not moments of the estimators. How important this distinction is remains to be seen. Further, very fruitful work that uses Edgeworth–Charlier series approximations to the moments and distributions of estimators has been reported by Sargan (1976).

Monte Carlo studies also have been employed in an effort to determine the finite-sample properties of alternative estimators (see Sowey 1973). Generally, these studies have been marred by an inadequate coverage of the high-dimensional parameter spaces associated with models, even simple two-equation supply and demand models that usually contain about ten or more parameters. Because risk functions of estimators usually intersect, failure to examine the entire parameter space can yield misleading and confusing results regarding the dominance of one estimator relative to another in terms of, for example, MSE. Thus, the results of Monte Carlo experiments that investigate the behavior of estimators over a limited number of points in the parameter space must be considered very cautiously. See Thornber (1967) for a valuable illustration of this point.

While much effort has been directed at determining the finite-sample properties of given, asymptotically justified estimators, relatively little work has been done on the problem of producing estimates that have a small-sample justification. Using approximate moment expressions, Nagar (1959) attempted to define an approximate minimal-MSE estimator for structural coefficients within the $\mathcal{K}$-class. Unfortunately, his "estimator" depends on parameters with unknown values that have to be estimated to operationalize his estimator. When these parameters are estimated, it appears that the "optimal" properties of his estimator are vitally affected. Nagar's work provides some evidence that use of a value of $\mathcal{K}$ less than one, the value that produces the 2SLS estimate, is probably better than the value of one. Analysis by Sawa (1972) provides the approximate MSE of a $\mathcal{K}$-class estimator for a structural parameter of a simple model and points in the same direction, namely, that finite-sample MSE usually is lower, and sometimes much lower, when a value of $\mathcal{K} < 1$ is employed. Sawa (1972) has also reported properties of estimators that are a linear combination of the 2SLS and the inconsistent CLS estimators. By appropriate choice of the weights, he has obtained approximately unbiased and approximate minimal-MSE estimators. These results do not appear relevant for cases in which the second moment of the 2SLS estimator does not exist, and the justification for considering a linear combination of a consistent and an inconsistent estimator is not apparent.

... Fuller (1977) has presented modified limited-information ML and modified fixed $\mathcal{K}$-class estimators that have finite moments. Restricting these modified estimators to have the same, but arbitrary, bias, he shows that to order T^{-2}, where T is the sample size, the modified ML estimator dominates in terms of approximate MSE.

In almost all the analytical finite-sample work on the sampling properties of estimates, problems with time series complications have not been analyzed, for example, estimates of parameters of models with lagged endogenous variables and/or serially correlated error terms. Relatively little effort has been devoted to obtaining good finite-sample estimates of error terms' covariance matrices ... [S]tatistical work by Perlman, Eaton, and others certainly seems relevant. It is highly probable that all, or almost all, the asymptotically justified estimators mentioned are inadmissible under quadratic loss and other loss functions over a wide range of conditions. This range of "Steinian" issues has received very little attention in connection with finite-sample work on structural parameter estimators' properties. The impact of pretesting on the finite-sample properties of the usual structural-coefficient estimators is relatively unexplored. For example, I have conjectured that the limited-information ML estimator's distribution subject to a favorable outcome of the rank test for identifiability will possess finite moments. In the simple case in which a structural parameter is the reciprocal of a reduced form regression coefficient, $\theta = 1/\pi$, it is easy to establish that the ML estimator $\hat{\theta} = 1/\hat{\pi}$, subject to the outcome of a t test that rejects $\pi = 0$, possesses finite moments, where $\hat{\pi}$ is the least squares (ML) estimate of π. Last and most basic, the relevance of sampling-theory criteria, such as unbiasedness, admissibility, and minimal MSE of estimators for the analysis of a given sample of data, has not been considered adequately in the econometric literature. Sampling properties of procedures seem relevant before we take the data in connection with design problems or in characterizing average properties of estimation procedures. The relevance of these average properties in analyzing a particular set of data is not clear. Further, as many, including Tiao and Box (1973), have emphasized, the computed value of an optimal point estimator can be a very bad representation of the information in a given set of data. Likelihood advocates emphasize the importance of studying properties of likelihood functions, while Bayesians emphasize the desirability of studying both likelihood functions and posterior distributions to understand the information content of a given sample for possible values of parameters of a model. For both likelihood advocates and Bayesians, a point estimate is just a summary measure that does not necessarily convey all or most of the information in a sample regarding parameters' probable values.

2.2.3 Bayesian estimation results [Since the mid-1960s], there has been a growing amount of research concerned with developing and applying the Bayesian approach to the problems of estimating values of parameters in SEMs and other econometric models, and elements of the Bayesian approach have appeared in econometric textbooks. As is well known, inferences about parameters' values, for example, elements of a parameter vector $\boldsymbol{\theta}$, are based on the posterior probability density function (pdf) for $\boldsymbol{\theta}$,

$$p(\boldsymbol{\theta} \mid D, I) = cp(\boldsymbol{\theta} \mid I)\ell(\boldsymbol{\theta} \mid D), \tag{2.1}$$

where $c = [\int p(\theta \mid I)\ell(\theta \mid D)d\theta]^{-1}$ is a normalizing constant, D denotes the data, I denotes the prior information, $p(\boldsymbol{\theta} \mid I)$ is the prior pdf, and

$\ell(\boldsymbol{\theta} \mid D)$ is the likelihood function. The following points are relevant particularly for analyses of SEMs:

1. The posterior pdf in (2.1) is an exact finite-sample pdf, and, hence, large-sample approximations, while sometimes convenient and useful, are, in principle, not needed. This statement applies to the analysis of all kinds of models, including static and dynamic SEMs.
2. Use of the prior pdf, $p(\boldsymbol{\theta} \mid I)$, enables an investigator to incorporate prior information in an analysis, as much or as little as he sees fit. Of course, if no sample information were available, as is the case in some low-income countries, prior information would be the only kind of information available. In connection with SEMs, prior information must be introduced in some form to identify structural parameters. In sampling-theory approaches, the identifying information has almost always been introduced as exact restrictions on parameter values (e.g. setting certain coefficients equal to zero, equivalent to using a degenerate or dogmatic prior pdf for these parameters in a Bayesian setting). Use of prior pdfs enables investigators to represent this required prior information more flexibly; see Drèze (1975) and Kadane (1975).
3. Use of Bayes' Theorem provides the complete posterior pdf for parameters of interest and not just a summary point estimate. If a point estimate is desired, however, it usually can be obtained readily. For example, for quadratic loss functions, it is well known that the mean of the posterior pdf, if it exists, is an optimal point estimate in the sense of minimizing posterior expected loss.
4. Generally, Bayesian estimates have very good sampling properties, because they minimize average risk when average risk is finite and are admissible.
5. In large samples under general conditions, the posterior pdf, $p(\boldsymbol{\theta} \mid D, I)$, assumes a normal form with mean vector equal to the ML estimate of $\boldsymbol{\theta}$ and covariance matrix equal to the inverse of the estimated Fisher information matrix. Thus, in large samples there is a dovetailing of Bayesian and sampling-theory numerical results; however, their interpretation is quite different.

I now turn from the general features of the Bayesian approach to a brief review of some Bayesian estimation results for the SEM. A representation of the linear (in the parameters) SEM is

$$\underset{n\times g}{Y}\,\underset{g\times g}{\Gamma} = \underset{n\times k}{X}\,\underset{k\times g}{B} + \underset{n\times g}{U}, \tag{2.2}$$

where Y is an $n \times g$ matrix of observations on g endogenous (or dependent) variables and X is an $n \times k$ matrix of observations on k predetermined variables, assumed of rank k. Predetermined variables include both exogenous (independent) and lagged endogenous variables. Γ is a

$g \times g$ structural parameter matrix, assumed non-singular, and B is a $k \times g$ matrix of structural parameters. U is an $n \times g$ matrix of disturbance or error terms. It will be assumed that the rows of U have been independently drawn from a g-dimensional normal distribution with zero mean vector and $g \times g$ positive definite symmetric (pds) covariance matrix Σ. Note that if $\Gamma = I_g$, the system in (2.2) is in the form of a multivariate regression model when X contains no lagged endogenous variables or in the form of a multivariate autoregressive system with input variables when X contains both exogenous (or independent) and lagged endogenous variables. In the special case $\Gamma = I_g$, analysis of (2.2) from the Bayesian point of view would proceed pretty much along multivariate regression lines if initial values for the lagged endogenous variables are taken as given (see Zellner 1971 and the references cited there).

The unrestricted reduced form (URF) system associated with the SEM in (2.2) is given by postmultiplying both sides of (2.2) by Γ^{-1} to yield:

$$Y = XB\Gamma^{-1} + U\Gamma^{-1} \tag{2.3}$$

or

$$\underset{n\times g}{Y} = \underset{n\times k}{X}\,\underset{k\times g}{\Pi} + \underset{n\times g}{V} \tag{2.4}$$

where

$$\Pi = B\Gamma^{-1} \quad \text{and} \quad V = U\Gamma^{-1}, \tag{2.5}$$

with the $k \times g$ matrix Π being the (URF) coefficient matrix and the $n \times g$ matrix V the URF disturbance or error matrix. The assumptions about the rows of U imply that the rows of V can be considered independently drawn from a g-dimensional normal distribution with zero mean vector and $g \times g$ pds covariance matrix $\Omega_g = (\Gamma^{-1})'\Sigma\Gamma^{-1}$.

Under the assumptions made earlier, the parameters Π and Ω_g of the URF system in (2.4) are identified and can be estimated by using Bayesian or non-Bayesian techniques whether or not the structural parameters in Γ, B, and Σ are identified. It has long been recognized that, under the assumptions made above, Γ, B, and Σ are not identified and that additional prior information must be added in order to identify these structural parameters. Identifying prior information can take various forms. Here we discuss only the case in which it involves restrictions that subsets of structural parameters assume zero values. In addition, it is necessary to adopt a normalization rule for elements of the Γ matrix. Here we let all diagonal elements of Γ be equal to 1. We shall write the system in (2.2) with identifying restrictions and normalization rule imposed as

$$Y\Gamma_r = XB_r + U. \tag{2.6}$$

Then the restricted reduced form system is given by

$$Y = XB_r\Gamma_r^{-1} + U\Gamma_r^{-1} \tag{2.7}$$

$$= X\Pi_r + V_r. \tag{2.8}$$

Obviously, the fundamental function of the restrictions is to reduce the number of free structural parameters and, by so doing, to provide a model in which the remaining free structural parameters in Γ_r and B_r are identified. Explicit statements of the conditions for identification of structural parameters are given in econometrics textbooks and other works. Since the free parameters in Γ_r, B_r, and Σ are identified, their number cannot exceed the number of parameters in the URF system in (2.4), namely, kg parameters in Π and $g(g+1)/2$ distinct parameters in the $g \times g$ RF covariance matrix Ω_g.

The likelihood function for the restricted structural system in (2.6) is

$$\ell(\Gamma_r, B_r, \Sigma \mid D) \propto \{\text{mod}\,|\Gamma_r|\}^n |\Sigma|^{-n/2} \cdot \exp\left\{-\tfrac{1}{2}\text{tr}\,(Y\Gamma_r - XB_r)'(Y\Gamma_r - XB_r)\Sigma^{-1}\right\} \tag{2.9}$$

where $\propto$ denotes proportionality, D denotes the data, and mod $|\Gamma_r|$ denotes the absolute value of the Jacobian determinant, $|\Gamma_r|$, for the transformation from the n rows of U to the n rows of Y in (2.6). If the system in (2.6) is autoregressive, (2.9) is the likelihood function conditional upon initial values (assumed given). Then, from (2.1), the posterior pdf for the free parameters in Γ_r, B_r, and Σ is given by

$$p(\Gamma_r, B_r, \Sigma \mid D, I) \propto p(\Gamma_r, B_r, \Sigma \mid I)\ell(\Gamma_r, B_r, \Sigma \mid D), \tag{2.10}$$

where $p(\Gamma_r, B_r, \Sigma \mid I)$ is the prior distribution and the prior information is denoted by I. Given a prior distribution and the likelihood function, the technical problems of analyzing properties of the posterior distribution, that is, obtaining its normalizing constant, its marginal distributions, and its moments, remain.

In the special case of a fully recursive SEM, Γ_r is in triangular form, implying that $|\Gamma_r| = 1$, and Σ is assumed to have a diagonal form. These assumptions simplify the likelihood function in (2.9) considerably and also simplify the analysis of the posterior pdf in (2.10) (see Zellner 1971 for details). The fully recursive case, however, is a very special case of a SEM. In the general case, work has concentrated on the analysis of (2.10) using the likelihood function in (2.9). In several studies, the posterior distribution in (2.10) has been computed for a few simple models. Drèze and Morales (1976), Harkema (1971), Morales (1971), and Richard (1973) have analyzed it by using several different informative prior distributions. Except for some approximate results provided by Zellner (1971) and

Zellner and Vandaele (1975), it is necessary to use numerical integration techniques to analyze features of the posterior distribution. These approximate results have been used by Mehta and Swamy (1978) to provide a ridgelike Bayesian estimate. Kloek and van Dijk (1976) have studied the application of Monte Carlo numerical-integration techniques in analyzing posterior distributions. Although more experience with applications, assessing and using various prior distributions, and computational procedures would be useful, past research has yielded results that will be valuable in obtaining better analyzes of given SEMs, particularly in small-sample cases.

Bayesian research also has focused on limited-information analyzes, that is, estimation of parameters of a single equation or of a subset of equations of a SEM. Complete posterior distributions for these problems have been obtained and analyzed by Drèze (1972, 1976), Morales (1971), Reynolds (1977), Rothenberg (1975), and Zellner (1971). A single equation of the system in (2.6), say the first, is given by

$$\underset{n\times 1}{\mathbf{y}_1} = \underset{n\times m_1}{Y_1}\ \underset{m_1\times 1}{\gamma_1} + \underset{n\times k_1}{X_1}\ \underset{k_1\times 1}{\beta_1} + \underset{n\times 1}{\mathbf{u}_1}, \tag{2.11}$$

where $\mathbf{y}_1$ and Y_1 are components of Y, that is, $Y = (y_1 \vdots Y_1 \vdots Y_0)$ with the variables in Y_0 not appearing in (2.11), X_1 is a submatrix of X, $X = (X_1 \vdots X_0)$ with the variables in X_0 not appearing in (2.11), and $\mathbf{u}_1$ is a subvector of U, $U = (u_1 \vdots U_0)$ and γ_1 and β_1 are parameter vectors to be estimated. The assumptions introduced about the rows of U imply that the elements of $\mathbf{u}_1$ have been independently drawn from a normal distribution with zero mean and variance σ_{11}. The URF equations for $\mathbf{y}_1$ and Y_1, a subset of the equations in (2.4), are

$$(\mathbf{y}_1 \vdots Y_1) = X(\pi_1 \vdots \Pi_1) + (\mathbf{v}_1 \vdots V_1). \tag{2.12}$$

On postmultiplying both sides of (2.12) by $(1 \vdots -\gamma'_1)'$ and comparing the result with (2.11), we achieve compatibility, given that

$$\pi_1 = \Pi_1\gamma_1 + \begin{pmatrix}\beta_1\\ \mathbf{0}\end{pmatrix}, \tag{2.13}$$

where the zero vector on the r.h.s. of (2.13) is $(k - k_1) \times 1$, *and* $u_1 = v_1 - V_1\gamma_1$. From (2.12) and (2.13), the estimation problem can be viewed as a restricted multivariate regression problem with the structural parameters γ_1 and β_1 involved in the restrictions on the elements of π_1 and Π_1. A necessary condition for the identification of γ_1, β_1, and σ_{11} is that $k - k_1 \geq m_1$. Note that (2.13) reflects restrictions arising from just the first equation (2.11) of a system. The information in restrictions similar to (2.13) associated with other structural equations is not taken into account in estimating γ_1, β_1, and σ_{11} and hence the nomenclature, *limited-information* or *single-equation* analysis.

Previous research has shown that ML, 2SLS, and 3SLS estimates are approximate means of posterior pdfs for structural parameters under special conditions. Of course, given complete posterior pdfs for parameters, optimal point estimates can be computed, that is, posterior means for quadratic loss functions and medians for absolute-error loss functions. A particularly simple optimal point estimate under a generalized quadratic loss function can be derived as follows. Upon multiplying both sides of (2.13) on the left by $X = (X_1 \vdots X_0)$, we obtain

$$X\pi_1 = X\Pi_1\gamma_1 + X_1\beta_1 = \bar{Z}_1\delta_1, \tag{2.14}$$

where $\bar{Z}_1 = (X\Pi_1 \vdots X_1)$ and $\delta'_1 = (\gamma'_1 \vdots \beta'_1)$. Take as loss function,

$$\begin{aligned} L &= (X\pi_1 - \bar{Z}_1\mathbf{d}_1)'(X\pi_1 - \bar{Z}_1\mathbf{d}_1) \\ &= (\delta_1 - \mathbf{d}_1)'\bar{Z}'_1\bar{Z}_1(\delta_1 - \mathbf{d}_1), \end{aligned} \tag{2.15}$$

a generalized quadratic loss function. Given a posterior pdf for the reduced form parameters π_1 and Π_1, the posterior expectation of L in the first line of (2.15) can be evaluated yielding $EL = \mathrm{E}\pi_1'\mathrm{X}'\pi_1 - 2\mathrm{d}_1' E\bar{Z}_1'X\pi_1 + \mathrm{d}_1' E\bar{Z}_1'\bar{Z}_1\mathrm{d}_1$, where E is the posterior expectation operator. Then the value of d_1 that minimizes expected loss, $\mathrm{d}_1{}^*$, termed a minimum-expected-loss (MELO) estimate, is given by Zellner (1978):

$$\mathrm{d}_1{}^* = (E\bar{Z}_1'\bar{Z}_1)^{-1}E\bar{Z}_1'X\pi_1. \tag{2.16}$$

When the system in (2.12) is analyzed under a diffuse prior for the regression coefficients π_1 and Π_1 and for the error covariance matrix, the marginal posterior pdf for $(\pi_1 \vdots \Pi_1)$ is in the matrix Student t form, and, hence, the expectations in (2.16) are readily available. In this case, Zellner (1978) has shown that $\mathrm{d}_1{}^*$ is in the form of a K-class estimate with a value of k that depends on the sample size and is less than one in finite samples. Also, $\mathrm{d}_1{}^*$ possesses at least first and second sampling moments (Zellner and Park 1979). Further, the optimal estimate of γ_1 is a matrix-weighted average of the 2SLS and CLS estimates with the weight on the CLS estimate going to zero as the sample size increases (Zellner 1976). In small samples, however, the optimal estimate of γ_1 and β_1 can be very close to or exactly equal to the CLS estimate. Thus, empirical workers who have persisted in their use of CLS estimates may not be very far from an optimal estimate in small samples. Further, this averaging of 2SLS and CLS estimates bears some resemblance to the work of Sawa, mentioned previously; however, the weights that Sawa uses and those associated with (2.16) are different. Last, this point-estimation approach has been applied to yield a MELO estimate of parameters appearing in all equations of a system, that is, $\boldsymbol{\delta}_i, \boldsymbol{\delta}_2, \ldots, \boldsymbol{\delta}_g$, where $\boldsymbol{\delta}'_i = (\boldsymbol{\gamma}'_i \vdots \boldsymbol{\beta}'_i)$. Also see Mehta and Swamy (1978) for some useful Bayesian results for obtaining point estimates of the $\boldsymbol{\delta}_i$s that are related to ridge-regression results.

Some additional issues regarding the Bayesian approach have been aptly summarized in the following remarks by Tukey (1976):

> It is my impression that rather generally, not just in econometrics, it is considered decent to use judgment in choosing a functional form, but indecent to use judgment in choosing a coefficient. If judgment about important things is quite all right, why should it not be used for less important ones as well? Perhaps the real purpose of Bayesian techniques is to let us do the indecent thing while modestly concealed behind a formal apparatus. If so, this would not be a precedent. When Fisher introduced the formalities of the analysis of variance in the early 1920s, its most important function was to conceal the fact that the data was being adjusted for block means, an important step forward which if openly visible would have been considered by too many wiseacres of the time to be "cooking the data." If so, let us hope the day will soon come when the role of "decent concealment" can be freely admitted.

2.3 *Hypothesis testing and SEMs*

Sampling-theory procedures used for testing hypotheses relating to structural coefficients' values have in the main been large-sample procedures, usually large-sample likelihood ratio tests or large-sample tests based on the Wald criterion, that is, for testing the rank condition for identifiability, over-identifying restrictions on structural parameters, and general linear hypotheses regarding structural parameters' values . . . [R]esearch (Berndt and Savin 1975; Savin 1976) has emphasized that asymptotically equivalent testing procedures can produce conflicting results when used in finite-sample situations with a given nominal significance level. Analysis of asymptotic power functions by Morgan and Vandaele (1974) has demonstrated that certain ad hoc testing procedures are dominated by standard large-sample testing procedures. Also, finite-sample approximations to the sampling distribution of the likelihood-ratio test statistic have received little, if any, attention in the econometric literature.

In special cases, the exact finite-sample distribution of a test statistic is available. In one such case the null hypothesis specifies the values of all coefficients of endogenous variables in an equation, for example, $\gamma_1 = \gamma_1^0$ and $\beta_{1i} = 0$ in $\mathbf{y}_1 = Y_1\gamma_1 + X_1\beta_1 + \mathbf{u}_1$. Conditional on $\gamma_1 = \gamma_1^0$, it is seen that $\mathbf{y}_1 - Y_1\gamma_1^0 = X_1\beta_1 + \mathbf{u}_1$ is in the form of a multiple regression given that X_1 does not contain lagged endogenous variables. In this special situation, test statistics that have exact t or F distributions are available; however, the requirement that the null hypothesis specify values for all elements of γ_1 is quite restrictive. Also, when the system is dynamic, that is, when X_1 contains lagged endogenous variables, only approximate large-sample test procedures are available. The quality of the approximation and finite-sample power functions for widely used large-sample approximate tests are relatively unexplored topics in econometric research.

Another topic that has received very little attention in econometric research is the effects of pretests on the properties of subsequent tests and on estimators' and predictors' properties. That pretesting can vitally affect properties of estimators is evident from consideration of simple cases, for example, $y_{1t} = \gamma y_{2t} + u_{1t}$ and $y_{2t} = \pi_2 x_t + u_{2t}$. The RF equations for this simple system are $y_{1t} = \pi_1 x_t + v_{1t}$ and $y_{2t} = \pi_2 x_t + v_{2t}$ with $\gamma = \pi_1/\pi_2$. The ML estimator for γ is $\hat{\gamma} = \hat{\pi}_1/\hat{\pi}_2$, where $\hat{\pi}_i = \Sigma x_t y_{it}/\Sigma x_t^2$, $i = 1, 2$. $\hat{\gamma}$ does not possess finite moments; however, the distribution of $\hat{\gamma}$ subject to the outcome of a pretest that rejects $\pi_2 = 0$, namely, $|\hat{\pi}_2| > c s_{\hat{\pi}_2} > 0$, where c is a critical value and $s_{\hat{\pi}_2}$ is the standard error associated with $\hat{\pi}_2$, does possess finite moments.

Work on Bayesian posterior odds ratios for selected hypotheses relating to SEMs' parameters' values is reported in Reynolds (1977). The posterior odds ratio, K_{12}, for two mutually exclusive hypotheses, H_1 and H_2, is given by

$$K_{12} = 0_{12} \times \int p_1(\boldsymbol{\theta})\ell_1(\boldsymbol{\theta} \mid \mathbf{y})d\boldsymbol{\theta} \Big/ \int p_2(\boldsymbol{\theta})\ell_2(\boldsymbol{\theta} \mid \mathbf{y})d\boldsymbol{\theta}, \qquad (2.17)$$

where 0_{12} is the prior odds ratio, and for $i = 1, 2$, $p_i(\boldsymbol{\theta})$ is the prior pdf, and $\ell_i(\boldsymbol{\theta} \mid \mathrm{y})$ is the likelihood function. If H_1 and H_2 are mutually exclusive and exhaustive, and H_1 is $\theta = \theta^0$, while H_2 is $\theta \neq \theta^0$, a pretest estimate that is optimal relative to quadratic loss is given by

$$\begin{aligned} \hat{\boldsymbol{\theta}} &= p_1\boldsymbol{\theta}^0 + p_2\hat{\boldsymbol{\theta}}_2 \\ &= \boldsymbol{\theta}^0 + (\hat{\boldsymbol{\theta}}_2 - \boldsymbol{\theta}^0)/(K_{12} + 1), \end{aligned} \qquad (2.18)$$

where $K_{12} = p_1/p_2 = p_1/(1 - p_1)$, where p_1 and p_2 are the posterior probabilities on H_1 and H_2, respectively, and $\hat{\boldsymbol{\theta}}_2$ is the posterior mean for $\boldsymbol{\theta}$ under H_2. The Bayesian pretest estimate, which also can be computed for other kinds of hypotheses, is a neat solution to the pretesting problem as it relates to estimation. Similar considerations apply in obtaining combined, optimal predictions from two or more alternative models.

2.4 *Prediction procedures for SEMs*

Several alternative methods for generating predictions from SEMs have been discussed in the literature. First, it has been recognized that the URF system $Y = X\Pi + V$ can be fitted by least squares and used to generate unrestricted reduced form predictions (URFPs). Such predictions will not generally be efficient because restrictions on structural coefficients that imply restrictions on the elements of Π are not reflected in URFPs. Second, from the restricted SEM, $Y\Gamma_r = XB_r + U$, we can obtain the restricted reduced form system $Y = XB_r\Gamma_r^{-1} + V_r$, and restricted reduced form predictions (RRFPs) can be obtained from $\hat{\mathbf{y}}'_f = \mathbf{x}'_f\hat{B}_r\hat{\Gamma}_r^{-1}$, where $\mathbf{x}'_f$ is a given vector and $\hat{B}_r$ and $\hat{\Gamma}_r$ are estimated restricted structural-coefficient estimates. Such predictions will be asymptotically efficient if $\hat{B}_r$ and $\hat{\Gamma}_r$ are asymptotically efficient estimates and if, of course, there are no specification errors. If B_r and Γ_r are estimated by inefficient but consistent methods, it is not always the case that a predictor based on them will be better in large samples than an URFP (Dhrymes 1973). Last, the partially restricted reduced form (PRRF) equations can be used to generate predictions, namely, $y_i = X\Pi_i\gamma_i + X_i\beta_1 + v_i, i = 1, 2, \ldots, g$. Estimates of Π_i, γ_i, and β_i along with given vectors $\mathbf{x}'_f$ and $\mathbf{x}'_{if}$ yield the PRRFPs $\hat{y}_i = \mathbf{x}'_f\bar{\Pi}_i\bar{\gamma}_i + \mathbf{x}'_{if}\bar{\beta}_i, i = 1, 2, \ldots, g$. Since the PRRFPs use more prior information than the URFPs in over-identified SEMs, they will have higher precision in large samples than URFPs. On the other hand, they will not generally be as precise as RRFPs in large samples when no specification errors are present in the SEM. Approximate expressions for the variance-covariance matrix of forecast error vectors are available in the literature for the prediction procedures mentioned previously. Further, it is apparent that specification errors can vitally affect relative large-sample properties of these predictors. Then, too, only limited attention has been given to the problems of predicting future values of the exogenous variables in X.

It has been pointed out in the literature that the RRF predictor, $\hat{\mathbf{y}}'_f = \mathbf{x}'_f\hat{B}_r\hat{\Gamma}_r^{-1}$, will not in general possess finite moments, whereas the other predictors mentioned will have finite moments in general for the URF predictor and in most situations for the PRRF predictor (Knight 1977). More thorough analyzes of alternative predictors' finite-sample properties would be most valuable; see Schmidt (1977) for Monte Carlo experimental evidence that led him to conclude that "The first main conclusion . . . is that inferences about forecasts are not terribly reliable, unless one's sample is fairly large" (p. 1004).

From the Bayesian point of view, the predictive probability function for the URF system, $Y = X\Pi + V$, is available. Its mean vector is an optimal point prediction relative to a quadratic loss function. Optimal multi-step predictions for the URF when it has autoregressive complications have been obtained by Chow (1973). Richard (1973) has studied predictive pdfs for the SEM and has applied some of his results that incorporate restrictions on structural coefficients in the analysis of small models. Further work to enlarge the range of prior pdfs used in these analyzes and to provide computer programs to perform calculations conveniently would be worthwhile.

Some other issues that arise in use of econometric models for forecasting are (a) procedures for using judgmental information and econometric models in making forecasts; (b) ways of combining forecasts from alternative models (Nelson 1972; Granger and Newbold 1977); (c) criteria for the evaluation of the accuracy of forecasts (Granger and Newbold 1973, 1977); (d) data quality and forecasting (Zellner 1958); and (e) seasonal adjustment and forecasting (Plosser 1976a, 1976b). Further, the relative forecasting performance of univariate auto-regressive-integrated-moving average (ARIMA) time series and econometric models has been the subject of much research (Leuthold *et al.* 1970; Cooper 1972; Nelson 1972; Christ 1975; also see section 3).

3 The SEMTSA approach

As mentioned before, much past econometric research has concentrated on the analysis of given models and yielded relatively little on formal methods for checking whether formulated models are consistent with information in sample data and for improving models. In addition, many time series aspects of econometric modeling have not been adequately treated. This is not to say that time series considerations were totally absent from econometric research, but rather that there was no systematic synthesis of econometric modeling and time series analysis.

Most important in stimulating some econometricians' interest in time series techniques was the good forecasting performance of simple, univariate time series models relative to that of large econometric models in the work of Cooper (1972) and Nelson (1972). Much earlier, Milton Friedman suggested that econometric models' forecasts be compared with those of simple, univariate "naive" models, a suggestion implemented by Christ (1951). The relatively good forecasting performance of simple univariate autoregressive (AR) or Box–Jenkins' ARIMA models surprised econometric model builders. In theory, a properly specified multi-equation econometric model should yield more precise forecasts than a univariate time series model, since the former incorporates much more sample and prior information. The reasonable conclusion, drawn by many from these forecasting studies, is that the econometric models considered in these studies probably contain serious specification errors (e.g. see Hickman 1972). For example, the econometric models may contain incorrect functional forms for relations, inappropriate lag structures, incorrect assumptions about the exogeneity of variables, incorrect assumptions about error terms' properties, and so forth. Because the relationship between econometric models and univariate ARIMA processes was not clearly understood, many econometricians considered simple time series models to be ad hoc, mechanical, alternative models. Further, it was not apparent how time series analysis could be used to improve properties of SEMs. These issues were taken up in an article by Zellner and Palm (1974) and have since been pursued in a number of other works such as Evans (1975, 1976, 1978); Palm (1976, 1977); Plosser (1976a, 1976b); Prothero and Wallis (1976), Trivedi (1975), Wallis (1976, 1977), and Zellner and Palm (1975).

This research on the SEMTSA approach has, first, emphasized that dynamic, linear (in the parameters) SEMs are a special case of multivariate or multiple time series processes, such as studied by Quenouille (1957) and others. Second, it has been shown that assuming variables to be exogenous places important restrictions on the parameters of a multiple time series process. Third, the transfer function (TF) equation system associated with a dynamic linear SEM has been derived and shown to be strongly restricted by structural assumptions. While the TF equation system had appeared in the econometric literature earlier under other names, its role in econometric model building had not been emphasized. Fourth, in the case of random exogenous variables generated by a multiple time series process, it is possible to derive the final equations (FEs), associated with the SEM, and individual FEs are in the form of ARIMA processes of the type studied by Box and Jenkins (1970) and others. Thus, as emphasized in the SEMTSA approach, the Box–Jenkins ARIMA processes are not ad hoc, alternative (to SEMs), mechanical models but are, in fact, implied by SEMs (see the studies cited previously for explicit examples). In addition, assumptions about structural equations' properties have strong implications for the forms of FEs and TFs that can be tested.

To make some of these considerations explicit, a multiple time series process for a $p \times 1$ vector of random variables z_t (assumed mean-corrected for convenience) is represented as follows (Quenouille 1957):

$$\underset{p\times p}{H(L)}\ \underset{p\times 1}{\mathbf{z}_t} = \underset{p\times p}{F(L)}\ \underset{p\times 1}{\mathbf{e}_t} \qquad t = 1, 2, \ldots, T, \tag{3.1}$$

where $H(L)$ and $F(L)$ are finite order matrix polynomials (assumed of full rank) in the lag operator L, and $\mathbf{e}_t$ is a vector of serially uncorrelated errors with zero mean vector and identity covariance matrix. If, for example,

$F(L)$ is of degree zero in L, that is, $F(L) = F_0$, with F_0 of full rank, then the error vector in (3.1) is $F_0\mathbf{e}_t$ with zero mean and only a non-zero contemporaneous covariance matrix, $EF_0\mathbf{e}_t\mathbf{e}_t'F_0' = F_0F_0'$. Other specifications of $F(L)$ allow for moving average error terms. For stationarity $|H(L)| = 0$ must have all its roots outside the unit circle, while for invertibility the roots of $|F(L)| = 0$ must lie outside the unit circle.

Upon multiplying both sides of (3.1) by the adjoint matrix associated with $H(L)$, denoted by $H^*(L)$, we obtain

$$|H(L)|\mathbf{z}_t = H^*(L)F(L)\mathbf{e}_t, \tag{3.2}$$

a set of FEs for the elements of $\mathbf{z}_t$. Each of the FEs in (3.2) is in autoregressive moving average (ARMA) form that is, $|H(L)|z_{it} = \alpha_i' e_t$, where $|H(L)|$ is an autoregressive polynomial, and α'_i, a $1 \times p$ vector of polynomial operators, is the *i*th row of $H^*(L)F(L)$. That $\alpha'_i\mathbf{e}_t$, a sum of moving average processes, can be represented as a moving average process in a single random variable has been proved in the literature. Thus even with the general multiple time series process in (3.1), processes on individual variables will be in the Box–Jenkins form.

In structural econometric modeling it is usually assumed that some of the variables in $\mathbf{z}_t$ are exogenous. Let $\mathbf{z}'_t = (\mathbf{y}'_t \vdots \mathbf{x}_t')$, where $\mathbf{y}_t$, a $p_1 \times 1$ vector, denotes the vector of endogenous variables and $\mathbf{x}_t$, a $p_2 \times 1$ vector, denotes the exogenous variables. Then (3.1) can be written as

$$\left(\begin{array}{c|c} H_{11} & H_{12} \\ \hline H_{21} & H_{22} \end{array}\right)\left(\begin{array}{c} \mathbf{y}_t \\ \hline \mathbf{x}_t \end{array}\right) = \left(\begin{array}{c|c} F_{11} & F_{12} \\ \hline F_{21} & F_{22} \end{array}\right)\left(\begin{array}{c} \mathbf{e}_{1t} \\ \hline \mathbf{e}_{2t} \end{array}\right), \tag{3.3}$$

where the partitioning of $H(L) = \{H_{ij}\}$, $F(L) = \{F_{ij}\}$, and $\mathbf{e}_t$ has been made to conform to that for $\mathbf{z}'_t = (\mathbf{y}'_t \vdots \mathrm{x}'_t)$. The assumption that $\mathbf{x}_t$ is exogenous places the following restrictions on the matrix lag operators in (3.3):

$$H_{21} \equiv 0, \quad F_{12} \equiv 0, \quad \text{and} \quad F_{21} \equiv 0. \tag{3.4}$$

On inserting (3.4) in (3.3), we have

$$H_{11}\mathbf{y}_t + H_{12}\mathbf{x}_t = F_{11}\mathbf{e}_{1t}, \tag{3.5}$$

and

$$H_{22}\mathbf{x}_t = F_{22}\mathbf{e}_t. \tag{3.6}$$

The equation system in (3.5) is the dynamic structural-equation system, while that in (3.6) is the multivariate ARMA process generating the exogenous variables in $\mathbf{x}_t$.

By multiplying both sides of (3.5) on the left by the adjoint matrix associated with H_{11}, denoted by $H_{11}{}^*$, we obtain the TF system,

$$|H_{11}|\,\mathbf{y}_t + H_{11}{}^* H_{12}\mathbf{x}_t = H_{11}{}^* F_{11}\mathbf{e}_{1t}. \tag{3.7}$$

Last, the FEs associated with (3.5)–(3.6) are obtained by multiplying both sides of (3.6) on the left by $H_{22}{}^*$, the adjoint matrix associated with H_{22}, to obtain

$$|H_{22}|\,\mathbf{x}_t = H_{22}{}^* F_{22}\mathbf{e}_{2t}, \tag{3.8}$$

and substituting for $\mathbf{x}_t$ in (3.7) from (3.8) to yield

$$|H_{11}|\,|H_{22}|\,\mathbf{y}_t = -H_{11}^* H_{12} H_{22}^* F_{22}\mathbf{e}_{2t} + |H_{22}|\, H_{11}^* F_{11}\mathbf{e}_{1t}. \tag{3.9}$$

Equations (3.8) and (3.9) are the FEs for the variables in $\mathbf{x}_t$ and $\mathbf{y}_t$, respectively. Each variable has an ARMA process, as mentioned before. Simple modifications of the analysis presented previously to take account of non-stochastic exogenous variables, such as time trends, and seasonal or other "dummy" variables, can easily be made.

In structural econometric modeling in the past, workers have concentrated attention on the SEM given in (3.5). Economic and other considerations have been employed to justify the classification of variables into the two categories, endogenous and exogenous. Further special assumptions regarding the matrices H_{11}, H_{12}, and F_{11} are required to achieve identification (e.g. see Hannan 1971). These assumptions place restrictions on lag patterns in equations, serial correlation properties of error terms, and on which variables appear with non-zero coefficients in equations of the system. If the resultant system is appropriately specified and estimated, it of course can be used for forecasting, control, and structural analysis, the traditional objectives of SEMs. It must be recognized, however, that a large number of specifying assumptions have to be made to implement the SEM in (3.5), and the probability that errors will be made in specifying an initial variant of (3.5) generally will be high. The solution to this problem is not to discard the initial variant of (3.5), which may contain much valuable information, but to pursue complementary analyses that can help to identify problems in the formulation of the initial variant and to suggest appropriate reformulation of specifying assumptions. Also, it is important that these complementary analyses yield useful results along the way toward obtaining a good SEM.

In the SEMTSA approach, it is suggested that workers use economic theory and other outside information to formulate an initial, tentative form for (3.5). The next step involves deducing algebraically the forms of the TF system in (3.7) and the FEs in (3.9). As is obvious from the forms of the TF and FE systems, assumptions regarding the SEM in (3.5) will result in a number of important restrictions on TFs and FEs that can be checked empirically. For example, from (3.7), (3.8), and (3.9), it is seen that the AR parts of the FEs and TFs will be identical when lag operators do not contain common factors. As pointed out in Zellner and Palm (1974), systems with special features, that is, fully recursive systems or systems in which H_{11} is block-diagonal, will lead to cancellation, and thus the AR parts of FEs and TFs will not be identical. Also, other special assumptions about the forms of H_{11} and H_{12} in (3.5) will result in TFs and FEs with different AR lag polynomials (see Zellner and Palm 1975 for an example). Work on examining the implications of specific SEMs for the forms of FEs and TFs is extremely important in enhancing understanding of SEMs. For example, the effect of changing a variable's classification from exogenous to endogenous on the forms of the TFs and FEs can be easily determined. Also, structural assumptions about lag structures, properties of structural error terms, and forms of policy-makers' control policies all result in strong restrictions on TFs and FEs. In addition, Quenouille (1957, ch. 5) has provided valuable analysis of the effects of incorrect inclusion or exclusion of variables, measurement errors, parameters varying with time, non-linearities, and so on.

When the forms of TFs and FEs associated with a SEM have been derived, the next step in the SEMTSA approach is to analyze data to determine or identify the forms of FEs and TFs to check that the empirically determined FEs and TFs are compatible with those implied by the tentatively formulated SEM. Of course, this work not only provides checks on a SEM but also estimates FEs that can be used for prediction and TFs that can be used for prediction and control. If the analysis of the FEs and TFs provides results compatible with the implications of the SEM, the SEMs' parameters can be estimated, and it can be used for prediction, control, and structural analysis. If, as is usually the case, the results of FE and TF analysis do not confirm the implications of an initial variant of a SEM, the SEM must be reformulated. This reformulation process is facilitated considerably by knowing the results of TF and FE analyses. That is, the latter analyses usually indicate specific deficiencies of the initial variant of a SEM, and many times recognition of these deficiencies is an important first step in finding remedies for them. When the initial variant of a SEM has been reformulated, its implications for the forms of FEs and TFs can be checked empirically. Also, the roots of FEs and TFs can be calculated, estimated, and examined for reasonableness.

The SEMTSA approach provides an operational and useful synthesis of traditional econometric and time series analysis techniques that can produce SEMs with fewer specification errors and better forecasting performance. As with traditional SEMs, however, some statistical problems associated with the SEMTSA approach require further research. First, there is the problem of determining the forms of the FEs from sample data. Box and Jenkins' well-known suggested techniques based on properties of estimated autocorrelation and partial autocorrelation functions are helpful in ruling out a number of forms for FEs; however, these techniques are rather informal. For nested FE models, large-sample likelihood-ratio tests can be employed to aid in discriminating among alternative FE models. For nested and non-nested models, Bayesian posterior odds ratios also are useful. For example, in discriminating between a white-noise process and a first order moving average process, Evans (1978) has shown that the posterior odds ratio is a function not only of the first order sample serial correlation coefficient, r_1, but also of higher order sample serial correlation coefficients, $r_2, r_3, \ldots$, the latter having weights that decline as the sample size increases. Because r_1 is not a sufficient statistic and because the r_is are highly correlated in small samples, a large-sample test using just r_1 does not use all the sample information and can lead to erroneous inferences. Posterior odds ratios also are useful in situations in which roots of AR polynomials lie on the unit circle, a situation in which it is known that usual large-sample likelihood-ratio tests based on χ^2 statistics are invalid. Geisel (1976) has reported work indicating that Bayesian posterior odds ratios performed better than variants of Box–Jenkins procedures in discriminating among alternative ARIMA schemes. Extensions of this work and the early work of Whittle (1951) on Bayesian hypothesis testing in time series analysis would be very valuable. This work also can shed light on the problem of determining the degree, if any, of differencing required to induce stationarity. Note that in formulating a posterior odds ratio, stationarity is not required. Stationarity is required for most uses of sample autocorrelation and partial autocorrelation functions.

Second, there is the problem of determining the forms of TFs. Important work on this problem for simple TFs has been reported by Box and Jenkins (1970, 1976), Haugh and Box (1977), Haugh (1972), Granger and Newbold (1977), and others. Also, the econometric work on distributed lag models is relevant (e.g. see Aigner 1971, Dhrymes 1971, Griliches 1967, and Nicholls, Pagan, and Terrell 1975) . . . [W]ork of Sims (1972, 1975), Skoog (1976), Pierce and Haugh (1977), Wu (1978), and others on tests for special recursive structures, along with procedures suggested by Box, Jenkins, Haugh, Granger, Newbold, and others, may be useful in checking the assumptions about input variables' properties. In TFs with several input variables, it may be advisable to reduce the number of free parameters by using some of the assumptions in the distributed-lag literature regarding coefficients of current and lagged input variables (e.g. see Shiller 1973). As with determining the forms of FEs, it is probable that posterior odds ratios will be found useful in discriminating among alternative forms for TFs and in obtaining posterior probabilities associated with alternative variants of TF's.

Third, there is the problem of obtaining good estimates of parameters in FEs, TFs, and SEMs. Currently, various asymptotically justified estimates are available, and some of these take account of random initial conditions and restrictions implied by the assumptions of stationarity and invertibility. The small-sample properties of these asymptotically justified estimates require much further investigation, a point also emphasized by Newbold (1976), who writes: "As regards estimation, I am not sure that uncritical use of maximum likelihood estimates is justified in small samples without some investigation of their sampling properties." As pointed out in section 2, ML estimators do not in general possess good finite-sample properties. These comments imply that more work to obtain good finite-sample estimates is required. Extensions of the valuable work of Box and Jenkins (1970, 1976), Newbold (1973), Tiao and Hillmer (1976), and others on Bayesian estimation of time series models seem to be possible and can provide additional good finite-sample estimation results.

Fourth, the problems associated with seasonality are important in formulating and analyzing SEMs and yet have received relatively little attention. Because seasonal variation accounts for a large fraction of the variation of many economic variables, a proper treatment of seasonality is critical. In much econometric work, seasonally adjusted variables are used with little or no attention to the procedures employed for seasonal adjustment and their possible effects on determination of lag structures and other features of SEMs. In the SEMTSA approach, Plosser (1976a, 1976b) and Wallis (1976) have provided valuable analyzes of seasonality in SEMs.

Fifth, the problem of measurement errors in economic time series requires much more attention. It is well known that a number of economic series are derived wholly or in part from sample surveys. Many statistical analyses of such data are based on the usually erroneous assumption of simple random sampling. Analyzes that take proper account of the designs of sample surveys, their sampling errors, and possible biases would be most welcome. Further work to consider SEMs subject to measurement error would also be valuable. For example, (3.5) could be formulated in terms of the true values of variables, $\mathbf{z}'_t = (\mathbf{y}'_t \vdots \mathbf{x}'_t)$. The measured values of variables $\mathbf{z}_t^{m'} = (\mathbf{y}_t^{m'}, \mathbf{x}_t^{m'})$ could be assumed given by $\mathbf{z}_t^m = R\mathbf{z}_t + \boldsymbol{\xi}_t$, where R is a matrix of coefficients reflecting systematic measurement errors and $\boldsymbol{\xi}_t$ is a vector of random measurement errors. In this form, the SEM becomes what engineers call a state-variable model. Perhaps use of results in the engineering literature would be useful in work with SEMs. Measurement problems are not insignificant: Initial and subsequently revised figures for GNP and other important quarterly economic series differ considerably, in some cases systematically, and provide different information regarding cyclical turning points (see Zellner 1958). Similar results have been obtained in current work with preliminary and revised figures for quarterly nominal GNP. Revisions in the preliminary estimates of quarterly GNP amounting to 5–10 billion dollars are common. For example, the first and subsequently revised figures (in billion current dollars) for GNP in the fourth quarter of 1954 are 361.0, 362.0, 367.1, 367.1, 367.7, and 373.4; for the fourth quarter of 1965, the preliminary and subsequently revised figures are 694.6, 697.2, 704.4, 708.4, 710.0, and 710.0. These figures illustrate an important measurement problem confronting econometric model builders and forecasters that has not been adequately treated in the literature.

Sixth, aggregation problems have received increased attention . . . Articles by Geweke (1976), Tiao and Wei (1976), Wei (1976), and Rose (1977) provide valuable results on temporal and other kinds of aggregation in the context of time series models. In work by Laub (1971, 1972), Peck (1973, 1974) and Levedahl (1976), attention has been focused on economic models for individual firms and consumers using panel data and the implications of these microanalyses for aggregate dividend, investment, and automobile expenditure functions. At the microlevel, discrete decisions, such as buy/not buy or change/don't change the dividend rate, are extremely important. Yet macroformulations of behavioral relationships that are incorporated in many SEMs do not properly take account of this discrete microbehavior and as a result are misspecified. Many estimated investment, dividend, and automobile expenditure functions that are based on partial adjustment models show long response lags that are spurious and are the result of aggregation over buyers and non-buyers or corporations that change and those that do not change the dividend rate in a particular quarter. Levedahl (1976) has shown analytically and empirically that the adjustment coefficient in a partial adjustment model for automobile expenditures is related to the proportion of consumers purchasing a car in a particular period. Because this proportion varies considerably over time, the adjustment coefficient is an unstable parameter, and models fitted under the assumption that it is stable have obvious problems in forecasting. These findings relating to defects of widely used partial adjustment equations have serious implications for SEMs that incorporate such equations. Further work on formulating macro-SEMs that takes better account of discrete elements in economic behavior seems very important in obtaining better models.

Seventh, forms of policy-makers' control may change and thus cause instability in lag parameters and other features of a model, a point emphasized by Lucas (1973). Analyzes using subsamples of data may indicate the empirical importance of this problem.

Last, time series analysts have identified relatively simple, low order ARIMA processes for economic variables appearing in SEMs. On the other hand, the ARIMA processes or FEs associated with most SEMs are usually complicated, high order schemes. To illustrate, Leuthold *et al.* (1970) formulated and estimated a SEM for analysis of hog markets with daily data. They also identified and fitted ARIMA processes for the daily price and quantity variables. Their time series model for price was found to be a simple random walk. As shown in Zellner (1974), the form of their SEM implies FEs for price and quantity with AR parts of at least third order, quite at variance with their random walk finding for the price variable. It seems that they forced a misspecified SEM on the data, one that involves the implicit assumption of unexploited profit opportunities in the hog markets. Indeed Muth (1961, p. 327, n. 11), in his pathbreaking paper on rational expectation models (i.e. models that do not imply unexploited profit opportunities), writes in connection with a general supply and demand model: "If the production and consumption flows are negligible compared with the speculative inventory level, the process [on price] approaches a random walk. This would apply to daily or weekly price movements of a commodity whose production lag is a year." Thus, economic theory provides some support for the empirical finding that daily hog prices follow a random walk and that the SEM for the hog markets is probably misspecified.

Similar considerations apply to the Hendry (1974) model of the UK analyzed in Prothero and Wallis (1976). The latter workers identified rather simple ARIMA processes for variables appearing in Hendry's SEM. The FEs associated with Hendry's estimated model have ninth order AR parts. Prothero and Wallis (1976, p. 483) apparently take this finding of a ninth order AR part of the FEs of Hendry's model seriously and attribute the relatively low orders of the empirically identified FEs to "relatively small coefficients of higher powers of L [the lag operator], which proved difficult to detect in our univariate analyzes." Also, they state that "the size of the available sample [forty-two quarterly observations] has clearly restricted our ability to detect subtle higher-order effects." Whether these subtle higher order effects are real or are results of specification errors present in the eight-equation Hendry model is a point that deserves further attention. In addition, the burgeoning literature on rational expectation economic models has important implications for the formulation and analysis of SEMs (for some examples, see

Ranson 1974, Evans 1975, Grossman 1975, Nelson 1975, Sargent and Wallace 1975, Wickens 1976, McCallum 1977, and Flood and Garber 1978).

4 Conclusions

This review of some of the research on SEMs has emphasized the following major points:

1. Substantial progress has been made in research on statistical methods for constructing, analyzing, and using econometric models.
2. There is a serious need for developing and vigorously applying additional statistical, mathematical, computer simulation, and economic diagnostic checks of properties of the SEMs.
3. For given SEMs, more work has to be done to develop and apply estimation, testing, and prediction procedures that have finite-sample justifications. In this connection, the present author and others believe that Bayesian procedures offer good solutions for many finite-sample problems.
4. More formal procedures for using prior information in the analysis of given SEMs are required, a problem area that can be approached most satisfactorily at present by use of the Bayesian approach.
5. Most serious is the need for formal, sequential statistical procedures for constructing SEMs.
6. The synthesis of traditional econometric model-building techniques and modern time series analysis techniques, called the SEMTSA approach previously, will probably lead to improved SEMs, a view of the present writer, Granger and Newbold (1975), and others.
7. Further use of existing economic theory, such as the theory of efficient markets (see Fama 1970 for a review of this theory), and rational expectations theory probably will yield better SEMs. Having SEMs consistent with elements of sound economic theory has long been emphasized in the econometric literature, and further attention to this point in current work with SEMs is critical.

So that this listing of research needs not be construed as misrepresenting the quality of current US macro-SEMs that are used to generate quarterly forecasts of important economic variables, such as GNP, unemployment, prices, and interest rates, it is relevant to consider Christ's (1975) thoughtful and relatively favorable review of the forecasting properties of such models. In the opening sentence of his article, he writes: "Econometric models of the U.S. economy have been developed to the point where forecasters who use them can forecast real and nominal GNP two or three quarters ahead with root mean square errors of less than one percent, and six quarters ahead with RMS errors of one to two percent." This rather optimistic summary statement fails to take account of the fact that population RMS errors have been estimated from rather small samples of forecast errors and hence are not very precise. A confidence interval at a reasonable level for the population RMS error would probably be rather broad. Also, the implication of a 1–2 percent error for nominal GNP that now exceeds 1.5 trillion (1,000 billion) dollars, is about a 15 to 30 billion dollar or greater error, which would be considered substantial by most analysts.

Further with respect to the very same models that yield RMS errors of 1 or 2 percent, Christ (1975) in the second paragraph of his article writes: "Though the models forecast well over horizons of four to six quarters, they disagree so strongly about the effects of important monetary and fiscal policies that they cannot be considered reliable guides to such policy effects, until it can be determined which of them are wrong in this respect and which (if any) are right." This statement clearly indicates that at least some, or perhaps all, of the models that Christ considered (Wharton; Data Resources Inc.; Bureau of Economic Analysis; St. Louis; Fair; Liu-Hwa; Hickman-Coen; and University of Michigan) probably contain serious specification errors.

Next, Christ (1975, p. 59) writes:

> In general, it appears that *subjectively adjusted* forecasts using *ex ante* exogenous values are better than the others. It is no surprise that subjective adjustment helps. It may surprise some that the use of actual exogenous values does not help, and sometimes hinders. But there is likely to be some interaction, in the sense that if a forecaster feels that the preliminary forecast turned out by his model is unreasonable, he may both adjust the model and change his *ex ante* forecast of the exogenous variables, in order to obtain a final forecast that he thinks is more reasonable (emphasis in the original).

Christ's conclusion that "*subjectively adjusted* forecasts . . . are better than the others" underlines the importance of using prior information carefully in preparing forecasts. His statement that use of the actual values, rather than the anticipated values, of exogenous variables "does not help" is indeed surprising. In this connection, it should be appreciated that the subjective adjustments often take the form of adjusting the values of intercept terms in equations of a model. Because equations often are formulated in terms of non-stationary variables and may be considered as local approximations, adjustments to intercept terms and slope coefficients will be needed when values of the variables move away from sample values. In such situations, thoughtful adjustment of intercept terms is a partial step in the direction of obtaining better results; but, because it is partial, there is no assurance that use of actual rather than anticipated values of exogenous variables will produce better results in general.

Last, from the information on models' forecast errors that Christ has assembled, it appears that SEMs for both nominal and real GNP outperform univariate ARIMA schemes for these variables in terms of estimated RMS errors. As pointed out before, many have recognized, implicitly or explicitly, that a correctly specified multi-equation SEM should, in theory, perform better in forecasting than a univariate ARIMA process. For example, Box and Jenkins (1976, p. 493) comment:

> If the question is whether a set of univariate [ARIMA] models of this kind which takes no account of relationships between the series describes a set of related time series better than the corresponding multivariate [econometric] model then predictably the answer must be "No." It is a sobering commentary on lack of expertise in the practical aspects of modeling that instances have occurred where well-built univariate models have done better than poorly built multivariate "econometric" ones.

In closing, it must be concluded from what has been presented and from Christ's remarks, that, while considerable progress has been made in work with SEMs, an econometric model as satisfactory as the Ford Model T has not as yet appeared.

BIBLIOGRAPHY

Aigner, D. J. (1971), "A compendium on estimation of the autoregressive moving average model from time series data," *International Economic Review* 12, 348–69

Anderson, T. W. (1977), "Asymptotic expansions of the distributions of estimates in simultaneous equations for alternative parameter sequences," *Econometrica* 45, 509–18

Berndt, E. and N. E. Savin (1975), "Conflict among criteria for testing hypotheses in the multivariate linear regression model," Discussion Paper 74–21 (rev.), Department of Economics, University of British Columbia

Box, G. E. P. and G. M. Jenkins (1970), *Time Series Analysis, Forecasting and Control* (San Francisco, Holden-Day) (rev. edn., 1976)

(1976), "Discussion of the paper by Dr. Prothero and Dr. Wallis," *Journal of the Royal Statistical Society*, Series A 139, 493–4

Chow, G. C. (1973), "Multiperiod predictions from stochastic difference equations by Bayesian methods," *Econometrica* 41, 109–18; reprinted in S. E. Fienberg and A. Zellner (eds.), *Studies in Bayesian Econometrics and Statistics in Honor of Leonard J. Savage* (Amsterdam, North-Holland 1975), 313–24

Christ, C. F. (1951), "A test of an econometric model for the United States, 1921–1947," in *Conference on Business Cycles*, New York, National Bureau of Economic Research, 35–107

(1975), "Judging the performance of econometric models of the US economy," *International Economic Review* 16, 54–74
Cooper, R. L. (1972), "The predictive performance of quarterly econometric models of the United States," in B. G. Hickman (ed.), *Econometric Models of Cyclical Behavior*, 2 (New York, Columbia University Press), 813–926
Dhrymes, P. J. (1971), *Distributed Lags: Problems of Estimation and Formulation* (San Francisco, Holden-Day)
(1973), "Restricted and unrestricted reduced forms: asymptotic distribution and relative efficiency," *Econometrica* 41, 119–34
Drèze, J. H. (1972), "Econometrics and decision theory," *Econometrica* 40, 1–17
(1975), "Bayesian theory of identification in simultaneous equation models," in S. E. Fienberg and A. Zellner (eds.), *Studies in Bayesian Econometrics and Statistics* (Amsterdam, North-Holland), 159–74
(1976), "Bayesian limited information analysis of the simultaneous equations model," *Econometrica* 44, 1045–76
Drèze, J. H. and J. A. Morales (1976), "Bayesian full information analysis of simultaneous equations," *Journal of the American Statistical Association* 71, 919–23
Evans, P. (1975), "Time series analysis of a macromodel of the US economy, 1880–1915," H. G. B. Alexander Research Foundation, Graduate School of Business, University of Chicago
(1976), "A time series test of the natural-rate hypothesis," H. G. B. Alexander Research Foundation, Graduate School of Business, University of Chicago
(1978), "Time-series analysis of the German hyperinflation," *International Economic Review* 19, 195–209
Fama, E. F. (1970), "Efficient capital markets: a review of theory and empirical work," *Journal of Finance* 25, 383–417
Flood, R. P. and P. M. Garber (1978), "An economic theory of monetary reform," Department of Economics, University of Virginia, unpublished manuscript
Fuller, W. A. (1977), "Some properties of a modification of the limited information estimator," *Econometrica* 45, 939–53
Geisel, M. S. (1976), "Box–Jenkins or Bayes?," Report of research to the Econometrics and Statistics Colloquium, University of Chicago
Geweke, J. (1976), "The temporal and sectoral aggregation of seasonally adjusted time series," Paper presented to the NBER–CENSUS Conference on Seasonal Analysis of Economic Time Series, September 1976; published in A. Zellner, *Seasonal Analysis of Economic Time Series*, Washington, DC: US Government Printing Office, 1978, 411–27
Granger, C. W. J. and P. Newbold (1973), "Some comments on the evaluation of economic forecasts," *Applied Economics* 5, 35–47
(1975), "The time series approach to econometric model building," Paper presented to the Seminar on New Methods in Business Cycle Research, Federal Reserve Bank of Minneapolis, November
(1977), *Forecasting Economic Time Series* (New York, Academic Press)
Griliches, Z. (1967), "Distributed lags – a survey," *Econometrica* 35, 16–49
Grossman, S. (1975), "Rational expectations and the econometric modeling of markets subject to uncertainty: a Bayesian approach," *Journal of Econometrics* 3, 255–72

Hale, C., R. S. Mariano, and J. G. Ramage (1978), "Finite sample analysis of misspecification in simultaneous equation models," Discussion Paper 357, Department of Economics, University of Pennsylvania

Hamilton, H. R., S. E. Goldstone, J. W. Milliman, A. L. Pugh III, E. R. Roberts, and A. Zellner (1969), *Systems Simulation for Regional Analysis: An Application to River-Basin Planning* (Cambridge, Mass., MIT Press)

Hannan, E. J. (1971), "The identification problem for multiple equation systems with moving average errors," *Econometrica* 39, 751–65

Harkema, R. (1971), *Simultaneous Equations: A Bayesian Approach* (Rotterdam, Rotterdam University Press)

Haugh, L. D. (1972), "The identification of time series interrelationships with special reference to dynamic regression," Department of Statistics, University of Wisconsin, Madison, unpublished doctoral dissertation

Haugh, L. D. and G. E. P. Box (1977), "Identification of dynamic regression (distributed lag) models connecting two time series," *Journal of the American Statistical Association* 72, 121–30

Hendry, D. F. (1974), "Stochastic specification in an aggregate demand model of the United Kingdom," *Econometrica* 42, 559–78

Hickman, B. F. (ed.) (1972), *Econometric Models of Cyclical Behavior*, 1 and 2 (New York, Columbia University Press)

Jeffreys, H. (1957), *Scientific Inference*, 2nd edn. (Cambridge, Cambridge University Press)

(1967), *Theory of Probability*, 3rd rev. edn. (London, Oxford University Press)

Kadane, J. B. (1975), "The role of identification in Bayesian theory," in S. E. Fienberg and A. Zellner (eds.), *Studies in Bayesian Econometrics and Statistics* (Amsterdam, North-Holland), 175–91

Kloek, T. and van Dijk, H. K. (1976), "Bayesian estimates of equation system parameters: an application of integration by Monte Carlo," Report 7622/E, Econometric Institute, Erasmus University, Rotterdam

Knight, J. L. (1977), "On the existence of moments of the partially restricted reduced-form estimators from a simultaneous-equation model," *Journal of Econometrics* 5, 315–21

Laub, P. M. (1971), "The dividend–earning relationship: a study of corporate quarterly panel data, 1947–65," Graduate School of Business, University of Chicago, unpublished doctoral dissertation

(1972), "Some aspects of the aggregation problem in the dividend–earning relationship," *Journal of the American Statistical Association* 67, 552–9

Leuthold, R. M., A. J. A. MacCormick, A. Schmitz, and D. G. Watts (1970), "Forecasting daily hog prices and quantities: a study of alternative forecasting techniques," *Journal of the American Statistical Association* 65, 90–107

Levedahl, J. W. (1976), "Predictive error of the stock adjustment model," H. G. B. Alexander Research Foundation, Graduate School of Business, University of Chicago

Lucas, R. E. (1973), "Econometric policy evaluation: a critique," Graduate School of Industrial Administration, Carnegie–Mellon University

McCallum, B. T. (1977), "Price-level stickiness and the feasibility of monetary stabilization policy with rational expectations," *Journal of Political Economy* 85, 627–34

Mehta, J. S. and P. A. V. B. Swamy (1978), "The existence of moments of some simple Bayes estimators of coefficients in a simultaneous equation model," *Journal of Econometrics* 7, 1–13
Morgan, A. and W. Vandaele (1974), "On testing hypotheses in simultaneous equation models," *Journal of Econometrics* 2, 55–66
Morales, J. A. (1971), *Bayesian Full Information Structural Analysis* (New York, Springer-Verlag)
Muth, J. F. (1961), "Rational expectations and the theory of price movements," *Econometrica* 29, 315–35
Nagar, A. L. (1959), "The bias and moment matrix of the general K-class estimators of the parameters in simultaneous equations and their small sample properties," *Econometrica* 27, 575–95
Nelson, C. R. (1972), "The prediction performance of the FRBMIT-Penn model of the US economy," *American Economic Review* 62, 902–17
(1975), "Rational expectations and the estimation of econometric models," *International Economic Review* 16, 555–61
Newbold, P. (1973), "Bayesian estimation of Box–Jenkins transfer function-noise models," *Journal of the Royal Statistical Society*, Series B 35, 323–36
(1976), "Discussion of the paper by Dr. Prothero and Dr. Wallis," *Journal of the Royal Statistical Society*, Series A 139, 490–1
Nicholls, D. F., A. R. Pagan, and R. D. Terrell (1975), "The estimation and use of models with moving average disturbance terms: a survey," *International Economic Review* 16, 113–34
Palm, F. C. (1976), "Testing the dynamic specification of an econometric model with an application to Belgian data," *European Economic Review* 8, 269–89
(1977), "On univariate time series methods and simultaneous equation econometric models," *Journal of Econometrics* 5, 379–88
Peck, S. C. (1973), "A test of alternative theories of investment using data from the electric utilities industry," Graduate School of Business, University of Chicago, unpublished doctoral dissertation
(1974), "Alternative investment models for firms in the electric utilities industry," *Bell Journal of Economics and Management Science* 5, 420–58
Pierce, D. A. and L. D. Haugh (1977), "Causality in temporal systems: characterizations and a survey," *Journal of Econometrics* 5, 265–94
Plosser, C. F. (1976a), "A time series analysis of seasonality in econometric models with application to a monetary model," Graduate School of Business, University of Chicago, unpublished doctoral dissertation
(1976b), "Time series analysis and seasonality in econometric models," Paper presented to the NBER–CENSUS Conference on Seasonal Analysis of Economic Time Series, September; published in A. Zellner (ed.), *Seasonal Analysis of Economic Time Series* (Washington, DC: US Government Printing Office 1978), 365–407; chapter 9 in this volume
Prothero, D. L. and K. F. Wallis (1976), "Modelling macroeconomic time series," *Journal of the Royal Statistical Society*, Series A 139, 468–86
Quenouille, M. H. (1957), *The Analysis of Multiple Time-Series* (New York, Hafner Publishing Co.)

Ranson, R. D. (1974), "Money, capital, and the stochastic nature of business fluctuations," Graduate School of Business, University of Chicago, unpublished doctoral dissertation

Reynolds, R. (1977), "Posterior odds for the hypothesis of independence between stochastic regressors and disturbances," H. G. B. Alexander Research Foundation, Graduate School of Business, University of Chicago

Richard, J. F. (1973), *Posterior and Predictive Densities for Simultaneous Equations Model* (New York, Springer-Verlag)

Rose, D. E. (1977), "Forecasting aggregates of independent ARIMA processes," *Journal of Econometrics* 5, 323–46

Rothenberg, T. (1975), "Bayesian analysis of simultaneous equations models," in S. E. Fienberg and A. Zellner (eds.), *Studies in Bayesian Econometrics and Statistics in Honor of Leonard J. Savage* (Amsterdam, North-Holland), 405–24

Sargan, J. D. (1976), "Econometric estimators and the Edgeworth approximation," *Econometrica* 44, 421–48

Sargent, T. J. and N. Wallace (1975), "Rational expectations, the optimal monetary instrument and the optimal money supply rule," *Journal of Political Economy* 83, 241–54

Savin, N. E. (1976), "Conflict among testing procedures in a linear regression model with autoregressive disturbances," *Econometrica* 44, 1303–15

Sawa, T. (1972), "Finite-sample properties of the K-class estimators," *Econometrica* 40, 653–80

(1973), "The mean square error of a combined estimator and numerical comparison with the TSLS estimator," *Journal of Econometrics* 1, 115–32

Schmidt, P. (1977), "Some small sample evidence on the distribution of dynamic simulation forecasts," *Econometrica* 45, 997–1005

Shiller, R. J. (1973), "A distributed lag estimator derived from smoothness priors," *Econometrica* 41, 775–88

Sims, C. A. (1972), "Money, income and causality," *American Economic Review* 62, 540–52

(1975), "Exogeneity and causal ordering in macroeconomic models," Paper presented to the Seminar on New Methods in Business Cycle Research, Federal Reserve Bank of Minneapolis, 13–14 November

Skoog, G. R. (1976), "Causality characterizations: bivariate, trivariate and multivariate propositions," Staff Report 14, Federal Reserve Bank of Minneapolis

Sowey, E. R. (1973), "A classified bibliography of Monte Carlo studies in econometrics," *Journal of Econometrics* 1, 377–95

Thornber, H. (1967), "Finite sample Monte Carlo studies: an autoregressive illustration," *Journal of the American Statistical Association* 62, 801–8

Tiao, G. C. and G. E. P. Box (1973), "Some comments on 'Bayes' Estimators," *The American Statistician* 27, 12–14; reprinted in S. E. Fienberg and A. Zellner (eds.), *Studies in Bayesian Econometrics and Statistics in Honor of Leonard J. Savage* (Amsterdam, North-Holland), 620–6

Tiao, G. C. and S. Hillmer (1976), "Seasonal adjustment: a Bayesian view," Paper presented to the 13th meeting of the NBER–NSF Seminar on Bayesian Inference in Econometrics, November

Tiao, G. C. and W. S. Wei (1976), "Effect of temporal aggregation on the dynamic relationship of two time series variables," *Biometrika* 63, 513–23

Trivedi, P. K. (1975), "Time series analysis versus structural models: a case study of Canadian manufacturing behavior," *International Economic Review* 16, 587–608

Tukey, J. W. (1976), "Discussion of Granger on seasonality," Paper presented at the NBER–CENSUS Conference on Seasonal Analysis of Economic Time Series, September; published in A. Zellner (ed.), *Seasonal Analysis of Economic Time Series* (Washington, DC, US Government Printing Office 1978), 50–3

Wallis, K. F. (1976), "Seasonal adjustment and multiple time series analysis," Paper presented to the NBER–CENSUS Conference on Seasonal Analysis of Economic Time Series, September; published in A. Zellner (ed.), *Seasonal Analysis of Economic Time Series* (Washington, DC, US Government Printing Office, 1978), 347–57

(1977), "Multiple time series analysis and the final form of econometric models," *Econometrica* 45, 1481–97

Wei, W. S. (1976), "Effects of temporal aggregation in seasonal time series models," Paper presented to the NBER–CENSUS Conference on Seasonal Analysis of Economic Time Series, September; published in A. Zellner (ed.), *Seasonal Analysis of Economic Time Series* (Washington, DC, US Government Printing Office), 433–44

Whittle, P. (1951), *Hypothesis Testing in Time Series Analysis* (Uppsala, Almqvist & Wiksells Boktryckeri AB)

Wickens, M. R. (1976), "Rational expectations and the efficient estimation of econometric models," Working Paper 35, Faculty of Economics, Australian National University

Wu, D. M. (1978), "Causality test and exogeneity test," Department of Economics, University of Kansas, unpublished manuscript

Zellner, A. (1958), "A statistical analysis of provisional estimates of gross national product and its components, of selected national income components, and of personal saving," *Journal of the American Statistical Association* 53, 54–65

(1970), "The care and feeding of econometric models," Selected Paper 35, Graduate School of Business, University of Chicago

(1971), *An Introduction to Bayesian Inference in Econometrics* (New York, John Wiley)

(1974), "Time series analysis and econometric model construction," Invited paper presented to the Conference on Applied Statistics, Dalhousie University, Halifax, Nova Scotia; published in R. Gupta (ed.), *Applied Statistics* (Amsterdam, North-Holland, 1975), 373–98

(1976), "A note on the relationship of minimum expected loss (MELO) and other structural coefficient estimates," H. G. B. Alexander Research Foundation, Graduate School of Business, University of Chicago, unpublished manuscript

(1978), "Estimation of functions of population means and regression coefficients including structural coefficients: a minimum expected loss (MELO) approach," *Journal of Econometrics* 8, 127–58

Zellner, A. and F. C. Palm (1972), "Time series analysis and simultaneous equation econometric models," Paper presented to the Olso Econometric Society Meeting; published in *Journal of Econometrics* 2, 1974, 17–54; chapter 1 in this volume

(1975), "Time series and structural analysis of monetary models of the US economy," *Sankhyā: The Indian Journal of Statistics*, Series C 37, 12–56; chapter 6 in this volume

Zellner, A. and S. B. Park (1979), "Minimum expected loss (MELO) estimators for functions of parameters and structural coefficients of econometric models," *Journal of the American Statistical Association* 74, 185–93

Zellner, A. and W. Vandaele (1975), "Bayes–Stein estimators for *k*-means, regression and simultaneous equation models," in S. E. Fienberg and A. Zellner (eds.), *Studies in Bayesian Econometrics and Statistics in Honor of Leonard J. Savage* (Amsterdam, North-Holland), 317–343

Comment (1979)

David A. Belsley and Edwin Kuh

Zellner's [chapter 2] marks a significant step in econometric writings, for it is one of the first pieces by a major econometrician that recognizes the full spectrum of the actual practice of econometric model building – including its trial-and-error (or iterative) aspects that are often viewed askance by econometric theorists. Implicit in Zellner's broad-gauged view is a realization that the classical statistical techniques, in their purist sense, provide a model-building methodology that is as constricting as it is beautiful. And it has been in an effort to make progress in the highly complex real world of quantitative economics that econometricians, against this formal statistical backdrop, have developed an informal, backroom set of procedures that have proved a necessary part of actual model building: Econometricians have long been classroom theorists and closet pragmatists. It takes courage to recognize openly that econometric model building is an iterative process. It takes courage because it gives recognition to a practice that is highly vulnerable to attack – and for good reasons. In particular, once one admits the legitimacy of modifying an errant hypothesis in light of its failure to account for the facts or for the investigator's implicit prior beliefs, then one must always fear, contrary to the classical philosophy, that the final hypothesis will be determined mainly by the data and become devoid of a rigorous interpretation. Although we all understand the rational import of such criticism, most of us, deep down, also realize

Originally published as an Invited paper; *Journal of the American Statistical Association* 74 (1979).

that enlightened and clever iteration in model building, combined with appropriate retesting with new information, can greatly defuse this criticism while helping to speed the process of applied econometric research – indeed often allowing it to occur at all. The problem comes in discovering means for making iterative model building enlightened and clever, and it is in part with such issues that Zellner is concerned.

Realizing that many informal econometric practices, practices that attain their effective legitimacy in the heat of battle rather than in the textbooks, are at variance with the various formal procedures, is it not better to recognize such practices, study them for what value they may have, and encourage research into means for correctly modifying and testing such models? We think so, and we feel that Zellner has provided a possible framework for such evolution. In his approach, we can see that such a growth in the discipline of econometrics can have its own theoretical foundations. Indeed, Zellner calls attention to many areas of frontier econometric research, in which theoretical development might provide the practicing econometrician with a more rigorous framework for iterative model-building techniques. These areas include Bayesian estimation, robust estimation, time-varying parametric structures, sequential hypothesis testing, and Zellner's structural econometric modeling time series analysis (SEMTSA), which joins traditional econometric structural modeling and the more mechanical application of the Box–Jenkins type of time series analysis. To this list we would like to add the diagnostic analysis of data, an analysis that is capable of assessing the suitability of the data for estimating specific models and for testing specific hypotheses. Such diagnostics are particularly useful in analyzing the nonexperimental data that typically arise in the social sciences. This is an important topic that Zellner does not discuss. We have little doubt that procedures for model building that draw properly on these topics, topics often considered peripheral to the accepted body of econometrics, could provide a theoretically sound iterative approach to model building that is more productive than a practice based solely on currently accepted theory.

Why does such a large gap exist between what econometricians learn in the classroom and what they practice? There are several reasons for this gap that could, in principle, be removed in a straightforward manner, which include a lack of appropriate software and a lack of adequate training, even in accepted theory, among some practitioners. Other reasons are less easily dealt with. Some modeling techniques, for example, even if well known and otherwise implementable, are simply considered too cumbersome or time-consuming. Proper construction of Bayesian priors affords a good example of this and, at present, constitutes one of the major practical drawbacks to the implementation of many Bayesian techniques. Another reason, perhaps the most important one, lies in the inability of econometrics, based on existing theory, to handle the complexities encountered in analyzing the real world – including the frailty of the human mind correctly to provide all important prior restrictions for a given model before "machine touches data." This last reason virtually ensures that models will, in fact, be built piecemeal. It is time for the econometrics profession to recognize this simple truth and to place various iterative procedures under its theoretical umbrella.

The status of econometrics today, as Zellner indicates, offers no grounds for complacency. It is true that the discipline of econometrics has come a long way [in the years since 1950] and even provides a body of knowledge that serves as a model for some other disciplines. Yet one may at times sense a large and increasing gap between the current orthodoxy of econometric theory and the research frontier in econometrics (and in related disciplines that have much to offer econometrics), a gap that is manifest in a widespread reluctance to consider novel procedures seriously. This is readily apparent when one considers the effect that the mention of terms such as the *Stein estimator*, *Kalman filtering*, or even *numerical analysis* can have. A similar effect could have been evoked only a few years ago by the mention of terms such as *Bayesian estimation*, *random coefficients*, or even *spectral analysis*, an effect that still exists to some extent today. There are obvious reasons why econometricians might resist the introduction of new techniques and new methodology. Econometricians can be justifiably proud of the techniques they have developed. Most have invested much effort in mastering their knowledge and skill, and it is only natural to resist new ideas, unless they can be proved to be substantially better. Also, new techniques often originate in other "seemingly unrelated" disciplines, usually with notational differences that are irksome. This apparently small problem has surely been one of the more important reasons why the efforts from various fields of engineering, numerical analysis, or data analysis, all of which have some useful things to teach us, have been slow to take hold.

We have little to add to what Zellner says on the potential roles and importance of the topics he mentions: Bayesian estimation, robust estimation, time-varying parameters, sequential testing, and SEMTSA. We would, however, like to add to the overall picture a few thoughts on the importance of the diagnostic analysis of data. Examination of the suitability of a particular data set for a specific econometric analysis has been widely ignored in actual economic practice. This happens, in part, because econometricians are effectively stuck with non-experimental data over which they typically have no control, and they therefore accept them without further question. In addition, little is known about data problems, such as errors in variables, collinearity, and outliers. What little is known presents such theoretical or practical headaches that an ostrich-like posture seems somehow justified. In fact, these problems with econometric data make an analysis of their suitability to a particular application even more imperative here than in other disciplines, such as in physics and in some branches of engineering, in which experiments constructed to be devoid of many problems are possible. Collinearity can clearly render a specific set of non-experimental data useless for many important tests of hypotheses based on the estimated coefficients of the collinear variates. Likewise, anomalous data points, arising perhaps through error or perhaps from a data structure not relevant to the model at hand, can dominate estimation or hypothesis testing, much to the detriment of correctly understanding the model being investigated. Therefore, reliable diagnostic techniques are required that allow the econometric practitioner – either before estimation or as a concomitant part of the estimation procedure – systematically to assess the suitability of the data for estimating the particular model or testing specific hypotheses. Initial efforts toward such diagnostics are given in Mosteller and Tukey (1977) and in Belsley, Kuh, and Welsch (1979), and, at least in the latter case, these diagnostics have indeed been presented by the authors as an initial stage in a more general iterative model-building scheme, such as that advocated by Zellner.

BIBLIOGRAPHY

Belsley, D. A., E. Kuh, and R. E. Welsch (1979), *Regression Diagnostics: Identifying Disparate Data and Sources of Collinearity* (New York, John Wiley)

Mosteller, F. and J. W. Tukey (1977), *Data Analysis and Regression* (Reading, Mass., Addison-Wesley)

Comment (1979)

Carl F. Christ

Both parts of Zellner's [chapter 2] are admirable. The first part is on the current state of the art of structural econometric modeling. The second is on the relation of that art to time series analysis *à la* Quenouille and Box–Jenkins. Zellner himself has made significant contributions to both these areas. In addition, he offers many thoughtful suggestions for future work.

The chapter's first main part, section 2, provides an excellent summary of the preconceptions and procedures involved in structural econometric

Originally published as an Invited paper; *Journal of the American Statistical Association* 74 (1979).

models and their specification, estimation, evaluation, revision, and use. The discussion is illustrated with the aid of the general linear-in-parameters additive-disturbances structural econometric model (SEM) displayed in (2.2) and its reduced form (2.3). The model will be useful to statisticians seeking an overview of econometrics, to those who seek to put specialized knowledge of econometrics into perspective, and to those seeking problems to solve.

An additional comment may be helpful regarding Zellner's entirely correct (but, perhaps, misleadingly brief) statement in subsection 2.2.2 that many consistent econometric estimators do not have finite means or variances and, therefore, have infinitive mean square errors (MSEs) and are inadmissible relative to quadratic loss functions. The discovery of this fact can have a traumatic effect on students. In subsections 2.2.2 and 2.3, Zellner offers one possible way to avoid it. He does not mention another way suitable for asymptotically normal estimators, which is to adopt a truncated quadratic loss function, thus: $L = (\hat{\theta} - \theta)^2$ when $|\hat{\theta}-\theta| \leqq k$, and $L = k^2$ otherwise, where L is the loss, θ and $\hat{\theta}$ are the parameter and its estimator, and k is some positive constant. This renders the expected loss finite and allows it to be approximated for large samples by the MSE of the limiting normal distribution that is associated with $\hat{\theta}$. The use of a bounded loss function is preferable to an unbounded one (such as the MSE) not only because of the foregoing but also because an unbounded objective function can lead to grossly unrealistic behavior prescriptions, as is well shown by the celebrated St. Petersburg Paradox.

The chapter's second main part, section 3, summarizes the relationship, elucidated in Zellner and Palm (1974), between SEMs and time series analysis (see (3.1)–(3.9)). Before the nature of this relationship was realized, it was possible to believe that time series analysis was a mindless technique for seeking empirical relations among observed variables, quite divorced from (and indeed the antithesis of) the use of economic theory to formulate SEMs. Now, however, it becomes clear (as Zellner points out) that the SEM in (2.2) or (2.6) is a special case of the Quenouille-type multiple time series model in (3.1) and that the maintained economic hypothesis specifying that certain variables are exogenous in the SEM has important testable implications for the parameters of the time series model. In my view, this is a neat and important discovery. Based on that, Zellner proposes a synthesis that he calls structural econometric modeling time series analysis (SEMTSA). The name is close enough to that of Senta, the heroine who, by her faithfulness, redeemed Wagner's Flying Dutchman from his fate of having to sail the oceans foreover, to inspire in me the fantasy that SEMSTA will, by its faithful application, redeem econometrics from its fate of being at sea forever for lack of suitable statistical testing. Zellner does not go that far, but he does believe (and I with him) that SEMSTA is a promising approach.

The final part of the chapter, section 4, . . . contains a summary and also some references to Christ (1975) concerning the accuracy of forecasts made with the aid of econometric models of the United States. Zellner gives a quotation from p. 59 of my article and then expresses surprise that the use of actual values of the exogenous variables (rather than *ex ante* forecasts thereof) does not help. I submit that the unquoted remainder of the quoted paragraph removes the surprise. Please read the quotation in Zellner's section 4, and then read the rest of the paragraph that is quoted here:

> This suggests that when unadjusted models are used, *actual* exogenous values should yield better forcasts than *ex ante* values. The two sets of forecasts from the Fair model bear this out. It also suggests that if *subjectively adjusted* models are used, *ex ante* exogenous values should yield better forecasts than actual values. A comparison of the EAF3 and EPF forecasts from the BEA model bears this out.

Zellner's chapter has an extensive bibliography. In sum, Zellner's chapter is excellent. The editors are to be congratulated for inviting him to contribute it.

BIBLIOGRAPHY

Christ, C. F. (1975), "Judging the performance of econometric models of the US economy," *International Economic Review* 16, 54–74

Zellner, A. and F. C. Palm (1974), "Time series and structural analysis of monetary models of the US economy," *Sankhyā: The Indian Journal of Statistics*, Series C 37, 12–56; chapter 6 in this volume

Comment (1979)

Peter M. Robinson

Zellner has treated us [in chapter 2] to an illuminating discussion of many of the tools available for the statistical analysis of some econometric models. I feel, however, that his remarks on the issue of finite-sample theory versus asymptotic theory need to be placed in perspective. Zellner reminds us that asymptotic properties may be inappropriate for use in

Research was supported by National Science Foundation Grant Soc. 78–05803.

Originally published as an Invited paper; *Journal of the American Statistical Association* 74 (1979).

statistical inferences based on the sample sizes often encountered in economics. He stresses the importance of examining finite-sample properties of asymptotically justifiable decision procedures and of developing procedures with desirable finite-sample properties. In commenting on the relevance of asymptotic theory and describing the progress that has been made in finite-sample theory, however, he does not mention that there often are considerable differences between the assumptions on which these theories rest. Some exact finite-sample properties (such as bias and risk) of relatively simple estimators can be obtained under mild conditions. Unfortunately, it is often difficult to investigate analytically the properties of more complicated estimators, let alone find their exact finite-sample distributions, without resorting to precise distributional assumptions. Moreover, to the extent that such properties can be obtained at all, the need for mathematical tractability sometimes presents us with little choice of parent distribution. Much of the finite-sample work Zellner describes assumes that the disturbances in the econometric model are independent, identically distributed (iid) normal random variables. Zellner explicitly makes this assumption in his discussion of the Bayesian analysis of the simultaneous equation model, in subsection 2.2.3, and, previously, takes it for granted when reporting some of the results in subsection 2.2.2. Therefore, application of the finite-sample theory might be accompanied by data analysis to assess the suitability of the normality assumption, and consideration might be also given to use of normality-inducing transformations.

Which conditions are sufficient for asymptotic properties – consistency, asymptotic normality, asymptotic efficiency – of various estimators (ML, 2SLS, 3SLS, Bayesian, etc.) for the simultaneous equations system? One needs some (often uncheckable) assumptions of the existence of certain limits involving the exogenous varibles, but not a precise description of the residuals. If these are martingale differences satisfying relatively mild moment and homogeneity assumptions, the estimators typically are consistent and have the same asymptotic normal distribution as they would have under the iid normal assumptions. Moreover, one can determine asymptotic properties under more dramatic departures from the classical assumptions. It often is possible to establish consistency and asymptotic normality (with a consistently estimable covariance matrix) when the residuals have an unknown serial correlation structure or are heteroskedastic. One also can construct estimators that are asymptotically efficient in these circumstances, using weighted least squares, autoregressive transformation, spectral methods, and so on, although these may require even larger sample sizes to justify using the asymptotics. Zellner describes . . . developments in finite-sample theory, but he tends to ignore those in asymptotic theory. I agree, of course, that application of asymptotics to small samples can produce misleading results and use of exact finite-sample theory, when appropriate, is desirable. If it is a question of deciding whether an asymptotic theory or a finite-sample theory provides the better approximation, I believe one should consider not only whether the sample size is "truly large" but also the different assumptions on the variables and any information on robustness to departures from these assumptions. My earlier remarks suggest that this choice may not be clear-cut. Another factor that might be considered is the relative ease of applying the central limit theorem compared with that of applying many finite-sample distributions.

For a number of econometric models, one has little or no power to choose between asymptotic and finite-sample properties or between asymptotically justified and finite-sample estimators. Zellner emphasizes closed form estimators of the linear structural system, involving endogenous variables with continuous distributions, and iid disturbances. He reports some progress in finite-sample work on time series models, but this seems limited, and the problems to be overcome in developing readily computable finite-sample estimators and properties for the range of models served by the current asymptotic theory seem great. (Note also that estimators of many time series models, based on an exact or approximate Gaussian likelihood and certain moment estimators, often have desirable asymptotic properties, even in the absence of Gaussianity.) For some econometric models, we have large-sample estimators and properties (some of which, admittedly, seem to rest on precise distributional assumptions), but little or nothing is known about finite-sample theory, such as the many regressions and structural systems that are non-linear in the parameters, the models containing discrete-valued or limited dependent variables, and the models for markets in disequilibrium. One can sympathize with Zellner's call for simplicity in model building without necessarily wishing to exclude all these latter models from consideration.

Section 3 of the chapter is concerned with Zellner's structural econometric modeling time series analysis (SEMTSA) method, which may well help to improve the quality of econometric models and forecasts. Although SEMTSA has a number of desirable features, I do not believe that a proposed econometric model should be too readily rejected on the basis of this type of analysis, because the introduction of the autoregressive moving average model (ARMA) representation ((3.6)) of the exogenous variables might produce misleading results. Of course, this specification is not involved in the usual estimates of the system $H_{11}\mathbf{y}_t + H_{12}\mathbf{x}_t = \mathbf{u}_t$ that is being judged. Zellner observes that his analysis also allows for non-stochastic $\mathbf{x}_t$, although it does not generally work if $\mathbf{x}_t$ is stochastic but not ARMA. It cannot always be taken for granted that an ARMA representation – at least one of manageably low order – will be sufficiently accurate to justify the ensuing analysis.

Zellner suggests that workers use prior knowledge to choose an initial specification of the "structural system" (3.5), but does not say how the remaining component of (3.1), the model (3.6) for $\mathbf{x}_t$, will arise. Presumably, univariate or multivariate ARMA identification procedures would be involved. The accuracy with which (3.6) is specified is clearly important in determining the form of the FE (3.8). Small differences in the ARMA orders of elements of $\mathbf{x}_t$ can correspond to large differences in the degrees of polynomials, such as $|H_{11}||H_{22}|$. Failure to confirm an initial transfer function of the form $-H_{11}^{-1}H_{12}$ might be due in part to poor testing power resulting from deviations from the ARMA specifications of $\mathbf{u}_t$ and $\mathbf{x}_t$. Zellner mentions that some high order ARMA coefficients in the FE may be so small as to make rejection of a low order model difficult. This notion seems plausible, because one can write $|H_{11}||H_{22}| = \Pi_j(1 - \theta_j L)$ and because $|\theta_j| < 1$ for all j, in a stationary model, the coefficients of high powers of L might be very small. Even when this is not the case, some methods for ARMA identification, such as those in which models are considered in increasing order of complexity, are capable of producing a model that incorrectly omits high order terms.

Comment (1979)

Thomas J. Rothenberg

Zellner has provided [in chapter 2] a thoughtful survey of some . . . research and unsolved problems in the statistical analysis of econometric models. He wisely restricts his attention to a limited class of structural equation models; to survey all the models used by economists would be an impossible task. Readers unfamiliar with the work of econometricians, however, should not think that multi-equation forecasting models exhaust the economist's repertoire. Single-equation models describing individual behavior, models of income and wealth distribution, unobserved-variable models, contingency-table analysis, Markov transition models, and so on, also are used widely. (Indeed, Zellner has been a major contributor to numerous areas of econometric theory that are not discussed in his chapter.) As indicated in Zellner's introductory comments, the

Originally published as an Invited paper; *Journal of the American Statistical Association* 74 (1979).

material surveyed is illustrative of the research and problems encountered by economists whenever they try to develop stochastic models of economic behavior.

In his survey, Zellner makes a number of critical comments on the state of current econometric modeling techniques and statistical procedures. I am in substantial agreement with his major points and would simply add the following remarks:

1. . . . [M]ost econometric inference procedures are justified on the basis of large-sample theory. The exact sampling distributions generally are very complicated and, hence, are approximated by asymptotic distributions based on the central limit theorem. Better approximations, usually based on the Edgeworth expansion, are generally available and have been developed for special cases in the work surveyed by Zellner. Additional . . . work by Anderson, Phillips, Sargan, and others treat more general cases. A large body of literature by mathematical statisticians also is relevant. Improved test statistics, which yield approximately the correct size, have been known since the late 1930s and are summarized by Bartlett (1954), Box (1949), and Lawley (1956). General methods for constructing estimators and tests that are optimal to a second order of approximation have been developed by Pfanzagl (1973), Ghosh and Subramanyam (1974), Efron (1975), and Pfanzagl and Wefelmeyer (1978a, 1978b).

   This research on improved statistical procedures based on second order asymptotic approximations has not yet been incorporated into the econometrics textbooks. Furthermore, the application of general statistical theorems to the specific models used by econometricians is still in progress. Improved approximations in time series models, for example, are still at an early stage of development. Zellner surely is correct in calling for more work in this area. But the foundations for a small-sample justification of inference procedures in econometrics now exist.

   In multi-parameter models, the algebra of deriving Bayesian posterior marginal densities and moments often is just as difficult as that of deriving exact sampling distributions. In practice, Bayesians are forced to assume unpleasantly simple conjugate prior distributions or to rely on large-sample approximations in order to obtain tractable posterior distributions. Of course, better approximations to posterior distributions can be found by using Edgeworth-type expansions. Thus, it seems that the development of better approximations to distribution functions will be important for both Bayesian and non-Bayesian econometricians.
2. Both Bayesians and sampling theorists assume a known likelihood function, typically one based on normally distributed errors. In

practice, we do not know the probability distribution for the errors. A solution, of course, is to embed our model in a larger one containing more parameters. Unfortunately, our ignorance is often so great that we are led to models with a large number of parameters, and the result is very imprecise measurement. The Bayesian approach that adds parameters with a fairly tight prior distribution is an attractive compromise. The classical theory of robust estimation is another alternative that needs to be extended to structural equation models.

3. The art of model building involves much trial and error, as indicated in Zellner's overview of the modeling process. It also involves the interaction of many different persons and many overlapping data sets. We base our models on all the past models we have read about, even if we only dimly remember them. It is extremely difficult to make this process formal. Indeed, one sometimes is led in despair to give up all hope of using formal statistical methods to describe our data-mining practices. A more optimistic position is taken in the pathbreaking monograph by Leamer (1978), in which sequential specification searches are modeled from a Bayesian point of view.
4. The synthesis of time series methods and structural equation modeling is an important . . . development and should prove helpful in improving our specification of dynamic relations. Of course, not all structural models involve time in an essential way. Much econometric work analyses cross-section data over a single time interval. Furthermore, if the major question to be answered involves the long-run effects of policy changes, it seems unlikely that improved specification of short-run dynamics will be of much help. But, as a tool for improving short-run forecasts, the SEMTSA approach holds much promise.

   One weakness of the approach should be pointed out. Any structural specification does indeed lead to an implication about the order of the autoregressive part of the final equation; however, tests of the order of an autoregressive process are likely to have very low power. Given any low order process, there exists a high order process (with very small coefficients, perhaps) that has approximately the same likelihood. The inability to reject the null hypothesis that the order is low need not be convincing evidence that the structural model is false. The specific numerical values of the final equation's coefficients implied by a given structure must be tested against the data.
5. With respect to the merits of the Bayesian approach, I have little to add to what I have said elsewhere (Rothenberg 1975a, 1975b). Unlike physical constants (like the weight of a coin, the length of a table, or the number of balls in an urn), the parameters of econometric models are subjective, hypothetical constructs. Likewise, the error terms in structural equation models represent a host of poorly understood effects,

due to various sorts of misspecification. It certainly is not unnatural to treat both the parameters and the errors symmetrically as random variables where the probability distributions represent the econometrician's uncertainty. For some purposes, I find it quite reasonable to treat some parameters and some error terms as having subjective prior probability distributions. I am not convinced that it is always reasonable. Ultimately, a Bayesian's credibility depends on his or her ability to justify the particular prior distribution used in the analysis. Most sensible statistical procedures are Bayes (or approximately Bayes) with respect to some prior distribution and some loss function. In reporting results of statistical analysis, perhaps the best we can do is to indicate the range of possible conclusions that arise from a reasonable set of priors. Again, Leamer (1978) has much of interest to say on this subject.

In addition to these remarks, some more general comments about the current state of econometrics are perhaps in order. I think it is useful to distinguish three different levels at which there are unsolved problems in econometrics. First, there are the very basic questions of methodology: What are the fundamental principles for modeling complex economic phenomena? Is traditional statistical theory at all relevant for econometric models? If so, is the Bayesian approach more satisfactory than sampling theory alternatives? Second, there are the technical and mathematical difficulties of deriving the statistical properties of estimators and test statistics, of calculating marginal and conditional distributions, and of constructing numerical algorithms for computing actual estimates and distributions. Third, there are the questions of art: For a given applied problem, what are reasonable assumptions to make about data, number of variables, number of equations, functional form, error distribution, prior distribution, and so forth?

Although the second class of problems is naturally of interest to statisticians and theoretical econometricians, it is important to emphasize that the basic methodological questions and the questions of art are critical. Indeed, one can argue that the technical and mathematical difficulties in econometrics are trivial when compared with the almost insurmountable problems in trying to capture an extraordinarily complex reality in manageable models. With perseverance and better statistical training, econometricians will produce the distribution theory and computer algorithms that are needed. There is much more doubt in my mind that in the near future we shall be able to produce economic models that are widely accepted. I do not find it surprising that macroeconometric models are unreliable guides to the effects of government policy. More surprising is the fact that they forecast as well as they appear to do. Econometrics is a field with much technique and as yet only modest accomplishment. Because many different models can explain the available data equally well, model formulation must be based on considerable intuition, institutional knowledge, and economic understanding. Unfortunately, these are very scarce commodities. We have not yet produced the econometric version of the Model T, and it is unlikely that better statistical theory alone will provide the technological breakthrough that is needed.

BIBLIOGRAPHY

Bartlett, M. S. (1954), "A note on the multiplying factors for various χ^2 approximations," *Journal of the Royal Statistical Society*, Series B 16, 296–8

Box, G. (1949), "A general distribution theory for class of likelihood criteria," *Biometrika* 36, 317–46

Efron, B. (1975), "Defining the curvature of a statistical problem (with applications to second-order efficiency)," *Annals of Statistics* 3, 1189–1242

Ghosh, J. K. and K. Subramanyam (1974), "Second-order efficiency of maximum likelihood estimators," *Sankhyā*, Series A 36, 325–58

Lawley, D. N. (1956), "A general method for approximating to the distribution of likelihood ratio criteria," *Biometrika* 43, 295–303

Leamer, E. (1978), *Specification Searches: Ad Hoc Inference With Nonexperimental Data* (New York, John Wiley)

Pfanzagl, J. (1973), "Asymptotically optimum estimation and test procedures," in J. Hajek (ed.), *Proceedings of the Prague Symposium on Asymptotic Statistics* I, 201–72

Pfanzagl, J. and W. Wefelmeyer (1978a), "An asymptotically complete class of tests," *Zeitschrift für Wahrscheinlichkeidtstheorie und verwandte Gebiete* 45, 49–72

(1978b), "A third-order optimum property of the maximum likelihood estimator," *Journal of Multivariate Analysis* 8, 1–29

Rothenberg, T. J. (1975a), "The Bayesian approach and alternatives in econometrics," in S. E. Fienberg and A. Zellner (eds.), *Studies in Bayesian Econometrics and Statistics in Honor of Leonard J. Savage* (Amsterdam, North-Holland), 55–67

(1975b), "Bayesian analysis of simultaneous equations models," in S. E. Fienberg and A. Zellner (eds.), *Studies in Bayesian Econometrics and Statistics in Honor of Leonard J. Savage* (Amsterdam, North-Holland), 405–24

Rejoinder (1979)

Arnold Zellner

I thank the discussants for their kind comments on my [chapter 2] and for the thoughtful points they have raised. I shall respond to the discussants in reverse alphabetical order, which I prefer for an obvious reason.

Originally published as an Invited paper; *Journal of the American Statistical Association* 74 (1979).

Rothenberg correctly points out that econometricians have analyzed many different problems by using a broad array of statistical models, some of which were mentioned briefly at the beginning of my chapter. Many of the statistical points made in the chapter are relevant for analyses of more complicated models, for example, non-linear models or models for discrete random variables, also mentioned by Robinson. With respect to improved second order asymptotic approximations, Takeuchi (1978) has reported on his and M. Akahira's work and on independent work by Pfanzagl and Wefelmeyer (1978) on the third order asymptotic efficiency of the maximum likelihood and Bayesian estimators with respect to unbounded loss functions (*pace* Christ, see below) and smooth prior distributions in the case of the Bayesian estimator. Takeuchi notes that the comparison of asymptotic efficiency is made possible only after estimators are adjusted for asymptotic bias or asymptotic median or modal bias and states (1978 p. 7) that "Before adjustments nothing can be said about the relative advantages of various estimators in terms of any 'loss' such as the mean square error, nor can we always gain in terms of MSE by adjusting for bias as was proved by Morimune [in the case of the single-equation LIML estimator]." This statement abstracts from the fact that Bayesian estimators relative to specific priors and loss functions have well-known finite-sample optimal sampling properties. Regarding improved sampling-theory testing procedures, Rothenberg has provided references relating to my remarks in section 2.3 on finite-sample approximations to distributions of test statistics. These references do not contribute to the many delicate and deep issues regarding alternative approaches for comparing and testing hypotheses. On the algebra of deriving Bayesian marginal posterior densities and the choice of prior distributions, I agree that good approximation procedures are useful and can extend the range of prior distributions that can be employed. Also, advances in numerical integration techniques have been, and will be, very useful. As Rothenberg and Robinson remark, everything in the world is not normal. It is thus fortunate that transformations that may induce normality can and have been used in econometrics. In addition, some analyses in the financial economic literature have been based on stable-Paretian, univariate Student t and mixture distributions, and analyses of multiple and multivariate regression models based on multivariate and matrix Student t distributions, respectively, have appeared (Tong 1976; Zellner 1976; Fraser and Ng 1978). On the SEMTSA approach, it seems relevant for cross-section data, because such data relate to time series processes for individuals in the cross-section. The time series aspect of cross-section data should not be overlooked and, in fact, is a primary reason for the growing emphasis on longitudinal or panel data for which time series considerations are extremely important. On testing the implications of a SEM, mentioned by both Rothenberg and Robinson, not only are the implications for the FEs testable but those for the TFs and structural equations are also obviously testable. On the ability to discriminate among alternative forms for the FEs, the hog market example discussed in my chapter is relevant. In the area of macroeconomics, in Ranson (1974) economic theory is developed that predicts random-walk or near-random-walk behavior for some macroeconomic variables, and time series analyses provide some support for his theory. Thus, in some cases economic theory and empirical analyses point in the direction of low order final equations. This topic needs much more research, as I suggested in my discussion of the Prothero–Wallis results. It is always possible to complicate a model by adding more parameters. The basic issue is whether addition of such parameters is useful or required in relation to the objectives of an analysis. On long-run versus short-run models, my experience (see Hamilton *et al.* 1969) is that the role of prior information is very important in long-run modeling and that the discussion relating to my figure (2.1) is quite relevant. Time series considerations also are relevant, for example, in deciding whether trends are stochastic and/or deterministic. On the issues of capturing an extraordinarily complex reality in manageable models, questions of art in modeling, physical constants and parameters in econometric models, and other philosophical issues, much could be said. At present, I shall just point to the early skeptics who refused to believe that there was any merit in applying mathematics and statistics in analyses of social science problems. I believe that subsequent experience has shown them to be wrong.

Robinson is right in pointing to different assumptions underlying asymptotic and finite-sample analyses. The critical point, in my opinion, is well expressed by Robinson, namely, "I agree, of course, that application of asymptotics to small samples can produce misleading results, and use of exact finite-sample theory, when appropriate, is desirable." Also, in agreement with Robinson, I emphasized the importance of diagnostic checking and robustness of statistical procedures. On the power to discriminate between the fruitfulness of finite-sample and asymptotic analyses, mathematical analyses and Monte Carlo experiments can be most helpful. Of course, in problems for which no finite-sample results are available, asymptotic approximations will have to be used, and Monte Carlo experimental results can be employed to check the quality of the asymptotic approximations. Use of the simplicity criterion in model building does not preclude, as Robinson apparently suggests, the use of a wide range of models. It does appear fruitful, however, to choose the simplest model compatible with achieving the objectives of a modeling project. On processes for the stochastic vector $\mathbf{x}_t$ in (3.3) and (3.6) of my chapter, these should be carefully identified from the data. If it is found that they are not ARMA processes, then, of course, one should not use ARMA processes. In such cases, one can determine the forms of these processes and use them, or one can go ahead conditionally on given values for the $\mathbf{x}_t$s and analyze the TF and structural equation systems. In forecasting, however, one has to generate future values of the $\mathbf{x}_t$ vector so that the problem of modeling the $\mathbf{x}_t$ process arises again. In a number of problems, for example, supply and demand models, an ARMA process for $\mathbf{x}_t$ may not be a bad approximation. On the other hand, when elements of $\mathbf{x}_t$ are subject to policy control, close attention must be paid to the nature of control processes.

Christ raises a very important point regarding the use of unbounded loss functions. In connection with existence of estimators' moments and associated infinite risks, in Zellner (1978, p. 154) I wrote in my conclusions, "ML and perhaps other estimators might perform better relative to performance criteria that are not sensitive to the existence or nonexistence of sampling moments," in part in response to a comment on my work made by George Barnard (1975) on the possible use of bounded loss functions. Use of bounded loss functions will affect the Gauss–Markov theorem, Stein's results, and use of posterior means as point estimates as well as any other estimates that are optimal relative to unbounded loss functions. Robustness of results to the form of the criterion or loss function is certainly an issue that deserves more research (see Zellner and Geisel 1968, Zellner 1973, and Varian 1975 for some work on this topic). Ideally, the form of the criterion or loss function should reflect serious subject-matter considerations.

Belsley and Kuh remark that "Proper construction of Bayesian priors . . . constitutes one of the major practical drawbacks to the implementation of many Bayesian techniques." This remark perhaps overlooks the well-known fact that many non-Bayesian estimation results can be produced by Bayesian methods based on diffuse priors or as means of conditional posterior distributions in which nuisance parameters are set equal to sample estimates. Thus, many non-Bayesian estimates in use can be given a Bayesian interpretation. Belsley and Kuh are right in saying that construction of operational, informative prior distributions is a difficult problem, akin to the problem of formulating good models for observations. Some . . . Bayesian research on formulating informative priors by Bernardo, Dickey, Kadane, Lindley, Novick, Press, Winkler, myself, and others may be helpful with respect to this problem. Last, I agree fully with Belsley and Kuh that "diagnostic analysis of data" is a crucial topic that deserves more attention in econometrics and elsewhere.

Clearly, what one gets out of an analysis depends on what one puts into it, and the data are an important input. In conclusion, I appreciate this opportunity to express my views on matters that appear to me to be critical for improving statistical analyses of econometric models. Among other useful results, improved statistical analyses can help to discover possible defects of current econometric models and prompt work to correct them.

BIBLIOGRAPHY

Barnard, G. (1975), "Comment," in "Report of the Tenth NBER–NSF Seminar on Bayesian Inference in Econometrics," University of Chicago, May 9–10

Fraser, D. A. S. and K. W. Ng (1978), "Multivariate regression: analysis with spherical error," University of Toronto, unpublished manuscript

Hamilton, H. R., S. E. Goldstone, J. W. Millman, A. L. Pugh III, E. R. Roberts, and A. Zellner (1969), *Systems Simulation for Regional Analysis: An Application to River-Basin Planning* (Cambridge, Mass., MIT Press)

Pfanzagl, J. and W. Wefelmeyer (1978), "A third order optimum property of the maximum likelihood estimator," *Journal of Multivariate Analysis* 8, 1–29

Ranson, R. D. (1974), "Money, capital, and the stochastic nature of business fluctuations," Graduate School of Business, University of Chicago, unpublished doctoral dissertation

Takeuchi, K. (1978), "Asymptotic higher order efficiency of ML estimators of parameters in linear simultaneous equations," Paper presented at the Kyoto Econometrics Seminar Meeting, University of Kyoto, June 27–30

Tong, K. (1976), "Bayesian analysis of multivariate regression with matrix-*t* error terms," H. G. B. Alexander Research Foundation, Graduate School of Business, University of Chicago, unpublished manuscript

Varian, H. R. (1975), "A Bayesian approach to real estate assessment," in S. E. Fienberg and A. Zellner (eds.), *Studies in Bayesian Econometrics and Statistics in Honor of Leonard J. Savage* (Amsterdam, North-Holland), 195–208

Zellner, A. (1973), "The quality of quantitative economic policy-making when targets and costs of change are misspecified," in W. Sellekart (ed.), *Selected Readings in Econometrics and Economic Theory: Essays in Honor of Jan Tinbergen* (London, Macmillan), 147–64

(1976), "Bayesian and non-Bayesian analysis of the regression model with multivariate student-*t* errors," *Journal of the American Statistical Association* 71, 400–5

(1978), "Estimation of functions of population means and regression coefficients including structural coefficients: a minimum expected loss (MELO) approach," *Journal of Econometrics*, 8, 127–158

Zellner, A. and M. S. Geisel (1968), "Sensitivity of control to uncertainty and the form of the criterion function," in D. G. Watts (ed.), *The Future of Statistics* (New York, Academic Press), 269–83

3 Structural econometric modeling and time series analysis: an integrated approach (1983)

Franz C. Palm

1 Introduction

An important and difficult part of econometric modeling is the specification of the model. Any applied econometrician knows how troublesome it can be to obtain a satisfactory specification of the model. While the problem of specification analysis has received increasing attention in econometric research in recent years, many of the existing econometric textbooks provide few guidelines on how to obtain a satisfactory specification. This is surprising as the specification of the model is necessary in order to justify the choice of an estimation or testing procedure among the large variety of existing procedures, the properties of which are well established given that the true model is known. The consequences of misspecification errors due to the exclusion of relevant explanatory variables are more extensively discussed in standard textbooks on econometrics. Misspecification tests such as the Durbin–Watson test belong to the tools of any empirical econometrician. Among the exceptions to what has been said about the treatment of specification analysis in textbooks, we should mention the book by Leamer (1978), in which he distinguishes six types of specification searches and presents solutions for each of them within a Bayesian framework. But the present state of econometric modeling leads us to stress once more Zellner's (1979, p. 640) conclusion concerning the research on structural econometric models (SEMs): "Most serious is the need for formal, sequential statistical procedures for constructing SEMs."

The computations reported in this chapter were carried out by David A. Kodde, I wish to thank him for his able assistance and his useful comments. I have benefitted from helpful discussions with and comments by Carl F. Christ, Jean-François Richard, and especially Christopher A. Sims and Arnold Zellner, and from useful remarks by the participants of the ASA–CENSUS–NBER Conference on Applied Time Series Analysis of Economic Data held in Washington and by members of the Econometrics Seminar at CORE, University of Louvain.

Originally published in A. Zellner (ed.), *Applied Time Series Analysis of Economic Data*, Proceedings of the Conference on Applied Time Series Analysis of Economic Data, October 13–15, 1981, Arlington, VA, Economic Research Report ER-5, Washington, DC, Bureau of the Census, US Department of Commerce, October 1983, 99–233.

In section 2 of this chapter, we review existing approaches to econometric modeling. We shall first briefly outline the traditional approach and the time series approach to dynamic econometric model building. Then we present the structural econometric modeling and time series analysis (SEMTSA, see Zellner 1979), which integrate the best features of econometric and time series techniques to analyze regression and structural equations in a framework of sequential testing of hypotheses.

In section 3, the SEMTSA will be applied to a multivariate dynamic model for seven Dutch quarterly macroeconomic variables for the period 1952–79. The initial model is an unrestricted vector autoregressive (VAR) model, which is assumed to be sufficiently general to include the data-generating process. Next, in a top-to-bottom approach, theoretically meaningful restrictions on the parameters of the VAR model are formulated and confronted with the information in the data. The dynamic properties of the restricted model are compared with those of the unrestricted model. Also, the postsample forecasting performance of the unrestricted and the restricted models is investigated. In this way, we try to iterate into a model that is a good parsimoniously parametrized approximation for the data-generating process.

In section 4, we shall draw some tentative conclusions concerning the application of the SEMTSA in general and the empirical results of our study in particular. We shall point to problems that remain to be solved.

The estimation and testing procedures used throughout this chapter are chosen on the basis of their large sample properties. Their finite sample properties are known for special models only.

2 Approaches to econometric model building

2.1 The traditional approach to econometric modeling

The methodology of traditional econometric modeling will be briefly outlined in this section. For a more detailed description and a schematic representation of model-building activities, the reader is referred to Zellner (1979).

Formally, one assumes that the model is given. The observations are used to estimate the parameters of the model. In practice, however, econometricians derive the model at least in part from the data. When specifying an initial model, the investigator makes use of economic theory, knowledge about institutional arrangements, and other subject-matter considerations. Sometimes a heavily – perhaps too strongly – restricted model is chosen as an initial model because the estimation of its parameters is straightforward.

The initial model is estimated using an estimation technique which is appropriate according to criteria such as unbiasedness, consistency, efficiency, . . . , provided the initial model is the true model. The estimation results of the model are judged on the basis of the algebraic t-values, the plausibility of the parameter estimates and their expected sign, the stability over time of the estimates, the serial correlation properties of the residuals, and the fit of the equations. When the initial model is not satisfactory as judged by one or more of these criteria, it is respecified and re-estimated. For instance, a significant Durbin–Watson test statistic has often led to fitting a regression model with first order autoregressive disturbances. Similarly, the finding that two-stage least squares [2SLS]estimates differ slightly from ordinary least squares [OLS]estimates is used as argument to ignore the simultaneity aspect. Certainly, in many situations the correct remedy has been applied to cure the model. However, as long as there is no systematic way to analyze the sample evidence, the diagnostic checking and reformulation of the initial model may be done quite differently by two independent investigators. That different final model specifications have been reported in the economic literature for similar data sets and observation periods is evidence for this statement.

The traditional approach to econometric modeling has certainly yielded very valuable results. These lines should not be interpreted as generally convicting econometricians of bad practice. Instead, we want to emphasize the need for a more systematic, formal approach to econometric modeling, in which the best elements of the traditional approach ought to be incorporated.

2.2 *Time series identification of dynamic econometric models*

Besides the progress made in modeling univariate time series during the [1970s], many contributions to formal modeling of regression equations, bivariate and multivariate models have been made by time series analyzes. (See Box and Jenkins 1970, chs. 10, 11: Granger and Newbold 1977: Haugh and Box 1977; and Jenkins and Alavi 1981 among many others.)

Similar to univariate ARIMA modeling, modern time series model building of vector processes consists of three stages: Identification, estimation and diagnostic checking. In contrast to the econometric approach, time series analysts explicitly rely on data to determine the model specification (or identification). More specifically, the time series analysis is directed towards finding a transformation of the data into a vector of innovations that are orthogonal to the lagged variables included in the model. Thereby, the aim of many time series analysts using a time domain approach is to find a parsimonious representation of the data-generating process.

Usually, the series to be modeled are made stationary and prewhitened. The cross-correlation function between the prewhitened series is used to check for the presence of feedback. When there is unidirectional Wiener–Granger "causality" (see Granger 1969) present, say from x to y only, the bivariate process for y and x can be modeled as a dynamic regression equation for y given x and a univariate (ARMA) model for the input x. The cross-correlation function for the prewhitened series ϵ_y and ϵ_x is used to determine the degree of distributed lag polynomials in the regression equation.

In vector time series models with feedback present, the autocorrelation and partial autocorrelation matrices are used to achieve a parsimonious parametrization of the model (see Jenkins and Alavi 1981, Tiao and Box 1981).

At this point, we shall make several comments on the time series approach to econometric model building:

(1) Usually the approach is applied to low dimensional vector processes. As most data in econometrics are non-experimental, dynamic econometric modeling has to account for the effects of the explanatory variables, which vary over the sample period. Therefore, there will usually be more than one explanatory variable included in an econometric equation, so that the specification of the lag structure using estimated cross-correlation functions becomes difficult, if not impossible in practice.

(2) The assumption that all the variables in the model are generated by a vector ARIMA process may be unrealistic. For instance, structural changes, which occur frequently in econometric models, can be modeled by expanding the set of explanatory variables, using dummy variables or products of explanatory variables and dummy variables. A structural change in the parameters of the ARMA process of x does not hamper the analysis of the regression function of y on x as long as the marginal process for x is of no direct interest in the analysis. Nevertheless, if one wants to transform the process of x into a white noise, the presence of a structural change in the process for x will complicate matters substantially. Special cases such as the effect of interventions on a given response variable in the form of changes in levels have been studied by Box and Tiao (1975).

(3) Mostly, the forms and the parameter values of the linear filters which prewhiten the variables are not known but have to be determined

empirically. Owing to the small samples available in many econometric studies, the estimates of the univariate ARIMA models are often not very precise and their use may crucially affect the results of the subsequent analysis.

(4) In tests on the cross-correlations of prewhitened series, the favorite null hypothesis is that of independence of the series. Under this hypothesis, the population correlation coefficients of the prewhitened series are zero and the asymptotic distribution of the sample cross-correlations is known. They are independently normally distributed with mean zero and variance equal to $(n - k)^{-1}$, with n being the sample size and k being the order of the cross-correlations. An asymptotic test of the null hypothesis of independent series is easily constructed. However, in economic applications, where economic theory indicates that there is a relationship between endogenous and exogenous variables, the hypothesis of independence of the series is not the most natural null hypothesis (see Hernández-Iglesias and Hernández-Iglesias 1980). Rather, econometricians often would like to find out what the shape of the lag distribution between y and x looks like, given that there exists a relationship between the series.

(5) Finally, for the use of autocorrelations and cross-correlations, stationary series are needed. In regression analysis, one can dispense with this requirement. In fact, the mean of the endogenous variable is assumed to vary with the explanatory variables. Also, the non-stationarity of the regressor variables may sometimes help to increase the precision of the regression coefficients estimates and forecasts.

Although the time series approach is not always appropriate in econometric applications, it can be very valuable when a bivariate or a low dimensional vector time series model constitutes the appropriate framework of analysis. For instance, when the aim of an application is to forecast an economic series, y, the use of a leading indicator, x, may increase forecasting precision. Similarly, when y has to be controlled through x, knowledge of the regression function for y can be useful if not requisite. Sometimes, economic theory implies testable restrictions on the parameters of a joint time series process, such as the absence of Wiener–Granger "causality" in one or both directions. Here, too, the usefulness of vector time series models has been demonstrated. From the discussion in this section, we conclude that in empirical work one has to combine the best features of the time series approach with existing econometric techniques. In the next subsection, we shall present the SEMTSA, which is a blend of econometric and time series methods.

2.3 *Structural econometric modeling and time series analysis*

Continuing research efforts during the 1970s . . . led to a combination of econometric and time series methods and their joint application in econometric modeling. Under the influence of modern time series analysis, the role of the data for the choice of a specification has become very important. Besides a large number of theoretical contributions, many empirical studies have been done. A more detailed survey can be found in Palm (1981).

In this subsection, we shall discuss the predominant features of the SEMTSA of regression models and behavioral equations.

2.3.1 Testing restrictions In the SEMTSA, economic theory and other subject-matter considerations such as institutional knowledge, relationships established empirically for similar data are used to specify a model and to formulate restrictions on the model. The restrictions and the assumptions underlying the model are formally confronted with the information in the data. Restrictions that are not contradicted by the sample information are incorporated in the model. Hypotheses regarding lag length, parameter stability, and exogeneity are tested.

Examples of restrictions originating from theoretical considerations are:

- A partial adjustment model, adaptive or rational expectations schemes
- Exclusion restrictions as a result of some causal mechanism
- The requirement of homogeneity of degree zero or 1 with respect to some or all explanatory variables, such as, for example implied by modern demand theory
- An "error correction" mechanism, such as introduced by Davidson *et al.* (1978), an "integral correction" term proposed and applied by Hendry and Von Ungern-Sternberg (1980).

The index models introduced by Sargent and Sims (1977) also include theoretically plausible restrictions. Dynamic econometric models based on more sophisticated optimizing behavior such as presented by Sargent (1981) obviously have an economic interpretation.

Among the restrictions that are easily imposed on the model without having necessarily an economic interpretation, we mention the Almon lag polynomials, which are equivalent to linear restrictions on distributed lag coefficients and the common factor restrictions leading to a regression model or a structural equation with autoregressive disturbances (e.g. Sargan 1964). However, testing the non-linear restrictions implied by the presence of common factors can create problems (see Sargan 1980).

2.3.2 From general to specific considerations In several . . . contributions to econometric modeling, the authors advocate – for very different reasons – starting the specification analysis with a fairly general model, that is, a model with a sufficient number of explanatory variables and lags for the true model to be nested within the initial model. For instance, Sims (1980a) argues that we generally do not have strong *a priori* knowledge (restrictions) to impose on the model. As an alternative to the traditional SEM, Sims uses multivariate autoregressive models that have a large number of unrestricted parameters.

In order to formulate an initial model, that includes the data-generating process, Zellner and Palm (1974) expand the small Haavelmo model by specifying finite-lag polynomials for the disturbances and for those variables, for which the dynamics were very uncertain. Their analysis is an example of a "bottom-up" or "specific-to-general" approach. (For a discussion of the advantages of a "bottom-up" approach compared with a "top-down" approach, see Zellner 1980.) A general initial model in a specification analysis can be obtained from a simple model by expanding the dynamics of those equations, for which the lag structure is *a priori* indeterminate.

Mizon (1977) and Mizon and Hendry (1980), among others, propose to start with a general model, specifying a uniquely ordered sequence of nested hypotheses and compare them using formal statistical tests.

Although starting with a loosely parametrized model implies a loss of degrees of freedom and possibly the presence of high multicollinearity between the regressors, it reduces the danger of analyzing models that are overly restricted. In agreement with Zellner and Palm (1974), rejecting the nested model, when it is true, will be a less serious error than using a restricted model, when the restrictions are not true. Similar considerations are sometimes put forward as an argument in favor of a specification analysis starting with a general model.

The general initial model can be used as a maintained hypothesis throughout the specification analysis, which aims at searching for the true model inside the initial model. As long as the true model is nested in the restricted model under the null hypothesis, H_o, the distribution of the test statistic under H_o is correct and the data can guide us towards the true model. Ideally, the investigator will formulate a sequence of nested hypotheses on the parameters of the initial model and test whether more restricted versions of the model are compatible with the data. Restrictions such as discussed in subsection 2.3.1 will be included in the sequence of hypotheses. Tests of specification in the form of a uniquely ordered nested sequence have desirable asymptotic properties. They are uniformly most powerful (see Anderson 1971, p. 263) in the class of unbiased tests. In practice, however, selecting a uniquely ordered sequence of hypotheses will be quite difficult, as several alternative sequences might be *a priori* reasonable. Therefore, the investigator will often carry along several alternative model specifications. In general, when a hypothesis in the sequence is not rejected by the data, it is imposed on the model. As a safeguard against misspecification, the serial correlation properties of the residuals of the restricted model should be checked. The sequence of tests stops when one hypothesis is rejected or when the last hypothesis cannot be rejected while the residuals of the most restricted model do not indicate any misspecification. Starting the specification analysis with a general model with serially uncorrelated disturbances has the following advantages:

(1) All the dynamics are incorporated in the systematic (explained) part of the equation instead of being left in the disturbance term. This enables the investigator to interpret the parameters more easily in terms of economic behavior.
(2) The conditions for the identification of the structural parameters in dynamic models with (vector) white-noise errors can be checked more easily in practice than those for models with serially correlated disturbances (see Hannan 1971).
(3) Many of the structural estimation methods and testing procedures designed for the static SEM can be applied to the dynamic SEM with white-noise disturbances. If the disturbances of an initial regression model are uncorrelated and homoskedastic, OLS has well-known optimal properties besides its obvious computational advantages, which can be important in a sequential testing setup. In a regression model with autocorrelated disturbances, but no lagged endogenous variables present, the OLS estimator is unbiased and consistent, but it is not efficient and the formula for the standard errors for OLS is no longer appropriate. Similarly, the *F*- and *t*-tests for linear and exclusion restrictions are no longer valid as such (see Kiviet 1979).

Notice that an initial finite order dynamic model with autoregressive disturbances can be transformed into a higher order finite distributed lag model with white-noise errors.

If the disturbances of the initial model are generated by a moving average process, the transformed model has infinite distributed lags and a finite order starting model can at best be considered as an approximation to the data-generating process. To limit the size of the approximation error, the number of lags included in the model will usually have to be large, which can lead to a substantial loss of degrees of freedom. Finally, although modeling the moving average process for the disturbances jointly with the regression or structural coefficients can be computationally cumbersome, it is necessary to achieve efficient estimation.

2.3.3 Model diagnostic checking Under the constant emphasis in the time series literature on residual correlation analysis and other diagnostic checks, econometricians have shifted their attention from the analysis of low order autoregressive and moving average processes to more general auto-correlation and cross-correlation schemes. Tests, such as the Box–Pierce (1970) test, have been very useful and have led to the development of many new tests for the presence of correlation in time series.

Diagnostic checking is synonymous with misspecification analysis. Given a model, one investigates whether more general models are more appropriate according to some criterion. It is going from specific to general, to use the terminology of Mizon and Hendry (1980) (see also Mizon 1977). Silvey's (1959) Lagrange multiplier and Rao's (1973) efficient score testing principle are well suited for misspecification analysis and many of the . . . tests developed [in the 1970s] are applications of these principles. (See Newbold 1981 for a survey of model checking tests.) Misspecification analysis is and has to be part of thorough econometric modeling. In the SEMTSA approach, the initial and the most restricted version of the model will have to be subjected to misspecification analysis.

2.3.4 Checking the overall consistency of the model Checking the overall consistency of the model is an important part of econometric modeling. An econometric model should be consistent with *a priori* knowledge and with the information in the data. Granger (1981) provides several examples of inconsistent models. Points, such as raised in his paper, should be taken into consideration when formulating a model. In addition, one should analyze the dynamic properties of the model, check the implications of the joint data-generating process for the associated marginal processes, and check the forecasting performance of the model.

One of the first questions asked by model builders is whether the different equations specified separately fit together. Common practice is to solve the complete model, either analytically, if the model is linear, or numerically, if the model is non-linear. Implausible values for the multipliers and for the solution of the model may lead to a reformulation of the model.

Subsequently, the implications of the restricted structural form for the properties of the transfer functions and the final equations ought to be checked along the lines proposed by Zellner and Palm (1974, 1975). The set of transfer functions associated with the structural form of a dynamic simultaneous equation model with vector moving average errors is the solution of the system which expresses each endogenous variable as a function of its own lagged values, of the current and lagged values of the exogenous variables, and of an error term which can be represented as a moving average in one variable. The lag length and the parameter values for the individual transfer function equations can be determined empirically and compared with those derived from the structural model. Under the additional assumption that the exogenous variables are generated by a multivariate ARMA model, the set of final equations for the endogenous (and exogenous) variables can be obtained after substitution for the exogenous variables. In the system of final equations, the endogenous variables are expressed as a set of restricted seemingly unrelated ARMA equations. The individual final equations can be analyzed, for example, along the lines proposed by Box and Jenkins (1970).

Any incompatibility between the results of the empirical analysis of the individual transfer functions and final equations and those derived from the tested structural form is an indication of a misspecification in at least one of these forms of the model and can be used to reformulate the model.

The role of the empirical analysis of the final equation and the transfer function forms for the structural form and for the properties of a simultaneous equation model has been discussed and illustrated by Zellner and Palm (1974, 1975). They also show how the model can be respecified when an incompatibility has been detected. The analysis of the final equations as a means for checking the dynamics of a simultaneous equation model has been pursued by Trivedi (1975), Prothero and Wallis (1976), and Wallis (1977), among others.

When the implications of the structural form of the model are in agreement with the results of the empirical analysis of the transfer functions and final equations, the model can be used to predict postsample observations. (See Christ 1975 on this point.)

If postsample data are available, the predictive performance of the structural form ought to be compared with that of the transfer functions and/or the final equations. If it predicts less well than the transfer functions or the final equations, there are good reasons for believing that the structural model is misspecified. If all three forms predict badly, the model is either misspecified or it has been subject to a structural change during the postsample period.

The predictive performance of the model can be formally checked using a test based on the distribution of the forecast errors – either assuming that the parameters of the model are known (see Hendry 1980) or that they have been estimated (see Dhrymes *et al.* 1972).

2.3.5 Some general remarks on SEMTSA The procedure outlined in the preceding subsections ought to be considered as a guideline for modeling systems of dynamic equations. Zellner (1979) discusses some of the statistical problems associated with the SEMTSA approach that require further research.

On many occasions, the data will not contain sufficient information to validate or reject all the assumptions underlying a simultaneous equation model, so that the tests will be inconclusive or the investigator has to rely on untested assumptions.

Also, before starting with the specification analysis, one has to decide whether a full information analysis of the complete initial model is feasible and desirable or whether one has to opt for an analysis under limited information (not necessarily through limited information maximum likelihood). Owing to the size of many simultaneous equation models used in practice, a full information analysis will hardly be feasible for most instances – except perhaps for models constructed for a small-scale purpose. In addition, one might expect an analysis under limited information to be robust against errors of misspecification in the remaining equations. With respect to the single-equation methods applied to a simultaneous equation model with autoregressive errors, Hendry (1974, p. 576) concludes that they pointed up the existence of misspecification and provided clues to its solution. About the disadvantages of testing subgroups of larger hypotheses, as will happen with a specification analysis under limited information, Darroch and Silvey (1963, p. 557) write: "Separate tests of h_1 and h_2 may induce a poor test of $h_1 \cap h_2$ because it is possible that for some Θ with high probability, $L(h_1)$ and $L(h_2)$ are both "near 1" while $L(h_1 \cap h_2)$ is small." For this reason, Byron (1974) suggests testing the restrictions on single structural equations first and, on the acceptance of all these tests, to test jointly for all over-identifying restrictions on the reduced form.

The computational intractability of an analysis under full information due to the size of the model has been put forward by Drèze (1976) as an argument in favor of limited information analysis in a Bayesian context . . . Malinvaud (1981) stressed this argument in a call for more research into estimation and testing procedures under limited information. However, limited information estimates and tests are usually not independent so that full information considerations are needed. In our application, we opt for an analysis under limited information and formulate the restrictions on the parameters of each equation separately.

3 An application of SEMTSA

3.1 *Specification of an initial model*

Given that a specification analysis starts with a fairly general model with possibly white-noise disturbances, we consider the following *p*th order VAR model

$$\underset{m\times 1}{\underline{z}_t} = \underset{m\times m}{A_1}\underset{m\times 1}{\underline{z}_{t-1}} + \ldots A_p\underline{z}_{t-p} + \underset{m\times 3}{B}\ \underset{3\times 1}{\underline{s}_t} + \underset{m\times 1}{\underline{\delta}} + \underset{m\times 1}{\underline{\gamma_t}} + \underset{m\times 1}{\underline{u}_t}\,, \tag{3.1}$$

where $\underline{z}_t$ is a vector of observed random variables, $A_i, i = 1, \ldots p$, and B are matrices of constant parameters, $\underline{s}_t$ is a vector of seasonal dummy variables denoted by s_{it}, $i = 1, 2, 3$, and being equal to 1 in the *i*th quarter and zero otherwise, $\underline{\delta}$ and $\underline{\gamma}$ are vectors of parameters and $\underline{u}_t$ is a vector of random disturbances assumed to be normally distributed, with mean zero and covariance matrices $E\,\underline{u}_t\underline{u}'_{t'} = \delta_{tt'}\,\Sigma$, where $\delta_{tt'}$ is the Kronecker delta.

In this section, we report the results of an empirical analysis of a vector autoregressive (VAR) model specified for seven quarterly seasonally unadjusted macroeconomic variables for the Netherlands. The VAR model serves as an initial model to which we subsequently apply a specification analysis, check for possible misspecification and investigate the overall consistency of the finally retained version of the model.

The sample covers the period 1952–79. Among the chosen variables, there are the major macroeconomic indicators: Aggregated gross national expenditures in constant prices (*Y*) and their price index (*P*), the unemployment rate (*U*) and a wage variable (*W*), nominal money balances (*M*) as measured by M_2 and a long-term interest rate (*R*) on government bonds, and an index of import prices (*PI*). Domestic variables are included in pairs of a real or a nominal variable and the associated price index. The index of import prices is introduced in order to take account of some of the effects of changes abroad on the open economy of the Netherlands. The choice of the variables is quite similar to that made by Sims (1980a). All the variables, except the interest rate, are expressed in natural logarithms. Quite obviously this multivariate AR process is not a complete model for the Dutch economy. If we assume a complete macroeconomic model for the Netherlands to be approximately log-linear with exogenous variables generated by a multivariate ARMA process with linear trend, the marginal process for the seven variables considered above will also be a multivariate ARMA model with linear trend. A VAR model is then an approximation for the marginal ARMA process, from which all other variables appearing in a complete model for the Dutch economy have been eliminated by integration. Although the marginal process is compatible with a larger model for the Dutch economy, one should keep in mind that its parameters may not be stable with respect to policy interventions, other than those in the form of a pure innovation. As will be seen below, after applying a fourth order VAR filter to the data, little correlation is left in the series. The VAR model is simple in the sense that its implications for the properties of the data-generating process are well understood. For instance, it implies that the long-run solution of the model does not depend on the time path of the variables. Nevertheless, when using an unrestricted VAR model, one usually violates the Principle of Parsimony. Our ultimate objective is to arrive at a more parsimoniously parametrized model that takes account of the correlation in the data and that can be interpreted in terms of economic behavior.

To detect possible structural changes, the empirical analysis has first been done for the subperiod 1952–73. Obviously, the choice of the year 1973 is not arbitrary. Two major developments are thought to have induced a structural change in the Dutch economy; the increase of the price of oil in 1973 and the change from a regime of fixed exchange rates to a system with partly flexible rates.

We first fit an unrestricted fourth order VAR model to the seven variables described above. The variables have been arranged in the following order: M, W, U, Y, P, R, PI. As expected, the estimates of the unrestricted reduced form parameters are not very precise. They do not exhibit any regular pattern. Many coefficients are not significantly different from zero. The estimates are not reproduced here. Notice also that by fitting a fourth order VAR model, we estimate the matrix of fourth order partial autocorrelations. Partial autocorrelations and stepwise vector autoregressions often play an important role in multiple time series modeling (see Tiao and Box 1981).

To investigate whether a four-period lag structure is sufficiently general, the residual correlation matrices for the unrestricted VAR model have been computed. The overall picture is that the estimated correlations are quite small. If we use $2T^{-1/2}$ (twice the approximate large sample standard error), with T being the sample size, as a yardstick for the precision of the estimates, very few residual correlations are significantly different from zero. We conclude that the fourth order VAR model is acceptable as a starting point for the specification analysis. Of course, the visual inspection of the residual correlations is not a perfect substitute for formal testing of the appropriateness of the starting model with respect to the lag length. A formal test such as implemented by Sims (1980a) requires that we extend the set of explanatory variables and is expensive in terms of degrees of freedom.

For the period 1952–73, none of the coefficients of the variables *M*, *W*, *P*, *R* and *PI* lagged four periods was significantly different from zero at conventional significance levels. This result led to the formulation of the null hypothesis of the coefficients of the five variables listed above. Using a large sample likelihood ratio test, we get $\chi^2(35) = 47.630$, which indicates that the fourth order lags on the nominal variables are not significantly different from zero at a 5 percent level ($\chi^2_{0.95}(35) = 49.57$). It is plausible that the lags for nominal variables are shorter than those for real variables.

A similar picture arises for the fourth order VAR model for the period 1952–79. In order to take account of possible structural changes, a dummy variable, denoted D74 and with value 1 in 1974–9 and zero elsewhere, has been included in the VAR model.

For the complete sample period, too, it was decided to use a VAR model excluding the fourth lag of the five nominal variables. The likelihood ratio test $\chi^2(35) = 55.08$ is significant at the 5 percent level. If we correct it for the loss of degrees of freedom, it will become insignificantly different from zero. The estimated residual correlations of these VAR models, which we will call the unrestricted models, are given in tables 3.8 and 3.10 (pp. 124, 128). Keeping the maximal lag for a given variable the same in all the equations of the model has the advantage that OLS estimates of the unrestricted reduced form equations separately will also be maximum likelihood estimates given initial conditions.

After the determination of the lag length, we next formulate restrictions on the parameters of the reduced form of the VAR model. Quite naturally one is interested in the exogeneity of *PI*, which implies block triangular reduced form matrices $A_i, i = 1, \ldots 4$, in (3.1). (See Geweke 1978 for a discussion of exogeneity in systems of equations.) For a small country under a regime of fixed exchange rates, import prices are often assumed to be exogenous. With pegged exchange rates, policy-makers attempting to stabilize the exchange rate might generate the exogeneity of the exchange rate with respect to the policy instruments (see Sims 1977). These restrictions are easily incorporated and tested.

A large sample Wald test that is equivalent to a likelihood ratio test yields a $\chi^2(20) = 95.95$ for 1952–1973 and a $\chi^2(20) = 115.10$ for 1952–79, which are significant at the level 0.005. For the hypothesis of the block exogeneity of *PI*, *R*, *P*, *Y*, and *U*, the test statistics $\chi^2(30) = 193.69$ for 1952–73 and $\chi^2(30) = 141.44$ are significant at the level 0.005 too.

In the monetary approach to the balance of payments under a regime of fixed exchange rates, the variables *PI*, *R*, *P*, *Y* (and *U*) are sometimes assumed to be jointly exogenous with respect to the reserves flow, which forms a component of the money supply. The exogeneity of the five variables with respect to M is stronger than what is needed in the monetary approach to the balance of payments. Rather surprisingly, there is little evidence in our data set pointing towards some block triangular structure of the VAR process. This also holds true for the joint exogeneity of the real sector with respect to the monetary sector.

3.2 *A limited information analysis of the initial model*

3.2.1 Introduction As the hypotheses of exogeneity formulated above do not seem to be supported by the data, we decide to use the unrestricted VAR model as a maintained hypothesis and to analyze it equation by equation. Our aim is to formulate restrictions that are meaningful in terms of economic behavior and that are not contradicted by the information in the data. For computational purposes, there is little need to restrict the number of parameters in the fourth order VAR model for seven variables, although the unrestricted VAR model seems to be highly over-parametrized. Regularities in and similarities between the VAR models for different countries and/or different sample periods show up in the moving average representation (MAR) of these models. Also, many economic series seem to follow a random-walk process or some other low degree univariate ARIMA model. For some series, in particular for financial data, these empirical findings have been explained and justified by economic theory (see Samuelson 1965). The regularities in the dynamics for many economic series point towards the existence of some common underlying structure. Therefore, testing restrictions on the parameters of the VAR model may be very useful for exploring, interpreting, and possibly understanding the dynamics of these models.

A limited information single-equation approach is clearly a second best strategy in terms of the power or the efficiency of the statistical procedures. However it is tractable and computationally less demanding compared with handling the complete model, possibly with non-linear restrictions. Given the size of our model, we could jointly analyze the complete model. By using an equation-by-equation approach, we also hope to get more insight into the performance of a limited information single-equation analysis, which will often be applied in larger models.

When modeling the equations separately we take the unrestricted VAR model as a maintained framework in which the alternative specifications for the single equations will be nested. In this way, the maximum lag length is determined and the list of predetermined variables needed for two-stage least squares (2SLS) is given. The following specifications were chosen using theoretical considerations and the information from the estimated unrestricted reduced form. Here, the inclusion of a variable in differenced form in a specification is equivalent to imposing a linear restriction on the coefficients of the lag polynomials. As we assume that all the dynamics are incorporated in the "systematic" part of the model, the equations can be consistently estimated by OLS or, when more than one current endogenous variable is present in an equation, by 2SLS. There is some evidence that 2SLS estimates are not "optimal" in finite samples (see Zellner and Park 1979). However, the method is easy to implement and it is also consistent when the structural equation to be estimated is actually a recursive form equation.

The specification of behavioral (or structural form) equations is a means of formulating over-identifying restrictions on the reduced form of the model. Historically, the structural form of the standard econometric model originates from the deterministic models used in economics, to which a disturbance term has been added. The structural coefficients usually have economic interpretations and they are assumed to be stable with respect to policy interventions. Statistically, the structural form corresponds to an over-parametrization of the model in order to subsequently reduce the number of parameters by means of identifying and over-identifying restrictions. Given that there are many predetermined variables in our unrestricted model, it is not difficult to identify the parameters in a behavioral equation and more importantly to generate restrictions on the reduced form parameters by imposing exclusion and other restrictions on a behavioral equation. Of course, one can argue that economic theory tells us which variables are of importance for the behavior of an economic agent and that theory is rather silent about the exclusion of other variables from a behavorial equation. To the extent that over-identifying restrictions generated in this way are tested and confronted with the information in the data, the danger of using false restrictions may be limited.

For the variables M, W, and P, it is possible in the framework of our VAR model to formulate relationships that have a behavioral interpretation. The equations for Y and R include the current endogenous variable P, but the restrictions imposed on the equations for these variables are basically data-instigated. Finally, for U and PI, we directly impose restrictions on the reduced form equations. These restrictions are suggested by patterns in the parameter estimates of the unrestricted reduced form equations. One can look at the restricted model as a system consisting of a set of behavioral equations analyzed under limited information to which one adds (restricted) reduced form equations to make the model complete. Any incompatibility with the data-generating process of the set of restrictions formulated in an equation-by-equation approach should show up in a joint test of all restrictions considered.

Table 3.1 *Nominal money balances (2SLS)*

	ΔM	
	1952–1973[a]	1952–1979[b]
ΔM_{-1}	0.214 (2.025)	0.408 (4.741)
$-V_{-1}$	−0.055 (4.195)	−0.045 (3.281)
ΔP_{-2}	−0.214 (1.875)	−0.059 (0.523)
ΔR_0	−0.012 (1.222)	0.003 (0.401)
$Y_0 - Y_{-2}$	0.071 (1.046)	0.069 (0.990)
S_1	0.027 (3.501)	0.037 (4.669)
S_2	0.040 (6.564)	0.044 (6.952)
S_3	0.025 (5.154)	0.019 (3.878)
C	−0.315 (4.284)	−0.264 (3.443)

Notes:
[a] SER = 0.013; DW = 1.888; GP = 0.075.
[b] SER = 0.015; DW = 1.879; GP = 0.609.

The results of the single equation analysis are reported in tables 3.1–3.7. The symbols Δ and C are used to denote the first difference operator and a constant term, respectively. The symbol D74 represents a dummy variable which is equal to 1 in the period 1974–9 and zero otherwise. A subscript indicates the number of lags. The variable V denotes the velocity of money, i.e. $V = P + Y - M$. Figures in parentheses are t-values (in absolute value). SER and 2SLS denote standard error of regression and two-stage least squares, respectively. DW denotes the Durbin–Watson statistic. GP denotes Godfrey's (1976) π-statistic for testing for first order disturbance serial correlation in an equation from a dynamic simultaneous equation system. This statistic has an approximate standard normal distribution. The data and the choice of the specifications for the individual equations deserve a short explanation.

The data are quarterly seasonally unadjusted observations on:

$M_t =$ total domestic money balances as measured by M_2 in the hands of the public averaged over the quarter (in million guilders)

$W_t =$ index of weekly wages, according to regulations, private and public sector, vacations, and other additional pay included, all adult employees: 1975 = 100

$U_t =$ quarterly average of unemployed males in percentage of total male employees

Y_t = gross national expenditures in quarter t, in million guilders per year, expressed in constant prices

P_t = price index of gross national expenditures; 1975 = 100

R_t = average of the interest rates on the three most recently issued long-term government bonds

PI_t = price index of all import goods; 1975 = 100.

3.2.2 Nominal money balances (M) The specification for nominal money balances can be interpreted as a demand for money function. The variables that are usually included in an empirical demand function for money appear as explanatory variables in the equation for M which has been retained after an extensive investigation into the shape of a demand for money function for the Netherlands (see Blommestein and Palm 1980). The relative change of nominal money balances is explained by the relative change of real total expenditures averaged over two quarters and of its price index, the change in the interest rate, and by the inverse of the velocity of money as perceived in period $t-1$. This last explanatory variable – also called error correction term (see Davidson *et al.* 1978) – takes account of the effect on the change in money balances of a disequilibrium in money holdings compared with total nominal expenditures in period $t-1$. As such, the specification describes the serial correlation properties of monetary balances very well. The steady state solution of the model implies a constant velocity of circulation. The value of the velocity depends on the rates of change prevailing in a given steady state. Alternative specifications, in which the interest rate level is included and which imply that the steady velocity of circulation also depends on the interest rate level, do not yield satisfactory results. The coefficient of the level of the interest rate was usually insignificant and had a "wrong" sign.

The amount of nominal balances is determined by the demand side. The main policy instruments of the Dutch Central Bank are the discount rate, credit regulations, which actually take the form of a penalty for excessive lending by private banks, and interventions in foreign currency markets to stabilize the exchange rate. To the extent that the determinants of the demand for money are included in the model, it is not necessary to include a behavioral equation for the Central Bank in order to complete the model.

The specification for the demand for money is not entirely stable over time. Some estimated coefficients change when the sample period is extended. In particular, the effect of a change in interest rates on the growth of money balances becomes positive although it is small and insignificant. In the 1970s, nominal balances measured by M_2 have been affected by heavy variations of short-term interest rates such as the interest rate on three-month interbank deposits, which induced substitution between time deposits and assets not included in M_2 (see den Butter and Fase 1979). Furthermore, the change in the exchange rates system has had its impact on the behavior of economic agents. Also the composition of M_2 has changed during the period of observation. The ratio of currency stock to demand and time deposits decreased in the 1960s and 1970s. An analysis of the effect of these changes requires a more disaggregated approach and naturally leads to an extension of the model or the use of more satisfactory monetary aggregates (see Barnett 1980), a line that will not be pursued in the present chapter. As the effect of ΔR is small, we retain the specification for both sample periods. Finally, we notice that when the lagged growth rate of money is left out of the specification, all parameter estimates have the expected sign. However, then there is much correlation left in the residuals, in particular for the complete sample period.

3.2.3 The wage equation (W) The wage equation is a Phillips-curve type specification in which the relative change in nominal wages is explained by the unemployment rate and the expected rate of inflation $P_t^* - P_{t-1}$, denoted as ΔP_t^*. We assume that expectations are rational, i.e. $P_t^* = E(P_t \mid \Phi_{t-1})$, where the expectations are taken, given the model and the set of variables up to the period $t - 1$, Φ_{t-1}. Following McCallum's (1976) proposal, the equation has been estimated by 2SLS after substitution of ΔP_t for ΔP_t^*. In this way, consistent estimates of the parameters of the wage equation are obtained. If we assume the "natural" rate of unemployment to be constant, the empirical finding that the coefficient of ΔP_t^* is not significantly different from one suggests that there is little or no long-run trade-off between inflation and unemployment. The constant term can be interpreted as being composed of the "natural" rate of unemployment and some "autonomous" wage rate change such as due to an increase of the productivity, the contributions to social security, and tax rates during the sample period. Notice also that there is some seasonality present in the equation. Phillips-curve type equations are used in macroeconomic models for the Netherlands (see Driehuis 1972). It should also be noted that in the reduced form of the restricted model, wages depend on P_{t-1} and P_{t-2} with coefficients summing to 0.9, which implies almost complete compensation for increases in the price index of total expenditures (table 3.2).

The specifications of the wage equation in macroeconomic models in which the wage sum per worker in enterprises is usually explained are, in general, more sophisticated than the specification retained here. Our choice of explanatory variables is limited through the size of the initial

Table 3.2 *Nominal wages (2SLS)*

	ΔW	
	1952–1973[a]	1952–1979[b]
U_{-1}	−0.0028 (0.694)	−0.0048 (1.801)
ΔP_t^*	0.874 (4.630)	0.713 (4.050)
S_1	0.017 (2.694)	0.020 (3.786)
S_2	0.0028 (0.472)	0.0047 (1.008)
S_3	0.0058 (1.162)	0.009 (2.071)
C	0.0063 (1.543)	0.0073 (1.944)

Notes:
[a] SER = 0.016; DW = 2.47; GP = −2.354.
[b] SER = 0.016; DW = 2.24; GP = −1.265.

VAR model. Single-equation modeling in the framework of a multivariate model naturally leads to an extension of the dimension of the model.

3.2.4 Unemployment (U) The specification for the unemployment rate ought to be interpreted as a restricted reduced form equation. The variables finally included in the specification have been selected because their coefficients were significant in the unrestricted reduced form. The numerical values of the unrestricted reduced form parameter estimates pointed towards restrictions that could easily be imposed on the parameters. The plausibility of the results in terms of the sign of the parameters, of the presence of some variables, also played a role in the formulation of the restrictions. For instance, the restricted equation is homogeneous of degree zero in all nominal variables.

However, when using a formal large sample chi-square test, the restrictions imposed on the unemployment equations reported in table 3.3 are significantly different from zero at conventional levels of significance. Presently, we retain the restricted version of the equation. In the analysis of the complete restricted model, we shall pay more attention to the specification of the unemployment equation.

3.2.5 Total expenditures (Y) For the aggregate expenditures in constant prices, an initial fairly general structural equation has been formulated, identified through exclusion restrictions and estimated by 2SLS. Next, variables for which the coefficients were not significantly different

Table 3.3 *Unemployment (OLS)*

	U	
	1952–1973[a]	1952–1979[b]
$(M-P)_{-1}$	−1.522 (3.192)	−1.361 (3.572)
$(W-P)_{-1}$	1.149 (2.814)	1.032 (3.083)
ΔY_{-2}	−0.680 (1.461)	−1.531 (3.906)
$(P-PI)_{-3}$	0.085 (0.285)	0.097 (0.432)
$\Delta^2 PI_{-1}$	−1.628 (3.576)	−1.127 (3.109)
U_{-1}	1.409 (13.60)	1.392 (16.46)
U_{-2}	−0.826 (4.665)	−1.130 (8.535)
U_{-3}	0.513 (2.958)	0.980 (7.415)
U_{-4}	−0.095 (0.894)	−0.266 (2.925)
S_1	0.008 (0.073)	0.215 (2.973)
S_2	−0.617 (5.937)	−0.279 (4.163)
S_3	−0.002 (0.017)	0.091 (1.050)
C	10.08 (3.273)	8.903 (3.605)
D 74		0.173 (2.739)

Notes:
[a] SER = 0.103; $R^2 = 0.966$; ln $L = 79.046$; DW = 2.065.
[b] SER = 0.109; $R^2 = 0.97$; ln $L = 93.224$; DW = 2.092.

from zero and for which the parameter estimates were implausible have been excluded. This procedure led to the finally retained specifications in table 3.4. It seems to be difficult to give the specifications in table 3.4 a behavioral interpretation, given that the variable Y is the total of the expenditures of all agents in the economy. In the restricted equation, total expenditures are explained by real money balances, the change in the unemployment rate, the domestic inflation rate, the foreign price level, and lagged expenditures.

The specifications in table 3.4 are not homogeneous in the nominal variables. Several alternative specifications, which were homogeneous of degree zero in nominal magnitudes or in which the effect of the level of the interest rate and of the unexpected component of the inflation rate were introduced, did not yield satisfactory empirical results. That *a priori* meaningful restrictions are apparently not supported by the sample information is possibly explained by the highly aggregate nature

Table 3.4 *Total expenditures (2SLS)*

	Y	
	1952–1973[a]	1952–1979[b]
$(M-P)_{-2}$	0.169 (2.304)	0.163 (2.573)
ΔU_{-1}	−0.028 (1.116)	−0.006 (0.380)
$P_0 - P_{-2}$	−0.087 (0.451)	−0.055 (0.296)
PI_{-2}	−0.165 (2.830)	−0.099 (4.055)
Y_{-1}	0.406 (3.368)	0.431 (4.208)
Y_{-2}	0.524 (3.945)	0.487 (4.400)
Y_{-3}	−0.302 (2.296)	−0.305 (2.858)
Y_{-4}	0.229 (1.876)	0.152 (1.507)
S_1	−0.057 (2.737)	−0.075 (4.871)
S_2	0.014 (0.747)	0.005 (0.304)
S_3	0.027 (1.246)	0.036 (2.152)
C	1.311 (1.511)	2.115 (2.467)
t	0.0006 (0.626)	0.002 (1.903)

Notes:
[a] SER = 0.022; DW = 2.002; GP = −0.609
[b] SER = 0.023; DW = 1.986; GP = −0.428.

of the variable Y. Again, quite naturally one is led to expand the model through disaggregation of Y into consumption, investment and government expenditures, variations in inventory holdings, and other expenditure categories.

3.2.6 The price index of aggregate expenditures (P) The rate of change of the total expenditures deflator is explained by the relative change in the wages, the import price, and total expenditures in constant prices. The rate of change in total expenditures has a negative impact on the rate of inflation. As the constant term was very small and insignificant, we opted for a homogeneous specification for the price in the period 1952–73. For the complete sample period, we include the dummy variable $D74$, defined above.

In the specification for the domestic price variable, variations in prices are explained by changes in the major cost components, wages, and imports, corrected for the variations in total expenditures. As such, the

Table 3.5 *The price equation (2SLS)*

	ΔP	
	1952–1973[a]	1952–1979[b]
ΔW	0.452 (9.375)	0.461 (9.026)
ΔW_{-1}	0.133 (2.883)	0.116 (2.636)
ΔY	−0.036 (1.984)	−0.035 (1.914)
ΔPI_{-1}	0.104 (1.848)	0.048 (1.300)
D 74		0.006 (2.615)

Notes:
[a] SER = 0.010; DW = 1.919; GP = −0.803.
[b] SER = 0.010; DW = 2.042; GP = −1.668.

equation is a generalized version of the full cost pricing. (For more details on the theoretical justification of aggregate price equations, see Driehuis 1972 and Nieuwenhuis 1980.) As the first price equation in table 3.5 is homogeneous, it has a static equilibrium solution. With the inclusion of a constant term, that could be interpreted as the effect of cost components which are not explicitly taken account of, the price equation would not have a static equilibrium solution. The coefficient estimates are actually not affected by the introduction of a constant term. Therefore, the solution of the homogeneous part of the complete model is insensitive too in this respect. As we shall see in subsection 3.4.3, the predictions for P_t could probably be improved by the inclusion of a constant term in the price equation. For the period 1974–9, it is consistent with a steady state solution of 8 percent per annum.

3.2.7 The interest rate (R) Several specifications have been fitted to the interest rate. The closed economy version of the Fisher equation stating that nominal interest rates equal the anticipated real rate of interest plus the expected rate of inflation is not very useful in this context. Furthermore, it requires a model for the *ex ante* real rate of interest which has apparently not been constant in the Netherlands during the period 1952–79. For our data, the closed economy version of the Fisher equation combined with alternative simple models for the *ex ante* real rate of interest did not yield standard errors of regression smaller than 2 percentage points.

An open economy version of the Fisher equation requires a two-regime model. For the period of fixed exchange rates, one ought to expect the domestic interest rates of a small economy to be closely linked to the

Table 3.6 *The interest rate (2SLS)*

	R	
	1952–1973[a]	1952–1979[b]
R_{-1}	0.954 (34.72)	0.928 (41.63)
$-V_{-1}$	−0.359 (1.012)	−0.953 (2.847)
$PI_{-1} - PI_{-3}$	2.426 (2.604)	2.080 (2.644)
ΔU_{-2}	−0.240 (1.425)	−0.424 (2.309)
$P_0 - P_{-2}$	3.039 (1.984)	0.819 (0.363)
S_1	−0.043 (0.416)	0.053 (0.450)
S_2	0.130 (0.769)	0.372 (2.187)
S_3	0.164 (1.106)	0.368 (2.460)
C	1.845 (0.977)	−5.103 (2.810)

Notes:
[a] SER = 0.197; DW = 1.621; GP = 1.579.
[b] SER = 0.289; DW = 1.665; GP = 1.649.

interest rates on international money markets. In a regime of flexible exchange rates, foreign interest rates and the spot and forward exchange rates are the major determinants of the domestic interest rates. For both regimes, the set of variables in the model has to be extended in order to get a theoretically satisfactory relation for the interest rate.

In this chapter we do not follow this line, but try to specify a parsimoniously parametrized equation for the nominal interest rate on government bonds. This is done along the lines of the approach that we applied to the total expenditures in constant prices. In table 3.6, the nominal interest rate is explained by the liquidity ratio, the rate of inflation of imports, the domestic inflation rate averaged over two quarters, the change in the unemployment rate, and the lagged nominal interest rate. The presence of a slight seasonal pattern shows up in the specification. The explanatory variables included in the equations of table 3.6 also appear in the interest rate equation of some macroeconomic models for the Netherlands. Notice that the rates of change are not expressed as percentage points but as fractions. The estimate of the coefficient of the domestic price change, which is also a consistent estimate of the coefficient of the expected inflation rate, differs from the value that it ought to take according to the Fisher equation. Finally, the results in table 3.6 suggest that the coefficient of R_{-1} is insignificantly different from 1, so

Table 3.7 *The price of imports (OLS)*

	PI	
	1952–1973[a]	1952–1979[b]
ΔM_{-2}	0.354 (2.635)	0.245 (2.077)
$(P-W)_{-2}$	0.206 (3.483)	0.176 (3.080)
ΔP_{-2}	0.391 (2.952)	0.360 (2.459)
Y_{-2}	0.161 (3.548)	0.080 (1.706)
PI_{-1}	0.899 (24.63)	0.987 (61.608)
S_1	0.018 (3.610)	0.017 (3.153)
S_2	0.015 (2.290)	0.018 (2.713)
S_3	0.011 (1.720)	0.008 (1.271)
C	−1.607 (3.078)	−1.049 (1.799)
t	0.0003 (0.410)	0.0010 (1.426)
$T89$		0.167 (8.563)

Notes:
[a] SER = 0.016; $R^2 = 0.94$; ln L = 236.047; DW = 1.37.
[b] SER = 0.018; $R^2 = 0.99$; ln L = 281.72; DW = 1.38.

that a specification in which the change in the interest rate is explained is in line with our empirical findings.

3.2.8 The import price (PI) One would have expected that this variable passed the exogeneity test for the period of fixed exchange rates, as the import price is the product of the exchange rate times the price of import goods expressed in foreign currency, which could be assumed to be exogenous. Given that we had to reject the exogeneity of the import prices, we decided to adopt a strategy of restricting the reduced form equation for the import price. The results of our analysis are reported in table 3.7, where the price of imports is explained by the change in money balances and domestic prices, the level of total expenditures, real wages, and lagged import prices. The estimated coefficient of the lagged import price is very close to 1, suggesting that the data support a specification in which the rate of change of import prices is the variable to be explained. A dummy variable, *T*89, has been included for the first quarter of 1974, when the shock of the oil price increase worked through in import prices. Notice that the variable $T89$ has not been included in the unrestricted initial model. As the import price series includes prices of primary, intermediate, and final import products, it is again difficult to give an interpretation in terms of economic behavior of the specification finally chosen. The signs of the estimated coefficients are in line with what one expects. For instance, an increase in the domestic rate of inflation may be expected to lead to an increase of the price of the competitive import commodities and/or to an increase in the rate of exchange.

To summarize, whenever possible, we fitted a specification with theoretically meaningful restrictions. Thereby, we limited ourselves to linear relationships among the seven variables listed above and with maximum lag equal to 4. When it was too difficult to formulate a behavioral relationship owing to the limited number of variables included in the model, we restricted the single structural or reduced form equations using the information in the data.

The model consisting of the equations reported in tables 3.1–3.7 will be called the "restricted model." It implies a plausible block recursive (Wold) structure for the systematic part (the disturbance covariance matrix is not block diagonal) of the model for the observable variables:

$$\text{Past} \rightarrow \left[\begin{array}{llll} W \rightarrow & P \rightarrow & R \rightarrow & M \\ U & \uparrow\downarrow & \nearrow & \\ PI & Y & & \end{array}\right. \tag{3.2}$$

The variables for the labor market and the import price are explained by predetermined variables only, whereas the remaining variables are jointly determined as indicated in the graph (figure 3.2, p. 135). It is not surprising that wages are not simultaneously determined with prices, as the correction of wages for price inflation usually takes place two times a year, i.e. with a delay of one quarter on average. Similarly, the nominal variables R and M are instantaneously affected by changes in aggregate expenditures or the price index of aggregate expenditures, whereas they influence the labor and commodity market variables with a lag of one quarter. One has to be careful when interpreting the restricted model. Only the equations for M, W, and P have a behavioral interpretation. For the variables R and Y, the structural form equations were purely instrumental in generating restrictions on the reduced form equations for the remaining variables in the system. The reduced form equations for these variables have been added to the behavioral equations of M, W, and P to obtain a complete system representing the marginal process for the seven variables considered in this chapter.

The number of parameters in the initial model (196 for 1952–73, 203 for 1952–79) has been reduced by more than two-thirds (64 for 1952–73,

67 for 1952–79) among which there are 18 seasonal parameters. Three additional parameters have been included to take account of the structural changes that occurred after 1973. Given that the restricted model is a low dimensional marginal process, from which variables (e.g. policy instruments) that could lead to a structural change in the parameters have been eliminated by integration, its parameters seem to be fairly stable over the sample period. Although 11 out of 40 coefficients other than constant terms, seasonal, and trend coefficients change by a factor more than two, only the estimates of the coefficient of U_{-3} in the unemployment equation, of the constant term in the interest rate equation, and the coefficient of PI_{-1} in the import equation change by more than two coefficient standard errors for 1952–73. A fully satisfactory treatment of the parameter instability requires an extension of the model. We have achieved a substantial reduction of the number of parameters, although we claim neither that our model is the most parsimonious parametrization of the VAR model nor that we have not imposed any false restriction. With exception of the dummy variable *T*89 in the import price equation for the complete sample period, the restricted model is nested in our starting model.

The estimates of the restricted model are not fully efficient as they are single-equation (limited-information) estimates. Owing to the nonlinearity of the restrictions implied by the rational expectations assumption and the still fairly large number of parameters in the restricted model, it is difficult to obtain fully efficient parameter estimates for testing all restrictions jointly, as suggested by Byron (1974). A likelihood ratio test is based on the quotient of the determinants of efficiently estimated reduced form disturbance covariance matrices for the unrestricted and restricted models, respectively. When comparing the determinant of the 2SLS residual covariance matrix with that of the estimated unrestricted VAR model, one obtains an upper bound for the likelihood ratio, i.e. a test statistic that is biased towards rejection of the restricted model. These bounds are 336 for the period 1952–73 and 324 for the period 1952–79 (or 224 and 242.6, respectively, if we correct for the loss of degrees of freedom as Sims 1980a suggests). Three-stage least squares (3SLS) estimates are not fully efficient asymptotically, as the non-linear restrictions implied by the rational expectations hypothesis are ignored. A joint test based on 3SLS estimates under the null hypothesis yields approximate likelihood ratio statistics of 133.3 for 1952–73 and 286.5 for 1952–79 (or 90.7 and 201.6, respectively, when we correct for the loss of degrees of freedom), which are asymptotically chi-square distributed with 133 and 137 degrees of freedom, respectively. When estimating the model jointly by 3SLS, we imposed the additional restriction that the coefficient of ΔP^* equals 1. For the period 1952–73, the restricted model is not rejected at conventional significance levels, whereas for the complete period we have to reject the restricted model, although one should keep in mind that a test based on 3SLS estimates (which are not exact maximum likelihood estimates) is still biased towards rejection of the null hypothesis. Also, when using Akaike's information criterion, the restricted model would be rejected at the margin only for the complete period. These results should be interpreted with care. Little is known about the power of these tests in small samples and on how to adjust the significance level when sample size increases. Schwarz's (1978) large sample approximation $\ln K_{01} = -{}^1/_2[\chi_q^2 - q\ln T]$ for the posterior odds ratio K_{01} of two alternative nested hypotheses, where χ_q^2 is the likelihood ratio statistic, q is the number of restrictions implied by the null hypothesis, and T is the sample size, strongly supports the restricted model for both sample periods. Notice that the posterior odds explicitly take account of the sample size.

Restricting the coefficient of ΔP^* to 1 yields a Phillips-curve with no trade-off present in the long run between unemployment and inflation. Interestingly, this restriction implies that the reduced form equations for ΔW and ΔP are identical, except for the coefficients of U_{-1} and the seasonal dummies, the constant term, and the error variance. Interpreted in this way, the restriction does not seem to be implausible. Also, with this restriction imposed, the 3SLS estimates of the parameters and disturbance correlations are very reasonable. Without the additional restriction, the 3SLS estimates seem to be somewhat ill-conditioned. This is not surprising, given that the number of parameters to be estimated is large. To conclude, the restrictions imposed on the VAR model are in accord with the information in the data for 1952–73. For the complete period, the restrictions are not entirely compatible with the sample evidence – at least, when 3SLS estimates are used as subsititutes for the exact likelihood estimates. This is probably due to a structural change induced by the oil price increase in 1973.

3.3 *Diagnostic checking*

To check the adequacy of the restricted model, we computed the residual correlation matrices. They are given in tables 3.8–3.11 for the two sample periods respectively. As the residuals of an equation do not necessarily sum to zero, the residuals have been taken in deviation from their sample mean.

The *i-j*th element of matrix θ in tables 3.8–3.11 is the sample correlation between $\hat{u}_{it+\theta}$ and $\hat{u}_{jt}$. If we use $2T^{-1/2}$ as a yardstick for the significance of the individual residual correlations, 21 and 27 among the 392 residual correlations are significantly different from zero in the periods 1952–73 and 1952–79, respectively. Among them, there are two and eight

Table 3.8 *Residual correlation matrices for the unrestricted reduced form, 1952–1979*

			$\theta = 0$			
1.00	−0.01	−16	−0.05	0.04	0.03	0.11
−0.01	1.00	0.03	0.13	0.53	0.21	0.08
0.16	0.03	1.00	−0.30	0.08	−0.01	−0.06
−0.05	0.13	−0.30	1.00	−0.19	0.03	0.11
0.04	0.53	0.08	−0.19	1.00	0.14	0.19
0.03	0.21	−0.01	0.03	0.14	1.00	0.39
0.11	0.08	−0.06	0.11	0.19	0.39	1.00
			$\theta = 1$			
0.02	−0.01	0.02	−0.04	0.08	−0.02	0.08
0.08	0.00	0.04	−0.00	0.03	0.03	0.11
0.04	0.05	0.05	0.02	−0.02	0.06	0.08
0.05	−0.00	−0.03	0.03	−0.03	−0.04	−0.07
0.03	0.02	0.00	−0.02	0.18	0.03	0.22
0.01	0.06	0.01	−0.07	0.03	0.04	0.14
0.01	0.06	0.03	−0.04	0.07	0.11	0.37
			$\theta = 2$			
−0.07	0.01	0.10	−0.06	0.05	−0.03	0.13
0.13	0.03	0.13	0.01	0.11	0.04	0.22
0.00	0.03	0.15	0.04	−0.01	0.01	0.12
0.09	−0.04	0.05	−0.06	−0.02	−0.00	−0.03
0.07	0.08	0.04	−0.03	0.16	0.09	0.26
0.06	0.09	−0.09	0.00	0.16	0.10	0.24
0.06	0.05	−0.03	−0.03	−0.09	0.05	0.09
			$\theta = 3$			
−0.04	−0.08	−0.06	−0.10	−0.00	0.03	0.24
−0.00	0.01	0.01	0.02	0.12	−0.02	0.11
−0.01	0.07	−0.02	−0.13	0.03	0.17	0.14
−0.02	−0.13	0.04	0.08	−0.01	0.01	−0.08
0.08	−0.06	0.02	−0.01	−0.07	−0.00	0.20
−0.12	−0.11	−0.05	0.03	−0.09	0.10	0.08
−0.05	−0.07	−0.06	0.04	−0.19	0.03	−0.02
			$\theta = 4$			
−0.00	0.20	0.15	−0.17	0.10	−0.05	0.12
0.04	−0.11	0.13	−0.02	0.11	−0.00	0.13
0.01	−0.03	−0.07	−0.01	−0.15	−0.08	0.12
−0.03	−0.17	0.06	−0.03	−0.02	0.01	0.00
−0.06	0.08	0.05	−0.06	0.11	0.06	0.18
−0.08	−0.07	−0.09	−0.12	−0.04	−0.13	−0.10
−0.08	−0.09	0.07	−0.13	0.01	−0.10	−0.12

Table 3.8 (*cont.*)

			$\theta = 5$			
−0.06	−0.07	−0.26	0.12	−0.03	0.05	0.02
0.08	−0.04	−0.07	−0.03	−0.04	−0.03	0.11
−0.15	0.01	−0.20	0.15	0.03	0.08	0.11
−0.03	−0.10	0.05	−0.09	−0.05	0.00	−0.11
0.13	0.10	0.04	0.01	−0.14	−0.02	0.15
0.01	−0.01	−0.03	0.08	−0.13	−0.12	−0.12
0.01	−0.12	0.07	−0.01	−0.14	−0.12	−0.15
			$\theta = 6$			
0.13	0.00	−0.04	0.12	−0.02	0.08	−0.04
0.17	−0.08	0.05	−0.02	−0.08	0.03	0.17
0.05	−0.12	−0.13	0.01	−0.14	−0.05	−0.12
−0.05	0.12	−0.05	0.16	0.07	0.07	0.10
0.23	−0.10	0.23	−0.16	−0.15	−0.12	0.01
0.02	0.01	0.02	0.09	0.02	−0.07	0.02
0.14	0.01	0.17	0.02	−0.04	−0.12	−0.10
			$\theta = 7$			
−0.04	0.08	−0.00	−0.18	0.11	−0.12	−0.11
−0.00	0.08	−0.09	−0.08	−0.04	−0.04	−0.04
0.04	−0.02	−0.19	0.06	−0.09	0.01	0.03
0.02	−0.05	−0.13	−0.08	0.01	0.00	−0.01
−0.02	−0.06	0.09	−0.06	−0.12	−0.15	−0.10
0.04	−0.08	0.07	0.05	−0.11	−0.26	−0.04
0.01	−0.11	0.10	−0.08	−0.17	−0.08	−0.16
			$\theta = 8$			
−0.02	−0.22	0.12	−0.08	−0.04	−0.16	−0.07
0.08	0.08	0.10	−0.03	0.06	0.04	−0.00
0.03	−0.08	0.10	0.03	−0.13	0.14	−0.09
−0.00	0.21	−0.06	−0.12	0.08	−0.01	−0.04
−0.07	0.03	0.15	−0.05	0.08	0.06	−0.02
0.24	0.12	0.15	0.18	−0.03	−0.04	−0.14
−0.04	−0.02	0.05	−0.05	−0.00	−0.02	−0.02

Note: The underlined figures (in absolute value) are greater than 2 approximate standard errors, $2T^{-1/2}$, $T = 108$.

Table 3.9 *Residual correlation matrices for the restricted reduced form, 1952–1979*

$\theta = 0$						
1.00	−0.10	0.14	−0.10	0.02	0.01	0.09
−0.10	1.00	0.01	0.12	−0.19	0.13	0.05
0.14	0.01	1.00	−0.25	0.02	−0.09	−0.02
−0.10	0.12	−0.25	1.00	−0.18	0.03	0.28
0.02	−0.19	0.02	−0.18	1.00	−0.10	0.07
0.01	0.13	−0.09	0.03	−0.10	1.00	0.25
0.09	0.05	−0.02	0.28	0.07	0.25	1.00
$\theta = 1$						
0.05	0.04	−0.19	−0.07	−0.04	0.11	−0.10
0.08	−0.12	0.05	0.04	0.12	−0.15	−0.05
−0.04	0.09	−0.05	0.07	0.08	−0.13	−0.13
0.17	−0.18	−0.16	0.01	0.13	0.11	0.12
0.09	−0.07	0.14	0.11	−0.06	0.05	0.21
−0.11	0.08	0.06	−0.15	0.08	0.16	0.03
0.07	−0.03	−0.11	0.02	0.29	0.19	0.28
$\theta = 2$						
0.00	0.17	0.13	−0.08	−0.19	0.16	0.07
0.08	0.00	0.10	−0.00	0.05	0.01	0.12
0.02	−0.10	0.18	0.17	−0.07	−0.13	−0.02
−0.14	0.11	−0.03	−0.09	0.06	0.07	0.05
−0.10	0.07	−0.27	0.16	−0.02	0.08	0.01
0.07	−0.10	0.02	0.05	0.15	−0.04	−0.03
0.13	−0.07	−0.15	−0.06	0.00	0.17	0.01
$\theta = 3$						
−0.07	0.05	0.11	−0.07	0.06	0.07	0.10
−0.08	0.04	−0.13	−0.12	0.08	−0.01	−0.05
−0.14	0.07	−0.06	0.03	0.06	0.25	0.04
−0.13	0.00	−0.09	0.12	0.09	0.06	−0.01
0.14	−0.16	0.11	−0.00	−0.14	0.02	−0.04
−0.03	−0.17	−0.07	0.04	0.01	−0.02	−0.07
−0.12	0.09	−0.22	−0.04	−0.04	0.29	−0.01
$\theta = 4$						
0.18	0.20	0.33	−0.15	−0.04	−0.13	0.08
0.06	0.03	0.07	−0.01	0.14	0.08	−0.01
−0.05	−0.03	0.14	0.00	−0.02	0.03	0.19
−0.04	−0.12	−0.03	0.06	0.15	−0.02	0.09
−0.10	0.15	−0.02	−0.15	0.12	−0.00	−0.14
−0.03	0.01	−0.07	−0.09	−0.07	−0.16	−0.30
−0.04	0.00	0.11	−0.20	0.09	−0.05	−0.02

Table 3.9 (*cont.*)

			$\theta = 5$			
−0.13	0.04	−0.17	0.17	0.09	−0.05	−0.18
0.05	−0.05	−0.02	−0.03	0.03	0.03	0.14
−0.15	0.06	−0.07	0.21	0.28	0.07	0.06
0.06	−0.09	0.02	−0.08	−0.02	−0.08	−0.00
0.01	0.20	0.04	0.09	−0.20	−0.09	0.02
−0.03	−0.04	0.10	0.05	−0.15	−0.07	−0.20
0.09	−0.05	0.06	0.01	0.06	−0.14	−0.03
			$\theta = 6$			
0.06	−0.01	0.11	−0.04	−0.06	0.06	−0.07
0.16	−0.00	0.05	−0.02	−0.04	0.03	0.04
0.03	−0.17	−0.05	0.07	0.06	0.00	0.06
0.03	0.04	−0.19	0.22	−0.07	−0.01	0.02
−0.02	−0.04	0.05	−0.01	0.05	−0.11	−0.15
−0.06	0.11	0.13	0.00	−0.02	−0.12	−0.03
0.16	0.10	−0.04	0.06	−0.06	−0.08	−0.04
			$\theta = 7$			
−0.15	0.15	0.06	−0.11	0.05	−0.05	0.05
−0.02	−0.00	−0.13	−0.07	−0.06	−0.08	−0.18
−0.11	−0.00	−0.16	0.08	0.08	0.10	0.10
0.13	−0.04	−0.05	−0.13	0.00	0.03	−0.07
−0.07	−0.10	0.07	0.04	−0.07	0.02	0.06
−0.00	−0.03	0.10	−0.06	0.12	−0.30	−0.16
0.01	0.06	0.01	−0.13	−0.15	−0.05	−0.25
			$\theta = 8$			
0.09	−0.11	0.22	−0.13	0.17	−0.15	0.07
0.00	0.21	0.13	0.02	0.00	0.07	−0.01
0.09	−0.04	0.19	0.10	−0.02	0.08	0.03
0.00	0.23	−0.06	−0.17	−0.05	−0.04	0.07
−0.15	0.14	0.06	−0.01	0.20	0.03	−0.02
0.10	0.09	0.16	0.22	−0.13	0.02	0.00
−0.04	−0.02	0.13	−0.11	−0.10	0.04	−0.07

Note: The underlined figures (in absolute value) are greater than 2 approximate standard errors, $2T^{-1/2}$, $T = 108$.

Table 3.10 *Residual correlation matrices for the unrestricted reduced form, 1952–1973*

			$\theta = 0$			
1.00	−0.14	0.15	−0.04	−0.07	−0.09	0.24
−0.14	1.00	−0.14	0.17	0.42	0.22	0.11
0.15	−0.14	1.00	−0.28	0.06	0.07	−0.09
−0.04	0.17	−0.28	1.00	−0.13	0.01	0.27
−0.07	0.42	0.06	−0.13	1.00	0.14	0.08
−0.09	0.22	0.07	0.01	0.14	1.00	0.31
0.24	0.11	−0.09	0.27	0.08	0.31	1.00
			$\theta = 1$			
0.01	0.01	0.01	−0.07	0.14	−0.06	0.14
0.08	−0.07	0.05	−0.06	0.03	0.03	0.10
0.02	0.01	0.00	−0.03	0.10	−0.02	0.09
0.15	0.02	−0.01	0.02	−0.04	0.00	0.01
−0.04	−0.03	−0.01	0.01	0.17	−0.04	0.17
0.04	0.04	0.03	−0.03	0.14	−0.07	0.12
0.05	−0.01	0.03	0.05	0.09	−0.02	0.16
			$\theta = 2$			
−0.17	0.03	0.14	−0.02	0.00	0.04	0.06
0.11	0.01	0.05	0.06	0.13	−0.08	0.11
−0.06	0.04	−0.06	0.04	0.13	−0.01	0.09
0.02	−0.15	0.03	−0.08	−0.06	−0.05	0.05
0.11	0.14	−0.02	0.08	0.17	0.08	0.23
−0.07	0.09	−0.09	0.16	0.03	0.03	0.04
−0.05	−0.07	−0.02	−0.01	−0.21	−0.05	−0.05
			$\theta = 3$			
0.09	0.06	0.07	−0.06	−0.01	−0.02	0.14
−0.06	−0.08	−0.07	0.06	0.06	0.08	0.04
−0.01	0.04	−0.05	−0.13	0.13	0.19	0.16
−0.03	−0.15	0.08	0.04	0.00	−0.15	−0.13
0.05	−0.11	−0.04	0.07	−0.22	0.12	0.18
0.04	−0.12	−0.05	−0.03	−0.07	0.18	−0.19
0.12	−0.07	0.09	0.04	−0.27	0.05	−0.07
			$\theta = 4$			
0.07	0.17	0.20	−0.18	0.05	−0.07	0.01
0.07	−0.16	0.18	0.05	0.01	−0.01	0.13
0.01	0.10	−0.11	−0.03	−0.10	−0.06	0.09
0.05	−0.10	0.20	−0.09	−0.04	0.00	−0.01
0.01	0.08	−0.01	0.16	−0.09	0.06	0.14
0.10	−0.12	−0.05	−0.12	−0.06	−0.21	−0.03
−0.02	−0.14	0.13	−0.08	−0.04	0.02	−0.04

Table 3.10 (*cont.*)

$\theta = 5$						
−0.19	−0.16	−0.05	0.07	−0.06	−0.07	0.01
0.02	0.03	0.00	0.09	−0.14	0.10	0.20
−0.26	0.05	−0.12	0.11	0.01	−0.06	−0.02
0.13	−0.06	0.14	−0.16	0.02	0.10	0.04
0.11	0.10	−0.05	0.14	−0.28	0.16	0.18
−0.02	0.14	−0.02	0.20	−0.02	0.06	0.08
−0.12	−0.08	0.04	−0.04	−0.10	0.09	0.01
$\theta = 6$						
0.16	−0.05	0.07	0.18	−0.16	0.13	0.06
0.05	−0.04	0.00	0.14	−0.11	−0.01	0.14
0.08	−0.17	−0.02	−0.05	−0.25	−0.06	−0.22
−0.02	0.14	−0.03	0.03	0.12	−0.06	0.18
0.17	−0.07	0.16	−0.03	−0.10	−0.12	−0.05
0.04	−0.01	−0.12	0.10	0.06	−0.14	0.07
0.01	−0.06	0.04	−0.00	−0.02	−0.01	0.02
$\theta = 7$						
−0.11	0.11	−0.11	−0.10	0.10	0.15	0.11
−0.04	−0.01	−0.12	0.02	−0.20	−0.11	−0.16
0.15	−0.06	−0.17	0.03	−0.16	−0.03	0.10
0.01	0.13	−0.05	0.00	0.04	0.11	−0.07
−0.08	−0.24	0.09	−0.04	−0.20	−0.16	−0.15
0.06	0.03	0.01	0.11	−0.14	−0.13	0.05
−0.07	−0.05	0.02	−0.15	−0.18	−0.15	−0.29
$\theta = 8$						
−0.18	−0.17	0.13	−0.10	−0.03	−0.16	−0.10
0.09	−0.09	0.07	−0.06	−0.18	−0.00	−0.13
−0.03	−0.13	0.16	−0.01	−0.17	0.12	0.00
−0.01	0.27	0.05	−0.20	0.07	0.10	0.05
−0.04	−0.07	0.18	−0.10	0.07	−0.06	−0.17
0.10	0.09	0.00	0.17	0.00	−0.08	−0.05
−0.23	0.10	−0.12	0.04	0.04	−0.01	−0.08

Note: The underlined figures (in absolute value) are greater than 2 approximate standard errors, $2T^{-1/2}$, $T = 84$.

Table 3.11 *Residual correlation matrices for the restricted reduced form, 1952–1973*

			$\theta = 0$			
1.00	−0.06	0.12	−0.08	−0.01	0.14	0.19
−0.06	1.00	−0.09	0.19	−0.16	0.12	0.11
0.12	−0.09	1.00	−0.25	0.13	−0.08	0.00
−0.08	0.19	−0.25	1.00	−0.15	−0.10	0.23
−0.01	−0.16	0.13	−0.15	1.00	−0.14	0.11
0.14	0.12	−0.08	−0.10	−0.14	1.00	0.19
0.19	0.11	0.00	0.23	0.11	0.19	1.00
			$\theta = 1$			
0.02	0.01	−0.02	−0.02	−0.01	0.01	−0.08
0.01	−0.26	−0.03	0.09	0.18	−0.13	−0.13
−0.11	0.22	−0.03	0.15	−0.00	−0.07	0.03
0.14	−0.23	−0.05	−0.01	0.09	−0.02	0.07
0.10	0.01	0.14	0.01	−0.01	0.09	0.09
−0.07	0.13	−0.11	0.12	0.03	0.19	0.05
0.04	0.00	0.00	0.16	0.29	−0.09	0.27
			$\theta = 2$			
0.10	0.14	0.11	−0.03	−0.17	0.02	0.01
0.00	−0.04	0.15	0.05	−0.02	0.15	0.20
−0.16	−0.08	0.04	0.12	−0.05	−0.13	0.02
−0.19	0.21	−0.01	−0.08	0.04	0.18	0.08
−0.09	0.01	−0.20	0.15	−0.02	−0.12	−0.01
0.26	−0.18	−0.06	0.12	0.09	0.12	0.06
0.11	−0.05	−0.10	−0.03	−0.01	0.02	−0.00
			$\theta = 3$			
−0.02	−0.04	0.04	−0.01	0.05	−0.07	−0.06
−0.15	0.05	−0.10	−0.12	0.11	0.04	−0.16
−0.06	−0.06	0.09	−0.04	0.07	0.14	0.16
−0.10	0.04	−0.10	0.09	0.11	−0.03	−0.07
0.22	−0.18	0.08	0.02	−0.16	−0.04	0.02
0.09	−0.10	−0.16	−0.08	0.10	0.15	−0.20
0.01	0.05	−0.23	−0.04	−0.02	0.20	−0.03
			$\theta = 4$			
0.17	0.23	0.25	−0.07	−0.05	−0.04	0.03
0.09	−0.08	0.07	−0.01	0.12	−0.13	−0.05
−0.09	0.00	0.11	−0.11	0.02	0.04	0.11
0.09	−0.14	0.04	0.06	0.14	0.00	0.13
−0.07	0.10	0.03	−0.16	0.06	0.05	−0.09
0.15	−0.08	−0.16	−0.09	−0.10	−0.12	−0.29
0.13	0.01	0.02	−0.17	0.10	0.13	−0.01

Table 3.11 (*cont.*)

			$\theta = 5$			
−0.16	−0.04	0.05	0.25	0.13	−0.12	−0.10
−0.02	−0.06	−0.02	−0.03	−0.04	0.04	0.07
−0.24	0.09	−0.02	−0.18	0.22	−0.08	0.03
0.13	−0.07	0.04	−0.08	−0.02	0.21	0.04
0.06	0.31	−0.04	0.12	−0.20	−0.02	0.02
0.07	−0.14	0.04	0.11	−0.12	−0.12	−0.17
0.09	−0.05	0.18	−0.04	0.08	−0.00	0.02
			$\theta = 6$			
0.06	−0.10	0.19	0.09	−0.04	0.08	0.00
0.10	0.02	0.02	0.02	−0.09	−0.03	0.00
0.05	−0.15	−0.01	−0.01	−0.01	0.08	−0.01
0.03	0.06	−0.16	0.12	0.03	−0.13	−0.02
−0.01	−0.07	0.04	0.07	0.05	−0.17	−0.13
0.04	0.11	0.01	0.03	0.01	−0.09	0.00
0.26	0.16	0.07	0.02	0.05	−0.11	−0.08
			$\theta = 7$			
−0.22	0.16	−0.02	0.01	0.02	0.01	0.07
−0.14	−0.05	−0.02	−0.03	−0.15	−0.17	−0.24
0.10	−0.08	−0.10	−0.06	0.10	0.06	0.15
0.07	−0.07	−0.00	−0.08	−0.03	0.03	−0.04
−0.02	−0.16	0.07	−0.05	−0.01	0.08	0.04
−0.03	0.15	0.05	0.15	−0.01	−0.18	−0.06
0.09	−0.01	0.01	−0.05	−0.13	−0.18	−0.30
			$\theta = 8$			
−0.08	−0.05	0.18	−0.07	0.15	−0.18	−0.01
0.06	0.17	0.05	−0.07	0.02	0.02	−0.02
0.09	0.02	0.20	0.06	−0.00	0.00	0.01
0.00	0.27	−0.04	−0.18	−0.06	0.04	0.11
−0.17	0.16	0.03	−0.02	0.27	−0.02	0.02
0.04	0.11	0.02	0.24	−0.10	−0.07	−0.02
−0.03	−0.04	0.07	−0.07	−0.07	−0.06	−0.04

Note: The underlined figures (in absolute value) are greater than 2 approximate standard errors, $2T^{-1/2}$, $T = 84$.

significant autocorrelations, respectively. Notice that many correlations are only marginally significant. Although we did not jointly test the vector white noise assumption of the disturbances, we conclude from the residual analysis that the restricted model performs fairly well in this respect. (Compare tables 3.8 and 3.10 with 3.9 and 3.11, respectively.) Few existing econometric models have been checked for the cross-equation residual correlations.

3.4 *Dynamic properties and forecasting performance of the model*

3.4.1 Solving the model After choosing a restricted specification, we now look into the dynamic properties of the model. Instead of solving the characteristic equation, which is a polynomial of degree 28, we apply the simulation approach used by Sims (1980a) to the unrestricted and the restricted versions of the model. In order to take account of all the correlation properties of the model while having orthogonal system innovations, we write the model in recursive form. The structural form of a pth-order VAR model with expectations variables, denoted by $\underline{z}_t^*$, can be written as

$$\begin{aligned} \underset{m\times m}{C_0}\ \underset{m\times 1}{\underline{z}_t} &= \underset{m\times m}{C_o^*}\ \underset{m\times 1}{\underline{z}_t^*} + C_1\underline{z}_{t-1} \\ &+ \cdots + C_p\underline{z}_{t-p} \\ &+ \underset{m\times 3}{D}\ \underset{3\times 1}{\underline{s}_t} + \underset{m\times 1}{\quad} + \underset{m\times 1}{\underline{\beta} t} + \underset{m\times 1}{\underline{\eta_t}}, \end{aligned} \tag{3.3}$$

where $C_i, i = 0, \ldots p, C_0^*, D, \underline{\alpha}$ and $\underline{\beta}$ are matrices and vectors of coefficients respectively, $\underline{z}_t$, $\underline{s}_t$, and t are defined in (3.1), and $\underline{\eta}_t$ is a vector of normally distributed disturbances with $E\underline{\eta}_t = 0$, $E\underline{\eta}_t\underline{\eta}'_{t'} = \delta_{tt'}\Omega$, with $\delta_{tt'}$ being the Kronecker delta. When $\underline{z}_t^* = E(\underline{z}_t|\underline{z}_{t-\theta}, \theta = 1, 2, \ldots)$, the matrices and vectors of the reduced form for the observable variables given in (3.1), satisfy the following relationships

$$\begin{aligned} &A_i = FC_i, i = 1, \ldots p, \text{ with} \\ &F = C_0^{-1}[C_0^*[C_0 - C_0^*]^{-1} + I] \\ &B = FD,\ \underline{\delta} = F\underline{\alpha},\ \underline{\gamma} = F\underline{\beta},\ \underline{u}_t = C_0^{-1}\underline{\eta_t} \text{ and} \\ &\Sigma = C_0^{-1}\Omega C_0^{\prime -1}. \end{aligned} \tag{3.4}$$

The covariance matrix Σ can be decomposed as $\Sigma = G\Lambda G'$, where G is a lower triangular matrix, with diagonal elements equal to 1, Λ is a diagonal matrix. Premultiplying (3.1) by G^{-1} yields one recursive form of the model.

We write the model in recursive form with the variables arranged in the following order M, W, U, Y, P, R, and PI, with the recursive and reduced form equations for PI being identical. The equation for M includes the current values of the remaining six variables in the model, the equation for W includes all current endogenous variables except M, etc. Given the openness of the Dutch economy, we expect that the variables PI, R, and P are strongly and quickly influenced by changes in the world economy, while the four remaining variables are also more strongly determined by changes in domestic economic conditions.

The solution of the model (excluding the seasonals and trend term) for the effect of a shock in the initial period equal to one standard error of regression is given in figures 3.1–3.7. This solution is the MAR of a VAR. It is given by the coefficients of the infinite matrix lag polynomial $[I - \sum_{i=1}^{p} A_i L^i]^{-1} G\Lambda^{1/2}$, with L being the lag operator. We solved the model for 120 quarters, but we report the response pattern for the first 60 quarters only. The solution of the homogeneous part of the model seems to be fairly stable. The inclusion of a time trend term in the reduced form equations takes up most of the instability owing to the sustained economic growth during almost the entire sample period. The value of the shock in the initial period is inferable from the figures. For instance, in figure 3.1, the shock of the restricted model for the period 1952–73 equals 0.012. (See the response of M to shock of M.)

With the exception of the response of M, W, P, and PI to shocks of R and the response of P to shocks of Y, the MAR of the unrestricted model is not very sensitive to the choice of the sample period. The empirical results indicate some instability over time of the impact of shocks of R on other nominal variables in the unrestricted VAR model. In general, the unrestricted and the restricted models exhibit a different dynamic behavior. The length of the period and the amplitude of the dominant cycle increase and the shape of the solution becomes much smoother when restrictions are imposed. This empirical finding clearly shows that the dynamic (interim) multipliers of a model can be very sensitive to imposing restrictions on the parameters of the model. Usually, the short-run response patterns of a variable to its own innovation are similar for the unrestricted and the restricted model. This result is not surprising given that the own lags are generously specified in most restricted equations. The response to own innovations seems to be slightly over-estimated in the restricted model. Differences between the restricted and unrestricted models show up in the cross-effects for some variables. In particular, the responses of the nominal variables W, P, and R to shocks of U are reversed when restricting the model, a finding that is a useful hint on how to improve the dynamic specification of the restricted model.

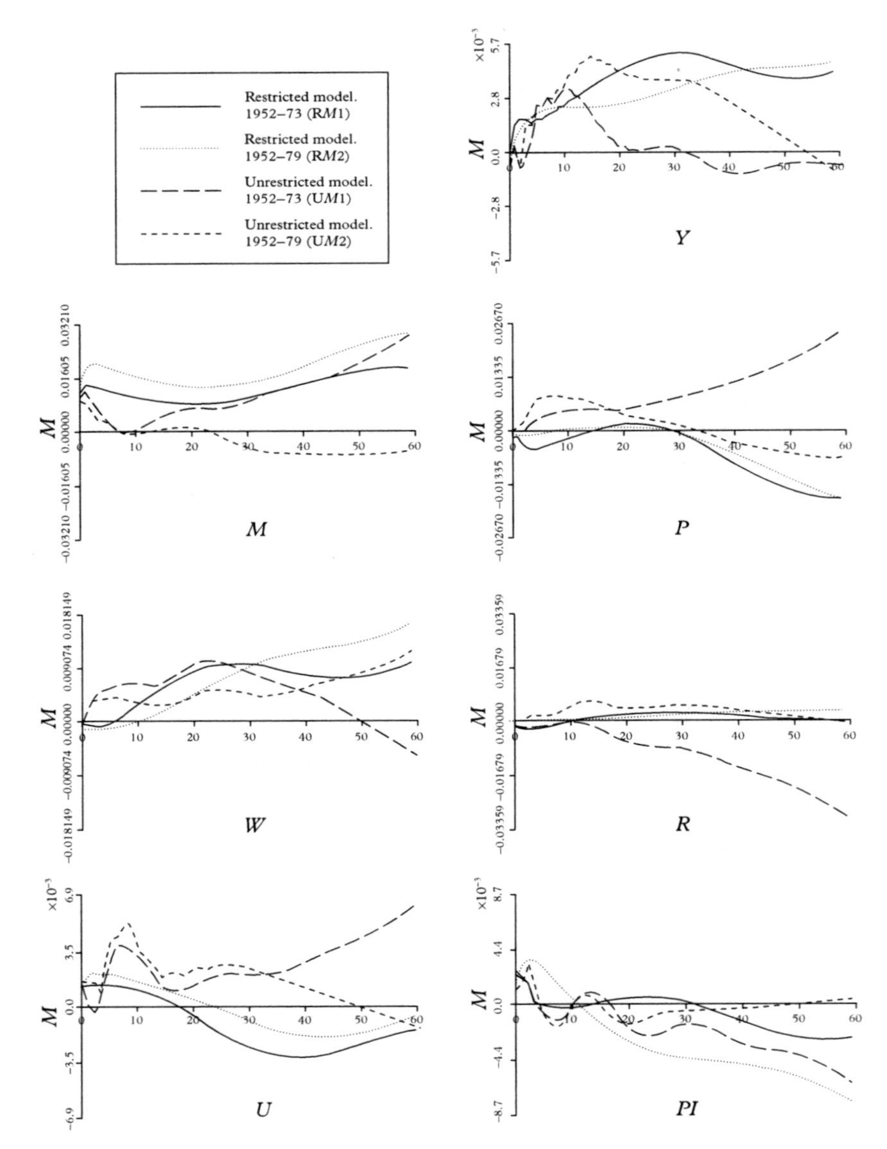

Figure 3.1 Responses of M to shocks of M, W, U, Y, P, R, PI

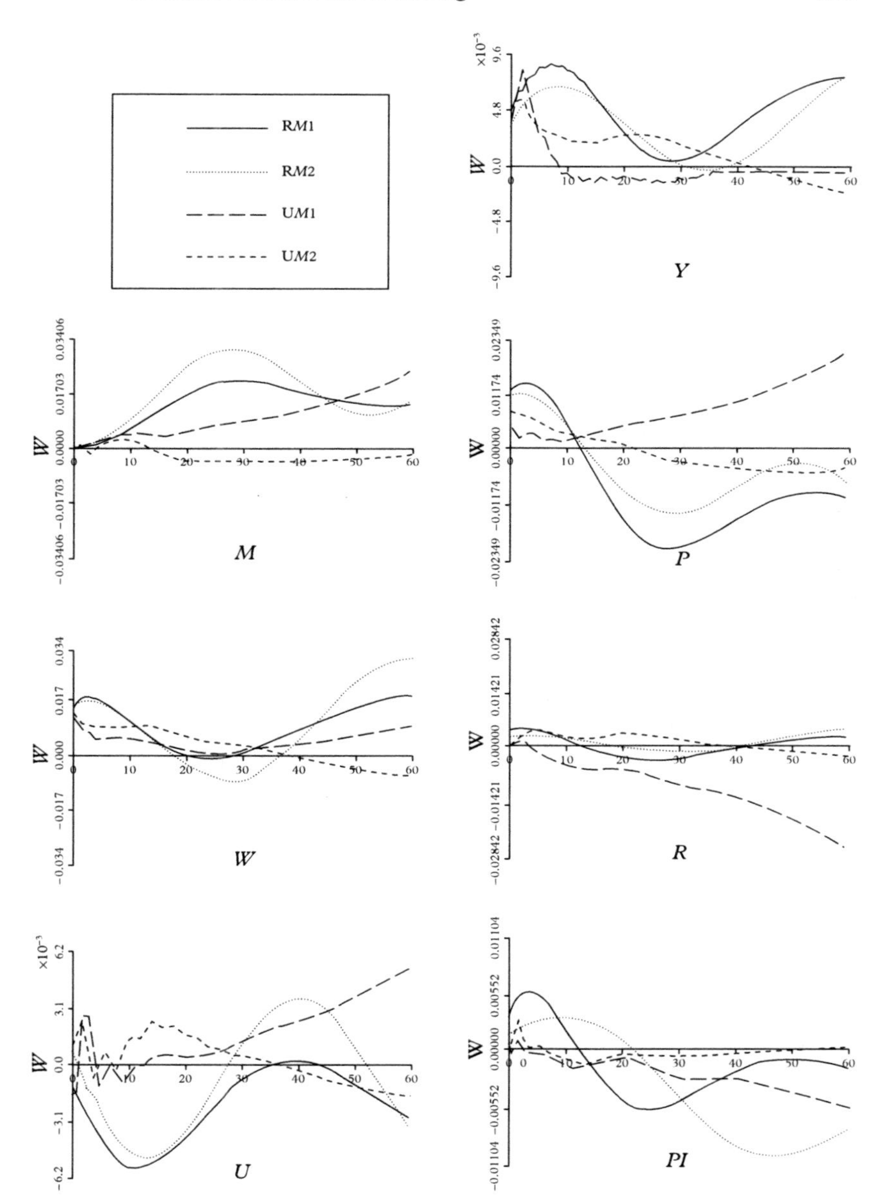

Figure 3.2 Responses of W to shocks of M, W, U, Y, P, R, PI

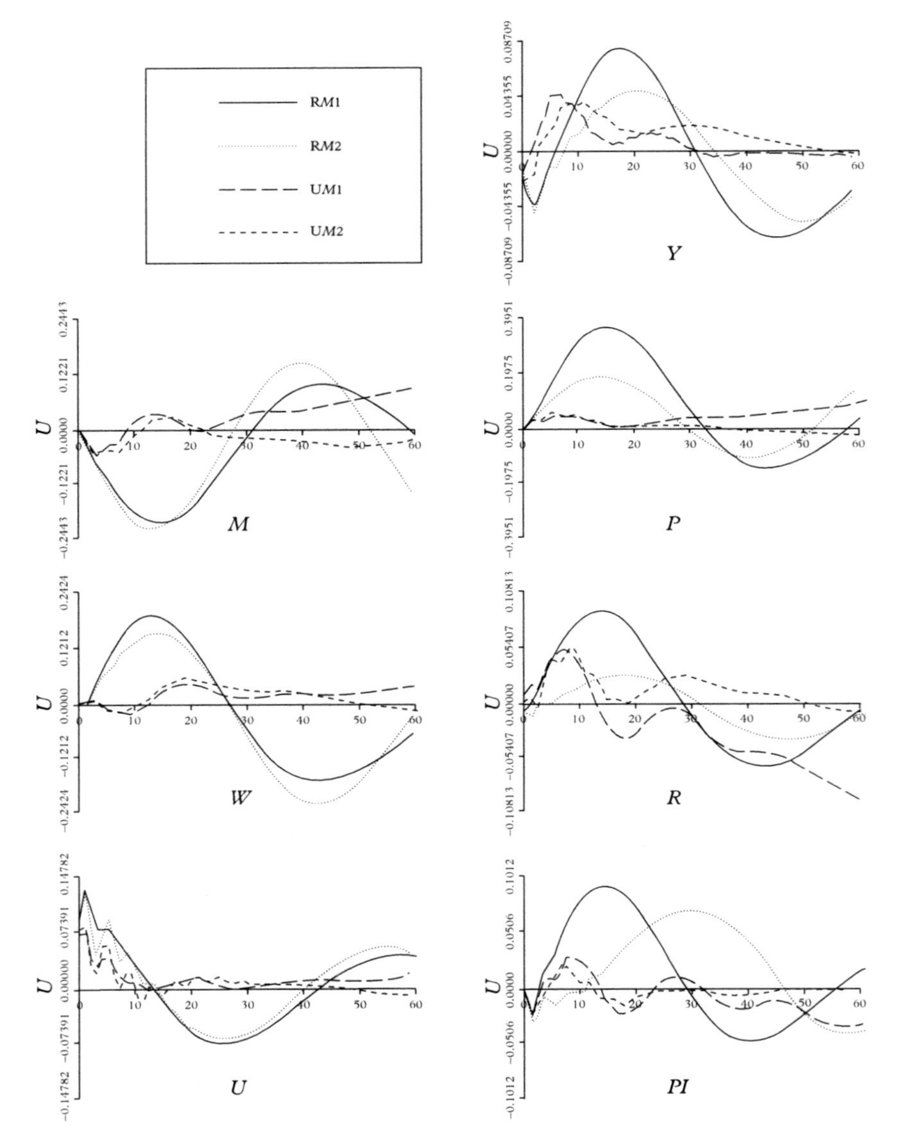

Figure 3.3 Responses of U to shocks of M, W, U, Y, P, R, PI

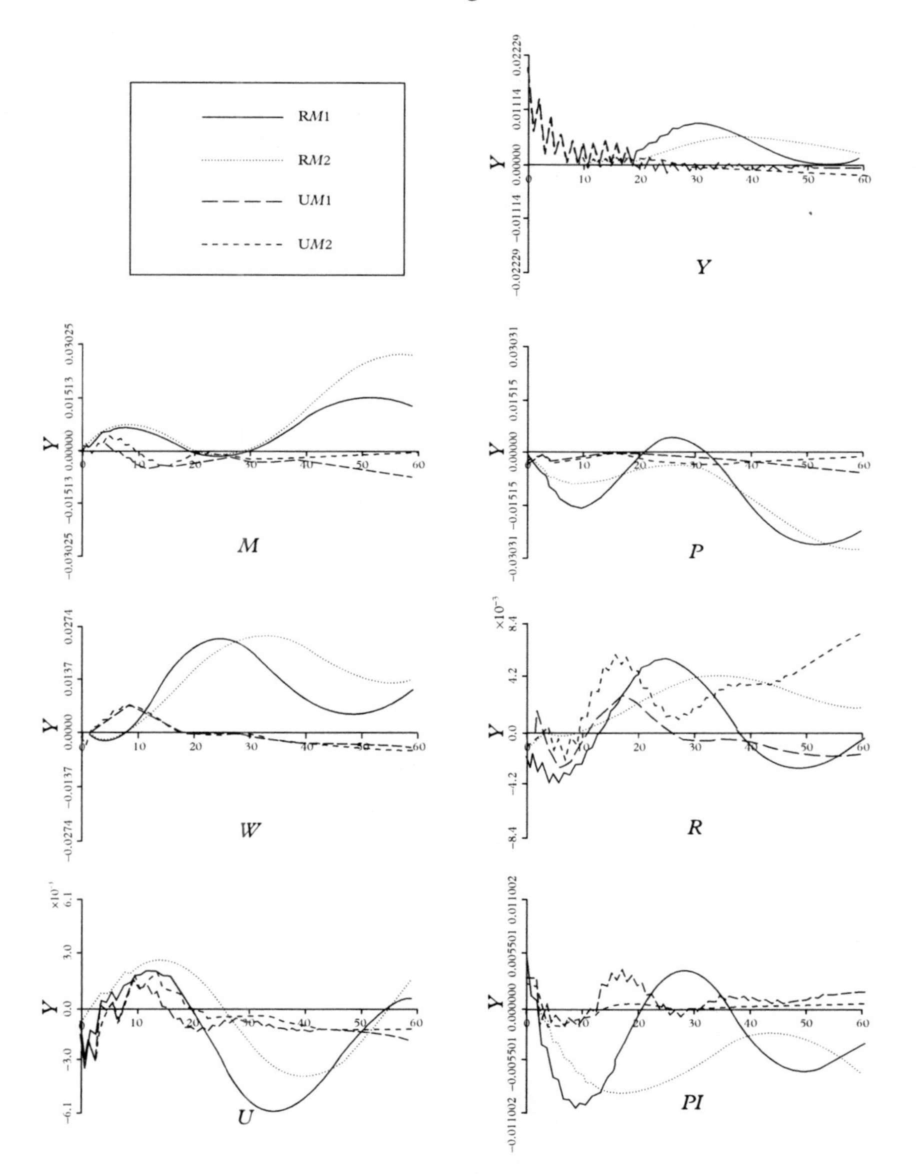

Figure 3.4 Responses of Y to shocks of M, W, U, Y, P, R, PI

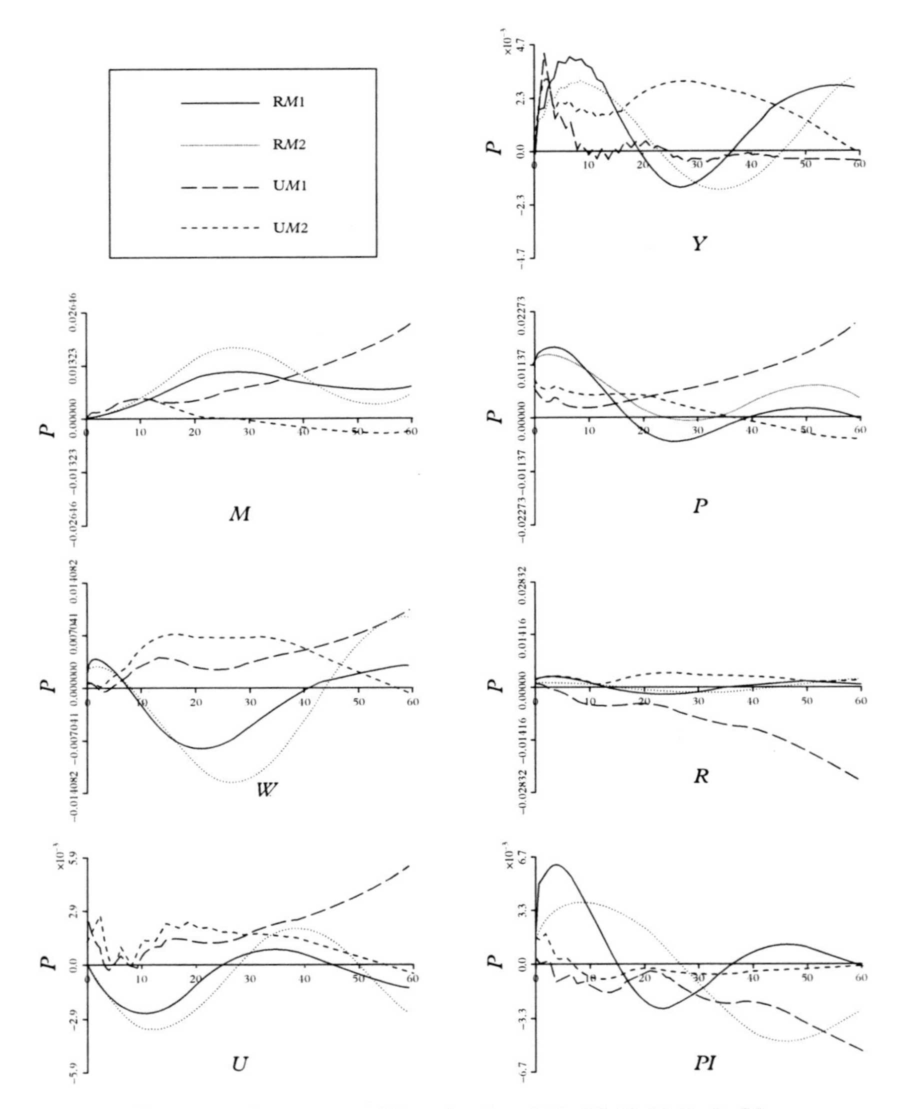

Figure 3.5 Responses of P to shocks of M, W, U, Y, P, R, PI

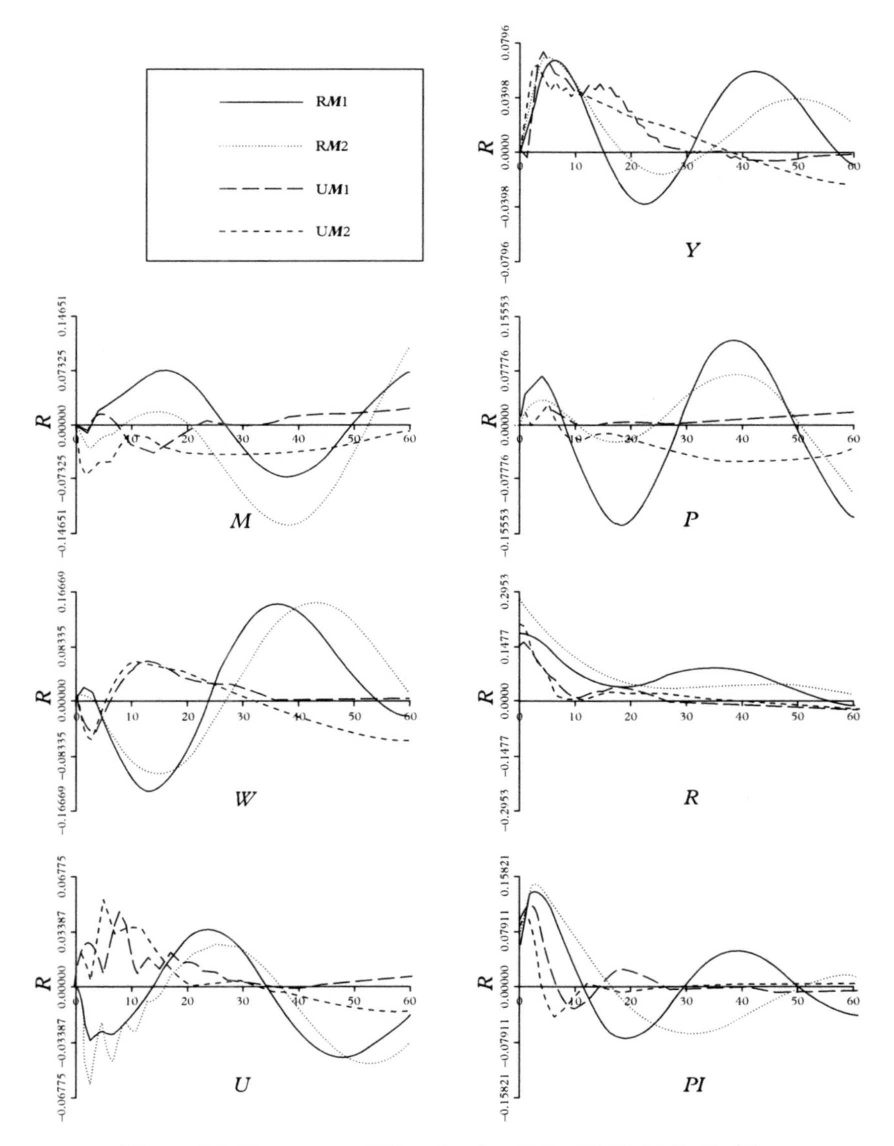

Figure 3.6 Responses of R to shocks of M, W, U, Y, P, R, PI

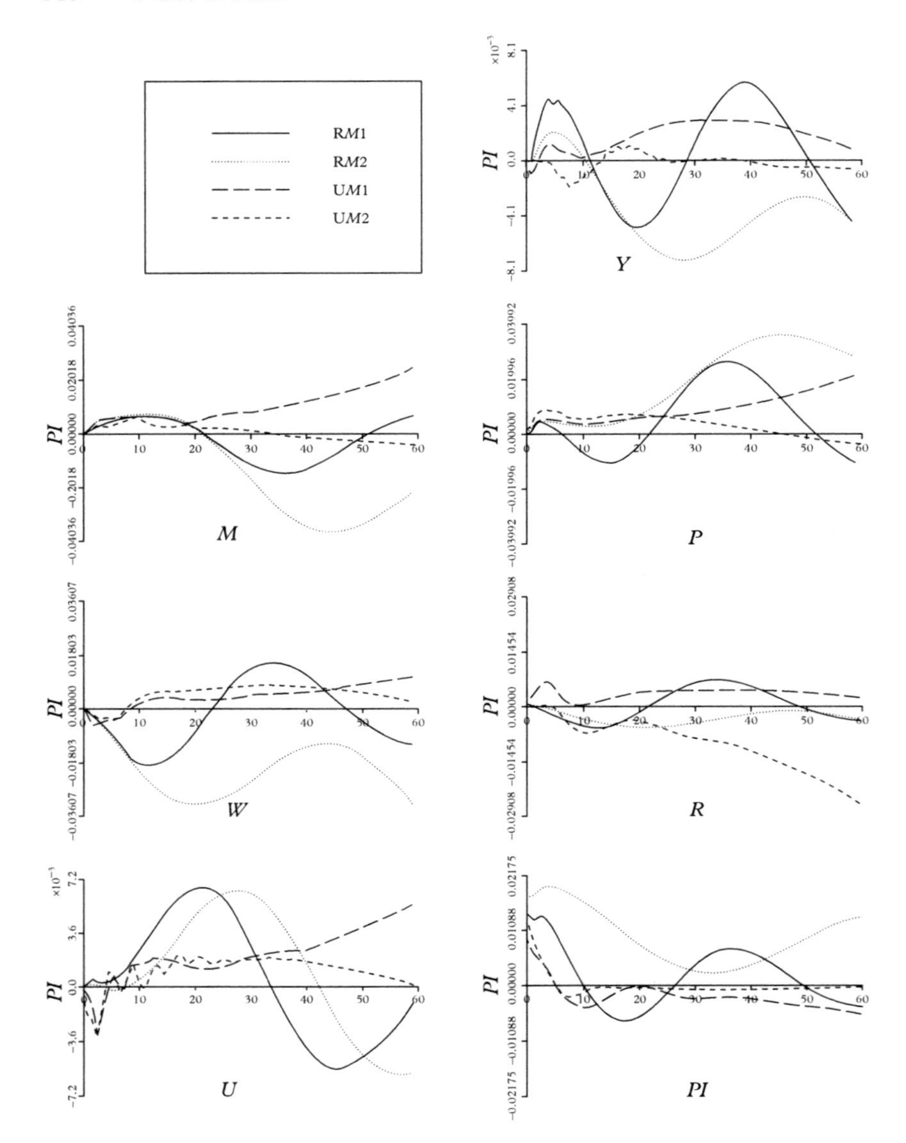

Figure 3.7 Responses of *PI* to shocks of *M*, *W*, *U*, *Y*, *P*, *R*, *PI*

Finally, the estimated step response function obtained by summation of the impact and interim multipliers is usually more robust with respect to imposing restrictions than the impulse response. This finding is important for the use of restricted dynamic models for medium- and long-term policy evaluation.

Similar to the results obtained by Sims (1980a) for the United States, money innovations have persistent effects on the nominal variables in the unrestricted model. For the restricted model, the reactions to money innovations are cyclical. Monetary shocks have some effects on real variables in the unrestricted model.

The response of U in the restricted model has the same shape for all seven innovations, whereas the phase of the cycles in the response is different. The unemployment innovation is followed by an accommodating monetary policy, a decrease in Y first and an increase in Y later on. The reactions of wages and import prices to an unemployment innovation (in the restricted model) are similar to those for the US data. They differ from the pattern obtained by Sims (1980a) for Germany.

The impact of wage shocks on real variables (U, Y) is small in the unrestricted model. It takes much longer than for German data, before the wage innovation has a negative effect on Y. The impact of a wage increase on unemployment becomes really perceptible after two-and-a half years. Prices and wages have similar reactions to shocks in all the variables. Their response to nominal variables is greater than that to real variables. An expenditure innovation is followed by wage and price increases, by a reduction in unemployment first and an increase after a lag of two–seven quarters. Its impact on interest rates is substantial, whereas that on import prices is negligible. The shape of the reaction of Y, W, and PI to an impulse in Y is the same for the Netherlands and for Germany. The impact of an initial price shock on nominal money balances is different for the two periods (unrestricted model). In the first period, a price impulse leads to an expansion of money balances, while for the complete period, the money supply finally reacts negatively to an initial price increase. Real variables are not too sensitive to a price shock. However, they are affected by an increase in the interest rates. The reaction of W, P, and PI to P has a similar shape for the Netherlands and the United States whereas the responses of M, Y, W, P, and PI to a price increase have similar patterns for Germany and the Netherlands. In the reaction of interest rates to a monetary impulse, the Keynesian liquidity effect lasts for two or three quarters in the period 1952–73. It lasts much longer in the period 1952–79. The Fisher effect, that increased liquidities lead to more (expected) inflation and therefore to higher nominal interest rates, is absent from the unrestricted model for 1952–79. It is interesting to note that for long-term US interest rates for the period 1952–71, the liquidity effect disappears after three–four quarters. (See Taylor 1972.) When money balances are measured by M_1, no liquidity effect shows up for US short-term interest rates in an unrestricted twelfth order VAR model for monthly postwar (1948–78) data. (See Sims 1980b.) Finally, import prices have a negative effect on unemployment, followed by a positive effect, and a reverse effect on expenditures. The shape of the reaction of *PI* to its own impulse in the unrestricted model is similar to that for the United States and for Germany.

If we assume that the money supply, the nominal wages, and the interest rates are the instruments for economic policy, controlling the money supply or the interest rates seems to be more effective than wage controls in reducing unemployment in the short run – at least according to the unrestricted model. When using the restricted model, monetary and wage policies seem to be more effective in fighting unemployment than an interest rate policy. They are equally effective in stabilizing the price level in the short run. Compared with the results for the United States and Germany given by Sims (1980a), the pattern of the response functions is much more erratic. This is probably due to the use of seasonally unadjusted data in our study. Notice finally that there is no indication in the simulation results of absence of Wiener–Granger "causality" (or of strong exogeneity, which is more stringent; see Engle, Hendry, and Richard 1980) for any variable in the system as each variable is affected by shocks in any of the seven variables in the model.

In conclusion, given that imposing restrictions on the parameters can substantially affect the dynamic behavior of a model, one should carefully investigate the consequences of these restrictions for the dynamics of the model. The MAR is a very useful means for verifying the stability over time of dynamic relationships, checking the dynamic properties of a model, comparing them with the dynamics of more general models, and investigating the exogeneity of a variable or a set of variables. Although the MAR is sometimes very sensitive to the choice of the variables in the model, there appear similarities in the MAR of models for different periods, countries, and data sets. These similarities point towards the existence of some common underlying structure that has to be disclosed and explained.

3.4.2 The final equation form Next, we check the properties of the final equations or the univariate ARIMA models associated with the VAR processes considered in this study. Premultiplying the system (3.1) by the adjoint matrix of the polynomial matrix operator

$$A(L) = [I - \sum_{i=1}^{p} A_i L^i],\ A^*(L),$$

we obtain the set of final equations associated with (3.1)

$$|A(L)|\underline{z}_t = A^*(L)[B\underline{s}_t + \underline{\delta} + \underline{\gamma} t] + A^*(L)\underline{u}_t, \tag{3.5}$$

where $|A(L)|$ is the determinant of $A(L)$. As there are no exogenous variables in these VAR processes, the analysis of the transfer functions outlined in subsection 2.3.4 cannot be done. From the results of the model simulations, we can conclude that the homogeneous part of our VAR model is stable. Taking annual changes. $I - L^4$, where L is the lag operator, eliminates a linear trend and seasonal dummy variables. This transformation of the endogenous variables is expected to yield stationary seasonally adjusted series, provided the assumptions on the VAR process hold true.

The empirical analysis of the single series confirmed our findings for the multivariate model. Very simple schemes are sufficient to model the seasonals in the series. (See Zellner 1978 for modeling seasonality.) In table 3.12, we report the estimated autocorrelation functions (ACF) and partial autocorrelation functions (PACF) of the differenced series, and in table 3.13, we give the estimated univariate ARIMA models[1] for the period 1952–73. Parsimoniously parametrized specifications have been chosen after an analysis of the estimated autocorrelation functions. One root of the characteristic equation associated with the univariate models for M and for P is slightly smaller than 1. All other roots are substantially smaller than 1. First differencing apparently induced stationarity, which confirms the results for the vector processes. We should mention that we computed the ACFs and PACFs for the complete sample period. The estimated ACFs and PACFs for W and for P are insensitive to the choice of the sample period. For Y, the ACFs and PACFs point towards a slight parameter instability, whereas for M, R, and PI, a structural change in 1974–9 clearly shows up in the ACFs and PACFs. As we had to use seasonal lag polynomials instead of dummy variables in the univariate models, it becomes difficult to check further implications of the unrestricted and restricted VAR models for the properties of the lag polynomials of the final equations. Still, as stated above, the univariate ARIMA models can be used as a standard of comparison for the forecasting properties of the VAR models.

3.4.3 The predictive performance of the models Next we compare the postsample forecasting properties of the unrestricted VAR, the restricted VAR, and the univariate ARIMA models. Each model has been estimated from the data up to 1973. The multi-step-ahead predictions, we

[1] The computations were performed using a computer program for non-linear least squares estimation developed by C. R. Nelson, University of Washington, Seattle.

Table 3.12 *Estimated autocorrelations (A) and partial (P) autocorrelations for the series, 1952–1973*

Series[1]	Filter	1	2	3	4	5	6	7	8	9	10	11	12	S.E.	
M	$1 - L^4$	A	0.83	0.62	0.41	0.23	0.15	0.09	0.03	0.03	0.08	0.15	0.18	0.17	0.11
		P	0.83	−0.24	−0.07	−0.08	0.17	−0.10	−0.04	0.09	0.18	0.05	−0.12	−0.02	0.11
W	$1 - L$	A	−0.10	−0.08	−0.05	0.23	−0.13	−0.09	−0.09	0.38	−0.04	−0.06	−0.06	0.31	0.11
		P	−0.10	−0.09	−0.07	0.22	−0.10	−0.08	−0.10	0.32	0.05	−0.00	−0.04	0.19	0.11
U	$1 - L^4$	A	0.90	0.70	0.46	0.22	0.05	−0.07	−0.14	−0.19	−0.24	−0.28	−0.33	−0.35	0.11
		P	0.90	−0.61	−0.10	−0.02	0.27	−0.31	0.07	−0.21	−0.04	−0.05	−0.08	0.09	0.11
Y	$1 - L^4$	A	0.73	0.55	0.31	0.07	−0.03	−0.10	−0.18	−0.27	−0.24	−0.24	−0.24	−0.14	0.11
		P	0.73	0.03	−0.23	−0.20	0.12	0.01	−0.19	−0.21	0.19	0.02	−0.23	0.08	0.11
P	$1 - L^4$	A	0.82	0.66	0.44	0.25	0.20	0.18	0.20	0.23	0.23	0.21	0.16	0.08	0.11
		P	0.82	−0.05	−0.26	−0.07	0.31	0.06	−0.03	0.02	0.01	−0.01	−0.05	−0.09	0.11
R	$1 - L$	A	0.31	0.07	0.03	−0.03	−0.05	−0.14	−0.17	−0.01	0.02	−0.13	−0.13	−0.12	0.11
		P	0.31	−0.03	0.02	−0.05	−0.03	−0.12	−0.10	0.08	0.01	−0.17	−0.07	−0.09	0.11
PI	$1 - L$	A	0.19	0.15	0.16	0.23	0.03	0.00	−0.18	0.10	0.05	−0.14	−0.07	−0.02	0.11
		P	0.19	0.11	0.12	0.18	−0.06	−0.06	−0.25	0.15	0.08	−0.13	0.01	−0.07	0.11

Series[1]	Filter		13	14	15	16	17	18	19	20	21	22	23	24	S.E.
M	$1 - L^4$	A	0.14	0.12	0.17	0.24	0.30	0.33	0.33	0.31	0.28	0.20	0.13	0.05	0.21
		P													
W	$1 - L$	A	−0.13	−0.04	−0.05	0.27	−0.11	−0.07	−0.17	0.20	−0.09	−0.05	0.00	0.28	0.13
		P													
U	$1 - L^4$	A	−0.33	−0.28	−0.19	−0.09	0.01	0.09	0.15	0.17	0.15	0.09	0.04	−0.01	0.24
		P													
Y	$1 - L^4$	A	−0.12	−0.07	−0.03	−0.05	−0.05	−0.04	−0.01	−0.00	−0.02	−0.09	−0.16	−0.16	0.20
		P													
P	$1 - L^4$	A	−0.01	−0.10	−0.17	−0.17	−0.12	−0.06	0.01	0.04	0.07	0.08	0.07	0.08	0.23
		P													
R	$1 - L$	A	0.05	0.09	0.08	−0.01	−0.01	0.02	−0.03	−0.03	0.02	0.02	−0.01	−0.00	0.13
		P													
PI	$1 - L$	A	−0.03	−0.02	−0.02	−0.05	−0.03	−0.01	0.01	0.00	0.00	0.08	0.06	0.11	0.13
		P													

Table 3.13 *Estimated univariate ARIMA models for the series, 1952–1973*

Series	The estimated model	RSS	DF	σ_ϵ^2 (back forecast residuals excl.)	Q_{12}	*DF*	Q_{24}	*DF*	Q_{36}	*DF*
M	$(1-\underset{(112.8)}{1.220}\mathrm{L}+\underset{(6.445)}{0.240}L^2)(1-\mathrm{L}^4)M_t=\underset{(0.745)}{0.0017}+(1-\underset{(6.445)}{0.666}L^4)\epsilon_t$	0.0164	80	0.00020	8.6	8	15.0	20	19.7	32
W	$(1-\underset{(2.372)}{0.250}\mathrm{L}^4(1-\mathrm{L})W_t=\underset{(4.911)}{0.015}+\epsilon_t$	0.0410	85	0.00048	13.7	10	26.8	22	33.5	34
U	$(1-\underset{(18.24)}{1.546}\mathrm{L}+\underset{(7.365)}{0.634}\mathrm{L}^2)(1-\mathrm{L}^4)U_t=-\text{‘}\underset{(0.242)}{0.001}+(1-\underset{(6.185)}{0.601}\mathrm{L}^4)\epsilon_t$	1.120	80	0.0134	7.5	8	11.5	20	20.8	32
Y	$(1-\underset{(6.396)}{0.721}\mathrm{L}-\underset{(0.215)}{0.024}\mathrm{L}^2)(1-\mathrm{L}^4)Y_t=\underset{(2.531)}{0.013}+\epsilon_t$	0.06693	81	0.00082	14.0	9	18.9	21	22.3	33
P	$(1-\underset{(7.812)}{0.864}\mathrm{L}+\underset{(0.091)}{0.010}\mathrm{L}^2)(1-\mathrm{L}^4)P_t=\underset{(2.219)}{0.0068}+\epsilon_t$	0.0178	81	0.00022	16.2	9	24.9	21	26.6	33
R	$(1-\underset{(3.043)}{0.313}\mathrm{L})(1-\mathrm{L})R_t=\underset{(1.481)}{0.034}+\epsilon_t$	3.864	85	0.0454	6.9	10	8.4	22	11.7	34
PI	$(1-\mathrm{L})PI_t=\underset{(0.005)}{0.00002}+(1+\underset{(2.111)}{0.214}\mathrm{L}+\underset{(3.456)}{0.317}\mathrm{L}^4)\epsilon_t$	0.0347	84	0.00039	18.2	9	21.1	21	26.7	33

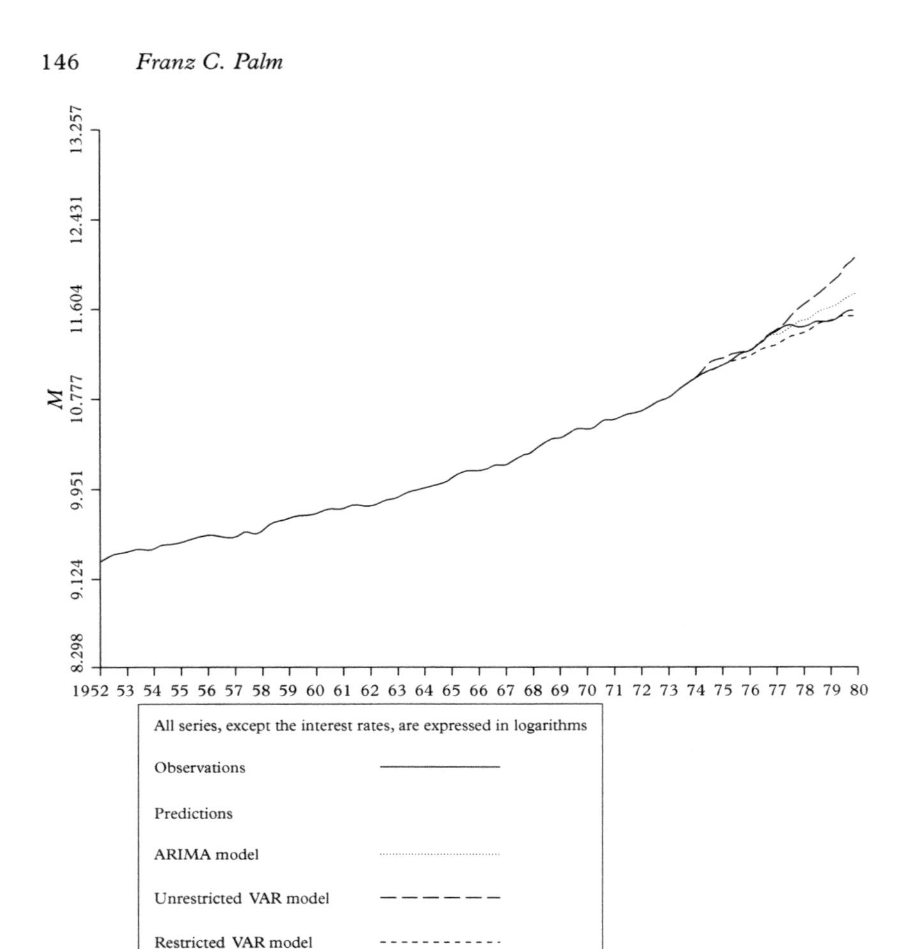

Figure 3.8 Nominal money balances (*M*), 1952–1980

compute, are minimum mean-square error forecasts of the (logarithms of the) series given the model, observations up to the original date and parameter estimates.

In figures 3.8–3.14, the observations (in natural logarithms, except *R*) for the period 1952–79 are plotted. We have also plotted the predictions generated in 1973 for the period 1974–9 using the univariate ARIMA processes, and the unrestricted and restricted VAR models. Except for *Y* and *PI*, the forecasts of the series using the unrestricted VAR model are above the realized values, those for the restricted model are below the true series. The forecasts for the univariate ARIMA models are usually close to the observed values. The medium-term forecasts of the unemployment

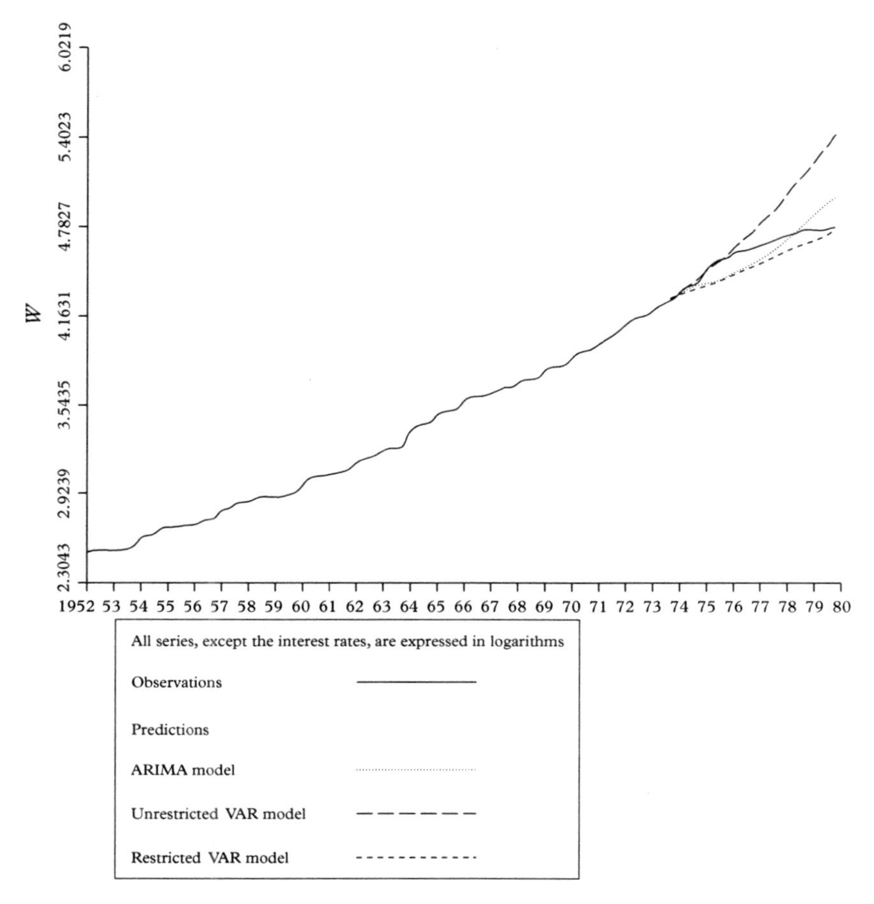

Figure 3.9 Wages (W), 1952–1980

rate using the restricted VAR model are rather inaccurate. The medium-term forecasts for U are very sensitive to the effect of the variable $(M-P)_{-1}$ (its coefficient has been rounded off to 1.5) in the restricted equation (see table 3.3), indicating once more that the specification of the restricted unemployment equation is not entirely satisfactory. Note that by choosing the year 1973 with the large increase of the oil price and the change in the exchange rate regime for the Netherlands to generate forecasts up to twenty-four quarters ahead, we investigate the predictive performance of the models under rather severe conditions.

To assess the forecast performance of the alternative models for various forecast horizons, we computed the ratio of the root mean-square

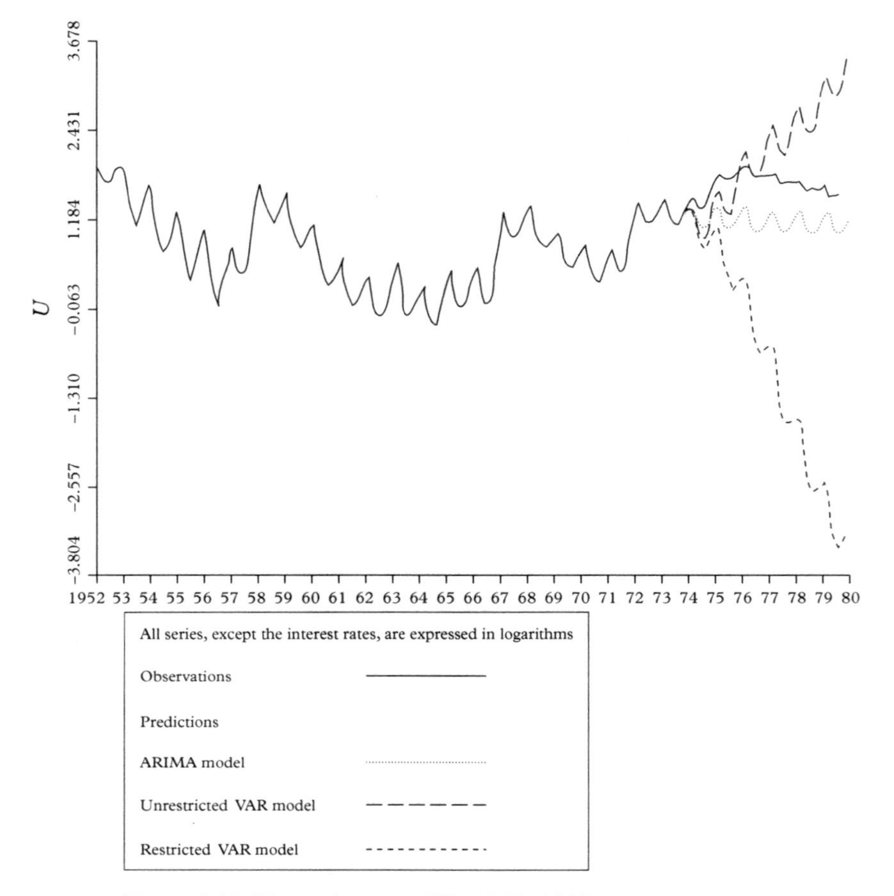

Figure 3.10 Unemployment (U), 1952–1980

prediction error to the root mean square of the realizations for the period 1974–9. These inequality coefficients denoted by

$$U_t(\ell) = \left[\frac{\sum_{t=t_0}^{T} [y_{it+\ell} - \hat{y}_{it}(\ell)]^2}{\sum_{t=t_0}^{T} y_{it-\ell}^2} \right]^{1/2},$$

where $y_{it+\ell}$ is the realization of (the logarithm of) variable i at time $t + \ell$ and $\hat{y}_{it}(\ell)$ is the ℓ-step-ahead forecast of variable i made at time t, are basically descriptive measures of the forecasting precision, although one

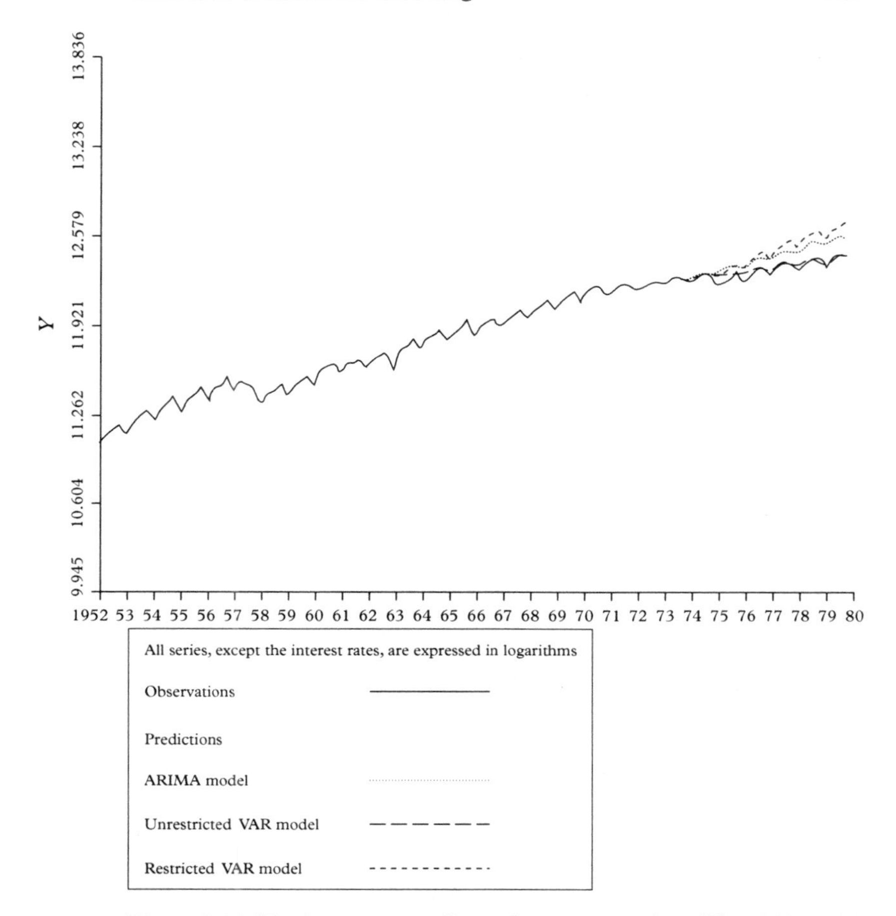

Figure 3.11 Total gross expenditures in constant prices (Y), 1952–1980

could design a decision problem that would require minimization of $U_i(\ell)$ computed for a single variable. In table 3.14, we report the inequality coefficients for ℓ-step-ahead predictions, $\ell = 1, \ldots 16$. The models have been estimated for the period 1952–73 and re-estimated for the period 1952–75. For each model, we generate $(25 - \ell)$ ℓ-step-ahead predictions, $\ell = 1, \ldots 16$, for the period 1974–9 and compute the inequality coefficients. The bias of the forecasts appearing in figures 3.8–3.14 for almost all models and variables disappears when the model is re-estimated using data beyond 1973, or when a different origin date is chosen. Although in practice a forecaster will probably re-estimate his models as soon as new observations become available, we have only once

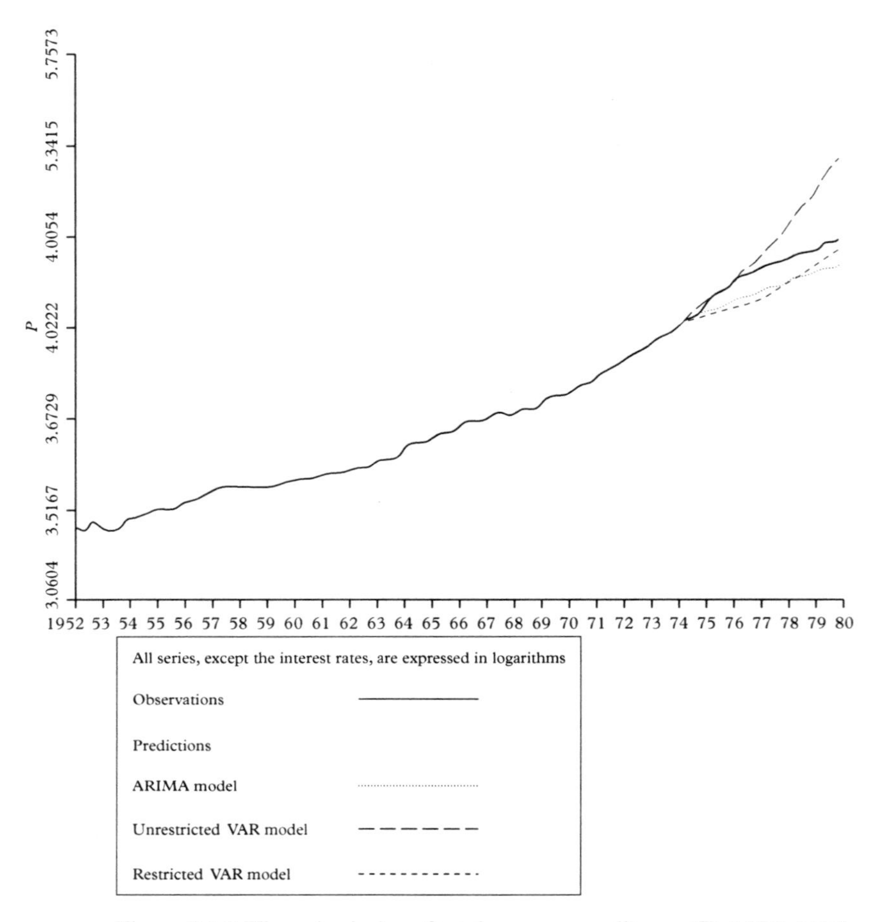

Figure 3.12 The price index of total gross expenditures (*P*), 1952–1980

re-estimated our models for the period 1952–75 in order to limit the computations.

In table 3.14, we also give the inequality coefficients for predictions obtained when smoothness restrictions are imposed on the reduced form parameters of the unrestricted VAR model. Litterman (1980) successfully used smoothness restrictions in forecasting US macroeconomic series. These restrictions can be interpreted as a shrinkage technique or as prior information in a Bayesian framework. We write the ith reduced form equation as

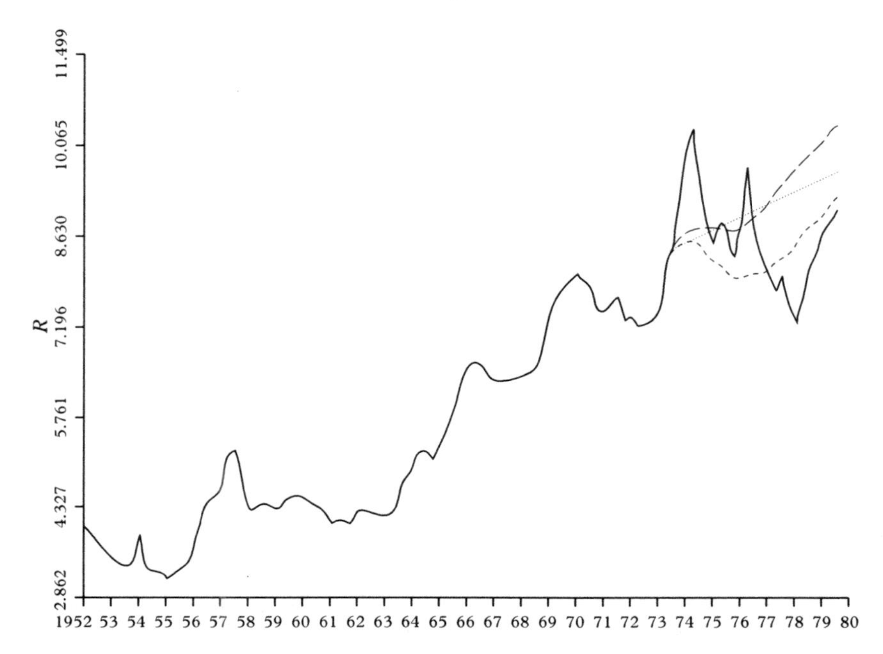

Figure 3.13 Interest rates on long-term government bonds (R), 1952–1980

$$z_{it} = \sum_{j=1}^{7}\sum_{\ell=1}^{4} \alpha_{ij\ell} z_{jt-\ell} + \sum_{k=1}^{3} b_{ik} s_{kt} + \delta_i + \gamma_i t + u_{it}. \tag{3.6}$$

Briefly, in line with Litterman's approach (for more details, refer to Litterman 1980), because a random walk model fits and predicts many economic series rather well, it is used to center the coefficient of z_{it-1} at one, i.e. $E(\alpha_{ii1}) = 1$, and all remaining coefficients at zero. All parameters, except the disturbance variance σ_i^2, b_{ik}, and δ_i, have an informative (normal) prior distribution. The prior standard deviations (SD) are assumed to decrease with ℓ:

$$SD(\alpha_{ij\ell}) = \lambda/\ell, \quad i = j, \quad SD(\gamma_i) = \lambda \tag{3.7a}$$

$$= \theta\lambda\partial_i/\ell\hat{\sigma}_j, \quad i \neq j, \tag{3.7b}$$

where λ and θ are parameters to be specified and $\hat{\sigma}_j$ is the standard error of regression j. The correlation between α_{ii1} and γ_i is assumed to be -0.7, whereas the remaining coefficients are assumed to be uncorrelated with each other. These restrictions can be summarized in a set of stochastic linear restrictions on the regression coefficients and lead to a

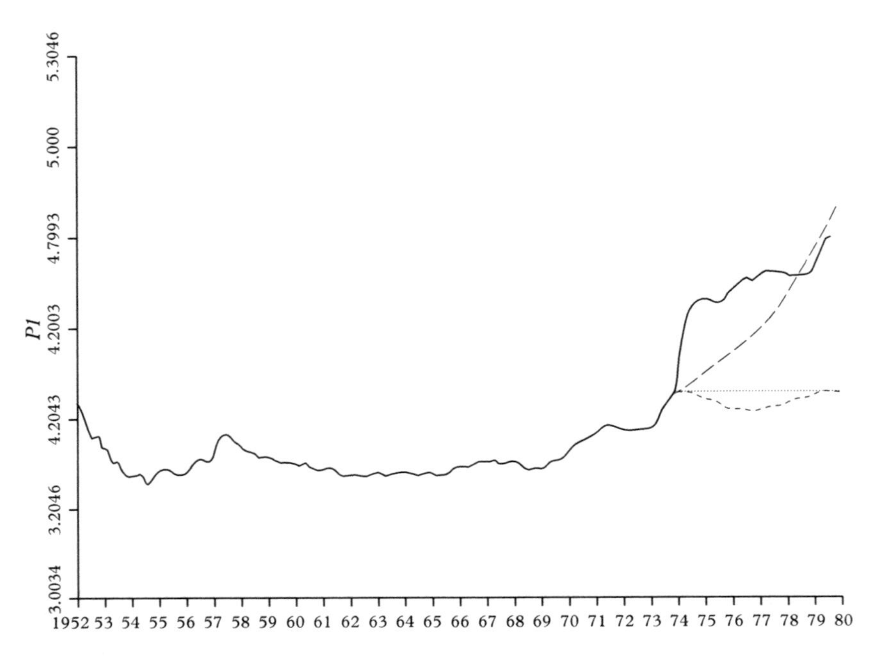

Figure 3.14 The price index of all import goods (*PI*), 1952–1980

"mixed estimator" that has been used to generate the forecasts for the period 1974–9. The parameter λ, which measures the precision of the smoothness restrictions, takes the values 0.5 and 0.1. When $\theta = 1$, own and cross-variables dynamics have equal prior weights. For $\theta = 0.2$, the variance of cross-variables effects is reduced (own weight) by a factor 25. Formulating smoothness restrictions as proposed by Litterman (1980) can be interpreted as going from specific to general; that is, the starting point is a random walk model for which one allows for some cross-variables interaction. Alternatively, as the smoothness restrictions force the coefficients of the (finite) AR representation to die out, they fulfill a role similar to that of an MA part, which implies that the higher order coefficients of the (infinite) autoregressive representation of an ARMA model follow a mixture of exponentials and damped sine waves. Restrictions originating from economic theory can also be implemented in a soft way ("soft theory") through smoothness restrictions. For a given reduced form equation, the explanatory variables are subdivided into variables that are judged important for explaining the endogenous variable, and the remaining predetermined variables. For the first group of explanatory variables, the coefficient standard deviation is given in (3.7a), whereas

Table 3.14 *Inequality coefficients $U_i(\ell)$ for t-step-ahead predictions, 1974–1979*

	Unrestricted VAR model							Smoothness prior restrictions													
								Equal weight: $\lambda = 0.5, \theta = 1$							Equal weight: $\lambda = 0.1, \theta = 1$						
ℓ	*M*	*W*	*U*	*Y*	*P*	*R*	*PI*	*M*	*W*	*U*	*Y*	*P*	*R*	*PI*	*M*	*W*	*U*	*Y*	*P*	*R*	*PI*
1	0.002	0.012	0.128	0.003	0.006	0.061	0.014	0.002	0.013	0.133	0.003	0.006	0.071	0.014	0.002	0.008	0.147	0.003	0.004	0.056	0.011
2	0.003	0.023	0.227	0.004	0.012	0.082	0.021	0.003	0.024	0.229	0.005	0.013	0.117	0.021	0.003	0.016	0.238	0.005	0.009	0.061	0.015
3	0.006	0.031	0.236	0.006	0.018	0.103	0.0226	0.006	0.033	0.249	0.006	0.018	0.154	0.027	0.006	0.024	0.217	0.006	0.014	0.071	0.020
4	0.009	0.040	0.225	0.005	0.023	0.132	0.030	0.009	0.041	0.257	0.005	0.023	0.171	0.030	0.008	0.031	0.200	0.006	0.019	0.081	0.024
5	0.012	0.048	0.269	0.002	0.029	0.152	0.034	0.013	0.049	0.322	0.002	0.029	0.181	0.033	0.011	0.038	0.244	0.005	0.023	0.087	0.027
6	0.015	0.056	0.340	0.004	0.035	0.160	0.034	0.016	0.055	0.410	0.005	0.034	0.184	0.034	0.014	0.044	0.302	0.006	0.027	0.094	0.029
7	0.019	0.064	0.366	0.006	0.041	0.156	0.033	0.020	0.061	0.463	0.005	0.039	0.179	0.033	0.017	0.049	0.318	0.006	0.030	0.097	0.030
8	0.023	0.073	0.382	0.006	0.047	0.140	0.032	0.024	0.068	0.494	0.006	0.044	0.165	0.031	0.020	0.055	0.325	0.006	0.033	0.092	0.029
9	0.027	0.081	0.393	0.006	0.053	0.124	0.030	0.028	0.075	0.512	0.005	0.049	0.152	0.030	0.023	0.060	0.325	0.003	0.037	0.088	0.029
10	0.032	0.090	0.407	0.008	0.060	0.112	0.029	0.032	0.082	0.516	0.007	0.055	0.143	0.028	0.026	0.066	0.335	0.005	0.049	0.087	0.028
11	0.036	0.099	0.413	0.010	0.067	0.108	0.029	0.036	0.089	0.510	0.008	0.062	0.138	0.028	0.030	0.071	0.352	0.005	0.043	0.086	0.026
12	0.040	0.109	0.398	0.011	0.075	0.112	0.031	0.040	0.097	0.485	0.009	0.063	0.138	0.030	0.033	0.077	0.357	0.005	0.047	0.086	0.024
13	0.045	0.119	0.384	0.011	0.083	0.119	0.034	0.044	0.105	0.435	0.009	0.077	0.137	0.033	0.036	0.083	0.331	0.003	0.052	0.083	0.023
14	0.050	0.131	0.430	0.013	0.093	0.131	0.039	0.048	0.114	0.390	0.010	0.087	0.140	0.039	0.039	0.089	0.285	0.006	0.057	0.083	0.021
15	0.055	0.143	0.551	0.014	0.104	0.146	0.045	0.053	0.124	0.444	0.010	0.096	0.151	0.046	0.043	0.096	0.307	0.006	0.062	0.078	0.020
16	0.861	0.156	0.678	0.014	0.116	0.156	0.052	0.057	0.133	0.502	0.010	0.106	0.149	0.052	0.046	0.102	0.306	0.007	0.068	0.083	0.018

Table 3.14 (*cont.*)

	Restricted VAR model (2SLS)							Own weight: $\lambda = 0.5, \theta = 0.2$							Own weight: $\lambda = 0.1, \theta = 0.2$						
ℓ	*M*	*W*	*U*	*Y*	*P*	*R*	*PI*	*M*	*W*	*U*	*Y*	*P*	*R*	*PI*	*M*	*W*	*U*	*Y*	*P*	*R*	*PI*
1	0.002	0.004	0.119	0.003	0.003	0.056	0.012	0.002	0.009	0.120	0.002	0.005	0.059	0.012	0.002	0.006	0.154	0.003	0.003	0.059	0.010
2	0.002	0.007	0.219	0.005	0.004	0.060	0.015	0.003	0.017	0.218	0.005	0.011	0.070	0.016	0.004	0.013	0.258	0.004	0.008	0.071	0.011
3	0.004	0.010	0.255	0.006	0.005	0.069	0.020	0.006	0.025	0.234	0.005	0.017	0.087	0.022	0.007	0.019	0.247	0.005	0.013	0.090	0.013
4	0.005	0.013	0.267	0.005	0.007	0.080	0.026	0.008	0.032	0.257	0.005	0.022	0.101	0.026	0.009	0.025	0.223	0.005	0.018	0.111	0.016
5	0.006	0.016	0.326	0.003	0.009	0.100	0.033	0.011	0.038	0.319	0.002	0.027	0.108	0.029	0.012	0.031	0.281	0.002	0.022	0.127	0.018
6	0.008	0.019	0.435	0.005	0.011	0.127	0.039	0.014	0.044	0.395	0.005	0.031	0.114	0.031	0.015	0.036	0.355	0.005	0.027	0.142	0.020
7	0.010	0.022	0.540	0.005	0.013	0.157	0.045	0.018	0.050	0.445	0.005	0.036	0.115	0.031	0.018	0.042	0.387	0.005	0.032	0.152	0.021
8	0.012	0.026	0.655	0.006	0.015	0.184	0.051	0.021	0.055	0.475	0.006	0.040	0.109	0.030	0.021	0.847	0.400	0.006	0.036	0.152	0.023
9	0.014	0.029	0.784	0.005	0.016	0.209	0.056	0.024	0.061	0.495	0.005	0.045	0.105	0.030	0.024	0.052	0.428	0.004	0.041	0.152	0.024
10	0.016	0.032	0.958	0.008	0.017	0.230	0.061	0.027	0.067	0.515	0.007	0.050	0.109	0.028	0.027	0.057	0.474	0.007	0.045	0.157	0.024
11	0.018	0.035	1.160	0.011	0.018	0.252	0.065	0.031	0.072	0.533	0.008	0.055	0.112	0.027	0.030	0.062	0.505	0.008	0.049	0.157	0.025
12	0.020	0.037	1.372	0.014	0.018	0.272	0.069	0.034	0.077	0.529	0.009	0.060	0.122	0.026	0.034	0.066	0.518	0.008	0.053	0.161	0.025
13	0.022	0.040	1.590	0.016	0.018	0.284	0.072	0.038	0.083	0.509	0.009	0.065	0.133	0.025	0.038	0.070	0.524	0.008	0.057	0.165	0.025
14	0.025	0.044	1.861	0.019	0.018	0.290	0.074	0.042	0.089	0.500	0.011	0.071	0.149	0.025	0.041	0.074	0.539	0.010	0.061	0.177	0.024
15	0.028	0.049	2.175	0.023	0.019	0.272	0.076	0.045	0.096	0.552	0.012	0.078	0.171	0.025	0.045	0.077	0.563	0.011	0.065	0.199	0.023
16	0.031	0.057	2.508	0.027	0.023	0.252	0.078	0.049	0.102	0.606	0.013	0.085	0.190	0.025	0.049	0.081	0.566	0.013	0.069	0.215	0.021

Table 3.14 (*cont.*)

	Univariate ARIMA models							"Soft theory": $\lambda = 0.5, \theta = 0.2$							"Soft theory": $\lambda = 0.1\ \theta = 0.2$						
ℓ	M	W	U	Y	P	R	PI	M	W	U	Y	P	R	PI	M	W	U	Y	P	R	PI
1	0.004	0.004	0.043	0.002	0.003	0.060	0.009	0.002	0.009	0.142	0.003	0.005	0.070	0.014	0.002	0.007	0.162	0.003	0.003	0.056	0.011
2	0.006	0.006	0.099	0.005	0.004	0.064	0.009	0.004	0.017	0.254	0.005	0.011	0.106	0.020	0.005	0.014	0.266	0.005	0.007	0.057	0.014
3	0.008	0.010	0.100	0.006	0.007	0.067	0.009	0.007	0.024	0.292	0.006	0.017	0.141	0.027	0.008	0.021	0.259	0.006	0.011	0.066	0.019
4	0.010	0.014	0.079	0.005	0.011	0.071	0.009	0.010	0.031	0.307	0.005	0.021	0.162	0.031	0.011	0.028	0.248	0.006	0.014	0.073	0.023
5	0.013	0.018	0.045	0.005	0.014	0.075	0.010	0.013	0.038	0.343	0.003	0.026	0.173	0.035	0.014	0.034	0.205	0.004	0.017	0.079	0.026
6	0.016	0.022	0.104	0.007	0.017	0.079	0.010	0.017	0.044	0.393	0.005	0.031	0.179	0.036	0.018	0.040	0.326	0.005	0.020	0.086	0.028
7	0.019	0.026	0.110	0.008	0.020	0.083	0.010	0.020	0.049	0.429	0.005	0.035	0.177	0.036	0.021	0.046	0.342	0.005	0.022	0.091	0.030
8	0.022	0.030	0.079	0.008	0.024	0.089	0.011	0.024	0.055	0.456	0.006	0.039	0.164	0.035	0.024	0.051	0.376	0.005	0.024	0.085	0.031
9	0.025	0.034	0.050	0.048	0.027	0.090	0.011	0.027	0.062	0.471	0.005	0.044	0.149	0.034	0.027	0.057	0.433	0.003	0.026	0.080	0.031
10	0.028	0.038	0.116	0.010	0.030	0.096	0.012	0.030	0.068	0.473	0.007	0.049	0.138	0.032	0.030	0.063	0.507	0.006	0.028	0.080	0.032
11	0.031	0.043	0.125	0.012	0.033	0.101	0.012	0.034	0.074	0.489	0.008	0.054	0.127	0.030	0.033	0.068	0.620	0.007	0.029	0.080	0.032
12	0.033	0.046	0.091	0.012	0.036	0.109	0.012	0.037	0.081	0.486	0.009	0.060	0.120	0.030	0.036	0.073	0.750	0.009	0.030	0.081	0.032
13	0.036	0.050	0.062	0.012	0.038	0.117	0.013	0.041	0.088	0.437	0.009	0.067	0.114	0.030	0.039	0.078	0.863	0.010	0.031	0.081	0.032
14	0.040	0.054	0.134	0.014	0.041	0.097	0.013	0.044	0.096	0.337	0.011	0.074	0.111	0.031	0.042	0.003	0.971	0.013	0.032	0.089	0.031
15	0.044	0.058	0.149	0.016	0.043	0.106	0.013	0.048	0.104	0.315	0.011	0.081	0.118	0.034	0.045	0.099	1.107	0.015	0.032	0.095	0.030
16	0.048	0.060	0.109	0.015	0.046	0.121	0.014	0.051	0.113	0.273	0.011	0.089	0.115	0.036	0.047	0.095	1.249	0.017	0.033	0.111	0.030

that for the second category of variables is given in (3.7b). The lagged values of the endogenous variable and the price variable P are assumed to be relevant in all equations. In addition, for M, the first group consists of Y and R; for W, the relevant variable is U; for U, the relevant variables are M, W, Y, and PI; for Y, these are M, U, and PI; for P, these are W, Y, and PI; for R, these are M, U, Y, and PI; and finally for PI, these are M, W, and PI. Inequality coefficients for models using "soft theory" are also given in table 3.14.

The results in table 3.14 are very interesting. Despite the fact that we re-estimated our models only once and that we analyzed the forecast performance for the period after the oil price increase, the forecasting performance is very reasonable, with the exception of the accuracy of the medium-term forecasts for the unemployment variable. The restricted VAR model predicts (the logarithms of) M and P better than any alternative model. For W, the restricted VAR model predicts as well as the univariate ARIMA model and it predicts better than the remaining models. Notice that we have been able to formulate behavioral equations for these three variables taking the unrestricted VAR model as a maintained hypothesis. As might be expected from the results in figure 3.10, the unemployment rate is rather inaccurately predicted over a horizon longer than four quarters when we use the restricted VAR model. Beyond a horizon of ten quarters, the predictions are worse than zero-level extrapolations. This also happens for 15- and 16-step-ahead forecasts of U using "soft theory" stochastic restrictions with $\lambda = 0.1$. For y, the short-run forecast performance of all the models is almost identical. Over a horizon of sixteen quarters, it is quite similar, whereas over a longer horizon, the equal weight prior with $\lambda = 0.1$ performs better than other models. Finally, for the import prices, the univariate ARIMA model and the model with equal weight prior restrictions and $\lambda = 0.1$ predict better than the other models. Contrary to the conclusion from the exogeneity test of PI, the results in table 3.14 seem to indicate an absence of Wiener–Granger "causality" from domestic variables to PI.

To conclude, the restricted VAR model seems to predict most of the variables better than the unrestricted VAR model does. Also, models with smoothness restrictions imposed generally forecast rather well. However, there is not a single set of parameter values λ and θ for which the forecasting precision is uniformly better over the different variables and horizons.

Of course, the results in table 3.14 should be interpreted with care. As we do not have the probability distribution of the inequality coefficients, we cannot use them to formally test the predictive properties. Approximate forecast intervals can be straightforwardly computed using the MAR of the model. Also, one could average the inequality coefficients, weighting them by a factor inversely related to the length of the forecast horizon, thereby ignoring the dependence between forecast errors for successive periods. More work on the forecasting performance of models analyzed here is in progress.

To see how smoothness restrictions affect the dynamics of the model, we computed the MAR for the models, when the smoothness restrictions are imposed on the reduced form parameter estimates for 1952–73. As an illustration, the graphs of the MAR for M and U are represented in figures 3.15 and 3.16. With equal weight or "soft theoretical" restrictions imposed, the MAR for all variables usually has the same shape as that of the unrestricted VAR model. Own weight restrictions sometimes heavily distort the shape of the MAR. The conclusion that models with "soft theoretical" restrictions seem to predict fairly well and have dynamics similar to those of the unrestricted VAR model is very interesting and indicates a sensible way of restricting densely parametrized models. The possible implications of this result for modeling have to be more extensively explored.

To summarize this section, we investigated the lag length and the structural stability of a VAR model for seven macroeconomic variables for the Netherlands. Next we tested for the exogeneity of PI and for that of PI, R, P, Y, and U jointly. Upon acceptance, these restrictions combined with the absence of instantaneous Wiener–Granger "causality" would imply a block recursive model, for which the transformation of the structural form into the reduced form would not affect the maximal lag length within a block. But, given that the exogeneity restrictions had to be rejected, we restricted the individual equations in the model using a limited information approach along the lines of traditional econometric modeling. Thereby, we used the unrestricted VAR model as a framework in which the true model is assumed to be nested. The outcome of the test of all restrictions jointly is not unambiguous, suggesting that some of the restrictions might not be correct. We did not find much evidence in the residual correlations that points towards a misspecification of the restricted and unrestricted models. The dynamic properties of the unrestricted VAR model were found to be different from those of the restricted one. The accuracy of the forecasts from the unrestricted and the restricted VAR models and from the univariate ARIMA schemes is different too. The use of restrictions, either exact or stochastic, often improves the forecast performance. Both VAR models were found to be consistent on a number of points with the properties of the univariate ARIMA schemes.

Finally, although the restricted VAR model can be improved in many ways, in particular the cross-variables interaction in some equations,

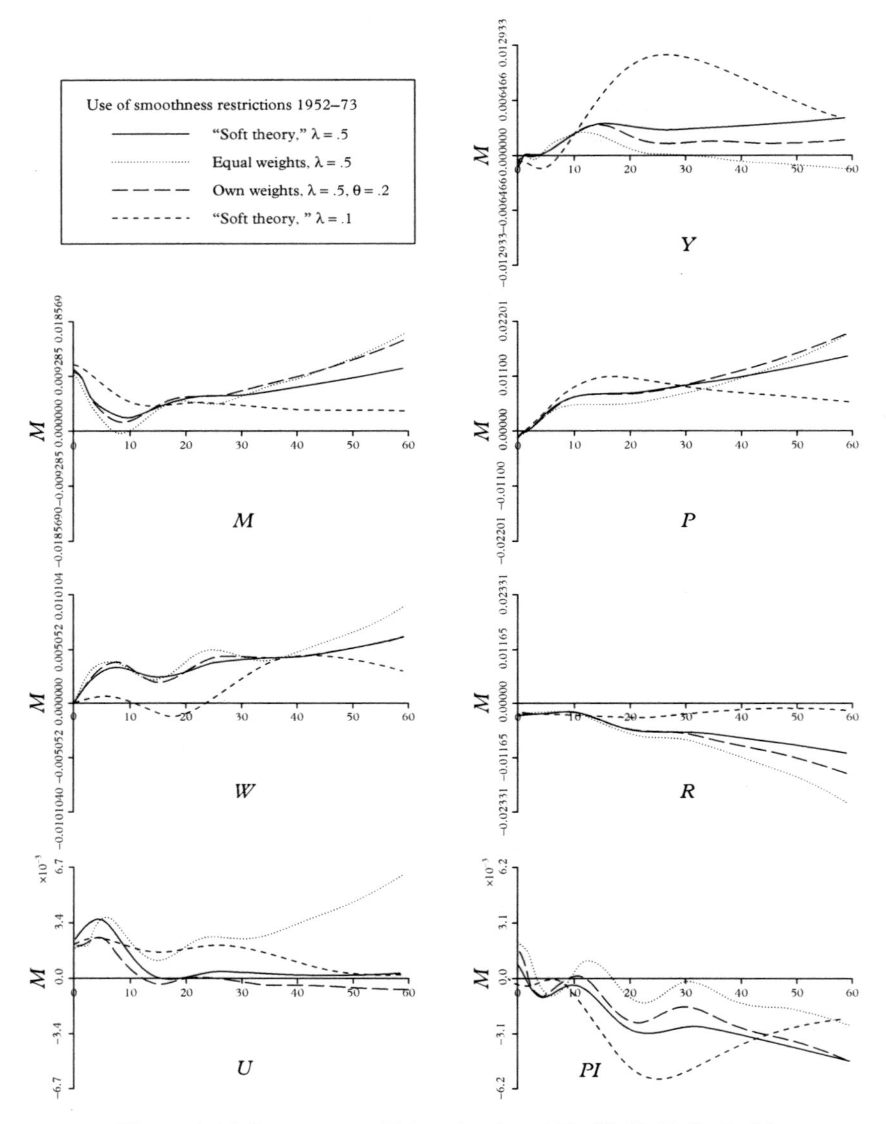

Figure 3.15 Responses of M to shocks of M, W, U, Y, P, R, PI

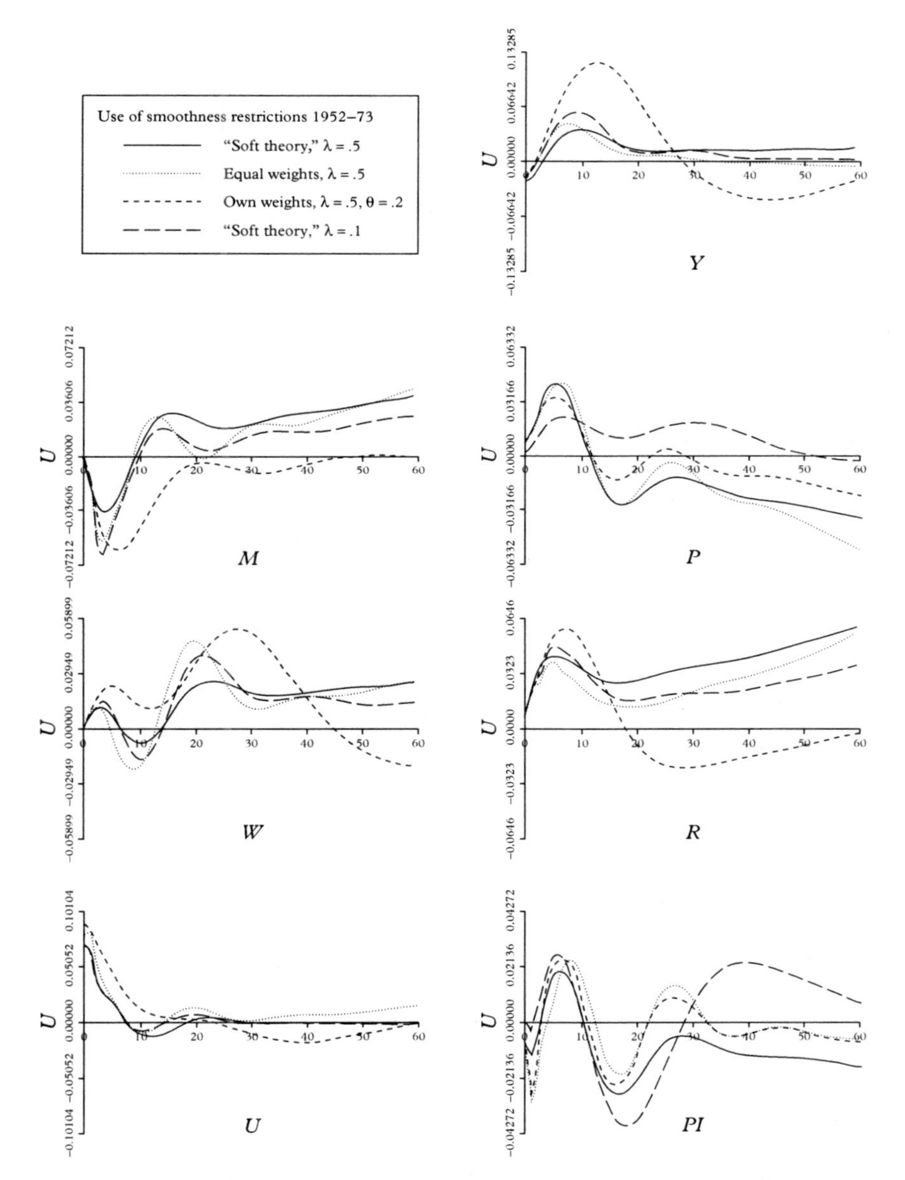

Figure 3.16 Responses of U to shocks of M, W, U, Y, P, R, PI

it does not seem to be very inferior to the unrestricted model, despite the substantial reduction of the number of parameters. Also, "soft" restrictions may be a useful alternative to dogmatic restrictions in econometric modeling.

4 Some tentative conclusions

In this chapter, we first presented the traditional approach to econometric modeling and several procedures proposed and applied in the time series literature on modeling bivariate and multivariate processes. Then we outlined the main features of the SEMTSA, which is an attempt to integrate econometric specification analysis and time series modeling that should be complements to rather than substitutes for each other. In the second part, we applied the SEMTSA to a VAR model for some of the main macroeconomic variables for the Netherlands. The restricted model obtained through SEMTSA has different dynamic properties than the unrestricted one, but it does not seem to be inferior to the unrestricted VAR model in terms of the results of diagnostic checking and forecasting properties. Certainly, the specification of the restricted model can be improved. A comparison of the MAR of the restricted and unrestricted model indicates where the cross-variables dynamics are affected by the restrictions. One should, of course, allow for sampling errors in the MAR estimates. For this purpose, it would be useful to compute approximate confidence intervals for the MAR parameters.

As already indicated earlier, the formulation of behavioral and theoretically meaningful restricted relationships often requires the introduction of additional variables in the model. For instance, the restricted reduced form equation for unemployment predicts rather badly. However, there is more information available on the medium-term development of the labor market than we used in our model. The change of the total labor force, the hiring for the public sector can be predicted fairly well over a horizon of two–three years. However, it is not possible to model this kind of detail in small dimensional VAR models. High dimensional VAR models with a rich lag structure still are computationally intractable. When modeling a larger number of economic variables using a top-to-bottom approach, the general initial model will have to be obtained from a restricted model by expanding the dynamics of those equations for which the lag structure is very uncertain. The alternative "bottom-up" approach, which consists of formulating a simple, parsimoniously parametrized model, usually does not suffer from a computational intractability owing to a large number of parameters.

However, it should be noted that a "bottom-up" approach often has elements in common with a "top-down" approach. For instance, the "bottom-up" approach proposed by Box and Jenkins (1970) (for multiple time series, see Tiao and Box 1981) starts with an analysis of the autocorrelations and partial autocorrelations, i.e. an analysis of a very dense parametrization of a second order stationary process. For the gas furnace data, Tiao and Box (1981, p. 813) show that the correct model structure could not be detected if only very low order vector autoregressive processes were considered as is sometimes done in a "bottom-up" approach.

Clearly, a better understanding of the interaction of economic variables in time is needed. We fully agree with the statement by Nerlove (1972, p. 277): "Without strong theoretical justification for a particular form of lag distribution, and perhaps even strong prior belief about the quantitative properties of that distribution and the factors on which those properties depend, it is generally impossible to isolate the lag distribution in any very definitive way from the sort of data generally available." However, we want to add that a theoretically justified dynamic model only lacks a confrontation with "hard facts," i.e. the empirical validation of the model.

Finally, a number of questions arise with the formal procedures for econometric modeling in general. The statistical properties of the procedures presented and applied in SEMTSA are only partially known. Quite often, one has to reject a set of restrictions tested at once or sequentially when the overall size is fixed at conventional levels (e.g. using the Bonferroni inequality or the Scheffé procedure). This happens for the large sample chi-square test even if one corrects it for the loss of degrees of freedom as Sims (1980a) does. More research into the finite sample properties such as the power of the sequential tests used in specification analysis is needed and it is expected to be very rewarding. The contributions to the field of pretest estimators may be very valuable, too, although some areas of application of pretesting which are relevant for SEMTSA are still relatively unexplored.

Instead of looking for the statistical properties of the modeling procedure as a whole, one can interpret it as a pursuit of consistency of the accepted model in its different forms with the information available such as *a priori* information on structural parameters and on multipliers, the conformity of the autocorrelations of the endogenous and exogenous variables and the residuals of the different forms with the properties of the autocorrelation functions implied by the finally accepted model. Many econometricians consider this as a minimum requirement.

APPENDIX DATA SOURCES

The series M, U, Y, P, and R have been collected at De Nederlandsche Bank NV, and were kindly provided to us by Professor Dr. M. M. G. Fase. The series W and PI are published in *Maandschrift van het C.B.S.* (Den Haag, Centraal Bureau voor de Statistiek).

BIBLIOGRAPHY

Anderson, T. W. (1971), *The Statistical Analysis of Time Series* (New York, John Wiley)

Barnett, W. A. (1980), "Economic monetary aggregates," *Journal of Econometrics* 14, 11–48

Blommestein, H. J. and F. C. Palm (1980), "Econometric specification analysis – an application to the aggregate demand for money in the Netherlands," Paper presented at the Econometric Society World Meeting, Aix-en-Provence

Box, G. E. P. and G. M. Jenkins (1970), *Time Series Analysis, Forecasting and Control* (San Francisco, Holden-Day)

Box, G. E. P. and D. A. Pierce (1970), "Distribution of residual autocorrelations in autoregressive integrated moving average time series models," *Journal of the American Statistical Association* 65, 1509–26

Box, G. E. P. and G. C. Tiao (1975), "Intervention analysis with applications to economic and environmental problems," *Journal of the American Statistical Association* 70, 70–9

Byron, R. P. (1974), "Testing structural specification using the unrestricted reduced form," *Econometrica* 42, 869–83

Christ, C. F. (1975), "Judging the performance of econometric models of the US economy," *International Economic Review* 16, 54–74

Darroch, J. N. and S. D. Silvey (1963), "On testing more than one hypothesis," *Annals of Mathematical Statistics* 34, 555–67

Davidson, J. E. H. *et al.* (1978), "Econometric modelling of aggregate time series relationship between consumers' expenditure and income in the United Kingdom," *Economic Journal* 88, 661–92

den Butter, F. A. G. and M. M. G. Fase (1979), "The demand for money in EEC countries," Research Report, De Nederlandsche Bank NV

Dhrymes, Ph. J., *et al.* (1972), "Criteria for evaluation of econometric models," *Annals of Economic and Social Measurement* 1/3, 291–324

Drèze, J. H. (1976), "Bayesian limited information analysis of the simultaneous equations model," *Econometrica* 44, 1045–76

Driehuis, W. (1972), *Fluctuations and Growth in a Near Full Employment Economy* (Rotterdam, Universitaire Pers Rotterdam)

Engle, R. F., D. F. Hendry and J. F. Richard (1980), "Exogeneity, causality and structural invariance in econometric modelling," CORE Discussion Paper 8038, University of Louvain

Geweke, J. (1978), "Testing the exogeneity specification in the complete dynamic simultaneous equation model," *Journal of Econometrics* 7, 163–86

Godfrey, L. G. (1976), "Testing for serial correlation in dynamic simultaneous equation models," *Econometrica* 44, 1077–84

Granger, C. W. J. (1969), "Investigating causal relationships by econometric models and cross-spectral methods," *Econometrica* 37, 424–38

(1981), "Some properties of time series data and their use in econometric model specification," *Journal of Econometrics* 16, 121–30

Granger, C. W. J. and P. Newbold (1977), *Forecasting Economic Time Series* (New York, Academic Press)

Hannan, E. J. (1971), "The identification problem for multiple equation systems with MA errors," *Econometrica* 39, 751–65

Haugh, L. D. and G. E. P. Box (1977), "Identification of dynamic regression (distributed lag) models connecting two time series," *Journal of the American Statistical Association* 72, 121–30

Hendry, D. F. (1974), "Stochastic specification in an aggregate demand model of the United Kingdom," *Econometrica* 42, 559–78

(1980), "Predictive failure and econometric modelling in macroeconomics: the transactions demand for money," in P. Ormerod (ed.), *Modelling the Economy* (London, Heinemann Educational)

Hendry, D. F. and Von Ungern-Sternberg, Th. (1980), "Liquidity and inflation effects on consumers' expenditure," in A. S. Deaton (ed.), *Essays in the Theory and Measurement of Consumers' Behaviour* (Cambridge, Cambridge University Press)

Hernández-Iglesias, C. and F. Hernández-Iglesias (1980), "Causality and the independence phenomenon: the case of the demand for money," *Journal of Econometrics* 15, 247–64

Jenkins, G. M. and A. S. Alavi (1981), "Some aspects of modelling and forecasting multivariate time series," *Journal of Time Series Analysis* 2, 1–48

Kiviet, J. (1979), "Bounds for the effects of ARMA disturbances on tests for regression coefficients," Report, University of Amsterdam

Leamer, E. (1978), *Specification Searches – Ad Hoc Inference With Nonexperimental Data* (New York, John Wiley)

Litterman, R. B. (1980), "A Bayesian procedure for forecasting with vector autoregressions," manuscript

Malinvaud, E. (1981), "Econometrics faced with the needs of macroeconomic policy," *Econometrica* 49, 1363–77

McCallum, B. T. (1976), "Rational expectations and the natural rate hypothesis: some consistent estimates," *Econometrica* 44, 43–52

Mizon, G. E. (1977), "Model selection procedures," in G. J. Artis and A. R. Nobay (eds.) *Studies in Modern Economic Analysis* (Oxford, Basil Blackwell)

Mizon, G. E. and D. F. Hendry (1980), "An empirical application and Monte Carlo analysis of tests of dynamic specification," *Review of Economic Studies* 47, 21–46

Nerlove, M. (1972), "Lags in economic behaviour," *Econometrica* 40, 221–51

Newbold, P. (1981), "Model checking in time series analysis," Paper presented at the ASA–CENSUS–NBER Conference on Applied Time Series Analysis of Economic Data, Washington, DC, October

Nieuwenhuis, A. (1980), "Aggregate price equations," Central Planning Bureau, The Hague, manuscript

Palm, F. C. (1981), "Structural econometric modelling and time series analysis – towards an integrated approach," Vrije Universiteit, Economic Faculty, R. M. 1981–4

Prothero, D. L. and K. F. Wallis (1976), "Modelling macroeconomic time series" (with discussion), *Journal of the Royal Statistical Society*, A 139, 468–500

Rao, C. R. (1973), *Linear Statistical Inference and its Applications*, 2nd edn. (New York, John Wiley)

Samuelson, P. A. (1965), "Proof that properly anticipated prices fluctuate randomly," *Industrial Management Review* 6, 41–9

Sargan, J. D. (1964), "Wages and prices in the United Kingdom: a study in econometric methodology," in P. E. Hart, G. Mills, and J. K. Wittaker (eds.), *Econometric Analysis for National Economic Planning* (London, Butterworth)

(1980), "Some tests of dynamic specification for a single equation," *Econometrica* 48, 879–98

Sargent, Th. J. (1981), "Interpreting economic time series," *Journal of Political Economy* 89, 213–48

Sargent, Th. J. and C. A. Sims (1977), "Business cycle modeling without pretending to have too much a priori economic theory," in C. A. Sims (ed.), *New Methods in Business Cycle Research: Proceedings From a Conference*, Federal Reserve Bank of Minneapolis, 45–109

Schwarz, G. (1978), "Estimating the dimension of a model," *Annals of Statistics* 6, 461–4

Silvey, S. D. (1959), "The Lagrange multiplier test," *Annals of Mathematical Statistics* 30, 389–407

Sims, C. A. (1977), "Exogeneity and causal ordering in macroeconomic models," in C. A. Sims, (ed.), *New Methods in Business Cycle Research: Proceedings From a Conference*, Federal Reserve Bank of Minneapolis, 23–43

(1980a), "Macroeconomics and reality," *Econometrica* 48, 1–48

(1980b), "Comparison of interwar and postwar business cycles: monetarism reconsidered," *American Economic Review* 70 (2), 250–7

Taylor, D. G. (1972), "The effects of monetary growth on the interest rate," unpublished PhD dissertation, University of Chicago

Tiao, G. C. and G. E. P. Box (1981), "Modelling multiple time series with applications," *Journal of the American Statistical Association* 76, 802–16

Trivedi, P. K. (1975), "Time series analysis versus structural models: a case of Canadian manufacturing behavior," *International Economic Review* 16, 587–608

Wallis, K. F. (1977), "Multiple time series analysis and the final form of econometric models," *Econometrica* 45, 1481–98

Zellner, A. (ed.) (1978), *Seasonal Analysis of Economic Time Series* (Washington, DC: US Department of Commerce, Bureau of the Census)

(1979), "Statistical analysis of econometric models," *Journal of the American Statistical Association* 74, 628–51; chapter 2 in this volume

(1980), "Statistical analysis of hypotheses in economics and econometrics," *Business and Economics Statistics Section, Proceedings of the American Statistical Association*, 199–203

Zellner, A. C. and F. Palm (1974), "Time series analysis and simultaneous equation econometric models," *Journal of Econometrics* 2, 17–54; chapter 1 in this volume

(1975), "Time series and structural analysis of monetary models of the US economy," *Sankhyā: The Indian Journal of Statistics*, Series C 37, 12–56; chapter 6 in this volume

Zellner, A. and S. Park (1979), "Minimum expected loss (MELO) estimators for functions of parameters and structural coefficients of econometric models," *Journal of the American Statistical Association* 74, 185–93

Comment (1983)

Carl F. Christ

In science, two opposite approaches are useful. One is to begin with a theory and test it against data. The other is to begin with data, look for regularities, and seek to build a theory to account for them.

Structural econometric modeling (SEM), in principle, begins with a theory and tests it against data. I agree with Professor Palm that the specification chosen for the theory is of crucial importance. Unfortunately, there is no systematic method for discovering a good theoretical specification. I agree also that it is desirable to begin with a model that is general enough to include the correct model as a special case, and then appeal to data to narrow the general model down to a more specific model. Unfortunately again, it's hard to begin that way, because models that are simple enough to be tractable are not necessarily general enough to include the correct model.

Time series analysis (TSA) seeks to discover empirical regularities that connect current and past observations of variables and that have purely random error terms. When I first encountered TSA, it struck me as a mindless method of data-mining, using no knowledge or theories about the subject matter being studied. However, SEM regards certain variables as exogenous, and supposes no knowledge of the processes that determine them. Hence, the proposal by Zellner and Palm (1974) to combine SEM for a set of variables designated as endogenous with TSA for a set of variables designated as exogenous appears promising.

The ARIMA equations obtained from TSA for exogenous variables need to be tested against future data, just as do the econometric equations obtained from SEM for endogenous variables given the exogenous variables. Do the equations and parameter values fitted to past data continue to fit future data as expected? If not, they are suspect. If so, we may have tentative confidence in them.

Palm tells us that if all methods predict the endogenous variables poorly, then either the model is misspecified or it has been subjected to a structural change. I suggest that an apparent structural change is a case of misspecification, in this sense: If we can explain the structural change, we can then specify a more general model that will give a unified explanation of events both before and after the apparent structural change.

Palm begins with brief descriptions of SEM and TSA. Then, he describes and recommends techniques that can be used to combine them, once an initial model has been chosen that is general enough to include the correct model as a special case nested within the general model.

In the remainder of [chapter 5], Palm illustrates the use of these techniques by means of a model of the Dutch economy. It is useful to distinguish several steps in his procedure.

First, he introduces us to his initial "unrestricted" seven-equation quarterly autoregressive dynamic model of the Dutch economy. We learn that the model is linear in seven endogenous variables and four exogenous variables. The seven endogenous variables are the interest rate R and the logarithms of the money stock, the wage rate, the unemployment rate, real income, the price level, and the import price level. (These logs are denoted by M, W, U, Y, P, and PI respectively.) The four exogenous variables include no economic variables: They are a linear time trend and three seasonal dummies. The unrestricted model contains current values and four lagged values of each endogenous variable. It has no identifying restrictions, and hence is unidentified. (Palm wisely does not attempt to estimate it.) Formally, its structural form can be represented by Palm's (3.3).

Second, consider the reduced form of this unrestricted model. It is represented by (3.1). Palm estimates it by ordinary least squares[OLS], for 1952–73 and for 1952–79. He doesn't show the estimates, but he tells us about them, and shows the residual correlation matrices in tables 3.8 and 3.10 . He concludes that this model "is acceptable as a starting point." As noted below, I am not ready to accept this conclusion.

Third, Palm examines the t-ratios of the reduced form coefficients, and drops the fourth lag of all variables except U and Y on grounds of insignificance. He tests and rejects the hypotheses that (a) import prices are exogenous to a subset of six equations, and (b) that the five variables PI, R, P, Y, and U are exogenous to a subset of two equations. He does this by testing for block triangularity of the reduced form matrices.

Fourth, Palm specifies a restricted structural model, nested within the original unrestricted model (except that he now includes two more dummy variables). He is frank to say that most of the equations of this restricted model do not have a clear behavioral derivation or interpretation. These equations are given in tables 3.1–3.7, along with 2SLS estimates for the same two sample periods. There is very little simultaneity in the coefficient matrix of these restricted equations. Three of them are, in effect, single-equation models, each determining the current value of a single variable, based on past data. These are the equations for unemployment (3.3), import prices (3.7), and the wage rate (3.2). (The latter contains price expectations, but these are assumed to be based on past data.) The only simultaneity in the matrix occurs next in the causal chain: Current income and prices are simultaneously determined by (3.4) and (3.5), based on the previously determined wage rate and past data. Then, (3.6) determines the current interest rate, based on the price level and past data. And, finally, (3.1) determines the money stock, based on income and the interest rate and past data. But Palm notes that the system is not block-recursive, because the covariance matrix of disturbances is not block-diagonal. I might add that when the equilibrium system is considered (obtained by setting all lagged variables equal to current values), the system's coefficient matrix is fully simultaneous, as determined by the neat method of McElroy (1978). It is a curious system from the economic point of view. All seven variables are determined without reference to any policy variables, or indeed any exogenous economic variables. The model ignores taxes, government expenditures, and Central Bank policy variables.

Fifth, Palm compares the structural estimates for the two periods, and reports that the parameters of the restricted model seem to be fairly stable over the sample period. He says this even though, of the 40 parameters other than constant terms and seasonal dummy coefficients, 11 change by more than a factor of 2 when the years 1974–9 are added to the sample. He also performs several tests of the restricted model, some of which reject it for the longer sample period (1952–79).

Sixth, Palm estimates the reduced form of the restricted model. Again the estimates are not presented, but the residual correlation matrices are shown in tables 3.9 and 3.11.

Seventh, for forecasting purposes, he forms a recursive system from the reduced form (3.1) by premultiplying it by the matrix G^{-1}, which diagonalizes the covariance matrix of disturbances. (See (3.4) and the paragraph following it.) The recursive system is used to obtain dynamic multipliers, plotted in figures 3.1–3.7 for both sample periods and for both the unrestricted and restricted models. The two models yield very different dynamic multipliers, as Palm notes. This means that it's very important to know which of the models (if either) is close to reality. Changing the sample period makes a substantial difference to the unrestricted model multipliers in about half of the 49 cases.

Eighth, Palm obtains the univariate ARIMA equations (the final equation forms), and exhibits them in table 3.13 . The autoregressive process order never exceeds 4, and in six of seven cases is 0 or 1 or 2. The moving average process order never exceeds 4, and in four of seven cases is zero.

Ninth, Palm presents and summarizes forecasts (table 3.14 and figures 3.8–3.14) made by the two recursive forms (of the unrestricted and restricted models) and the univariate ARIMA models. There are some very bad forecasting errors, especially for unemployment via the recursive forms (errors of 38 percent–150 percent for forecasts two to four years ahead). Four years ahead, the ARIMA forecasts are either best or second best among the three for every one of the seven variables. That doesn't speak very well for either of the structural models.

The imposition of smoothness restrictions on the reduced form of the unrestricted model makes a modest improvement in its forecasting ability for income and the interest rate. (See table 3.14.)

Let us reconsider the unrestricted seven-equation model described above, from which Palm's analysis began. He tells us that it "is assumed to be sufficiently general to include the data-generating process." I regard this model as a useful vehicle for illustrating the techniques he has in mind. But surely the "data-generating process" for the Dutch economy is not captured by such a simple linear quarterly difference equation model of seven endogenous variables with four lags. Where is monetary policy represented in this model? Fiscal policy? Foreign exchange policy? International trade and investment? The stock of productive resources? Productivity?

Palm concedes that "quite obviously" his fourth order multivariate autoregressive reduced form (3.1) is not a complete model of the Dutch economy. He claims, however, that it can be considered as an approximation to the reduced form of a complete model of the Dutch economy, from which all but his seven variables have been eliminated by integration. (This claim is weaker than the assumption that the seven-equation model is "sufficiently general to include the data-generating process.") The claim may be correct. It has testable implications about the behavior of variables that belong in a complete model of the Dutch economy but do not appear in Palm's seven-equation model. It would be desirable to test these implications empirically before accepting Palm's claim.

In conclusion, the statistical testing methods illustrated here are interesting and promising. Structural econometric model builders will do well to learn more about them. But the method of obtaining the unrestricted and restricted models, and the seven equation models so obtained, are seriously flawed, in my judgment.

BIBLIOGRAPHY

McElroy, F. W. (1978), "A method of causal ordering," *International Economic Review* 19, 1–23

Zellner, A. and F. C. Palm (1974), "Time series analysis and simultaneous equation econometric models," *Journal of Econometrics* 2, 17–54; chapter 1 in this volume.

Comment (1983)

Christopher A. Sims

There are very few examples in the econometric literature of economy-wide models estimated with serious attention to their properties as multivariate stochastic processes. The model Palm estimates in his [chapter 3] is one of these examples, and it deserves imitation and extension. I believe that what Palm is doing here is important, and I have only small disagreements with the way he specifies, estimates, and presents his model.

Since there is so much work underlying the chapter, I cannot mention every aspect of it that merits special commendation or criticism. Let me begin by drawing attention to Palm's extensive use of plots of the multivariate moving average representations of models with different specifications as ways of comparing them. Like the autocovariance function, the spectral density, and the autoregressive representation, the moving average representation is a complete summary of the second order properties of the model. All these ways of summarizing a model's properties are also connected to the model's forecast accuracy in a natural way – if a sequence of models has forecast error variances converging to the forecast error variance of the true model, it must also have moving average representations (and autocorrelation functions, and spectral densities, and autoregressive representations) converging at least pointwise to their true values. In this, these summaries differ from some other apparently natural ones. Lag lengths, parameter values, and characteristic equation roots in a sequence of ARIMA models can stay arbitrarily far away from those of the true model while forecast errors converge to those of the true model. Thus, models that appear to be very different in terms of their ARIMA parameterizations may have similar second order properties. It is worthwhile, therefore, to compare their moving average representations to see whether and how they differ.

As compared to the other summaries which connect naturally to forecast error, the moving average representation has the advantage that it can be interpreted as a set of simulations of responses to typical "shocks."

These shocks, in turn, often invite behavioral economic interpretation. (Some would even argue that they make such interpretation misleadingly natural.) It appears to be more difficult to interpret the other summaries in most circumstances. The spectral density is valuable for intepreting models of seasonal data, despite the appearance of imaginary numbers in its multivariate form. The Fourier transform of the moving average representation might combine some advantages of the spectral density and the moving average representation; John Geweke of the University of Wisconsin has . . . [made] use of functions of the Fourier transform of the moving-average representation in interpreting multivariate economic time series models. The coefficients of the autoregressive representation, when treated as real valued functions of the lag, will have highest power (largest absolute values of their Fourier transforms) at precisely those frequencies where the spectral density is smallest. This means that for smooth economic time series, the coefficients of the autoregressive representation are likely to be oscillatory and erratic, making them difficult to interpret directly.

I wish Palm had gone a bit further (of course this is unfair, considering how much work he has already done) in using the differences in the moving average representations (MARs) and other diagnostic devices he employs to determine what it is about his restricted model that gives it such a strong predictive advantage over all the other models for M and P and over all the other multivariate models for W. Correspondingly, he might have gone further to discover why all the multivariate models do so much worse than the univariate model for U. He suggests that, since the restricted model does nearly worst of all the models at predicting U, the restrictions on that equation deserve special scrutiny. But it is not necessarily true that the restrictions on the U equation are primarily responsible for the model's poor performance in predicting U. It could even be that the same restrictions which help predict M are making the U forecasts deteriorate. That it is not mainly the U equation restrictions which are to blame is borne out by the fact that all the multivariate models are more similar to each other in their poor forecasts of U than they are to the much better univariate model forecasts for this variable.

Palm formally follows the standard textbook prescription for generating linear restrictions on his simultaneous equations model from economic theory. Some of the equations are explicitly acknowledged to be reduced form equations, and we are given no explanation for how *a priori* theory leads to restrictions on these equations, though restrictions are imposed. Two more, those for R and Y, are equations containing more than one current endogenous variable, but are explicitly labeled as not having any behavioral intepretation. For these equations also there is no explanation of how *a priori* theory leads to restrictions. The M equation is claimed to be a demand for money equation, and theory clearly is behind at least one of the restrictions on it – particularly the omission of level effects of the interest rate because they proved to be of the wrong sign. But since the interest rate equation has no behavioral interpretation, how is it meant to be distinguished from the demand for money equation? Would it not have been simpler to treat this equation, as well as the Y equation, as straightforward reduced forms? Perhaps there is a case to be made for introducing simultaneity into the part of the model that has been specified on a purely empirical basis, but the case is not made in the chapter. The W and P equations are claimed to be behavioral, but the claims are weak in my opinion.

On the whole, Palm's procedure for introducing restrictions seems to me reasonable, but in fact nearly entirely an empirical simplification of the model rather than an imposition of any *a priori* known information. It is hard to see in what application the restricted model might be hoped to be structural in the sense of invariant under a policy intervention, when the unrestricted model was not.

As I have already mentioned, it is appealing to make certain "natural" identifications of the disturbances of multivariate time series models and to use these to interpret the moving average representation. Particularly attractive, and no doubt often reasonable, is the practice of treating innovations in policy variables as being generated by policy choice, so that the moving average coefficients on these innovations represent responses to policy-induced changes in the policy variables. One must always bear in mind, however, the possibility that policy variables may have moved historically in response to non-policy influences, so that the response to innovations in these variables is not appropriately interpreted as a response to a policy-induced change in the variable. For money, wages, and interest rates, it is certainly possible that some of the historical variation was generated by the private sector, not by policy. Palm's conclusions about the effectiveness of various policies from examination of the MARs should, therefore, probably have been put forth more cautiously.

My own research interests and philosophy run so close to those displayed in this chapter that I could continue at some further length with discussion of technical fine points. This seems a good point to close these formal comments, however, with the hope that the profession will be seeing more empirical work along these lines and displaying these high standards of thoroughness and integrity.

Response to the discussants (1983)

Franz C. Palm

I would like to thank the discussants for the thoughtfulness of their comments. They have raised several interesting issues. In my response, I shall consider three themes.

1 Marginalization, exogenous variables, and parameter stability

One of the points raised by Professor Christ is concerned with the absence from the model of (exogenous) economic variables such as taxes, government expenditures, international trade, and investment. On many occasions (for instance, in the presence of a closed loop policy), policy variables can be treated as partly endogenous and stochastic. Similarly, exogenous variables are often stochastic. These variables can then be eliminated from the model by integration (or substitution). However, the parameters of the marginal, multivariate or univariate, model are functions of the parameters in the broader model. They remain stable as long as the parameters of the latter model are stable. It is legitimate to restrict the analysis to the marginal process provided its parameters are approximately stable. One can think of three reasons why a multivariate marginal model could be flawed: (a) A log-linear dynamic specification is too simple to be a reasonable approximation for the data-generating process; (b) some exogenous variables are non-stochastic and should, therefore, appear in the model; (c) the parameters have changed over the sample period. In my opinion, a log-linear specification for the model and the assumption of stochastic exogenous variables are reasonable working hypotheses in the present context. Concerning the last point, I hope that I have carefully analyzed the parameter stability and, in particular, the implications for the period 1974–9 of a model that has been specified and estimated from data up to 1973, a year for which a structural change in the economy of the Netherlands is very likely. Also, I hope that I have been sufficiently cautious in my conclusions about the presence of structural changes in the sample period.

Finally, from my own experience and in line with a remark by Professor Sims, the estimates of the (restricted) autoregressive representation are less stable than those of the moving average representation (MAR). Therefore, the MAR is better suited for an analysis of the structural stability than the autoregressive representation. As stated in the chapter, it is the response of several variables in the model to shocks in the interest rate that seems to be most strongly affected by parameter instability and for which the specification could be improved accordingly.

2 Simultaneity and economic interpretation

The restricted multivariate model can be interpreted as a set of structural equations that has been completed by adding reduced form equations. The model is statistically complete, but it is not a complete behavioral model (think, for instance, of limited-information maximum likelihood). The remark that "It is a curious system from an economic point of view" comprises an interpretation which the model can and should not be given.

The question of whether the interest rate equation is actually a demand for money equation can be answered by stating what the determinants of the demand for money are. I explain money balances by income, prices, and interest rates only (besides the seasonal dummies). These variables have been selected on the basis of *a priori* considerations, although the dynamics of the equation have been at least in part derived from the observations. Exclusion of other variables from the equation implies the identification of the parameters of the demand function.

Whether it would have been simpler to directly specify reduced form equations for Y and R, as Professor Sims suggests, is not obvious. The reader should realize that specifying a relationship with more than one current endogenous variable is a straightforward way of incorporating restrictions in the reduced form. For instance, the following relationship

$$y_{1t} + \beta y_{2t} + \gamma x_{1t} = u_{1t}, \quad \text{with } Eu_{1t} = 0,$$

implies that the expectations of y_{1t}, y_{2t}, and x_{1t} lie in a hyperplane or alternatively that the reduced form equation of y_{1t} is proportional to that of y_{2t}, except for the coefficient of x_{1t}. The specifications for Y and R should be interpreted in this way.

3 Forecasting

First, I would like to emphasize that the forecasts have been computed from the reduced form and not from the recursive form, as Professor Christ states.

When comparing the forecasting performance of the vector models with that of the univariate ARIMA models, one should keep in mind that differencing of the series not only eliminates the linear trend and the seasonal dummy variables but also transforms the stochastic structure of the disturbances of the univariate ARIMA models. If the disturbances in the univariate models for the original series are non-stationary and generated by a random walk, differencing will induce a white-noise disturbance term. Assuming that the factors that affect the unemployment rate are (especially after 1973) non-stationary could possibly imply non-stationary disturbances in the univariate model for U. It may also explain why the forecasting performance of the univariate model for $(1 - L^4)U$ is superior to that of the vector models.

Finally, I am glad to see that the discussants do not have major objections against the approach proposed in my chapter and that they are sympathetic to the statistical methods that I used.

4 Time series analysis, forecasting, and econometric modeling: the structural econometric modeling, time series analysis (SEMTSA) approach (1994)

Arnold Zellner

In this chapter an account of our experiences in modeling and forecasting the annual output growth rates of 18 industrialized countries is presented. A *s*tructural *e*conometric *m*odeling, *t*ime-*s*eries *a*nalysis (SEMTSA) approach is described and contrasted with other approaches. Theoretical and applied results relating to variable and model selection and point and turning-point forecasting are discussed. A summary of results and directions for future research concludes the chapter.

1 Introduction

In this chapter we shall provide an account of some of our experiences in modeling, forecasting, and interpreting time series data. Since the literature on these topics is so extensive, a comprehensive survey would require one or probably more volumes. Thus, we have decided to describe our approach, the experience that we have had with it, and its relation to a part of the statistical and econometric time series literature.

Obtaining good macroeconomic and microeconomic and other time series models is important since they are useful in explanation, prediction, and policy-making or control. The basic issue that is addressed in this chapter is how to produce such good time series models. Our SEMTSA approach (see, e.g., Zellner and Palm 1974, 1975; Plosser 1976, 1978; Zellner 1979, 1984, 1991; Palm 1983; Wallis 1983; Webb 1985; Manas-Anton 1986; Hong 1989; Min 1991) will be briefly compared to several other approaches that have emerged in the literature. Rather than just present theoretical procedures that may be useful in producing good

This research was financed by income from the H. G. B. Alexander Endowment Fund, Graduate School of Business, University of Chicago, and by a grant from the National Science Foundation.

Originally published in the *Journal of Forecasting* 13 (1994), 215–33. CCC 0277-6693/94/020215–19.

models, an account will be given of both theoretical and applied results in what follows. Obviously, the issue of how well procedures work in serious applications is of first order importance.

The plan of the chapter is as follows. In section 2, a description of the SEMTSA approach and some brief comparisons with other approaches are provided. Also, our experience in applying it in the analysis of macroeconomic time series data is described. In section 3 theoretical and applied results on variable and model selection procedures are presented. Section 4 takes up some aggregation issues in model formulation as they relate to forecasting performance treated . . . by Espasa and Matea (1990) and others. The final section [5] contains a summary of results and some conclusions about future work.

2 Background on the SEMTSA approach and its applications

In the SEMTSA approach the first step involves a description of main objectives, as emphasized in Zellner (1979). In our present modeling work, our objectives are to produce models which (1) forecast well, (2) are useful in explanation, and (3) serve policy-makers' needs adequately. As will be seen, in our work we do not attempt to achieve all three goals at once, as is the case in some structural econometric modeling approaches (see, for example, Hickman 1972; Fair 1991). Rather, we proceed sequentially by approaching the forecasting goal first, then the explanatory goal, and finally the policy or control goal using many sets of data, in our case data for many countries.

As regards the forecasting goal, we indicate the variable or variables which are of major concern. It is recognized that these variables can be modeled in a multiple or multivariate time series model, including vector autoregressions (VARs) as a special case, or in a multi-equation structural econometric model. However, as recognized by Keynes, Friedman, Christ, Box, Tiao, Sims, this author, and many others, in such multivariate models, there are many parameters, which usually makes tests of model specification not very powerful and estimates and predictions not very precise. Further, there is a high probability that errors may occur in formulating equations of such models which can lead to models with unusual properties (see, e.g., the results of simulation experiments with large-scale econometric models reported in Adelman and Adelman 1959; Hickman 1972; Zellner and Peck 1973). Further, . . . these multiple equation models are [often] non-linear in variables and parameters, a fact which makes understanding them somewhat difficult even with the results of simulation experiments available. For example, it is difficult to establish whether such large, non-linear models have unique or many solutions, what their global dynamic properties are, and what are the finite sample properties of estimation, testing, and prediction techniques, particularly when there is considerable use of the data in pretesting to determine forms of models. Add to this the problems of systematic and random measurement errors in the variables employed and one begins to understand the serious problems that arise in attempting to model a set of time series variables in a "one-shot" attempt to get a satisfactory multivariate time series or structural econometric model.

In the SEMTSA approach, we formulate the components of a model, using as much sound background information as possible, and then establish that the components work well in forecasting a good deal of out-of-sample data. Then, using relevant subject-matter theory and background information, we attempt to put the components together to form a sensible explanatory model. The explanatory model so formulated can then be tested further using as much new data as possible, that is, data for additional time periods going forward or backward in time and for additional economic entities (say, countries). As will be seen, the formulation of the component forecasting relations, usually transfer functions, is conditioned by certain information relating to properties of an overall multivariate model for the set of variables. Finally, adapting the model to make it useful not only for explanation and prediction but also for policy making will take additional work.

To make some of the above points explicit, suppose that we initially tentatively entertain a linear multiple time series model as put forward many years ago by Quenouille (1957) and now often referred to as a multivariate autoregressive, moving average (MVARMA) process:

$$H(L)z_t = F(L)e_t, \tag{2.1}$$

where z_t and e_t are $m \times 1$ vectors of random variables, the former observable and mean-corrected and the latter a white-noise error vector. $H(L)$ and $F(L)$ are $m \times m$ matrix lag operators with L being the lag or backshift operator. If $F(L)$ is of zero degree, the system is a VAR. Note that if m and the degree of $H(L)$ are large, this VAR will contain a very large number of parameters which, as pointed out above, will present problems. Now, if $H(L)$ is invertible, that is, $H^{-1}(L) = H(L)^*/|H(L)|$, where $H(L)^*$ is the adjoint matrix associated with $H(L)$, we can write (2.1) as follows:

$$|H(L)|z_t = H(L)^* F(L)e_t \tag{2.2a}$$

or

$$|H(L)|z_{it} = a_i' e_t, \tag{2.2b}$$

where z_{it} is the ith element of z_t and a_i' denotes the ith row of $H^*(L)F(L)$. Equation (2.2b) indicates that individual elements of z_t have processes in the univariate ARMA form. If $H(L)$ is large in dimension, that is, if m is large, and of high degree, the autoregressive polynomial $|H(L)|$ in (2.2) will be of high degree unless there are some common roots on both sides of (2.2) which cancel. Given a tentatively restricted form for (2.1), it is possible to derive the implied ARMA processes for individual variables, as shown in (2.2) and check their forms against forms determined from the data (see references cited above for examples illustrating this procedure). Note, however, that in order to get fixed parameter, stationary processes in (2.2), a number of rather strong assumptions must be made about the process in (2.1) which may not be satisfied in practice. For example, some of the variables in z_t may not be covariance stationary and/or some of the elements of $H(L)$ and $F(L)$ may be time-varying.

Further, in connection with (2.1), if we are willing to assume, as many model builders do, that a subvector x_t' of $z_t' = (y_t', x_t')$ is exogenous, then we can write system (2.1) as follows:

$$\begin{pmatrix} H_{11} & H_{12} \\ 0 & H_{22} \end{pmatrix} \begin{pmatrix} y_t \\ x_t \end{pmatrix} = \begin{pmatrix} F_{11} & 0 \\ 0 & F_{22} \end{pmatrix} \begin{pmatrix} e_{1t} \\ e_{2t} \end{pmatrix} \tag{2.3a}$$

or

$$\begin{aligned} H_{11} y_t &= -H_{12} x_t + F_{11} e_{1t} \\ H_{22} x_t &= F_{22} e_{2t}, \end{aligned} \tag{2.3b}$$

where $H(L)$ and $F(L)$ have been partitioned in conformity with the partitioning of z_t and for convenience, the submatrices' dependence on L is not explicitly shown. The assumption that x_t is exogenous implies that the following submatrices of $H(L)$ and $F(L)$ are identically zero, namely H_{21}, F_{12}, and F_{21}. The first line of (2.3b) is in the form of a linear dynamic econometric model while the second provides a tentative multiple time series process for the exogenous variables in x_t.

From (2.3b), given that H_{11} is invertible, the transfer function (TF) system is given by:

$$|H_{11}| y_t = -H_{11}^* H_{12} x_t + H_{11}^* F_{11} e_{1t}, \tag{2.4a}$$

and a single TF equation (say, the ith) is

$$|H_{11}| y_{it} = b_i' x_t + c_i' e_{1t}, \tag{2.4b}$$

where H_{11}^* is the adjoint matrix associated with H_{11}, b_i' is the ith row of $-H_{11}^* H_{12}$, and c_i' is the ith row of $H_{11}^* F_{11}$. It is seen from (2.4b), that the polynomial operator hitting each element of y_t is the same, barring the cancellation of common roots in particular equations of (2.4a). Also, the use of (2.4) does not necessarily imply that the input or exogenous variables are covariance stationary. Further, if we are willing to put some restrictions on the parameters of the structural equations given in (2.3b), these will imply restrictions on the TF system in (2.4a), which can be tested (see Zellner and Palm 1975, for explicit examples with applications to checking the properties of a small structural model of the US economy using monthly data). Here we started with a restricted version of (2.3b), the structural equations and assumptions regarding x_t, as many other model builders have done. Unfortunately, several variants of the structural equation system, with adaptive or rational expectations assumptions, were found not to be completely supported by the information in the data. This is an example of an attempt to obtain a multivariate model in a "one-shot" approach. For a description of other failures of this approach, see the evaluation by McNees (1986) of a number of structural econometric models. Also, as shown by Litterman (1980) and McNees (1986), attempts to use unrestricted VARs to model quarterly macroeconomic variables for the US economy have not been successful. Use of prior distributions on autoregressive parameters, as in Litterman (1980, 1986), Highfield (1986), and others' work which center processes for individual variables at random walks with drift, tend to improve on unrestricted VARs' performance in forecasting but were not entirely satisfactory in forecasting financial variables and recent turning points. Perhaps centering processes for individual variables at random walks is inappropriate since random-walk models do not always perform well in forecasting (see below for further discussion and empirical results bearing on this issue). Further, as pointed out above, a VAR usually implies marginal ARMA processes for individual variables which have very high order AR and MA parts (see (2.2)). Finally, for ten or so macroeconomic variables and data sets generally available, an unrestricted multiple time series approach based on (2.1) does not seem fruitful in view of the large number of parameters and model uncertainty present.

In view of the failures of "one-shot" approaches to modeling macroeconomic variables, including our own, the problem is how to proceed in order to obtain reliable models. In research for the paper (Garcia-Ferrer *et al.* 1987) we argued as follows. First, since many had tried one-shot approaches and failed, we decided to take a component-by-component or a variable-by-variable approach. We selected an important, key variable, the real total output of an economy as measured by real gross national or domestic product (data available for many countries in the IMF computerized database at the University of Chicago). Most "one-shot" attempts to model this variable along with many others have failed. As shown by Christ (1951), Cooper (1972), and Nelson

(1972), many large-scale models' forecasts were not as good as those of very simple univariate naive models (e.g. random walks, low order AR models, or simple ARMA models). Also, in the exchange rate area, Meese and Rogoff (1983) showed that random-walk models performed better in forecasting than did three structural exchange rate models (see also Wolff 1985, 1987, who improved the structural exchange rate models' performance by use of time-varying parameter state-space versions but not to an extent that they entirely dominated random-walk models' predictive performance). In interpreting such results, years ago Friedman (1951), who was influential in having such forecasting tests performed, suggested that if a large model could not perform better in forecasting than a univariate naive model then it was probably faulty and needed reformulation.

The above considerations, along with Jeffreys' (1967) advice to consider all variation random unless shown otherwise and his "simplicity postulate" which suggests that simpler models will probably work better than complicated models, led us in Garcia-Ferrer *et al.* (1987) to start with a relatively simple autoregressive model of order three for the rate of growth of real annual GNP (RGNP) for a country. That is, with $y_{it} = \ln \mathrm{RGNP}_{it} - \ln \mathrm{RGNP}_{it-1}$, the growth rate of the ith country in the tth year, we entertained the following autoregressive model of order three, denoted by AR(3):

$$y_{it} = \beta_{0i} + \beta_{1i} y_{i,t-1} + \beta_{2i} y_{i,t-2} + \beta_{3i} y_{i,t-3} + u_{it} \quad \begin{array}{l} i = 1, 2, \ldots, N \\ t = 1, 2, \ldots, T, \end{array} \tag{2.5}$$

where u_{it} was assumed to be a scalar white-noise error term. Now it may be asked, why did we first difference and choose an AR(3)? We first differenced the log of RGNP to obtain the rate of growth because there is a great interest in this variable. Also, first differencing is helpful in reducing the effects of certain types of constant or time-varying systematic measurement biases, and it is, of course, a procedure for possibly inducing stationarity, although we were not too sanguine on this point. Further, by using an AR(3) process, we allowed for the possibility of having complex conjugate roots giving rise to an oscillatory component and a real root producing a local trend. Indeed, in subsequent analyses by Geweke (1988) and Hong (1989), it was shown empirically using data for many countries that in each case there are two complex roots and one real root, giving rise to damped cycles and non-explosive local trends. Note that the Nelson–Plosser (1982) (0, 1, 1) ARIMA process for US RGNP does not admit the possibility of complex roots and associated cyclical components. In this regard, see also Harvey and Todd (1983), who favored

(2, 1, 1) or (2, 1, 2) models. Finally, Cooper (1972) showed that a simple AR process performed as well or better in forecasting than did complicated multi-equation models.

In view of the above considerations, we decided to begin our analyzes with an AR(3) model for the growth rate of real annual output as shown in (2.5) and annual data for eight European countries and the United States, 1951–73 for fitting and 1974–81 for out-of-sample one-step-ahead forecasting. The results of these forecasting experiments produced what Thomas Huxley has called an "ugly fact," namely a fact that destroyed the *a priori* arguments mentioned in the previous paragraph. Our AR(3) models for the nine countries did not forecast any better than various naive random walk models (see reported RMSEs in Garcia-Ferrer *et al.* 1987). Given the simplicity of the AR(3) model, it was not difficult to determine the reason for poor forecasting performance. The model was missing badly in forecasting downturns and upturns, generally showing over-shooting and under-shooting. With the nature of the problem clear, it was not too hard to remember Burns' and Mitchell's (1946) fundamental work on business cycles using pre-Second World War data relating to several economies. They found that stock prices and money generally led aggregate economic activity in business fluctuations. Also, Moore, Zarnowitz, and others had emphasized the value of using such leading indicator variables in forecasting even though many were skeptical. Stock prices reflect quickly all kinds of news events, policy changes, etc. affecting an economy while the real economy usually responds to such events with a lag. Also, money supply changes can operate to affect consumer and producer demands as well as reflect information that policy-makers may have that is not available to the general public. Further, since world events probably affect individual economies, we decided to introduce a "common world effect" into each country's equation. Thus (2.5) was embellished to incorporate these considerations to produce the following third order autoregressive-leading indicator (AR(3)LI) model:

$$\begin{aligned} y_{it} &= \beta_{0i} + \beta_{1i} y_{it-1} + \beta_{2i} y_{it-2} + \beta_{3i} y_{it-3} + \beta_{4i} SR_{it-1} \\ &\quad + \beta_{5i} SR_{it-2} + \beta_{6i} GM_{it-1} + \beta_{7i} WSR_{t-1} + u_{it} \\ &= \mathbf{x}'_{it}\beta_i + u_{it}, \end{aligned} \tag{2.6}$$

where SR_{it} and GM_{it} are the rates of change of real stock prices and real money, respectively, and WSR_t is the world stock return, the median of the SR_{it}s in year t. Specifically, $SR_{it} = (1 - L)\ln(SP_{it}/P_{it})$ and $GM_{it} = (1 - L)\ln(M_{it}/P_{it})$, where SP_{it} is a stock price index, P_{it} is a price index, and M_{it} is nominal money (M_1) holdings at the end of year t.

With respect to the trend-stationary (TS) versus difference-stationary (DS) issue, we note that it has been considered in the literature in terms of models, AR(1)s, etc. which have not been shown to perform well in forecasting. Thus results based on these models may not be dependable. In (2.6), the output variable, y_{it}, is clearly affected by the input leading indicator variables which may be generated by non-stationary processes and can induce varied cyclical and trend components in the output variable, y_{it}. As indicated below, the AR(3)LI model in (2.6) performs reasonably well in point and turning point forecasting.

When (2.6) was implemented by simply using least squares estimation with annual data 1951–73 for nine countries[1] and one-year-ahead forecasts were made for the years 1974–81, it was found that there was a decided improvement in forecasting performance *vis-à-vis* use of AR(3) models and several random-walk models. Further, an interesting feature of the error terms was noted when we employed the common effect variable, lagged world stock returns, WSR_{t-1}, namely the contemporaneous error terms for individual countries were not very highly correlated. Also, there was little indication that the error terms are autocorrelated (see Zellner, Hong, and Gulati 1990 for the fitted relations and measures of autocorrelation). Also, in Garcia-Ferrer *et al.* (1987) there are some comparisons of the forecasting performance of (2.6) to that of large-scale OECD models combined with judgmental adjustments.

Further, in Garcia-Ferrer *et al.* (1987) additional computations were performed to check the effects of using two types of Stein-like shrinkage techniques in forecasting which generally produced better overall results. These calculations exploited the fact that the coefficient vectors for different countries are not too different in value. Using vector and matrix notation, we can write (2.5) for the ith country as follows: $y_i = X_i\beta_i + u_i$, $i = 1, 2, \ldots, N$. Then, assuming along with Swamy (1971), Lindley and Smith (1972), and others, $\beta_i = \theta + v_i$, where θ is the common mean of the β_is and v_i is an error vector, it is not difficult to combine the information in the data with the information in the distribution of the β_is to obtain estimates of the β_is. As we specified the system, the estimate of β_i is an average of the least squares estimate $\hat{\beta}_i = (X_i'X_i)^{-1}X_i'y_i$ and an estimate of θ, the mean of the β_is. On using such an estimate for each country, the nine annual, one-step-ahead forecasts, 1974–81, were generally improved relative to using just the least squares or diffuse prior Bayes' forecasts, with RMSE used as a criterion of forecasting performance.

[1] All data are taken from the IMF IFS data bases.

Also, we analyzed the data using a time-varying parameter model, namely, $y_{it} = x_{it}'\beta_{it} + u_{it}$, and $\beta_{it} = \beta_{it-1} + \varepsilon_{it}$, a vector random walk. Parameters were assumed time varying to reflect possible effects of wars, policy changes, aggregation, and so on. On implementing the above time-varying parameter model there were improvements relative to use of fixed-parameter models for many countries. Recall that the period 1951–81 includes at least two wars, the Korean and Vietnamese, two oil crises, changes from fixed exchange rates to floating exchange rates in the early 1970s for many countries, etc. In performing our estimation and forecasting calculations, no points were omitted nor were any dummy or intervention variables employed. It appears that the leading indicator variables alone or coupled with shrinkage or time-varying parameters produced reasonably good forecasting results for the nine countries.

Since there is a possibility that in some sense the sample of nine countries and/or the period employed were "special" it was considered extremely important to check previously obtained results with an expanded sample of countries and a longer time period. Thus in Hong (1989) and Zellner and Hong (1989), data for eighteen countries, European countries including Spain, as well as for the United States, Canada, Japan, and Australia, 1951–84 were employed. This expanded data-base included data for the sharp 1982 recession which were not included in earlier analyzes. Generally, the one-year-ahead forecasting results for 1974–84 were similar to those found earlier; results reported in Zellner and Hong (1989). Further improvements were made by including a second general effect variable in each country's equation, namely the median of the countries' growth rates for year t, denoted by w_t. That is, the equation for each country shown in (2.6) was modified as follows:

$$y_{it} = \alpha_i w_t + x_{it}'\beta_i + u_{it} \tag{2.7a}$$

and the following equation was assumed for w_t:

$$\begin{aligned} w_t &= \delta_0 + \delta_1 w_{t-1} + \delta_2 w_{t-2} + \delta_3 w_{t-3} + \delta_4 MSR_{t-1} \\ &\quad + \delta_5 MGM_{t-1} + v_t, \end{aligned} \tag{2.7b}$$

an AR(3)LI with the following leading indicator variables, MGM_t and MSR_t, the medians of the eighteen countries' real money and real stock price growth rates (see figure 4.1 for plots of these data). We denote the model in (2.7) an ARLI/WI model. Use of it along with complete shrinkage led to improved forecasting results for most countries (see Zellner and Hong 1989). Also, Hong (1989) showed that the AR(3)LI models performed better in one-year-ahead forecasting, 1974–84, for eighteen countries than the Nelson–Plosser (1982) (0, 1, 1) ARIMA model and

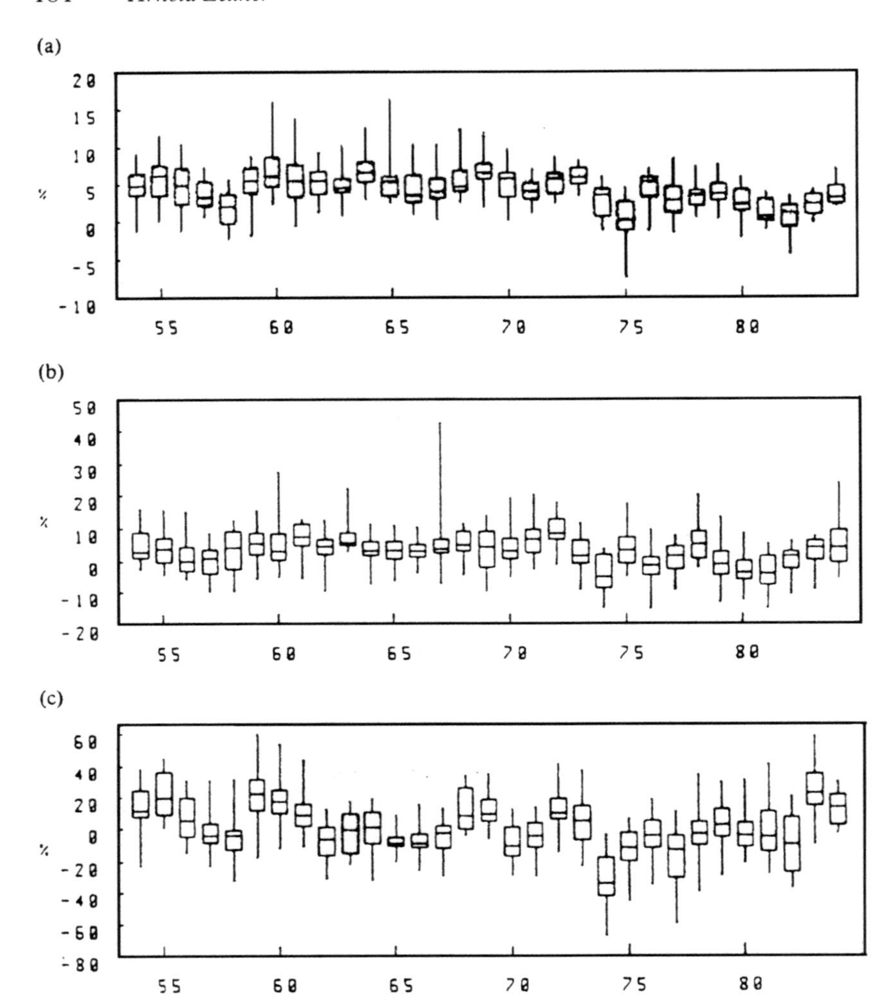

Figure 4.1 Boxplots of data for eighteen countries, 1954–1984. For each year the horizontal line in the box is the median growth rate and the height of the box is the interquartile range of the eighteen countries' growth rates. The end points of the lines extending from each box indicate the highest and lowest growth rates in the given year. Shown are annual growth rates of (a) real output, (b) real money, and (c) real stock prices

a version of the Barro (1978) "money surprise" model. Further, Hong (1989) computed the roots of the ARLI process for each country and showed that in each instance there was a high posterior probability that there are two complex roots and one real root with amplitudes less than one. Plots of the posterior distributions of the periods indicated cyclical periods of roughly four–six years' duration. Finally, Otter (1990) re-analyzed some of the Zellner–Hong (1989) data using a canonical correlation approach and Hankel matrix identification procedures which require strong stationarity assumptions. While he claimed to get better forecasting results, he did not make comparisons with the Zellner–Hong shrinkage forecasting results which are better than his and the results that he did report in his confused table 4.1, when unscrambled reveal no forecasting improvement as measured by median RMSE of forecast, 2.41 for the AR(3)LI model, and 2.94 and 2.44 for two variants of his canonical correlation approach.[2]

To check further the forecasting properties of the ARLI and ARLI/WI models, a Bayesian decision theoretic methodology for forecasting turning points in the rates of growth of real GNP was developed building on earlier work of Wecker (1979) and Kling (1987). In Zellner and Hong (1989), Zellner *et al.* (1990), and Zellner, Hong, and Min (1991) this methodology was developed and applied to forecast turning points in eighteen countries' growth rates. In the last reference, using a variety of models including fixed- and time-varying parameter models with and without pooling, approximately 70 percent or more of 158 turning points, 1974–87, were correctly forecasted. The procedures employed involved a definition of a turning point and use of a predictive density to compute the probability of a downturn or of an upturn. Having these probabilities calculated and a 2×2 loss structure, it is possible to choose a forecast that minimizes expected loss. For example, if the 2×2 loss structure is symmetric, then one forecasts downturn when the probability of a downturn is greater than $1/2$ and no-downturn otherwise. In figure 4.2 and table 4.1 are shown the results of these probability calculations for eighteen countries year by year. The turning point forecasts, based on these computed probabilities, were compared with those of some naive models using Breier scores and were found to be much superior to all naive models used. That the models performed so well in forecasting turning points (approximately 70 percent of 158 turning points correctly forecasted) was indeed a pleasant surprise.

[2] Further, Otter (1990) excluded data for several countries, made some incorrect comparisons, and did not employ exactly the same data as employed in our work.

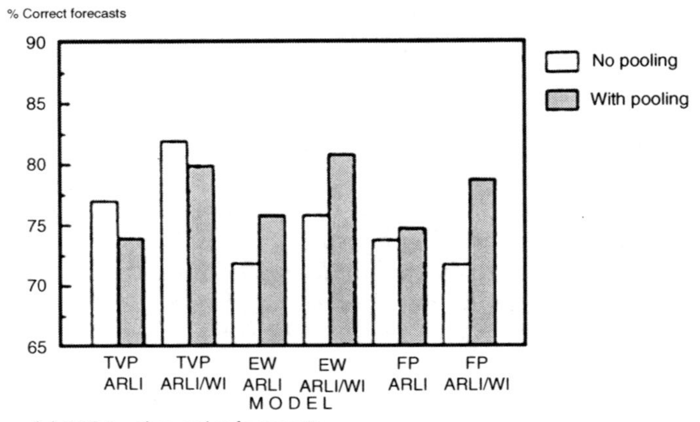

(a) 158 turning-point forecasts

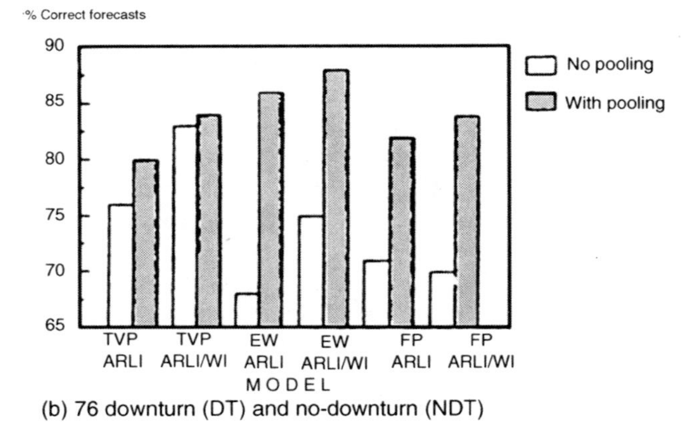

(b) 76 downturn (DT) and no-downturn (NDT)

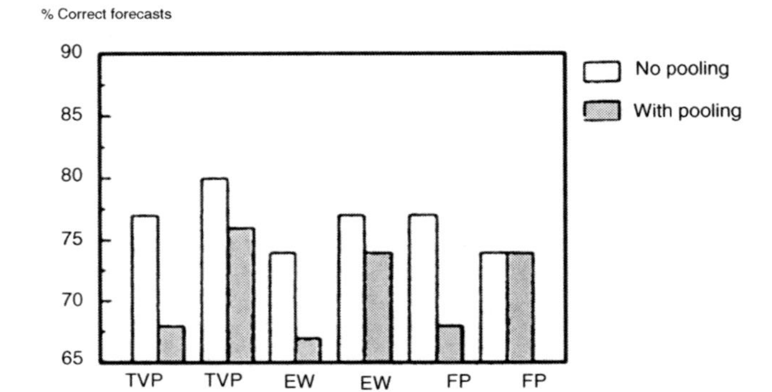

(c) 82 upturn (UT) and non-upturn (NUT) forecasts

Figure 4.2 Percentages of correct forecasts, 1974–1986.
See Zellner, Hong, and Gulati (1990) for additional information regarding methods and models

Table 4.1 *Computed probabilities of downturns by year and country from the pooled TVP/ARLI/WI model*[a], *1974–1986*

	Year												
Country	1974	1975	1976	1977	1978	1979	1980	1981	1982	1983	1984	1985	1986
Australia				0.68*			0.83*					0.94*	
Austria				0.77*			0.95*				0.27*		0.36*
Belgium	0.82*			0.90*				0.89*				0.54*	
Canada	0.94*			0.91*							0.37–	0.88*	
Denmark				0.95*			0.80*			0.65*		0.33–	0.36*
Finland						0.13–	0.93*						0.69*
France				0.88*						0.44*			0.51–
Germany	0.87*			0.92*			0.85*				0.27–	0.64*	
Ireland					0.74*				0.92*			0.60*	
Italy	0.81*			0.87*			0.81*					0.66*	
Japan				0.83–	0.55*							0.67*	
Netherlands	0.88*			0.84*			0.70*				0.23–	0.58*	
Norway		0.98*		0.95*			0.78*				0.58–	0.66*	
Spain					0.51*						0.44–	0.31–	0.37–
Sweden	0.68*					0.12–	0.80*				0.37–	0.80*	
Switzerland					0.47*		0.74–	0.94*				0.57–	0.77*
United Kingdom	0.99*			0.85*		0.31*				0.32–	0.54*		0.44*
United States	0.86*			0.85*		0.52*					0.42–	0.96*	

Notes:
[a] An asterisk (*) indicates that a downturn occurred while a dash (–) indicates that a downturn did not. The definition of a downturn is two consecutive growth rates below a third and the fourth below the third, that is, $y_{T-2}, y_{t-1} < y_T > y_{T+1}$. See Zellner, Hong, and Min (1991) for additional results and information about models and techniques employed.

Table 4.2 *RMSEs for pooled and unpooled ARLI/WI models' forecasts, by country, 1974–1987*[a]

RMSE(%)	Countries									Freq.	Prop.
(a) Pooled TVPM											
1.00–1.49	FRN	GER	NET	SPN						4	0.22
1.50–1.99	AUR	BEL	CAN	FIN	ITY	NOR	SWD	UKM	USA	9	0.50
2.00–2.49	AUL	DEN	JAP	SWZ						4	0.22
2.50–2.99	IRE									1	0.06
3.00–3.49	–									0	0.00
	Median = 1.74			Minimum = 1.17			Maximum = 2.53			18	1.00
(b) Unpooled TVPM											
1.00–1.49	UKM									1	0.06
1.50–1.99	BEL	FRN	GER	NET	SPN	SWD				6	0.33
2.00–2.49	AUR	USA								2	0.11
2.50–2.99	CAN	DEN	ITY	NOR						4	0.22
3.00–3.49	AUL	FIN	IRE	JAP	SWZ					5	0.28
	Median = 2.37			Minimum = 1.39			Maximum = 3.32			18	1.00
(c) Pooled FPM											
1.00–1.49	NOR	SPN								2	0.11
1.50–1.99	AUR	BEL	CAN	FIN	FRN	GER	NET	SWD	UKM	9	0.50
2.00–2.49	AUL	DEN	IRE	ITY	JAP	SWZ	USA			7	0.39
2.50–2.99	–									0	0.00
3.00–3.49	–									0	0.00
	Median = 1.86			Minimum = 1.21			Maximum = 2.48			18	1.00
(d) Unpooled FPM											
1.00–1.49	–									0	0.00
1.50–1.99	BEL	NET	NOR	UKM	USA					5	0.28
2.00–2.49	FRN	SPN	SWD							3	0.17
2.50–2.99	AUR	CAN	GER	IRE						4	0.22
3.00–3.49	AUL	DEN	FIN	JAP	SWZ					5	0.28
3.50–3.99	ITY									1	0.06
	Median = 2.60			Minimum = 1.50			Maximum = 3.68			18	1.01

Note:
[a] See Min and Zellner (1990) for explicit specification of the models employed in these calculations, namely variants of the model in (2.7) of the chapter.

In table 4.2, taken from Min and Zellner (1993), there is a significant demonstration of the effects of pooling on forecasting performance for two models. Note that pooling results in eighteen countries' forecast RMSEs being highly concentrated around a low value whereas without pooling there is much more dispersion in countries' forecast RMSEs and their median value is higher than in the pooled case. That pooling has such a significant effect on the precision of forecasts is indeed remarkable.

Having provided an overview of selected aspects and results of the SEMTSA approach, in the next section results provided by use of Bayesian model and variable selection techniques will be presented.

3 Model and variable selection procedures and results

Here selected Bayesian model and variable selection procedures will be described and applied. As explained above, the issue of whether fixed-parameter (FP) models or time-varying parameter (TVP) models are more appropriate for modeling economic time series is a central one. If parameter values change because of Lucas effects, aggregation, wars, strikes, and other causes, FP models may not be adequate for modeling and forecasting. To approach this problem, in Garcia-Ferrer *et al.* (1987) both FP and TVP models were estimated and used in forecasting. The results favored somewhat the TVP models, the AR(3)LI model in (2.6) with the parameter vector assumed generated by a vector random walk. In Min and Zellner (1993), posterior odds for FP versus TVP models were derived and evaluated year by year for the period 1974–87 for eighteen countries. The prior densities used in forming the posterior odds for 1974 were the posterior densities based on annual data, 1954–73, with initial priors rather diffuse. In tables 4.3(a) and 4.3(b) some of these posterior odds are presented. It is seen that for a number of countries the odds favour TVP models. Also, in table 4.4 it is seen that when odds favor TVP models, the TVP models tend to have lower RMSEs of forecast than do FP models and vice versa. Thus the posterior odds appear to be useful in screening models for forecasting purposes. Further, they can be used in a decision theoretic framework to choose between or among alternative forecasts or to combine forecasts of alternative models (see Min and Zellner 1993; Palm and Zellner 1992, for theoretical and empirical analyses illustrating these points). Important in this context is the issue of whether or not to combine models when the models considered do or do not constitute an exhaustive set (see also Diebold 1989). That Bayesian posterior odds are useful in these problems is indeed fortunate.

As regards variable selection, the approach described in Zellner and Siow (1980) and Zellner (1984, ch. 3.7) will be used here to determine what variables to include in a FP ARLI model. This approach to selecting or testing alternative leading indicator and other variables to include in a model has also been employed in several studies reported in Poirier (1991) and is different from the approach employed by Stock and Watson (1991). Here we are interested in comparing the AR(3)LI model in (2.6) with some narrower and some broader models. Thus we consider a broad model containing a constant term, c, and eight input variables, namely

Table 4.3(a) *Posterior odds for fixed versus time-varying parameter models computed from annual data, 1973–1987*[a]

	Models			
	Unpooled		Pooled	
Country	ARLI[b]	ARLI/WI[c]	ARLI[d]	ARLI/WI[e]
Australia	0.16	0.34	1.01	0.94
Austria	0.01	0.07	0.21	1.02
Belgium	1.64	0.45	0.40	1.57
Canada	0.41	2.21	0.57	1.72
Denmark	0.01	0.39	0.05	0.99
Finland	0.78	1.68	0.36	0.76
France	0.01	1.30	0.47	1.42
Germany	0.00	0.01	0.16	1.43
Ireland	6.18	3.04	1.87	1.50
Italy	0.01	0.06	0.32	1.21
Japan	0.45	0.48	6.81	0.91
Netherlands	0.41	2.32	0.10	1.92
Norway	6.60	4.95	1.36	1.09
Spain	0.27	0.75	0.57	1.35
Sweden	0.25	0.34	0.04	0.58
Switzerland	0.71	4.25	0.21	0.91
United Kingdom	0.22	8.67	0.13	0.82
United States	9.18	17.31	0.24	0.81

Notes:

[a] Posterior distributions for models' parameters, computed using annual data 1954–72, were used to form Bayes factors for the period 1973–87. Prior odds were set 1:1 in all cases. The entries in the table are odds in favor of fixed parameter (FP) models. See Min and Zellner (1993) for derivations and other results.

[b] Unpooled FP/ARLI versus unpooled TVP/ARLI.

[c] Unpooled FP/ARLI/WI versus unpooled TVP/ARLI/WI.

[d] Pooled FP/ARLI versus pooled TVP/ARLI.

[e] Pooled FP/ARLI/WI versus pooled TVP/ARLI/WI.

three lagged values of the output growth rate, denoted here by y_1, y_2, and y_3 as well as rates of growth of real stock prices and of real money, lagged one or two years and denoted by S_1 and S_2 and M_1 and M_2, respectively, and the lagged median growth rate of real stock prices, denoted by W_1. With all models containing a constant term, c, there is just one model with all eight input variables, eight models containing any seven variables, 28 models containing any six variables, 56 models containing any five variables, 70 models containing any four variables, and so on with a total

Table 4.3(b) *Number of countries, by values of posterior odds, fixed versus time-varying parameter models, annual data, 1973–1987*[a]

	Models			
	Unpooled		Pooled	
Posterior odds, FP versus TVP	ARLI ARLI	ARLI/WI ARLI/WI	ARLI ARLI	ARLI/WI ARLI/WI
0–0.49	12	8	12	0
0.5–0.99	2	1	2	8
1.00–1.99	1	2	3	10
≥ 2.00	3	7	1	0

Notes:
[a] See Min and Zellner (1993) for derivation of posterior odds based on 1:1 prior odds.

Table 4.4 *Number of countries by posterior odds and root mean-squared error (RMSE) of forecast, by type of model, annual output growth rates, 1974–1987*[a]

		ARLI Lower RMSE for:			ARLI/WI Lower RMSE for:		
Posterior odds favour		FPM	TVPM	Tot.	FPM	TVPM	Tot.
(a) Unpooled models							
	FPM	3	1	4	5	4	9
	TVPM	3	11	14	1	8	9
	Tot.	6	12	18	6	12	18
(b) Pooled models							
	FPM	4	0	4	3	6	9
	TVPM	0	14	14	1	8	9
	Tot.	4	14	18	4	14	18

Note:
[a] Taken from Min and Zellner (1993).

of 256 possible models. Thus the AR(3)LI is just one of 256 possible linear models involving the eight input variables and a constant. Testing it against 255 alternatives raises the issue of "selection" effects. That is, it may be that with so many alternative models, one may fit the data just by chance or due to over-fitting. This issue will be discussed further below.

Table 4.5 *Model selection using posterior odds, mean-squared error and out-of-sample forecast RMSEs*

Model[a]	1954–1973 Odds[b]	1954–1973 MSE[c]	1954–1987 Odds[b]	1954–1987 MSE[c]	Out-of-sample, 1974–1987, RMSE of forecast[d]
1. AR(3)LI $cy_1y_2y_3S_1S_2M_1W_1$	2.80	1.79	6.19	2.70	2.31
2. $cy_3M_1M_2W_1$	17.42	1.57	2.74	3.46	2.74
3. $cy_3S_1M_1M_2W_1$	15.67	1.53	14.29	2.88	2.44
4. $cy_3S_1M_1W_1$	8.61	1.81	26.34	2.87	2.30
5. $cy_3S_1S_2M_1W_1$	12.88	1.59	59.92	2.56	2.22
6. $cS_1S_2M_1M_2W_1$[e]	2.18	2.27	1.68	3.44	2.59
7. $cy_1S_1M_1W_1$[f]	2.11	2.39	2.55	3.48	2.54
8. AR(3) $cy_1y_2y_3$	0.04	5.41	0.01	5.76	2.69
9. Random walk for ln GNP with drift, c	0.06	4.89	0.02	5.75	2.68
10. General model $cy_1y_2y_3S_1S_2M_1M_2W_1$	1.00	1.81	1.00	2.80	2.47

Notes:

[a] The general model in line 10 is given by $y_t = c + \beta_1 y_{t-1} + \beta_2 y_{t-2} + \beta_3 y_{t-3} + \beta_4 SR_{t-1} + \beta_5 SR_{t-2} + \beta_6 GM_{t-1} + \beta_7 GM_{t-2} + \beta_8 W_{t-1} + \varepsilon_t$ and denoted by $cy_1y_2y_3S_1S_2M_1M_2W_1$. Other models are particular cases of the general model.

[b] Posterior odds are computed for each specific model versus the general model in line 9 with prior odds in each case taken 1:1. See Zellner (1984, ch. 3.7) for expressions for odds.

[c] Within-sample mean-squared error $= \sum_{t=1}^{N} \hat{\varepsilon}_t^2/v$, where $v =$ degrees of freedom, or number of observations minus number of estimated parameters.

[d] One-step-ahead forecast errors, Δ_t were employed to compute RMSEs given by

$$\text{RMSE} = \sqrt{\sum_{1}^{14} \Delta_t^2/14}$$

[e] This is the model selected by Mittnik (1990), based on data 1953–73 for the United States.

[f] This is the model selected by Otter (1990), based on data 1954–73 for the United States.

Here we shall just note that variable selection will initially be performed using the data for 1954–73 and the preferred models will be evaluated further in out-of-sample forecasting using data for 1974–87, fourteen annual one-step-ahead forecasts. Then the odds are recalculated using all the data, 1954–87. The results of these calculations are presented in table 4.5.

It is seen from table 4.5, based on data for the United States, that the posterior odds on the basic AR(3)LI model in (2.6) versus the general model containing a constant c and eight input variables, shown in line 9, is 2.80:1.00. This posterior odds is based on prior odds of 1:1, data 1954–73, and the procedures in Zellner (1984, ch. 3.7). Also, the within-sample MSE $= \sum_{i=1}^{20} \hat{\varepsilon}_i^2/v$, where $v =$ degrees of freedom, is 1.79 for the AR(3)LI model in line 1 and 1.81 for the general model in line 9. Further, the RMSEs of forecast for out-of-sample forecasts, 1974–87 (fourteen forecasts) are 2.31 and 2.47 percentage points, respectively. Finally, the recomputed odds for the entire period 1954–87 is 6.19:1.00 for the AR(3)LI model versus the general model. Thus in this comparison, the ARLI(3)LI model fares well.

In lines 8 and 9 of table 4.5 are shown results for a random walk with drift and an AR(3) model. The posterior odds, MSEs, and forecast RMSEs overwhelmingly favor the AR(3)LI relative to these models, as in previous work. For example, the RMSEs of forecast are 2.31, 2.69, and 2.68 respectively, for these three models.

With respect to the period 1954–73, the models in lines 2–5 are the ones most favored by the posterior odds and/or the MSE criteria after examining all 256 models. The posterior odds reported in table 4.5 are all based on equal prior probabilities on all 256 models. There is a question as to whether this is appropriate given that we selected models from a large number, 256. If p_A is the probability associated with the AR(3)LI model and p_B is the probability associated with the other 255 models and each of them is given an equal probability, the prior odds for any pair of models is $p_A/(p_B/255) = 255.\ p_A/p_B$ will usually not be equal to one. For example, if $p_A = 1/16$ and $p_B = 15/16$, then the prior odds $= 255/15 = 17$, that is, 17:1 on the AR(3)LI model versus any other of the 255 alternative models. In this calculation, we have assumed the 256 models are exhaustive, that is, $p_A + p_B = 1$. Of course, this is usually not the case. We could take $p_A = 1/16$ and $p_B = 5/16$, in which case the odds would be equal to $255/5 = 51$. These and other assumptions (e.g. see Jeffreys 1967, p. 254) indicate that when we are considering a single model, here an AR(3)LI model versus a large number of alternatives, to avoid "selection effects" it is not wise to use 1:1 prior odds but odds which favor the basic or null model, here an AR(3)LI model.

With the above proviso regarding prior odds, we see that with a correction for selection, the odds on the models in lines 2–5 *vis-à-vis* the AR(3)LI model in line 1 are not high. For example, with a conservative factor of 10 to guard against selection effects, the odds on the AR(3)LI model versus that in line 2 is $(10)(2.80)/(17.42) = 1.6$ in favor of the AR(3)LI model. Note that the forecast RMSEs for these models are 2.31 for the AR(3)LI model versus 2.74 for the model in line 2. Similar comparisons can be made with the models in lines 3–5. Finally, we note that only y_3 (that is, y_{t-3}) appears in these models but not y_1 and y_2 (that is, y_{t-1} and y_{t-2}). It is the case that $y_t + \beta y_{t-3} = 0$, with $0 < \beta < 1$ yields a damped oscillatory solution with period equal to six years. Thus inclusion of only y_{t-3} is a parsimonious way of including a six-year cyclical component in a model.

The model in line 6 of table 4.5 is the one identified for the United States using Hankel matrix identification procedures in Mittnik (1990). As can be seen, it is not favored by the odds or by a MSE comparison *vis-à-vis* the AR(3)LI model. Also, its RMSE of forecast 2.59 is somewhat larger than that of the AR(3)LI model, namely 2.31. Similar conclusions relate to the Otter (1990) model in line 7 versus the AR(3)LI model in line 1. Of course, it would be desirable to extend these comparisons using data for other countries as Mittnik (1990) did in his careful study.

4 Aggregation and non-linearity

In this section a brief discussion of and some preliminary empirical results relating to aggregation and non-linearity will be presented.

With respect to aggregation, one main concern is whether an aggregate variable is better modeled and forecasted using a model for it or whether it is better to model its components, forecast them, and use their sum as the forecast of the aggregate variable. Of course, if there is interest in forecasting the components, they will have to be modeled. However, there is still the issue, raised in personal conversation with Espasa in 1990, whether to add the component forecasts to forecast the aggregate or to forecast it directly, an issue treated in Espasa and Matea (1990).

In terms of the vector y_t in (2.4a), an aggregate can be defined as $Y_t^A = \sum_{i=1}^{N} y_{it} = \iota' y_t$ where $\iota' = (1\ 1 \ldots 1)$, a vector with all elements equal to 1. Thus from (2.4a)

$$\begin{aligned} |H_{11}| Y_t^A &= -\iota' H_{11}^* H_{12} x_t + \iota' H_{11}^* F_{11} e_{1t} \\ &= \phi'(L) x_t + \theta'(L)\, e_{1t}, \end{aligned} \tag{4.1}$$

where $\phi'(L) \equiv -\iota' H_{11}^* H_{12}$ and $\theta'(L) \equiv c' H_{11}^* F_{11} e_{1t}$. It is seen that the polynomial hitting Y_t^A, the aggregate variable, is the same one hitting each component of y_t, as shown in (2.4), given no cancellation of common roots. This point is being checked empirically in our current work with the components of real GNP.

Also, note that in (2.4), each y_{it} can depend on specific subsets of the elements of x_t, leading indicator variables specific to individual components of real GNP. On the other hand, in (2.4), the relation for the aggregate, probably all the elements of x_t appear, which makes it hard to implement. However, aggregate leading indicator variables may be used as an approximation. Preliminary theoretical analyzes indicate that forecasting components using specific input variables for each and adding forecasts of them to forecast an aggregate is the preferred approach under many conditions.

As regards non-linearities, non-linear effects can enter the AR(3)LI model in (2.6) through the leading indicator variables. That is, for example, erratic, chaotic movements in the money supply variable will induce such movements in the output variable. Also, with TVP models, parameter values can move to produce non-linear effects. While these and other statistical models (for example, threshold autoregressions, etc.) can produce non-linear effects, it would be desirable to have an explicit subject-matter justification for non-linearities, as, for example, in the asymmetric cost of adjustment literature. In addition, we have considered certain generalized production functions which have variable returns to scale and associated U-shaped average cost functions put forward by Zellner and Revankar (1969). One such function is $\log Y_t + \theta Y_t = x_t'\beta$, where $Y_t =$ output or RGNP, θ is a positive parameter, and x_t' is a row vector of input variables. On first differencing this relation we obtain

$$y_t = \log Y_t / Y_{t-1} = \frac{1}{1 + \theta Y_{t-1}} \Delta x_t' \beta + u_t, \tag{4.2}$$

where u_t is an additive white-noise error term and the approximation $\log Y_t / Y_{t-1} = (Y_t - Y_{t-1})/Y_{t-1}$ has been employed. On tentatively relating $\Delta x_t'$ to lagged values of y_t, leading indicator variables, and a constant, we obtain

$$\begin{aligned} y_t = {} & \frac{1}{1 + \theta Y_{t-1}} [\beta_0 + \beta_1 y_{t-1} + \beta_2 y_{t-2} + \beta_3 y_{t-3} + \beta_4 SR_{t-1} \\ & + \beta_5 S_{t-2} + \beta_6 GM_{t-1} + \beta_7 W_{t-1}] + u_t, \end{aligned} \tag{4.3}$$

a non-linear version of our ARLI model. Note that with $\theta > 0$, when Y_{t-1} is near full employment, the coefficients on the input variables are smaller in absolute value than they are when Y_{t-1} has a lower value.

When (4.3) was estimated using annual data for the United States and used to produce one-year-ahead forecasts, the results shown in table 4.6 were obtained. It is seen that using a value of θ different from zero has produced somewhat lower RMSEs of forecast. Also, it should be noted that with the use of generalized production functions, demand functions for labor and capital assume simple non-linear forms, somewhat similar in form to (4.2). Taking account of such non-linearities as well as those associated with asymmetric costs of adjustment may well indeed improve the explanatory and forecasting performance of our models.

Table 4.6 *Forecasting errors of real output growth rates for the United States, 1974–1990*[a]

Real output growth rates		Forecast errors	(Forecasts – actual values)		
		Value of θ			
		0 (ARLI)	0.001	0.005	0.01
1974	−0.54	0.94	0.85	0.86	0.87
1975	−1.27	1.02	1.30	1.45	1.48
1976	4.77	−3.52	−3.93	−4.03	−4.04
1977	4.56	0.81	−0.19	−0.45	−0.50
1978	5.16	−0.84	−1.19	−1.31	−1.34
1979	2.44	−0.07	−0.50	−0.60	−0.62
1980	−0.16	−0.51	−0.81	−0.82	−0.82
1981	1.92	−2.13	−2.88	−2.99	−3.01
1982	−2.58	3.93	3.85	3.83	3.82
1983	3.51	3.25	1.63	1.25	1.19
1984	6.43	−2.50	−3.07	−3.26	−3.29
1985	3.43	1.10	0.19	−0.03	−0.06
1986	2.81	3.64	1.77	1.35	1.28
1987	3.31	2.64	1.45	1.14	1.08
1988	4.32	−2.60	−3.60	−3.79	−3.82
1989	2.48	0.41	−0.60	−0.81	−0.84
1990	0.96	1.56	0.05	−0.25	−0.29
RMSE (74–87)		2.30	2.08	2.09	2.09
RMSE (74–90)		2.22	2.09	2.11	2.12

Note:
[a] Updated data (August 1991) were used. The forecasting model for the output growth rate, y_t, is:

$$y_t = \frac{1}{(1+\theta Y_{t-1})}\{\beta_0 + \beta_1 Y_{t-1} + \beta_2 Y_{t-2} + \beta_3 Y_{t-3} + \beta_4 SR_{t-1} + \beta_5 SR_{t-2} + \beta_6 GM_{t-1} + \beta_7 MSR_{t-1}\} + u_t$$

where $y_t = \ln(Y_t/Y_{t-1})$ with Y_t = real GNP in year t.

5 Summary and conclusions

Selected theoretical and empirical results have been presented illustrating the SEMTSA approach to model building. As can be seen from what has been presented, it is an operational approach that has already yielded useful results. Fixed-parameter and time-varying parameter models which forecast growth rates of real output for eighteen countries have been formulated and tested in out-of-sample forecasting tests, with encouraging results. Bayesian shrinkage forecasting techniques have been shown to be effective in reducing RMSEs of forecast. Bayesian decision theoretic procedures for forecasting turning points have been formulated and applied with considerable success, with about 70 percent of 158 turning points for eighteen countries correctly forecasted. Model and variable selection procedures have been formulated and successfully applied. Preliminary results on procedures for forecasting components of aggregates and adding them up to obtain a forecast of an aggregate have been briefly mentioned and are under study. Also, selected results for a class of nonlinear models have been presented which are encouraging. Works that link these results to economic theory have been cited.

To conclude, in the present problem area the SEMTSA approach has already provided a number of sound, tested reliable procedures and results. Future work to extend them and to integrate them with and to improve economic theory will lead to even more such results and, hopefully, models that are dependable in explanation, prediction, and control.

BIBLIOGRAPHY

Adelman, I. and F. L. Adelman (1959), "The dynamic properties of the Klein–Goldberger model," *Econometrica* 27, 569–625

Barro, R. J. (1978), "Unanticipated money, output and the price level in the United States," *Journal of Political Economy* 86, 549–80

Burns, A. F. and W. C. Mitchell (1946), *Measuring Business Cycles*, (New York, National Bureau of Economic Research

Christ, C. F. (1951), "A test of an econometric model for the United States, 1921–1947," in *Conference on Business Cycles* (New York, National Bureau of Economic Research, 35–107

Cooper, R. L. (1972), "The predictive performance of quarterly econometric models of the United States," in B. G. Hickman (ed.), *Econometric Models of Cyclical Behavior* (New York, Columbia University Press), 813–926

Diebold, F. X. (1989), "Forecast combination and encompassing, reconciling two divergent literatures," *International Journal of Forecasting* 5, 589–92

Espasa, A. and M. L. Matea (1990), "Underlying inflation in the Spanish economy: estimation and methodology," Bank of Spain, manuscript

Fair, R. C. (1991), "How might the debate be resolved?," Paper presented at a conference on the business cycle held at the Federal Reserve Bank of St. Louis, October

Friedman, M. (1951), "Comment on 'A test of an econometric model for the United States, 1921–1947', by Carl Christ," in *Conference on Business Cycles* (New York, National Bureau of Economic Research), 107–14

Garcia-Ferrer, A., R. A. Highfield, F. C. Palm, and A. Zellner (1987), "Macroeconomic forecasting using pooled international data," *Journal of Business and Economic Statistics* 5, 53–67; chapter 13 in this volume

Geweke, J. (1988), "The secular and cyclical behavior of real GNP in nineteen OECD countries, 1957–83," *Journal of Business and Economic Statistics* 6, 479–86

Harvey, A. C. and P. H. J. Todd (1983), "Forecasting economic time series with structural and Box–Jenkins models: a case study," *Journal of Business and Economic Statistics* 1, 299–307

Hickman, B. G. (ed.) (1972), *Econometric Models of Cyclical Behavior* (New York, Columbia University Press)

Highfield, R. A. (1986), *Forecasting with Bayesian State Space Models*, Doctoral dissertation, Graduate School of Business, University of Chicago

Hong, C. (1989), *Forecasting Real Output Growth Rates and Cyclical Properties of Models: A Bayesian Approach*, Doctoral dissertation, Department of Economics, University of Chicago

Jeffreys, H. (1967), *Theory of Probability*, 3rd edn. (Oxford, Oxford University Press)

Kling, J. L. (1987), "Predicting the turning points of business and economic time series," *Journal of Business* 60, 201–38

Lindley, D. V. and A. F. M. Smith (1972), "Bayes estimates for the linear model," *Journal of the Royal Statistical Society*, Series B 34, 1–41

Litterman, R. B. (1980), "A Bayesian procedure for forecasting with Bayesian vector autoregressions," Department of Economics, MIT, Cambridge, Mass., unpublished manuscript

(1986), "A statistical approach to economic forecasting," *Journal of Business and Economic Statistics* 4, 1–4

Manas-Anton, L. A. (1986), *Empirical Regularities in Short-Run Exchange Rate Behavior*, Doctoral dissertation, Department of Economics, University of Chicago

McNees, S. K. (1986), "Forecasting accuracy of alternative techniques: a comparison of US macroeconomic forecasts," *Journal of Business and Economic Statistics* 4, 5–15

Meese, R. and K. Rogoff (1983), "The out-of-sample failure of empirical exchange rate models: sampling error or misspecification?," in J. A. Frenkel (ed.), *Exchange Rates and International Economics* (Chicago, University of Chicago Press)

Min, C. (1991), *Economic Analysis and Forecasting of International Growth Rates Using a Bayesian Approach*, Thesis proposal, Department of Economics, University of Chicago

Min, C. and A. Zellner (1993), "Bayesian and non-Bayesian methods for combining models and forecasts with applications to forecasting international growth rates," *Journal of Econometrics, Annals* 56, 89–118; chapter 17 in this volume

Mittnik, S. (1990), "Macroeconomic forecasting using pooled international data," *Journal of Business and Economic Statistics* 8, 205–8

Nelson, C. R. (1972), "The prediction performance of the FRB–MIT–Penn model of the US economy," *American Economic Review* 62, 902–17

Nelson, C. R. and C. I. Plosser (1982), "Trends and random walks in macroeconomic time series," *Journal of Monetary Economics* 10, 139–62

Otter, P. W. (1990), "Canonical correlation in multivariate time series analysis with an application to one-year-ahead and multiyear-ahead macroeconomic forecasting," *Journal of Business and Economic Statistics* 8, 453–7

Palm, F. C. (1983), "Structural econometric modeling and time series analysis: an integrated approach," in A. Zellner (ed.), *Applied Time Series Analysis of Economic Data* (Washington, DC, Bureau of the Census, US Department of Commerce), 199–233; chapter 3 in this volume

Palm, F. C. and A. Zellner (1992), "To combine or not to combine? Issues of combining forecasts," *Journal of Forecasting* 11, 687–701

Plosser, C. I. (1976), *A Time Series Analysis of Seasonality in Econometric Models with an Application to a Monetary Model*, Doctoral dissertation, Graduate School of Business, University of Chicago

(1978), "A time series analysis of seasonality in econometric models," in A. Zellner (ed.), *Seasonal Analysis of Economic Time Series* (Washington, DC, Bureau of the Census, US Department of Commerce, 365–407; chapter 9 in this volume

Poirier, D. J. (ed.) (1991), "Bayesian Empirical Studies in Economics and Finance," Journal of Econometrics, Annals 49

Quenouille, M. H. (1957), *The Analysis of Multiple Time Series* (London: C. Griffin & Co.)

Stock, J. H. and M. W. Watson (1991), "Predicting recessions," Paper presented at Conference on Recent Advances in Time Series Analysis and Their Impact on Economic Forecasting, Madrid

Swamy, P. A. V. B. (1971), *Statistical Inference in Random Coefficient Regression Models* (New York, Springer-Verlag)

Wallis, K. F. (1983), "Models for X-11 and X-11 forecast procedures for preliminary and revised seasonal adjustments," in A. Zellner (ed.), *Applied Time Series Analysis of Economic Data* (Washington, DC, Bureau of the Census, US Department of Commerce), 3–11

Webb, R. I. (1985), "The behavior of speculative prices and the consistency of economic models," *Journal of Econometrics* 27, 123–30; chapter 10 in this volume

Wecker, W. E. (1979), "Predicting the turning points of a time series," *Journal of Business* 52, 35–50

Wolff, C. C. P. (1985), *Exchange Rate Models, Parameter Variation and Innovations: A Study on the Forecasting Performance of Empirical Models of Exchange Rate Determination*, Doctoral dissertation, Graduate School of Business, University of Chicago

(1987), "Time-varying parameters and the out-of-sample forecasting performance of structural exchange rate models," *Journal of Business and Economic Statistics* 5, 87–97

Zellner, A. (1979), "Statistical analysis of econometric models," *Journal of the American Statistical Association* 74, 628–51; chapter 2 in this volume

(1984), *Basic Issues in Econometrics* (Chicago, University of Chicago Press), paperback edn., 1987

(1991), "Comment on Ray C. Fair's thoughts on 'How might the debate be resolved?'," Paper presented at a conference on the business cycle held at the Federal Reserve Bank of St. Louis, October

Zellner, A. and C. Hong (1989), "Forecasting international growth rates using Bayesian shrinkage and other procedures," *Journal of Econometrics, Annals* 40, 183–202; chapter 14 in this volume

Zellner, A., C. Hong, and G. M. Gulati (1990), "Turning points in economic time series, loss structures, and Bayesian forecasting," part VI in S. Geisser, J. S. Hodges, S. J. Press, and A. Zellner (eds.), *Bayesian and Likelihood Methods in Statistics and Econometrics: Essays in Honor of George A. Barnard* (Amsterdam, North-Holland), 371–93; chapter 15 in this volume

Zellner, A., C. Hong, and C. Min (1991), "Forecasting turning points in international output growth rates using Bayesian exponentially weighted autoregression, time-varying parameter, and pooling techniques," *Journal of Econometrics* 49, 275–304; chapter 16 in this volume

Zellner, A. and F. C. Palm (1974), "Time series analysis and simultaneous equation econometric models," *Journal of Econometrics* 2, 17–54; chapter 1 in this volume

(1975), "Time series and structural analysis of monetary models of the US economy," *Sankhyā. The Indian Journal of Statistics*, Series C 37, 12–56; chapter 6 in this volume

Zellner, A. and S. Peck (1973), "Simulation experiments with a quarterly macroeconometric model of US economy," in A. A. Powell and R. A. Williams (eds.), *Econometric Studies of Macro and Monetary Relations* (Amsterdam, North-Holland), 149–68; reprinted in A. Zellner, *Basic Issues in Econometrics* (Chicago, University of Chicago Press)

Zellner, A. and N. S. Revankar (1969), "Generalized production functions," *Review of Economic Studies* 36 (2), 241–50

Zellner, A. and A. Siow (1980), "Posterior odds ratios for selected regression hypotheses," in J. M. Bernardo *et al.* (eds.), *Bayesian Statistics* (Valencia, University Press), 585–603

5 Large-sample estimation and testing procedures for dynamic equation systems (1980)

Franz C. Palm and Arnold Zellner

1 Introduction

In this chapter we consider large-sample estimation and testing procedures for parameters of dynamic equation systems with moving average error terms that are frequently encountered in econometric work (see, e.g., Quenouille 1957 and Zellner and Palm 1974). As pointed out in Zellner and Palm (1974), three-equation systems that are particularly relevant in econometric model building are (1) the final equations (FEs), (2) the transfer functions (TFs), and (3) the structural equations (SEs). In the present work, we specify these equation systems and develop large-sample "joint" or "system" estimation and testing procedures for each system of equations. These "joint" or "system" estimation procedures are iterative. They provide asymptotically efficient estimates of the parameters at the second step of iteration. The maximum likelihood (ML) estimator is obtained by iterating until convergence. The "joint" estimation methods provide parameter estimates that are more precise in large samples than those provided by single-equation procedures and the "joint" testing procedures are more powerful in large samples than those based on single-equation methods.

The aim of the chapter is to present a unified approach for estimating and testing FE, TF, and dynamic SE systems. In the chapter we use the results of previous work on the asymptotic properties of the ML estimator of the parameters of a dynamic model. We extend the recent work on efficient two-step estimation of dynamic models (e.g. Dhrymes and Taylor 1976, Hatanaka 1976, Reinsel 1976, 1977, Palm 1977a). It is interesting to note that the development of estimation methods for dynamic models is

Research financed by National Science Foundation Grants GS 40033 and SOC 7305547, income from the H. G. B. Alexander Endowment Fund, Graduate School of Business, University of Chicago, and the Belgian National Science Foundation. Some of the results in this chapter [were] presented in an earlier draft completed in 1974.

Originally published in the *Journal of Econometrics* 12 (1980), 251–83.

very similar to that for static models where iterative estimation procedures such as ML estimation under limited and full information developed at the Cowles Commission were also followed by asymptotically equivalent but computationally simpler methods such as the two-stage least squares [2SLS], three-stage least squares [3SLS], and linearized ML methods, respectively.

Previous work related to the present work includes that of Hannan (1969, 1971), Deistler (1975, 1976), and Hatanaka (1975), who have considered the identification problem for dynamic SE systems with moving average error terms. Maximum likelihood estimation of dynamic SEMs with moving average errors has been considered by Byron (1973), Phillips (1966) and Wall (1976) in the time domain, and for dynamic SEMs with stationary errors by Espasa (1977) in the frequency domain. Spectral estimation methods for static SEMs with stationary errors have been proposed by Hannan and Terrell (1973) and by Espasa (1977). Among many other workers, Durbin (1959), Box and Jenkins (1970), Maddala (1971), Pierce (1972), Akaike (1973), Pesaran (1973), Wilson (1973), Anderson (1975), Hannan (1975), Kang (1975), Nelson (1976), Nicholls (1976), Osborn (1976), and Reinsel (1976) have considered estimation of parameters of single-equation or multi-equation ARMA and transfer function (TF) models. The problem of TF estimation in a single-equation context has been extensively studied in the "distributed-lag" area. Closely related to our approach for FEs is the work of Nelson (1976) who considered joint estimation of a special FE system with diagonal MA matrices.

For a system of TFs, Wilson (1973) proposes an iterative procedure leading to a ML estimator. With respect to ML methods for TFs (e.g. Wilson 1973) and dynamic SEMs (e.g. Byron 1973, Phillips 1966, and Wall 1976), our approach is computationally more convenient to implement while having similar asymptotic properties. Many of the spectral methods apply to more general models, in the sense that the authors assume a stationary error process. In the light of Espasa's conclusion that treating the errors simply as a stationary process when they are generated by an AR or an MA model can lead, in the presence of lagged endogenous variables, to a loss of statistical efficiency, we parameterize the errors as a multivariate MA process. For an extensive review of the literature, the reader is referred to Aigner (1971), Nicholls, Pagan, and Terrell (1975), and Åström and Bohlin (1966). Finally, estimation methods for dynamic models with autoregressive errors, which have a long tradition in econometrics, are reviewed by Hendry (1976).

In what follows we shall specify the FE system that we consider in section 2 and then go on to develop estimation and testing procedures for parameters of the FE system. In section 3, a TF system is specified and inference procedures for it are developed, while in section 4 the SE system is presented and procedures for analyzing it are developed. Section 5 is devoted to a summary and discussion of the results with particular emphasis on relating them to the structure of econometric estimation procedures and on pointing to problems that remain to be analyzed.

2 Specification of and estimation and testing procedures for final equations

Let $\boldsymbol{z}_t' = (z_{1t}, z_{2t}, \ldots, z_{pt})$ be a vector of p observable random variables generated by the following multivariate autoregressive moving average (ARMA) process such as studied by Quenouille (1957):

$$\underset{p\times p}{H(L)}\ \underset{p\times 1}{\boldsymbol{z}_t} = \underset{p\times 1}{\bar{\boldsymbol{c}}} + \underset{p\times p}{F(L)}\ \underset{p\times 1}{\boldsymbol{e}_t}, \qquad t = 1, 2, \ldots, T, \tag{2.1}$$

where $\bar{\boldsymbol{c}}$ is a $p \times 1$ vector of constants, L is a lag operator such that $L^n \boldsymbol{z}_t = \boldsymbol{z}_{t-n}$, $H(L) = \{h_{ij}(L)\}$ and $F(L) = \{f_{ij}(L)\}$ are $p \times p$ matrix lag operators with typical elements being finite degree polynomials in L, namely $h_{ij}(L)$ and $f_{ij}(L)$, respectively, and $\boldsymbol{e}_t$ is a $p \times 1$ random error vector. We assume that $\boldsymbol{e}_t$ is normally distributed with

$$E\boldsymbol{e}_t = 0 \quad \text{and} \quad E\boldsymbol{e}_t\boldsymbol{e}_{t'}' = \delta_{tt'} I_p, \tag{2.2}$$

for all t and t', where $\delta_{tt'}$ is the Kronecker delta. Contemporaneous and serial correlation as well as different variances for the error process in (2.2) can be introduced through appropriate specification of $F(L)$. We further assume that the inverse of $H(L)$, $H^{-1}(L) = H^*(L)/|H(L)|$, exists, where $H^*(L)$ is the adjoint matrix associated with $H(L)$ and $|H(L)|$ is the determinant of $H(L)$ that is a scalar polynomial of finite degree in L with roots lying outside the unit circle.

The "final equations" (FEs) associated with (2.1), obtained by multiplying both sides of (2.1) on the left by $H^*(L)$ and normalizing the system, are given by

$$\theta(L)\boldsymbol{z}_t = \boldsymbol{c} + A(L)\boldsymbol{e}_t, \tag{2.3}$$

where

$$\theta(L) = d^{-1}|H(L)| = (1 - \theta_1 L - \cdots - \theta_n L^n)$$

is a scalar polynomial in L, d is a normalizing constant,

$$c = d^{-1} H^*(L)\bar{\boldsymbol{c}}$$

is a vector of constants, and

$$A(L) = d^{-1}H^*(L)F(L)$$

is a matrix lag operator of degree m.

In order to identify the system (2.3), we assume among other things that the roots of $|A(L)|$ are outside the unit circle, and that both sides of (2.3) do not have common factors. As pointed out in previous work, Zellner and Palm (1974, 1975), the AR polynomial $\theta(L)$ operates on each element of $\boldsymbol{z}_t$. Unless there is canceling, the AR parts of the equations in (2.3) should be of identical order and have the same parameters. Since it is often of interest to test that the AR parameters are the same in different equations and also for greater generality, we shall take up the problem of estimating parameters of the following system:

$$\theta_i(L)z_{it} = c_i + u_{it}, \quad i = 1, 2, \ldots, p, \tag{2.4}$$

where

$$\theta_i(L) = 1 - \theta_{i1}L - \theta_{i2}L^2 - \cdots - \theta_{in_i}L^{ni},$$

with n_i given, $i = 1, 2, \ldots, p$, and u_{it} is the ith element of the vector $\boldsymbol{u}_t = A(L)\boldsymbol{e}_t$.

In connection with convenient estimation of the parameters in (2.4), we express the error vector $\boldsymbol{u}_t$ as

$$\begin{aligned} \boldsymbol{u}_t &= A_0\boldsymbol{e}_t + A_1\boldsymbol{e}_{t-1} + \cdots + A_m\boldsymbol{e}_{t-m} \\ &= \boldsymbol{v}_t + G_1\boldsymbol{v}_{t-1} + \cdots + G_m\boldsymbol{v}_{t-m}, \end{aligned} \tag{2.5}$$

where $G_i = A_iA_0^{-1}$, $i = 1, 2, \ldots, m$, A_0 is assumed to be non-singular, and $\boldsymbol{v}_t = A_0\boldsymbol{e}_t$ is normally distributed with $E\boldsymbol{v}_t = \boldsymbol{0}$, and

$$E\boldsymbol{v}_t\boldsymbol{v}_t' = A_0A_0' = \Omega_v \quad \text{and} \quad E\boldsymbol{v}_t\boldsymbol{v}_{t'}' = 0, \quad t \neq t'. \tag{2.6}$$

A typical element of $\boldsymbol{u}_t$, say u_{it}, may be represented as a moving average in one random variable (see, e.g., Ansley, Spivey and Wroblenski 1977, Palm 1977b, or Granger and Morris 1976),

$$u_{it} = v_{it} + \lambda_{i1}v_{it-1} + \cdots + \lambda_{im}v_{it-m}, \tag{2.7}$$

where the λ_{ih}s are such that they reproduce the autocorrelation structure of u_{it}, i.e.,

$$\omega_{ii}\sum_{h=0}^{m-j}\lambda_{ih+j}\lambda_{ih} = \sum_{h=0}^{m-j}\boldsymbol{a}_{ih+j}'\boldsymbol{a}_{ih}, \quad j = 0, 1, \ldots, m,$$

with $\lambda_{i0} = 1, \omega_{ii}$ being the $i \times i$ element of Ω_v as defined in (2.6), and $\boldsymbol{a}_{ih}'$ being the ith row of A_h in (2.5). The disturbances v_{it} on the r.h.s. of

(2.7) are normally and independently distributed, each with zero mean and common variance ω_{ii}.

Each FE may be estimated separately using a single-equation non-linear least squares or single-equation ML procedure. Joint estimation of the parameters in the system shown in (2.4) and (2.5) will now be considered. In doing so, we ignore possible restrictions implied by the underlying structural model (2.1). For an example of joint ML estimation of a set of FEs, the reader is referred to Wallis (1977). We write the system of FEs as

$$\begin{bmatrix} \theta_1(L) & 0 & \dots & 0 \\ 0 & \theta_2(L) & & \vdots \\ \vdots & & \ddots & 0 \\ 0 & \dots & 0 & \theta_p(L) \end{bmatrix} \begin{bmatrix} z_{1t} \\ z_{2t} \\ \vdots \\ z_{pt} \end{bmatrix} = \boldsymbol{c} + \boldsymbol{v}_t + \sum_{h=1}^{m} G_h \boldsymbol{v}_{t-h}, \quad (2.8)$$

or alternatively as

$$\boldsymbol{z}_t = \boldsymbol{c} + W_{1t}\boldsymbol{\theta} + \boldsymbol{u}_t, \quad (2.9)$$

with

$$\underset{p\times k}{W_{1t}} = \begin{bmatrix} z_{1t-1} & z_{1t-2} & \dots & z_{1t-n_1} & 0 & \dots & \dots & \dots & \dots\dots & 0 \\ 0 & \dots & \dots & 0 & z_{2t-1} & z_{2t-2} & \dots & z_{2t-n_2} & 0 \dots & 0 \\ \vdots & & & & & & & & & \vdots \\ 0 & \dots & \dots & \dots & \dots & \dots & \dots & \dots & \dots\dots & z_{pt-n_p} \end{bmatrix}$$

$$\underset{1\times k}{\boldsymbol{\theta}'} = (\theta_{11}, \theta_{12}, \theta_{13}, \dots, \theta_{1n_1}, \theta_{21}, \dots, \theta_{2n_2}, \dots, \theta_{pn_p}),$$

$$k = \sum_{i=1}^{p} n_i,$$

and for a sample of T observations

$$\underset{Tp\times 1}{\boldsymbol{z}} = \underset{(Tp\times p)}{W_0} \underset{(p\times 1)}{\boldsymbol{c}} + \underset{(Tp\times k)}{W_1} \underset{(k\times 1)}{\boldsymbol{\theta}} + \underset{(Tp\times 1)}{\boldsymbol{u}}, \quad (2.10)$$

where

$$\begin{aligned} \boldsymbol{z}' &= (\boldsymbol{z}_1', \boldsymbol{z}_2', \dots, \boldsymbol{z}_T'), \\ W_0' &= (I_p, I_p, \dots, I_p), \\ W_1' &= (W_{11}', W_{12}', \dots, W_{1T}'), \\ \boldsymbol{u}' &= (\boldsymbol{u}_1', \boldsymbol{u}_2', \dots, \boldsymbol{u}_T'). \end{aligned}$$

Assuming initial conditions to be zero, the vector $\boldsymbol{u}$ may be expressed in terms of $\boldsymbol{v}$,

$$\boldsymbol{u} = M\boldsymbol{v},$$

where

$$M = \begin{bmatrix} I_p & 0 & \dots & \dots & \dots & \dots & 0 \\ G_1 & I_p & 0 & \dots & \dots & \dots & \vdots \\ G_2 & G_1 & I_p & 0 & \dots & \dots & \vdots \\ \vdots & & & & & & \vdots \\ G_m & G_{m-1} & \dots & \dots & I_p 0 & \dots & \vdots \\ 0 & G_m & \dots & \dots & \dots & \dots & \vdots \\ 0 & \dots & \dots & \dots & G_m & \dots & I_p \end{bmatrix},$$

and

$$\boldsymbol{v}' = (\boldsymbol{v}'_1, \boldsymbol{v}'_2, \dots, \boldsymbol{v}'_T).$$

The assumption of zero initial values is basically made for the purpose of simplicity. One can also "backforecast" the values of the initial conditions for a set of FEs,[1] as Box and Jenkins (1970) do for single-equation ARMA models, or treat the initial conditions as unknown parameters (e.g. Phillips 1966). Whether backforecasting improves the properties of estimators under all conditions is not known. The treatment of the initial conditions generally does not affect the asymptotic properties of the estimators presented in this chapter, but may be very important in small-sample situations – see, e.g., Kang (1975) and Osborn (1976). Under zero initial conditions, the likelihood function is

$$L(\boldsymbol{\theta}, \mathbf{c}, M, \Omega_v, \boldsymbol{z}) \propto |\Omega_v|^{-T/2} \exp{(S)}, \tag{2.11}$$

where

$$S = -\tfrac{1}{2}(\boldsymbol{z} - W_0\mathbf{c} - W_1\theta)'M'^{-1}(I_T \otimes \Omega_v^{-1})M^{-1}(\boldsymbol{z} - W_0\mathbf{c} - W_1\boldsymbol{\theta}).$$

The first order conditions for a maximum of the log-likelihood function (see, e.g., Palm 1977a) are

$$\partial S/\partial \boldsymbol{\beta} = W'M'^{-1}(I_T \otimes \Omega_v^{-1})M^{-1}\boldsymbol{u} = \boldsymbol{0}, \tag{2.12}$$
$$\beta' = (\mathbf{c}', \boldsymbol{\theta}', \gamma'),$$
$$\gamma = \text{vec}\,[G_1\, G_2 \dots G_m]',$$

[1] For example for a vector MA (1) model $\boldsymbol{x}_t = \varepsilon_t - A\varepsilon_{t-1}$, the backward version is given by $\boldsymbol{x}_t = \boldsymbol{e}_t - \boldsymbol{B}\boldsymbol{e}_{t+1}$, where ε_t and $\boldsymbol{e}_t$ are both white noise with covariance matrix Ω. From the autocovariance function of $\boldsymbol{x}_t$, we have $\Omega + A\Omega A' = \Omega + B\Omega B'$ and $-A\Omega = -\Omega B'$. Given consistent parameter estimates of the forward version $\hat{A}$ and $\hat{\Omega}$, $\hat{B} = \hat{\Omega}\hat{A}'\hat{\Omega}^{-1}$ is a consistent estimate of B, that can be used to backforecast $\boldsymbol{x}_t$ using the backward version of the model.

with "vec" denoting the operation of vectorizing a matrix, stacking column after column,

$$W = [W_0\ W_1\ W_2],$$

with W_2 being a $T_p \times p^2 m$ matrix of disturbances,

$$W_2' = \left[\boldsymbol{v}_{ij}^{l'}\right], \quad i = 1, \ldots, p, \quad l = 1, \ldots, m, \quad j = 1, \ldots, p,$$

and with typical row

$$\boldsymbol{v}_{ij}^{l'} = [\underbrace{0, 0, \ldots, 0}_{lp \text{ times}}\ \underbrace{0 \ldots 0 v_{j1}\, 0 \ldots 0}_{p \text{ elements}}\ \underbrace{0 \ldots 0 v_{j2}\, 0 \ldots 0}\ \underbrace{0 \ldots 0 v_{jT-l}\, 0 \ldots 0}].$$

(ith position)

For given Ω_v, the set of equations in (2.12) is non-linear in the parameters of M. The solution of (2.12) requires an iterative procedure.

An alternative to the exact ML solution of (2.12) is provided by approximating the first order condition (2.12). Using a lemma given by Dhrymes and Taylor (1976),[2] a two-step estimator of β with the same asymptotic properties as the ML estimator is given by

$$\hat{\hat{\beta}} = \hat{\beta} - \Gamma^{-1}(\hat{\beta}) \frac{\partial S}{\partial \beta}(\hat{\beta}), \tag{2.13}$$

where $\Gamma(\hat{\beta})$ is non-singular matrix such that

$$\operatorname*{plim}_{T \to \infty} \frac{1}{T} \Gamma(\hat{\beta}) = \operatorname*{plim}_{T \to \infty} \frac{1}{T} \frac{\partial^2 S}{\partial \beta \partial \beta'}(\beta_0),$$

β_0 is the true parameter value and $\hat{\beta}$ is a consistent estimator of β_0 such that $T^{\frac{1}{2}}(\hat{\beta} - \beta_0)$ has some limiting distribution. The matrix Γ and the vector $\partial S / \partial \beta$ in (2.13) also depend on the unknown parameters of Ω_v.

The lemma given by Dhrymes and Taylor (1976) applies to all the parameters in the likelihood function (i.e. to $\boldsymbol{\vartheta}' = (\beta', \omega')$, where ω is the vector of unknown parameters in Ω_v). As the information matrix is block-diagonal for β and Ω_v, the use of a block-diagonal matrix Γ to approximate the Hessian matrix of the log-likelihood function with respect to $\boldsymbol{\vartheta}$ yields expression (2.13) for the subvector of parameters β. The use of consistent but not efficient estimates for the unknown elements of Ω_v in (2.13) will be sufficient for $\hat{\hat{\beta}}$ to have the same asymptotic properties as the ML estimator.

[2] For an earlier discussion of approximations to the ML solution, the reader is referred to Fisher (1925, ch. 9), Kendall and Stuart (1961, pp. 48–51), and Rothenberg and Leenders (1964).

Expression (2.13) defines a class of two-step estimators. A member of this class can be characterized by a particular choice for the matrix Γ. If we use the Hessian matrix for Γ in (2.13), we implement the second step of the Newton–Raphson algorithm.[3] Provided one starts with consistent parameter estimates an approximate solution of the system (2.12), which satisfies the requirements for (2.13) is also obtained at the second step of the Gauss–Newton algorithm. The function S in (2.11) can be written as an inner product of vectors.

$$\boldsymbol{v}'\left(I_T \otimes \Omega_v^{-1}\right)\boldsymbol{v} = \varepsilon'\varepsilon, \tag{2.14}$$

where

$$\varepsilon = P\boldsymbol{v} = (\boldsymbol{e}_1' \, \boldsymbol{e}_2' \ldots \boldsymbol{e}_T')',$$

and $P = (I_T \otimes A_0^{-1})$ is the matrix obtained from the decomposition of the positive definite matrix $(I_T \otimes \Omega_v^{-1}) = P'P$. The derivative of ε with respect to $\boldsymbol{\beta}$ is

$$\frac{\partial \varepsilon}{\partial \boldsymbol{\beta}} = -W'M'^{-1}P'. \tag{2.15}$$

The second step of the Gauss–Newton procedure can be written as

$$\hat{\hat{\beta}} = \hat{\beta} - \left\{ \left[\frac{\partial \varepsilon}{\partial \boldsymbol{\beta}} \frac{\partial \varepsilon'}{\partial \boldsymbol{\beta}} \right]^{-1} \frac{\partial \varepsilon}{\partial \boldsymbol{\beta}} \varepsilon \right\}_{\beta = \hat{\beta}}, \tag{2.16a}$$

$$= \hat{\beta} + [\hat{W}'\hat{M}'^{-1}(I_T \otimes \hat{\Omega}_v^{-1})\hat{M}^{-1}\hat{W}]^{-1}\hat{W}'\hat{M}'^{-1}(I_T \otimes \hat{\Omega}_v^{-1})\hat{\boldsymbol{v}}, \tag{2.16b}$$

where the carats above the symbols denote that the quantities are evaluated at consistent parameter estimates, for example $\hat{W} = (W_0 \, W_1 \, \hat{W}_2)$. The second order derivative of S with respect to β_i and β_j is

$$\frac{\partial^2 S}{\partial \beta_i \partial \beta_j} = -\left[\frac{\partial^2 \varepsilon'}{\partial \beta_i \partial \beta_j} \right] \varepsilon - \left[\frac{\partial \varepsilon'}{\partial \beta_i} \right] \left[\frac{\partial \varepsilon}{\partial \beta_j} \right]. \tag{2.17}$$

Under the assumptions underlying our model, it can be shown by using the strong law for martingales (see, e.g., Feller 1966, p. 238) that the first r.h.s. term in (2.17) converges to zero in probability, so that the two-step estimator in (2.16) implements expression (2.13) and therefore the two-step Gauss–Newton procedure (2.16) is one member of a class of estimators which are asymptotically equivalent to the ML estimator, defined as the exact solution of expression (2.12). Further, expression (2.16) can be calculated either using the analytical derivatives given in

[3] The reader, who is not familiar with numerical procedures to solve systems of non-linear equations, is referred to Goldfeld and Quandt (1972).

(2.15), evaluating them at consistent parameter estimates, as has been done, e.g., by Nelson (1976), or by numerical calculation of the partial derivatives in (2.16a) of ε with respect to β, as is proposed by Box and Jenkins (1970) for univariate models.

The two-step Gauss–Newton estimator (2.16) can be interpreted as a "residual-adjusted" Aitken estimator. Adding the quantity $\sum_{h=1}^{m} G_h \boldsymbol{v}_{t-h}$ to both sides of (2.9), we get for the sample period

$$\begin{aligned} \boldsymbol{y} &= W_0 \boldsymbol{c} + W_1 \boldsymbol{\theta} + W_2 \gamma + M\boldsymbol{v} \\ &= W\beta + M\boldsymbol{v}, \end{aligned} \tag{2.18}$$

where

$$\boldsymbol{y}' = (\boldsymbol{y}_1', \ldots, \boldsymbol{y}_t', \ldots, \boldsymbol{y}_T') \quad \text{and} \quad \boldsymbol{y}_t = \boldsymbol{z}_t + \sum_{h=1}^{m} G_h \boldsymbol{v}_{t-h}.$$

Application of generalized least squares to the system (2.18), after evaluation of the regressand, the regressors, and the disturbance covariance matrix at consistent parameter estimates denoted by the carats, yields

$$\hat{\hat{\beta}} = \left[\hat{W}'\hat{M}'^{-1}(I_T \otimes \hat{\Omega}_v^{-1})\hat{M}^{-1}\hat{W}\right]^{-1}\left[\hat{W}'\hat{M}'^{-1}(I_T \otimes \hat{\Omega}_v^{-1})\hat{M}^{-1}\hat{\boldsymbol{y}}\right], \tag{2.19}$$

which is equivalent to (2.16b).

The two-step estimator in (2.19) is similar to those proposed by Reinsel (1976) for other models and has the same asymptotic properties as the ML estimator under fixed and known initial values and ignoring possible restrictions on the FE parameters implied by the specification (2.1). The estimator $\hat{\hat{\beta}}$ is consistent, asymptotically normally distributed and efficient, with a large sample covariance matrix consistently estimated by

$$\hat{V}(\hat{\hat{\beta}}) = \left[\hat{W}'\hat{M}'^{-1}(I_T \otimes \hat{\Omega}_v^{-1})\hat{M}^{-1}\hat{W}\right]^{-1}. \tag{2.20}$$

Joint estimation of the parameters in the system (2.4) and (2.5) involves the following steps:

(1) Using the error representation shown in (2.7), estimate the parameters of each equation separately using, for example, the Box–Jenkins (1970) non-linear least squares approach or a univariate ML procedure. The estimates so obtained will be consistent but not efficient. They will be asymptotically efficient within the class of estimation methods for the univariate ARMA representation of $\boldsymbol{z}_{it}$. As the first-step estimator plays a crucial role in a two-step estimation procedure, it is preferable to use a non-linear least squares or a single-equation maximum likelihood method to estimate the parameters of each FE separately instead of using, for example, an instrumental variables

approach. The main objective of this first step is to obtain an estimate of $\boldsymbol{v}_t$, say $\hat{\boldsymbol{v}}_t$, $t = 1, 2, \ldots, T$.

(2) Use the $\hat{\boldsymbol{v}}_t$s to form an estimate of the covariance matrix $\Omega_v = E\boldsymbol{v}_t\boldsymbol{v}_t'$, namely

$$\hat{\Omega}_v = \sum_{t=1}^{T} \hat{\boldsymbol{v}}_t\hat{\boldsymbol{v}}_t'/T. \qquad (2.21)$$

(3) Express the ith equation of the system (2.4) as

$$z_{it} = c_i + \boldsymbol{w}_{1it}'\boldsymbol{\theta}_i + \boldsymbol{q}_t'\gamma_i + v_{it}, \qquad \begin{matrix} i = 1, 2, \ldots, p, \\ t = 1, 2, \ldots, T, \end{matrix} \qquad (2.22a)$$

where

$$\begin{aligned} \boldsymbol{w}_{1it}' &= (z_{it-1}, z_{it-2}, \ldots, z_{it-n_i}), \\ \boldsymbol{\theta}_i' &= (\theta_{i1}, \theta_{i2}, \ldots, \theta_{in_i}), \\ \boldsymbol{q}_t' &= (\boldsymbol{v}_{t-1}', \boldsymbol{v}_{t-2}', \ldots, \boldsymbol{v}_{t-m}'), \\ \gamma_i' &= (\gamma_{i1}', \gamma_{i2}', \ldots, \gamma_{im}'), \end{aligned}$$

with γ_{ij}' being the ith row of G_j, $j = 1, 2, \ldots, m$.

Expressing (2.22a) for all t, we have

$$\boldsymbol{z}_i = c_i\boldsymbol{\iota} + W_{1i}\boldsymbol{\theta}_i + Q\gamma_i + \boldsymbol{v}_i, \quad i = 1, 2, \ldots, p, \qquad (2.22b)$$

where $\boldsymbol{w}_{1it}'$ and $\boldsymbol{q}_t'$ are typical rows of W_{1i} and Q, respectively, and $\boldsymbol{\iota}$ is a $T \times 1$ vector with elements equal to one.

We then apply ordinary least squares to each equation (2.22a), after replacing Q by $\hat{Q}$, a matrix of the first-step residuals $\hat{\boldsymbol{v}}_t$, to obtain consistent estimates of γ_i.

(4) Compute expression (2.19). Iteration of steps 1–4 yields the ML estimator[4] given known and fixed initial conditions. In small samples it is not clear that the iterated estimator for β is to be preferred to $\hat{\beta}$. For example, it is well known that ML estimators for parameters of many models have poor finite sample properties relative to usually employed loss functions.[5] Nelson (1976) provides Monte Carlo results pertaining to a system similar to a particular set of FEs that

[4] Other iterative algorithms that may be computationally more efficient can be employed to compute the ML estimate (see, e.g., Chow and Fair 1973) who considered a dynamic system with AR errors.

[5] See Zellner (1971a) for some results relating to ML estimation of parameters of the log-normal distribution. For static simultaneous equation models, ML estimators frequently are found to possess no finite moments and hence have unbounded risk relative to a quadratic loss function. Last, Stein's well-known results indicate that ML estimators are often inadmissible relative to a quadratic loss function (see references and analysis in Zellner and Vandaele 1974).

indicate a substantial gain of efficiency of the joint estimators with respect to univariate procedures. The finite sample properties of $\hat{\hat{\beta}}$ in (2.19) and estimators obtained by iteration are as yet not established.

The system of FEs (2.10) can also be written as

$$\boldsymbol{z} = W_0\boldsymbol{c} + W_1\boldsymbol{\theta} + W_2\gamma + \boldsymbol{v}. \tag{2.23}$$

Generalized least squares applied to (2.23) after replacing the lagged error terms in W_2 by their sample estimates $\hat{W}_2$ and Ω_v by a consistent estimate $\hat{\Omega}_v$ leads to

$$\hat{\hat{\beta}}_{\mathrm{GLS}} = \left[\hat{W}'\left(I_T \otimes \hat{\Omega}_v^{-1}\right)\hat{W}\right]^{-1}\hat{W}'\left(I_T \otimes \hat{\Omega}_v^{-1}\right)\boldsymbol{z}. \tag{2.24}$$

Expression (2.24) gives a consistent joint estimator for $\boldsymbol{\beta}$, but it usually is not efficient. From a comparison with (2.19), it is obvious that the estimator in (2.24) is not a solution to the first order conditions for a maximum of the likelihood function, so that iterative solution of (2.24) will not yield the ML estimator. It rather gives a solution of the first order conditions for a maximum of the likelihood function with respect to $\boldsymbol{\beta}$ under the condition that $W_2 = \hat{W}_2$, implying that $\partial S/\partial\boldsymbol{\beta}$ is linear in $\boldsymbol{\beta}$ (see, e.g., Maddala 1971 for a similar discussion of single-equation models).

As mentioned above, FEs are often encountered in which the $\boldsymbol{\theta}_i$ coefficient vectors in (2.22) are all the same, that is $\boldsymbol{\theta}_1 = \boldsymbol{\theta}_2 = \boldsymbol{\theta}_3 = \cdots = \boldsymbol{\theta}_p = \boldsymbol{\theta}^{(r)}$. In such cases, the restricted matrix W_1 in (2.10) takes the form

$$\underset{Tp\times n}{W_1^{(r)}} = \begin{bmatrix} \boldsymbol{z}_0 & \boldsymbol{z}_{-1} & \cdots & \boldsymbol{z}_{-n+1} \\ \boldsymbol{z}_1 & \boldsymbol{z}_{-2} & & \boldsymbol{z}_{-n+2} \\ \vdots & & & \vdots \\ \boldsymbol{z}_{T-1} & \cdots & \cdots & \boldsymbol{z}_{T-n} \end{bmatrix}. \tag{2.25}$$

We then write the system (2.10) as

$$\boldsymbol{z} = W_0\boldsymbol{c} + W_1^{(r)}\boldsymbol{\theta}^{(r)} + \boldsymbol{u}. \tag{2.26}$$

Then the approximate ML estimators for $\boldsymbol{\beta}^{(r)'} = (\boldsymbol{c}', \boldsymbol{\theta}^{(r)'}, \gamma')'$ are given by

$$\begin{bmatrix} \hat{\hat{\boldsymbol{c}}} \\ \hat{\hat{\gamma}} \end{bmatrix}^{(r)} = \left[(W_0\,\hat{W}_2)'\hat{\Omega}^{-1}(W_0\,\hat{W}_2)\right]^{-1}(W_0\,\hat{W}_2)\hat{\Omega}^{-1}\left(\hat{\boldsymbol{y}} - W_1^{(r)}\hat{\hat{\boldsymbol{\theta}}}^{(r)}\right), \tag{2.27a}$$

$$\hat{\hat{\boldsymbol{\theta}}}^{(r)} = \left[W_1^{(r)'}RW_1^{(r)}\right]^{-1}W_1^{(r)'}R\hat{\boldsymbol{y}}, \tag{2.27b}$$

with

$$R = \hat{\Omega}^{-1} - \hat{\Omega}^{-1}(W_0\,\hat{W}_2)[(W_0\,\hat{W}_2)'\hat{\Omega}^{-1}(W_0\,\hat{W}_2)]^{-1}(W_0\,\hat{W}_2)'\hat{\Omega}^{-1},$$

where $\hat{\Omega}$ is an estimate of $\Omega = [M(I_T \otimes \Omega_v)M']$ and $\hat{W}_2$ is formed using lagged residuals. The large sample covariance matrix for the restricted estimators in (2.27), $\hat{\hat{\beta}}^{(r)}$, denoted V_r, is consistently estimated by

$$\hat{V}_r = [\hat{W}^{(r)\prime}\hat{\Omega}^{-1}\hat{W}^{(r)}]^{-1} \quad \text{where} \quad \hat{W}^{(r)} = \left[W_0 \, W_1^{(r)} \, \hat{W}_2\right]. \tag{2.28}$$

In case of general linear restrictions on the elements of β in (2.18), say $C\beta = a$, where C is a given matrix with q linearly independent rows of rank q, and $\boldsymbol{a}$ is a given $q \times 1$ vector, an estimator of β satisfying the restriction is given by

$$\tilde{\beta} = \hat{\hat{\beta}} - (\hat{W}'\hat{\Omega}^{-1}\hat{W})^{-1}C'[C(\hat{W}'\hat{\Omega}^{-1}\hat{W})^{-1}C']^{-1}(C\hat{\hat{\beta}} - \boldsymbol{a}), \tag{2.29}$$

with $\hat{\hat{\beta}}$ as shown in (2.19), $\hat{\Omega} = [\hat{M}(I_T \otimes \hat{\Omega}_v)\hat{M}']$, and large sample covariance matrix, $V(\tilde{\beta})$, consistently estimated by

$$\hat{V}(\tilde{\beta}) = (\hat{W}'\hat{\Omega}^{-1}\hat{W})^{-1} - (\hat{W}'\hat{\Omega}^{-1}\hat{W})^{-1}C'[C(\hat{W}'\hat{\Omega}^{-1}\hat{W})^{-1}C']^{-1}C(\hat{W}'\hat{\Omega}^{-1}\hat{W})^{-1}. \tag{2.30}$$

While (2.29) and (2.30) are relevant for the case of general linear restrictions, it should be appreciated that the matrices involved in the expressions are quite large from a numerical point of view for systems of even moderate size.[6]

To test the restriction that $\boldsymbol{\theta}_1 = \boldsymbol{\theta}_2 = \cdots = \boldsymbol{\theta}_p = \boldsymbol{\theta}^{(r)}$, an $n \times 1$ vector, introduced in connection with (2.26), we consider the following residual sums of squares:

$$SS_r = \left(\hat{\boldsymbol{y}} - \hat{\hat{W}}^{(r)}\hat{\hat{\beta}}^{(r)}\right)' \hat{\hat{\Omega}}^{-1}\left(\hat{\boldsymbol{y}} - \hat{\hat{W}}^{(r)}\hat{\hat{\beta}}^{(r)}\right), \tag{2.31}$$

and

$$SS_u = (\hat{\boldsymbol{y}} - \hat{\hat{W}}\hat{\hat{\beta}})'\hat{\hat{\Omega}}^{-1}(\hat{\boldsymbol{y}} - \hat{\hat{W}}\hat{\hat{\beta}}), \tag{2.32}$$

where the carats denote that the quantities are computed using the second-step estimates of $\beta^{(r)}$ and β, respectively. Thus, the approximate likelihood ratio,

$$N \log(SS_r/SS_u), \tag{2.33}$$

[6] With respect to the large matrices that are encountered in joint estimation procedures and that will usually lead to a multicollinearity problem, it is worthwhile to mention the use of approximate Bayes estimates such as considered by Zellner and Vandaele (1974). In a Monte Carlo study of the small sample ($T = 20$) properties of several estimators for a dynamic model with first order autoregressive errors, Swamy and Rappoport (1978) conclude that in terms of mean-square errors the ridge regression and the approximate minimum mean-square error estimates of the regression coefficients are significantly better than alternative estimates such as ML or Hatanaka's (1974) residual adjusted estimates.

is distributed as χ^2_m in large samples where $N = Tp$ and $m = n(p-1)$, the number of restrictions involved in $\boldsymbol{\theta}_1 = \boldsymbol{\theta}_2 = \cdots = \boldsymbol{\theta}_p = \boldsymbol{\theta}^{(r)}$. Thus (2.33) provides a large sample χ^2 test of a frequently encountered hypothesis in model construction. In a similar fashion, large sample χ^2 tests of the general linear hypothesis $C\boldsymbol{\beta} = \boldsymbol{a}$ can be constructed.

The computation of the likelihood ratio (LR) requires parameter estimates for the restricted and the unrestricted versions of the model and can therefore be laborious. For the purpose of computational convenience, it is preferable to use the Wald-test (W) (see, e.g., Sargan 1975) when it is easier to estimate the unrestricted model than to estimate the restricted model and to use the Lagrange multiplier (LM)-test (see, e.g., Breusch and Pagan 1978) or the efficient score (ES)-test (see, e.g., Rao 1973) when the restricted model is easily estimated. All four test statistics have the same asymptotic properties.

Imposing the restrictions $\boldsymbol{\theta}_i = \boldsymbol{\theta}^{(r)}$, $i = 1, 2, \ldots, p$, leads to a reduction of the size of the matrices involved in the joint estimation of a set of FEs, so that the LM test may be preferred. The approximate LM (and ES) test statistic is obtained by evaluation of the first derivatives of the log-likelihood function with respect to $\boldsymbol{\beta}$ in (2.12) at the restricted parameter estimates $\hat{\hat{\boldsymbol{\beta}}}^{(r)}$ in (2.27) and forming the quadratic form

$$\phi_{LM} = \{\boldsymbol{u}'\Omega^{-1} W_1 [W_1' R W_1]^{-1} W_1' \Omega^{-1} \boldsymbol{u}\}\bigg|_{\substack{\phi_i = \hat{\phi}^{(r)} \\ \boldsymbol{c} = \hat{\boldsymbol{c}}^{(r)} \\ \gamma = \hat{\gamma}^{(r)}}}, \tag{2.34}$$

where R is defined in (2.27b). Under the null hypothesis $\boldsymbol{\theta}_i = \boldsymbol{\theta}^{(r)}$, $i = 1, 2, \ldots, p$, ϕ_{LM} is distributed as χ^2_m in large samples where $m = n(p-1)$.

In order to test non-linear restrictions on the parameters $\boldsymbol{\beta}$ – for example, restrictions on $\boldsymbol{\beta}$ implied by the underlying structural form (2.1) – the W-test will usually be preferred on the basis of its computational convenience. Suppose that we want to test a set of m non-linear differentiable restrictions on $\boldsymbol{\beta}$, $h_i(\boldsymbol{\beta}) = 0$, $i = 1, \ldots, m$. Under this null hypothesis, the approximate W-statistic given by

$$\phi_W = \left\{ \boldsymbol{h}(\boldsymbol{\beta})' \left[\frac{\partial \boldsymbol{h}}{\partial \boldsymbol{\beta}} (W'\Omega^{-1} W)^{-1} \frac{\partial \boldsymbol{h}'}{\partial \boldsymbol{\beta}} \right]^{-1} \boldsymbol{h}(\boldsymbol{\beta}) \right\}_{\boldsymbol{\beta} = \hat{\hat{\boldsymbol{\beta}}}}, \tag{2.35}$$

where $\boldsymbol{h}(\boldsymbol{\beta})$ is an $m \times 1$ vector with typical element $h_i(\boldsymbol{\beta})$ and $\partial \boldsymbol{h}/\partial \boldsymbol{\beta}$ is the matrix of first derivatives of $\boldsymbol{h}$ with respect to $\boldsymbol{\beta}$, is in large samples distributed as χ^2_m. Work is needed to establish the finite-sample properties of the asymptotic test statistics discussed above.

3 Specification of and estimation and testing procedures for sets of transfer functions

To specify a set of transfer functions (TFs), we partition the vector $\boldsymbol{z}_t$ in (2.1) as follows, $\boldsymbol{z}_t' = (\boldsymbol{y}_t' \boldsymbol{x}_t')$, where $\boldsymbol{y}_t$ is a $p_1 \times 1$ vector of endogenous variables and $\boldsymbol{x}_t$ is a $p_2 \times 1$ vector of exogenous variables with $p_1 + p_2 = p$. With $\boldsymbol{z}_t$ so partitioned, the system in (2.1) becomes

$$\begin{bmatrix} H_{11}(L) & H_{12}(L) \\ H_{21}(L) & H_{22}(L) \end{bmatrix} \begin{bmatrix} \boldsymbol{y}_t \\ \boldsymbol{x}_t \end{bmatrix} = \begin{bmatrix} \bar{\boldsymbol{c}}_1 \\ \bar{\boldsymbol{c}}_2 \end{bmatrix} + \begin{bmatrix} F_{11}(L) & F_{12}(L) \\ F_{21}(L) & F_{22}(L) \end{bmatrix} \begin{bmatrix} \boldsymbol{e}_{1t} \\ \boldsymbol{e}_{2t} \end{bmatrix}. \quad (3.1)$$

The assumption that $\boldsymbol{x}_t$ is exogenous gives rise to the following restrictions on the system in (3.1):

$$H_{21}(L) \equiv 0, \quad F_{12}(L) \equiv 0, \quad F_{21}(L) \equiv 0. \quad (3.2)$$

With the restrictions in (3.2) imposed on (3.1), we have

$$H_{11}(L)\boldsymbol{y}_t + H_{12}(L)\boldsymbol{x}_t = \bar{\boldsymbol{c}}_1 + F_{11}(L)\boldsymbol{e}_{1t}, \quad (3.3a)$$

$$H_{22}(L)\boldsymbol{x}_t = \bar{\boldsymbol{c}}_2 + F_{22}(L)\boldsymbol{e}_{2t}. \quad (3.3b)$$

The system in (3.3a) is in the form of a set of linear, dynamic simultaneous equations while that in (3.3b) is a set of ARMA equations for the exogenous variables.

The TFs associated with (3.3a), obtained by multiplying both sides of (3.3a) by $H_{11}^*(L)$, the adjoint matrix associated with $H_{11}(L)$ and normalizing the system, are

$$\phi(L)\boldsymbol{y}_t = \boldsymbol{c}_1 + \Delta(L)\boldsymbol{x}_t + B(L)\boldsymbol{e}_{1t} \quad (3.4a)$$

$$= \boldsymbol{c}_1 + \Delta(L)\boldsymbol{x}_t + K(L)\varepsilon_t, \quad (3.4b)$$

where

$$\phi(L) = g^{-1}|H_{11}(L)|,$$

with g being the normalizing constant and $|H_{11}(L)|$ being the determinant of $H_{11}(L)$;

$$\boldsymbol{c}_1 = g^{-1} H_{11}^*(L)\bar{\boldsymbol{c}}_1$$

is a $p_1 \times 1$ vector of constants,

$$\Delta(L) = -g^{-1} H_{11}^*(L) H_{12}(L) = \sum_{i=0}^{r} \Delta_i L^i,$$

and

$$B(L) = g^{-1} H_{11}^*(L) F_{11}(L) = \sum_{j=0}^{q} B_j L^j.$$

The coefficients of the matrix lag operator $K(L)$ in (3.4b) are given by $K_i = B_j B_0^{-1}$, $j = 1, 2, \ldots, q$, and $\varepsilon_t = B_0 \boldsymbol{e}_{1t}$ is a $p_1 \times 1$ disturbance vector with $E(\varepsilon_t) = \mathbf{0}$, $E(\varepsilon_t \varepsilon_t') = B_0 B_0' = \Omega_\varepsilon$, and zero serial correlations. In order to identify (3.4), we assume that $|H_{11}(L)|$ and $|K(L)|$ have their roots outside the unit circle and that the r.h.s. and the l.h.s. of (3.4) have no factors in common. Just as with the FEs, the AR polynomial $\phi(L)$ is the same in each equation if no cancelling occurs. To allow for possibly different $\phi(L)$ in different equations, we shall write the TF system as

$$\phi_i(L) y_{it} = c_{1i} + \boldsymbol{\delta}_i'(L) \boldsymbol{x}_t + \boldsymbol{k}_i'(L) \varepsilon_t, \quad i = 1, 2, \ldots, p_1, \tag{3.5}$$

where

$$\phi_i(L) = 1 - \phi_{i1} L - \phi_{i2} L^2 - \phi_{im_i} L^{m_i},$$

with m_i assumed known, and $\boldsymbol{\delta}_i'(L)$ is the ith row of $\Delta(L)$ and $\boldsymbol{k}_i'(L)$ is the ith row of $K(L)$. Since the error terms in (3.5) have a structure similar to those in (2.4), the representation presented in (2.7) is relevant here.

Each TF can be estimated separately using single-equation non-linear least squares or the single-equation ML procedure. Joint estimation of the parameters of the set of TFs (3.5) will now be considered. We write the system (3.5) as

$$\boldsymbol{y}_t = Y_t \boldsymbol{\phi} + \boldsymbol{c}_1 + X_t \boldsymbol{\delta} + \boldsymbol{u}_{1t}, \tag{3.6}$$

where

$$\underset{p_1 \times \sum_{i=1}^{p1} m_i}{Y_t} = \begin{bmatrix} y_{1t-1} & y_{1t-2} & \cdots & y_{1t-m_1} & 0 & \cdots & \cdots & \cdots & 0 \\ 0 & & & 0 & y_{2t-1} & \cdots & y_{2t-m_2} & \cdots & 0 \\ \vdots & & & & & & & & 0 \\ 0 & \cdots & \cdots & \cdots & \cdots & \cdots & \cdots & y_{p_1 t - m_{p_1}} & \end{bmatrix},$$

$$\boldsymbol{\phi}' = [\phi_{11}\ \phi_{12} \ldots \phi_{1m_1}\ \phi_{21} \ldots \phi_{2m_2} \ldots \phi_{p_1 m_{p_1}}],$$

and X_t is a $p_1 \times p_1 p_2 (r+1)$ matrix with typical row, say the jth, given by

$$\boldsymbol{x}_{jt}' = [\underbrace{0, 0, \ldots, 0}_{(j-1)(r+1)p_2 \text{ elements}}, \underbrace{x_{1t}, x_{1t-1}, \ldots, x_{1t-r}, x_{2t}, \ldots, x_{p_2 t-r}}_{(r+1)p_2 \text{ elements}}, \underbrace{0, 0, \ldots, 0}_{(p_1 - j)(r+1)p_2 \text{ elements}}],$$

$\delta = \text{vec}\,[\Delta_0\ \Delta_1\ \ldots\ \Delta_r]'$ and, provided the values of the initial conditions are zero, $\boldsymbol{u}_1 = \boldsymbol{K}\varepsilon$ where $\varepsilon' = [\varepsilon'_1, \varepsilon'_2, \ldots, \varepsilon'_T]$ and

$$K = \begin{bmatrix} I_{p_1} & 0 & \cdots & \cdots & 0 \\ K_1 & \ddots & & & \vdots \\ \vdots & & \ddots & & \vdots \\ K_q & & & \ddots & \vdots \\ 0 & & & & \vdots \\ 0 & \cdots & K_q & K_1 & I_{p_1} \end{bmatrix}.$$

Of course some of the columns of the matrix X_t will be deleted when we have exclusion restrictions on the vector $\boldsymbol{\delta}$.

For a sample of T observations, we write the system (3.6) as

$$\begin{aligned} \boldsymbol{y} &= Y\phi + X_0\boldsymbol{c}_1 + X\delta + \boldsymbol{u}_1 \\ &= Z_1\boldsymbol{\lambda}_1 + \boldsymbol{u}_1, \end{aligned} \tag{3.7}$$

with

$$Z_1 = (Y, X_0, X), \quad \boldsymbol{\lambda}'_1 = (\phi', \boldsymbol{c}'_1, \delta').$$

As in the preceding section, the likelihood function for the unrestricted system of TFs, conditional on zero starting values,[7] can be written as

$$L(\boldsymbol{\lambda}_1, \Omega_\varepsilon, K, \boldsymbol{y}) \propto |\Omega_\varepsilon|^{-T/2}\exp(S), \tag{3.8}$$

where

$$S = -\tfrac{1}{2}(\boldsymbol{y} - Z_1\boldsymbol{\lambda}_1)'K'^{-1}\left(I_T \otimes \Omega_\varepsilon^{-1}\right)K^{-1}(\boldsymbol{y} - Z_1\boldsymbol{\lambda}_1).$$

The first order conditions for a maximum of the log-likelihood function are

$$\frac{\partial S}{\partial \lambda} = Z'K'^{-1}\left(I_T \otimes \Omega_\varepsilon^{-1}\right)K^{-1}\boldsymbol{u}_1 = \boldsymbol{0}, \tag{3.9}$$

where

$$\boldsymbol{\lambda}' = (\boldsymbol{\lambda}'_1, \boldsymbol{\lambda}'_2),$$

with $\boldsymbol{\lambda}_2 = \text{vec}[K_1, \ldots, K_q]'$, and

$$Z = [Z_1, Z_2],$$

[7] As with the set of FEs, the starting values may be "backforecasted" using single TF equations. They may also be considered as unknown parameters to be estimated (e.g. Phillips 1966).

with Z_2 being a $Tp_1 \times p_1^2 q$ matrix of disturbances,

$$Z_2' = \left[\varepsilon_{ij}^{l'}\right], \quad i = 1, 2, \ldots, p_1, \quad l = 1, 2, \ldots, q, \quad j = 1, 2, \ldots, p_1,$$

and with typical row

$$\varepsilon_{ij}^{l'} = [\underbrace{0, 0, \ldots, 0}_{lp_1 \text{ times}}, \underbrace{0, \ldots, 0\varepsilon_{j1}0 \ldots 0}_{p_1 \text{ elements}} \ldots \underbrace{0 \ldots 0\varepsilon_{jT-1}0 \ldots 0}_{p_1 \text{ elements}}] .$$

(ith position)

As in the preceding section, a fully[8] efficient two-step estimator of $\boldsymbol{\lambda}$ is obtained using expression (2.13) to yield

$$\hat{\hat{\boldsymbol{\lambda}}} = \left[\hat{Z}'\hat{K}'^{-1}\left(I_T \otimes \hat{\Omega}_\varepsilon^{-1}\right)\hat{K}^{-1}\hat{Z}\right]^{-1}\left[\hat{Z}'\hat{K}'^{-1}\left(I_T \otimes \hat{\Omega}_\varepsilon^{-1}\right)\hat{K}^{-1}\hat{\boldsymbol{w}}\right], \tag{3.10}$$

where the carats denote that the unobserved quantities are computed at consistent estimates of $\boldsymbol{\lambda}$, $\hat{\boldsymbol{\lambda}}$, and the sample residuals obtained from the estimates, and

$$\boldsymbol{w}' = \left(\boldsymbol{w}_1', \ldots, \boldsymbol{w}_T'\right) \quad \text{where} \quad \boldsymbol{w}_t = \boldsymbol{y}_t + \sum_{h=1}^{q} K_h \varepsilon_{t-h}.$$

The remarks made in the preceding section concerning expressions (2.13) and (2.19) also apply to (3.10). It should be noticed that the requirement on the limit of Γ in (2.13) has also to be satisfied in TF context, where it leads to conditions on the limiting values of the second moments of $\boldsymbol{x}_t$. The steps in obtaining joint estimates of the parameters in (3.5) are as follows:

(1) Fit individual equations of (3.5) to obtain consistent estimates of the contemporaneous residuals, $\hat{\varepsilon}_t$, where $\varepsilon_t = B_0 \boldsymbol{e}_{1t}$.

(2) Use the residuals to form a consistent estimate of the contemporaneous covariance matrix of ε_t, Ω_ε, namely

$$\hat{\Omega}_\varepsilon = \sum_{t=1}^{T} \hat{\varepsilon}_t \hat{\varepsilon}_t' / T. \tag{3.11}$$

(3) For T observations, write the ith equation of (3.5) as

$$\begin{aligned} \boldsymbol{y}_i &= c_{1i}\boldsymbol{\iota} + Y_i \boldsymbol{\phi}_i + X_i \boldsymbol{\delta}_i + Q\boldsymbol{\kappa}_i + \varepsilon_i \\ &= \mathcal{J}_i \boldsymbol{\eta}_i + \varepsilon_i, \quad i = 1, 2, \ldots, p_1, \end{aligned} \tag{3.12}$$

[8] This means, that the estimator is as efficient as the ML estimator for the parameters of the TF form, ignoring restrictions coming from the underlying structural form.

where $\boldsymbol{y}_i$ has typical element y_{it}, Y_i has typical row $(y_{it-1}, y_{it-2}, \ldots, y_{it-m_i})$, X_i has typical row $(x_{1t}, x_{1t-1}, \ldots, x_{1t-r_{1i}}, \ldots, x_{p_2t}, x_{p_2t-1}, \ldots, x_{p_2t-r_{p2i}})$, Q has typical row $(\varepsilon'_{t-1}, \varepsilon'_{t-2}, \ldots, \varepsilon_{t-q})$, which does not depend on i.

Imposing restrictions on the error serial correlations could lead to a different matrix Q_i for different $i = 1, 2, \ldots, p_1$. Further, ε_i has typical element ε_{it}, $\boldsymbol{\phi}'_i = (\phi_{i1}, \phi_{i2}, \ldots, \phi_{im_i})$, $\boldsymbol{\delta}'_i = (\delta_{i10}, \delta_{i11}, \ldots, \delta_{i1r_{1i}}, \ldots, \delta_{ip_20}, \delta_{ip_21}, \ldots, \delta_{ip_2r_{p2i}})$ is the vector of coefficients in the elements of vector $\boldsymbol{\delta}'_i(L)$. And $\boldsymbol{\kappa}'_i = (\boldsymbol{\kappa}'_{i1}, \boldsymbol{\kappa}'_{i2}, \ldots, \boldsymbol{\kappa}'_{iq})$ with $\boldsymbol{\kappa}'_{ij}$ being a typical row of K_j, $j = 1, 2, \ldots, q$, $\boldsymbol{\iota} = (1, 1, \ldots, 1)'$, $T \times 1$, $\mathcal{J}_i = (\boldsymbol{\iota}, Y_i, X_i, Q)$, $\boldsymbol{\eta}'_i = (c_{1i}, \boldsymbol{\phi}'_i, \boldsymbol{\delta}'_i, \boldsymbol{\kappa}'_i)$.

We then apply ordinary least squares to each equation (3.12), after replacing ε_i by $\hat{\varepsilon}_i$, a matrix of the first step residuals $\hat{\varepsilon}_t$ to obtain consistent estimates of Q.

(4) Compute expression (3.10), using the residuals obtained in step 2 and the consistent parameter estimates $\hat{\Omega}_\varepsilon$ and $\hat{\boldsymbol{\eta}}_i$ to evaluate the unknown quantities in (3.10). Iteration of (3.10) yields the ML estimator given known and fixed initial conditions. To compute the inverse of $\hat{K}$, one ought to exploit the block-triangular structure of this matrix. This reduces the inversion of a Tp_1 matrix to addition and multiplication of $p_1 \times p_1$ matrices.

The large sample covariance matrix of the estimator proposed in (3.10) is consistently estimated by

$$\widehat{\text{var}}(\hat{\hat{\boldsymbol{\lambda}}}) = \left[\hat{Z}'\hat{K}'^{-1}\left(I_T \otimes \hat{\Omega}_\varepsilon^{-1}\right)\hat{K}^{-1}\hat{Z}\right]^{-1}. \tag{3.13}$$

In considering TF estimation, it is important to realize that the number of parameters in each TF can be large when there are several input variables in the vector $\boldsymbol{x}_t$ and lags relating to them are long. In such cases it will be expedient to consider reducing the number of free parameters to be estimated by making assumptions regarding the forms of lagged responses as is done in the distributed lag literature (see, e.g., Almon 1965, Dhrymes 1971, Zellner 1971b, ch. 7, and Shiller 1973). And of course, introducing the restriction that $\phi_1(L) \equiv \phi_2(L) \equiv \cdots \equiv \phi_{p_1}(L) \equiv \phi(L)$ in (3.5), when warranted, will lead to fewer free parameters to be estimated. A large-sample χ^2 test of the hypothesis that $\phi_1(L) \equiv \phi_2(L) \equiv \cdots \equiv \phi_{p_1}(L) \equiv \phi(L)$ can be constructed as in the case of FEs where a similar hypothesis was considered. Linear and non-linear restrictions on the TF parameters can be tested along the lines of the testing procedures proposed for the system of FEs. An example of a test of non-linear restrictions is J. D. Sargan's test on common factors in the polynomials

$\phi_i(L)$ and $\delta_i'(L)$ in the TF equation (3.5) which, if the common factor hypothesis is not rejected, leads to an ARMA representation of the disturbances in (3.5).

4 Estimation and testing procedures for structural equations

The structural equations (SEs), shown explicitly in (3.3a), will now be considered. We shall assume that a sufficient number of zero restrictions has been imposed on the parameters of the system such that the remaining free parameters are identified (see, e.g., Hannan 1971 and Hatanaka 1975). In what follows, we shall first take up "single-equation" estimation techniques for parameters in individual SEs and then go on to develop a "joint estimation" procedure that can be employed to estimate parameters appearing in a set of SEs.

4.1 Single-equation estimation procedure

The ith SE of the system in (3.3a) is given by

$$\sum_{j=1}^{p_1} h_{ij}(L)y_{jt} + \sum_{j=p_1+1}^{p} h_{ij}(L)x_{jt} = \bar{c}_{1i} + \sum_{j=1}^{p_1} f_{ij}(L)e_{jt}, \quad t = 1, 2, \ldots, T. \tag{4.1}$$

On imposing identifying zero restrictions and a normalization rule, $h_{ii0} = 1$ in $h_{ii}(L) = h_{ii0} + h_{ii1}L + \cdots$, the remaining free parameters of (4.1) can be estimated utilizing the techniques described below.

As shown in connection with (2.7) above, we can write

$$\sum_{j=1}^{p_1} f_{ij}(L)e_{jt} = \phi_i(L)\varepsilon_{it}, \tag{4.2}$$

where

$$\phi_i(L)\varepsilon_{it} = \varepsilon_{it} + \phi_{i1}\varepsilon_{it-1} + \phi_{i2}\varepsilon_{it-2} + \cdots + \phi_{iq}\varepsilon_{it-q_i},$$

where $q_i \leqq \max_j q_{ij}$, with q_{ij} the degree of $f_{ij}(L)$, and ε_{it} is a non-autocorrelated, normally distributed disturbance term with zero mean and constant finite variance, σ_ε^2, for all t. On substituting from (4.2) in (4.1) we have

$$\sum_{j=1}^{p_1} h_{ij}(L)y_{jt} + \sum_{j=p_1+1}^{p} h_{ij}(L)x_{jt} = \bar{c}_{1i} + \phi_i(L)\varepsilon_{it}, \quad t = 1, 2, \ldots, T, \tag{4.3}$$

with property that x_{jt} and $\varepsilon_{it'}$ are independent for all j, t and t'.

Since more than one current endogenous variable appears in (4.3), along with lagged endogenous variables, and since the disturbance terms are serially correlated, it is well known that usual estimation techniques such as two-stage least-squares, etc. yield inconsistent structural coefficient estimates.

Similarly, non-linear techniques for minimizing $\sum_{t=1}^{T} \varepsilon_{it}^2$ with respect to the parameters of (4.3) yield inconsistent coefficient estimates because of the "simultaneous equation" complication of correlation between disturbances $\phi_i(L)\varepsilon_{it}$ and the endogenous variables among the explanatory variables in (4.3).

To get consistent estimates of the parameters in (4.3), one can use an instrumental variables method using as instruments for the current and lagged endogenous variables, the current and lagged exogenous variables. One can also use the y_{jt-q_i-l}, $j = 1, \ldots, p_1, l = 1, 2, \ldots,$ as instruments for the current and lagged endogenous variables, as these instruments are independent of the error term $\phi_i(L)\varepsilon_{it}$. The use of lagged endogenous variables as instruments has been proposed by Phillips (1966). On the basis of the instrumental variable estimates, $\hat{\bar{c}}_{1i}$ and $\hat{h}_{ij}(L), j = 1, 2, \ldots, p$, in (4.3), one can compute the residuals

$$\hat{\eta}_{it} = \sum_{j=1}^{p_1} \hat{h}_{ij}(L)y_{jt} + \sum_{j=p_1+1}^{p} \hat{h}_{ij}(L)x_{jt} - \hat{\bar{c}}_{1i},$$

and then fit a q_ith order MA model to the residuals to get consistent estimates of the ϕ_{il}s, $l = 1, 2, \ldots, q_i$.

Alternatively, as explained in Zellner and Palm (1974), one may use the FEs (or TFs) to substitute for current endogenous variables appearing in (4.3) with coefficients with unknown values, that is $y_{jt}, j = 1, 2, \ldots, p_1$, for $j \neq i$. For example, the FEs for the y_{jt} given in (2.4) are

$$y_{jt} = c_j + \bar{\theta}_j(L)y_{jt} + u_{it}, \tag{4.4}$$

where $\bar{\theta}_j(L)$ the homogeneous part of $\theta_j(L)$. On substituting from (4.4) in (4.3) for $y_{jt}, j = 1, 2, \ldots, p_1$, for $j \neq i$, we obtain

$$\begin{aligned} y_{it} + \sum_{\substack{j=1 \\ j\neq i}}^{p_1} h_{ij0}[c_j + \bar{\theta}_j(L)y_{jt}] + \sum_{j=1}^{p_1} \bar{h}_{ij}(L)y_{jt} + \sum_{j=p_1+1}^{p} h_{ij}(L)x_{jt} \\ = \bar{c}_{1i} + \psi_i(L)\xi_{it}, \end{aligned} \tag{4.5}$$

where

$$\psi_i(L)\xi_{it} = \phi_i(L)\varepsilon_{it} - \sum_{\substack{j=1 \\ j\neq i}}^{p_1} h_{ij0}u_{it},$$

with $\psi(L)$, usually being a polynomial of degree m, and $\bar{h}_{ij}(L)$ is the homogeneous part of $h_{ij}(L)$. The error terms, $\xi_{it}, t = 1, 2, \ldots, T$, are normally and independently distributed each with zero mean and finite, common variance, $\sigma^2_{\xi_i}$.

Note that for given values of c_j and the parameters in $\bar{\theta}_j(L)$, (4.5) is in the form of a TF that is linear in the parameters. In view of this our estimation approach involves analyzing (4.5) as a TF with $c_j = \tilde{c}_j$ and $\bar{\theta}_j(L) = \tilde{\bar{\theta}}_j(L)$, where $\tilde{c}_j$ and $\tilde{\bar{\theta}}_j(L)$ are consistent estimates obtained from estimation of the FEs in (4.4).

Given that these consistent estimates of c_j and $\theta_j(L)$ are inserted in (4.5), a non-linear computational algorithm, e.g. Marquardt's, can be utilized to obtain consistent estimates of the remaining free parameters of (4.5) by minimizing the residual sum of squares $S = \sum_{t=1}^{T} \tilde{\xi}^2_{it}$ with respect to the free parameters, where ξ_{it} is given by

$$\tilde{\xi}_{it} = y_{it} - \bar{c}_{1i} + \sum_{\substack{j=1 \\ j\neq i}}^{p_1} h_{ij0}[\tilde{c}_j + \tilde{\bar{\theta}}_j(L)y_{jt}]$$

$$+ \sum_{j=1}^{p_1} \bar{h}_{ij}(L)y_{jt} + \sum_{j=p_1+1}^{p} h_{ij}(L)x_{jt} + \sum_{j=1}^{m} \psi_{ij}\tilde{\xi}_{it-j}. \tag{4.6}$$

Of course, one has to be cautious that the regressor matrix in (4.6) does not become singular, as one substitutes linear combinations of lagged endogenous variables for the current endogenous variables. The inverse of the Hessian matrix of S, evaluated at the consistent estimates, provides large-sample standard errors. These results in conjunction with the large-sample normal distribution of the estimates provide a basis for performing large sample tests of hypotheses.

The above procedure for estimating parameters of (4.6) can be applied for $i = 1, 2, \ldots, p_1$, to obtain "single-equation" parameter estimates and residuals, $\hat{\xi}_{it}, i = 1, 2, \ldots, p_1, t = 1, 2, \ldots, T$. Since

$$\xi_{it} = \varepsilon_{it} - \sum_{\substack{j=1 \\ j\neq i}}^{p_1} h_{ij0}v_{it},$$

where v_{it} is a FE disturbance defined in (2.7),

$$\hat{\varepsilon}_{it} = \hat{\xi}_{it} + \sum_{\substack{j=1 \\ j \neq i}}^{p_1} \hat{h}_{ij0} \hat{v}_{it}, \quad i = 1, 2, \ldots, p_1, \tag{4.7}$$

is a consistent estimate of ε_{it} in (4.2). Also from (4.2), $\varepsilon_t = \boldsymbol{F}_{110} \boldsymbol{e}_{1t}$ where $\varepsilon_t = (\varepsilon_{1t}, \varepsilon_{2t}, \ldots, \varepsilon_{p1t})$, so that

$$\begin{aligned} F_{11}(L)\boldsymbol{e}_{1t} &= F_{110}\boldsymbol{e}_{1t} + F_{111}\boldsymbol{e}_{1t-1} + \cdots + F_{11q}\boldsymbol{e}_{1t-q} \\ &= \varepsilon_t + F_1 \varepsilon_{t-1} + \cdots + F_q \varepsilon_{t-q}, \end{aligned} \tag{4.8}$$

with $F_j = F_{11j} F_{110}^{-1}$, $j = 1, 2, \ldots, q$. The error vector ε_t in (4.8) is normally distributed with mean zero and covariance matrix $E(\varepsilon_t \varepsilon_t') = F_{110} F_{110}' = \Sigma_\varepsilon$ and zero serial correlations. Thus from (4.7), it is possible to compute $\hat{\varepsilon}_t$ once all equations of the system are estimated. The $\hat{\varepsilon}_t$s thus computed will play a role in the joint estimation procedure to be described in the next section.

4.2 *Joint estimation of a set of structural equations*

We now consider (4.1) for $i = 1, 2, \ldots, p_1$,

$$\sum_{l=0}^{s} H_{11l} \boldsymbol{y}_{t-l} + \sum_{j=0}^{r} H_{12j} \boldsymbol{x}_{t-j} = \bar{\boldsymbol{c}}_1 + \sum_{h=0}^{q} F_{11h} \boldsymbol{e}_{1t-h}, \tag{4.9}$$

where the diagonal elements of H_{110} are equal to one. For a sample of T observations, we can write the system (4.9) as

$$\boldsymbol{y} = Z_1 \boldsymbol{\eta}_1 + Y_1 \boldsymbol{\eta}_2 + F\varepsilon, \tag{4.10}$$

where

$$\begin{aligned} \boldsymbol{y}' &= (\boldsymbol{y}_1', \boldsymbol{y}_2', \ldots, \boldsymbol{y}_T'), \\ Z_1 &= (X_0 X_1), \quad p_1 T \times p_1(1 + p_1 s + p_2 r + p_2), \\ X_0' &= [I_{p_1} I_{p_1} \ldots, I_{p_1}], \quad p_1 \times p_1 T, \\ Y_1 &= \begin{bmatrix} I_{p_1} \otimes \ldots \boldsymbol{y}_1' \\ \ldots \\ I_{p_1} \otimes \ldots \boldsymbol{y}_T' \end{bmatrix}, \quad p_1 T \times p_1^2, \\ X_1 &= \begin{bmatrix} I_{p_1} \otimes (\boldsymbol{y}_0' \boldsymbol{y}_1' \ldots \boldsymbol{y}_{-s+1}' \boldsymbol{x}_1' \boldsymbol{x}_0' \ldots \boldsymbol{x}_{-r+1}') \\ I_{p_1} \otimes (\boldsymbol{y}_1' \boldsymbol{y}_0' \qquad \boldsymbol{x}_{-r+2}') \\ I_{p_1} \otimes (\boldsymbol{y}_{T-1}' \ldots \boldsymbol{y}_{T-s}' \boldsymbol{x}_T' \ldots \quad \boldsymbol{x}_{T-r}') \end{bmatrix}, \\ & \qquad p_1 T \times p_1(p_1 s + p_2 r + p_2), \end{aligned}$$

with some of the columns of Z_1 and Y_1 being deleted when the exclusion restrictions are imposed in order to identify the structural parameters,

$$\underset{p_1T\times p_1T}{F} = \begin{bmatrix} I_{p_1} & 0 & \cdots & \cdots & \cdots & \cdots & 0 \\ F_1 & \ddots & & & & & \vdots \\ \vdots & \ddots & \ddots & & & & \vdots \\ F_q & & \ddots & \ddots & & & \vdots \\ 0 & \ddots & & \ddots & \ddots & & \vdots \\ \vdots & & \ddots & & \ddots & \ddots & 0 \\ 0 & \cdots & 0 & F_q & \cdots & F_1 & I_{p_1} \end{bmatrix},$$

and

$$\varepsilon' = (\varepsilon_1' \ldots \varepsilon_T'),$$

$\varepsilon_t = F_{110}\boldsymbol{e}_{1t}$ is defined in (4.8).

Since the ε_ts are assumed to be normally distributed, the likelihood function can be written as

$$L(\boldsymbol{y}, Z_1, \boldsymbol{\eta}_1, \boldsymbol{\eta}_2, \Sigma_\varepsilon, F) \propto |H_{110}|^T |\Sigma_\varepsilon|^{-T/2} \times \exp\{-\tfrac{1}{2}[(\boldsymbol{y} - Z_1\boldsymbol{\eta}_1 - Y_1\boldsymbol{\eta}_2)' F'^{-1}(I_T \otimes \Sigma_\varepsilon^{-1}) F^{-1}(\boldsymbol{y} - Z_1\boldsymbol{\eta}_1 - Y_1\boldsymbol{\eta}_2)]\}. \tag{4.11}$$

In order to keep the block-triangular structure of the matrix F, we proceed in a way slightly different from Reinsel (1977) and write the first order conditions for a maximum of the likelihood function as

$$\frac{\partial \ln L}{\partial \boldsymbol{\eta}_1} = Z_1' F'^{-1}(I_T \otimes \Sigma_\varepsilon^{-1}) F^{-1}\boldsymbol{u} = \boldsymbol{0}, \tag{4.12a}$$

$$\frac{\partial \ln L}{\partial \boldsymbol{\eta}_2} = -T\text{vec}(H_{110})^{-1} + Y_1' F'^{-1}(I_T \otimes \Sigma_\varepsilon^{-1}) F^{-1}\boldsymbol{u} = \boldsymbol{0}, \tag{4.12b}$$

$$\frac{\partial \ln L}{\partial \boldsymbol{\beta}} = X_2' F'^{-1}(I_T \otimes \Sigma_\varepsilon^{-1}) F^{-1}\boldsymbol{u} = \boldsymbol{0}, \tag{4.12c}$$

where

$$\boldsymbol{\beta} = \text{vec}[F_1 F_2 \ldots F_q]',$$

$$X_2' = [\varepsilon_{ij}^{l'}], \quad i = 1, 2, \ldots, p_1, \quad l = 1, 2, \ldots, q, \quad j = 1, 2, \ldots, p_1,$$

with a typical row

$$\varepsilon_{ij}^{l'} = [\underbrace{0, 0 \ldots 0}_{p_1\ l \text{ times}}, \underbrace{0 \ldots \varepsilon_{j1} \ldots 0}_{p_2 \text{ elements}}, \underbrace{0 \ldots \varepsilon_{j2} \ldots 0}_{} \ldots 0 \varepsilon_{jT-l} 0],$$

ith position

and

$$\boldsymbol{u} = F\varepsilon.$$

The first r.h.s. term of (4.12b) may be written as

$$-\text{vec}\big(\hat{\Sigma}_{\varepsilon}^{-1} V' V H_{110}^{-1}\big), \tag{4.13}$$

where

$$\hat{\Sigma}_{\varepsilon} = \frac{1}{T} \sum_{t=1}^{T} \varepsilon_t \varepsilon_t' = \frac{1}{T} V' V,$$

with

$$\underset{T \times p_1}{V} = \begin{bmatrix} \varepsilon_{11} & \varepsilon_{p_1 1} \\ \varepsilon_{1T} & \varepsilon_{p_1 T} \end{bmatrix},$$

and $VH_{110}^{-1} = W$ is the $T \times p_1$ matrix of reduced form disturbances. Also,

$$-\text{vec}\big(\hat{\Sigma}_{\varepsilon}^{-1} V' W\big) = -\big(W' \otimes \hat{\Sigma}_{\varepsilon}^{-1}\big)\text{vec}(V') = -(W' \otimes I_{p_1})\big(I_T \otimes \hat{\Sigma}_{\varepsilon}^{-1}\big)\varepsilon.$$

We can write the set of first order conditions for a maximum of the likelihood function with respect to $\boldsymbol{\eta}' = (\boldsymbol{\eta}_1', \boldsymbol{\eta}_2', \boldsymbol{\beta}')$ as

$$\frac{\partial \ln L}{\partial \boldsymbol{\eta}} = Z' F'^{-1} \big(I_T \otimes \hat{\Sigma}_{\varepsilon}^{-1}\big) F^{-1} \boldsymbol{u} = 0, \tag{4.14}$$

where

$$Z = (X_0, X_1, Y_1 - F(W \otimes I_{p_1}), X_2).$$

As discussed by Reinsel (1977), neglecting terms which, divided by T, have zero probability limit as $T \to \infty$, we have

$$-\underset{T \to \infty}{\text{plim}} \frac{1}{T} \frac{\partial^2 \ln L}{\partial \boldsymbol{\eta} \partial \boldsymbol{\eta}'} = \underset{T \to \infty}{\text{plim}} \frac{1}{T} Z' F'^{-1} \big(I_T \otimes \hat{\Sigma}_{\varepsilon}^{-1}\big) F^{-1} Z.$$

Using the results given in (2.13), the following two-step estimator for $\boldsymbol{\eta}$ has the same asymptotic distribution as the ML estimator:

$$\hat{\hat{\boldsymbol{\eta}}} = \hat{\boldsymbol{\eta}} - \Gamma(\hat{\boldsymbol{\eta}})^{-1} \frac{\partial \ln L}{\partial \boldsymbol{\eta}}(\hat{\boldsymbol{\eta}}), \tag{4.15}$$

where $\hat{\eta}$ is a consistent estimator of η such that $T^{\frac{1}{2}}(\hat{\eta} - \eta_0), \eta_0$ being the true parameter value, has some limiting distribution and the matrix $\Gamma(\hat{\eta})$ is such that

$$\operatorname*{plim}_{T\to\infty} \frac{1}{T}\Gamma(\hat{\eta}) = \operatorname*{plim}_{T\to\infty} \frac{1}{T}\frac{\partial^2 \ln L}{\partial\eta\partial\eta'}(\eta_0),$$

where the unknown matrix Σ_ε is replaced by a consistent estimate

$$\hat{\Sigma}_\varepsilon = \frac{1}{T}\sum_{t=1}^{T} \varepsilon_t(\hat{\eta})\varepsilon_t'(\hat{\eta}).$$

Since the information matrix is block-diagonal with respect to η and Σ_ε, it is sufficient to substitute a consistent estimate of Σ_ε in order to efficiently estimate η.

Applying (4.15) to the present problem yields

$$\hat{\hat{\eta}} = \hat{\eta} + [\hat{Z}'\hat{F}'^{-1}(I_T \otimes \hat{\Sigma}_\varepsilon^{-1})\hat{F}^{-1}\hat{Z}]^{-1}\hat{Z}'\hat{F}'^{-1}(I_T \otimes \hat{\Sigma}_\varepsilon^{-1})\hat{F}^{-1}\hat{\boldsymbol{u}}, \tag{4.16}$$

where the carats denote that the unknown quantities are evaluated at consistent parameters estimates. The first-step consistent estimates can be obtained using one of the single-equation estimation methods proposed in section 4.1. Since we can write the system in (4.10) as

$$\bar{\boldsymbol{y}} = [\boldsymbol{y} - F(W \otimes I_{p_1})\eta_2 + \boldsymbol{u} - \varepsilon] = Z\eta + \boldsymbol{u} = Z\eta + F\varepsilon, \tag{4.17}$$

we may apply generalized least squares to (4.17) after having evaluated the regressand $\bar{\boldsymbol{y}}$, the regressors Z and the disturbance covariance matrix $F(I_T \otimes \Sigma_\varepsilon)F'$ at consistent parameter estimates – this is in fact one way of computing the two-step estimator in (4.16) and it shows that the two-step estimator (4.16) can also be interpreted as a residual-adjusted estimator. Reinsel (1977) derives a slightly different estimator to which he gives an instrumental variables interpretation. It is obvious that the computation of the two-step estimator (4.16) which, if iterated until convergence, yields the ML estimator given fixed and known initial conditions, involves the inverse of the $p_1T \times p_1T$ disturbance covariance matrix $F(I_T \otimes \Sigma_\varepsilon)F'$. In the way we have analyzed the problem, this involves the inversion of F which is a blockband triangular matrix. As shown by Palm (1977a), it only requires multiplication and addition of matrices of order $p_1 \times p_1$.

As already discussed in section 2, the approximation in (4.16) to the second step of the Newton–Raphson algorithm, is in fact the second step of the Gauss–Newton algorithm starting from consistent parameter estimates. The large sample covariance matrix of $\hat{\hat{\eta}}$ is consistently estimated by

$$\hat{V}(\hat{\hat{\eta}}) = [\hat{Z}'\hat{F}'^{-1}(I_T \otimes \hat{\Sigma}_\varepsilon^{-1})\hat{F}^{-1}\hat{Z}]^{-1}. \tag{4.18}$$

Since $\hat{\hat{\eta}}$ will be approximately normally distributed in large samples, approximate tests of hypotheses can be constructed along the lines discussed in section 2.

As in the discussion of TF estimation, it is important to emphasize that (4.16) involves rather large matrices when the dimensionality of η is large. The situation is similar to that encountered in three-stage least squares but here in addition to structural coefficients, there are also parameters of the MA disturbance process to estimate. As with three-stage least squares, the estimation approach described above can be applied to subsets of the structural equations.

4.3 *Single-equation structural estimation reconsidered: two-step LIML*

Given that full information methods usually involve complicated computations and that the complete system is not always fully specified, we consider in this section single-equation methods from a ML point of view.

Consider a structural equation, assumed to be identified by exclusion restrictions, of the system (4.1), say the first,

$$Y_{(1)}\boldsymbol{\eta}_{(1)} + X_{(1)}\boldsymbol{\beta}_{(1)} = \boldsymbol{u}_1, \tag{4.19}$$

where $Y_{(1)} = (\boldsymbol{y}_1 \; Y_1)$ is the $T \times m_{(1)}$ matrix of observations on the current endogenous variables included in the first equation, with $m_{(1)} = m_1 + 1$; $X_{(1)}$ is the matrix of observations on included lagged endogenous, included current and lagged exogenous variables and a column of 1s for the constant term; $\boldsymbol{\eta}_{(1)}$ and $\boldsymbol{\beta}_{(1)}$ are vectors of the non-zero structural coefficients in the first equation; and $\boldsymbol{u}_1' = (u_{11} \; u_{12} \; . \; . \; . \; u_{1T})$.

We write the unrestricted reduced form for $Y_{(1)}$ as

$$Y_{(1)} = X\Pi_{(1)} + V_{(1)} = X_{(1)}\Pi_{1.} + X_{(0)}\Pi_{0.} + V_{(1)}, \tag{4.20}$$

where $\Pi_{(1)}' = (\Pi_{1.}' \vdots \Pi_{0.}')$ and $X_{(0)}$ denotes the $T \times k_0$ matrix of predetermined variables excluded from the first equation.

Postmultiplying (4.20) by $\boldsymbol{\eta}_{(1)}$ and comparing the result with (4.19) yields the following restrictions:

$$\Pi_{0.}\boldsymbol{\eta}_{(1)} = 0, \quad \Pi_{1.}\boldsymbol{\eta}_{(1)} = \boldsymbol{\beta}_{(1)}. \tag{4.21}$$

From the assumptions on the model, the rows of $V_{(1)}$ are normally distributed, with zero mean and common covariance matrix $\Omega_{(1)}$. Each row of $V_{(1)}$ can be represented as a MA of order $q_{(1)}$. In order to get a simple structure for the disturbance term covariance matrix, we vectorize the model (4.20) as follows:

$$\mathrm{vec}(Y'_{(1)}) = W_1 \,\mathrm{vec}(\Pi_{1.}) + W_0 \,\mathrm{vec}(\Pi_{0.}) + \mathrm{vec}(V'_{(1)}), \qquad (4.22)$$

with

$$W_j = \begin{bmatrix} I_{m_{(1)}} \otimes \boldsymbol{x}'_{j1} \\ \cdots \\ I_{m_{(1)}} \otimes \boldsymbol{x}'_{jT} \end{bmatrix},$$

with $j = 1, 0$ and $\boldsymbol{x}'_{jt.}$ being the tth row of $X_{(j)}$.

We write (4.22) as

$$\boldsymbol{y}_{(1)} = W_1\boldsymbol{\pi}_1 + W_0\boldsymbol{\pi}_0 + \boldsymbol{v}_{(1)}, \qquad (4.23)$$

and the MA representation of $\boldsymbol{v}_{(1)}$ as

$$\boldsymbol{v}_{(1)} = F_{(1)}\varepsilon_{(1)}, \qquad (4.24)$$

where

$$F_{(1)} = \begin{bmatrix} I_{m_{(1)}} & 0 & \cdots & & 0 \\ F_{(1)1} & \ddots & & & \vdots \\ F_{(1)q_{(1)}} & & \ddots & & \\ 0 & \ddots & & \ddots & 0 \\ 0 & 0 & F_{(1)q_{(1)}} & F_{(1)1} & I_{m_{(1)}} \end{bmatrix}, \qquad (4.25)$$

and $\varepsilon_{(1)}$ is normally distributed with covariance matrix $\Omega_{(1)}$ and zero serial correlations. The likelihood function may then be written as

$$L(\boldsymbol{y}_{(1)}, W_1, W_0, \boldsymbol{\pi}_1, \boldsymbol{\pi}_0, \Omega_{(1)}, F_{(1)}) \propto |\Omega_{(1)}|^{-T/2} \exp(S), \qquad (4.26)$$

where

$$S = -\tfrac{1}{2}(\boldsymbol{y}_{(1)} - W_1\boldsymbol{\pi}_1 - W_0\boldsymbol{\pi}_0)'\Omega^{-1}(\boldsymbol{y}_{(1)} - W_1\boldsymbol{\pi}_1 - W_0\boldsymbol{\pi}_0),$$

and

$$\Omega = F_{(1)}(I_T \otimes \Omega_{(1)})F'_{(1)}.$$

We define the LIML estimator in a way slightly different from the usual definition, as the estimator which maximizes (4.26) with respect to $\boldsymbol{\pi}_1$, $\boldsymbol{\pi}_0$, $F_{(1)}$, and $\boldsymbol{\eta}_{(1)}$ subject to $\boldsymbol{\Pi}_{\mathbf{0}.}\boldsymbol{\eta}_{(1)} = \mathbf{0}$. In terms of asymptotic properties of the LIML estimator it does not matter whether we maximize the likelihood function concentrated with respect to $\Omega_{(1)}$ or use a consistent estimate for $\Omega_{(1)}$ in the first order conditions for a maximum of the likelihood function with respect to the remaining parameters. The restrictions may also be vectorized as

$$[\boldsymbol{\eta}'_{(1)} \otimes I_{k_0}]\boldsymbol{\pi}_0 = \mathbf{0}. \tag{4.27}$$

The Lagrangean expression is

$$\boldsymbol{z} = -\frac{T}{2}\log|\Omega_{(1)}| + S - \boldsymbol{\lambda}'[\boldsymbol{\eta}'_{(1)} \otimes I_{k_0}]\boldsymbol{\pi}_0, \tag{4.28}$$

where $\boldsymbol{\lambda}$ is the $k_0 \times 1$ vector of Lagrange multipliers.

The set of first order conditions for a maximum is

$$\partial \boldsymbol{z}/\partial \boldsymbol{\pi}_1 = W'_1\Omega^{-1}\boldsymbol{v}_{(1)} = \mathbf{0} \tag{4.29a}$$

$$\partial \boldsymbol{z}/\partial \boldsymbol{\pi}_0 = W'_0\Omega^{-1}\boldsymbol{v}_{(1)} - (\boldsymbol{\eta}_{(1)} \otimes I_{k_0})\boldsymbol{\lambda} = \mathbf{0} \tag{4.29b}$$

$$\partial \boldsymbol{z}/\partial \boldsymbol{\phi} = W'_2\Omega^{-1}\boldsymbol{v}_{(1)} = \mathbf{0} \tag{4.29c}$$

where

$$\boldsymbol{\phi} = \text{vec}[F_{(1)1}, F_{(1)2}, \ldots, F_{(1)q_{(1)}}]'$$

$$W'_2 = \left[\varepsilon^{l'}_{ij}\right], \quad i = 1, 2, \ldots, m_{(1)}, \quad l = 1, 2, \ldots, q_{(1)}, \quad j = 1, 2, \ldots, m_{(1)},$$

and

$$\varepsilon^{l'}_{ij} = [\underbrace{0, 0 \ldots 0}, \underbrace{0 \ldots \varepsilon_{j1}, 0 \ldots 0}_{lm_{(1)} \text{ times, } m_{(1)} \text{ elements}}, \underbrace{0 \ldots \varepsilon_{j2}, 0 \ldots 0, , \ldots 0 \ldots 0\, \varepsilon_{jT-l} 0 \ldots 0}_{i\text{th position}}].$$

$$\partial \boldsymbol{z}/\partial \boldsymbol{\eta}_1 = -(I_{m_{(1)}} \otimes \boldsymbol{\lambda}')\boldsymbol{\pi}_0 = \mathbf{0}, \tag{4.29d}$$

$$\partial \boldsymbol{z}/\partial \boldsymbol{\lambda} = -[\boldsymbol{\eta}'_{(1)} \otimes I_{k_0}]\boldsymbol{\pi}_0 = \mathbf{0}. \tag{4.29e}$$

We can solve (b) for $\boldsymbol{\pi}_0$, to get

$$\boldsymbol{\pi}_0 = \left(W'_0\Omega^{-1}W_0\right)^{-1}\left[W'_0\Omega^{-1}(\boldsymbol{y}_{(1)} - W_1\boldsymbol{\pi}_1) - (\boldsymbol{\eta}_{(1)} \otimes I_{k_0})\boldsymbol{\lambda}\right]. \tag{4.30}$$

Substituting (4.30) into (4.29e) and solving for $\boldsymbol{\lambda}$ gives

$$\begin{aligned}\boldsymbol{\lambda} =& [(\boldsymbol{\eta}'_{(1)} \otimes I_{k_0})[W'_0\Omega^{-1}W_0]^{-1}(\boldsymbol{\eta}_{(1)} \otimes I_{k_0})]^{-1} \\ &\times [(\eta'_{(1)} \otimes I_{k_0})(W'_0\Omega^{-1}W_0)^{-1}W'_0\Omega^{-1}(\boldsymbol{y}_{(1)} - W_1\boldsymbol{\pi}_1)].\end{aligned} \tag{4.31}$$

The set of first order conditions for a maximum in (4.29) is clearly nonlinear in the parameters. We can approximate the solution by a two-step Newton–Raphson procedure, as has been done in (2.13), starting with consistent estimates for $\boldsymbol{\pi}_0$, $\boldsymbol{\pi}_1$, $\boldsymbol{\phi}$, and $\boldsymbol{\eta}_{(1)}$, computing $\boldsymbol{\lambda}$ from expression (4.31) and evaluating

$$\hat{\hat{\theta}} = \hat{\theta} - \Gamma(\hat{\theta})^{-1} [\partial z / \partial \theta]_{\theta = \hat{\theta}}, \tag{4.32}$$

where

$$\theta = (\pi'_1, \phi', \pi'_0, \eta'_{(1)}, \lambda')',$$

$\hat{\theta}$ is a consistent estimate of θ satisfying the requirement in (2.13), and

$$\Gamma(\theta) = \begin{bmatrix} W'_1\Omega^{-1}W_1 & W'_1\Omega^{-1}W_2 & W'_1\Omega^{-1}W_0 & 0 & 0 \\ W'_2\Omega^{-1}W_1 & W'_2\Omega^{-1}W_2 & W'_2\Omega^{-1}W_0 & 0 & 0 \\ W'_0\Omega^{-1}W_1 & W'_0\Omega^{-1}W_2 & W'_0\Omega^{-1}W_0 & -(I_{m_{(1)}} \otimes \lambda) & -(\eta_{(1)} \otimes I_{k_0}) \\ 0 & 0 & -(I_{m_{(1)}} \otimes \lambda') & 0 & -\Pi'_{0.} \\ 0 & 0 & -(\eta'_{(1)} \otimes I_{k_0}) & -\Pi_{0.} & 0 \end{bmatrix} \tag{4.33}$$

The unknown elements of $\Omega_{(1)}$ in $\Omega = F_{(1)}(I_T \times \Omega_{(1)})F'_{(1)}$ are replaced by consistent estimates. The probability limit of the matrix $(1/T)\Gamma(\hat{\theta})$ in (4.33) usually is the matrix $(1/T)$ $(\partial^2 z/\partial\theta\partial\theta')(\theta_0)$ where θ_0 is the vector of true parameter values of θ. Of course, one can iterate the expression (4.32) to get the exact solution of the first order conditions for a maximum of the likelihood function, which is the limited-information ML estimator given fixed and known initial conditions. In terms of asymptotic efficiency, it is not necessary to continue the iteration after the second step.

5 Some concluding remarks

(1) In this chapter, we have presented several estimators for the three forms of a dynamic SEM with moving average disturbances and discussed their asymptotic properties. The results essentially rely upon:

(a) the asymptotic properties of the ML estimator of the parameters of dynamic models, and

(b) a result given by Fisher (1925), Kendall and Stuart (1961), Rothenberg and Leenders (1964) and later by Dhrymes and Taylor (1976) concerning the asymptotic properties of a two-step iteration of the first-order conditions for a maximum of the likelihood function.

Of course, the starting values for the iteration and the matrix Γ approximating the matrix of second order derivatives of the log-likelihood function have to satisfy some conditions (see, e.g., (2.13)), which we give in the text, but which we do not verify explicitly for the estimation problems considered. It ought to be clear that these requirements, such as stated in (2.13), have to be checked in practical situations.

(2) Computation of the estimators presented above generally involves operations on large matrices. For example, in each case one has to compute the inverse of the covariance matrix of a vector-MA process. The

estimation methods presented here open an immense field of application for good numerical matrix inversion procedures exploiting the special features of the covariance matrix of an MA process.

(3) Despite the fact that the field of application of the methods presented is probably limited to small models, the results of the chapter clarify a number of questions concerning the asymptotic properties of estimators for dynamic and static models. For example, if the disturbances of the TF system in (3.6) are not correlated, i.e. $K_h = 0, h = 1, 2, \ldots, q$, then the two-step estimator given in (3.10) specializes to Zellner's estimator for seemingly unrelated regressions.

As a second example, assume that $H_{11}(L)$ in (3.3a) is an unimodular matrix, i.e. $|H_{11}(L)| =$ constant, then the expression given in (3.10) specializes to an expression with $\hat{\boldsymbol{Z}} = (X_0, X, \hat{\boldsymbol{Z}}_2)$ and the covariance matrix of the estimator in (3.10) will be asymptotically a block-diagonal matrix as $\text{plim}_{T\to\infty}(1/T)[(X'_0X)'\hat{Z}_2] = 0$ under suitable conditions. Therefore it will be sufficient to have consistent estimates of $\boldsymbol{\lambda}_2$ to efficiently estimate $(\boldsymbol{c}'_1, \boldsymbol{\delta}')$ in (3.7). A similar result has been established by Amemiya (1973).

(4) It is to be expected that the estimation results can, at least for samples of the size encountered in applied work, be improved by using two-step estimators approximating the first order conditions for a maximum of the exact likelihood function. One step in the direction of using the exact likelihood function is to "backforecast" the values of the initial conditions for FE, TF, or structural equation systems. This aspect however deserves additional work.

(5) The discussion has been in terms of large-sample properties of the estimators and test statistics for dynamic models. Small-sample properties of the estimators and test statistics have to be investigated. However, the Monte Carlo results obtained by Nelson (1976) justify some optimism about improving estimation precision in small samples by use of the joint estimates which we have considered.

BIBLIOGRAPHY

Aigner, D. J. (1971), "A compendium on estimation of the autoregressive-moving average model from time series," *International Economic Review* 12, 348–71

Akaike, H. (1973), "Maximum likelihood identification of Gaussian autoregressive-moving average models," *Biometrika* 60, 255–65

Almon, S. (1965), "The distributed lag between capital appropriations and expenditures," *Econometrica* 33, 178–96

Amemiya, T. (1973), "Generalized least squares with estimated autocovariance matrix," *Econometrica* 14, 723–32

Anderson, T. W. (1975), "Maximum likelihood estimation of parameters of autoregressive processes with moving average residuals and other covariance matrices with linear structure," *The Annals of Statistics* 3, 1283–1304

Ansley, C. F., W. A. Spivey, and W. J. Wroblenski (1977), "On the structure of moving average processes," *Journal of Econometrics* 6, 121–34
Åström, K. J. and T. Bohlin (1966), "Numerical identification of linear dynamic systems from normal operating records," in Ph.H. Hammond (ed.), *Theory of Self-Adaptive Control Systems* (New York, Plenum Press)
Box, G. E. P. and G. M. Jenkins (1970), *Time Series Analysis, Forecasting and Control* (San Francisco, Holden-Day)
Breusch, T. S. and A. R. Pagan (1978), "The Lagrange multiplier test and its applications to model specification in econometrics," CORE Discussion Paper
Byron, R. P. (1973), "The computation of maximum likelihood estimates for linear simultaneous systems with moving average disturbances," Australian National University, Sydney, mimeo
Chow, G. C. and R. C. Fair (1973), "Maximum likelihood estimation of linear equation systems with autoregressive residuals," *Annals of Economic and Social Measurement* 2, 17–28
Deistler, M. (1975), "Z-transform and identification of linear econometric models with autocorrelated errors," *Metrika* 22, 13–25
(1976), "The identifiability of linear econometric models with autocorrelated errors," *International Economic Review* 17, 26–46
Dhrymes, P. J. (1971), *Distributed Lags* (San Francisco, Holden-Day)
Dhrymes, P. J. and J. B. Taylor (1976), "On an efficient two-step estimator for dynamic simultaneous equations models with autoregressive errors," *International Economic Review* 17, 362–76
Durbin, J. (1959), "Efficient estimation of parameters in moving-average models," *Biometrika* 46, 306–16
Espasa, A. (1977), *The Spectral Maximum Likelihood Estimation of Econometric Models with Stationary errors* (Göttingen, Vandenhoeck & Ruprecht)
Feller, W. (1966), *An Introduction to Probability Theory and its Applications*, 2, 2nd edn. (New York, John Wiley)
Fisher, R. A. (1925), *Statistical Methods for Research Workers* (Edinburgh, Oliver & Boyd)
Goldfeld, S. M. and R. E. Quandt (1972), *Non-Linear Methods in Econometrics* (Amsterdam, North-Holland)
Granger, C. W. and M. J. Morris (1976), "Time series modeling and interpretation," *Journal of the Royal Statistical Society*, A 139, 246–57
Hannan, E. J. (1969), "The identification of vector mixed autoregressive-moving average systems," *Biometrika* 56, 223–5
(1971), "The identification problem for multiple equation systems with MA errors," *Econometrica* 39, 751–65
(1975), "The estimation of ARMA models," *The Annals of Statistics* 3, 975–81
Hannan, E. J. and R. D. Terrell (1973), "Multiple equation system with stationary errors," *Econometrica* 41, 299–320
Hatanaka, M. (1974), "An efficient two-step estimator for the dynamic adjustment model with autoregressive errors," *Journal of Econometrics* 2, 199–220
(1975), "On global identification of the dynamic simultaneous equations model with stationary disturbances," *International Economic Review* 16, 545–54

(1976), "Several efficient two-step estimators for the dynamic simultaneous equations model with autoregressive disturbances," *Journal of Econometrics* 4, 189–204

Hendry, D. F. (1976), "The structure of simultaneous equations estimators," *Journal of Econometrics* 4, 51–88

Kang, K. M. (1975), "A comparison of estimators for moving average processes," Unpublished paper, Australian Bureau of Statistics, Canberra

Kendall, M. G. and A. Stuart (1961), *Advanced Theory of Statistics*, 2 (London, Griffin & Co.)

Kmenta, J. and R. F. Gilbert (1968), "Small sample properties of alternative estimates of seemingly unrelated regressions," *Journal of the American Statistical Association* 63, 1180–1200

Maddala, G. S. (1971), "Generalized least squares with estimated variance covariance matrix," *Econometrica* 39, 23–33

Nelson, C. R. (1976), "Gains in efficiency from joint estimation of systems of autoregressive-moving average processes," *Journal of Econometrics* 4, 331–48

Nicholls, D. F. (1976), "The efficient estimation of vector linear time series models," *Biometrika* 63, 381–90

Nicholls, D. F., A. R. Pagan, and R. D. Terrell (1975), "The estimation and use of models with moving average disturbance terms: a survey," *International Economic Review* 16, 113–34

Osborn, D. R. (1976), "Maximum likelihood estimation of moving average processes," *Annals of Economic and Social Measurement* 5, 75–87

Palm, F. C. (1977a), "On efficient estimation of the final equation form of a linear multiple time series process," *Cahiers du Centre d'Etudes de Recherche Opérationnelle* 19, 297–308

(1977b), "On univariate time series methods and simultaneous equation models," *Journal of Econometrics* 5, 379–88

Pesaran, M. H. (1973), "Exact maximum likelihood estimation of a regression with a first-order moving average error," *Review of Economic Studies* 41, 529–36

Phillips, A. W. (1966), "Estimation of systems of difference equations with moving average disturbances," Paper read at the Econometric Society Meeting in San Francisco; reprinted in A. R. Bergstrom, A. J. L. Catt, M. H. Peston, and B. D. J. Silverstone (eds.), *Stability and Inflation* (New York, John Wiley, 1978)

Pierce, D. A. (1972), "Least squares estimation in dynamic-disturbance time series models," *Biometrika* 59, 73–8

Quenouille, M. H. (1957), *The Analysis of Multiple Time Series* (London, Griffin & Co.); 2nd edn. (1968)

Rao, C. R. (1973), *Linear Statistical Inference and Its Applications*, 2nd edn. (New York, John Wiley)

Reinsel, G. (1976), "Maximum likelihood estimation of vector autoregressive moving-average models," Department of Statistics, Carnegie–Mellon University, Pittsburgh, mimeo

(1977), "FIML estimation of the dynamic simultaneous equations model with ARMA disturbances," University of Wisconsin, Madison, mimeo

Rothenberg, T. J. and C. T. Leenders (1964), "Efficient estimation of simultaneous equation system," *Econometrica* 32, 57–76
Sargan, J. D. (1975), "A suggested technique for computing approximations to Wald criteria with application to testing dynamic specification," Discussion Paper, London School of Economics
Shiller, R. J. (1973), "A distributed lag estimator derived from smoothness priors," *Econometrica* 41, 775–88
Swamy, P. A. V. B. and P. N. Rappoport (1978), "Relative efficiencies of some simple Bayes estimators of coefficients in a dynamic equation with serially correlated errors – II," *Journal of Econometrics* 7, 245–58
Wall, K. D. (1976), "FIML estimation of rational distributed lag structural models," *Annals of Economic and Social Measurement* 5, 53–62
Wallis, K. F. (1977), "Multiple time series analysis and the final form of econometric models," *Econometrica* 45, 1481–98
Wilson, G. T. (1973), "The estimation of parameters in multivariate time series models," *Journal of the Royal Statistical Society* B, 76–85
Zellner, A. (1971a), "Bayesian and non-Bayesian analysis of the log-normal distribution and log-normal regression," *Journal of the American Statistical Association* 66, 327–30
(1971b), *An Introduction to Bayesian Inference in Econometrics* (New York, John Wiley)
Zellner, A. and W. Vandaele (1974), "Bayes–Stein estimators for *k*-means, regressions and simultaneous equation models," in S. E. Fienberg and A. Zellner (eds.), *Studies in Bayesian Econometrics and Statistics* (Amsterdam, North-Holland)
Zellner, A. and F. C. Palm (1974), "Time series analysis and simultaneous equation econometric models," *Journal of Econometrics* 2, 17–54; chapter 1 in this volume
(1975), "Time series and structural analysis of monetary models of the US economy," *Sankhyā: The Indian Journal of Statistics*, Series C 37, 12–56; chapter 6 in this volume

Rejoinder (1981)

Franz C. Palm and Arnold Zellner

1 Introduction

In Palm and Zellner (1980), hereafter referred to as P–Z, we have presented an asymptotically efficient two-step and iterative estimation procedure for the parameters of the final equation (FE), transfer function (TF) and structural form of a dynamic simultaneous equation model (SEM)

Originally published in the *Journal of Econometrics* 17 (1981), 131–8.

with vector moving average (MA) disturbances. Asymptotic efficiency of the two-step estimator requires the use of consistent initial parameter estimates. The procedure outlined in P–Z to generate the initial estimates usually does not yield consistent estimates.

We are grateful to J. McDonald and J. Darroch for pointing out (in the preceding comment, hereafter denoted as Mc–D, and in Darroch and McDonald 1981) that the procedure to compute the initial estimates of the MA parameters of the disturbances and a method to get initial estimates of the structural coefficients generally are not consistent. The reason for this is that the single-equation innovations which play a crucial role in our estimation procedure have correlation properties that are different from those of the innovations in the system. In fact, each single-equation innovation is a weighted sum of all current and past innovations of the system.

McDonald and Darroch indicate how consistent initial estimates can be obtained. The procedures that they suggest are computationally demanding. There is a need for computationally less cumbersome procedures. In this note, we shall show how consistent initial estimates can be obtained in a simpler way by modifying the procedures suggested in P–Z.

In section 2, we shall discuss consistent estimation of vector MA processes. Section 3 is devoted to single structural equation estimation. We conclude [in section 4] with some final remarks, one of which is on consistent estimation of a single reduced form equation when the disturbances of the system are generated by a vector MA process.

2 Consistent estimation of a vector MA process

In order to outline how consistent estimates for the parameters of a vector MA process can be obtained, we consider the following vector MA model (see also (2.5) in P–Z and (1) in Mc–D:

$$\underset{p\times 1}{\boldsymbol{u}_t} = \underset{p\times p}{G(L)}\ \underset{p\times 1}{\boldsymbol{v}_t}\ , \tag{2.1}$$

where $\boldsymbol{u}_t$ is a vector of stationary random variables, $\boldsymbol{v}_t$ is the vector of normally distributed system innovations, with mean zero and covariance matrices $E\boldsymbol{v}_t\boldsymbol{v}'_{t'} = \delta_{tt'}\Omega_v$, $\delta_{tt'}$ being the Kronecker delta. The $\boldsymbol{u}_t$s may be observable random vectors or unobservable disturbances in a system of equations. In the latter case, the residuals obtained by consistently estimating the system can be used. The matrix

$$G(L) = \sum_{j=0}^{q} G_j L^j, \quad G_0 = I_p$$

is an invertible matrix operator, whose elements are finite polynomials of degree q in the lag operator L.

A typical element of $\boldsymbol{u}_t$, say u_{it}, can be represented as an MA of degree $q_i \leqq q$ in one variable, v_{it}^*, called the univariate or single-equation innovation, which is a normally distributed white noise. Then, the vector $\boldsymbol{u}_t$ can be written as

$$\boldsymbol{u}_t = \Lambda(L)\boldsymbol{v}_t^*, \tag{2.2}$$

where $\Lambda(L)$ is a diagonal matrix polynomial with $\Lambda(0) = I_p$ (see also (3) in Mc–D). The univariate innovations can be obtained by applying a rational matrix lag operator to the $\boldsymbol{v}_t$s, i.e., $\boldsymbol{v}_t^* = \Lambda^{-1}(L)\boldsymbol{u}_t = \Lambda^{-1}(L)G(L)\boldsymbol{v}_t$ (provided $\Lambda(L)$ is invertible). Therefore, each element v_{it}^* usually depends on the current and all lagged values of u_{it} or equivalently on the current and lagged values of v_{jt}, $j = 1, \ldots, p$. Also the $\boldsymbol{v}_t^*$s will usually be cross-correlated. Premultiplying (2.1) by the adjoint matrix of $G(L)$, $G^*(L)$, we obtain

$$G^*(L)\boldsymbol{u}_t = |G(L)|\boldsymbol{v}_t, \tag{2.3}$$

where $|G(L)|$ is the determinant of $G(L)$. As the system (2.1) has been normalized through the requirement $G_0 = I_p$, the off-diagonal elements in $G^*(L)$ will be homogeneous polynomials in L, so that the ith equation in (2.3) is in the form of a transfer function equation with an MA process in the system innovation v_{it}. Notice that the MA parameters in (2.3) are the same in different equations.

The estimation procedure presented by P–Z (pp. 209, 217, 221), can be modified as follows to yield consistent initial estimates of the parameters in $G(L)$ and Ω_v.

Step 2: Estimation of the system innovations and of Ω_v. After substitution of consistent estimates of the final equations', transfer functions' or structural equations' disturbances, say $\hat{\boldsymbol{u}}_t$, obtained in Step 1 by instrumental variables estimation or derived from the univariate innovations using (2.2), the transfer function equations in (2.3) can be estimated separately using, e.g., non-linear least squares. The residuals in (2.3) are consistent estimates of the system innovations and can be used to estimate Ω_v by

$$\hat{\Omega}_v = \frac{1}{T}\sum_{t=1}^{T}\hat{\boldsymbol{v}}_t\hat{\boldsymbol{v}}_t'.$$

Step 3: Estimation of the matrices $G_j, j = 1, \ldots, q$. The parameters in (3) are functions of the parameters of the G_js. Consistent estimates of some, possibly all, parameters in (2.1) can be obtained from the single-equation estimates of (2.3). In any case, they can be obtained from the regressions of the $\hat{u}_{it}$s on $\hat{\boldsymbol{v}}_{t-1}, \hat{\boldsymbol{v}}_{t-2}, \ldots, \hat{\boldsymbol{v}}_{t-q}$. The G_js can also be estimated by the method described in Step 3 of P–Z using the system innovations.

The estimates obtained in Steps 2 and 3 are used in Step 4 to generate fully efficient estimates of the system parameters. The procedure outlined in Steps 2 and 3 has computational advantages compared with the traditional method of solving the set of non-linear equations

$$\frac{1}{T}\sum_{t=1}^{T} \hat{\boldsymbol{u}}_t \hat{\boldsymbol{u}}'_{t-i} = \sum_{j=i}^{q} G_j \Omega_v G'_{j-i}, \quad i = 0, 1, \ldots, q, \tag{2.4}$$

for the elements in Ω_v, $G_1, \ldots, G_q$, using, e.g., the algorithm developed by Wilson (1969), and then solving (2.1) to get estimates of the $\boldsymbol{v}_t$s. It will also be more easily implemented than frequency domain methods, such as those presented by Hannan (1970, pp. 383–8, 1975) and suggested by Reinsel (1979).

As an illustration, consider the following bivariate process:

$$\begin{bmatrix} u_{1t} \\ u_{2t} \end{bmatrix} = \begin{bmatrix} 1 + g_1 L & g_2 L \\ g_3 L & 1 + g_4 L \end{bmatrix} \begin{bmatrix} v_{1t} \\ v_{2t} \end{bmatrix}, \tag{2.5}$$

or

$$\boldsymbol{u}_t = (I + G_1 L)\boldsymbol{v}_t,$$

which satisfies the assumptions made for (2.1). The univariate innovations are given by $v^*_{it} = (1 - \lambda_i L)^{-1} u_{it}, i = 1, 2$, where λ_i is the root of $\rho_i(1)\lambda_i^2 + \lambda_i + \rho_i(1) = 0$, satisfying $|\lambda_i| < 0$, with $\rho_i(1)$ being the first order autocorrelation coefficient of u_{it}. A single-equation innovation can usually be expressed as a sum of infinite MAs in the system innovations, e.g. $v^*_{1t} = (1 - \lambda_1 L)^{-1} \times (1 + g_1 L)v_{1t} + (1 - \lambda_1 L)^{-1} g_2 L v_{2t}$.

The model (2.5) can be written in the form of (2.3),

$$\begin{aligned} (1 + g_4 L)u_{1t} - g_2 L u_{2t} &= [(1 + g_1 L)(1 + g_4 L) - g_2 g_3 L^2] v_{1t}, \\ (1 + g_1 L)u_{2t} - g_3 L u_{1t} &= [(1 + g_1 L)(1 + g_4 L) - g_2 g_3 L^2] v_{2t}. \end{aligned} \tag{2.6}$$

Consistent estimates of the g_is, $i = 1, \ldots, 4$, and of the disturbances v_{1t} and v_{2t} are obtained through fitting separately the transfer function equations in (2.6) with second order MA errors. Consider now the model (2.5) with the restriction $g_3 = 0$. Then in (2.5), u_{2t} is written as a univariate MA with $\lambda_2 = -g_4$ and $v^*_{2t} \equiv v_{2t}$. Substitution of $u_{2t} = (1 + g_4 L)v_{2t}$ into the first equation in (2.6) leads to a simplified transfer function equation for u_{1t} with one "explanatory variable," $v_{2t} \equiv v^*_{2t}$, and a first order MA disturbance,

$$u_{1t} - g_2 v_{2t-1} = (1 + g_1 L)v_{1t}, \tag{2.7}$$

from which g_1, g_2 and v_{1t} can be estimated. Similarly, if $g_2 = g_4 = 0$, $u_{1t} = (1 + g_1 L)v_{1t}$, with $\lambda_1 = -g_1$ and $v_{1t} \equiv v^*_{1t}$. The second equation in (2.6) becomes

$$u_{2t} - g_3 v^*_{1t-1} = v_{2t}, \tag{2.8}$$

so that g_3 and v_{2t} can be consistently estimated in a linear regression of u_{2t} on $\hat{v}^*_{1t-1}$.

This analysis illustrates the usefulness in some situations of single-equation innovations for the estimation of MA parameters in systems of equations. At present, the relevance of univariate innovations in time series modeling is widely recognized (see, e.g., Haugh and Box 1977).

Obviously, for models with diagonal MA matrices the single-equation innovations are identical with the system innovations, and the single-equation parameter estimates and residuals can be used straightforwardly in the expressions for the asymptotically efficient estimators. Among others, Nelson (1976) assumes diagonal MA matrices, an assumption which leads to a substantial reduction of the number of parameters to be estimated. Notice also, that any vector ARMA model can be transformed into a restricted autoregressive model with diagonal MA matrices through premultiplication as in (2.3) by the adjoint matrix associated with the MA part. In the transformed model, single-equation and system innovations are identical.

3 Single structural equation estimation

In the notation of P–Z a single structural equation, say the ith one, of a dynamic SEM with vector MA disturbances can be written as

$$y_{it} + \sum_{j=1}^{p_1} h_{ij}(L)y_{jt} + \sum_{j=p_1+1}^{p} h_{ij}(L)x_{jt} = c_i + u_{it}, \tag{3.1}$$

where the $h_{ij}(L) = \sum_{l=0}^{r_{ij}} h_{ijl}L^l$ are scalar polynomials in L of degree r_{ij}, $h_{ii}(0) = 0$, the y_{jt}s are endogenous variables, the x_{jt}s are exogenous variables, and u_{it} is the ith element of a vector of disturbances generated by a $p_1 \times 1$ vector MA such as defined in (2.1). The variables x_{jt}s are assumed strictly exogenous, i.e., x_{jt} and $u_{it'}$ are independent, all j, t, t'. On imposing identifying exclusion restrictions the remaining parameters in (3.1) can be consistently estimated by an instrumental variables method as described in P–Z (p. 270).

As an alternative Zellner and Palm (1974) (see also P–Z, pp. 221–2) proposed to substitute the estimated FEs or TFs for the current endogenous variables in (3.1) (except for y_{it}) and then to apply non-linear least squares to obtain estimates for the parameters in the $h_{ij}(L)$s.

However, as shown by Darroch and McDonald (1981) when non-linear least squares is applied, the structural disturbance is represented as an MA in one innovation that depends on all current and past values of the system innovations v_{it}, so that this single-equation innovation and some of the explanatory variables are correlated. Therefore, the non-linear least squares method applied to a structural equation, after substitution as described above, fails to produce consistent estimates of the structural coefficients. OLS applied to (3.1) after substitution of fitted values $\hat{y}_{jt-k}$ for $y_{jt-k} = \hat{y}_{jt-k} + \hat{w}_{jt-k}$, $j = 1, \ldots, p_1, k = 0, 1, \ldots, q$, $k = 0, 1, \ldots, q$, (except for y_{it}),

$$y_{it} + \sum_{j=1}^{p_1} \left[\sum_{l=0}^{q} h_{ijl} \hat{y}_{jt-l} + \sum_{l=q+1}^{r_{ij}} h_{ijl} y_{jt-l} \right] + \sum_{j=p_1+1}^{p} h_{ij}(L) x_{jt}$$

$$= c_i + \left[u_{it} - \sum_{j=1}^{p_1} \sum_{l=0}^{q} h_{ijl} \hat{w}_{jt-l} \right], \tag{3.2}$$

where we assume without loss of generality that $q \leq$ all r_{ij}, will yield consistent estimates of the parameters in $h_{ij}(L)$, provided the l.h.s. explanatory variables are orthogonal to the disturbance term between brackets on the r.h.s. One can obtain $\hat{y}_{jt-k}$s that satisfy this requirement by OLS applied to a regression of y_{jt-k} on $\boldsymbol{z}_t$,

$$y_{jt-k} = \boldsymbol{z}_t' \gamma + w_{jt-k}, \tag{3.3}$$

where the vector $\boldsymbol{z}_t$ consists of all the variables appearing on the l.h.s. of (3.1), except y_{jt-l}, $j = 1, \ldots, p_1, l \leq q$, and includes enough additional variables that are uncorrelated with u_{it} (e.g. other lagged exogenous variables not appearing in (3.1)) for the inverse of the OLS estimator applied to (3.2) to exist.

The estimation procedure can be applied to all p_1 structural equations. The residuals can be computed and used as described in section 2 to obtain consistent estimates of the system MA parameters. When $q = 0$, the structural disturbances are uncorrelated and this estimation procedure leads to the truncated two-stage least squares estimator (see Brundy and Jorgenson 1974) where, instead of using all predetermined variables of the model in the OLS regression of the first stage, one uses a selected subset. Notice the similarity with the model with uncorrelated disturbances. As in the model with MA errors all current and past endogenous variables up to lag q are correlated with the disturbance term u_{it}, they have to be treated as "dependent" variables for which an instrumental variable has to be used or a substitution made. Also, none of them can be included as a regressor in $\boldsymbol{z}_t$. Despite this fact, there are many ways to obtain consistent estimates of the structural parameters in the $h_{ij}(L)$s.

It should be noted that non-linear least squares estimation of (3.2), implementing GLS, is inconsistent as there will be correlation between some explanatory variables and the innovations associated with (3.2). In the final form or the TF form each endogenous variable in the system of equations consisting of (3.1), $i = 1, \ldots, p_1$, is expressed as a rational distributed lag on all exogenous variables and an error term, which is a sum of ARMA processes in the v_{it}s. Substitution of the estimated final form or TF equation for the current and all lagged endogenous variables in (3.1) leads to a linear regression equation with an error term that can be represented as a univariate ARMA process with an innovation that is independent of the regressors. As stated in Darroch and McDonald (1981) non-linear least squares estimation will be consistent. It is expected to be more efficient than OLS estimation, which is also consistent.

Concluding remarks

(a) To conclude, through a modification of Steps 2 and 3 of the procedure presented in P–Z, consistent estimates of the MA parameters and of the system innovations can be obtained. They can be used to generate fully efficient estimates along the lines outlined by P–Z. In some cases, the single-equation innovations will be very useful to get initial consistent estimates of the vector MA parameters. These procedures are expected to be computationally less demanding than the procedure proposed in Mc–D.

(b) There are many ways to obtain consistent estimates of the structural parameters of a dynamic SEM with vector MA errors. One can use instrumental variables or substitute for the "dependent" variables in the structural equation. After appropriate substitution OLS will be consistent whereas non-linear least squares (or GLS) may not be consistent.

(c) The points discussed in this note are also relevant for single-equation reduced form estimation, when lagged endogenous variables are present and the disturbances are generated by a vector MA process. Although the disturbance term of the single reduced form equation can be represented by an MA in one variable, neither OLS nor non-linear least squares estimates of a single reduced form equation are consistent. In both cases there will be correlation between explanatory variables and the disturbance or the innovation. But the methods presented in section 3 and appropriately specialized are consistent.

BIBLIOGRAPHY

Brundy, J. M. and D. W. Jorgenson (1974), "Consistent and efficient estimation of systems of simultaneous equations by means of instrumental variables," in P. Zarembka (ed.), *Frontiers in Econometrics* (New York, Academic Press), 215–44

Darroch, J. and J. McDonald (1981), "Sums of moving average processes: some implications for single equation structural estimation," mimeo

Hannan, E. J. (1970), *Multiple Time Series* (New York, John Wiley)

(1975), "The estimation of ARMA models," *The Annals of Statistics* 3, 975–81

Haugh, L. D. and G. E. P. Box (1977), "Identification of dynamic regression (distributed lag) models connecting two time series," *Journal of the American Statistical Association* 72, 121–30

McDonald, J. and J. Darroch (1981), "On large sample estimation and testing procedures for dynamic equation systems" [*Journal of Econometrics* 17, pp. 131–8]

Nelson, C. R. (1976), "Gains of efficiency from joint estimation of systems of autoregressive-moving average processes," *Journal of Econometrics* 4, 331–48

Palm, F. C. and A. Zellner (1980), "Large-sample estimation and testing procedures for dynamic equation systems," *Journal of Econometrics* 12, 251–83; chapter 5 in this volume

Reinsel, G. (1979), "FIML estimation of the dynamic simultaneous equations model with ARMA disturbances," *Journal of Econometrics* 9, 263–82

Wilson, G. T. (1969), "Factorisation of the covariance generating function of a pure moving average process," *SIAM Journal of Numerical Analysis* 6, 1–7

Zellner, A. and F. C. Palm (1974), "Time series analysis and simultaneous equation econometric models," *Journal of Econometrics* 2, 17–54; chapter 1 in this volume

Part II

Selected applications

6 Time series and structural analysis of monetary models of the US economy (1975)

Arnold Zellner and Franz C. Palm

1 Introduction

In previous work, Zellner and Palm (1974), an approach for building and analyzing dynamic econometric models was presented that is a blend of recently developed time series techniques and traditional econometric methods. This approach was applied in analyzing dynamic variants of a small Keynesian macroeconometric model formulated by Haavelmo (1947). In the present chapter, we apply our approach in the analysis of variants of a dynamic monetary model formulated by Friedman (1970, 1971).

We commence our present analysis by presenting the structural equations of an initial variant of Friedman's model, denoted S^0, that is viewed as a starting point for our analyses. That is, as in previous work we set forth a number of testable implications of S^0, in particular the implications of S^0 for the forms of the final and transfer equations for the variables of S^0. Using monthly data for the US economy, 1953–72, and time series analysis, the implications of S^0 are checked against the information in the data. As will be seen, some of S^0's implications do not square with the information in the data. This leads us to consider other variants of the model whose implications can be checked with the data. In this way we attempt to iterate in on a variant of the model that is in accord with the information in the data. When a variant has been obtained that is in accord with the data information, it can be checked further with new sample information.

In considering possible variants of the initial model S^0, we shall be concerned with, among others, the following issues: (1) rational vs. other representations of the formation of anticipations, (2) open versus closed

Research supported by National Science Foundation Grant GS-40033, H. G. B. Alexander Endowment Fund, Graduate School of Business, University of Chicago, and the Belgian National Science Foundation.

Originally published in *Sankhyā: The Indian Journal of Statistics*, Series C 37 (1975), 12–56.

loop control policies, (3) relationship of "nominal" and "real" sectors of the economy, (4) lag structures and other features of behavioral equations, and (5) serial correlation properties of disturbance terms. As will be seen, these issues can be analyzed by comparing the theoretical and empirical properties of final equation and transfer equation systems.

The plan of the chapter is as follows. In section 2, we present and discuss the structural, transfer, and final equations of the initial variant of Friedman's model. We then turn to the results of empirical analyses designed to check the consistency of the initial variant with the information in our data. After summarizing the empirical findings, we go on in section 3 to describe and analyze variants of the initial model. In section 4, we consider properties of a variant of the model that we believe is consistent with the information in our data. Additional empirical analyses and tests of our final variant are reported. A summary of results and some concluding remarks are presented in section 5.

2 Analysis of an initial variant, S^0, of a monetary model

2.1 *Structural equations of* S^0

In this section we describe the structural equations of a simple monetary model that we regard as a good starting point in our search for a formulation that is consistent with the information in our data. If it is found that the simple model is inconsistent with the data, alternative variants will be considered. The equations of the initial variant include (1) a money demand equation, (2) a money supply equation, (3) a money market clearing relationship, (4) the Fisher equation, and (5) an anticipation formation equation. In the initial variant of the model, the anticipation formation equation is formulated as a "partial adjustment" equation and possible lag structures in the structural equations are purposely suppressed. The equations of the model are first presented in deterministic form and then transformed to a stochastic difference equation representation.

The first equation of the model is a money demand equation that we write as follows:

$$Y_t = Ae^{\gamma_1 i_t} M_t^D \quad A, \gamma_1 > 0, \tag{2.1}$$

where A and γ_1 are constant parameters, the subscript t denotes the value of a variable in the t-th time period, Y_t = nominal income, M_t^D = nominal money balances demanded, and i_t = nominal interest rate. Following Friedman, we have assumed that (2.1) is homogeneous of degree one in M_t^D and P_t, the price level.[1] Further, we have included a particular functional form for the dependence of money demand on the interest rate.[2] Last, we have intentionally excluded any lags in the money demand relations, a point that will receive attention below.

The second relationship of the model is a money supply function, formulated as follows:

$$M_t^S = Be^{\gamma_2 i_t} H_t \quad B, \gamma_2 > 0, \tag{2.2}$$

where B and γ_2 are constant parameters, M_t^S = nominal supply of money, and, H_t = "high-powered" money or the monetary base (currency plus bank deposits with the Federal Reserve System).

We assume that H_t is an exogenous variable, an assumption whose implications will be explored below.

Given that the money market clears each period, we have

$$M_t^D = M_t^S = M_t \tag{2.3}$$

where M_t is the actual stock of money.

The next equation of the model is the Fisher equation,

$$i_t = \rho + r_{p_t}^*, \tag{2.4a}$$

where ρ = the real rate of interest, assumed constant, and $r_{p_t}^* = (\Delta \log P_t)^* = (\Delta P_t/P_{t-1})^*$, the anticipated rate of inflation. Given that $(\Delta \log Y_t)^* = (\Delta \log P_t)^* + (\Delta \log y_t)^*$, where $r_{Y_t}^* = (\Delta \log Y_t)^*$ is the anticipated rate of change of nominal income and $g_t^* = (\Delta \log y_t)^*$ is the anticipated rate of change of real income, y_t, we have $r_{p_t}^* = r_{Y_t}^* - g_t^*$. If we, along with Friedman, assume that $g_t^* = g_t = g$, a constant,[3] where $g_t = \Delta \log y_t$, the actual rate of change of real income, (2.4a), can be rewritten as,

$$i_t = \rho - g + r_{Y_t}^* = c + r_{Y_t}^*, \tag{2.4b}$$

with $c = \rho - g$, assumed constant.

The last equation of the model is an "adaptive expectations" equation that has been employed in many studies, namely,

$$r_{Y_t}^* - r_{Y_{t-1}}^* = \beta(r_{Y_{t-1}} - r_{Y_{t-1}}^*) \quad \beta > 0, \tag{2.5}$$

where $r_{Y_{t-1}} = \Delta \log Y_{t-1}$, the lagged rate of growth of nominal income.

[1] Some empirical results of Laidler (1966) suggest that this approximation may be acceptable.

[2] Other functional forms could be employed. However, some alternatives lead to a model that is non-linear in the variables.

[3] Below, we relax this assumption.

Upon substituting from (2.3) in (2.1) and (2.2), taking logarithms of (2.1) and (2.2) and differencing them and (2.4b), the resulting equations of the initial model S^0 are:

$$r_{Y_t} = \gamma_1 \Delta i_t + r_{M_t}, \tag{2.6}$$

$$r_{M_t} = \gamma_2 \Delta i_t + r_{H_t}, \tag{2.7}$$

$$\Delta i_t = \Delta r^*_{Y_t}, \tag{2.8}$$

$$\Delta r^*_{Y_t} = \beta(r_{Y_{t-1}} - r^*_{Y_{t-1}}). \tag{2.9}$$

Since (2.9) can be expressed as $[1 - (1-\beta)L]r^*_{Y_t} = \beta L r_{Y_t}$, where L is a lag operator ($Lx_t = x_{t-1}$), we have $[1-(1-\beta)L](1-L)r^*_{Y_t} = \beta(1-L)Lr_{Y_t}$. Then on noting that (2.8) is $(1-L)r^*_{Y_t} = (1-L)i_t$, we can replace (2.8) and (2.9) by

$$[1-(1-\beta)L]\Delta i_t = \beta L(1-L)r_{Y_t}. \tag{2.10}$$

Thus the three equations of the model involving the three observable endogenous variables r_{Y_t}, Δi_t, and r_{M_t} and the observable exogenous variable r_{H_t} are given by (2.6), (2.7), and (2.10). These three equations can be expressed in matrix form with the addition of random disturbance terms, u_{1t}, u_{2t} and u_{3t}, as follows:

$$\begin{bmatrix} 1 & -1 & -\gamma_1 \\ 0 & 1 & -\gamma_2 \\ -\beta L(1-L) & 0 & 1-(1-\beta)L \end{bmatrix} \begin{bmatrix} r_{Y_t} \\ r_{M_t} \\ \Delta i_t \end{bmatrix} = \begin{bmatrix} 0 \\ 1 \\ 0 \end{bmatrix} r_{H_t} + \begin{bmatrix} u_{1t} \\ u_{2t} \\ u_{3t} \end{bmatrix}, \tag{2.11a}$$

or in more general form,

$$H_{11}\boldsymbol{y}_t = \boldsymbol{\alpha} r_{H_t} + \boldsymbol{u}_t, \tag{2.11b}$$

where H_{11} denotes the matrix on the l.h.s. of (2.11a), $\boldsymbol{y}'_t = (r_{Y_t}, r_{M_t}, \Delta i_t)$ $\boldsymbol{\alpha}' = (0, 1, 0)$, and $\boldsymbol{u}'_t = (u_{1t}, u_{2t}, u_{3t})$.

We shall now derive the transfer functions (TFs) and final equations (FEs) associated with the structural equations S^0 in (2.11) and establish properties of the TF and FE systems.

2.2 *Transfer functions (TFs) for* S^0, *(2.11)*

Solving the system (2.11) for r_{Y_t}, r_{M_t}, and Δi_t in terms of r_{H_t} by premultiplying (2.11) by the adjoint matrix H^*_{11} associated with H_{11}, we get the TFs

$$|H_{11}|\boldsymbol{y}_t = H^*_{11}\boldsymbol{\alpha} r_{H_t} + H^*_{11}\boldsymbol{u}_t, \tag{2.12}$$

where $|H_{11}|$, the determinant of H_{11}, is given by $|H_{11}| = 1 - (1-\beta)L - \beta(\gamma_1 + \gamma_2)L(1-L)$, which is a second degree polynomial in L, and the adjoint matrix H_{11}^* is given by:

$$H_{11}^* = \begin{pmatrix} 1-(1-\beta)L & 1-(1-\beta)L & \gamma_1+\gamma_2 \\ \gamma_2\beta L(1-L) & 1-(1-\beta)L - \gamma_1\beta L(1-L) & \gamma_2 \\ \beta L(1-L) & \beta L(1-L) & 1 \end{pmatrix}. \quad (2.13)$$

Explicitly, the TF system in (2.12) is then:

$$[1-(1-\beta)L - \beta(\gamma_1+\gamma_2)L(1-L)] \begin{bmatrix} r_{Y_t} \\ r_{M_t} \\ \Delta i_t \end{bmatrix} = \begin{bmatrix} 1-(1-\beta)L \\ 1-(1-\beta)L - \gamma_1\beta L(1-L) \\ \beta L(1-L) \end{bmatrix} r_{H_t} + \begin{bmatrix} v_{1t} \\ v_{2t} \\ v_{3t} \end{bmatrix}, \quad (2.14a)$$

with $\boldsymbol{v}_t = H_{11}^* \boldsymbol{u}_t$, where $\boldsymbol{v}_t' = (v_{1t}, v_{2t}, v_{3t})$.

From examination of the l.h.s. of (2.14a), it is seen that S^0 implies that each of the three TFs has an autoregressive (AR) part that is of second order and that the parameters of the AR parts of the three TFs are identical. Further the implied lags on r_{H_t}, the rate of growth of high-powered money, in the TFs are shown in the following table:

TF for	Order of lag on r_{H_t}
r_{Y_t}	1
r_{M_t}	2
Δi_t	2

In addition to these implications of S^0 for the forms of the TFs, it should be noted that (2.14) involves some strong restrictions on the parameters of the TFs that can be appreciated by rewriting (2.14a) as follows:

$$[1-(1-\beta+\eta)L + \eta L^2] \begin{bmatrix} r_{Y_t} \\ r_{M_t} \\ \Delta i_t \end{bmatrix} = \begin{bmatrix} 1-(1-\beta)L \\ 1-(1-\beta+\eta_1)L + \eta_1 L^2 \\ \beta L - \beta L^2 \end{bmatrix} r_{H_t} + \begin{bmatrix} v_{1t} \\ v_{2t} \\ v_{3t} \end{bmatrix}, \quad (2.14b)$$

where $\eta = (\gamma_1 + \gamma_2)\beta$ and $\eta_1 = \gamma_1\beta$. The following are restrictions on the parameters of (2.14b) implied by S^0:

(i) The sum of the coefficients of $r_{Y_{t-1}}$ and $r_{Y_{t-2}}$ equals the coefficient of $r_{H_{t-1}}$ in the TF for r_{Y_t}.
(ii) The coefficient of r_{H_t} should equal one in the TF for r_{Y_t} and r_{M_t} and be equal to zero in the TF for Δi_t.
(iii) In each of the three TFs, the sum of the AR parameters is equal to β, the coefficient of $r_{H_{t-1}} - r_{H_{t-2}}$ in the TF for Δi_t.
(iv) In the TF for Δi_t, the sum of the coefficients of $r_{H_{t-1}}$ and $r_{H_{t-2}}$ is zero.
(v) The sum of the coefficients of $r_{H_{t-1}}$ and $r_{H_{t-2}}$ in the TF for r_{M_t} equals the coefficient of $r_{H_{t-1}}$ in the TF for r_{Y_t}.

The restrictions (i)–(v) can be tested in empirical analysis of the TF system in (2.14). Further, the properties of the error vector $\boldsymbol{v}_t$ in (2.14) can also be investigated. Since $\boldsymbol{v}_t = H_{11}^* \boldsymbol{u}_t$, with H_{11}^* shown in (2.13), it is clear that assumptions about the serial correlation properties of $\boldsymbol{u}_t$ imply testable implications regarding the serial correlation properties of $\boldsymbol{v}_t$. For example, if the $\boldsymbol{u}_t$s are assumed to be serially uncorrelated, then the elements of $\boldsymbol{v}_t$, given by

$$\boldsymbol{v}_t = \begin{pmatrix} v_{1t} \\ v_{2t} \\ v_{3t} \end{pmatrix} = \begin{pmatrix} [1-(1-\beta)L]u_{1t} + [1-(1-\beta)L]u_{2t} + (\gamma_1+\gamma_2)u_{3t} \\ \gamma_2 \beta L(1-L)u_{1t} + [1-(1-\beta)L - \gamma_1\beta L(1-L)]u_{2t} + \gamma_2 u_{3t} \\ \beta L(1-L)u_{1t} + \beta L(1-L)u_{2t} + u_{3t} \end{pmatrix}, \tag{2.15}$$

will be autocorrelated. In fact, under the assumption that the u_{it}s are serially uncorrelated, the moving average (MA) processes on the v_{it}s will have the following properties given that $\gamma_1, \gamma_2 \neq 0$ and $0 < \beta < 1$:

Error term	Order of MA process
v_{1t}	1
v_{2t}	2
v_{3t}	2

In empirical analyses, these implications regarding the orders of the MA error processes, that are lower bounds, can be tested empirically.

2.3 *Final equations for* S^0

We assume that the exogenous variable r_{H_t} is generated independently[4] of the other variables by the following autoregressive moving average (ARMA) process:

$$\phi_p(L) r_{H_t} = \theta_q(L) u_{4t}, \tag{2.16}$$

[4] Below we consider the implications of a breakdown of this independence assumption.

Table 6.1 *Selected properties of FEs*

Variable	Order of AR	Order of MA error[a]
r_{Y_t}	$2+p$	$\max(1+q, p+n_1)$
r_{M_t}	$2+p$	$\max(2+q, p+n_2)$
Δi_t	$2+p$	$\max(2+q, p+n_3)$
r_{H_t}	p	q

Note:
[a] n_i, $i = 1, 2, 3$, denotes the order of the MA processes for v_{it}, $i = 1, 2, 3$, respectively. Since, as shown above, the $n_i > 0$, $i = 1, 2, 3$, the orders of the MA errors in the FEs are bounded from below.

where $\phi_p(L)$ and $\theta_q(L)$ are polynomials in L of finite degrees, p and q, respectively, and u_{4t} is a non-autocorrelated error term with zero mean, constant variance that is distributed independently of the structural disturbance terms appearing in (2.11a). Properties of the process in (2.16), including the values of p and q, will be determined from the data on r_{H_t}.

If we premultiply the system in (2.14a) by $\phi_p(L)$ and substitute $\phi_p(L)r_{H_t} = \theta_q(L)u_{4t}$, from (2.16), we obtain the FEs for endogenous variables r_{Y_t}, r_{M_t} and Δi_t, namely:

$$[1-(1-\beta)L-\beta(\gamma_1+\gamma_2)L(1-L)]\phi_p(L)\begin{pmatrix} r_{Y_t} \\ r_{M_t} \\ \Delta i_t \end{pmatrix}$$
$$= \begin{pmatrix} 1-(1-\beta)L \\ 1-(1-\beta)L-\gamma_1\beta L(1-L) \\ \beta L(1-L) \end{pmatrix}\theta_q(L)u_{4t} + \phi_p(L)\begin{pmatrix} v_{1t} \\ v_{2t} \\ v_{3t} \end{pmatrix}, \tag{2.17}$$

It is seen from (2.17) that each FE is in ARMA form with *identical* AR parts, an implication that will be tested with our data. In table 6.1, we summarize the features of the FEs in (2.16) and (2.17).

Given that we determine the forms and estimate the parameters of the FEs from our data, it is possible to assess whether the information in our data, is consistent with the implications of S^0 that have been set forth above.

2.4 *Empirical analyses of final equations of the initial model* S^0

In this subsection we report the results of analyses of the FEs (2.16) and (2.17) first employing Box–Jenkins (1970) techniques and then utilizing likelihood ratio tests and posterior odds. The latter techniques, used previously in Zellner and Palm (1974), appear to us to be less judgmental than Box–Jenkins (hereafter, BJ) techniques.

BJ (1970) suggest differencing a series until it is stationary and then computing and studying estimates of the autocorrelation and partial autocorrelation functions in order to determine the orders of the AR and MA parts of the final equations.[5] For the monthly series[6] under consideration – that is, those for nominal personal income, high-powered money, and nominal money balances – the proportionate rate of growth of each of these variables seems to be generated by a stationary process while the first difference of the monthly nominal interest rates also appears to be stationary (see figure 6.1).

In figure 6.2, the estimated autocorrelation and partial autocorrelation functions for the series Δi_t, the first difference of the monthly market interest rate on three-month treasury bills (3MTB) are presented.[7] The bands represent a large sample $\pm 2\hat{\sigma}$ confidence interval for the autocorrelation parameters where $\hat{\sigma}$ is a large-sample standard error associated with the sample estimates of the autocorrelation parameters. The estimates of the first, sixth, and seventh autocorrelation coefficients lie outside the $\pm 2\hat{\sigma}$ band. Using the first order autocorrelation coefficient, which appears to be quite different from zero, as a cut-off of the autocorrelation function, the results suggest a first order MA process. With respect to the estimated partial autocorrelation function, only the estimate of the first order partial autocorrelation coefficient lies outside the $\pm 2\hat{\sigma}$ interval. If the first order partial autocorrelation coefficient is deemed significantly different from zero while all higher order coefficients are assumed equal to zero, the error process in the FE for Δi_t would be a first order AR process. Estimation of a (1, 1, 1) process[8] produced the following result using 240 monthly observations on i_t, 1953–72:

$$(1 - \underset{(0.147)}{0.124}L)\Delta i_t = \underset{(0.020)}{0.012} + (1 + \underset{(0.139)}{0.334}L)e_t \quad s^2 = 0.0534, \tag{2.18}$$

where e_t is a non-autocorrelated error term with zero mean and constant variance, s^2 is the residual sum of squares divided by the number

[5] See Box and Jenkins (1970), Nelson (1973), and Zellner and Palm (1974) for further discussion and applications of these techniques.

[6] The data are discussed in the appendix at the end of the chapter (p. 286).

[7] The computations were performed using a computer program developed by C. R. Nelson and S. Beveridge, Graduate School of Business, University of Chicago.

[8] In Box and Jenkins (1970, terminology a (p, d, q) process for a variable denotes a process that is stationary in the dth difference of the variable with AR part of order p and MA error process of order q.

of degrees of freedom, and the figures in parentheses are large-sample standard errors.

Analyses similar to those with the 3MTB rate were performed with the interest on four–six-month prime commercial paper (4–6PCP). Again the process suggested by the autocorrelation and partial autocorrelation functions' estimates is a (1, 1, 1) process that was estimated using 240 monthly observations on i_t, 1953–1972, with the following result:

$$(1 - \underset{(0.100)}{0.573}L)\Delta i_t = \underset{(0.012)}{0.006} + (1 - \underset{(0.122)}{0.0569}L)e_t \quad s^2 = 0.0378. \tag{2.19}$$

The first differences of ln M_1, where M_1 is currency plus demand deposits, appear to be subject to a slight trend indicating an increasing proportionate rate of change in nominal balances. The autocorrelation function for the first differences, shown in figure 6.2, falls off slowly while the partial autocorrelation function has values significantly different from zero at lags 1, 2, and 3. If we consider lag 3 as a cut-off, the underlying process for $\Delta \ln M_{1t}$ is a pure third order AR process that was estimated with the following result:

$$(1 - \underset{(0.0641)}{0.240}L - \underset{(0.0647)}{0.204}L^2 - \underset{(0.0642)}{0.235}L^3)\Delta \ln M_{1t} = \underset{(0.00027)}{(0.00095)} + e_t$$
$$s^2 = 0.00000757. \tag{2.20}$$

Notice that all coefficients in (2.20) are significantly different from zero at a reasonable significance level.

With respect to monthly personal income, Y_t, the first difference of $\ln Y_t$ appears to be stationary, see figure 6.1. Except for lag 4 and lag 9, the estimates of the autocorrelation coefficients lie within the $\pm 2\hat{\sigma}$ band and those of the partial autocorrelation coefficient all lie within or close to the $\pm 2\hat{\sigma}$ band (see figure 6.3). If we take the value of the autocorrelation function at lag 4 to be a cut-off of the function, the process for $\Delta \ln Y_t$ may be a fourth order MA with zero coefficients for e_{t-1}, e_{t-2} and e_{t-3}. Estimation of this scheme led to:

$$\Delta \ln Y_t = \underset{(0.0004)}{0.0052} + (1 + \underset{(0.0656)}{0.144}L^4)e_t \quad s^2 = 0.00002. \tag{2.21}$$

Last, the first differences of $\ln H_t$ show a slight trend that can be eliminated by second differencing. For the levels ln H_t, the autocorrelation function falls off slowly while the partial autocorrelation function has a value significantly different from zero at lag 1. For $\Delta \ln H_t$, the estimated partial autocorrelation function has some values up to lag 6 that lie outside the $\pm 2\hat{\sigma}$ band while the autocorrelation function has significantly

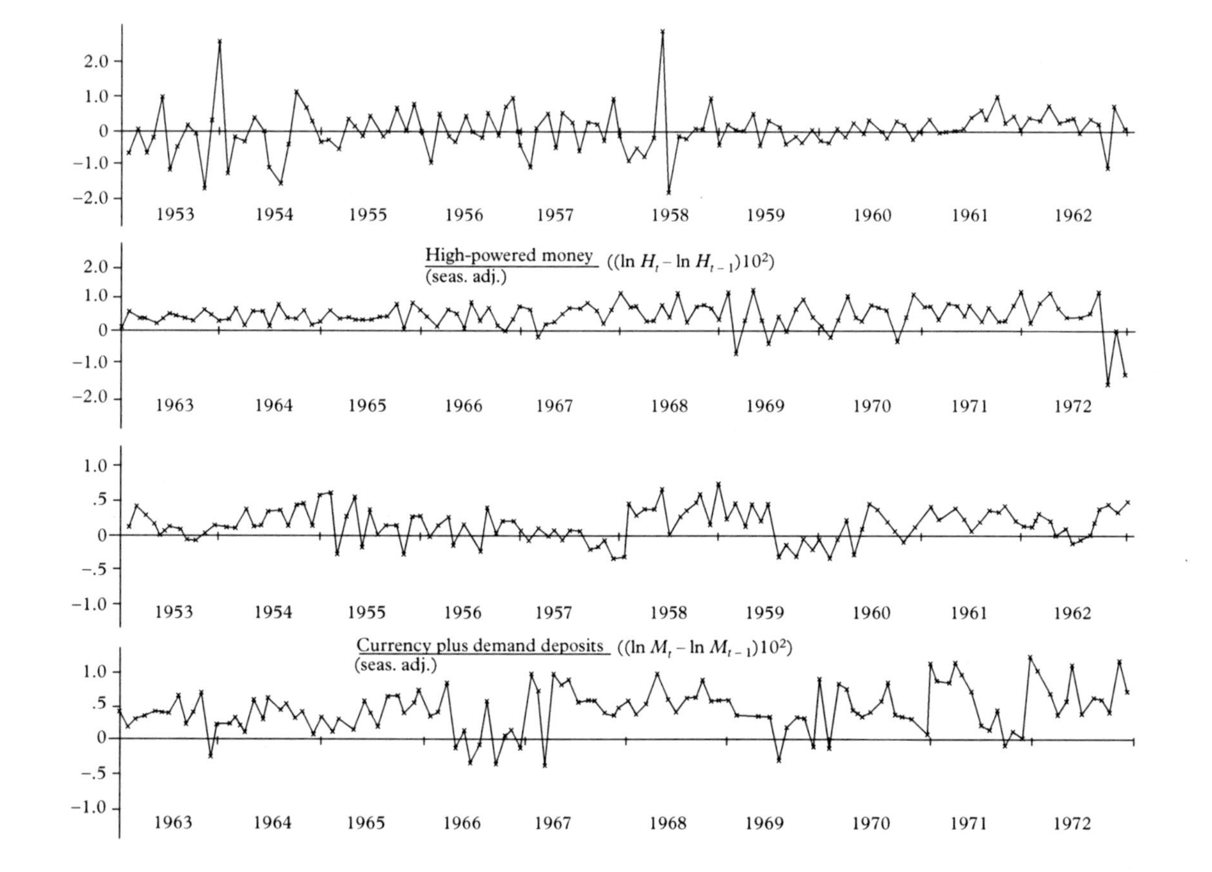

2.0
1.0
0
−1.0
−2.0
1953
1954
1955
1956
1957
1958
1959
1960
1961
1962
High-powered money ((ln H_t − ln H_{t-1})10^2)
(seas. adj.)
1963
1964
1965
1966
1967
1968
1969
1970
1971
1972
1.0
.5
0
−.5
−1.0
Currency plus demand deposits ((ln M_t − ln M_{t-1})10^2)
(seas. adj.)

Figure 6.1

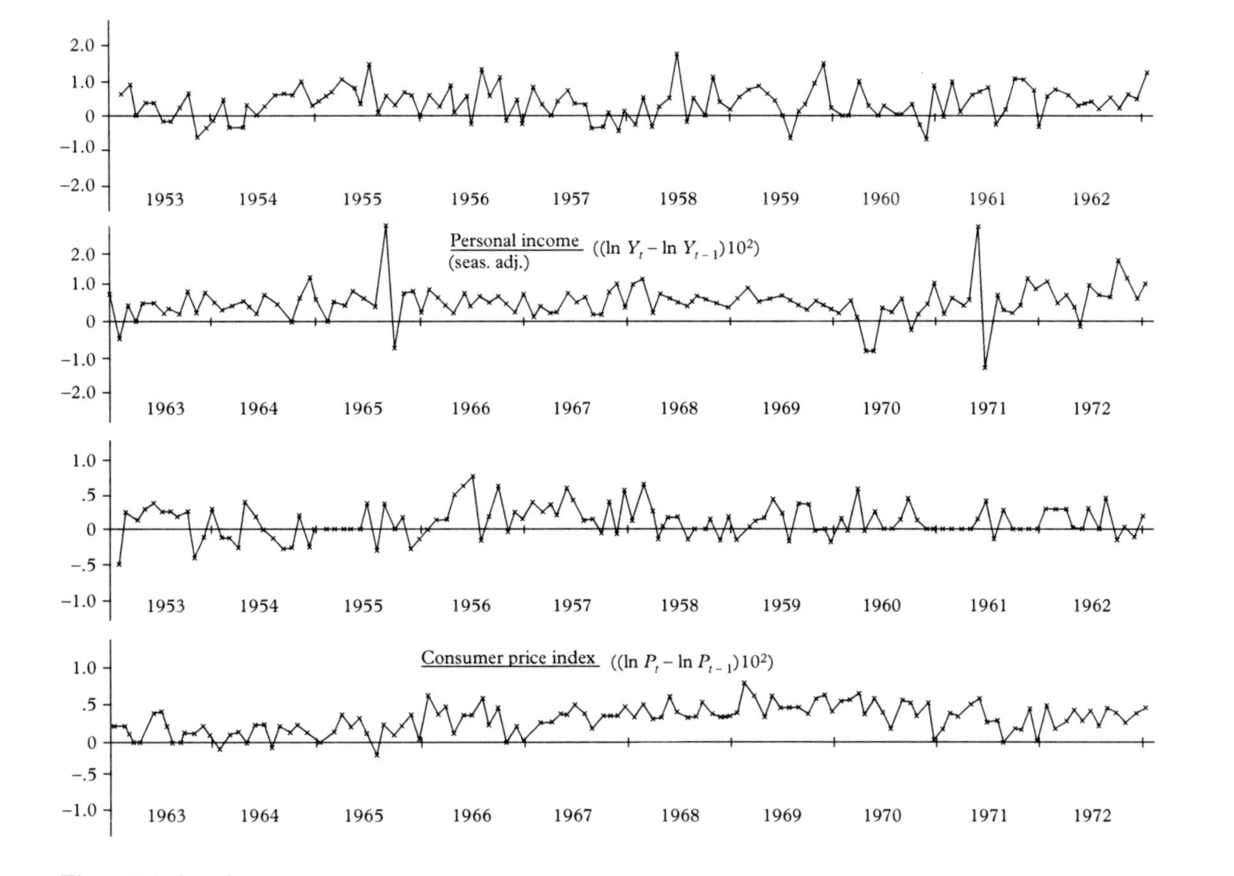

Figure 6.1 (*cont.*)

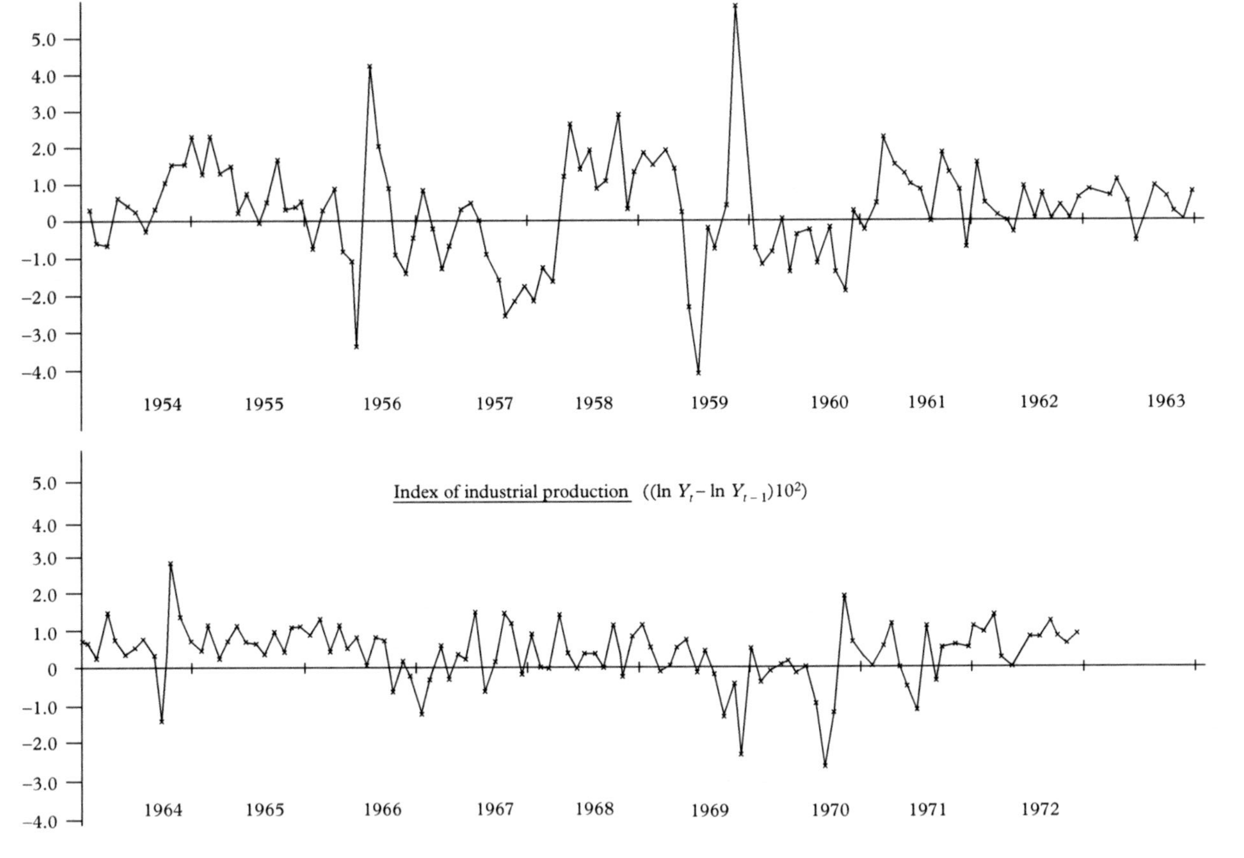

Figure 6.1 (*cont.*)

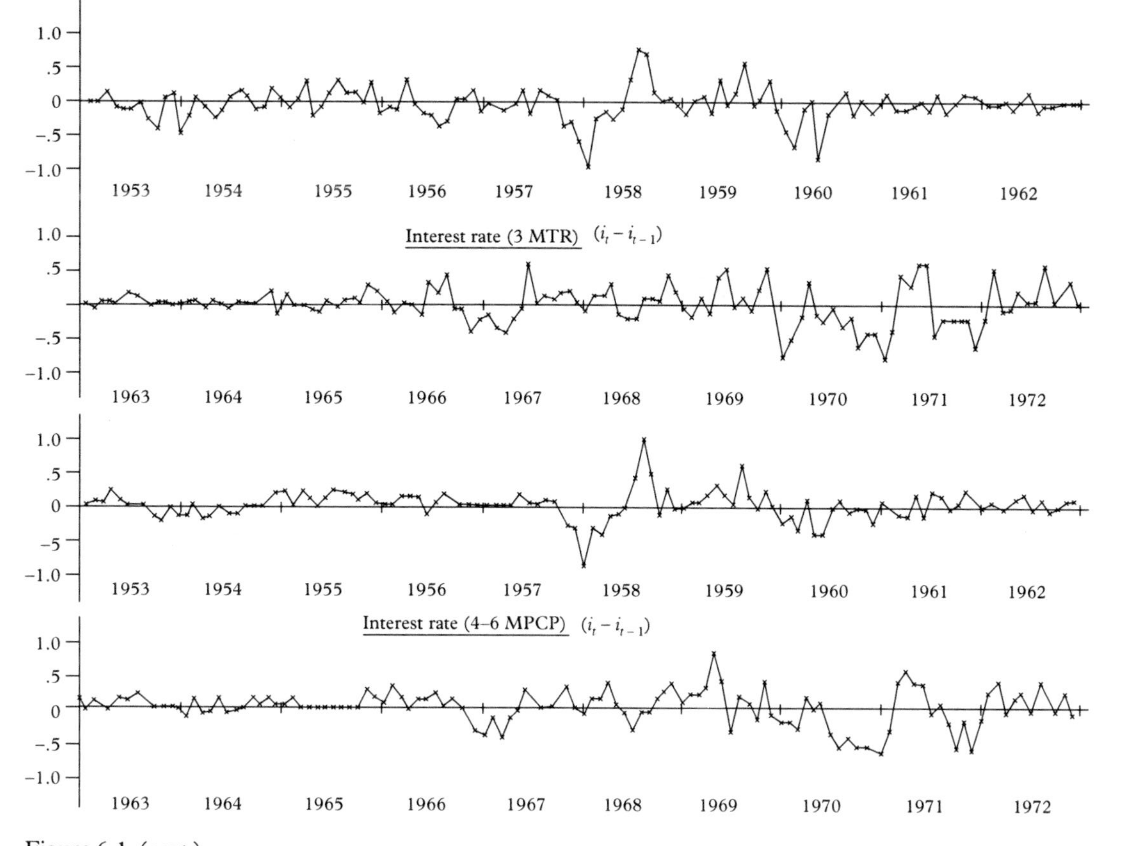

Figure 6.1 (*cont.*)

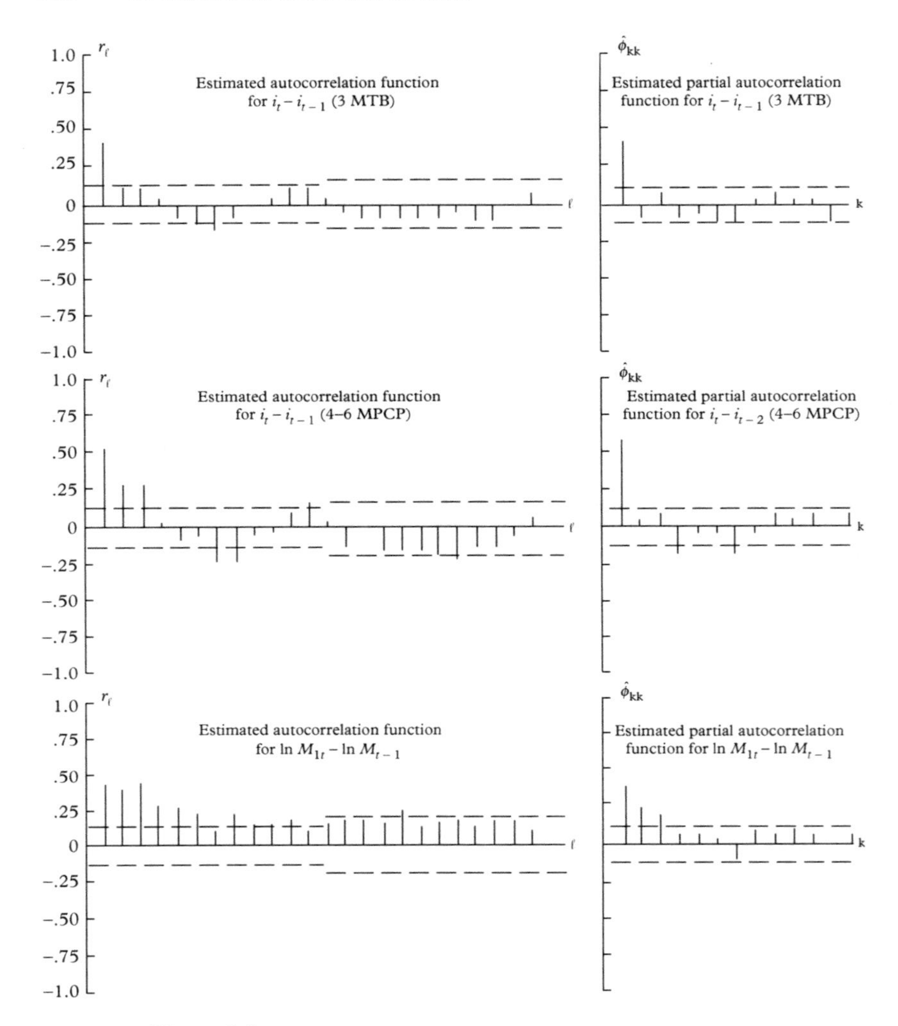

Figure 6.2
Note: Figures below abscissa should be read as negative.

positive autocorrelations for a large number of lags (see figure 6.3). In view of these results, the choice of an underlying scheme is not obvious. On fitting a (1, 1, 1) scheme, the result is:

$$(1 + \underset{(0.403)}{0.615}L)\Delta \ln H_t = \underset{(0.0012)}{0.004} + (1 - \underset{(0.372)}{0.683}L)e_t$$

$$s^2 = 0.0000326. \qquad (2.22)$$

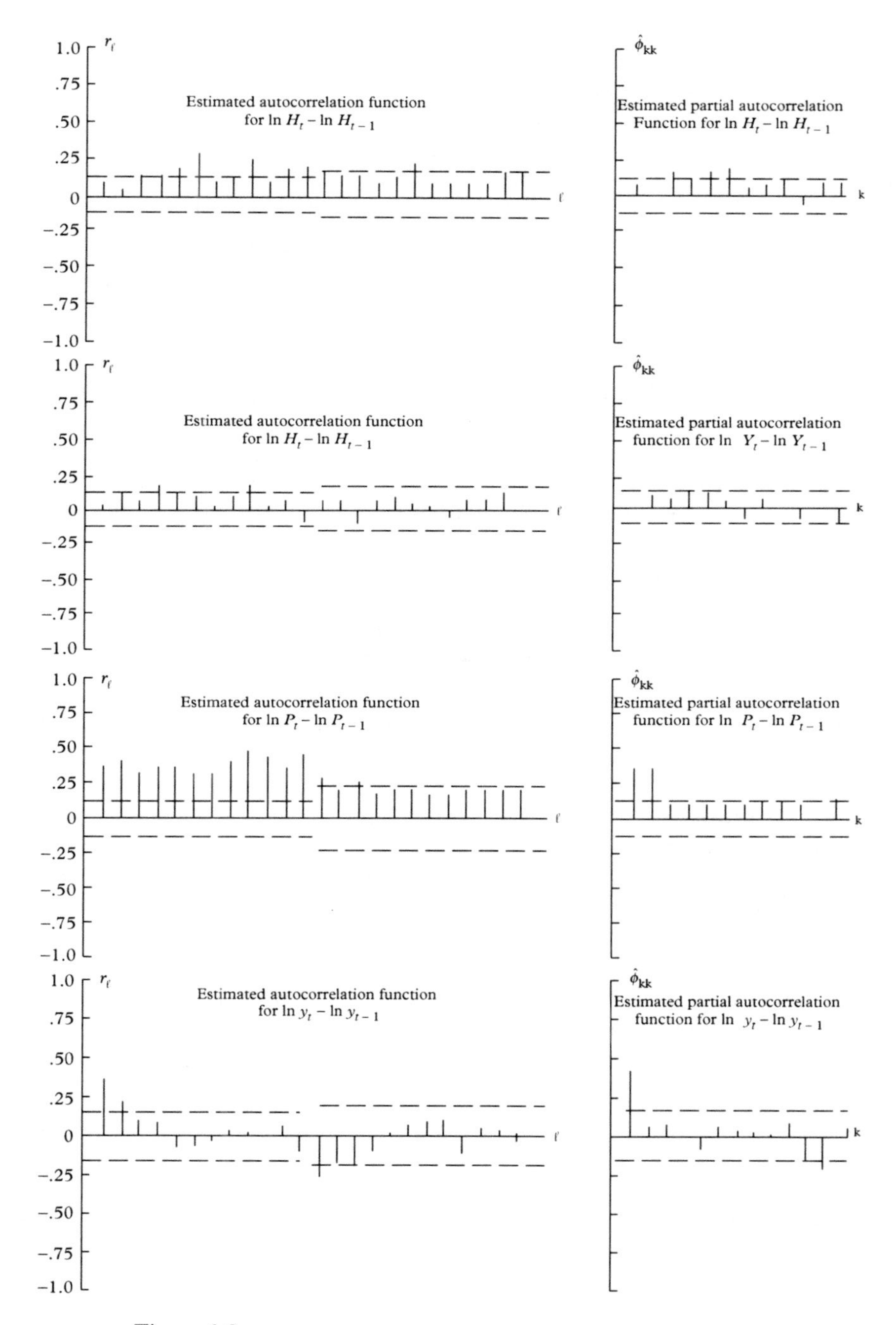

Figure 6.3
Note: Figures below abscissa should be read as negative.

Table 6.2 *Results of Box–Jenkins analyzes of FEs*

Variable	ARMA process
i_t : (3MTB)	(1, 1, 1)
i_t : (4-6PCP)	(1, 1, 1)
$\ln M_{1t}$	(3, 1, 0)
$\ln Y_t$	(0, 1, 4)
$\ln H_t$	(1, 1, 1)

It is seen that neither the AR parameter nor the MA parameter is significantly different from zero. If both are taken equal to zero, $\ln H_t$ would follow a random walk.

The analyses reported above provide a tentative identification of the processes generating Δi_t, $\Delta \ln M_{1t}$, $\Delta \ln Y_t$ and $\Delta \ln H_t$, with the results summarized in table 6.2.

As indicated in the previous subsection, AR parts of the FEs for the endogenous variables, Δi_t, $\Delta \ln M_{1t}$ and $\Delta \ln Y_t$ should be identical and of order $p + 2$, where p is the order of the AR part of the FE for $\ln H^t$, namely $p = 1$ as shown in table 6.2. On comparing the results in table 6.2 with the implications of the model S^0, shown in table 6.1, we see that the AR part of the FE for $\Delta \ln M_{1t}$ has order 3, consistent with the requirement that it be $p + 2 = 3$. However, the empirically determined orders of the AR parts of the FEs for Δi_t and for $\Delta \ln Y_t$, 1 and 0, respectively, are not consistent with the implication of S^0 that they be of order $p + 2 = 3$. Further, the empirically determined orders of the MA error processes of the FEs for $\Delta \ln M_{1t}$ and Δi_t are, respectively, 0 and 1, which is inconsistent with the implication of S^0 that they be equal to or greater than $1 + q = 2$, where q is the order of the MA error process in the FE for $\Delta \ln H_t$.

It is clear that the empirical analyses of the FEs have produced findings apparently inconsistent with the implications of model S^0. However, before considering modifications of S^0, we shall consider the FEs' specifications employing large sample likelihood ratio tests and posterior odds ratios.

2.5 *Empirical analyses of final equations using likelihood ratio tests and posterior odds ratios*

In this subsection, we compare alternative specifications of the FEs using large-sample likelihood ratios that approximate posterior odds ratios.[9] In

[9] That is if λ = ratio of maximized likelihood functions under hypotheses H_1 and H_2, the posterior odds ratio, K_{12}, in large-samples is given approximately by $K_{12} = (\pi_1/\pi_2)\lambda$,

addition to the variables appearing explicitly in model S^0, we also analyze processes for $\ln y_t$ and $\ln P_t$, where y_t is the monthly index of industrial production and P_t is the monthly consumer price index. These latter variables will appear in variants of S^0 considered below.

The results of estimating alternative schemes for the variables are reported in tables 6.3–6.9. The variants of the processes considered have been suggested by analyses reported in the previous subsection and also by the usual practice of considering somewhat broader schemes that may be supported by the information in the data. Large-sample χ^2 tests have been employed to determine whether schemes somewhat broader than those presented in the previous subsection are supported by the information in the data. Information regarding the application of the χ^2 tests is presented in tables 6.10–6.12.

For the 3MTB interest rate, i_t, a (2, 1, 3) model appears significantly better than either a (1, 1, 1) or a (2, 1, 2) model while the (3, 1, 3) model seems to be more in accord with the information in the data than either (1, 1, 1), (2, 1, 2), or (3, 2, 1) models. Since the estimation results provide a third order AR coefficient significantly different from zero at a reasonable significance level, we conclude that the data favor a (3, 1, 3) model for the 3MTB interest rate.

As regards the 4–6MPCP rate, (2, 1, 3), (3, 1, 2), and (3, 1, 3) are significantly different from (1, 1, 1) or (2, 1, 2) models at the significance levels indicated in table 6.3. However, it does not appear to be possible to discriminate among the (2, 1, 3), (3, 1, 2), and (3, 1, 3) models. Even though the (2, 1, 3) model is nested in the (3, 1, 3) model, the likelihood ratio is very close to 1. From the estimates of the (2, 1, 3) and (3, 1, 2) models it is difficult to discriminate between them and thus we shall tentatively carry along both variants.

According to the results of the likelihood ratio tests, a (3, 1, 3) model for $\ln M_{1t}$ is more in accord with the information in the data than (3, 1, 0), (2, 1, 2), (2, 1, 3), or (3, 1, 2) models. The estimation results for the (3, 1, 3) model suggest that all its parameters are significantly different from zero at a reasonable significance level.

With respect to $\ln Y_t$, the logarithm of nominal income, the (0, 1, 4) model suggested by the BJ identification techniques performed very well relative to alternative and is thus retained.

For $\ln H_t$, the logarithm of high-powered money, a random-walk model, (0, 1, 0), performs as well as more complicated models on the basis of the large-sample χ^2 tests and thus is retained at the present stage.

where π_1/π_2 is the prior odds ratio for the two hypotheses. See Lindley (1961), Palm (1973), and Zellner and Palm (1974, p. 22) for derivation and discussion of this approximation and Zellner (1971, ch. 10) for discussion of posterior odds ratios.

Table 6.3 *Estimated FEs for 3MTB market rate, 1953–1972*

Model (p, d, q)	RSS, residual sum of squares	DF	RSS/DF	Estimates of the AR part			Estimates of the MA part			Constant
				AR1	AR2	AR3	MA1	MA2	MA3	
1. (2, 0, 0)	13.089	237	0.0552	1.226 (0.011)	−0.244 (0.014)					0.0638 (0.033)
2. (1, 1, 1)	12.603	236	0.0534	0.124 (0.147)			−0.334 (0.139)			0.012 (0.020)
3. (2, 1, 2)	12.528	234	0.0535	0.234 (1.015)	0.133 (0.224)		−0.220 (1.015)	0.161 (0.335)		0.0087 (0.0199)
4. (2, 1, 3)	11.681	233	0.05013	1.514 (0.0243)	−0.942 (0.0258)		1.091 (0.0194)	−0.391 (0.0198)	−0.395 (0.009)	0.0052 (0.0100)
5. (3, 1, 3)	11.642	232	0.05018	1.401 (0.0502)	−0.767 (0.0767)	−0.111 (0.049)	0.995 (0.0269)	−0.248 (0.0303)	−0.491 (0.0115)	0.0055 (0.0108)
6. (3, 1, 2)	12.377	233	0.05312	−1.135 (0.2796)	−0.179 (0.189)	0.220 (0.112)	−1.607 (0.275)	−0.773 (0.280)		0.0287 (0.0506)

Table 6.4 *Estimated FEs for four–six-month interest rate on prime commercial paper, 1953–1972*

Model (p, d, q)	RSS, residual sum of squares	DF	RSS/DF	Estimates of the AR part			Estimates of the MA part			Constant
				AR1	AR2	AR3	MA1	MA2	MA3	
1. (2, 0, 0)	9.719	237	0.0410	1.263 (0.0054)	− 0.277 (0.0087)					0.058 (0.029)
2. (1, 1, 1)	8.917	236	0.0378	0.573 (0.100)			0.0569 (0.122)			0.006 (0.012)
3. (2, 1, 2)	8.818	234	0.0377	0.0678 (0.507)	0.356 (0.274)		− 0.470 (0.508)	0.147 (0.137)		0.0081 (0.017)
4. (2, 1, 3)	8.365	233	0.0359	1.422 (0.0872)	− 0.808 (0.081)		0.934 (0.103)	− 0.312 (0.0960)	− 0.376 (0.0645)	0.00535 (0.00925)
5. (3, 1, 3)	8.365	232	0.0361	1.494 (0.233)	− 0.923 (0.361)	0.0664 (0.190)	1.001 (0.228)	− 0.403 (0.280)	− 0.324 (0.138)	0.00508 (0.0089)
6. (3, 1, 2)	7.864	233	0.0338	− 0.384 (0.0378)	− 0.399 (0.0365)	0.607 (0.0337)	− 0.959 (0.0198)	− 0.986 (0.0061)		0.0185 (0.0347)

Table 6.5 *Estimated FEs for* $\ln M_1$*, 1953–1972*

Model (p, d, q)	RSS, residual sum of squares	DF	RSS/DF	Estimates of the AR part			Estimates of the MA part			Constant
				AR1	AR2	AR3	MA1	MA2	MA3	
1. (3, 1, 0)	0.001778	235	0.00000757	0.240 (0.0641)	0.204 (0.0647)	0.235 (0.0642)				0.00095 (0.00027)
2. (1, 0, 0)	0.023391	238	0.00009828	0.966 (0.00097)						0.173 (0.005)
3. (2, 1, 2)	0.001788	234	0.00000764	1.193 (0.484)	−0.318 (0.425)		0.976 (0.471)	0.353 (0.288)		0.000374 (0.00023)
4. (2, 1, 3)	0.001776	233	0.00000762	0.508 (0.546)	0.244 (0.481)		0.275 (0.542)	0.113 (0.354)	− 0.151 (0.0815)	0.00074 (0.00036)
5. (3, 1, 2)	0.001776	233	0.00000762	0.224 (0.451)	0.289 (0.397)	0.207 (0.098)	−0.0107 (0.455)	0.091 (0.309)		0.00084 (0.00041)
6. (3, 1, 3)	0.001711	232	0.00000737	− 0.273 (0.0356)	0.080 (0.033)	0.905 (0.024)	− 0.549 (0.0513)	−0.277 (0.059)	0.601 (0.031)	0.00087 (0.0003)

Table 6.6 *Estimated FEs for the* ln *of personal income, 1953–1972*

Model (p, d, q)	RSS, residual sum of squares	DF	RSS/DF	Estimates of the AR part: AR1	AR2	AR3	Estimates of the MA part: MA1	MA2	MA3	Constant
1. (0, 1, 4)	0.0058	237	0.00002						− 0.144 (0.0656)	0.0052 (0.0004)
2. (1, 0, 0)	0.0605	238	0.00025	0.971 (0.00098)						0.179 (0.0063)

Table 6.7 *Estimated FEs for the* ln *of high-powered money, 1953–1972*

Model (p, d, q)	RSS, residual sum of squares	DF	RSS/DF	Estimates of the AR part: AR1	AR2	AR3	Estimates of the MA part: MA1	MA2	MA3	Constant
1. (0, 1, 0)	0.007759	238	0.0000326							0.0025 (0.00037)
2. (1, 1, 0)	0.007738	237	0.0000326	0.0522 (0.0645)						0.0024 (0.0004)
3. (0, 1, 1)	0.007737	237	0.0000326				− 0.054 (0.0645)			0.0025 (0.00039)
4. (1, 1, 1)	0.007700	236	0.0000326	− 0.615 (0.403)			0.683 (0.372)			0.004 (0.0012)

With respect to $\ln y_t$, the logarithm of the monthly index of industrial production, BJ analysis and likelihood ratio tests both support a (1, 1, 0) model.[10]

Last, various processes for the logarithm of the monthly consumer price index, $\ln P_t$, appear better supported by the information in the data than the (2, 1, 0) model suggested by BJ techniques.[11] Since it does not seem possible to discriminate well among more complicated variants, we shall choose the simplest, a (2, 1, 1) model.

In summary, the results of our analyses using monthly data, 1953–72, suggest tentatively the findings reported in table 6.13 with regard to the processes that probably generated the observations on our variables. Also shown in table 6.13 are results using data for 1953–62 and 1963–72 subperiods that indicate little change in the orders of the lag polynomials for the processes considered.

Viewing the results in table 6.13 for the 1953–72 period in terms of the implications of S^0 for the FEs set forth in table 6.1, the finding for $\ln H_t$, a (0, 1, 0) process, indicates that $p = 0$ and $q = 0$. Given that $p = 0$,

[10] See figure 6.3 for estimates of the autocorrelation and partial autocorrelation functions for $\Delta \ln y_t$.

[11] See figure 6.3 for estimates of the autocorrelation and partial autocorrelation functions for $\Delta \ln P_t$.

Table 6.8 *Estimated FEs for the* ln *of industrial production index, 1953–1972*

Model (p, d, q)	RSS, residual sum of squares	DF	RSS/DF	Estimates of the AR part			Estimates of the MA part			Constant
				AR1	AR2	AR3	MA1	MA2	MA3	
1. (1, 0, 0)	0.0566	226	0.000250	0.968 (0.0016)						0.138 (0.007)
2. (1, 1, 2)	0.024234	223	0.000108	0.479 (0.333)			0.0999 (0.340)	0.0149 (0.148)		0.0019 (0.001)
3. (1, 1, 1)	0.024233	224	0.000108	0.4497 (0.153)			0.0710 (0.170)			0.002 (0.00085)
4. (2, 1, 1)	0.024223	223	0.000103	0.556 (2.377)	− 0.042 (0.935)		0.177 (2.372)			0.0018 (0.0054)
5. (1, 1, 0)	0.024250	225	0.000108	0.389 (0.061)						0.0023 (0.00072)

Table 6.9 *Estimated FEs for the* ln *of consumer price index, 1953–1972*

model (p, d, q)	RSS, residual sum of squares	DF	RSS/DF	Estimates of the AR part			Estimates of the MA part			Constant
				AR1	AR2	AR3	MA1	MA2	MA3	
1. (1, 0, 0)	0.01298	238	0.000054	0.961 (0.001)						0.178 (0.0048)
2. (2, 1, 0)	0.00097	236	0.000004	0.243 (0.0604)	0.308 (0.0604)					0.00087 (0.000185)
3. (2, 1, 1)	0.00090	235	0.00000385	0.964 (0.0493)	0.0135 (0.0486)		0.844 (0.0442)			0.00004 (0.000036)
4. (2, 1, 2)	0.000899	234	0.00000384	0.156 (0.241)	0.802 (0.233)		0.0618 (0.268)	0.652 (0.227)		0.000081 (0.000067)
5. (3, 1, 2)	0.000895	233	0.00000384	0.489 (0.501)	0.577 (0.443)	−0.094 (0.080)	0.359 (0.502)	0.439 (0.409)		0.000053 (0.000054)
6. (3, 1, 1)	0.000898	234	0.00000384	0.987 (0.0369)	0.0837 (0.082)	−0.088 (0.0680)	0.871 (0.042)			0.000033 (0.000029)

Table 6.10 *Results of the large-sample likelihood ratio test applied to FEs, 1953–1972*

Models compared		$\lambda = \frac{L(X \mid H_1)}{L(X \mid H_0)}$	$2 \ln \lambda$	r	Critical points for χ_r^2		
					$\alpha = 0.05$	$\alpha = 0.10$	$\alpha = 0.20$
1. *Market rates* 3*MTB*							
H_0 : (1, 1, 1) vs.	H_1 : (2, 1, 2)	2.04665	1.43241	2	5.99	4.60	3.22
(1, 1, 1)	(2, 1, 3)	9104.01	18.2329	3	7.82	6.25	4.64
(1, 1, 1)	(3, 1, 3)	13599.7	19.0356	4	9.49	7.78	5.99
(1, 1, 1)	(3, 1, 2)	8.76987	4.34265	3	7.82	6.25	4.64
(2, 1, 2)	(2, 1, 3)	4448.33	16.8006	1	3.84	2.71	1.64
(2, 1, 2)	(3, 1, 3)	6644.64	17.6031	2	5.99	4.60	3.22
(2, 1, 2)	(3, 1, 2)	4.28479	2.91014	1	3.84	2.71	1.64
(2, 1, 3)	(3, 1, 3)	1.49368	0.80249	1	"	"	"
(3, 1, 2)	(3, 1, 3)	1550.68	14.6929	1	"	"	"
2. *Interest rates* 4–6MPCP							
H_0 : (1, 1, 1) vs.	H_1 : (2, 1, 2)	3.81787	2.67938	2	5.99	4.60	3.22
(1, 1, 1)	(2, 1, 3)	2139.45	15.3366	3	7.82	6.25	4.64
(1, 1, 1)	(3, 1, 3)	2139.45	15.3366	4	9.49	7.78	5.99
(1, 1, 1)	(3, 1, 2)	$3540. \times 10^3$	30.1593	3	7.82	6.25	4.64
(2, 1, 2)	(2, 1, 3)	560.350	12.6571	1	3.84	2.71	1.64
(2, 1, 2)	(3, 1, 3)	560.350	12.6571	2	5.99	4.60	3.22
(2, 1, 2)	(3, 1, 2)	$9272. \times 10^2$	27.4799	1	3.84	2.71	1.64
(2, 1, 3)	(3, 1, 3)	1	0	1	"	"	"
(3, 1, 2)	(3, 1, 3)	$\lambda < 1$	–	1	"	"	"

Table 6.11 *Results of the large-sample likelihood ratio test applied to FEs, 1953–1972*

Models compared		$\lambda = \frac{L(X \mid H_1)}{L(X \mid H_0)}$	$2 \ln \lambda$	r	Critical points for χ_r^2		
					$\alpha = 0.05$	$\alpha = 0.10$	$\alpha = 0.20$
3. *Currency and demand deposits* : M_1							
H_0 : (3, 1, 0) vs.	H_1 : (3, 1, 2)	1.14450	0.269929	2	5.99	4.60	3.22
(3, 1, 0)	(3, 1, 3)	100.411	9.21855	3	7.82	6.25	4.64
(2, 1, 2)	(2, 1, 3)	2.24336	1.61595	1	3.84	2.71	1.64
(2, 1, 2)	(3, 1, 2)	2.24336	1.61595	1	"	"	"
(2, 1, 2)	(3, 1, 3)	196.813	10.5645	2	5.99	4.60	3.22
(3, 1, 2)	(3, 1, 3)	87.7229	8.34837	1	5.99	4.60	3.22
(2, 1, 3)	(3, 1, 3)	87.7229	8.94837	1	3.84	2.71	1.64
4. *Consumer price index*							
H_0 : (2, 1, 0) vs.	H_1 : (2, 1, 1)	8006.92	17.9761	1	3.84	2.71	1.64
(2, 1, 0)	(2, 1, 2)	9150.49	18.2431	2	5.99	4.60	3.22
(2, 1, 0)	(3, 1, 2)	15625.1	19.3133	3	7.82	6.25	4.64
(2, 1, 0)	(3, 1, 1)	10457.0	18.5100	2	5.99	4.60	3.22
(2, 1, 1)	(2, 1, 2)	1.14267	0.266728	1	3.84	2.71	1.64
(2, 1, 1)	(3, 1, 2)	1.95114	1.33683	2	5.99	4.60	3.22
(2, 1, 1)	(3, 1, 1)	1.30594	0.533844	1	3.84	2.71	1.64
(2, 1, 2)	(3, 1, 2)	1.70758	1.07015	1	"	"	"
(3, 1, 1)	(3, 1, 2)	1.	0	1	"	"	"

Table 6.12 *Results of the large-sample likelihood ratio test applied to FEs, 1953–1972*

Models compared		$\lambda = \frac{L(X\|H_1)}{L(X\|H_0)}$	2 ln λ	r	Critical points for χ_r^2 $\alpha = 0.05$	$\alpha = 0.10$	$\alpha = 0.20$
5. *High-powered money*							
H_0 : (0, 1, 0) vs.	H_1 : (1, 1, 0)	1.38423	0.650288	1	3.84	2.71	1.64
(0, 1, 0)	(0, 1, 1)	1.40588	0.681328	1	"	"	"
(0, 1, 0)	(1, 1, 1)	2.49906	1.83183	2	5.99	4.60	3.22
(1, 1, 0)	(1, 1, 1)	1.80518	1.18132	1	3.84	2.71	1.64
(0, 1, 1)	(1, 1, 1)	1.77744	1.15035	1	"	"	"
6. *Industrial production index*							
H_0 : (1, 1, 0) vs.	H_1 : (1, 1, 2)	1.08239	0.158334	2	5.99	4.60	3.22
(1, 1, 0)	(1, 1, 1)	1.08772	0.168169	1	3.84	2.71	1.64
(1, 1, 0)	(2, 1, 1)	1.14293	0.267185	2	5.99	4.60	3.22
(1, 1, 1)	(2, 1, 1)	1	0	1	3.84	2.71	1.64

Table 6.13 *Models suggested by large-sample likelihood ratio tests and estimation results*

Variable	Period of analysis 1953–1972	1953–1962	1963–1972
i_t : (3MTB)	(2, 1, 3) or (3, 1, 3)	(2, 1, 3) or (3, 1, 3)	(2, 1, 3) or (3, 1, 3)
i_t : (4-6PCP)	(2, 1, 3) or (3, 1, 2)	(2, 1, 3) or (3, 1, 2)	(2, 1, 3) or (3, 1, 2)
$\ln M_{1t}$	(3, 1, 3)	(2, 1, 2) or (3, 1, 3)	(2, 1, 3) or (3, 1, 3)
$\ln Y_t$	(0, 1, 4)	(0, 1, 4)	(0, 1, 4)
$\ln H_t$	(0, 1, 0)	(0, 1, 0) or (2, 1, 0)	(0, 1, 0)
$\ln y_t$	(1, 1, 0)	(1, 1, 0)	(1, 1, 0) or (1, 1, 1)
$\ln P_t$	(2, 1, 1) or (1, 1, 1)	(2, 1, 0) or (1, 1, 1)	(2, 1, 1) or (1, 1, 1)

model S^0 implies that the AR parts of the FEs for i_t, $\ln M_{1t}$ and $\ln Y_t$ should all be second order (see table 6.1). The findings reported in table 6.13 contradict this implication in that the orders of the AR parts of the FEs for i_t (3MTB) and for $\ln M_{1t}$ are 3 while that for $\ln Y_t$ is 0. Thus some major implications of S^0 are apparently in conflict with the information in the data and there is a need to consider variants of the initial model.[12]

[12] If instead of expressing S^0 in terms of the growth rate of nominal income, we expressed it in terms of the rate of growth of the price level, r_{p_t}, and if we continue to assume that the growth rate of real income or output is constant, this would be equivalent to substituting r_{p_t} and $r^*_{p_t}$ for r_{Y_t} and $r^*_{Y_t}$, respectively, where $r^*_{p_t}$ is the anticipated rate of inflation. Then the empirical result for $\ln P_t$ in table 6.13 is compatible with the implication of this variant of S^0 but there is still an incompatibility with respect to the empirical findings relating to the FE for $\ln M_{1t}$.

3 Formulation and analysis of variants of the initial model S^0

The findings reported in table 6.13 indicate a need to reformulate S^0. In particular, the finding that real output, $\ln y_t$, follows a (1, 1, 0) process may mean that the approximation of a constant growth rate for real output embedded in S^0 may be inadequate. Reformulation of S^0 to permit $r_{y_t} = \Delta \ln y_t$ to be variable along with a relaxation of the assumption that the income elasticity of demand for real balances is equal to 1 leads to a variant of S^0 that we shall refer to as S^1. The equations of S^1 are given in (3.1a–3.1d):

$$r_{p_t} + \alpha r_{y_t} = \gamma_1 \Delta i_t + r_{M_t} \tag{3.1a}$$

$$r_{M_t} = \gamma_2 \Delta i_t + r_{H_t} \tag{3.1b}$$

$$\Delta i_t = \Delta r^*_{p_t} \tag{3.1c}$$

$$r^*_{p_t} - r^*_{p_{t-1}} = \beta (r_{p_{t-1}} - r^*_{p_{t-1}}), \tag{3.1d}$$

where variables are defined as above in connection with the development of (2.6)–(2.9), $r_{y_t} = \Delta \ln y_t$, and α is the income elasticity of demand for real balances.

On expressing the unobservable anticipated rate of inflation $r^*_{p_t}$ in terms of observables the system S^1 in (3.1) can be expressed as follows:

$$\begin{bmatrix} 1 & -1 & -\gamma_1 \\ 0 & 1 & -\gamma_2 \\ -\beta L(1-L) & 0 & 1-(1-\beta)L \end{bmatrix} \begin{bmatrix} r_{p_t} \\ r_{M_t} \\ \Delta i_t \end{bmatrix} = \begin{bmatrix} 0 & -\alpha \\ 1 & 0 \\ 0 & 0 \end{bmatrix} \begin{bmatrix} r_{H_t} \\ r_{y_t} \end{bmatrix} + \begin{bmatrix} u_{1t} \\ u_{2t} \\ u_{3t} \end{bmatrix}, \tag{3.2}$$

where u_{it}, $i = 1, 2, 3$, are disturbance terms. Initially, we shall go forward under the assumption that r_{H_t} and r_{y_t} are exogenous variables.

Solving the system in (3.2) for r_{p_t}, r_{M_t}, and Δi_t, as in (2.14) above, we obtain the TFs associated with S^1 in (3.2):

$$[1-(1-\beta)L + \beta(\gamma_1+\gamma_2)L(1-L)] \begin{bmatrix} r_{p_t} \\ r_{M_t} \\ \Delta i_t \end{bmatrix} = \begin{bmatrix} 1-(1-\beta)L & -\alpha[1-(1-\beta)L] \\ 1-(1-\beta)L - \gamma_1\beta L(1-L) & -\alpha\gamma_2\beta L(1-L) \\ \beta L(1-L) & -\alpha\beta L(1-L) \end{bmatrix} \times \begin{bmatrix} r_{H_t} \\ r_{y_t} \end{bmatrix} + \begin{bmatrix} v_{1t} \\ v_{2t} \\ v_{3t} \end{bmatrix}, \tag{3.3}$$

Table 6.14 *Degrees of lag polynomials in (3.3)*

Variable	AR	MA for r_{H_t}	MA for r_{y_t}	MA for error[a]
r_{p_t}	2	1	1	$1\leq$
r_{M_t}	2	2	2	$2\leq$
Δi_t	2	2	2	$2\leq$

Note:
[a] "$1 \leq$" indicates that the order of the MA error process is at least 1 or greater and similarly for "$2 \leq$."

where $\boldsymbol{v}_t = H_{11}^* \boldsymbol{u}_t$, $\boldsymbol{v}_t' = (v_{1t}, v_{2t}, v_{3t})$, H_{11}^* is given in (2.13), and $\boldsymbol{u}_t' = (u_{1t}, u_{2t} u_{3t})$ are the structural disturbances in (3.2).

Notice that the AR parts of the TFs for S^1 in (3.3) are identical to those for S^0 in (2.14a) and thus should be of order 2 with identical parameters. If $\alpha = 1$, the MA polynomials for r_{H_t} and r_{y_t} in the first and third equations sum to 0. With $\alpha \neq 1$, this restriction is no longer satisfied. Further, the MA polynomial for r_{y_t} in the second equation is proportional to that for r_{y_t} in the third equation.[13] Table 6.14 provides a summary of degrees of the various polynomials appearing in (3.3). It is seen from the table that the degrees of the polynomials are rather low and the same in a number of instances, a point that can be checked empirically.

If we assume that the exogenous variables are generated independently of the other variables by the following ARMA schemes,

$$\phi_1(L) r_{H_t} = \theta_1(L) u_{4t} \tag{3.4}$$

$$\phi_2(L) r_{y_t} = \theta_2(L) u_{5t}, \tag{3.5}$$

where $\phi_i(L)$ and $\theta_i(L)$, $i = 1, 2$, are finite polynomials in L of degrees p_i and q_i, respectively, we can solve for the FEs of S^1 by substituting from (3.4)–(3.5) in (3.3). Note, however, that this involves assuming that the real growth rate, r_{y_t}, is exogenous to the monetary sector and that policy-makers, assumed to control r_{H_t}, the rate of growth of high-powered money, have adopted an "open-loop" control strategy.[14] With these qualifications in mind, we present the FEs of S^1 in (3.6):

[13] There are other restrictions on the parameters of (3.3) similar to those discussed in connection with (2.14) above.

[14] The latter assumption excludes possible feedback effects on the policy-makers' actions. See Sargent (1973) and Sargent and Wallace (1973) where such feedback effects are considered.

Table 6.15 *Degrees of lag polynomials in FEs for variables in (3.6)*

Variable	Degree of AR polynomial	Degree of MA polynomial
r_{P_t}	$2 + p_1 + p_2$	$1 + p_1 + p_2,\ 1 + p_2 + q_1$ and $1 + p_1 + q_2 \leq$
r_{M_t}	$2 + p_1 + p_2$	$2 + p_1 + p_2,\ 2 + p_2 + q_1$ and $2 + p_1 + q_2 \leq$
Δi_t	$2 + p_1 + p_2$	$2 + p_1 + p_2,\ 2 + p_2 + q_1$ and $2 + p_1 + q_2 \leq$

$$
\begin{aligned}
&[1-(1-\beta)L+\beta\gamma L(1-L)]\phi_1(L)\phi_2(L)\begin{bmatrix} r_{P_t} \\ r_{M_t} \\ \Delta i_t \end{bmatrix} \\
&= \begin{bmatrix} 1-(1-\beta)L & -\alpha[1-(1-\beta)L] \\ 1-(1-\beta)L-\gamma_1\beta L(1-L) & -\alpha\gamma_2\beta L(1-L) \\ \beta L(1-L) & -\alpha\beta L(1-L) \end{bmatrix} \\
&\times \begin{bmatrix} \phi_2(L)\theta_1(L)u_{4t} \\ \phi_1(L)\theta_2(L)u_{5t} \end{bmatrix} + \phi_1(L)\phi_2(L)\begin{bmatrix} v_{1t} \\ v_{2t} \\ v_{3t} \end{bmatrix}, \qquad (3.6)
\end{aligned}
$$

where $\gamma \equiv \gamma_1 + \gamma_2$. If no cancelling occurs, the degrees of the polynomials in the ARMA schemes given in (3.6) are as shown in table 6.15.

With respect to table 6.15, the empirical results for the FEs in table 6.13 indicate that $p_1 = q_1 = 0$, $p_2 = 1$, and $q_2 = 0$. Thus, if we retain a (3, 1, 3) or a (3, 1, 2) model for i_t and a (3, 1, 3) model for $\ln M_{1t}$, we have compatibility with the requirements set forth in table 6.15. However if no cancelling occurs, there is an incompatibility with respect to the degrees of the AR and MA polynomials in the empirically determined process for $\ln P_t$, namely (2, 1, 1). In addition, on viewing the estimates of the AR parameters in the (3, 1, 3) FE for $\ln M_{1t}$ and those for the (3, 1, 3) FEs for i_t, it is seen that they are far from being identical as required by the form of (3.6). Thus there appear to be fundamental problems with the S^1 formulation of the model.

In considering reformulation of S^1, the following points were considered:

(1) α in the money demand function (3.1a) could be a polynomial in L, $\alpha(L)$. This would bring current and lagged values of r_{y_t} into (3.1a). However, such a modification of S^1 leads to implications for the FEs that are incompatible with the empirical findings reported in table 6.13. This is also the case if other lagged effects were incorporated in the structural equations (3.1).

(2) The rate of growth of real output, r_{y_t}, considered to be exogenous in S^1, might be an endogenous variable. However, this assumption

leads to implications for the FEs at variance with the empirical results in table 6.13.

(3) The rate of growth of high-powered money, r_{H_t}, might be subject to closed loop control, that is dependent on current and/or lagged endogenous variables. This assumption has implications for the FEs properties that are not in agreement with the empirical findings for the FEs.

(4) Some cancelling may occur in the FE for r_{p_t} in (3.6). For example, if $\phi_2(L)$ were proportional to $1-(1-\beta)L$, cancelling would occur and (3.6) would be compatible with the empirical results insofar as the degrees of AR and MA polynomials are considered. However, there does not appear to be any obvious rationale for assuming that the AR polynomial in the FE for r_{y_t}, namely $\phi_2(L)$ in (3.5), is proportional to $1-(1-\beta)L$. Further, even if this were assumed, there is still the problem that estimates of corresponding parameters in the 3rd degree AR polynomials in the FEs for r_{y_t} and Δi_t appear to be quite dissimilar.

(5) It may be that use of Δi_t in (3.1b), the money supply relation, and/or the Cagan expectation formation equation, (3.1d), are inappropriate. Below we consider these possibilities.

As mentioned in (3.5), problems with the S^1 formulation of the model may be due to inadequacies in formulating the money supply relation (3.1b) and the expectation equation (3.1d). As the model is shown in (3.1), expectations are not rational in the sense of Muth (1961) . . . Sargent and Wallace (1973) have provided a model, similar in some respects to those considered above, within which the Cagan expectation formation process is rational. A variant of the Sargent–Wallace model, denoted S^2, will now be considered.

The equations of the Sargent–Wallace model, employing our notation with u_{1t} and u_{2t} structural disturbance terms, are:

$$r_{M_t} = r_{p_t} + \alpha_1(r^*_{p_t} - r^*_{p_{t-1}}) + (1-L)u_{1t} \quad \alpha_1 < 0 \tag{3.7}$$

$$r^*_{p_t} = \frac{(1-\lambda)}{1-\lambda L} r_{p_t} \tag{3.8}$$

$$r_{M_t} = \frac{1-\lambda}{1-\lambda L} r_{p_t} + u_{2t}. \tag{3.9}$$

Equation (3.7) is obtained by differencing a log-log demand equation for real balances with real income assumed constant. Sargent employs $r^*_{p_t} - r^*_{p_{t-1}}$ rather than our Δi_t; however, given that the Fisher equation (2.4a) holds with a constant real rate of interest,[15] this formulation is

[15] Alternatively, we could assume that the real rate is variable and follows a random walk.

equivalent to ours in (3.1a). Equation (3.8) represents Cagan's expectation formation process and is equivalent to our (3.1d) except that our equation involves using $r_{p_{t-1}}$ rather than r_{p_t} in (3.8). Equation (3.9) is a money supply equation that differs from ours in that r_{M_t} is made to depend on $r^*_{p_t} = \frac{1-\lambda}{1-\lambda L} r_{p_t}$, the anticipated rate of inflation that is assumed generated by (3.8) for the monetary authorities as well as for the demanders of money balances. If $\ln H_t$ is assumed exogenous and follows a (0, 1, 0) process, as appears to be the case in our data, $r_{H_t} = \Delta \ln H_t = \alpha_0 + e_t$ could be considered to be included in u_{2t}. Thus the major modifications embedded in Sargent's model *vis-à-vis* our S^0 is a timing change in the expectation equation and the inclusion of $r^*_{p_t}$ in the rate of change of money supply.

On substituting from (3.8) and (3.9) in (3.7), the FE for r_{p_t} is obtained which when substituted in (3.9) yields the FE for r_{M_t}.[16] Given that u_{1t} and u_{2t} are non-autocorrelated, these FEs are a (1, 1, 2) process for $\ln P_t$ and a (1, 1, 1) process for $\ln M_t$. Both of these processes are in conflict with the empirical findings reported in table 6.13 above, namely a (1, 1, 1) or a (2, 1, 2) process for $\ln P_t$ and a (3, 1, 3) process for $\ln M_{1t}$. Thus the "rational" Cagan model in (3.7)–(3.9) does not appear to be consistent with the information in our data. Even if we relax the assumption that real income is constant in (3.7)–(3.9) by introducing r_{y_t} as a variable in (3.7), we still find the model incompatible with our empirical results.

The last variant of the model to be considered, S^3, involves introducing a rational expectations relation to replace Cagan's [1956] assumption. In this formulation, our equations are:

$$\phi_0 r_{p_t} + \phi_1 r_{y_t} = \gamma_1 \Delta i_t + \phi_4 r_{M_t} + u_{1t}, \tag{3.10}$$

$$\phi_2 r_{M_t} = \phi_3 r_{p_t} + \gamma_2 \Delta^i_t + \delta r_{H_t} + u_{2t}, \tag{3.11}$$

$$\Delta i_t = \Delta r^*_{p_t} + u_{3t}, \tag{3.12}$$

$$r^*_{p_t} = \underset{t}{E}(r_{p_t} \mid .), \tag{3.13}$$

where the ϕ_is are polynomial lag operators, (3.10) is the money demand equation, (3.11) the money supply equation, (3.12) the Fisher equation,[17] and (3.13) the rational hypothesis where $\underset{t}{E}(r_{p_t} \mid .)$ denotes the

[16] The FEs are given explicitly below:

$$\phi(L) r_{p_t} = (1 - \lambda L) u_{2t} - (1 - \lambda L)(1 - L) u_{1t}$$

$$\phi(L) r_{M_t} = [1 - \lambda + \phi(L)] u_{2t} - (1 - \lambda)(1 - L) u_{1t},$$

where $\phi(L) \equiv \lambda + \alpha_1(1 - \lambda) - [\lambda + \alpha_1(1 - \lambda)]L$. Note that the factor $1 - \lambda L$ has been canceled in obtaining these FEs.

[17] If the real rate of interest is variable, its first difference can be assumed incorporated in the structural disturbance u_{3t}.

conditional expectation of r_{p_t} as of time t given the equations of the model and past information.

On substituting for the endogenous variables Δi_t and r_{M_t} in (3.10), we have

$$\begin{aligned} r_{p_t} &= \phi_0^{-1}[-\phi_1 r_{y_t} + \gamma_1 \Delta i_t + \phi_4 r_{M_t} + u_{1t}] \\ &= \phi^{-1}\big[-\phi_1 r_{y_t} + (\gamma_1 + \phi_4\phi_2^{-1}\gamma_2)\Delta r^*_{p_t} + \phi_4\phi_2^{-1}\phi_3 r_{p_t} \\ &\quad + \phi_4\phi_2^{-1}\delta r_{H_t}\big] + \phi_0^{-1}\big[u_{1t} + \phi_4\phi_2^{-1}u_{2t} + (\gamma_1 + \phi_4\phi_2^{-1}\gamma_2)u_{3t}\big]. \end{aligned} \tag{3.14}$$

Then, under the simplifying assumption that ϕ_0, ϕ_1, ϕ_3, γ_1, and γ_2 are constant parameters, ϕ_2 and ϕ_4 are polynomials of the same degree with the same roots we have:

$$\begin{aligned} \underset{t}{E}(r_{p_t} \mid .) &= \phi_0^{-1}\big[-\phi_1 E r_{y_t} + (\gamma_1 + \phi_4\phi_2^{-1}\gamma_2)\Delta r^*_{p_t} \\ &\quad + \phi_4\phi_2^{-1}\phi_3 E r_{p_t} + \phi_4\phi_2^{-1}\delta E r_{H_t}\big] \\ &\quad + \phi_0^{-1}\big[E u_{1t} + \phi_4\phi_2^{-1} E u_{2t} + (\gamma_1 + \phi_4\phi_2^{-1}\gamma_2) E u_{3t}\big], \end{aligned} \tag{3.15}$$

where all expectations on the r.h.s. of (3.15) denote conditional expectations at time t given past information. On multiplying both sides by $\phi_0\phi_2$, using (3.13), and rearranging terms, we have:

$$\begin{aligned} &[\phi_0\phi_2 - \phi_2\gamma_1\Delta - \phi_3\phi_4 - \phi_4\gamma_2\Delta]\Delta r^*_{p_t} \\ &\quad = -\phi_1\phi_2\Delta E r_{y_t} + \phi_4\delta\Delta E r_{H_t} + v_t, \end{aligned} \tag{3.16}$$

where

$$v_t = \phi_2 \Delta E u_{1t} + \phi_4 \Delta E u_{2t} + (\phi_2\gamma_1 + \phi_4\gamma_2)\Delta E u_{3t}.$$

Now if the variables r_{H_t} and r_{y_t} are generated by the following ARMA processes,

$$\begin{aligned} \theta_1(L) r_{H_t} &= w_1(L) u_{4t} \\ \theta_2(L) r_{y_t} &= w_2(L) u_{5t}, \end{aligned} \tag{3.17}$$

the conditional expectations of r_{H_t} and r_{y_t} can be written as

$$\begin{aligned} E r_{H_t} &= -\theta_1'(L) r_{H_t} + w_1'(L) u_{4t} \\ E r_{y_t} &= -\theta_2'(L) r_{y_t} + w_2'(L) u_{5t}, \end{aligned} \tag{3.18}$$

where $\theta_i'(L)$ and $w_i'(L)$ are the homogeneous parts of $\theta_i(L)$ and $w_i(L)$, respectively, $i = 1, 2$.

Substituting from (3.18) in (3.16) and using (3.12), we have

$$\phi(L)\Delta i_t = \psi_1(L)r_{H_t} + \psi_2(L)r_{y_t} + v'_t, \tag{3.19}$$

where

$$\begin{aligned}
\phi(L) &= (\phi_0\phi_2 - \phi_2\gamma_1\Delta - \phi_3\phi_4 - \phi_4\gamma_2\Delta)\\
\psi_1(L) &= -\phi_4\Delta\delta\theta'_1(L),\\
\psi_2(L) &= \Delta\phi_1\phi_2\theta'_2(L),
\end{aligned}$$

and

$$v'_t = v_t - \phi_1\phi_2\Delta w'_2 u_{5t} + \phi_4\delta\Delta w'_1 u_{4t} + \phi(L)u_{3t}.$$

In matrix form the system in (3.10)–(3.13) becomes

$$\begin{bmatrix} \phi_0 & -\phi_4 & -\gamma_1 \\ -\phi_3 & \phi_2 & -\gamma_2 \\ 0 & 0 & \phi \end{bmatrix}\begin{bmatrix} r_{p_t} \\ r_{M_t} \\ \Delta i_t \end{bmatrix} = \begin{bmatrix} 0 & -\phi_1 \\ \Delta & 0 \\ \psi_1 & \psi_2 \end{bmatrix}\begin{bmatrix} r_{H_t} \\ r_{y_t} \end{bmatrix} + \begin{bmatrix} u_{1t} \\ u_{2t} \\ v'_t \end{bmatrix}. \tag{3.20}$$

The determinant of the matrix on the l.h.s. of (3.20), denoted det H_{11}, is

$$\det H_{11} = \phi(\phi_0\phi_2 - \phi_3\phi_4), \tag{3.21}$$

while its adjoint matrix, H^*_{11}, is:

$$H^*_{11} = \begin{bmatrix} \phi_2\phi & \phi_4\phi & \gamma_1\phi_2 + \gamma_2\phi_4 \\ \phi_3\phi & \phi_0\phi & \gamma_1\phi_3 + \gamma_2\phi_0 \\ 0 & 0 & \phi_0\phi_2 - \phi_3\phi_4 \end{bmatrix}. \tag{3.22}$$

Then the transfer functions associated with (3.21) are

$$\begin{aligned}
\phi(\phi_0\phi_2 - \phi_3\phi_4)\begin{bmatrix} r_{p_t} \\ r_{M_t} \\ \Delta i_t \end{bmatrix} &= \begin{bmatrix} \delta\phi_4\phi + \psi_1(\gamma_1\phi_2 + \gamma_2\phi_4) \\ \delta\phi_0\phi + \psi_1(\gamma_1\phi_3 + \gamma_2\phi_0) \\ \psi_1(\phi_0\phi_2 - \phi_3\phi_4) \end{bmatrix} r_{H_t} \\
&\quad + \begin{bmatrix} -\phi_1\phi_2\phi + \psi_2(\gamma_1\phi_2 + \gamma_2\phi_4) \\ -\phi_1\phi_3\phi + \psi_2(\gamma_1\phi_3 + \gamma_2\phi_0) \\ \psi_2(\phi_0\phi_2 - \phi_3\phi_4) \end{bmatrix} r_{y_t} \\
&\quad + \begin{bmatrix} \phi_2\phi & \phi_4\phi & \gamma_1\phi_2 + \gamma_2\phi_4 \\ \phi_3\phi & \phi_0\phi & \gamma_1\phi_3 + \gamma_2\phi_0 \\ 0 & 0 & \phi_0\phi_2 - \phi_3\phi_4 \end{bmatrix}\begin{bmatrix} u_{1t} \\ u_{2t} \\ v'_t \end{bmatrix}.
\end{aligned} \tag{3.23}$$

If in (3.23) we substitute for r_{H_t} and r_{y_t} from (3.17), we obtain the final equations for r_{p_t}, r_{M_t}, and Δi_t. The empirical results for the processes on r_{H_t} and r_{y_t} indicate that θ_1 and w_1 in (3.17) are each of degree 0 while

θ_2 is of degree 1 and w_2 is of degree 0. These findings imply that $\psi_1 \equiv 0$ and that the final equations are given explicitly by

$$\phi(\phi_0\phi_2 - \phi_3\phi_4)\theta_2 \begin{bmatrix} r_{p_t} \\ r_{M_t} \\ \Delta i_t \end{bmatrix} = \begin{bmatrix} \delta\phi_4\phi \\ \delta\phi_0\phi \\ 0 \end{bmatrix} \theta_2 u_{4t} + \begin{bmatrix} -\phi_1\phi_2\phi + \psi_2(\gamma_1\phi_2 + \gamma_2\phi_4) \\ -\phi_1\phi_3\phi + \psi_2(\gamma_1\phi_3 + \gamma_2\phi_0) \\ \psi_2(\phi_0\phi_2 - \phi_3\phi_4) \end{bmatrix} u_{5t} + \theta_2 H_{11}^* \begin{pmatrix} u_{1t} \\ u_{2t} \\ v_t' \end{pmatrix}, \tag{3.24}$$

with H_{11}^* given in (3.22).

For the FEs in (3.24) to be compatible with our empirical findings in table 6.13, some cancelling has to occur. In particular, under the simplifying assumption made in (3.15) that ϕ_2 and ϕ_4 have the same degree and

$$\phi_2 \propto \phi_4, \quad \text{or} \quad \phi_4 = \lambda\phi_2, \tag{3.25}$$

we can eliminate the common factor ϕ_2 and (3.24) becomes

$$\begin{bmatrix} \phi'(\phi_0 - \phi_3\lambda)\theta_2 r_{p_t} \\ \phi(\phi_0 - \phi_3\lambda)\theta_2 r_{M_t} \\ \phi'\theta_2 \Delta i_t \end{bmatrix} = \begin{bmatrix} \delta\phi'\lambda \\ \delta\phi_0\phi' \\ 0 \end{bmatrix} \theta_2 u_{4t} + \begin{bmatrix} -\phi_1\phi' + \psi_2'(\gamma_1 + \gamma_2\lambda) \\ -\phi_1\phi_3\phi' + \psi_2'(\gamma_1\phi_3 + \gamma_2\phi_0) \\ \psi_2' \end{bmatrix} u_{5t} + \theta_2 \begin{bmatrix} \phi' & \phi'\lambda & (\gamma_1 + \gamma_2\lambda) \\ \phi_3\phi' & \phi_0\phi' & (\gamma_1\phi_3 + \gamma_2\phi_0) \\ 0 & 0 & 1 \end{bmatrix} \begin{bmatrix} u_{1t} \\ u_{2t} \\ v_t'' \end{bmatrix}, \tag{3.26}$$

where $\phi' = \phi/\phi_2$, $\psi_2' = \psi/\phi_2$ and $v_t'' = v_t'/\phi_2$.

We assume that

$\gamma_1, \gamma_2, \phi_1, \delta, \phi_0$ and ϕ_3 are each of degree 0;
ϕ is of degree 2 and thus ϕ' is of degree 1; and
ψ_2' is of degree 2,

and if the u_{it}s, $i = 1, 2, \ldots, 5$, are serially uncorrelated, $v_t'' = \phi'(L)u_{3t}$, where $\phi'(L)$ is of degree 1. Under these conditions, the FEs in (3.26) have the following properties:

$$\begin{aligned} i_t &: (2, 1, 2) \\ \ln M_t &: (3, 1, 3) \\ \ln P_t &: (2, 1, 2). \end{aligned}$$

On comparing the above properties of the FEs with the empirical findings reported in table 6.13, it is seen that there is a high degree of compatibility.

Explicit forms of the structural equations, (3.10)–(3.13), embodying the conditions given in (3.25) and those below equation (3.26), and the empirically determined final equations for $\ln y_t$ and $\ln H_t$ are given below

$$r_{M_t} = \phi_4^{-1}[\beta_1 r_{P_t} + \alpha r_{y_t} - \gamma_1 \Delta i_t - u_{1t}] \tag{3.10'}$$
$$r_{M_t} = \lambda\phi_4^{-1}[\beta_2 r_{P_t} + \gamma_2 \Delta i_t + \delta r_{H_t} + u_{2t}] \tag{3.11'}$$
$$\Delta i_t = \Delta r^*_{P_t} + u_{3t} \tag{3.12'}$$
$$r^*_{P_t} = \underset{t}{E}(r_{P_t} \mid .) \tag{3.13'}$$
$$r_{y_t} = \alpha_0 + \theta_2^{-1} u_{4t} \tag{3.17a'}$$
$$r_{H_t} = \alpha_0' + u_{5t}, \tag{3.17b'}$$

where we have taken $\phi_1 \equiv \alpha$, $\phi_0 \equiv \beta_1$, $\phi_3 \equiv \beta_2$, $\phi_4 = \lambda\phi_2$ in line with (3.25), and the λ, αs, γs and βs are scalar parameters. It is seen that variables *and* disturbances on the r.h.s. of the money demand equation, (3.10′), and the money supply equation, (3.11′) are "smoothed" by the same polynomial. Of course, it may be that their parameters are slightly different in equations (3.10′) and (3.11′) and that we have not picked up such differences in our empirical analyses.[18] However, the system presented above is compatible with our empirical findings for the FEs.

The next step in our work will involve analysis of the TFs associated with the compatible structural equation system presented above. From (3.23), the TFs are:

$$\begin{aligned} \phi'(\phi_0 - \phi_3\lambda) r_{P_t} &= \delta\lambda\phi' r_{H_t} + [-\phi_1\phi' + \psi_2'(\gamma_1 + \gamma_2\lambda)] r_{y_t} \\ &\quad + \phi' u_{1t} + \phi' u_{2t} + (\gamma_1 + \gamma_2\lambda) v_t'' \end{aligned} \tag{3.27}$$

$$\begin{aligned} \phi'(\phi_0 - \phi_3\lambda) r_{M_t} &= \delta\phi_0\phi' r_{H_t} + [-\phi_1\phi_3\phi' + \psi_2'(\gamma_1\phi_3 + \gamma_2\phi_0)] r_{y_t} \\ &\quad + \phi_3\phi' u_{1t} + \phi_0\phi' u_{2t} + (\gamma_1\phi_3 + \gamma_2\phi_0) v_t'' \end{aligned} \tag{3.28}$$

$$\phi' \Delta i_t = \psi_2' r_{y_t} + v_t''. \tag{3.29}$$

[18] On comparing estimates of the AR parameter in (3.17a) with estimates obtained from other FEs, we find them very similar.

Table 6.16 *Degrees of polynomials in TFs (3.27)–(3.29)*

Variable	AR	MA for r_{H_t}	MA for r_{y_t}	MA for error[a]
r_{p_t}	1	1	2	$1 \leq$
r_{M_t}	2	1	2	$1 \leq$
Δi_t	1	–	2	$1 \leq$

Note:
[a] If the u_{i_t}s, $i = 1, 2, 3$ are serially correlated we will have strict inequalities.

Under the assumptions made in (3.15), the polynomials in the TFs have the degrees shown in table 6.16, implications that will be checked empirically.[19]

Of course the implications of the TFs may or may not be found consistent with the information in our data. For example, if r_{y_t} is not exogenous, this will affect estimates of the TFs based on the assumption that r_{y_t} is an exogenous variable and will probably lead to incompatibilities. Further, if r_{y_t} is an endogenous variable and if the real sector is modeled along with the monetary sector, our empirically determined FEs imply that any such system must be characterized by cancelling of polynomials in order to have its FEs compatible with the empirically determined FEs presented above. These are some of the issues that will receive theoretical and empirical attention in future work.

4 Empirical analysis of the transfer functions of S^3

We now turn to the analysis of the TFs presented in (3.27), (3.28) and (3.29). If model S^3 has generated the data, the results in table 6.16 should empirically be verified. The empirical analysis of the TFs is done along the lines suggested by Box–Jenkins (1970) and by Zellner and Palm (1974). Alternative specifications for the TFs are compared with those implied by the model S^3 using a large-sample likelihood ratio (LR) test. The results of estimating these alternatives are reported in tables 6.17–6.20. Equation M_2 in each of the tables is compatible as far as the order of the polynomials is concerned with S^3 (see table 6.16). The transfer function M_3 is compatible with S^1. Alternative specifications are reported in tables 6.17–6.20 and compared using the LR test in table 6.21. The estimation results deserve some comments. The parameter estimates are in general rather imprecise. This is somewhat disquieting as the sample size is

[19] See Zellner and Palm (1974) for empirical analyses of TFs associated with a small dynamic Keynesian model.

Table 6.17 *Estimated TFs for 3MTB market rate, 1954–1972*

Model	RSS, residual sum of squares	DF	RSS/DF	Estimates of the AR and MA parts of r_{H_t}	Estimates of the AR and MA parts of r_{y_t}	Estimates of the AR and MA parts of the error process	Constant
M1	11.1252	215	0.0523		$\frac{\underset{(1.458)}{5.010} - \underset{(1.721)}{1.154}L + \underset{(1.780)}{2.624}L^2 + \underset{(1.495)}{3.074}L^3}{1 - \underset{(0.068)}{0.621}L + \underset{(0.058)}{0.941}L^2}$	$1 + \underset{(0.068)}{0.408}L + \underset{(0.068)}{0.031}L^2$	$\underset{(0.023)}{-0.009}$
M2	11.1843	219	0.0511		$\frac{\underset{(1.455)}{4.551} - \underset{(1.944)}{1.648}L + \underset{(1.747)}{0.155}L^2}{1 - \underset{(0.130)}{0.818}L}$	$1 - \underset{(0.062)}{0.393}L$	$\underset{(0.031)}{-0.043}$
M3	11.0750	214	0.0518	$\frac{\underset{(2.969)}{0.155} - \underset{(2.914)}{2.130}L + \underset{(3.522)}{3.690}L^2}{1 - \underset{(0.948)}{0.319}L}$	$\frac{\underset{(1.480)}{4.419} - \underset{(1.978)}{1.329}L - \underset{(1.795)}{0.143}L^2}{1 - \underset{(0.137)}{0.821}L}$	$1 - \underset{(0.069)}{0.401}L + \underset{(0.069)}{0.009}L^2$	$\underset{(0.037)}{-0.047}$
M4	11.2141	220	0.0509		$\frac{\underset{(1.117)}{3.871} + \underset{(1.599)}{3.095}L}{1 + \underset{(0.220)}{0.1696}L - \underset{(0.154)}{0.719}L^2}$	$1 + \underset{(0.062)}{0.397}L$	$\underset{(0.028)}{-0.037}$
M5	11.1181	219	0.0510		$\frac{\underset{(1.444)}{4.590} + \underset{(1.860)}{1.622}L}{1 - \underset{(0.107)}{0.824}L}$	$\frac{1 + \underset{(0.170)}{0.363}L}{1 - \underset{(0.158)}{0.035}L}$	$\underset{(0.031)}{-0.042}$
M6	11.11379	219	0.0507	$\underset{(2.922)}{0.064} - \underset{(2.865)}{3.418}L$	$\frac{\underset{(1.442)}{4.488} + \underset{(1.852)}{1.386}L}{1 - \underset{(0.106)}{.816}L}$	$1 + \underset{(0.062)}{0.400}L$	$\underset{(0.032)}{-0.0332}$
M7	11.2686	222	0.0508	$\underset{(2.729)}{1.802}$	$\frac{\underset{(1.076)}{3.638}}{1 - \underset{(0.093)}{0.767}L}$	$1 + \underset{(0.061)}{0.397}L$	$\underset{(0.028)}{-0.044}$
M8	11.2907	223	0.0506		$\frac{\underset{(1.073)}{3.707}}{1 - \underset{(0.092)}{0.765}L}$	$1 + \underset{(0.062)}{0.394}L$	$\underset{(0.027)}{+0.040}$

Table 6.18 *Estimated TFs for four–six-month interest rate on prime commercial paper, 1954–1972*

Model	RSS, residual sum of squares	DF	RSS/DF	Estimates of the AR and MA parts of r_{H_t}	Estimates of the AR and MA parts of r_{y_t}	Estimates of the AR and MA parts of the error process	Constant
M1	7.89821	215	0.03674		$\dfrac{\underset{(1.251)}{3.74} + \underset{(1.294)}{3.488}L - \underset{(1.711)}{0.778}L^2 + \underset{(1.509)}{0.691}L^3}{1+\underset{(0.097)}{0.106}L - \underset{(0.093)}{0.784}L^2}$	$1 + \underset{(0.068)}{0.477}L + \underset{(0.068)}{0.062}L^2$	$-\underset{-(0.0300)}{0.0642}$
M2	8.11412	219	0.03705		$\dfrac{\underset{(1.237)}{3.817} - \underset{(1.570)}{0.075}L - \underset{(1.447)}{0.352}L^2}{1-\underset{(0.066)}{0.853}L}$	$1 - \underset{(0.061)}{0.440}L$	$\underset{(0.030)}{0.068}$
M3	7.93732	213	0.03726	$\underset{(2.495)}{0.244} - \underset{(2.669)}{0.936}L - \underset{(2.577)}{2.426}L^2$	$\dfrac{\underset{(1.267)}{3.772} - \underset{(1.613)}{0.310}L - \underset{(1.474)}{0.026}L^2}{1-\underset{(0.080)}{0.847}L}$	$\dfrac{1-\underset{(0.378)}{0.010}L - \underset{(0.188)}{0.068}L^2}{1-\underset{(0.368)}{0.463}L}$	$-\underset{(0.036)}{0.054}$
M4	7.98409	220	0.03629		$\dfrac{\underset{(0.855)}{3.214} - \underset{(0.850)}{4.206}L}{1+\underset{(0.067)}{0.136}L - \underset{(0.063)}{0.792}L^2}$	$1 + \underset{(0.060)}{0.459}L$	$-\underset{(0.0263)}{0.0611}$
M5	8.00874	219	0.03657		$\dfrac{\underset{(1.233)}{3.814} - \underset{(1.448)}{0.355}L}{1-\underset{(0.064)}{0.848}L}$	$\dfrac{1+\underset{(0.151)}{0.144}L}{1-\underset{(0.145)}{0.321}L}$	$-\underset{(0.033)}{0.066}$
M6	8.11971	219	0.03708	$-\underset{(2.503)}{0.5387} + \underset{(2.455)}{0.134}L$	$\dfrac{\underset{(1.225)}{3.795} - \underset{(1.449)}{0.225}L}{1-\underset{(0.0518)}{0.840}L}$	$1 + \underset{(0.061)}{0.439}L$	$-\underset{(0.029)}{0.062}$
M7	8.13533	222	0.03665	$-\underset{(2.290)}{0.146}$	$\dfrac{\underset{(0.783)}{3.639}}{1-\underset{(0.049)}{0.837}L}$	$1 + \underset{(0.061)}{0.436}L$	$-\underset{(0.027)}{0.064}$
M8	8.13547	223	0.03648		$\dfrac{\underset{(0.778)}{3.635}}{1-\underset{(0.049)}{0.837}L}$	$1 + \underset{(0.061)}{0.436}L$	$-\underset{(0.027)}{0.065}$

Table 6.19 *Estimated TFs for r_{p_t}, 1954–1972*

Model	RSS, residual sum of squares	DF	RSS/DF	Estimates of the AR and MA parts of r_{H_t}	Estimates of the AR and MA parts of r_{y_t}	Estimates of the AR and MA parts of the error process	Constant
M1	0.000741	212	0.00000350	$\frac{\underset{(0.025)}{0.003} + \underset{(0.033)}{0.016}L - \underset{(0.025)}{0.005}L^2}{1-\underset{(0.009)}{0.984}L}$	$\frac{-\underset{(0.012)}{0.030} - \underset{(0.019)}{0.0137}L - \underset{(0.013)}{0.007}L^2 - \underset{(0.013)}{0.025}L^8}{1+\underset{(0.454)}{0.135}L}$	$1 + \underset{(0.069)}{0.103}L + \underset{(0.069)}{0.192}L^2$	$\underset{(0.00045)}{0.00035}$
M2	0.000807	217	0.00000372	$\frac{-\underset{(0.026)}{0.00146} + \underset{(0.027)}{0.0157}L}{1-\underset{(0.007)}{0.984}L}$	$-\underset{(0.013)}{0.030} - \underset{(0.013)}{0.015}L - \underset{(0.012)}{0.009}L^2$	$1 + \underset{(0.070)}{0.074}L$	$\underset{(0.0004)}{0.0003}$
M3	0.000796	220	0.00000362	$\frac{\underset{(0.014)}{0.0048}}{1-\underset{(0.022)}{0.993}L}$	$\frac{\underset{(0.004)}{0.006}}{1-\underset{(0.008)}{0.997}L}$	$1 + \underset{(0.066)}{0.138}L + \underset{(0.066)}{0.202}L^2$	$-\underset{(0.0007)}{0.003}$
M4	0.000794	218	0.00000364	$\frac{-\underset{(0.025)}{0.016} + \underset{(0.025)}{0.029}L}{1-\underset{(0.008)}{0.986}L}$	$\frac{-\underset{(0.012)}{0.029}L}{1+\underset{(0.394)}{0.037}L}$	$1 + \underset{(0.068)}{0.108}L + \underset{(0.068)}{0.181}L^2$	$\underset{(0.0005)}{0.0003}$
M5	0.000739	211	0.00000350	$\frac{-\underset{(0.026)}{0.0074} + \underset{(0.033)}{0.0188}L - \underset{(0.033)}{0.024}L^2 + \underset{(0.026)}{0.028}L^3}{1-\underset{(0.011)}{0.980}L}$	$\frac{-\underset{(0.012)}{0.029} - \underset{(0.019)}{0.014}L - \underset{(0.013)}{0.005}L^2 - \underset{(0.013)}{0.024}L^3}{1+\underset{(0.461)}{0.117}L}$	$1 + \underset{(0.069)}{0.010}L + \underset{(0.069)}{0.186}L^2$	$\underset{(0.00040)}{0.00066}$
M6	0.000796	221	0.00000360	$\frac{\underset{(0.0063)}{0.0132}}{1-\underset{(0.008)}{0.986}L}$	$-\underset{(0.012)}{0.0299}$	$1 + \underset{(0.067)}{0.120}L + \underset{(0.066)}{0.202}L^2$	$\underset{(0.00044)}{0.00016}$
M7	0.000791	217	0.00000365	$\frac{-\underset{(0.025)}{0.009} + \underset{(0.025)}{0.024}L}{1-\underset{(0.009)}{0.982}L}$	$\frac{-\underset{(0.013)}{0.028} - \underset{(0.024)}{0.014}L}{1+\underset{(0.712)}{0.230}L}$	$1 + \underset{(0.068)}{0.106}L + \underset{(0.068)}{0.169}L^2$	$\underset{(0.0004)}{0.0003}$
M8	0.000758	214	0.00000354	$\frac{\underset{(0.025)}{0.0032} + \underset{(0.032)}{0.0085}L + \underset{(0.025)}{0.0028}L^2}{1-\underset{(0.009)}{0.983}L}$	$\frac{-\underset{(0.012)}{0.030} - \underset{(0.015)}{0.037}L - \underset{(0.012)}{0.009}L^2}{1+\underset{(0.025)}{0.946}L}$	$1 + \underset{(0.069)}{0.108}L + \underset{(0.069)}{0.169}L^2$	$\underset{(0.0004)}{0.0003}$

Table 6.20 *Estimated TFs for r_{M_t}, 1954–1972*

Model	RSS	DF	RSS/DF	Estimates of the AR and MA parts of r_{H_t}	Estimates of the AR and MA parts of r_{y_t}	Estimates of the AR and MA parts of the error process	Constant
M1	0.001585	210	0.00000755	$\frac{\underset{(0.036)}{0.071} - \underset{(0.047)}{0.089}L + \underset{(0.037)}{0.047}L^2}{1-\underset{(0.015)}{0.972}}$	$\frac{\underset{(0.018)}{0.008} + \underset{(0.018)}{0.032}L + \underset{(0.028)}{0.020}L^2 + \underset{(0.020)}{0.012}L^3}{1+\underset{(0.728)}{0.244}}$	$\frac{1-\underset{(0.209)}{0.468}L + \underset{(0.215)}{0.103}L^2}{1-\underset{(0.087)}{0.6242}}$	$\underset{(0.000686)}{0.000162}$
M2	0.001663	216	0.00000770	$\frac{\underset{(0.037)}{0.087} - \underset{(0.0375)}{0.0738}L}{1-\underset{(0.004)}{0.994}L}$	$\frac{\underset{(0.0181)}{0.0148} + \underset{(0.0219)}{0.0318}L + \underset{(0.0347)}{0.0080}L^2}{1-\underset{(0.721)}{0.192}L}$	$1+\underset{(0.068)}{0.168}L$	$\underset{(0.0006)}{-0.0011}$
M3	0.001628	214	0.00000761	$\frac{\underset{(0.037)}{0.080} - \underset{(0.046)}{0.106}L + \underset{(0.036)}{0.051}L^2}{1-\underset{(0.011)}{0.977}L}$	$\frac{\underset{(0.018)}{0.012} + \underset{(0.020)}{0.031}L + \underset{(0.032)}{0.012}L^2}{1-\underset{(0.697)}{0.134}L}$	$1+\underset{(0.068)}{0.153}L+\underset{(0.069)}{0.125}L^2$	$\underset{(0.0006)}{0.0003}$
M4	0.001682	218	0.00000771	$\frac{\underset{(0.036)}{0.070} - \underset{(0.036)}{0.044}L}{1-\underset{(0.0096)}{0.979}L}$	$\frac{\underset{(0.016)}{0.029}}{1-\underset{(0.351)}{0.469}L}$	$1+\underset{(0.068)}{0.142}L+\underset{(0.068)}{0.134}L^2$	$\underset{(0.00071)}{0.00024}$
M5	0.001632	211	0.00000773	$\frac{\underset{(0.039)}{0.084} - \underset{(0.047)}{0.098}L + \underset{(0.047)}{0.062}L^2 - \underset{(0.039)}{0.017}L^3}{1-\underset{(0.012)}{0.972}L}$	$\frac{\underset{(0.018)}{0.011} + \underset{(0.019)}{0.037}L + \underset{(0.026)}{0.025}L^2 + \underset{(0.019)^3}{0.013}L^6}{1+\underset{(0.586)}{0.364}L}$	$1+\underset{(0.069)}{157}L+\underset{(0.069)^2}{0.130}L^2$	$\underset{(0.00062)}{0.00003}$
M6	0.001710	219	0.00000781	$\frac{\underset{(0.036)}{0.068} - \underset{(0.035)}{0.038}L}{1-\underset{(0.011)}{0.971}L}$	$\frac{\underset{(0.016)}{0.031}}{1-\underset{(0.352)}{0.430}L}$	$1+\underset{(0.068)}{0.162}L$	$\underset{(0.00054)}{0.00004}$
M7	0.001695	220	0.00000770	$\frac{\underset{(0.0093)}{0.0277}}{1-\underset{(0.011)}{0.974}L}$	$\frac{\underset{(0.016)}{0.030}}{1-\underset{(0.340)}{0.451}L}$	$1+\underset{(0.068)}{0.135}L+\underset{(0.068)}{0.124}L^2$	$\underset{(0.00049)}{0.00027}$
M8	0.001705	221	0.00000772	$\frac{\underset{(0.0099)}{0.0296}}{1-\underset{(0.0119)}{0.970}L}$	$\underset{(0.0175)}{0.0239}$	$1+\underset{(0.068)}{0.139}L+\underset{(0.068)}{0.129}L^2$	$\underset{(0.00044)}{0.00048}$

Table 6.21 *Results of the large-sample likelihood ratio test applied to the TFs 1954–1972*

Models compared		$\lambda = \dfrac{L(\boldsymbol{x}\mid H_1)}{L(\boldsymbol{x}\mid H_0)}$	2 ln λ	r	Critical points for χ_r^2	
					$\alpha = 0.05$	$\alpha = 0.20$
1. *Market rates*: 3MTB						
H_0 : M2 vs.	H_1 : M1	1.829	1.208	3	7.82	4.64
M4	M1	2.478	1.815	3	7.82	4.64
M8	M1	5.383	3.366	5	11.07	7.23
M8	M2	2.943	2.159	2	5.99	3.22
M2	M3	3.064	2.239	5	11.07	7.23
M6	M3	1.490	0.798	4	9.44	5.90
M7	M3	7.211	3.951	6	12.59	9.56
M8	M3	9.016	4.398	7	14.07	9.82
M8	M4	2.173	1.552	2	5.99	3.22
M8	M5	5.790	3.512	2	5.99	3.22
M7	M6	4.840	3.154	2	5.99	3.22
M8	M6	6.051	3.600	3	7.82	4.64
M8	M7	1.250	0.446	1	3.84	1.64
2. *Interest rates*: 4–6MPCP						
H_0 : M2 vs.	H_1 : M1	21.669	6.152	3	7.82	4.64
M4	M1	3.437	2.469	3	7.82	4.64
M8	M1	29.094	6.741	5	11.07	7.23
M8	M2	1.343	0.590	2	5.99	3.22
M2	M3	1.		5	11.07	7.23
M6	M3	1.088	0.169	4	9.44	5.90
M7	M3	1.343	0.590	6	12.59	8.56
M8	M3	1.343	0.590	7	14.07	9.82
M8	M4	8.465	4.272	2	5.99	3.22
M8	M5	5.927	3.559	2	5.99	3.22
M7	M6	1.234	0.421	2	5.99	3.22
M8	M7	1.		1	3.84	1.64
M8	M6	1.234		3	7.82	4.64
3. *Rate of change in consumer price index*						
H_0 : M2 vs.	H_1 : M1	16761.3	19.454	4	9.44	5.90
M3	M1	3506.0	16.324	5	11.07	7.23
M4	M1	2631.85	15.750	4	9.44	5.90
M6	M1	3506.02	16.324	5	11.07	7.23
M7	M1	1709.40	14.887	3	7.82	4.64
M8	M1	13.27	5.171	1	3.84	1.64
M6	M3	1.				
M3	M4	1.332	0.573	1	3.84	1.64
M6	M4	1.332	0.573	2	5.99	3.22
M1	M5	1.361	0.616	1	3.84	1.64
M2	M5	22809.56	20.070	5	11.07	7.23
M3	M5	4771.16	16.941	6	12.59	8.56
M4	M5	3581.55	16.367	5	11.07	7.23

Table 6.21 (*cont.*)

Models compared		$\lambda = \frac{L(\boldsymbol{x} \mid H_1)}{L(\boldsymbol{x} \mid H_0)}$	2 ln λ	r	Critical points for χ_r^2 $\alpha = 0.05$	$\alpha = 0.20$
H_0 : M6 vs.	H_1 : M5	4771.16	16.941	7	14.07	9.82
M7	M5	2326.23	15.504	4	9.44	5.90
M8	M5	18.06	5.787	2	5.99	3.22
M3	M7	2.05	1.436	2	5.99	3.22
M4	M7	1.54	0.864	1	3.84	1.64
M2	M8	1262.68	14.282	3	7.82	4.64
M3	M8	264.12	11.153	4	9.44	5.90
M4	M8	198.27	10.579	3	7.82	4.64
M6	M8	264.12	11.153	5	11.07	7.29
M7	M8	128.77	9.716	2	5.99	3.22
4. *Rate of change of the stock of money* (*M*1)						
H_0 : M2 vs.	H_1 : M1	238.99	10.953	4	9.44	5.90
M3	M1	21.15	6.103	3	7.82	4.64
M4	M1	872.62	13.543	5	11.07	7.29
M6	M1	5730.86	17.307	6	12.59	8.56
M7	M1	2098.99	15.298	6	12.59	8.56
M8	M1	4104.40	16.640	6	12.59	8.56
M6	M2	23.980	6.354	2	5.99	3.22
M2	M3	11.301	4.850	2	5.99	3.22
M4	M3	41.263	7.440	3	7.82	4.64
M6	M3	270.991	11.204	4	9.44	5.90
M7	M3	99.253	9.195	4	9.44	5.90
M8	M3	194.079	10.537	5	11.07	7.29
M6	M4	6.567	3.764	1	3.84	1.64
M7	M4	2.405	1.755	1	3.84	1.64
M8	M4	4.703	3.096	2	5.99	3.22
M2	M5	8.543	4.290	4	9.44	5.90
M6	M5	204.860	10.645	6	12.59	8.56
M7	M5	75.032	8.636	6	12.59	8.56
M8	M5	140.900	9.896	7	14.07	9.82
M8	M7	1.955	1.340	1	3.84	1.64

relatively large. Most of the alternatives are obtained by dropping some of the non-significant parameter estimates and lead to simpler schemes. The ideal procedure would be to fix upper limits to the degrees of the polynomials in the TFs and then to explore systematically the parameter space by changing the degrees inside the fixed limits and retaining the model with the highest RSS. Such a procedure would have ML justification. However it would be cumbersome to implement. Selecting a few alternatives may lead to the same results. At least it may lead us to reject the TF of a theoretical model so far inadequately entertained.

For the market rate for 3 MTBs none of the more complicated models $M_1, M_2, M_3, M_4, M_6, M_7$ does better than M_8 for which all the parameters except the constant are significant at the 5 percent level. Only M_5 is significantly different from M_8 at the level of 0.20. If we retain model M_5, it is not compatible with the TFs of S^3.

For the 4–6 MPCP rate, M_1 is preferred to M_2, and M_4 or M_5 are preferred to M_8 at the 0.2 level. If we retain M_4 or M_5, we have structures similar to those for 3MTB. Notice that most of the coefficients in M_4 and M_5 are significant at the 0.05 level. Also the parameter estimates for both series are similar. A few remarks are now appropriate. Although the results suggest that the TFs in accord with the information in the data are simpler than those implied by S^1 or S^3, we cannot entirely reject model S^3. The theoretically meaningful model S^3 implies that the variable r_{H_t} does not affect the interest rate. Bringing the variable r_{H_t} into the TF for 3MTB and 4–6 MPCP does not lead to an important gain in RSS. In addition no coefficient of r_{H_t} is significant at the 0.05 level. This finding supports some implications of S^3. The RSS do not vary significantly by increasing the number of lags. The likelihood function seems to be rather flat and thus does not permit sharp discrimination among the estimated models.

For the rate of change in the consumer price index, models M_1 and M_5 are preferred to $M_2, M_3, M_4, M_6, M_7, M_8$, and M_8 is more in accord with the data than M_2, M_3, M_4, M_6, M_7. It is reasonable to accept tentatively TF M_1. However few coefficients of M_1 are significant at the 0.05 level.

For the proportionate rate of change in the stock of money, model M_1 is preferred to M_2, M_4, M_6, M_7, M_8 at the 0.05 level, whereas it is preferred to M_3 at the 0.20 level. M_3 is significantly different from the models M_2, M_4, M_6, M_7, and M_8 at the 0.20 level. For r_{M_t}, model S^3 is empirically validated. Notice that one of the roots of the AR part for the three variables r_{p_t}, r_{M_t}, and Δi_t is close to 1, suggesting that perhaps second order differencing of the endogenous variables is adequate. Again parameter estimates are not very precise. If M_1 is retained, two roots of the AR part for r_{p_t} and r_{M_t} are very similar, giving additional evidence that the variables are generated by some joint process. Whether M_1 or M_5 is retained for the interest rates, the AR part has one root close to 1 and in common with the AR part for r_{p_t} and r_{M_t}.

5 Concluding remarks

At a first glance, the results of the TFs' analysis are not compatible with model S^3. At least for the interest rate series, they do not lead to a relevant choice of the dynamic structure. The schemes retained from the empirical analysis are of higher order than those summarized in table 6.16. For the interest rates the RSS does not vary much from model to model. The likelihood function seems to be very flat in the neighborhood of its maximizing parameter values. The sample is not powerful enough to discriminate among alternative specifications and we have to rely on external information (e.g. from economic theory) to choose the structure. The assumptions underlying S^3 (i.e. the "rational expectation" hypothesis and the variable $\ln H_t$ generated by a random walk) are not ruled out by the information in the data. Model S^1 is not compatible with the empirical findings which do not show any dependence of the interest rates on the money base.

A factor which may have disturbed the analysis and which has to be studied in future work is seasonality. It is obvious from the figures 6.2 and 6.3 that seasonal effects have not completely been eliminated from the seasonally adjusted variables. For example for the variable r_{p_t}, the twelfth order autocorrelation is significantly different from 0. For personal income and for the index of industrial production, the twelfth order partial autocorrelation is significant at the 5 percent level. The seasonally unadjusted interest rate series shows a clear presence of seasonal effects. The twelfth order autocorrelation is significantly different from 0 at the 5 percent level. In the price series, seasonality seems to be very strong. Both, the autocorrelation and the partial autocorrelation of order 12 are significantly different from 0. The seasonal effects show up again in the TFs' analysis. The estimated autocorrelation function of the disturbances of all the TFs has a significant value at lag 12.

In summary, the main findings of the present statistical analysis are:

(1) Although the results of the likelihood ratio tests appear to favor more complicated schemes, they do not lead to a systematic rejection of model S^3 for the US data for the period 1953–72.

(2) The data indicate that the variable $\ln H_t$ follows a random walk.

(3) The variables and the disturbances of the money demand equation and the money supply equation are smoothed by the same first order lag polynomial. (See 3.10′–3.11′). A surprising feature is that the degrees of the lag polynomials in the structural form are rather low.

(4) The structure of the model is stable over the twenty-year period covered by the sample. Splitting the sample into two parts and estimating the same schemes for the subperiods leads to very similar results and nearly identical parameter estimates for the series, except for the process $\ln H_t$ where an upward shift occurred in the drift parameter around the year 1963.

APPENDIX DATA SOURCES

1. Three-Month Treasury Bill market rates (3MTB) are averages computed from daily closing bid prices. Data were obtained from the *Federal Reserve Bulletin.*
2. Four–six-month Prime Commercial Paper rates (4–6MPCP), are averages of daily offering rates of dealers and were obtained from the *Federal Reserve Bulletin.*
3. Seasonally adjusted M1 (currency plus demand deposits) data were obtained as follows: (1) 1953–8 figures were obtained from the December 1970, *Federal Reserve Bulletin*, pp. 895–909; (2) 1959–72 figures were obtained from February 1973 *Federal Reserve Bulletin*, pp. 72–3. The monthly data are averages of daily figures.
4. The Consumer Price Index (CPI), Bureau of Labor Statistics index for city wage-earners and clerical workers was obtained from 1971 *Business Statistics* for 1953–70. For 1971–2 data were obtained from the *Federal Reserve Bulletin.*
5. Personal income (PI) data, seasonally adjusted, were obtained from 1971 *Business Statistics* for 1953–70. For 1971–2 data were obtained from the *Federal Reserve Bulletin.*
6. Index of Industrial Production (IIP) data, seasonally adjusted, were obtained from 1971 *Business Statistics* for 1953–70. For 1971–2 data were obtained from the *Federal Reserve Bulletin.*
7. High-powered money (H) data, seasonally adjusted, were taken as the sum of the average of daily figures for currency outside the Treasury, Federal Reserve Banks and vaults of all commercial banks, plus the average of daily figures for the total reserves of all member banks. The data were obtained as follows: (1) 1953–8 figures were obtained from the *Federal Reserve Bulletin*; (2) 1959–72 figures were obtained from the February 1973 *Federal Reserve Bulletin*, pp. 72–9.

BIBLIOGRAPHY

Box, G. E. P. and G. M. Jenkins (1970), *Time Series Analysis, Forecasting and Control* (San Francisco: Holden-Day)

Cagan, P. (1956), "The monetary dynamics of hyperinflation," in M. Friedman (ed.), *Studies in the Quantity Theory of Money* (Chicago, University of Chicago Press), 25–120

Friedman, M. (1956), "The quantity theory of money – a restatement," in M. Friedman (ed.), *Studies in the Quantity Theory of Money* (Chicago, University of Chicago Press), 3–24

(1970), "A theoretical framework for monetary analysis," *Journal of Political Economy* 78, 193–238

(1971), "A monetary theory of nominal income," *Journal of Political Economy* 79, 323–37

Haavelmo, T. (1947), "Methods of measuring the marginal propensity to consume," *Journal of the American Statistical Society* 42, 105–22; reprinted in W. Hood and T. C. Koopmans (eds.), *Studies in Econometric Methods* (New York, John Wiley, 1953)

Laidler, D. (1966), "The rate of interest and the demand for money, some empirical evidence," *Journal of Political Economy* 74, 545–55

Lindley, D. V. (1961), "The use of prior probability distributions in statistical inference and decision," in J. Neyman (ed.), *Proceedings of the Fourth Berkeley Symposium on Mathematical Statistics and Probability*, I (Berkeley, CA, University of California Press), 453–68

Muth, J. F. (1961), "Rational expectations and the theory of price movements," *Econometrica* 29, 315–35

Nelson, C. R. (1973), *Applied Time Series Analysis for Managerial Forecasting* (San Francisco, Holden-Day)

Palm, F. C. (1973), "On the Bayesian approach to comparing and testing hypotheses when 'Knowing Little,'" University of Chicago, manuscript

Sargent, T. J. (1973), "'Rational' Expectations, the Real Rate of Interest and the 'Natural' Rate of Unemployment," University of Minnesota, mimeo (Berkeley, CA, University of California Press)

Sargent, T. J. and N. Wallace (1973), "Rational expectations and the dynamics of hyperinflation," *International Economic Review* 14, 328–50

Zellner, A. (1971), *An Introduction to Bayesian Inference in Econometrics* (New York, John Wiley)

Zellner, A. and F. C. Palm (1974), "Time series analysis and simultaneous equation models," *Journal of Econometrics*, **2**, 17–54; chapter 1 in this volume

7 Time series versus structural models: a case study of Canadian manufacturing inventory behavior (1975)

Pravin K. Trivedi

1 Time series and structural dynamic models

The purpose of this chapter is to present some results which throw light on the relative strengths of time series models of the type popularized by Box and Jenkins (Box and Jenkins 1970; Naylor, Seans, and Wichern 1972) and structural models of inventory investment behavior in Canadian manufacturing.[1] The choice of inventory investment as a testing ground is motivated by the dual considerations of its importance in the short-run behavior of national income and in its extremely volatile behavior which makes its prediction especially difficult. Furthermore, even when considerable care is given to formulation and estimation of inventory at varying levels of aggregation (see Courchene 1967; Hirsch and Lovell 1969; Trivedi 1969), the results tend not to be robust so that for production purposes there may be some justification in resorting to mechanical ("naive") devices.[2] The comparison carried out in this chapter, however, does not solely concern the prediction problem. It is also concerned with the inter-relationships between time series models and structural models. The following aspects are considered:

(a) The restrictions (parametric or general) placed by structural models of inventory behavior on specification of the corresponding ARMA models

I am indebted to David Wilton of Statistics Canada for his encouragement and advice in writing this chapter, to Philip Smith for his helpful comments on section 3, and to Gloria Glaubowitz and Margaret Howes for assistance with data-gathering and estimation. The work reported here was started when I was at Statistics Canada.

Originally published in the *International Economic Review* 16 (1975), 587–608.

[1] A straight comparison of time series and structural model is odious since it is possible that the latter contains all information while the former contains none. However, this has not prevented some investigators from doing so. This exercise is, presumably, meaningful when the structural specification of the model is not robust or even approximately correct.

[2] The class of ARIMA models will be regarded here as the major alternative to structural model building. This choice is reasonable since it is the most well developed of the mechanical forecasting models. Also the principles on which it is based have been clearly expounded by Box and Jenkins.

(b) The characteristics of the structural model which lead to an unstable ARMA model of aggregate inventory investment.

As an illustration consider the following simple structural dynamically stable model;

$$\alpha(B)y_t = \beta(B)x_t + \theta(B)\eta_t, \tag{1.1}$$

where $\alpha(B)$, $\beta(B)$, and $\theta(B)$ are finite order polynomials in the lag-operator B, η_t is a stochastic disturbance, and x_t, is an explanatory variable. Suppose now that the generation of x_t may be described by the autoregressive- moving average model of the form[3]

$$\omega(B)x_t = \mu(B)\varepsilon_t \tag{1.2}$$

where $\omega(B)$, $\mu(B)$ are polynomials in the lag operator B, and ε_t is a stochastic term which for convenience is assumed to be NID $(0, \sigma_\varepsilon^2)$. Substituting (1.2) in (1.1), and multiplying by $\omega(B)$

$$\omega(B)\alpha(B)y_t = \beta(B)\mu(B)\varepsilon_t + \omega(B)\theta(B)\eta_t, \tag{1.3}$$

which is seen to be an autoregressive-moving average model.[4] (1.3) is of course simply the final form[5] of the structural model consisting of (1.1) and (1.2), but it differs from an unrestricted ARMA model in that it is subject to restrictions imposed on $\alpha(B)$, $\beta(B)$, $\theta(B)$, $\omega(B)$, and $\mu(B)$, by the structural model. That is, knowledge of restrictions on polynomials $\alpha(B)$, $\beta(B)$, $\theta(B)$, $\omega(B)$, and $\mu(B)$ leads to a particular ARMA model for the endogenous variable.

The foregoing argument suggests that when the time series of explanatory variable(s) in the structural dynamic model can be represented as a realization from an ARMA model, the reduced form of the model is also an ARMA model. In such a case it may be convenient to use it as a basis for forecasting, though, by proceeding directly to an unrestricted reduced form, the structural restrictions are neglected.[6] On the other hand, where structural models are themselves misspecified, or based on data with

[3] The assumption that the exogenous variable x is generated by an ARMA process is clearly restrictive and is one of convenience only. A preliminary time series analysis may help to test the appropriateness of the assumption.

[4] The polynomials on both sides of the equation.

$$\omega(B)\alpha(B)y_t = \beta(B)\mu(B)\varepsilon_t$$

are subject to certain restrictions. In particular, we assume that they have no common roots (an identification condition) and that coefficients associated with the highest powers of B in each of the polynomials $\omega(B)$, $\alpha(B)$, $\beta(B)$, $\mu(B)$ are non-zero.

[5] In using the term "final form" I am following Zellner and Palm (1974).

[6] The structural restrictions are often complex since they include not only the usual zero-type restrictions but also qualitative information about relative magnitudes of coefficients – information which is somewhat difficult to incorporate formally in

large measurement errors, or are not robust for other reasons, the corresponding restricted reduced forms are not appropriate for forecasting purposes.

If the structural model contains more than one explanatory variable, the final form of the model will still have the ARMA form provided that it is appropriate to express all the exogenous variables in the ARMA form. Similarly, the presence of additional endogenous variables poses no problems provided a complete model is specified.

The implications of the underlying structural model for the final form model may be examined somewhat more generally by rewriting the model (1.3) in terms of the roots of the polynomials. Thus

$$\prod_{i=1}^{p_1}(1-\omega_i^* B)\prod_{i=1}^{p_2}(1-\alpha_i^* B)y_t = \prod_{i=1}^{q_1}(1-\beta_i^* B)\prod_{i=1}^{r_1}(1-\mu_i^* B)\varepsilon_t + \prod_{i=1}^{p_1}(1-\omega_i^* B)\prod_{i=1}^{r_2}(1-\theta_i^* B)\eta_t, \tag{1.4}$$

where $\omega_i^*, \alpha_i^*, \beta_i^*, \mu_i^*$, and θ_i^* are, respectively, p_1, p_2, q_1, r_1, and r_2 roots of the polynomials $\omega(B)$, $\alpha(B)$, $\beta(B)$, $\mu(B)$, and $\theta(B)$. Imposing the necessary restrictions on the roots to secure identification, it is seen that the degree of the autoregressive part of the model is not greater than $(p_1 + p_2)$ and that of the moving average part is not greater than $\max(q_1 + r_1, p_1 + r_2)$. In practice, if some of the roots of the polynomials are close to zero, it may be possible to find an approximation with fewer autoregressive or moving average terms. Clearly one of the attractions of an ARMA model lies in the possibility that after neglecting small roots one may obtain a parsimonious representation. Neglecting the possibility of small roots, however, it is easy to see that knowledge of the underlying structural model will provide useful implications about the final ARMA model. For instance, if q is of high order and if most of the roots of $\beta(B)$ are non-negligible in size, it may not be easy to find a parsimonious moving average representation.

One final point regarding the r.h.s. of (1.4) is worth noting. The sum of the two moving average terms here may be replaced by another moving average by the following argument. Let $K = \max(q_1 + r_1, p_1 + r_2)$. Let us assume that the sum of the moving average errors defines a discrete time series stochastic process with a finite number of autocovariances. If

an estimation procedure. It may often be the case that the time series model cannot make use of such structural information very easily, even when it is available.

v_t denotes the composite error term and its K autocovariances exist, then the spectral density of v_t is

$$f_v(w) = \left(\frac{1}{2\pi}\right)\sum_{-K}^{K}(c_j e^{-ijw}), \tag{1.5a}$$

where $c_j = Ev_t v_{t-j}$ defines the jth autocovariance of v_t; $f_v(w)$ can be uniquely factorized as

$$f_v(w) = \left(\frac{1}{2\pi}\right)\left|\sum_{j=0}\delta_j e^{ijw}\right|^2, \tag{1.5b}$$

by choosing δ_0 to be real and requiring that the roots of $\delta_{k-1}z^{k-1} + \cdots + \delta_o = 0$ lie on or outside the unit circle. Thus the composite error term may be thought of as a simple K term moving average (see Hannan 1970, ch. 16).

A possible attraction of time series models arises from the difficulties of specifying structural models which, for example, involve unobservable variables, such as "expected sales" in inventory models, for which it is difficult to specify a generation mechanism at all precisely. Although this is only one illustration, it emphasizes that there are many obstacles to structural estimation which, in general, uses much more information. However, in the concluding section of this chapter, it is indicated that time series and structural estimation may be complementary, rather than competing, approaches.

Although the basic idea behind this chapter is a comparison of ARMA models and structural models of inventory investment in durables and non-durables groups of Canadian manufacturing, there are two useful by-products. First we obtain some new estimates of inventory equations for different categories of assets. Second we obtain some information regarding the effect of aggregation over types of assets on the specification of inventory equations.

The data are described in section 2, the fitted time series models in section 3, and the structural models in section 4. The concluding section [5] compares the results of sections 3 and 4.

2 Data

The data used in this study are obtained from the Statistics Canada publications *Inventories, Shipments and Orders in Manufacturing Industries* (Cat. 31–001) and *Indexes of Real Domestic Product by Industry* (Cat. 61–005). They are quarterly seasonally unadjusted, and relate to the period

1961(I)–1973(I). All figures are estimated values in millions of dollars in current prices and relate to the following variables:

S: Shipments
N: New orders received during quarter
U: Unfilled orders at end of quarter
RM: Total inventory held (raw materials) at end of quarter
GIP: Total inventory held (goods in process) at end of quarter
FG: Total inventory held (finished products) at end of quarter
TIH: Total inventory held at end of quarter
TIO: Total inventory owned (durables)
IHBNO: Inventory held but not owned.

The components of total inventory held (TIH) are RM, GIP, and FG.[7]

The data are available for non-durable and durable groups; we identify the two groups by the use of prefix ND and D, respectively.[8]

It certainly would have been possible to utilize data at a more disaggregated level, though this alternative has not been pursued in view of the problems of data accuracy. The greatest reservation one has regarding the data concerns the use of value rather than volume figures which must lead to biases of unknown magnitude. On the other hand, it is hard to see why, in spite of this limitation, the study should fail to provide at least some insights.

A well-established line of disaggregation in the literature is between the industries which produce to order (PTO) and those which produce for stocks (PFS) (see Courchene 1967; Beisley 1969; Trivedi 1970). The classification adopted in the chapter cuts across the PTO–PFS distinction in that each contains certain groups of industries which produce both to order and for stocks (see Courchene 1967, pp. 331–2). However, there are sufficient differences between the two groups to justify the present chapter.

[7] TIH differs from TIO in that it excludes the value of progress payments received by manufacturers for partially completed items such as aircraft, ships, or structures using fabricated steel. "As work proceeds on such items, total inventory values increase until the time of delivery. But deducting the balance or progress payments on manufacturers' books from total inventory, the value of manufacturers' investment inventory is derived" (see *Inventories, Shipments and Orders in Manufacturing Industries*, Concepts and Methods, Statistics Canada). The method of progress payments is important only for the durables group. For non-durables, TIH and TIO are equal. Thus for durables we have two distinct concepts of total inventory and the choice of any one must depend upon the purpose of the investigation.

[8] An additional variable which was used unsuccessfully is the real domestic product (RDP) of each broad sector of the manufacturing industry.

3 Time series models

First consider the results of fitting to each dependent variable an autoregressive-moving average model along the lines of Box and Jenkins (1970). The class of models considered may be written in their notation in the form

$$\begin{aligned}(1- B^{s})^{D}(1 - B)^{d}(1 - \phi_1 B - \cdots - \phi_p B^{P})\\ \times (1 - \psi_1 B^{s} - \psi_2 B^{2s} - \cdots - \psi_p B^{Ps})(y_t - \phi_0)\\ = (1 - \theta_1 B - \theta_2 B^{2} \cdots - \theta_q B^{q})\\ \times (1 - \xi_1 B^{s} - \xi_2 B^{2s} - \cdots - \xi_Q B^{Qs})(\varepsilon_t - \theta_0),\end{aligned} \tag{3.1}$$

where the successive polynomials in the lag operator B may be called the non-stationary seasonal part, the regular non-stationary part, the regular autoregressive part, the seasonal autoregressive part, the regular moving average part, and the seasonal moving average part, respectively. Each model belonging to this general class may be characterized by seven parameters p, d, q, P, D, Q, and s which denote the following: p is the number of regular autoregressive parameters $(\phi_1, \ldots, \phi_p)$, d is the number of regular differences, q is the number of regular moving average parameters $(\theta_1, \theta_2, \ldots, \theta_q)$, P is number of seasonal autoregressive parameters $(\psi_1, \psi_2, \ldots, \psi_p)$, D is the number of seasonal differences, Q is the number of seasonal moving average parameters, and s, the order of the seasonal. s equals four in our case since we use quarterly data throughout. The parameters ϕ_0 and θ_0 allow for the presence of a constant term in the model.

In line with the Box–Jenkins procedure we compute the simple and partial autocorrelation functions for each time series. The simple ones are presented in tables 7.1 and 7.2. In each case we present the first eleven autocorrelation coefficients. These are intended to suggest the choice of the appropriate (p, d, q) $(P, D, Q)^{s}$ model in each case. Two points are worth noting here. First the task of choosing the appropriate model may be considerably simplified by calculating many more autocorrelation coefficients than I have chosen to calculate. (My choice was dictated by the consideration that in a time series of about 45 observations this is roughly the number that can be computed reasonably precisely.) Second it may be difficult to identify uniquely the appropriate $(p, d, q)(P, D, Q)$ model on the basis of the study of the autocorrelation function alone, so that an extensive use of diagnostic checks is highly desirable after the time series models have been fitted.

Table 7.1 *Autocorrelation functions: manufacturing durables*

Category	Lag										
	1	2	3	4	5	6	7	8	9	10	11
TIH 1	0.642*	0.348*	0.448*	0.514*	0.247	0.009	0.235	0.460*	0.293*	0.235	0.472*
2	0.614*	0.335*	−0.005	−0.477*	−0.436*	−0.539*	−0.441*	−0.182	−0.143	−0.150	0.273
TIO 1	0.579	0.271*	0.467*	0.618*	0.276*	−0.008	0.236	0.474*	0.260	0.164	0.426*
2	0.589*	0.421*	0.157	−0.375*	−0.380*	−0.606*	−0.580*	−0.290	−0.281	0.043	0.208
RM 1	0.605*	0.358*	0.427*	0.518*	0.261	0.015	0.138	0.403*	0.300*	0.160	0.332*
2	0.510*	0.403*	0.220	−0.319*	−0.343*	−0.474*	−0.523*	−0.319*	−0.122	−0.023	0.161
GIP 1	0.587*	0.344*	0.248	0.206	0.134	0.087	0.096	0.275*	0.245	0.208	0.300*
2	0.476*	0.124	−0.141	−0.512*	−0.304*	−0.184	−0.217	−0.062	−0.077	0.039	0.172
FG 1	0.081	−0.684*	0.082	0.797*	0.030	−0.643*	0.109	0.711*	0.014	−0.551*	0.163
2	0.229	0.039	−0.098	−0.506*	−0.183	−0.133	−0.144	−0.033	0.020	0.078	0.210

Note:
Asterisks denote coefficients significantly different from zero.

Table 7.2 *Autocorrelation functions: manufacturing non-durables*

	Lag										
Category	1	2	3	4	5	6	7	8	9	10	11
TIH 1[a]	0.359*	0.224	0.210	0.631	0.237	0.087	0.112	0.530*	0.282*	0.192	0.243
2[a]	0.348*	0.241	0.028	−0.418*	−0.350*	−0.378*	−0.348*	−0.124	0.015	0.133	0.293
RM 1	0.118	−0.173	0.016	0.624*	0.152	−0.181	−0.045	0.485*	0.166	−0.096	−0.017
2	−0.130	0.021	0.215	−0.257	−0.073	−0.073	−0.124	−0.141	0.079	0.051	−0.187
GIP 1	0.411*	0.082	0.415*	0.586*	0.265	0.008	0.295*	0.572*	0.295*	0.071	0.357
2	0.222	0.153	0.111	−0.477*	−0.139	−0.209	−0.206	0.106	−0.081	0.076	0.132
FG 1	0.463*	0.312*	0.266	0.458*	0.216	0.090	0.134	0.477*	0.268	0.276*	0.323*
2	0.428*	0.210	−0.071	−0.587*	−0.328*	−0.375*	−0.298*	0.016	−0.006	0.193	0.289

Notes:
[a] Row 1: No seasonal differencing; row 2: one seasonal difference.
Asterisks denote coefficients significantly different from zero.

Tables 7.1 and 7.2 give computed autocorrelation coefficients for seasonally unadjusted inventory investment (row 1) and annual changes in inventory investment (row 2). From row 2 note that in case of manufacturing *durables* the coefficients are high not only at the low lags of order 1 and 2, but also at higher lags of order 4, 5, 6, and 7. The raw materials' component of aggregate inventory shares this characteristic more clearly than the two other. The finished goods component shows very little systematic behavior. For the manufacturing non-durables, the total inventories held category also shows presence of large (significantly different from zero) coefficients at lags of order 1, 4, 5, 6, and 7, though the only component that shares this pattern is the finished goods component. From this, it seems reasonable to draw the preliminary conclusion that the raw materials' component dominates the aggregate behavior in case of manufacturing durables, whereas the finished goods component does so for non-durables, total inventory.

In tables 7.3 and 7.4, are listed the various $(p, d, q)(P, D, Q)^s$ models which are fitted to the component series.[9] In addition to the estimated model, the standard errors of all coefficients, the sample size, residual variances, and goodness of fit statistics are also presented. The main diagnostic check used is based on the comparison of

$$\dot{Q} = n \sum_{k=1}^{J} \hat{r}_k^2,$$

where $\hat{r}_k$ is the kth order autocovariance of the residuals from the fitted model, J being the number of autocovariances computed, and n the sample size. A chi-square test based on the statistic Q provides a useful overall check of goodness of fit (see Pierce 1971). In a certain number of cases where the choice of the appropriate model was not immediately obvious, and more than one model was plausible, these were all fitted but have not necessarily been included in the tables.

The major features of these fitted models are as follows:

(i) With one exception of IHBNO the fitted models all have $d = 1, D = 1, Q = 1$.

(ii) Especially for the durables category it is a little difficult to choose "the" model. On one hand the $(0, 1, 6)(0, 1, 1)^4$ model for D/TIO and D/RM seems over-parameterized. On the other hand, other,

[9] In fitting the time series models I have used the program "TYMPAC" provided by Queen's University, Kingston, Ontario. This uses an algorithm described in Box and Jenkins (1970), which sets preperiod residuals equal to zero.

more parsimonious alternatives yield higher residual variances and significantly non-zero residual autocorrelations, especially at lags 6 and 7. We have presented both sets of results since there are grounds for choosing the parsimonious representation even where residual variance is greater.

(iii) For the durables group we observe a lack of homogeneity in the type of model fitted. That is, even where the model is the same, numerical coefficients are different. The "best" model for D/RM, D/GIP, and D/FG all have different forms, suggesting somewhat different underlying structures. In case of the non-durables group this is not true to the same extent. Such differences are plausible and to be expected given what is known about the differential responses of components of aggregate inventory to their determinants (see, for example Trivedi and Rowley 1975, ch. 6). One is led to the conclusion that the aggregation problem may exist for the pure time series models as it may for structural econometric models. The presence of significant aggregation biases would cast doubt on the temporal stability of the time series models in this area because the components are known to differ considerably in their volatility.[10]

(iv) Finally note that the fitted time series models indicate that in most cases at least the seasonal pattern of the inventory investment series is stable, that is, the coefficient of B^4 in the seasonal part of the model is not significantly different from unity. This point suggests that in the structural models the use of seasonal dummies in the usual way may be appropriate. But this conclusion should be treated with caution

[10] Consider, for example, the following coefficients of variation for the components and the aggregate,

	Coefficient of variation (percent)	
Component	*Durables*	*Non-durables*
TIO	43.1 (190.8)	74.4 (137.4)
TIH	50.6 (206.8)	74.4 (137.4)
RM	56.3 (80.0)	201.0 (57.6)
GIP	>100.0 (73.5)	66.9 (24.7)
FG	301.00 (55.0)	131.00 (55.04)

where the figures in parenthesis are the sample mean values in million dollars. This suggests that the success of time series models for prediction purposes in this area may be limited. Additional reasons for such skepticism are contained in the final section of this chapter, where I compare time, series models with their econometric counterparts.

Table 7.3 *Fitted time series models: durables*

Category	Model*	(p, d, q)	$(P, D, Q)^s$	Error variance	$\hat{Q}$	"Large" residual autocorrelations
TIO $n = 44$	$y_t = (1 - \underset{(0.1639)}{0.4615}B - \underset{(0.1565)}{0.6641}B^2 - \underset{(0.1676)}{0.5785}B^3 - \underset{(0.1735)}{0.6770}B^4 - \underset{(0.1637)}{0.5810}B^5 + \underset{(0.1754)}{0.0829}B^6)(1 + \underset{(0.1753)}{0.6616}B^4)e_t$	(0, 1, 6)	$(0, 1, 1)^4$	16950.0	13.80	
TIO $n = 44$	$y_t = (1 - \underset{(0.3265)}{0.5373}B - \underset{(0.3640)}{0.3593}B^2 - \underset{(0.3496)}{0.4818}B^3) \times (1 + \underset{(0.3978)}{0.4125}B^4)e_t$	(0, 1, 3)	$(0, 1, 1)^4$	20960	21.79	$r_6 = -0.41$; $r_7 = -0.37$
TIH $n = 43$	$1 - (\underset{(0.3678)}{0.4403}B)y_t = (1 - \underset{(0.3947)}{0.2059}B - \underset{(0.2539)}{0.2257}B^2 - \underset{(0.2219)}{0.2676}B^3)(1 + \underset{(0.1367)}{0.7953}B^4)e_t$	(1, 1, 3)	$(0, 1, 1)^4$	26300.0	15.60	
TIO $n = 44$	$(1 - \underset{(0.0587)}{1.054}B)y_t = (1 - \underset{(0.1688)}{0.5329}B - \underset{(0.1611)}{0.3879}B^2 - \underset{(0.1829)}{0.4446}B^3)(1 + \underset{(0.1997)}{0.5033}B^4)e_t$	(1, 1, 3)	$(0, 1, 1)^4$	21250.0	22.17	$r_6 = -0.39$; $r_7 = -0.38$

RM $n = 44$	$y_t = (1 - \underset{(0.1752)}{0.4241}B - \underset{(0.1754)}{0.4895}B^2 - \underset{(0.1957)}{0.9235}B^3 - \underset{(0.1781)}{0.5587}B^4 - \underset{(0.2012)}{0.4999}B^5 - \underset{(0.1995)}{0.03780}B^6)(1 + \underset{(0.1556)}{0.7558}B^4)e_t$	(0, 1, 6)	$(0, 1, 1)^4$	3781.0	8.18	
	$(1 + \underset{(0.2650)}{0.6121}B)y_t = (1 + \underset{(0.2131)}{0.8287}B^4)e_t$	(1, 1, 0)	$(0, 1, 1)^4$	5093	14.61	
GIP $h = 42$	$y_t = (1 - \underset{(0.2831)}{0.4438}B)(1 + \underset{(0.1870)}{0.8568}B^4)$	(0, 1, 1)	$(0, 1, 1)^4$	11830	11.50	
FG $n = 44$	$y_t = (1 - \underset{(0.1539)}{0.1577}B)(1 + \underset{(0.1133)}{0.7180}B^4)e_t$	(0, 1, 1)	$(0, 1, 1)^4$	6030.0	13.21	
IHBNO $n = 47$	$y_t = (1 + \underset{(0.1356)}{0.4074}B)e_t$	(0, 1, 1)	$(0, 1, 0)^4$	6940.0	29.32	$r_{13} = 0.38$

Notes:
* Numbers in parenthesis indicate standard errors.
The goodness of fit test based on the statistic $\hat{Q}$ (mentioned above) may be carried out at a chosen level of significance by comparing the calculated value with the tabulated chi-square value with $(\mathcal{J} - p - q - P - Q)$ degrees of freedom. In all our calculations $\mathcal{J}$ was fixed at 20 which is possibly too large in view of the small sample available. On applying this test the only cases in which $\hat{Q}$ approaches the critical value are the following two cases for D/TIO: $(0, 1, 3)(0, 1, 1)^4$ and $(1, 1, 3)(0, 1, 1)^4$. The relevant critical values are, respectively, 26.3 and 25.0. In the remaining cases in this table and table 7.4 a poor fit does not seem to be a major problem.

Table 7.4 *Fitted time series models: non-durables*

Category	Model	(p, d, q)	$(P, D, Q)^s$	Error variance	$\hat{Q}$	"Large" residual autocorrelations
TIH $n = 43$	$(1 - \underset{(0.1273)}{0.5913}B)y_t = (1 + \underset{(0.1232)}{0.6912}B + \underset{(0.1701)}{0.0426}B^2$ $(1 + \underset{(0.2125)}{0.5187}B^3)(1 + \underset{(0.1156)}{0.9145}B^4)e_t$	(1, 1, 3)	$(0, 1, 1)^4$	12380.0	12.96	
RM $n = 44$	$y_t = (1 + \underset{(0.3423)}{0.4742}B^4)e_t$	(0, 1, 0)	$(0, 1, 1)^4$	7107.0	13.05	
GIP	$y_t = (1 - \underset{(0.3079)}{0.2910}B)(1 + \underset{(0.2307)}{0.7696}B^4)e_t$	(0, 1, 1)	$(0, 1, 1)^4$	556.1	11.50	
FG $n = 42$	$(1 + \underset{(0.2934)}{0.4330}B)y_t = (1 + \underset{(0.1936)}{0.8645}B^4)e_t$	(1, 1, 0)	$(0, 1, 1)^4$	3414.0	12.61	

in view of a deficiency of the numerical algorithm that was used (see Kang 1973).

4 Econometric models of inventory investment

It is not possible in the space available here to provide detailed justification for the general form of specification that is used. The interested reader should consult Childs (1967), Beisley (1969), Hirsch and Lovell (1969), and Trivedi and Rowley (1975), on this point. However, the following comments may serve as a brief background.

The most common feature of specification is the use of the flexible accelerator model incorporating an adjustment process relating actual and desired inventories, the latter being a linear function of a small number of variables. *Expected demand* (or orders) is taken to be the most important determinant of desired stock. Expectations may relate to one or more future periods. This simple model may be modified in a number of ways. First, in industries producing primarily to order (PTO group) it is more pertinent to use a measure of established demand rather than expected demand because factors which encourage PTO are those which also make production time-consuming and storage of finished goods an uneconomic activity. A second modification is the introduction of a *buffer stock variable* which represents an error in forecasting sales to explain the unanticipated reductions in inventories. In a number of formulations– Lovell (1961), Lovell and Darling (1965), Hirsch and Lovell (1969), and Trivedi (1970) – this variable is either the current or one-period lagged rate of change of sales or shipments. It typically has a negative coefficient. There are other alternative specifications of the buffer stock variable such as that in Helliwell *et al.* (1972) which rely on economywide models to achieve a distinctive specification. The importance of a buffer stock variable is an empirical matter, but *a priori* reasoning suggests that its importance will be greater in those industries in which production for stocks (PFS) is common.

In addition to those factors mentioned above, the other main consideration explicitly introduced in the models is capacity utilization, or rather a proxy variable for it.

In a more detailed analysis a number of other factors would be considered. These include the role of production smoothing, speculative motive, and financial factors. However, previous studies have shown that the first of these may be largely seasonal and hence adequately captured by prior seasonal adjustment or explicit inclusion of dummy variables, whereas the latter two are not of major importance so that their omission may not be serious.

4.1 *Raw (or purchased) materials inventory*

4.1.1 Durables The appropriate decision variable to look at in this context is the *purchases* made by the firm – but published data usually relate to the observed change in materials inventories, $R_t - R_{t-1}$, which represents the difference between additions, A_t, and withdrawals from stocks, M_t; that is

$$R_t - R_{t-1} = A_t - M_t. \tag{4.1}$$

The variables A_t and M_t may be eliminated by use of auxiliary relations such as

$$A_t = \beta_0 + \beta_1 R_t^* + \beta_2 R_{t-1} + \beta_3 A_{t-1} \tag{4.2}$$

and

$$M_t = \beta_4 Q_t \tag{4.3}$$

or

$$M_t = \sum_{i=0}^{L} \delta_i N_{t-i}, \tag{4.3a}$$

where R_t^* represents the desired level of materials stocks. The logic behind (4.2) is simply the gradual adjustment of additions to stock to a desired level which is itself determined by either past commitments or future expectations. The choice of the appropriate variables determining R_t^* and M_t depends on the industry in question. Thus, for instance, in dealing with the PTO case, it seems realistic to substitute (4.3a) and

$$R_t^* = \sum_{i=0}^{K} \alpha_i N_{t-i} \tag{4.4}$$

in (4.2).

(The integer K (like L) is unknown.) Substitution into (4.2) and (4.1) yields

$$\Delta R_t = \beta_0 + \beta_1 \sum_{i=0}^{K} \alpha_i N_{t-i} - \beta_4 \sum_{i=0}^{L} \delta_i N_{t-i} + \beta_2 R_{t-1} + \beta_3 A_{t-1}. \tag{4.5}$$

When dealing with PFS industries modifications to (4.5) are needed, though not necessarily of a fundamental nature.

Some of the empirical estimates available in the literature are variants of (4.5); in general, non-availability of data on placement of orders leads to A_{t-1} being omitted. (In the specific context of (4.5), note that identification of α_i and δ_j ($i = 1, \ldots, K$; $j = 1, \ldots, L$) poses a problem since some of the terms such as $(\beta_1\alpha_1 - \beta_4\delta_1)$, $(\beta_1\alpha_2 - \beta_4\delta_2)$ could be close to zero.)

4.1.2 Non-durables In case of the non-durables group N_{t+j} is replaced by its anticipated value $\hat{N}_{t+j}$. Let $\hat{N}_t$ and $\hat{N}_{t+1}$ denote forecasts of orders receivable in the current and the following periods. Suppose that firms do not look beyond this horizon in forming their inventory plans. Then the relevant variant of (4.5) is (4.6)

$$\Delta R_t = \gamma_0 + \sum_{i=0}^{1} \gamma_{i+1} \hat{N}_{t+i} - \gamma_3 N_t - \gamma_4 R_{t-1}, \tag{4.6}$$

where the term $-\gamma_3 N_t$ represents the extent to which inventories are drawn down by current usage. Since we are dealing with quarterly seasonally unadjusted data, it is assumed that orders forecasts are generated recursively by

$$\hat{N}_t = \theta_1 N_{t-1} + \theta_4 N_{t-4} - \theta_1\theta_4 N_{t-5} \tag{4.7}$$

$$\hat{N}_{t+1} = \theta_1 \hat{N}_t + \theta_4 N_{t-3} - \theta_1\theta_4 N_{t-4}, \tag{4.8}$$

which when substituted in (4.6) yields

$$\begin{aligned}\Delta R_t = {} & \gamma_0 + \theta_1(\gamma_1 + \gamma_2\theta_1) N_{t-1} + \gamma_2\theta_4 N_{t-3} + \gamma_1\theta_4 N_{t-4} \\ & - \theta_1\theta_4(\gamma_1 + \gamma_2\theta_1) N_{t-5} - \gamma_3 N_t - \gamma_4 R_{t-1}.\end{aligned} \tag{4.9}$$

4.2 *Finished goods inventory*

4.2.1 Durables If the production behavior of this sector is dominated by industries which produce to order, it is not appropriate to regard the flexible accelerator hypothesis as very relevant for the explanation of finished goods inventory investment. The motives which otherwise encourage firms to maintain a stable relation between stocks and orders are absent here. Indeed Childs (1967) suggests that it is more appropriate to regard the backlog of unfilled orders as the appropriate decision variable in this case. This point is now well established in the literature. For this reason, we should not expect the kind of model applied to raw materials inventory to work very well for finished goods. I have nevertheless fitted this model as a check on this type of *a priori* reasoning. The finished goods equation analogous to (4.5) is referred to as (4.10).

4.2.2 Non-durables The model used is similar to that for raw materials. Beginning with

$$\Delta FG_t = \delta_0 + \delta_1 \hat{N}_t + \delta_2 \hat{N}_{t+1} + \delta_3(\hat{N}_t - N_t) - \delta_4 FG_{t-1}, \quad (4.10)$$

which incorporates the buffer stock variable $(\hat{N}_t - N_t)$ to capture the effect of errors of expectation and using the forecasting (4.7) and (4.8), the last equation can be reduced to

$$\begin{aligned}\Delta FG_t = {} & \delta_0 - \delta_3 N_t + \left((\delta_1 + \delta_3)\theta_1 + \delta_2\theta_1^2\right) N_{t-1} - \delta_2\theta_1\theta_2 N_{t-3} \\ & - (\delta_1 + \delta_3)\theta_2 N_{t-4} - \left((\delta_1 + \delta_3)\theta_1\theta_2 + \delta_2\theta_1^2\theta_2\right) N_{t-5} \\ & -\delta_4 FG_{t-1}. \end{aligned} \quad (4.11)$$

Note that the sign restrictions on this equation are the same as before.

4.3 Goods in progress

4.3.1 Durables Once again the production behavior of this group is assumed to be determined by the commitment to deliver goods for which there is an established demand. Hence a distributed lag function on past new orders is used as a proxy for expected shipments. Introducing the rate of change of unfilled orders as a measure of capacity utilization, we obtain an equation of the same form as raw materials:

$$\begin{aligned} GIP_t - GIP_{t-1} = {} & \mu_0 + \sum_{i=1}^{m} \mu_i N_{t-i+1} + \mu_{m+1}(U_t - U_{t-1}) \\ & -\mu_{m+2} GIP_{t-1}. \end{aligned} \quad (4.12)$$

Despite the apparent plausibility of this model, there are good theoretical reasons why we cannot expect it to work satisfactorily in practice. The main limitation is the basic inability of the flexible accelerator model to account for the complicated dynamics which characterize the behavior of goods in progress. This point is expanded below in dealing with the non-durables groups. A second point is the problem of valuation of goods in process which makes any estimate of changes in its stock subject to extremely wide margins of error, thereby complicating further the problem of assessing how well the given model fits the data.

4.3.2 Non-durables Previous econometric studies of work in progress behavior have shown insufficient recognition of the fact that the period of production, $\bar{p}$, is not time invariant and time profile of the consumption of inputs is highly variable in the short run.[11]

Two other issues concern the validity of a stock adjustment-type model and the importance of the buffer stock motive. Accelerator mechanism implies that investment in work in progress takes place at a rate determined by the rate of change of output or sales. However, empirical evidence has not always provided strong support for the lagged adjustment form of the accelerator.[12] Lastly, mention must be made for the need to incorporate sales expectations in this context; for when we admit the possibility of accumulating work in progress between stages of reproduction, it is possible that its behavior is akin to that of finished goods, though its actual importance is an empirical matter.

As a first approximation (which is hard to improve upon in the present state of data) I consider for goods in process the same model as for finished goods inventory (4.11). For convenience of reference it is written explicitly in the form

$$\begin{aligned}\Delta GIP = {} & \nu_0 - \nu_3 N_t + \left((\nu_1 + \nu_3)\theta_1 + \nu_2\theta_1^2\right) N_{t-1} \\ & + \nu_2\theta_1\theta_2 N_{t-3} - (\nu_1 + \nu_3)\theta_2 N_{t-4} \\ & - \left[(\nu_1 + \nu_3)\theta_1\theta_2 + \nu_2\theta_1^2\theta_2\right] N_{t-5} - \nu_4 GIP_{t-1},\end{aligned} \tag{4.13}$$

where the coefficients ν_i $(i = 0, 4)$ take place of the δ_i $(i = 0, 4)$ in (4.11). This formulation incorporates the buffer stock role and the importance of sales anticipations, but leaves out the role of other factors.

4.3.3 Stochastic specification To complete the specification of various equations we shall assume that the stochastic term in the equation has either moving average or autoregressive representation, of finite but

[11] Specific illustrations of this are provided by shortening of $\bar{p}$ by working overtime or by operating more machines: both possibilities retard the operation of the accelerator mechanism, $\bar{p}$ may vary cyclically and may be systematically related to capacity utilization. Not enough is known to us from theoretical models to suggest what factors underlie such variation. However, if it is present, it tends to vitiate the mechanical operation of the accelerator. Empirical research on the time shape of production processes throws some indirect light on this question.

[12] Abramovitz (1950, pp. 160–77, 380–8) reported that his findings for the continuous process industries in the United States showed work in progress investment to be related to the rate of change of output without a lag, and with slight lead in the discontinuous process industries. Stanback's (1962) study lends further support to these findings. Econometric investigation of the issue is hampered by unavailability of reliable disaggregated data. Although Lovell's (1961) results give some support to the flexible accelerator, these relate to the sum of work in progress and raw materials inventories, rather than to one component alone. Further result provided by Courchene (1967) and Trivedi (1970) provide at best only weak support for the lagged adjustment hypothesis, leaving open a distinct possibility that various other factors such as changes in composition of goods and the time profile of production dominate the mechanical role of the the accelerator.

unknown order. The assumption of autocorrelated residuals is realistic and may be interpreted as representing either inherent properties of economic disturbances or as a logical outcome of algebraic transformations, or assumptions about expectations formation, or misspecification of the relationship. There are considerable difficulties in distinguishing between the two possibilities (see Hendry 1974), and in choosing between autoregressive and moving average representation of the error.[13]

4.3.4 Seasonality Inclusion of seasonal dummies in the model may be regarded as an implicit allowance for production smoothing (see Darling and Lovell 1971).

Also the sales forecasting assumption we have used above for the non-durables group implies that seasonality considerations are built into formation of sales expectation and hence additional inclusion of seasonal dummy variables is not called for (see Trivedi 1970). If so, inclusion of seasonals provides a check on the specification. All specifications reported below were estimated with and without seasonal dummies. The latter were introduced in two ways – the usual zero–one type of variables and the product of the zero–one dummy with the lagged dependent variable on each estimated equation. This second variant allows for the possibility that the rate of adjustment to the desired stock varies seasonally.[14]

5 Results

The results of estimating (4.5), (4.10), and (4.12) for durables and (4.9), (4.11), and (4.13) for non-durables are given in tables 7.5–7.7. The last column of these tables indicates what stochastic specification was chosen. The only difference between tables 7.6 and 7.7 is that the latter pertains to specifications for non-durables which include seasonals. Each table also includes an equation, numbered (4.14)–(4.16), explaining the variation of *total* (sum of the three components) inventory investment with the same general form as other component equations. This gives some idea of the magnitude of aggregation effects on coefficients as well as goodness of fit.

[13] The estimation procedure used for fitting models with these stochastic specifications uses non-linear optimization algorithms. The details of these may be found in Trivedi (1970) and Hendry (1974).

[14] The use of seasonal dummy variables which interact with the lagged stock of inventories is interpreted as a way of taking account of an adjustment rate which varies seasonally. Alternative interpretations are also available – e.g. systematic seasonal changes in composition of inventories.

Table 7.5 *Manufacturing durables**

No.	Dependent variable	N_{t-1}	N_{t-2}	N_{t-3}	ΔU_{t-1}	Lagged dependent variable	Q_1	Q_2	Q_3	Constant	$\hat{\sigma}_{\varepsilon}^2$	Stochastic specification
(4.5)	D/ΔRM	0.0502 (1.54)	0.1375 (3.15)	0.0831 (1.92)	0.0802 (2.89)	−0.2129 (3.68)	0.0234 (4.12)	−0.0074 (1.07)	−0.0331 (2.61)	29.96	1717	
(4.10)	D/ΔFG	−0.03741 (1.51)	0.0191 (0.52)	0.1228 (3.66)	0.0662 (2.04)	−0.2201 (4.21)	0.0224 (2.25)	0.1150 (9.40)	0.0852 (8.25)	100.86	2842.	2MA
(4.12)	D/ΔGIP	0.1411 (0.1018)	0.0286 (0.39)	−0.0332 (0.35)	0.1278 (1.71)	−0.5095 (2.06)	−0.0006 (0.06)	−0.0024 (0.35)	−0.0259 (2.21)	303.3	8626	1AR
(4.14)	D/ΔTIO	0.0909 (2.13)	0.2937 (1.43)	0.1823 (3.80)	0.1186 (2.13)	−0.2712 (7.60)	0.0280 (4.50)	0.0046 (0.94)	0.0131 (2.40)	235.1	6762	
(4.15A)	D/ΔTIO	0.0918 (1.09)	0.2049 (1.99)	0.2887 (2.70)	0.1726 (2.28)	−0.2538 (6.47)	237.71** (2.91)	119.38** (1.83)	−109.22** (1.51)	287.54	12319.	

Equation Stochastic specification

(4.10) $U_t = \varepsilon_t - \underset{(2.90)}{0.5184}\varepsilon_{t-1} - \underset{(3.12)}{0.5401}\varepsilon_{t-2}$

(4.12) $U_t = \underset{(2.79)}{0.6220}U_{t-1} + \varepsilon_t$

Notes:

* The term in brackets below each coefficient is the associated *t*-ratio.

** These are coefficients of simple quarterly dummy variables.

Table 7.6 *Manufacturing non-durables*

No.	Dependent variable	N_t	N_{t-1}	N_{t-3}	N_{t-4}	N_{t-5}	Lagged dependent	Constant	$\hat{\sigma}_{\varepsilon}^{2}$	Stochastic specification
(4.9)	ND/ΔRM	−0.0869 (0.60)	0.3092 (1.97)	0.1040 (0.72)	−0.0291 (0.1744)	0.0514 (0.27)	−0.5322 (2.25)	559.8	5828	4MA
(4.11)	ND/ΔFG	−0.3598 (7.16)	0.3868 (6.17)	0.1859 (3.24)	−0.2087 (2.86)	−0.2451 (2.60)	−0.2444 (1.66)	172.5	2325	3MA
(4.13)	ND/ΔGIP	0.0646 (1.39)	0.1222 (2.55)	0.0951 (2.85)	−0.0367 (2.85)	−0.1280 (2.33)	−0.0646 (1.39)	5.92	498	2MA

Equation stochastic specification

(4.9) $U_t = \varepsilon_t - \underset{(0.33)}{0.0835}\varepsilon_{t-1} - \underset{(1.04)}{0.239}\varepsilon_{t-2} + \underset{(2.49)}{0.5387}\varepsilon_{t-3} + \underset{(2.79)}{0.6220}\varepsilon_{t-4}$

(4.11) $U_t = \varepsilon_t + \underset{(3.23)}{0.8352}\varepsilon_{t-1} + \underset{(1.25)}{0.3903}\varepsilon_{t-2} - \underset{(1.90)}{0.4987}\varepsilon_{t-3}$

(4.13) $U_t = \varepsilon_t + \underset{(1.43)}{0.3190}\varepsilon_{t-1} - \underset{(3.68)}{0.7896}\varepsilon_{t-2}$

Table 7.7 *Manufacturing non-durables*

No.	Dependent variable	N_t	N_{t-1}	N_{t-3}	N_{t-4}	N_{t-5}	Q_1	Q_2	Q_3	Lagged dependent	Constant	$\hat{\sigma}_\varepsilon^2$	Stochastic specification
(4.9a)	ND/ΔRM	0.1023 (0.87)	−0.0721 (0.38)	0.3887 (1.43)	−0.0485 (0.14)	−0.3784 (2.01)	0.0220 (0.95)	−0.0114 (0.55)	−0.0790 (3.11)	0.0127 (0.14)	37.16	4304	1MA
(4.11a)	ND/ΔFG	−0.2332 (2.75)	0.1698 (1.43)	−0.0736 (0.50)	0.4839 (4.18)	−0.0229 (0.1619)	0.0031 (0.29)	0.0357 (4.11)	0.0057 (0.36)	−0.4680 (2.50)	306.12	2464	2MA
(4.13a)	ND/ΔGIP		0.0511 (1.61)	−0.0053 (−0.09)	0.1491 (2.52)	−0.1752 (3.60)	−0.0564 (2.88)	−0.0079 (0.53)	0.0011 (0.10)	−0.0796 (0.91)	5.74	275	
(4.16a)	ND/ΔTIH	−0.3600 (1.11)	0.3792 (1.51)	0.4513 (1.09)	0.4313 (1.04)	−0.8719 (2.67)	0.0531 (2.46)	0.02672 (1.43)	0.0543 (3.21)	−0.0678 (0.750)	191.87	11455	

Equation stochastic specification

(4.9a) $u_t = \varepsilon_t - \underset{(3.19)}{0.6004}\varepsilon_{t-1}$

(4.11a) $u_t = \varepsilon_t + \underset{(6.27)}{1.073}\varepsilon_{t-1} + \underset{(4.34)}{0.7333}\varepsilon_{t-2}$

From Table 7.5, which contains results for durables, the following points emerge:

(i) The *a priori* sign restrictions are satisfied in case of (4.5), (4.10), (4.12), (4.14a), (4.15a), (4.15b) (with minor exceptions in (4.10) and (4.12) where N_{t-1}, N_{t-2} and N_{t-3} have the "wrong" sign). This is a relatively minor dent in the basic model as all the coefficients in question are not significantly different from zero. Furthermore, the model was not expected to be quite suitable for durables finished goods inventory.

(ii) For both durables and non-durables the fit is closest for ΔRM and worst for ΔGIP, with ΔFG in between. We also obtain this ordering in time series models. In both cases the ordering conforms to the ordering of the coefficient of variation of the dependent variable. The overall performance of the basic model in explaining the variation in the dependent variable is satisfactory. For durables, the level of past orders and the rate of change of unfilled orders contributes strongly to the explanation, D/N_{t-2} has a larger coefficient than D/N_{t-1} and D/N_{t-3}. The effect of an increase in the rate of change of D/U_{t-1}, interpreted here as an increase in the rate of change of capacity utilization, is to increase stocks. As in many previous studies the significant role of the stock adjustment process is confirmed by the results. For D/ΔGIP and D/ΔFG the preceding remarks still do not apply with the same force, since the flexible accelerator model appears inadequate for explaining D/ΔGIP. This is a worrying shortcoming in view of the high quarterly average value of this component. On the other hand, the performance of the model in explaining D/ΔFG is quite satisfactory though the seasonals dominate somewhat.[15] To test whether the estimated equations adequately reflect the complexity of the distributed lag, diagnostic checks provided by the empirical autocorrelation function of residuals and cross-correlation function between regression residuals and N_t were used. The calculated autocorrelation coefficients for residuals from (4.5) and (4.10) are now given.

Lag	1	2	3	4	5	6	7	8	9	10
Equation:										
(4.5)	0.088	0.109	0.073	−0.099	0.085	−0.286	−0.173	0.045	0.130	−0.148
(4.10)	0.26	0.077	−0.024	−0.063	0.100	0.035	0.135	−0.056	−0.170	−0.194

[15] The durables category also includes some industries for which it is not sensible to think of ΔFG as a decision variable.

Only the value at lag 6 borders on statistical significance. For non-durables it is more important to explain ND/ΔRM and ND/ΔFG than ND/ΔGIP, since the last item is on average about $2\frac{1}{2}$ times smaller. Table 7.6 shows that the model is very satisfactory for ND/ΔFG in the sense that all coefficients are well determined. The buffer stock variable and expected sales play their hypothesized role in determining actual inventory investment. The addition of interactive seasonals *does not* significantly improve the explanation though this conclusion should be interpreted with caution since equation (4.11) is *not* nested in (4.11a). On theoretical grounds (4.11) seems preferable. For ND/ΔRM the chosen model is not appropriately judged either by results in table 7.6 or table 7.7. Finally, for ND/ΔGIP the model is only marginally worse than for ND/ΔFG. Here, the inclusion of interactive seasonals improves the fit somewhat, the comparison being once again obscured by the fact that (4.13) is not nested in (4.13a).

(iii) The two preceding sections emphasized the heterogeneity of behavior of components of aggregate inventory; the reasons given suggest why a single explanatory model cannot be expected to perform uniformly well. To a certain extent an inability to explain the aggregate satisfactorily will reflect such difficulties both in the structural–econometric and in the time series framework. On the other hand, if the disaggregated series are individually subject to relatively greater random variation (including variation due to measurement errors) than the aggregate, then the aggregate equation provides a useful overall check on the empirical validity of the model. Considered in this way the results of (4.14), (4.15a), (4.16) and (4.16a) provide additional support for the basic model.

Equations (4.14) and (4.15a) differ in the choice of dependent variable and in the use of interactive seasonals in place of simple seasonals in the latter. Equation (4.14) uses the concept of total inventory owned (TIO) which differs from the total held (TIH) in that it excludes the value of progress payments. (See section 2, p. 291.) Thus it is to be expected, and table 7.5 confirms, that our model would provide a somewhat tighter fit when D/ΔTIO is the dependent variable. (In case of non-durables there is no conflict since ΔTIO = ΔTIH.) Both equations, while confirming the empirical validity of the model, suffer from the same limitation; that is, the autocorrelation function of the residuals shows high values at lags 6, 7, and 9: −0.314, −0.348, and −0.316, respectively. The time series models (see table 7.3), do not share this characteristic, but the residual variance for both D/ΔTIO and D/ΔTIH is considerably larger than for the structural models.

The above comments apply to non-durables as well, with the qualifications that the chosen stochastic specifications appear to deal with the problem of autocorrelation quite satisfactorily; on the other hand, the coefficients are less well determined. The residual variance of the structural model is once again smaller than that of the corresponding time series model, 11455 (table 7.7) compared with 12380 (table 7.4).

(iv) The structural models estimated seem to suggest that at least for the durables, the polynomial $\beta(B)$ (see (1.3)), is of a high order, perhaps 6 or 7, whereas, $\alpha(B)$ and $\eta(B)$ are low order polynomials. Thus when we also take account of $\omega(B)$ and $\mu(B)$ it seems unlikely that ARIMA models for durables inventory investment will be characterized by a low degree of parameterization. This problem is likely to be less serious for the non-durable group.

6 Conclusion

Structural analysis provides additional insight into the workings of time series (forecasting) models of inventory behavior. In particular, it appears that three difficulties which plague econometric research also affect time series models. These are: the problem of aggregation over different types of inventories, the complexity of lag distributions on explanatory variables, and the presence of unobservable variables on the structural model. Some might argue that these pose no difficulties at all for those who want to ignore all structural information and predict unrestricted ARIMA reduced forms. The present chapter, it seems, has provided some grounds for skepticism regarding the superiority of this alternative over econometric modeling.

From another point of view time series and structural models need not be thought of as competing approaches. A possible way of integrating them in the present context is to regard the expected orders or expected sales as the systematic part of a random variable whose generation may be described by an ARMA process. The decomposition into random and systematic components may be carried out by a preliminary fitting of an ARMA model to the sales series; the predictions generated by the fitted model can then be used as an explanatory variable in a structural model. This two-stage procedure is based on somewhat restrictive assumptions regarding the joint distribution of errors on the two relevant equations. Under somewhat more general stochastic assumptions one would fit the ARMA and the structural models simultaneously, utilizing any across-equation constraints that arise. Though computationally this is more difficult, statistically it may be more efficient. However, the issues arising from these considerations would take us beyond the scope of this chapter.

BIBLIOGRAPHY

Abramovitz, M. (1950), *Inventories and Business Cycles* (New York, National Bureau of Economic Research)
Beisley, D. A. (1969), *Industry Production Behaviour: The Order-Stock Distinction* (Amsterdam: North-Holland)
Box, G. E. P. and G. M. Jenkins (1970), *Time Series Analysis, Forecasting and Control* (San Francisco, Holden-Day)
Childs, G. L. (1967), *Unfilled Orders and Inventories* (Amsterdam, North-Holland)
Courchene, T. J. (1967), "Inventory behaviour and the stock-order distinction. An analysis by industry and stage of fabrication with empirical application to the Canadian manufacturing sector," *Canadian Journal of Economics and Political Science* 33, 325–57
Darling, P. G. and M. C. Lovell (1971), "Inventories, production smoothing and the flexible accelerator," *Quarterly Journal of Economics* 85, 357–62
Hannan, E. J. (1970), *Multiple Time Series* (New York, John Wiley)
Helliwell, J. F. *et al.* (1972), *The Structure of RDX2*, Bank of Canada Staff Research Studies 7, Ottawa, Ontario
Hendry, D. F. (1974), "Stochastic specification in an aggregate demand model of the UK," *Econometrica* 42, 559–74
Hirsch A. A. and M. C. Lovell (1969), *Sales Anticipations and Inventory Behavior* (New York, John Wiley)
Kang, K. (1973), "A comparison of least squares and maximum likelihood estimation for moving average processes," Paper presented at the Third National Congress of Australian and New Zealand Economists, Adelaide
Lovell, M. C. (1961), "Manufacturers' inventories, sales expectations, and the acceleration principle," *Econometrica* 29, 293–314
Lovell, M. C. and P. G. Darling (1965), "Factors influencing investment in inventories," in J. Duesenberry *et al.* (eds.), *The Brookings Quarterly Model of the US* (Chicago, Rand-McNally)
Naylor, T. H., T. G. Seaks, and D. W. Wichern (1972), "Box–Jenkins methods: an alternative to econometric models," *International Statistical Review* 40, 123–37
Pierce, D. A. (1971), "Distribution of residual autocorrelations in regression models with ARMA errors," *Journal of Royal Statistical Society*, Series B 33(1), 140–6
(1972), "Residual correlations and diagnostic checking in dynamic-disturbance time series models," *Journal of American Statistical Association* 47, 636–40
Stanback, T. M. (1962), *Postwar Cycles in Manufacturers' Inventories* (New York, National Bureau of Economic Research)
Trivedi, P. K. (1969), *An Econometric Study of Inventory Behaviour in the UK Manufacturing Sector*, 1956–67, unpublished PhD dissertation, University of London

(1970), "Inventory behaviour in UK manufacturing, 1956–67," *Review of Economic Studies* 37, 517–36

(1973), "Retail inventory investment behaviour," *Journal of Econometrics* 1, 61–80

Trivedi, P. K. and J. C. R. Rowley (1975), *Econometrics of Investment* (London, John Wiley)

Zellner A. and F. C. Palm (1974), "Time series analysis and simultaneous equation econometric models," *Journal of Econometrics* 2, 17–54; chapter 1 in this volume

8 Time series analysis of the German hyperinflation (1978)

Paul Evans

1 Introduction

Historical studies of periods of rapid and sustained inflation as well as empirical studies of the demand for and supply of money have convinced monetarists that a strong link exists between money and prices. Indeed, Friedman (1968) has argued that movements in the money supply dominate movements in the price level.

One study that monetarists have often cited as strong evidence for this link between money and prices is Phillip Cagan's [1956] study of hyperinflations. With data on the money supplies and price levels of six countries in the throes of hyperinflation, Cagan finds that the hyperinflations were apparently caused by the pressure of a rapidly growing and exogenous money supply against a stable demand for real money balances. Unfortunately, the statistical procedures available to Cagan did not enable him to test the specification of his model adequately. In particular, he tested neither his specification of the mechanism generating expected inflation rates nor his specification of an exogenous money supply. Indeed, he failed even to test for serial correlation of the error terms.[1] The Monte Carlo experiments of Granger and Newbold (1974) amply demonstrate that the goodness of fit of a regression is often greatly over-stated when serial correlation is present in the error term. It is therefore desirable to reassess Cagan's study.

In this chapter, I apply the technique advocated by Zellner and Palm (1974, 1975) to test three specifications of the dynamics of the German hyperinflation:

I have benefited from discussions with Arnold Zellner and from the comments of Milton Friedman and other participants of a Money and Banking Workshop at the University of Chicago. Of course, all errors are my sole responsibility.

Originally published in the *International Economic Review* 19 (1978), 195–209.

[1] Eden (1974) provides the Durbin–Watson statistics for Cagan's regressions. They indicate strong positive serial correlation in the error terms. Note, however, that Cagan's statistical techniques were sophisticated for the time at which he wrote his thesis.

- Model C, Cagan's original model;
- Model MC, a modification of Cagan's original model in which expectations of inflation are rational in the sense of Muth (1961); and
- Model S&W, named after Sargent and Wallace (1973), in which expectations are rational and the money supply is endogenous.

I find that none of these models is consistent with the data. In particular, expectations are not formed adaptively *à la* Cagan and the money supply is not exogenous. One can, however, conclude that the demand function was stable during the German hyperinflation.

2 Three models of hyperinflation

Model C can be written in discrete time as[2]

$$M_t = \Pi_t - aD\Pi_t^* + c(L)u_t, \quad a > 0; \tag{2.1}$$

$$\Pi_t^* = b\Pi_{t-1}^* + (1-b)\Pi_t, \quad 0 < b < 1; \tag{2.2}$$

$$DM_t = f(L)v_t; \tag{2.3}$$

where M_t is the exogenous rate of growth of the money supply (= money demand); Π_t is the rate of inflation; Π_t^* is the rate of inflation expected in period t to take place between periods t and $t+1$; D is the difference operator (e.g. $DM_t = M_t - M_{t-1}$); L is the lag operator (e.g. $Lv_t = v_{t-1}$); $c(L) = 1 - c_1L - \cdots - c_qL^q$ and $f(L) = 1 - f_1L - \cdots - f_rL^r$; a, b, $c_1, \cdots c_q, f_1, \ldots, f_r$ are parameters; and u_t and v_t are independently and identically distributed random variables with zero means and finite variances. Equation (2.1) is the demand function for real balances. It is written in first differences because real income, one of the real variables collapsed into the disturbance term $c(L)u_t$ is likely to be non-stationary in its levels.[3] This disturbance term is posited to be a qth order moving average in the independently and identically distributed error term u_t. Equation (2.2) is a standard adaptive expectations equation. It has the implication that the expected rate of inflation adapts by the fraction $1 - b$ of the departure of the actual rate of inflation from its expectation in the previous period. The growth rate of the money supply is a stochastic variable. As such, it can under weak conditions, be represented as an autoregressive, integrated, moving average (ARIMA) process of order

[2] I have suppressed all constant terms in the theoretical models of the paper.

[3] Let

$$\log(m_t/P_t) = a_0 + a_1 \log Y_t - a\Pi_t^* + U_t,$$

where m_t is the money supply, P_t is the price level, Y_t is real income, and U_t is a stationary disturbance term. If $\log Y_t$ is a difference-stationary process, then $U_t + a_1 \log Y_t$ is non-stationary while $D(U_t + a_1 \log Y_t)$ is stationary. If one identifies $c(L)u_t$ with $D(U_t + a_1 \log Y_t)$, differencing the above equation yields (2.1).

(p,d,r)[4] with p, d, and r finite (and, one must hope, small) positive integers. Equation (2.3), which will be found to be consistent with the data, is a special case of the general ARIMA model, where $p = 0$ and $d = 1$. Note that M_t is an exogenous variable because u_t is independent of v_t.

One can obtain the stochastic process of Π_t for model C by multiplying (2.1) by $1 - bL$ and substituting from (2.2):

$$(1 - bL)M_t = (1 - bL)\Pi_t - a(1 - b)D\Pi_t + (1 - bL)c(L)u_t,$$

or

$$\{[1 - a(1 - b)] - [b - a(1 - b)]L\}\Pi_t = (1 - bL)[M_t - c(L)u_t].$$

Now, differencing this equation and substituting for M_t from (2.3) gives

$$\begin{aligned}\{[1 - a(1 - b)] - [b - a(1 - b)]L\}D\Pi_t \\ = (1 - bL)[f(L)v_t - c(L)Du_t]\end{aligned} \tag{2.4}$$

According to Granger's Lemma, the sum of any number of moving averages can be written as a moving average in a single random variable (see Anderson 1971). The order of this moving average will typically equal the order of the longest constituent moving average. Therefore, (2.4) can be written as

$$g(L)D\Pi_t = h(L)e_t, \tag{2.5}$$

where

$$g(L) \equiv 1 - \left[\frac{b - a(1 - b)}{1 - a(1 - b)}\right]L;$$

$$h(L) \equiv 1 - h_1 L - \cdots - h_{q'}L^{q'};$$

$q' = \max(q + 2, r + 1)$; and e_t is an independently and identically distributed random variable. One can check whether Π_t follows this ARIMA process whose order is $(1, 1, q')$ by using Box–Jenkins techniques. This will be done in the next section.

One frequent objection to the expectations mechanism (2.2) is that it uses only the past history of the inflation rate in predicting its future evolution.[5] In fact, rational and fully informed economic actors would

[4] A variable x_t follows an ARIMA process of order (p,d,r) if it must be differenced at least d times to be stationary, its autoregressive part is of order p, and its moving-average part is of order r.

[5] Equation (2.2) may be written as

$$\Pi_t^* = \left(\frac{1 - b}{1 - bL}\right)\Pi_t = (1 - b)\sum_{i=0}^{\infty} b^i \Pi_{t-i}.$$

Thus, Π_t^* is a geometrically weighted average of current and all past rates of inflation.

incorporate all relevant information – not just the history of inflation rates – into Π_t^*. Moreover, Muth [1961] has shown that, even if only the history of inflation rates is known, the mechanism (2.2) is equivalent to the least squares predictor of Π_{t+1} if, and only if, Π_t obeys the stochastic process

$$D\Pi_t = (1 - bL)e_t.$$

Since $g(L)$ in (2.5) is of degree one, model C cannot obtain unless expectations are not rational *à la* Muth.

Sargent and Wallace have proposed that (2.2) be replaced by

$$\Pi_t^* = E_t\Pi_{t+1}, \tag{2.6}$$

where E_t denotes an expectation conditional on knowledge of Π_t, $\Pi_{t-1}, \ldots, M_t, M_{t-1}, \ldots$ and the structure of the model (2.1), (2.6), and (2.3) (hereafter called model MC). For time $t+1$, (2.1) and (2.6) imply that

$$\Pi_{t+1} = M_{t+1} + a(E_{t+1}\Pi_{t+2} - E_t\Pi_{t+1}) - c(L)u_{t+1}. \tag{2.7}$$

Now, applying the operator E_t to (2.7) and rearranging yields

$$E_t\Pi_{t+1} = \left(\frac{1}{1+a}\right)E_t[M_{t+1} - c(L)u_{t+1}] + \left(\frac{a}{1+a}\right)E_t\Pi_{t+2}. \tag{2.8}$$

Applying the same procedure iteratively enables one to obtain

$$E_t\Pi_{t+1} = \left(\frac{1}{1+a}\right)\sum_{i=0}^{\infty}\left(\frac{a}{1+a}\right)^i E_t[M_{t+i+1} - c(L)u_{t+i+1}], \tag{2.9}$$

where the end condition

$$\lim_{i\to\infty}\left(\frac{a}{1+a}\right)^i E_t\Pi_{t+i+1} = 0$$

has be imposed.[6] Substituting (2.9) into (2.6) and the result into (2.1) gives

$$\mathrm{M}_t - \Pi_t = -\left(\frac{a}{1+a}\right)\sum_{i=0}^{\infty}\left(\frac{a}{1+a}\right)^i DE_t[M_{t+i+1} - c(L)u_{t+i+1}] + c(L)u_t. \tag{2.10}$$

In order to obtain the ARIMA process generating Π_t, one first looks at the term

$$\left(\frac{a}{1+a}\right)\sum_{i=0}^{\infty}\left(\frac{a}{1+a}\right)^i E_t[\mathrm{M}_{t+i+1} - c(L)u_{t+i+1}]. \tag{2.11}$$

Equation (2.3) implies that

$$\begin{aligned} E_t\mathrm{M}_{t+1} &= \mathrm{M}_t - f_1 v_t - f_2 v_{t-1} \cdots - f_r v_{t-r-1} \\ \mathrm{E}_t M_{t+2} &= \mathrm{E}_t M_{t+1} - f_2 v_t - f_3 v_{t-1} - \cdots - f_r v_{t-r-2} \\ &= \mathrm{M}_t - (f_1 + f_2)v_t - (f_2 + f_3)v_{t-1} \cdots - f_r v_{t-r-1} \\ &\ \ \vdots \\ E_t\mathrm{M}_{t+k} &= M_t - (f_1 + f_2 + \cdots + f_r)v_t - (f_1 + f_2 + \cdots + f_{r-1}) \\ &\quad v_{t-1} - \cdots - f_r v_{t-r-1}, \quad k \geq r. \end{aligned}$$

Similarly,

$$\begin{aligned} E_t c(L)u_{t+1} &= -c_1 u_t - c_2 u_{t-1} - \cdots - c_q u_{t-q+1} \\ E_t c(L)u_{t+2} &= -c_2 u_t - c_3 u_{t-1} - \cdots - c_q u_{t-q+2} \\ &\ \ \vdots \\ E_t c(L)u_{t+k} &= 0, \quad k \geq q. \end{aligned}$$

Therefore, the expression (2.11) can be written as

$$aDM_t - F_1 Dv_t - \cdots - F_r Dv_{t-r+1} + C_1 Du_t + \cdots + C_q Du_{t-q+1}, \tag{2.12}$$

where $F_1, \ldots, F_r$ and $C_1, \ldots, C_q$ are linear combinations of the parameters $f_1, \ldots, f_r$ and $c_1, \ldots, c_q$, respectively. Substituting (2.12)

[6] This condition is necessary for stability since its failure at any time t would lead to an immediate explosion of the price level. It is satisfied so long as households anticipate that the rate of inflation will grow at a rate less rapid than $1/a$ per period. Consequently, households will continue to hold some real cash balances at any positive, but constant, rate of inflation, however great it might be. Note that stability does not depend on the rate at which households adjust their expectations to current conditions so long as this adjustment is rational *à la* Muth.

into (2.10) and rearranging gives us

$$\Pi_t = M_t + aDM_t - F_1 Dv_t - \cdots - F_r Dv_{t-r+1}$$
$$+ C_1 Du_t + \cdots + C Du_{t-q+1} - c(L)u_t.$$

Finally, differencing both members of this equation and substituting from (2.3) yields

$$D\Pi_t = (1 + a - aL) f(L)v_t - c(L) Du_t + D^2 [C_1 u_t + \cdots + C_q u_{t-q+1} - F_1 v_t - \cdots F_r v_{t-r+1}]. \tag{2.13}$$

By Granger's Lemma, the right-hand member of (2.13) is a moving average of order $q'' = \max[r + 1, q + 1]$. Therefore, Π_t follows an ARIMA process of order $(0, 1, q'')$. This implication of model MC can also be tested with Box–Jenkins techniques. This, too, will be done in the next section.

Sargent and Wallace point out that the German government printed money during the hyperinflations largely as an expedient for raising revenue. They argue that the government would have followed a rule for expanding the money supply that would on average have maintained its real command over resources. They then propose

$$M_t = \left(\frac{1 - b}{1 - bL}\right) \Pi_t + v_t \tag{2.14}$$

as such a rule. Note that the rate of growth of the money supply is endogenous since it depends on the current rate of inflation. Following Sargent and Wallace, I assume that the disturbance term v_t, which subsumes all other influences on monetary policy, is independent of u_t and serially independent.

In order to obtain the stochastic process of Π_t for the model (2.1), (2.6), and (2.14) (hereafter called model S&W), substitute (2.6) and (2.14) into (2.1):

$$\left(\frac{1 - b}{1 - bL}\right) \Pi_t + v_t - \Pi_t = aDE_t \Pi_{t+1} + u_t,$$

or

$$\left(\frac{b}{1 - bL}\right) \Pi_t = -aDE_t \Pi_{t+1} + v_t - u_t, \tag{2.15}$$

where I further assume that $c(L) = 1$. To solve this model, I first guess that

$$E_t D\Pi_{t+1} = \left(\frac{1 - b}{1 - bL}\right) \Pi_t, \tag{2.16}$$

and then confirm that the guess is correct. Substituting (2.16) into (2.15) yields

$$\left(\frac{b}{1-bL}\right) D\Pi_t = -a\left(\frac{1-b}{1-bL}\right) D\Pi_t + v_t - u_t,$$

or

$$D\Pi_t = (1-bL)(v_t - u_t)/[b + a(1-b)]. \tag{2.17}$$

Hence,

$$\begin{aligned} E_t\Pi_{t+1} &= \Pi_t - b(v_t - u_t)/[b + a(1-b)] \\ &= \Pi_t - \left[\frac{b}{b+a(1-b)}\right]\left[\frac{b+a(1-b)}{1-bL}\right]\Pi_t \\ &= \Pi_t - \left(\frac{b}{1-bL}\right) D\Pi_t = \left(\frac{1-b}{1-bL}\right)\Pi_t. \end{aligned}$$

The guess (2.16) is therefore correct, and the process (2.17) generates Π_t. Consequently, Box–Jenkins procedures should demonstrate that Π_t is generated by a (0,1,1) process if model S&W holds. Moreover, differencing (2.14) and substituting from (2.17) yields

$$DM_t = Dv_t + (1-b)(v_t - u_t)/[b + a(1-b)]. \tag{2.18}$$

Because the right-hand member of (2.18) is a first order moving average, M_t is generated by an ARIMA process of order (0,1,1) in model S&W. These implications will be checked in the next section.

3 Time series analysis

The techniques of Box and Jenkins (1970) are used in order to identify and fit the stochastic processes generating Π_t and M_t for the German hyperinflation.[7] Accordingly, the first step is to examine the sample autocorrelations in table 8.1.

Many non-stationary economic series can be transformed into stationary series by differencing enough times. I assume that Π_t and M_t has such a property. According to Box and Jenkins, one can determine the minimal degree of differencing necessary to induce stationarity by examining the sample autocorrelations of the series.[8] For example, in table 8.1 the

[7] See section 6 for a description of the data. To keep my study comparable with those of Cagan and Sargent and Wallace, I have used Cagan's data. I have also analyzed data from the original hyperinflation sources, obtaining similar results.

[8] Of course, nowadays one would decide this question with a battery of unit-root tests.

Table 8.1 *Sample autocorrelations for Π, M, M − Π, and various differences*[a]

	Lags											
Variable	1	2	3	4	5	6	7	8	9	10	11	12
Π	0.47	0.28	0.39	0.45	0.20	0.23	0.25	0.16	0.13	0.18	0.13	0.07
$D\Pi$	−0.25	−0.26	0.10	0.08	−0.17	0.03	0.01	−0.06	−0.03	0.06	0.09	0.10
$D^2\Pi$	−0.50	−0.15	0.25	−0.02	−0.15	0.13	−0.02	−0.05	0.01	−0.02	0.01	0.12
M	0.84	0.78	0.70	0.60	0.39	0.30	0.18	0.08	0.01	−0.04	−0.07	−0.10
DM	−0.25	0.00	−0.18	0.25	−0.25	−0.04	−0.07	−0.01	−0.01	−0.00	0.07	−0.00
D^2M	−0.52	0.17	−0.21	0.22	−0.15	0.01	−0.04	0.02	−0.01	−0.02	0.08	−0.00
M − Π	0.16	−0.23	0.08	0.04	−0.22	−0.13	−0.06	−0.11	0.01	0.14	0.14	0.03
D(M − Π)	−0.23	−0.37	0.29	−0.02	−0.18	0.07	−0.02	−0.13	0.02	0.04	0.09	0.10

Note:
[a] The approximate standard errors are 0.12 for the entries in the first three rows and 0.18 for the other entries.

first few sample autocorrelations of M_t have the usual pattern for a non-stationary series because they are large and decay slowly.[9] By contrast, the sample autocorrelations for DM_t hover around zero, suggesting that it is stationary. Examining the first few sample autocorrelations of D^2M_t, which are large and negative, strengthens this conclusion. They indicate that D^2M_t is over-differenced; that is, DM_t need not be differenced again to be stationary.

The most important distinguishing feature of a non-stationary series is its failure to decay at early lags. Consequently, the sample autocorrelations of Π_t suggest non-stationarity. By contrast, those of $D\Pi_t$ rapidly approach insignificance as the lag increases, suggesting stationarity.[10] The large negative sample autocorrelations of $D^2\Pi_t$ also indicate over-differencing. It is therefore reasonable to claim that Π_t is not stationary but that $D\Pi_t$ is.

After finding the minimal degree of differencing required for stationarity, the sample autocorrelations are useful for identifying the process generating the serial correlation of the differenced series. For example, the first two sample correlations of $D\Pi_t$ are significant but follow no apparent pattern. According to Box and Jenkins, this behavior suggests that $D\Pi_t$ is a second order moving average and hence that Π_t is an ARIMA (0,1,2)

[9] The first few sample autocorrelations of a non-stationary series tend to be larger and to decay more slowly, the larger is the sample size.

[10] In a large sample with n observations, the sample autocorrelations are distributed as $NID(0, n^{-1/2})$under the null hypothesis of no serial correlation. It may be misleading to base identifications on large-sample results for samples that are not in fact large.

Table 8.2 *Fitted processes for $D\Pi$*[a]

Order	AR1	MA1	MA2	MA3	Constant	S.E.	R^2	Q(12)	Q(24)
(0,1,1)		0.782 (0.074)			0.00613 (0.00488)	0.180	0.230	11.0	15.8
(0,1,2)		0.535 (0.123)	0.303 (0.123)		0.00623 (0.00383)	0.176	0.276	4.9	10.0
(0,1,3)		0.566 (0.130)	0.322 (0.148)	−0.073 (0.131)	0.00626 (0.00420)	0.176	0.280	4.6	9.1
(1,1,0)	−0.272 (0.121)				0.0132 (0.0234)	0.198	0.068	17.4	24.9
(1,1,1)	0.366 (0.120)	0.965 (0.017)			0.00354 (0.00136)	0.178	0.258	7.7	14.1
(1,1,2)	−0.167 (0.434)	0.389 (0.408)	0.408 (0.318)		0.00731 (0.00550)	0.176	0.279	4.7	9.3

Note:
[a] Approximate standard errors appear in parentheses below each parameter estimate. The Q(n) statistic is distributed in large sample as $\chi^2(n-k)$ where n is the order of the Q statistic and k is the number of parameters fitted.

Table 8.3 *Fitted processes for DM*[a]

Order	AR1	MA1	MA2	Constant	S.E.	R^2	Q(12)	Q(24)
(0,1,0)				0.0113 (0.0147)	0.0831	0.076	5.1	6.0
(0,1,1)		0.300 (0.175)		0.0103 (0.0104)	0.0825	0.076	5.1	6.1
(0,1,2)		0.303 (0.199)	−0.010 (0.230)	0.0102 (0.0107)	0.0839	0.076	5.1	6.1
(1,1,0)	−0.295 (0.192)			0.0131 (0.0148)	0.0827	0.073	5.1	6.1
(1,1,1)	−0.368 (0.748)	0.267 (0.692)		0.0106 (0.0130)	0.0839	0.076	5.1	6.1

Note:
[a] Approximate standard errors appear in parentheses below each parameter estimate. The Q(n) statistic is distributed in large samples as $\chi^2(n-k)$ where n is the order of the Q statistic and k is the number of parameters fitted.

process. Similarly, M_t would be identified as a (0,1,0) process because the sample autocorrelations are neither significant at the 0.10 level nor patterned.

Tables 8.2 and 8.3 contain the results of fitting these and some other processes for Π_t and M_t. Table 8.4 gives the test statistics[11] for comparing some of the models fitted in tables 8.2 and 8.3. The tentative identifications appear to be acceptable, but the process (0,1,1) for Π_t and M_t also

[11] Minus twice the logarithm of the likelihood ratio is distributed as $\chi^2(k)$ in large samples, where k is the number of restrictions tested.

Table 8.4 *Comparison of statistical models*

Variable	Models compared	Test statistic
Π	(0,1,1) versus (0,1,2)	2.05
	(0,1,2) versus (0,1,3)	0.23
	(0,1,1) versus (1,1,1)	1.22
	(0,1,2) versus (1,1,2)	0.17
M	(0,1,0) versus (0,1,1)	3.56[a]
	(0,1,1) versus (0,1,2)	0.00
	(0,1,0) versus (1,1,0)	3.47[a]
	(0,1,0) versus (1,1,1)	0.00

Note:
[a] Statistically significant at the 0.10 level.

Table 8.5 *Implied orders of the ARIMA processes for Π and M*[a]

	Π	M
Model C	$(1,1,q+2)$	(0,1,0) or (0,1,1)
Model MC	$(0,1,q+1)$ or $(0,1,\max[2,q+1])$	(0,1,0) or (0,1,1)
Model S&W	(0,1,1)	(0,1,1)
Data	(0,1,2) or (0,1,1)	(0,1,0) or (0,1,1)

Note:
[a] Rows one and two are based on the assumption that M_t is a process of order (0,1,0) or (0,1,1).

appears to be consistent with the data. The data, however, appear to rule out the model (1,1,2) for Π_t.

Table 8.5 presents the implications of the three models and the data for the orders of the ARIMA processes followed by Π_t and M_t. Model C does not appear to be consistent with the data. Models MC and S&W, however, are consistent with the observed orders of the ARIMA processes for Π_t and M_t.

We can eliminate model S&W from consideration by taking another tack, however. Equation (2.18) can be rewritten as

$$DM_t = (1 - \theta L)n_t,$$

where θ is a parameter approximately equal to 0.7[12] and n_t is a serially independent random variable with a zero mean and finite variance. Furthermore, the parameter θ is insensitive to the choice of b, $a(1 - b)$ and var u_t/var v_t. Therefore, a belief in model S&W can be formulated in a prior density function for θ with, say, a mean of 0.7 and a standard deviation of 0.15. In section 7, I derive the functional form of the posterior odds ratio for comparing the null hypotheses H_0 and $\theta = 0$ with the alternative hypothesis H_1 that θ is distributed with a probability density function $f(\theta)$, $-1 < \theta < 1$. In this problem, choosing $f(\theta)$ to be the beta distribution function

$$0.003262(1 + \theta)^{17}(1 - \theta)^2$$

allows one to impose a mean of 0.71 and a standard deviation of 0.15 (see Zellner 1971, pp. 371–3). The sample information then converts prior odds of 1:1 in favor of H_0 into posterior odds of 219:1. Consequently, one can reject model S&W.[13]

4 Further time series analysis of model MC

According to table 8.5, the implications of model MC for the orders of the ARIMA processes generating Π_t and M_t are consistent with the data if $q = 0$ or 1 and $r = 0$. Therefore, if model *MC* obtains, (2.1) can be rewritten as

$$M_t - \Pi_t = -aDE_t\Pi_{t+1} + u_t - c_1u_{t-1}, \tag{4.1}$$

[12] It is easy to show by equating the first order autocorrelations for (2.18) to $-\theta/(1 + \theta^2)$ that

$$\theta = 1 + \frac{x}{2} - \frac{x}{2}\sqrt{1 + \frac{4}{x}},$$

where

$$x \equiv \frac{(1 - b)^2(1 + \operatorname{var} u_t / \operatorname{var} v_t)}{[1 + a(1 - b)][b + a(1 - b)]}.$$

Because of (2.17), 0.787 in table 8.2 estimates b. Choosing $a(1 - b) = 0.75$, the average of Cagan's "reaction indices" ([1956, p. 69]), one finds that $x = 0.0166(1 + \operatorname{var} u_t/\operatorname{var} v_t)$. For variance ratios ranging from 1/5 to 5, θ ranges from 0.73 to 0.87. If either b or $a(1 - b)$ were smaller, one could rationalize somewhat smaller θs. Most reasonable values of b, $a(1 - b)$, and var u_t/var v_t nevertheless imply that θ is near 0.7. A standard deviation of 0.15 also seems to be consistent with these considerations.

[13] The Bayesian procedure for comparing hypotheses is superior to the informal Box–Jenkins and sampling theory methods used in the chapter. It would therefore be desirable to compare models C and MC by computing a posterior odds ratio. Unfortunately, obtaining an expression for the posterior odds ratio for this comparison is a non-trivial problem.

where the parameter c_1 could actually be zero. Equation (2.9) implies that

$$E_t\Pi_{t+1} = \left(\frac{1}{1+a}\right)\sum_{i=0}^{\infty}\left(\frac{a}{1+a}\right)^i E_t(\mathrm{M}_{t+i+1} - u_{t+i+1} + c_1 u_{t+i})$$
$$= \mathrm{M}_t + \left(\frac{c_1}{1+a}\right)u_t.$$

This equation, (2.3) with $r = 0$, and (4.1) lead to

$$\mathrm{M}_t - \Pi_t = -a\left[DM_t + \left(\frac{c_1}{1+a}\right)Du_t\right] + u_t - c_1 u_{t-1}$$
$$= -av_t + \left[\frac{1+a(1-c_1)}{1+a}\right]u_t - \left(\frac{c_1}{1+a}\right)u_{t-1}. \qquad (4.2)$$

According to (4.2), $\mathrm{M}_t - \Pi_t$ is a first order moving average unless $c_1 = 0$, in which case it is white noise. The sample autocorrelations for $\mathrm{M}_t - \Pi_t$ and $D(\mathrm{M}_t - \Pi_t)$ in table 8.1 suggest that $\mathrm{M}_t - \Pi_t$ is white noise. Nevertheless, it proves to be better represented by the first order moving average process

$$\mathrm{M}_t - \Pi_t = -\underset{(0.0520)}{0.0661} + (1 + \underset{(0.168)}{0.458}L)\hat{w}_t \qquad (4.3)$$
$$\text{S.E.} = 0.207,\ R^2 = 0.078,\ Q(12) = 4.0,\ Q(24) = 7.6,$$

where w_t is supposed to be a serially independent disturbance term with a zero mean and finite variance. Furthermore, at the 0.10 significance level, the first order moving average (4.3) cannot be rejected in favor of the second order moving average[14]

$$M_t - \Pi_t = -\underset{(0.0368)}{0.0638} + (1 + \underset{(0.191)}{0.321}L - \underset{(0.198)}{0.294}L^2)\hat{w}_t \qquad (4.4)$$
$$\text{S.E.} = 0.203,\ R^2 = 0.138,\ Q(12) = 2.3,\ Q(24) = 5.3.$$

These findings provide strong evidence that the disturbance term in the money demand function (2.1) was stationary. In this sense, then, money demand was stable during the German hyperinflation. This stability held notwithstanding the non-stationarity of the growth rate of the money supply.

Differencing (4.2), substituting from (2.3) with $r = 0$, and rearranging yields

$$D\Pi_t = v_t + aDv_t - \left[\frac{1+a(1-c_1)}{1+a}\right]Du_t + \left(\frac{c_1}{1+a}\right)Du_{t-1}. \qquad (4.5)$$

[14] Minus twice the logarithm of the likelihood ratio is 2.24.

Table 8.6 *Sample cross-correlations between the prewhitened M and Π series*[a]

Lag												
−6	−5	−4	−3	−2	−1	0	+1	+2	+3	+4	+5	+6
−0.14	−0.17	0.35	−0.25	−0.04	0.32	0.65	−0.43	0.14	−0.06	0.18	−0.15	−0.04

Note:
[a] The approximate standard error for each entry is 0.18. Positive lags imply that money leads prices, and negative lags imply that prices lead money.

By Granger's Lemma, Π_t has the representation

$$D\Pi_t = (1 - \theta_1 L - \theta_2 L^2) n_t, \tag{4.6}$$

where θ_1 and θ_2 are parameters and n_t is a serially independent disturbance term with a zero mean and finite variance. Furthermore, it is clear from (4.5) and (4.6) that the cross-correlations between v_t and n_{t+i} are non-zero only at the lags $i = 0$ and 1. By contrast, model C implies that they are cross-correlated at every lag $i \geq 0$ and not cross-correlated at negative lags, while model S&W implies that they are cross-correlated only at $i = -1$ and 0.[15]

Table 8.6 presents the sample cross-correlations between the residuals from the (0,1,0) process for M_t and residuals from the (0,1,2) process for Π_t.[16] Those at lags −1, 0, and +1 are significant at the 0.10 level, indicating that money not only affects prices but also is affected by them. Therefore, all three models are inconsistent with the data. Note, however, that in constructing his series for the Germany money supply, Cagan obtains his figures for the middle of the month by interpolating between end-of-month figures. The series on M_t could therefore appear to be correlated with Π_{t-1} even if the "true" M_t series was exogenous *vis-à-vis* Π_t. Nevertheless, the substantial cross-correlations at lags −4 and −3 make the conclusion that prices affect money inescapable. One may therefore rule out model MC.

5 Conclusions

None of the models considered in this chapter is consistent with Cagan's data from the German hyperinflation. In particular, Cagan's simple adaptive-expectations mechanism appears not to have described the formation of expectations. Moreover, the growth rate of the money supply

[15] See (2.4), (2.5), (2.17), and (2.18).
[16] See Haugh (1976) for a discussion of the method used below.

appears to have depended on past inflation rates. The growth rate of the German money supply did strongly and contemporaneously affect the inflation rate, however, a finding completely consistent with the existence of a stable demand function for money. There is also no evidence against the hypothesis that the demand for money depended on rationally expected future inflation rates.

6 Data

The basic data from which I calculated M_t, the rate of growth of the money supply, and Π_t, the rate of inflation, appear in Cagan (1956, pp. 102–3). The series Π_t is available from June 1917 to November 1923. I excluded the observations for July 1923–November 1923 for both series because preliminary analysis indicated that the error terms of their ARIMA processes are not covariance stationary when I include any of these observations. The demand for money may thus have been unstable in the last few months of the hyperinflation.

Note that the data are not ideal for the purposes of this [chapter] since Cagan calculated some of the observations by interpolation. It is well known that interpolation may disguise the stochastic properties of a series.

7 Bayesian analysis of the first order moving average process

Consider the first order moving average

$$y_t = u_t - \theta u_{t-1}, \tag{7.1}$$

where u_t is NID$(0,\sigma^2)$. The likelihood function for $y \equiv (y_1, y_2, \ldots, y_n)'$, a vector of n observations on y_t, is

$$(2\pi)^{-n/2}\sigma^{-n}|V(\theta)|^{-1/2}\exp\left[-\frac{y'V^{-1}(\theta)y}{2\sigma^2}\right], \tag{7.2}$$

where $V(\theta)$ is the $n \times n$ banded matrix

$$V(\theta) \equiv \begin{bmatrix} 1+\theta^2 & -\theta & 0 & \cdots & 0 \\ -\theta & 1+\theta^2 & -\theta & \cdots & 0 \\ 0 & -\theta & 1+\theta^2 & \cdots & 0 \\ \vdots & \vdots & \vdots & \ddots & \vdots \\ 0 & 0 & 0 & \cdots & 1+\theta^2 \end{bmatrix}. \tag{7.3}$$

The null hypothesis is

$$H_0 : \theta = 0 \quad \text{and} \quad p(\theta) \propto 1/\sigma, \quad 0 < \sigma < \infty. \tag{7.4}$$

The improper prior $p(\sigma) \propto 1/\sigma$ is frequently chosen when one is completely agnostic about the variance σ^2. The alternative hypothesis

$$H_1 : p(\theta, \sigma) \propto f(\theta)/\sigma, \quad -1 < \theta < 1, \quad \theta \neq 0, \quad 0 < \sigma < \infty, \tag{7.5}$$

is agnostic about the variance but informative about θ since $f(\theta)$ is a proper probability density function, which I assume to be bounded, continuous and positive over the entire parameter space. Following Zellner (1971, p. 298), one may therefore write the posterior odds ratio as

$$\begin{aligned} K_{01} &= \left[\frac{p(H_0)}{p(H_1)}\right] \\ &\times \left\{\frac{(2\pi)^{-n/2}|V(0)|^{-1/2}\int_0^\infty \sigma^{-n-1}\exp[-y'V^{-1}(0)y/2\sigma^2]d\sigma}{(2\pi)^{-n/2}\int_{-1}^1 f(\theta)|V(\theta)|^{-1/2}\int_0^\infty \sigma^{-n-1}\exp[-y'V^{-1}(\theta)y/2\sigma^2]d\sigma d\theta}\right\} \\ &= \left[\frac{p(H_0)}{p(H_1)}\right]\left\{\frac{\int_0^\infty \sigma^{-n-1}\exp(-y'y/2\sigma^2)d\sigma}{\int_{-1}^1 f(\theta)|V(\theta)|^{-1/2}\int_0^\infty \sigma^{-n-1}\exp[-y'V^{-1}(\theta)y/2\sigma^2]d\sigma d\theta}\right\}, \end{aligned}$$

or

$$K_{01} = \left[\frac{p(H_0)}{p(H_1)}\right] \Bigg/ \int_{-1}^{1} f(\theta)|V(\theta)|^{-1/2}[y'V^{-1}(\theta)y/y'y]^{-n/2}d\theta, \tag{7.6}$$

since V(0) is the $n \times n$ identity matrix. In (7.6), $p(H_0)/p(H_1)$ is the prior odds ratio.

One can show by multiplying the matrix below by $V(\theta)$ that

$$V^{-1}(\theta) \equiv \begin{bmatrix} C_{n-1}C_0 & \theta C_{n-2}C_0 & \theta^2 C_{n-3}C_0 & \cdots & \theta^{n-1}C_0C_0 \\ \theta C_{n-2}C_0 & C_{n-2}C_1 & \theta C_{n-3}C_1 & \cdots & \theta^{n-2}C_0C_1 \\ \theta^2 C_{n-3}C_0 & \theta C_{n-3}C_1 & C_{n-3}C_2 & \cdots & \theta^{n-3}C_0C_2 \\ \vdots & \vdots & \vdots & \ddots & \vdots \\ \theta^{n-1}C_0C_0 & \theta^{n-2}C_0C_1 & \theta^{n-3}C_0C_2 & \cdots & C_0C_{n-1} \end{bmatrix}, \tag{7.7}$$

where

$$C_i \equiv \frac{1-\theta^{2(i+1)}}{1-\theta^2}, \quad i = 1, \ldots, n, \tag{7.8}$$

is the determinant of the $i \times i$ banded matrix analogous to $V(\theta)$. Note that $V^{-1}(\theta)$ is symmetric and each entry v^{ij} above the diagonal equals

$\theta^{j-i}C_{n-j}C_{i-1}$. The diagonal entry v^{ii} is $C_{n-i}C_{i-1}$. Therefore,

$$y'V^{-1}(\theta)y = C_n^{-1}\sum_{i=1}^{n} C_{n-i}C_{i-1}y_i^2 + 2C_n^{-1}\sum_{i=1}^{n}\sum_{\substack{j=1\\ j>i}}^{n}\theta^{j-i}C_{n-j}C_{i-1}y_iy_j$$

$$= S(\theta) + 2\sum_{k=1}^{n-1}\theta^k R_k(\theta), \tag{7.9}$$

where

$$S(\theta) \equiv \sum_{i=1}^{n}\left[\frac{1-\theta^{2(n+1-i)}}{1-\theta^{2(n+1)}}\right]\left[\frac{1-\theta^{2i}}{1-\theta^2}\right]y_i^2, \tag{7.10}$$

and

$$R_k(\theta) \equiv \sum_{j=k+1}^{n}\left[\frac{1-\theta^{2(n+1-j)}}{1-\theta^{2(n+1)}}\right]\left[\frac{1-\theta^{2(j-k)}}{1-\theta^2}\right]y_jy_{j-k}, \quad k = 1, \ldots, n-1. \tag{7.11}$$

Substituting (7.9) into (7.6) yields

$$K_{01} = \left[\frac{p(H_0)}{p(H_1)}\right] \Big/ \int_{-1}^{1} \mathrm{f}(\theta) \times \left\{\frac{(1-\theta^2)y'y}{[1-\theta^{2(n+1)}]\left[S(\theta) + 2\sum_{k=1}^{n-1}\theta^k R_k(\theta)\right]}\right\}^{n/2} d\theta \tag{7.12}$$

since $|V(\theta)| = C_n = [1-\theta^{2(n+1)}]/(1-\theta^2)$.

BIBLIOGRAPHY

Anderson, T. W. (1971), *The Statistical Analysis of Time Series* (New York, John Wiley)

Box, G. E. P. and G. M. Jenkins (1970), *Time Series Analysis* (San Francisco, Holden-Day)

Cagan, P. (1956), "The monetary dynamics of hyperinflation," in M. Friedman (ed.), *Studies in the Quantity Theory of Money* (Chicago, University of Chicago Press)

Eden, B. (1974), "On the specification of the demand for money: some theoretical and empirical results," University of Chicago, unpublished manuscript

Friedman, M. (1968), "The role of monetary policy," *American Economic Review* 56, 1–17

Granger, C. W. J. and P. Newbold (1974), "Spurious regressions in economics," *Journal of Econometrics* 2, 111–20
Haugh, L. D. (1976), "Checking the independence of two covariance-stationary time series: a univariate residual cross-correlation approach," *Journal of the American Statistical Society* 71, 378–85
Muth, J. F. (1961), "Rational expectations and the theory of price movements," *Econometrica* 29, 315–35
Sargent, T. J. and N. Wallace (1973), "Rational expectations and the dynamics of hyperinflation," *International Economic Review* 15, 328–50
Zellner, A. (1971), *Introduction to Bayesian Econometrics* (New York, John Wiley)
Zellner, A. and F. C. Palm (1974), "Time series analysis and simultaneous equation econometric models," *Journal of Econometrics* 2, 17–54; chapter 1 in this volume
(1975), "Time series and structural analysis of monetary models of the US economy," *Sankhyā: The Indian Journal of Statistics*, Series C 37, 12–56; chapter 6 in this volume

9 A time series analysis of seasonality in econometric models (1978)

Charles I. Plosser

1 Introduction

The traditional literature on seasonality has mainly focused attention on various statistical procedures for obtaining a seasonally adjusted time series from an observed time series that exhibits seasonal variation. Many of these procedures rely on the notion that an observed time series can be meaningfully divided into several unobserved components. Usually, these components are taken to be a trend or cyclical component, a seasonal component, and an irregular or random component. Unfortunately, this simple specification, in itself, is not sufficient to identify a unique seasonal component, given an observed series. Consequently, there are difficult problems facing those wishing to obtain a seasonally adjusted series. For example, the econometrician or statistician involved in this adjusting process is immediately confronted with several issues. Are the components additive or multiplicative? Are they deterministic or stochastic? Are they independent or are there interaction effects? Are they stable through time or do they vary through time? Either explicitly or implicitly, these types of questions must be dealt with before one can obtain a seasonally adjusted series.

One approach to answering some of these questions would be to incorporate subject-matter considerations into the decision process. In particular, economic concepts may be useful in arriving at a better understanding of seasonality. Within the context of an economic structure (e.g. a simple

This work has been financed, in part, by the National Science Foundation under Grant GS 40033 and the H. G. B. Alexander Research Foundation, Graduate School of Business, University of Chicago. The author is grateful to Arnold Zellner for his helpful comments and encouragement. J. M. Abowd, R. E. Lucas, H. V. Roberts, G. W. Schwert, and W. E. Wecker also provided helpful suggestions. All remaining errors, however, remain the sole responsibility of the author.

Originally published in A. Zellner (ed.), *Seasonal Analysis of Economic Time Series*, Proceedings of the Conference on the Seasonal Analysis of Economic Time Series, Washington, DC, September 9–10, 1976, Economic Research Report ER-1, Washington, DC: Bureau of the Census, US Department of Commerce, December 1978, 365–407.

supply and demand model), the seasonal variation in one set of variables, or in one market, should have implications for the seasonal variation in closely related variables and markets.[1] For example, the seasonality in the amount of labor supplied in non-agricultural labor markets is not independent of the labor demanded in agricultural labor markets. Consequently, knowledge of the economic structure can provide one with a great deal of understanding about the seasonal variation of different variables, such as where it comes from and what might cause it to vary through time.

The purpose of this chapter is to suggest and investigate an approach that involves the incorporation of seasonality directly into an economic model.[2] Analyzing the problem from this perspective has two important implications. First, if an adjusted series is the objective, an economic model that incorporates seasonality may provide an analyst with a better understanding of the source and type of seasonal variation, as indicated in the previous paragraph. This understanding, in turn, may aid in the development of improved adjustment procedures. Second, including seasonality in an economic model avoids the necessity of using a seasonally adjusted data base in estimating an economic model and subsequent concern over whether the seasonal adjustment procedure itself may be causing distortions of the economic analysis and the interpretation of the model.[3] For example, although many economic time series are available in adjusted form, there are some series that are not adjusted at all (e.g. interest rates). Wallis (1974) shows how the use of adjusted and unadjusted data in the same model can lead to spurious dynamic relationships between variables where dynamic relationships do not otherwise exist.

Furthermore, to the extent model builders do not take seasonality into account in the specification of a model because they believe that using seasonally adjusted data has eliminated that need, they could be led into model misspecification, misleading inferences about parameter values, and poor forecasts. Such problems would naturally arise if the adjustment procedure did not effectively eliminate the seasonal variation in the data. Consequently, the adjustment procedure may have the effect of

[1] Kuznets (1933) was concerned with how seasonal movements worked their way through various markets. Fundamental to this approach is the idea of *induced or derived seasonal variation*. That is, seasonality is induced into some markets because of seasonality in other markets. However, Kuznets first obtained what he called the seasonal component of an observed series and proceeded to compare these seasonal components in related markets.

[2] An example of how an economic model can be built to generate seasonal or periodic behavior can be found in US Department of the Interior (1962).

[3] Laffer and Ranson (1971) were concerned with this problem and made use of seasonal dummies in an attempt to avoid the dependence on the seasonally adjusted data.

inducing properties on a series that are spurious concerning the model under consideration.

In Zellner and Palm (1974), techniques were developed for analyzing dynamic econometric models that combined traditional econometric modeling with the time series techniques developed by Box and Jenkins (1970). In a subsequent work, Zellner and Palm (1975), these techniques were applied to the analysis of several monetary models of the US economy. Using monthly data for 1953–72, the information in the data was checked against the implications derived from these models. They pointed out, in the conclusion of their work, that even though they were using seasonally adjusted data, effects of seasonality seemed to be present in the autocorrelation structure of some of the variables, as well as in the residuals of the transfer functions. These complications might be expected from data that are smoothed in the same manner, regardless of the underlying stochastic process or economic mechanism at work.

Finally, if the data being used to test and estimate a model are inappropriate for the particular model, the model is likely to produce poor forecasts. Even in the case of forecasting univariate time series, the effects of seasonal adjustment may cause poor predictions. This lack of prediction accuracy may arise from the fact that the adjustment procedures periodically undergo revision, such that the form of the filter and the weights employed are changing through time. That is, the raw data are being passed through a filter that may vary considerably over a particular sample period. The result would be to introduce an instability in the stochastic properties of the adjusted data that may not exist in the raw or unadjusted data.

Figures 9.1 and 9.2 provide an illustrative example of the type of prediction problem suggested in the preceding paragraph. Using the methodology of Box and Jenkins (1970), a univariate time series model was built for the unadjusted money stock (M1). The model was identified using monthly data for January 1953–December 1962 and then used to forecast unadjusted M1 through 1963 (i.e. forecasting up to twelve steps ahead). Subsequently, the model was updated with actual data through December 1963 and then used to forecast M1 for 1964. This process was repeated through 1972. The results of this exercise are presented graphically in figure 9.1. These are the plots of the actual and the predicted series as well as a set of 95 percent prediction intervals. As can be seen, the model seems to do rather well with the actual series coming close to being outside the prediction interval in 1967 and again in 1969. Even at the twelve-step-ahead forecast, the error is rarely more than 1 to 2 percent.

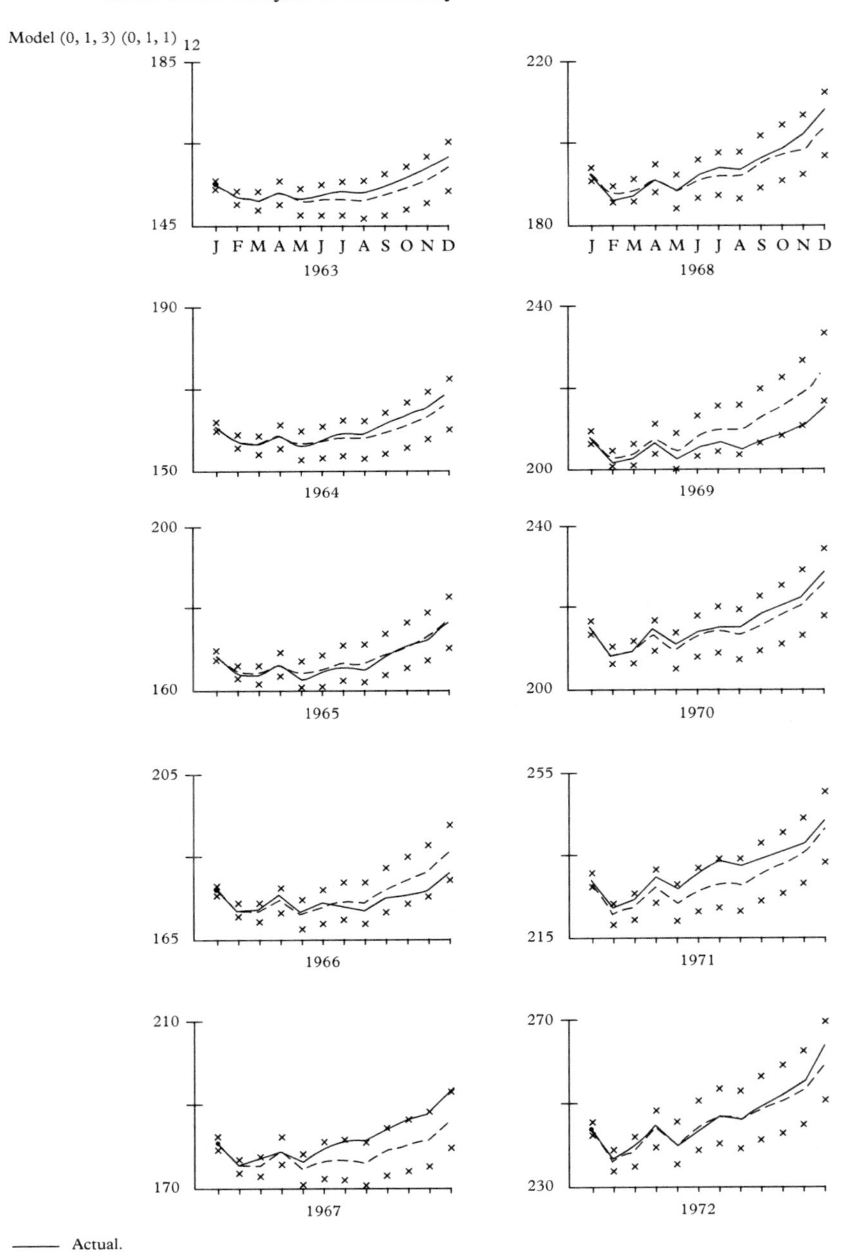

Figure 9.1 Updated predictions for M1, unadjusted, 1963–1972, billion dollars

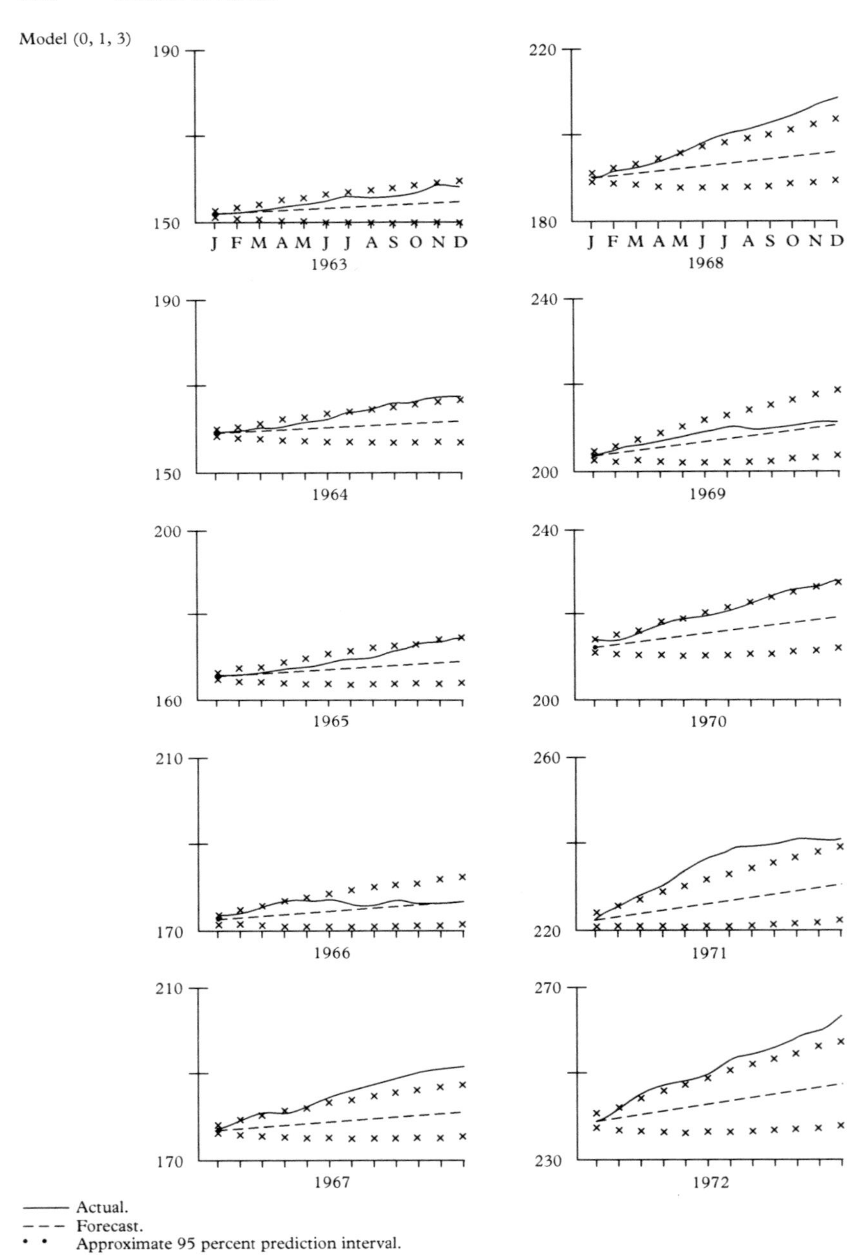

Figure 9.2 Updated predictions for M1, adjusted, 1963–1972, billion dollars

In contrast to this is a model developed using the same techniques for the seasonally adjusted money supply. The same updating and prediction procedure was performed with the model, and the results are shown in figure 9.2. Notice the relatively larger prediction errors at the twelve-step-ahead forecast. More important is the observation that the actual series often wanders outside the prediction interval. It is, of course, very difficult to compare the results in figures 9.1 and 9.2 directly because, in fact, the models are predicting two different series. A complete analysis of the findings presented in figures 9.1 and 9.2 would constitute a study in and of itself, but such an analysis is not the intention of this work. However, these simple results should be sufficient to cause one to ask questions concerning the role of adjustment – and, perhaps, its usefulness in forecasting.[4]

The organization of the remainder of this chapter is as follows: section 2 is a methodological section that includes a brief discussion of the analysis of linear dynamic econometric models as developed in Zellner (1975) and Zellner and Palm (1974), as well as some of the theoretical aspects involved in modeling seasonal time series. Suggestions are then made concerning the way one might go about building seasonality into a model and how to check the consistency of the specification of the model with data. In section 3, a simple economic model is proposed with explicit assumptions regarding the manner in which seasonality enters the system. This is followed by a detailed discussion of the implications of the model for the properties of the stochastic processes for the endogenous variables. In particular, consideration is given to how the effects of changes in the values of structural parameters and of properties of the processes for exogenous variables would lead to changes in the seasonal properties of the output variables of the model. Section 4 presents the results of an empirical analysis of the model, and section 5 provides a discussion of the results and implications for future research.

2 Methodology for analyzing seasonal economic models

In this section, a methodology is suggested for analyzing seasonal economic models. In . . . subsection [2.1] on the analysis of linear dynamic econometric models, a brief discussion is provided of the analysis of linear dynamic econometric models as developed by Zellner (1975) and Zellner and Palm (1974). In . . . subsection [2.2] on seasonality in time

[4] There are certainly alternative explanations for this observed phenomenon. However, these results are only meant to be suggestive, and not conclusive evidence of the distortions that may be caused by seasonal adjustment. The reader who is interested in the details of the development of the exact models used for this example is referred to Plosser (1976).

series data, several approaches to modeling data with seasonal variation are discussed. Finally, in . . . subsection [2.3] on an approach to the analysis of seasonality in structural models, the methodology developed in . . . subsection [2.1] . . . and [2.2] . . . is utilized to illustrate several ways of incorporating seasonality in an econometric model and techniques for checking the model's specification against information in the data.

2.1 *Analysis of linear dynamic econometric models*

As indicated by Quenouille (1968) and Zellner and Palm (1975), a linear multiple time series (MTS) process can be written as follows:

$$\underset{p\times p}{H(L)}\underset{p\times 1}{\underline{z}_t} = \underset{p\times p}{F(L)}\underset{p\times 1}{\underline{e}_t}, \quad \text{for } t = 1, 2, \ldots, T \tag{2.1}$$

where $\underline{z}_t$ is a vector of p observable variables (in this case written as deviations from their respective means), $\underline{e}_t$ is a $p \times 1$ vector of unobservable random errors, L is the lag operator such that $L^k x_t = x_{t-k}$, and $H(L)$ and $F(L)$ are $p \times p$ matrices of full rank having elements that are finite polynomials in L. In addition, the error vector $\underline{e}_t$ is assumed to have the following properties:

$$E\underline{e}_t = \underline{0} \quad \text{for all } t, t^* \quad E\underline{e}_t\underline{e}'_{t^*} = \delta_{tt^*} I_p, \tag{2.2}$$

where δ_{tt^*} is the Kronecker delta and I_p is a $p \times p$-unit matrix. Note that contemporaneous and serial correlations between errors are introduced through $F(L)$.

This general MTS model includes the linear dynamic simultaneous equation model as a special case. Assume that prior information, in particular economic theory, suggests that certain elements of $\underline{z}_t$ can be treated as being endogenous and others as being exogenous. The system (2.1) can then be written as follows:

$$\begin{bmatrix} H_{11} & H_{12} \\ H_{21} & H_{22} \end{bmatrix} \begin{bmatrix} \underline{y}_t \\ \underline{x}_t \end{bmatrix} = \begin{bmatrix} F_{11} & F_{12} \\ F_{21} & F_{22} \end{bmatrix} \begin{bmatrix} \underline{e}_{1t} \\ \underline{e}_{2t} \end{bmatrix}. \tag{2.3}$$

Given that $\underline{y}_t$ represents a vector of endogenous variables and $\underline{x}_t$ a vector of exogenous variables, the following restrictions are implied:

$$H_{21} \equiv 0, \quad F_{12} \equiv 0 \quad \text{and} \quad F_{21} \equiv 0; \tag{2.4}$$

with these restrictions imposed, the usual structural equations from (2.3) are given in (2.5):

$$H_{11}\underline{y}_t + H_{12}\underline{x}_t = F_{11}\underline{e}_{1t}, \tag{2.5}$$

and

$$H_{22}\underline{x}_t = F_{22}\underline{e}_{2t} \tag{2.6}$$

represents an autoregressive moving average process generating the exogenous variables.[5]

If it is now assumed that the roots of the determinantal equations $|H_{11}(\xi)| = 0$ and $|H_{22}(\xi)| = 0$ lie outside the unit circle, the system (2.5) can be rewritten in two forms that can be of use in analyzing the model. The first form represents a system of final equations (FEs) for the endogenous variables. They are obtained by substituting for $\underline{x}_t$ in (2.5) the expression

$$\underline{x}_t = H_{22}^{-1} F_{22}\underline{e}_{2t} \tag{2.7}$$

and then premultiplying both sides of the resulting expression by the adjoint of H_{11} that yields

$$|H_{11}|\underline{y}_t + H_{11}^{*} H_{12} H_{22}^{-1} F_{22}\underline{e}_{2t} = H_{11}^{*} F_{11}\underline{e}_{1t}, \tag{2.8}$$

or

$$|H_{11}||H_{22}|\underline{y}_t = -H_{11}^{*} H_{12} H_{22}^{*} F_{22}\underline{e}_{2t} + |H_{22}| H_{11}^{*} F_{11}\underline{e}_{1t}, \tag{2.9}$$

where $|H_{ij}|$ denotes the determinant and H_{ij}^{*} the adjoint matrix of H_{ij}. This representation implies that each endogenous variable can be written in the form of an autoregressive integrated moving average (ARIMA) model of the type developed and analyzed by Box and Jenkins (1970). Thus, as emphasized by Zellner and Palm, those who utilize the Box and Jenkins models for forecasting are not making use of a technique that is necessarily distinct from standard econometric models. In fact, they are utilizing a very specialized reduced form, the FE, that is well suited for forecasting but may or may not be very informative for structural analysis. However, this representation of the model can provide insights into the stochastic structure of the endogenous variables in the system. For example, if one is interested in seasonality, the autocorrelation coefficient at the seasonal lag can be analyzed with respect to changes in structural

[5] If one or more of the elements of $\underline{x}_t$ is deterministic, it can not be handled in this fashion but must be analyzed through the transfer functions, a discussion of which will follow.

parameters or changes in the processes generating the exogenous variables. Furthermore, this type of analysis is helpful in understanding what type of adjustment procedure may be suggested by the model.

Upon inspection, several things can be noted about (2.9). First, since the assumption is made that all the elements of $F(L)$ and $H(L)$ are finite polynomials in L, then if no cancellation takes place, it is apparent that each and every endogenous variable in the system will have identical autoregressive (AR) polynomials and they will be of order equal to or greater than the AR polynomials for the elements of $\underline{x}_t$. This theoretical restriction might be one means of testing the model against information obtained from the data. In addition, there are restrictions placed on the form of the moving average (MA) polynomial in (2.9). However, there are possible reasons why these theoretical restrictions on the AR and MA polynomials may not be observed in the data even when the model is true. One problem, mentioned by Zellner and Palm, is the possibility of cancellation. This will occur if there are common roots in the AR and MA portions. Depending upon the complexity of the structural model, this may or may not be noticed by the analyst but if not recognized could lead to estimated FEs that do not appear to satisfy the restrictions implied on the polynomials by the model.[6]

The second set of equations derived from the system (2.5) that can be of value in testing assumptions about the structural model is the set of transfer functions (TFs). These equations can be obtained from (2.5) by multiplying both sides by H_{11}^*; this yields

$$|H_{11}|\underline{y}_t = -H_{11}^* H_{12}\underline{x}_t + H_{11}^* F_{11}\underline{e}_{1t}, \tag{2.10}$$

or, alternatively,

$$\underline{y}_t = \frac{-H_{11}^* H_{12}}{|H_{11}|}\underline{x}_t + \frac{H_{11}^* F_{11}}{|H_{11}|}\underline{e}_{1t}, \tag{2.11}$$

As noted by Kmenta (1971), Pierce and Mason (1971), and Zellner and Palm (1974), this form expresses the current values of endogenous variables as functions of the current and past values of the exogenous variables and is restricted in form. Formally, (2.11) is a set of rational distributed lag (RDL) equations (Jorgenson (1966) and Dhrymes (1970), or a system of multiple-input transfer functions (MITF) of the type described by Box and Jenkins (1970).

This form of the model is useful for prediction and control. In particular, it is useful for assessing the response, over time and in total,

[6] Of course, if the model is incorrect or misspecified, then these restrictions will also fail to hold.

of endogenous variables to changes in exogenous variables. Notice that, here too, there are strong restrictions on the form of the TFs under the assumptions of a specific model. For instance, if no cancellation occurs and if all the elements of H_{11}^{*} and H_{12} are finite polynomials, then all of the inputs have the same denominator polynomial. There are also restrictions on the form of the error process in (2.11). Other tests that could be carried out concern testing the assumptions of the exogeneity of the $\underline{x}_t$s. By estimating and analyzing (2.11) and comparing the results with the restrictions implied by a specific structural model, it is felt that many useful insights can be obtained concerning the adequacy of the specification of the structure. In particular, interest here will focus on the specification of the seasonal aspects of the model.

2.2 *Seasonality in time series data*

Before discussing how one would incorporate seasonality in a structural model, it will be useful to review briefly several approaches to modeling data that have seasonal properties. The two approaches discussed here are the traditional concept of seasonality that treats an observed series as the sum of three components – a trend or cyclical component, a seasonal component, and a noise component[7] – and the multiplicative times series model as developed by Box and Jenkins (1970).

One of the more common approaches to seasonality within the framework of the aforementioned traditional model is the dummy variable model. The general form of such a model is

$$y_t = y_t^c + \sum_{i=1}^{s} \alpha_i d_{it} + \epsilon_t, \tag{2.12}$$

where y_t^c is the trend or cyclical component, ϵ_t is an error term, and the dummy variables d_{it} are used to represent the seasonal component of the series. (Often . . . , y_t^c is represented by a polynomial in t, time.) If monthly data were under consideration, one might use a dummy variable for each month representing a series with a fixed periodic or seasonal component. The estimate of α_i would represent the estimated mean for the ith month. If such a system is presumed to be the true model, it is then straightforward to obtain a seasonally adjusted series by just subtracting the seasonal component that yields

$$y_t^a = y_t - y_t^s = y_t^c + \epsilon_t, \tag{2.13}$$

[7] As noted earlier, it is in this conceptual framework that the idea that a series can be decomposed into a seasonal component and a seasonally adjusted series arises.

where

$$y_t^s = \sum_{i=1}^{s} \alpha_i d_{it}, \tag{2.14}$$

Another approach, also using this traditional decomposition, is the Census Bureau X–11 program (see US Department of Commerce 1967). The basic idea of this approach is to eliminate the seasonal component y_t^s through the application of symmetric moving average filters. That is, a seasonally adseries is obtained by passing the unadjusted data through a filter of the form

$$y_t^a = \sum_{i=-k}^{k} \beta_i y_{t-i} = B(L)y_t, \tag{2.15}$$

where the β_is are fixed weights such that $\beta_i = \beta_{-i}$ and $\sum_{i=-k}^{k} \beta_i = 1$, and L is the lag operator. In terms of the traditional components model, this filter is chosen such that the seasonal component is taken out and the trend or cyclical component is unaffected. (That is, $B(L)y_t^c = y_t^c$, and $B(L)y_t^s = 0$.)

Another class of models that is of special interest and that contains the dummy variable approach as a special case, is the multiplicative seasonal time series models of Box and Jenkins (1970). These models are of the general form:

$$\Gamma_P(L^s)\phi_p(L)\Delta_s^D \Delta^d z_t = \Omega_Q(L^s)\Theta_q(L)a_t, \tag{2.16}$$

where s is the length of the seasonal period (e.g. 12 for monthly data), $\Delta_s^D = (1 - L^s)^D$, $\Delta^d = (1 - L)^d$, Γ and Ω are seasonal polynomials in L^s of degree P and Q respectively, ϕ and θ are polynomials in L of degree p and q, respectively, and a_t is a white-noise error term. It is also assumed that the roots of $\Gamma(\xi) = 0$ and $\phi(\xi) = 0$ lie outside the unit circle so that the process is stationary and the roots of $\Omega(\xi) = 0$ and $\theta(\xi) = 0$ lie on or outside the unit circle. Box and Jenkins refer to this as a model of order (p,d,q) $(P,D,Q)_s$.

Consider the process $(0,1,1)$ $(0,1,1)_{12}$ as a simple example. It can be written as

$$(1 - L^{12})(1 - L)z_t = (1 - \Theta_1 L)(1 - \Omega_1 L^{12})a_t. \tag{2.17}$$

Let $w_t = (1 - L)$ $(1 - L^{12})z_t$ (i.e. let w_t equal the seasonal differences of the changes in z_t). Now the moving average process governing w_t is easily seen by multiplying out the polynomials on the r.h.s. of (2.17), yielding

$$w_t = (1 - \Theta_1 L - \Omega_1 L^{12} + \Theta_1 \Omega_1 L^{13})a_t. \tag{2.18}$$

Therefore, this multiplicative model can be interpreted as an ordinary MA process of order 13. The distinction is that the multiplicative formulation restricts the weights on lags 2 through 11 to be 0 and on lag 13 to be the product of the weights for lags 1 and 12.

In general, these multiplicative seasonal models cannot be decomposed or interpreted within the traditional unobserved components framework without precise definitions of the components and some further identifying restrictions.[8] However, there is one special case of (2.16) that has an interpretation as the dummy variable case described earlier. Assume that observations were taken quarterly on some variable z_t. In addition, assume that the true processes generating the z_ts were such that each quarter had a different mean but otherwise the series was just a random, non-autocorrelated variable, a_t. Such a process could be written as

$$z_t = \alpha_1 d_{1t} + \alpha_2 d_{2t} + \alpha_3 d_{3t} + \alpha_4 d_{4t} + a_t, \tag{2.19}$$

where d_{it} is a dummy variable that takes on the value 1 in the ith quarter and 0 elsewhere. The estimates of the α_is would represent the mean of the ith quarter. If one were to seasonally difference this process, then the remaining process would be

$$(1 - L^4)z_t = (1 - L^4)a_t. \tag{2.20}$$

The effect of seasonal differencing is to eliminate a constant, deterministic seasonal pattern. The process in (2.20) indicates that under the particular model in (2.19), the seasonal differences of z_t obey a first order seasonal moving average process with a parameter value of 1. Alternatively, if the as were considered non-autocorrelated and the model were found to have a first order seasonal moving average parameter of less than 1, then the implication would be that the seasonal pattern is changing through time. That is, the seasonal means are changing through time.[9]

The multiplicative model will be used in this work because of its flexibility in describing not only certain types of additive or deterministic seasonal patterns but also seasonal patterns that might not be constant through time. In addition, it readily fits into the framework of analysis of this chapter.

[8] See Cleveland (1972). He . . . proposed an underlying stochastic process for which the Census X–11 is nearly optimal from the standpoint of conditional expectation. He argues that, for processes very near this, the X–11 does quite well, but, when departures occur, the appropriateness of the X–11 decomposition is thrown into doubt.

[9] These models are the first satisfactory models for forecasting seasonal series with changing seasonal patterns. For a more complete development and discussion of these models see Box and Jenkins (1970, ch. 9).

2.3 *An approach to the analysis of seasonality in structural models*

One question with which this work is concerned is how seasonality enters a structural econometric model. The primary focus is on testing the assumption that seasonality enters the system through exogenous forces. That is, can the seasonal fluctuations of the endogenous variables of the system be explained by the seasonality of the exogenous variables? There are, of course, other possibilities, such as certain parameters in the structure that fluctuate seasonally and, therefore, induce a seasonal pattern in the endogenous variables even when the exogenous variables are non-seasonal.

One approach that might be put forward combines the traditional concept of seasonality and seasonal adjustment with the concepts and methodology presented in . . . subsection [2.1] . . . Assume that the endogenous variables of the system, denoted by $\underline{y}_t$, and the exogenous variables of the system, denoted by $\underline{x}_t$, can be written as follows:

$$\begin{aligned} \underline{y}_t &= \underline{y}_t^c + \underline{y}_t^s + \underline{v}_t \\ \underline{x}_t &= \underline{x}_t^c + \underline{x}_t^s + \underline{u}_t, \end{aligned} \tag{2.21}$$

where no superscript on x or y indicates an observed variable, a superscript c denotes the trend or cyclical component, s denotes the seasonal component, and $\underline{v}_t$ and $\underline{u}_t$ are noise components. In addition, assume that one believes the true economic relationship is in terms of the trend components. In the notation of . . . subsection [2.1] . . . , the model can be written as

$$H_{11}\underline{y}_t^c + H_{12}\underline{x}_t^c = F_{11}\underline{e}_{1t}. \tag{2.22}$$

Substitution yields

$$H_{11}\left(\underline{y}_t - \underline{y}_t^s - \underline{v}_t\right) + H_{12}\left(\underline{x}_t - \underline{x}_t^s - \underline{u}_t\right) = F_{11}\underline{e}_{1t}, \tag{2.23}$$

or

$$\underline{y}_t = \underline{y}_t^s - \frac{H_{11}^* H_{12}}{|H_{11}|}\underline{x}_t + \frac{H_{11}^* H_{12}}{|H_{11}|}\underline{x}_t^s + \frac{H_{11}^* F_{11}}{|H_{11}|}\underline{e}_{1t} + \frac{H_{11}^* H_{12}}{|H_{11}|}\underline{u}_t + \underline{v}_t. \tag{2.24}$$

It is clear that if (2.22) is the true model, then the model builder must be very concerned about how the trend component is obtained from the observed or unadjusted data. On the other hand, such a theory could be tested using the unadjusted data and the seasonal components, using (2.24). For example, a restriction implied by (2.22) on (2.24) is that the coefficient of $\underline{y}_t^s$ is 1 and the coefficient of $\underline{x}_t^s$ is the negative of the coefficient on $\underline{x}_t$.

Assume, on the other hand, that one believes that seasonal fluctuations in the exogenous variables work their way through the system like all other fluctuations in the exogenous variables. In addition, suppose interest is focused on the ability of the seasonal fluctuations in the exogenous variables to explain the seasonality in the endogenous variables. Under these conditions, (2.25) would have to hold

$$\underline{y}_t^s + \frac{H_{11}^* H_{12}}{|H_{11}|} \underline{x}_t^s = 0. \tag{2.25}$$

This restriction arises from the fact that the true economic model exists between the observed series, and, therefore, the seasonal portion of $\underline{x}_t$ should explain the seasonal portion of $\underline{y}_t$.

However, this approach still suffers from the problems of defining and obtaining an optimal adjustment and/or appropriate decomposition.

As indicated earlier, the approach taken in this chapter is slightly different. The structural model is written in a manner which presumes that its form holds for the observed data and not only the trend component

$$H_{11} y_t + H_{12} \underline{x}_t = F_{11} \underline{e}_{1t}. \tag{2.26}$$

The hypothesis to be tested is that seasonality enters the system through the process generating the exogenous variables. That is, the process generating $\underline{x}_t$, (2.6), is written as a multiplicative seasonal time series model. By doing this, it is hoped to broaden the model by allowing a slightly greater flexibility with regard to the form of the seasonal fluctuation.

Since one of the objectives is to avoid choosing an arbitrary decomposition prior to developing an adequate model, a means must be devised by which conclusions can be drawn concerning the ability of the exogenous variables to account for the seasonality in the endogenous variables. Fortunately, there is a straightforward method of doing this. Since the process generating the $\underline{x}_t$s will be associated, in general, with both seasonal AR and seasonal MA polynomials, it is possible to trace these polynomials through the analysis to determine their impact on the TFs and FEs of the system. Once the TFs and FEs have been obtained, they can be estimated and the results compared with the implications of the theory used in writing the structural model. To the extent that the estimated models are in agreement regarding the behavior of these seasonal polynomials, the hypothesis of exogenous seasonality will be accepted.

Proceeding in the manner previously described yields some interesting insights into the type of stochastic properties that are likely to be exhibited by the endogenous variables. Assume that (2.26) is written as a multivariate-multiplicative seasonal time series process.

$$H_{22}S_{22}\underline{x}_t = F_{22}T_{22}\underline{e}_{2t}, \tag{2.27}$$

where it is assumed that S_{22} and T_{22} are matrices having elements that are polynomials is L^s, where s is the seasonal period. For simplicity, consider the case where the exogenous variables are independent so that H_{22}, S_{22}, T_{22}, and T_{22} are all diagonal. This is sufficient to enable each and every exogenous variable to be written as a strictly multiplicative seasonal time series process.

Given (2.27) as the process generating $\underline{x}_t$, a set of FEs can be obtained by substituting (2.27) in to (2.26) with the following result:

$$|H_{11}||H_{22}||S_{22}|y_t = -H_{11}^*H_{12}S_{22}^*H_{22}^*F_{22}T_{22}\underline{e}_{2t} + H_{11}^*|H_{22}||S_{22}|\underline{e}_{1t}. \tag{2.28}$$

Inspection of (2.28) reveals that the AR portion of the processes for the endogenous variables will, in general, be in the form of the multiplicative seasonal model. However, the MA portion of (2.28) does not factor, in general, into the multiplicative form. Consequently, one might not, in general, expect to find the endogenous variables to be strictly multiplicative seasonal processes (i.e. multiplicative in both the AR and MA portions). It would seem that the MA term would have characteristics of both multiplicative and additive seasonal variation. This implication will be investigated further in the economic model analyzed in the next section, and it will be seen that if certain restrictions are placed on the structure and on the processes generating the exogenous variables, (2.28) will become strictly multiplicative.

Although no mention has been made, up to this point, of constant terms or intercept terms, it is straightforward to see how they can be handled in the framework that has been discussed. If these intercept terms are considered constants, they can be carried along as deterministic elements of $\underline{x}_t$, or, if they are considered random and generated by a process, perhaps seasonal, they can again be considered as elements of $\underline{x}_t$. In either case, the inclusion of these intercepts is a simple extension of the methodology outlined in this section.

To summarize, the approach that will be applied in the following sections is to:

1. Construct an economic model with an explicit specification of seasonality
2. Derive the implied TFs and FEs of the model noting where the seasonal specification places restrictions on the form of these equations
3. Empirically check these restrictions against the data

4. Utilize the empirical results to suggest alternative specifications of the model if the model under consideration proves deficient.

3 Analysis of an economic model

3.1 Model formulation

In this section, a simple monetary model is formulated and analyzed to illustrate how the techniques outlined in the previous sections might be helpful in gaining insights about seasonality and its role in an economic model.

The economic model contains five variables: Equilibrium money stock, a measure of real income or wealth, nominal interest rate, price level, and the monetary base. The model is written to allow for various types of lag structures having form and length that are to be inferred from the data. In addition, no restrictions are placed on the theoretical elasticities and the growth rate of real output is allowed to vary. Expectations in this model are generated rationally in the sense of Muth (1961). That is, expectations are formed, based on information in the past history of the exogenous variables and the structure of the model. Finally, the monetary base and real income (output) are treated as exogenous or independently determined, and seasonality is assumed to enter the system only through these variables.

Obviously, in a simple model, such as this, there are many possible sources of specification error. However, this study focuses on two important aspects of the model. First, the assumption of the exogeneity of the monetary base and real income may not be an adequate representation. For instance, as specified, the model assumes that an open loop control strategy has been adopted by the policy-makers with regard to the creation of the monetary base. The alternative is, of course, some sort of closed loop control scheme, whereby the authorities respond to changes in the price level or interest rate in determining the growth of the base. The exogeneity of real income assumes the absence of a Phillips-curve relationship or feedback from the monetary sector to the real sector.[10] Therefore, it is of interest to investigate the adequacy of the exogeneity assumptions in light of these other possible specifications of the model.

Secondly, seasonality is assumed to enter the model only through the exogenous variables. It may be that there are separate seasonal effects that enter directly through the money demand or money supply equations that

[10] See Lucas (1973) and Sargent (1973) for a more thorough treatment of the issues surrounding this phenomenon.

are different from those induced by the seasonal influence of real income and the base. Such effects may be due to seasonally varying parameters in the structure. If this is the case, the empirical results would be at variance with the implications of the model.

The equations of the model include (1) a money demand equation, (2) a money supply equation, (3) a money market equilibrium condition, (4) the Fisher equation, and (5) a rational expectations equation. We can write these equations as follows:

$$M_t^D = L(Y_t, i_t, P_t) \tag{3.1}$$

$$M_t^S = S(B_t) \tag{3.2}$$

$$M_t = M_t^D = M_t^S \tag{3.3}$$

$$i_t = \rho_t^* + \pi_t^* \tag{3.4}$$

$$\pi_t^* = \underset{t}{E}(\pi_t \mid \cdot), \tag{3.5}$$

where

M_t^D = nominal money demand at time t
M_t^S = nominal money supply at time t
Y_t = real income (output) at time t
i_t = nominal interest rate at time t
P_t = price level at time t
B_t = net source base at time t
ρ_t^* = anticipated real interest rate as of time t
π_t^* = anticipated rate of inflation as of time t

Equation (3.5) builds the rational expectations hypothesis into the model, and $\underset{t}{E}(\pi_t \mid \cdot)$ denotes a conditional expectation of inflation given the equations of the model and past information.[11]

It will be assumed that (3.1) can be written as

$$r_{M_t} = \alpha_1 r_{Y_t} - \beta_1 \Delta i_t + y_1 r_{P_t} + u_{1t}, \tag{3.6}$$

[11] Given the previous structure, there are many other issues that could also be raised. For example, most economists agree that permanent income, or possibly wealth, is a more appropriate income measure for the money demand function than real output. One might also consider an adjustment process rather than require market clearing at each time t. Finally, a more complicated money supply relationship might be considered to allow for changing reserve ratios, or changing interest rates that would affect the money multiplier. Clearly, a thorough examination of this model would have to consider these alternatives. However, the objective of this chapter is somewhat less ambitious. Here, the intent is to gain a better understanding of the techniques and the issues surrounding seasonality.

and that the money multiplier is non-autocorrelated so that (3.2) can be written as

$$r_{M_t} = \alpha_2 r_{B_t} + u_{2t} \tag{3.7}$$

where $\Delta = (1 - L)$ is the difference operator; hence, $r_k = \Delta \ell n(k)$ is the rate of growth of k. The coefficients are, in general, unrestricted in that they can be interpreted as polynomials in the lag operator. However, as a starting point, it will be assumed that they are constants, and the empirical results will be utilized to suggest alternative lag structures.[12] For convenience, both u_{1_t} and u_{2_t} will be considered independent, non-autocorrelated disturbance terms.

The remainder of the model involves the Fisher equation and the rational expectations hypothesis. Since $\pi_t = \ell n\, P_{t+1} - \ell n\, P_t = \Delta \ell n\, P_{t+1} = r_{P_{t+1}}$

$$i_t = \rho_t^* + r^*_{P_{t+1}}, \tag{3.8}$$

and the expectation can be written as

$$r^*_{P_{t+1}} = \underset{t}{E}(r_{P_{t+1}} \mid \cdot). \tag{3.9}$$

At this point, some assumption must be made about ρ_t^*, the anticipated real rate of interest. In order to keep this analysis from becoming unduly complicated, the anticipated real rate will be considered a random variable with a constant expected value. Therefore,

$$i_t = r^*_{P_{t+1}} + u_{3t}, \tag{3.10}$$

where u_{3_t} may have a non-zero mean. Of course, if the anticipated real rate were autocorrelated, then u_{3_t} would also be autocorrelated. In addition, u_{3_t} will be considered independent of u_{1_t} and u_{2t}.

Utilizing the assumption that the monetary base and real income are exogenous, the system can be completed by writing down the processes generating these variables.

$$\phi_B(L)\Gamma_B(L^{12})\Delta_{12} r_{B_t} = \Theta_B(L)\Omega_B(L^{12})u_{4t} \tag{3.11}$$

$$\phi_Y(L)\Gamma_Y(L^{12})\Delta_{12} r_{Y_t} = \Theta_Y(L)\Omega_Y(L^{12})u_{5t}, \tag{3.12}$$

where ϕ_i, Γ_i, Θ_i, and Ω_i are polynomials in the lag operator having roots that satisfy the stationarity and invertibility conditions, and Γ_B, Γ_Y, Ω_B, and Ω_Y represent the seasonal polynomials that are to be traced through the model.

[12] Whether differencing is appropriate for these structural relationships is not a real issue. The result of over-differencing would be to induce moving average complications into the error structure that can be handled in the estimation procedure. (See Plosser and Schwert 1976.)

Now that the model has been developed, the system represented by equations (3.6), (3.7), (3.9), and (3.10) can be rewritten in the form of a system of simultaneous equations as shown in (2.5), yielding[13]

$$\begin{bmatrix} 1 & -\upsilon_1 & -\beta_1\Delta \\ 1 & 0 & 0 \\ 0 & 0 & \Phi\Delta_{12} \end{bmatrix} \begin{bmatrix} r_{M_t} \\ r_{P_t} \\ i_t \end{bmatrix} + \begin{bmatrix} 0 & -\alpha_1 \\ -\alpha_2 & 0 \\ -\Psi_1\Delta_{12} & -\Psi_2\Delta_{12} \end{bmatrix} \begin{bmatrix} r_{B_t} \\ r_{Y_t} \end{bmatrix} = \begin{bmatrix} u_{1t} \\ u_{2t} \\ \Delta_{12}v_t \end{bmatrix}, \tag{3.13}$$

where

$$\begin{aligned} \Phi &= (\gamma_1 - \beta_1) \\ \Psi_1 &= \alpha_2 \sum_{j=0}^{\infty} \left(\frac{-\beta_1}{\gamma_1 - \beta_1} \right)^j \pi_{j+1}^{(B)}(L) \\ \Psi_2 &= -\alpha_1 \sum_{j=0}^{\infty} \left(\frac{-\beta_1}{\gamma_1 - \beta_1} \right)^j \pi_{j+1}^{(Y)}(L) \\ \Delta_{12}v_t &= \Delta_{12} \sum_{j=0}^{\infty} \left(\frac{-\beta_1}{\gamma_1 - \beta_1} \right)^j \\ &\quad (\underset{t}{E} u_{2t+j+1} - \beta_1 \underset{t}{E} u_{3t+j+1} - \underset{t}{E} u_{1t+j+1}) + \Phi\Delta_{12}u_{3t}, \end{aligned} \tag{3.14}$$

u_{1t}, u_{2t}, and u_{3t} are non-autocorrelated and independent disturbance terms. For convenience, let

$$H_{11} = \begin{bmatrix} 1 & -\gamma_1 & -\beta_1\Delta \\ 1 & 0 & 0 \\ 0 & 0 & \Phi\Delta_{12} \end{bmatrix}, \tag{3.15}$$

and

$$H_{12} = \begin{bmatrix} 0 & -\alpha_1 \\ -\alpha_2 & 0 \\ -\Psi_1\Delta_{12} & -\Psi_2\Delta_{12} \end{bmatrix}. \tag{3.16}$$

Through some simple algebraic manipulations, both the TFs and FEs can be written down, and the following analysis highlights some of the more interesting properties of the TFs and FEs.[14]

[13] For the mathematical derivation of (3.13), the reader is referred to appendix A.
[14] See appendix A for the derivations.

3.2 *Analysis of the transfer functions*

The TFs of the system are easily obtained by premultiplying both sides of (3.13) by H_{11}^{*}. Writing the resulting system of equations, one by one, the TF of each endogenous variable can be analyzed in greater detail.

The TF for the nominal money stock can be simplified to

$$\Delta_{12} r_{M_t} = \alpha_2 \Delta_{12} r_{B_t} + \Delta_{12} u_{2t}. \tag{3.17}$$

The money supply is seen to be a function only of the base and real income does not enter as an input. Under the assumption that α_2 is a constant coefficient, (3.17) is simply a regression model with moving average errors. If α_2 is a polynomial in the lag operator, it is a distributed lag model. In either case, note that α_2 can be directly estimated, using nonlinear techniques. In addition, if all of the seasonality in M is explained by the base (B), then the only evidence of seasonal autocorrelation should appear in the noise process as a seasonal moving average polynomial of order 1 and parameter value of 1. Alternatively, there might be seasonal fluctuations in the money multiplier. As was noted earlier, the model has implicitly assumed that the multiplier is non-autocorrelated. However, to the extent that the Federal Reserve Board offsets changes in the money multiplier by either increasing or decreasing the amount of currency as it deemed appropriate, the result would be to force the first order seasonal moving average parameter (SMA) away from 1 and to induce downward bias into the estimated value of α_2. In fact, if the Fed followed a policy of no money growth and sought only to offset the multiplier exactly, the estimate of α_2 and of the first order SMA parameter would be near zero.

The TF for prices is somewhat more complicated than the one describing money but, by that very fact, turns out to have interesting interpretations. Through some algebraic manipulations, the following expression is obtained:

$$\Delta_{12} r_{P_t} = \left(\frac{\alpha_2 \Phi - \beta_1 \Delta \Psi_1}{\gamma_1 \Phi} \right) \Delta_{12} r_{B_t} + \left(\frac{-\alpha_1 \Phi - \beta_1 \Delta \Psi_2}{\gamma_1 \Phi} \right) \Delta_{12} r_{Y_t} + \Delta_{12}(-\Phi u_{1t} + \Phi u_{2t} - \beta_1 \Delta v_t) \tag{3.18}$$

The analysis of this expression will depend, to a large degree, on what can be said about the form of the distributed lag on $\Delta_{12} r_{B_t}$ and $\Delta_{12} r_{Y_t}$. Fortunately, several interesting observations can be made. Consider the case where all the structural parameters in the model are polynomials of zero degree in L. Under these circumstances, the only polynomials in L (other than the difference operators) arise from the terms Ψ_1 and Ψ_2. Note from (3.14), where Ψ_1 and Ψ_2 are defined, and equations (A.19) and (A.20) in appendix A (p. 376), that, in general, Ψ_1 and Ψ_2

will be polynomials which are infinite in length. The implication is that even though there are no lagged relationships specified in the structural model, due to the expectational aspect of the model there exists an infinite distributed lag relationship between the exogenous variables and the endogenous variables of the system. Consequently, estimating this transfer function would, most likely, result in a rational distributed lag (RDL) model as a means of parsimoniously representing such a relationship.

Secondly, in appendix A it is shown that the expressions for Ψ_1 and Ψ_2 involve the summation of varying powers and cross-products of the parameters in the seasonal and non-seasonal polynomials that are generating the exogenous variables. It is possible that the data would not indicate a need for seasonal parameters (i.e. specific coefficients at the seasonal lags) in the RDL formulation. If this is true, then the only evidence of seasonal autocorrelation appears in the error term as a seasonal moving average polynomial of order 1 with parameter value of 1. The presence of Ψ_1 and Ψ_2 also indicate that, even though $\Delta_{12} r_{B_t}$ and $\Delta_{12} r_{Y_t}$ may be seasonal, the existence of an expectations mechanism has a smoothing effect on the output variable $\Delta_{12} r_{P_t}$. This smoothing effect arises out of the infinite distributed lag relationship between the inputs and the output variable $\Delta_{12} r_{P_t}$. In other words, $\Delta_{12} r_{P_t}$ will be a weighted average of all past values of $\Delta_{12} r_{B_t}$ and $\Delta_{12} r_{Y_t}$.

An additional point of interest is how this model can simplify under alternative assumptions about the structural model. For example, if the classical quantity theory of money were true, then β_1 would equal zero, and γ_1 would equal 1, allowing (3.18) to reduce to

$$\Delta_{12} r_{P_t} = \alpha_2 \Delta_{12} r_{B_t} - \alpha_1 \Delta_{12} r_{Y_t} + \Delta_{12}[-u_{1t} + u_{2t}]. \tag{3.19}$$

In a similar manner, the TF for the nominal interest rate can be written as

$$\Delta_{12} i_t = \left(\frac{\Psi_1}{\Phi}\right) \Delta_{12} r_{B_t} + \left(\frac{\Psi_2}{\Phi}\right) \Delta_{12} r_{Y_t} + \left(\frac{\Delta_{12}}{\Phi}\right) v_t. \tag{3.20}$$

Notice that here too, the distributed lags on $\Delta_{12} r_{B_t}$ and $\Delta_{12} r_{Y_t}$ will in general be infinite in length and, therefore, more easily modeled as a RDL even when the structural parameters indicate only contemporaneous relationships. As was pointed out, this is due to the expectations aspect of the model. In addition, if Ψ_1 and Ψ_2 do not display strong seasonal properties, the only evidence of seasonality that one would expect to find, if the model is correct occurs in the error term of the form Δ_{12}. Once again, it is worthy of note that because of Ψ_1 and Ψ_2 and the smoothing effect they have on $\Delta_{12} i_t$, the interest rate most likely would not display seasonal movements that are visually striking.

Table 9.1 *Summary of TFs*

General formulation	Simplified formulation ($\beta_1 = 0$ and $\gamma_1 = 1$)
$\Delta_{12} r_{M_t} = \alpha_1 \Delta_{12} r_{B_t} + \Delta_{12} u_{2t}$	$\Delta_{12} r_{M_t} = \alpha_2 \Delta_{12} r_{B_t} + \Delta_{12} u_{2t}$
$\Delta_{12} r_{P_t} = \left(\frac{\alpha_2 \Phi - \beta_1 \Delta \Psi_1}{\gamma_1 \Phi}\right) \Delta_{12} r_{B_t} + \left(\frac{\alpha_1 \Phi - \beta_1 \Delta \Psi_2}{\gamma_1 \Phi}\right) \Delta_{12} r_{Y_t} + \Delta_{12}(-\Phi u_{1t} + \Phi u_{2t} - \beta_1 \Delta v_t)$	$\Delta_{12} r_{P_t} = \alpha_2 \Delta_{12} r_{B_t} - \alpha_1 \Delta_{12} r_{Y_t} + \Delta_{12}(-u_{1t} + u_{2t})$
$\Delta_{12} i_t = \left(\frac{\Psi_1}{\Phi}\right) \Delta_{12} r_{B_t} + \left(\frac{\Psi_2}{\Phi}\right) \Delta_{12} r_{Y_t} + \left(\frac{\Delta_{12}}{\Phi}\right) v_t$	$\Delta_{12} i_t = \alpha_2 \left(1 - \frac{\phi_B \Gamma_B}{\Theta_B \Omega_B}\right) F \Delta_{12} r_{B_t} - \alpha_1 \left(1 - \frac{\phi_Y \Gamma_Y}{\Theta_Y \Omega_Y}\right) F \Delta_{12} r_{Y_t} + \Delta_{12} u_{3t}$

Table 9.1 summarizes the transfer functions (TFs) for the model under consideration. Both a general formulation and a simplified formulation suggested by the classical quantity theory of money, as previously discussed, are presented for comparison.

3.3 *Analysis of the final equations*

The next set of equations to be analyzed are the final equations (FEs) (table 9.2). They can be obtained, as indicated, in . . . subsection [2.1] . . . In deriving these equations, it is important to recognize that (3.11) and (3.12) are rewritten as

$$\Delta_{12} \begin{bmatrix} r_{B_t} \\ r_{Y_t} \end{bmatrix} = \begin{bmatrix} \phi_B \Gamma_B & 0 \\ 0 & \phi_Y \Gamma_Y \end{bmatrix}^{-1} \begin{bmatrix} \Theta_B \Omega_B & 0 \\ 0 & \Theta_Y \Omega_Y \end{bmatrix} \cdot \begin{bmatrix} u_{4t} \\ u_{5t} \end{bmatrix}$$
$$= H_{22}^{-1} F_{22} \begin{bmatrix} u_{4t} \\ u_{5t} \end{bmatrix}. \qquad (3.21)$$

This presumes the independence of u_{4t} and u_{5t}, but such a restriction is not necessary. An alternative specification might allow the (1,2) and (2,1) elements of F_{22} in (3.21) to be non-zero. This would allow for a dynamic relationship among the inputs.

As derived in appendix A, the FE for the equilibrium money stock (M) can be written as follows:

$$\gamma_1 \Phi \phi_B \phi_Y \Gamma_B \Gamma_Y \Delta_{12} r_{M_t} = \gamma_1 \Phi \alpha_2 \phi_Y \Gamma_Y \Theta_B \Omega_B u_{4t} + \phi_B \phi_Y \Gamma_B \Gamma_Y v_1 \Phi \Delta_{12} u_{2t}. \qquad (3.22)$$

Table 9.2 *Summary of FEs*

General formulation	Simplified formulation ($\beta_1 = 0, \gamma_1 = 1, \phi_B = \phi_Y = \Gamma_B = \Gamma_Y = 1$ and $\Omega_B \doteq \Omega_Y \doteq (1 - L^{12})$)
$\phi_B \Gamma_B \Delta_{12} r_{M_t} = \alpha_2 \Theta_B \Omega_B u_{4t} + \phi_B \Gamma_B \Delta_{12} u_{2t}$	$\Delta_{12} r_{M_t} = \Delta_{12}[\alpha_2 \Theta_B u_{4t} + u_{2t}]$
$\gamma_1 \Phi \phi_B \phi_Y \Gamma_B \Gamma_Y \Delta_{12} r_{P_t} = [\Phi \alpha_2 - \beta_1 \Delta \Psi_1] \phi_Y \Gamma_Y \Theta_B \Omega_B u_{4t} + [-\Phi \alpha_1 - \beta_1 \Delta \Psi_2] \phi_B \Gamma_B \Theta_Y \Omega_Y u_{5t} + \phi_B \phi_Y \Gamma_B \Gamma_Y \Delta_{12}[-\Phi u_{1t} + \Phi u_{2t} - \beta_1 \Delta v_t]$	$\Delta_{12} r_{P_t} = \Delta_{12}[\alpha_2 \Theta_B u_{4t} - \alpha_1 \Theta_Y u_{5t} - u_{1t} + u_{2t}]$
$\Phi \phi_B \phi_Y \Gamma_B \Gamma_Y \Delta_{12} i_t = \Psi_1 \phi_Y \Gamma_Y \Theta_B \Omega_B u_{4t} + \Psi_2 \phi_B \Gamma_B \Theta_Y \Omega_Y u_{5t} + \phi_B \phi_Y \Gamma_B \Gamma_Y \Delta_{12} v_t$	$\Delta_{12} i_t = \alpha_2(\Theta_B \Omega_B - 1) F u_{4t} - \alpha_1(\Theta_Y \Omega_Y - 1) F u_{5t} + \Delta_{12} u_{3t}$

Notice that $\gamma_1 \Phi \phi_Y \Gamma_Y$ can be factored out of both sides, leaving

$$\phi_B \Gamma_B \Delta_{12} r_{M_t} = \alpha_2 \Theta_B \Omega_B u_{4t} + \phi_B \Gamma_B \Delta_{12} u_{2t}. \tag{3.23}$$

The FE for the money stock is a function of the structural parameter α_2, the error term u_{2t}, and process generating the monetary base (B). More important is that, by introducing seasonality by way of the exogenous variables, seasonality is induced on the endogenous variable (M) and, in fact, on each and every endogenous variable in the system, as will be pointed out in subsequent analyses.

It is known (e.g. see Anderson 1971) that the sum of two moving average processes is representable as a single invertible linear process in one random variable. Consequently, given that u_{2t} and u_{4t} are independent due to the assumption that the monetary base is exogenous, the order of this moving average polynomial will be equivalent to the order of the expression $\alpha_2 \Theta_B \Omega_B$ or $\phi_B \Gamma_B \Delta_{12}$, whichever is greater.

The FE for prices (P) is shown in appendix A to be

$$\begin{aligned}\gamma_1 \Phi \phi_B \phi_Y \Gamma_B \Gamma_Y \Delta_{12} r_{P_t} = {} & [\Phi \alpha_2 - \beta_1 \Delta \Psi_1] \phi_Y \Gamma_Y \Theta_B \Omega_B u_{4t} \\ & + [-\Phi \alpha_1 - \beta_1 \Delta \Psi_2] \phi_B \Gamma_B \Theta_Y \Omega_Y u_{5t} \\ & + \phi_B \phi_Y \Gamma_B \Gamma_Y \Delta_{12} [-\Phi u_{1t} + \Phi u_{2t} - \beta_1 \Delta v_t].\end{aligned} \tag{3.24}$$

Once again, seasonality is seen to be induced on an endogenous variable only as a result of exogenous seasonality. This fact is evident from the presence of the Δ_{12} operator and the seasonal polynomials Γ_B, Γ_Y, Ω_B, and Ω_Y. As occurred in the FE for money, the AR side of (3.24) is in the form of the multiplicative seasonal time series model, and the MA portion is not. In fact, the MA portion appears to border on the unintelligible. However, some insights can be obtained from this representation.

In order to gain some understanding of (3.24), suppose $\phi_B = \phi_Y = \Gamma_B = \Gamma_Y = 1$ and that $\Omega_B = \Omega_Y = (1 - L^{12})$, then (3.24) can be rewritten as:

$$\begin{aligned}\gamma_1 \Phi \Delta_{12} r_{P_t} = {} & \Delta_{12} [(\Phi \alpha_2 - \beta_1 \Delta \Psi_1) \Theta_B u_{4t} \\ & + (-\Phi \alpha_1 - \beta_1 \Delta \Psi_2) \Theta_Y u_{5t} - \Phi u_{1t} + \Phi u_{2t} - \beta_1 \Delta v_t].\end{aligned} \tag{3.25}$$

Equation (3.25) now appears to be in the terms of the general multiplicative time series model. However, it is not, because both Ψ_1 and Ψ_2 are expressions involving seasonal polynomials and are, in general, of infinite length. Therefore, it is convenient to consider two possible cases for this expression, when $\beta_1 = 0$ and $\beta_1 \neq 0$.

Suppose that the classical quantity theory of money were to be considered. In that case, (3.25) reduces to

$$\Delta_{12} r_{P_t} = \Delta_{12}[\alpha_2 \Theta_B u_{4t} - \alpha_1 \Theta_Y u_{5t} - u_{1t} + u_{2t}], \tag{3.26}$$

which is obtained by allowing the exogenous variables to have no AR polynomials and for the seasonality to approach the seasonal means problem as well as having $\beta_1 = 0$ and $\gamma_1 = 1$ (i.e. restricting the interest rate elasticity of the demand for money to zero and requiring demand for real cash balances to the homogeneous of degree zero in the price level). Notice that, once again, as the economic model is simplified, so is the implied stochastic structure of the output variables of the system.

The implication of (3.26) is that the seasonally differenced rate of inflation would be a pure MA process. It would be in the form of the multiplicative seasonal model with the seasonal moving average polynomial of order 1 and parameter value close to 1.

As was noted previously, the model has been carried through under the assumption that the *u*s in (3.26) are independent of one another. Under such an assumption the order of the monthly MA process would be of the order of $\alpha_2 \Theta_B$ or Θ_Y, whichever is larger. However, u_{4_t} and u_{5_t} may not be independent either contemporaneously or through time, and similarly for u_{1_t} and u_{2t}.

Neither of these complications would alter the basic economics of the model but could affect the orders of the MA portions of the FEs. Therefore, if the classical quantity theory of money is true, one might expect to observe an ARIMA model for the natural log of prices to be of the form (0,1,q) $(0,1,1)_{12}$, where q is determined by Θ_Y, Θ_B and the covariance structure between the error terms.

The second case of (3.25) to be considered allows β_1 to be different from zero. In order to gain insight into this case, it is necessary to analyze the expressions for Ψ_1 and Ψ_2 in greater detail. Rewriting (3.25) yields

$$\begin{aligned}\gamma_1 \Phi \Delta_{12} r_{P_t} = {} & \Delta_{12}[\Phi \alpha_2 \Theta_B u_{4t} - \Psi \alpha_1 \Theta_Y u_{5t} - \Phi u_{1t} \\ & + \Phi u_{2t} - \beta_1 \Delta v_t] - \Delta_{12} \beta_1 \Delta [\Psi_1 \Theta_B u_{4t} + \Psi_2 \Theta_Y u_{5t}],\end{aligned} \tag{3.27}$$

or

$$\begin{aligned}\gamma_1 \Phi \Delta_{12} r_{P_t} = {} & W_t - \Delta_{12} \beta_1 \Delta \sum_{j=0}^{\infty} \left(\frac{-\beta_1}{\gamma_1 - \beta_1} \right)^j \left[\alpha_2 \pi_{j+1}^{(B)}(L) \Theta_B u_{4t} \right. \\ & \left. - \alpha_1 \pi_{j+1}^{(Y)}(L) \Theta_Y u_{5t} \right].\end{aligned} \tag{3.28}$$

Now, under the assumption that the structural coefficients are just constants, W_t is a finite MA polynomial of order equal to the maximum order of Δ, Θ_B or Θ_Y, with seasonal polynomial Δ_{12}. The second term is more complicated.

The expressions $\pi_{j+1}^{(B)}(L)$ and $\pi_{j+1}^{(Y)}(L)$ merely represent the weighting scheme applied to the infinite past history of $\Delta_{12}r_{B_t}$ and $\Delta_{12}r_{Y_t}$, respectively, to obtain the forecast of these variables at time t, for time $t+j+1$. This would imply that the FE for prices would involve an infinite MA polynomial. It is very difficult to evaluate the form of this polynomial for anything except the most trivial cases. However, if either Θ_B, Θ_Y, Ω_B, or Ω_Y are of degree greater than zero, then the polynomial will be of infinite length. In finite samples, this infinite MA model may be indistinguishable from a more parsimonious AR representation. If the decay of this infinite MA is very slow, then one might even be led into differencing the series or estimating an AR polynomial that had a root close to the unit circle. It is even more interesting to note that the presence of Ψ_1 and Ψ_2 is due to the necessity of generating expectations and has an apparent smoothing effect on the autocorrelation structure of $\Delta_{12}r_{P_t}$, resulting in the seasonality in prices that appears much less pronounced.

The last FE to be considered is the one implied for the nominal interest rate (i). It can be written as follows:

$$\begin{aligned}\gamma_1\Phi\phi_B\phi_Y\Gamma_B\Gamma_Y\Delta_{12}i_t &= \gamma_1\Psi_1\phi_Y\Gamma_Y\Theta_B\Omega_B u_{4t} + \gamma_1\Psi_2\phi_B\Gamma_B\Theta_Y\Omega_Y u_{5t} \\ &\quad + \phi_B\phi_Y\Gamma_B\Gamma_Y\gamma_1\Delta_{12}v_t. \end{aligned} \tag{3.29}$$

As has occurred for money and prices, seasonality has occurred in the nominal interest rate. In addition, the r.h.s. of (3.29) does not indicate that a multiplicative time series model is the correct representation of the data if the model is true but that some mixture of the multiplicative and additive models would be more appropriate. However, if it is assumed that $\Gamma_B = \Gamma_Y = 1$ and $\Omega_B = \Omega_Y = \Delta_{12}$, then (3.29) can be rewritten as

$$\Phi\phi_B\phi_Y\Delta_{12}i_t = \Delta_{12}[\Psi_1\phi_Y\Theta_B u_{4t} + \Psi_2\phi_B\Theta_Y u_{5t} + \phi_B\phi_Y v_t], \tag{3.30}$$

or allowing $\phi_B = \phi_Y = 1$, as

$$\Phi\Delta_{12}i_t = \Delta_{12}[\Psi_1\Theta_B u_{4t} + \Psi_2\Theta_Y u_{5t} + v_t]. \tag{3.31}$$

Notice that the terms Ψ_1 and Ψ_2 appear here as they did in the FE for prices. Consequently, if $\beta_1 \neq 0$, then the data may indicate the need for an AR polynomial for $\Delta_{12}i_t$. In addition, if Ψ_1 and Ψ_2 imply weights that decline very slowly, then $\Delta_{12}i_t$ may appear non-stationary in finite samples. Similarly, the presence of Ψ_1 and Ψ_2, most likely, indicates that the seasonality in the interest rate is greatly attenuated.

Alternatively, simplifications of the economic model naturally lead to a simplification of the stochastic structure of $\Delta_{12}i_t$. If $\beta_1 = 0$, i.e. the classical quantity theory is true, with some algebraic manipulation, (3.3) reduces to

$$\Delta_{12}i_t = \alpha_2(\Theta_B\Omega_B - 1)Fu_{4t} - \alpha_1(\Theta_Y\Omega_Y - 1)Fu_{5t} + \Delta_{12}u_{3t}, \tag{3.32}$$

where u_{3t} is obtained from the expression for $\Delta_{12}v_t$ in (42) and F is the forward shift operator so that $F^j z_t = z_{t+j}$. Therefore, the univariate model for the nominal interest rate might well be expected to follow something similar to a $(1,0,q)\ (0,1,1)_{12}$ or $(0,1,q')\ (0,1,1)_{12}$ process, where q and q' would be determined by Θ_B and Θ_Y and the covariance between u_{4t} and u_{5t}.

A summary of the FEs discussed in this section are presented in table 9.2. For comparison, both the general and the simplified versions are presented.

3.3.1 FEs and the census X–11 adjustment procedure In light of the work done by Cleveland (1972), who found a stochastic model for which the X–11 procedure is nearly optimal in the sense of conditional expectation, it is interesting to analyze the stochastic structure implied by the economic model to see if and when the model might imply a structure for which the X–11 method, for example, is appropriate. The model developed by Cleveland is

$$\begin{aligned}(1 - L)(1 - L^{12})y_t = {} & (1 - 0.28L + 0.27L^2 + 0.24(L^3 + \cdots + L^8) \\ & + 0.23L^9 + 0.22L^{10} + 0.16L^{11} - 0.50L^{12} \\ & + 0.34L^{13} + 0.07L^{14})c_t,\end{aligned} \tag{3.33}$$

where c_t is a white-noise error term.[15] This suggests that for data having an autoregressive structure $(1 - L)\ (1 - L^{12})$ and having a moving average structure of length 14 and similar to that specified in (61), the X–11 procedure may do a fairly accurate job of decomposition.

Consider, for example, the FE for the money stock. From (3.33) and (3.23), it can be seen that, if the economic model is correct and if ϕ_B and

[15] That is, for stochastic processes very similar to the one he derives, the seasonally adjusted data created by the X–11 can be considered approximately equal to the conditional expectation of a trend component, given the observed series.

Γ_B are identically equal to 1, then

$$\Delta_{12} r_{M_t} = \alpha_2 \Theta_B \Omega_B u_{4t} + \Delta_{12} u_{2t}, \tag{3.34}$$

or

$$(1 - L)(1 - L^{12})\text{-}nM_t = T(L)u_t, \tag{3.35}$$

where T is at least of order 12 and maybe higher depending on the order of $\alpha_2 \Theta_B \Omega_B$. Equation (3.35) suggests that the X–11 procedure may provide a satisfactory decomposition of $\ell n\, M_t$ under some restrictions on the behavior of the exogenous variables. Though $T(L)$ is not likely to conform exactly to the MA process described in (3.33), the AR position is identical. On the other hand, if ϕ_B or Γ_B are not one, i.e. if the exogenous variables display autoregressive properties, the X–11 procedure could produce grossly inaccurate results.

This analysis can also be done with the FE for prices and the interest rate. Consider (3.25) as the FE for prices. If $\gamma_1 \Phi = \gamma_1(\gamma_1 - \beta_1)$ is not a constant (i.e. contains a lag structure), the economic model would be indicating AR polynomials and, hence, a departure from the type of process for which the X–11 procedure is considered appropriate.

3.3.2 FEs and Box–Jenkins multiplicative seasonal model An additional point of interest is that the AR portion of all the FEs are already in the form of the multiplicative seasonal time series model, discussed in . . . subsection [2.3] . . . However, the MA portions do not appear to factor into seasonal and non-seasonal polynomials. In fact, the models, in general, imply a mixture type of model that contains some aspects of a multiplicative nature and others of an additive nature. This suggests that the properties of this type of mixed model should be investigated as a starting point for developing methods of adjustment. It would be of interest, however, to determine a set of conditions under which the theory would predict the multiplicative model. For the FE for the money stock, a sufficient set of conditions is to let $\Omega_B \doteq \Delta_{12} = (1 - L^{12})$ and $\Gamma_B = 1$, which yields, from (3.23),

$$\phi_B \Delta_{12} r_{M_t} = \Omega_B(\alpha_2 \Theta_B u_{4t} + \phi_B u_{2t}). \tag{3.36}$$

Finally, if $\phi_B = 1$, (3.36) reduces to a very simple pure seasonal moving average model

$$\Delta_{12} r_{M_t} = \Omega_B(\alpha_2 \Theta_B u_{4t} + u_{2t}). \tag{3.37}$$

These assumptions are equivalent to stating that the process generating the monetary base has no autoregressive polynomials associated with it, neither seasonal nor non-seasonal, and that the seasonality in the base is very close to following the seasonal means model. (See . . . subsection

[2.3] . . .) Recalling that u_{4t} and u_{2t} are assumed independent, and, considering the case where α_2 is just a constant, the r.h.s. of (3.28) reduces to a monthly MA polynomial having a degree that is equal to the degree of Θ_B and a seasonal polynomial of first degree and parameter value of approximately 1. Under such circumstances, the model implied for the natural log of money would be written as (0,1,q) $(0,1,1)_{12}$, where q depends on the properties of Θ_B.

Similarly, (3.26) represents a multiplicative formulation for the FE for the price variable. In this case, both Ω_B and Ω_Y need to approximately equal $(1 - L^{12})$, Γ_B, and Γ_Y equal to 1, and, in addition, the quantity theory of money must hold so that $\beta_1 = 0$ and $\gamma_1 = 1$.

3.3.3 FEs and dependence of seasonality on structural assumptions
Because the FE for the money stock is reasonably simple, it is instructive to investigate it further. In particular, consider the effects on key aspects of the autocorrelation structure of $\Delta_{12} r_{M_t}$ under some different assumptions about the polynomials and parameters on the r.h.s. of (3.34).

$$\Delta_{12} r_{M_t} = w_t = \alpha_2 \Theta_B \Omega_B u_{4t} + \Delta_{12} u_{2t}. \tag{3.38}$$

Assume that the base is truly exogenous, i.e. the model is correct so that

$$E(u_{4t} u_{2t-k}) = 0 \text{ for all } k. \tag{3.39}$$

By assumption,

$$E(u_{4t} u_{4t-k}) = \begin{cases} \sigma_4^2 & \text{if } k = 0 \\ 0 & \text{if } k \neq 0 \end{cases}. \tag{3.40}$$

Although it has been assumed, so far, that u_{2t} is serially uncorrelated, it is interesting to relax this assumption somewhat. Recall that in this model u_{2t} incorporates changes in the money multiplier. Now, the money multiplier may have seasonal properties that are unspecified here. In order to keep the problem manageable, assume that changes in the money multiplier are random except for a seasonal effect. That is, assume that

$$E(u_{2t} u_{2t-k}) = \begin{cases} \sigma_2^2 & \text{if } k = 0 \\ \gamma_{12}^{(2)} & \text{if } k = 12 \\ 0 & \text{otherwise,} \end{cases} \tag{3.41}$$

which implies that changes in the multiplier follow a seasonal MA(1) process. If the multiplier were non-autocorrelated, then, of course, $\gamma_{12}^{(2)} = 0$. Finally, assume that

$$\Theta_B = (1 - \Theta L), \tag{3.42}$$

and

$$\Omega_B = (1 - \Omega L^{12}). \tag{3.43}$$

Under these assumptions, the variance of $\Delta_{12} r_{M_t}$ or w_t can be shown to be

$$\gamma_0^{(w)} = \alpha_2^2(1 + \Theta^2)(1 + \Omega^2)\sigma_4^2 + 2\left(\sigma_2^2 - \gamma_{12}^{(2)}\right). \tag{3.44}$$

A convenient method of getting an idea of how different assumptions affect seasonality is to investigate the autocorrelation coefficient of w_t at lag 12. The autocovariance of w_t at lag 12 is simply

$$\gamma_{12}^{(w)} = -\alpha_2^2 \Omega(1 + \Theta^2)\sigma_4^2 - \sigma_2^2 + 2\gamma_{12}^{(2)}, \tag{3.45}$$

and the autocorrelation coefficient

$$\begin{aligned}\rho_{12}^{(w)} = \frac{\gamma_{12}^{(w)}}{\gamma_0^{(w)}} &= \frac{-\alpha_2^2 \Omega(1 + \Theta^2)\sigma_4^2 - \sigma_2^2 + 2\gamma_{12}^{(2)}}{\alpha_2^2(1 + \Theta^2)(1 + \Omega^2)\sigma_4^2 + 2\left(\sigma_2^2 - \gamma_{12}^{(2)}\right)} \\ &= \frac{-\alpha_2^2 \Omega(1 + \Theta^2)h - \left(1 + 2\rho_{12}^{(2)}\right)}{\alpha_2^2(1 + \Theta^2)(1 + \Omega^2)h + 2\left(1 - \rho_{12}^{(2)}\right)},\end{aligned} \tag{3.46}$$

where $h = \sigma_4^2/\sigma_2^2$ and $\rho_{12}^{(2)} = \gamma_{12}^{(2)}/\sigma_2^2$.

If $\rho_{12}^{(2)} = 0$ and $\Omega = 1$, then it is clear that $\rho_{12}^{(w)}$ is known with certainty, since the process for the money stock is simply the dummy variable case. That is,

$$\rho_{12}^{(w)} = -\frac{1}{2} \cdot \frac{\alpha_2^2(1 + \Theta^2)h + 1}{\alpha_2^2(1 + \Theta^2)h + 1} = -\frac{1}{2}. \tag{3.47}$$

However, if there is seasonality in the multiplier, meaning $\rho_{12}^{(2)} \neq 0$, then the implied value of $\rho_{12}^{(w)}$ is

$$\rho_{12}^{(w)} = -\frac{1}{2} \cdot \frac{\alpha_2^2(1 + \Theta^2)h + 1 - 2\rho_{12}^{(2)}}{\alpha_2^2(1 + \Theta^2)h + 1 - \rho_{12}^{(2)}}, \tag{3.48}$$

which, for $\rho_{12}^{(2)} > 0$, is greater than $-\frac{1}{2}$(or $|\rho_{12}^{(w)}| < \frac{1}{2}$), even though $\Omega = 1$.

Assume that the Fed. was interested in creating the simplest seasonal pattern possible in the money supply. If they knew the parameter α_2 and the stochastic structure of the money multiplier (σ_2^2 and γ_{12}^2 in this case), then values of Θ, Ω, and σ_4^2 could be chosen to obtain a $\rho_{12}^{(w)}$ of $-\frac{1}{2}$, which would imply that the seasonal pattern in the money supply was merely a stable seasonal mean. It would then be straightforward to either adjust the money supply or, for the Fed., to design an optimal control scheme to effectively eliminate seasonality in the money supply.

3.4 Summary

In this section, the basic framework of a simple monetary model was postulated. Explicit assumptions were made regarding several important aspects of the model. First, the assumptions were made that the monetary base and real income are exogenous inputs to the system. This places theoretical restrictions on the covariance matrix between these variables and the endogenous variables of the system that can be checked against the data. Another issue of importance is the question of whether the economic structure generates seasonal fluctuations or acts only as a transmitter of seasonality. In order to shed light on this issue, it was hypothesized that seasonality enters the system only through the exogenous variables. This approach would be consistent with the system transmitting only seasonality. It was shown that this resulted in seasonality being induced into each and every endogenous variable and the FEs and, more importantly, the TFs obtained from the model display restrictions concerning the location and magnitude of certain seasonal parameters and polynomials. An important point to make concerning the FE is that, due to cancellation, the AR portion of the endogenous variables is not identical. Therefore, the estimated univariate models should not be restricted to have the same AR polynomials in the empirical work.

In addition, the theory suggests that, in general, the multiplicative seasonal model is not implied by the structure. Instead, a more general structure is suggested that contains both additive and multiplicative characteristics. The model was then investigated in order to ascertain a set of assumptions sufficient to allow the theory to predict that a multiplicative seasonal model would be adequate in describing the FEs. It was found that, as the seasonality in the exogenous variables approached the simple seasonal means case and as the economic structure approached the classical quantity theory of money, the FEs approach a special case of the multiplicative seasonal model, or the seasonal means case. These results indicate that decomposition schemes, based on the general multiplicative time series model, would be inappropriate, since they are not suggested by the economic structure. In fact, it is clear that the multiplicative seasonal model will not, in general, result from linear models.

Another point investigated in this section was when the economic model implied that the stochastic behavior of the output variables would be of a form, similar to that suggested by Cleveland (1972), which might be appropriate for decomposition by the X–11 procedure. Finally, it was shown how an economic model can explain explicitly why seasonality in interest rates and prices does not appear to be important. The existence of an expectations mechanism has an attenuating effect on the seasonality and the autocorrelation structure of these series.

4 Empirical results

The purpose of this section is to demonstrate how one might utilize available data to test a theoretical economic model, such as the one outlined and analyzed in the previous section.

4.1 Analysis of the univariate time series

In this section, the results from the analysis of the univariate time series properties of the raw or unadjusted data for each variable in the model are reported and compared with the implications of the FEs, as discussed in previous sections. The techniques used are essentially those developed by Box and Jenkins (1970) for the analysis of time series data as well as several other techniques, including likelihood ratio tests and posterior odds ratios, as utilized by Zellner and Palm (1974, 1975) and Zellner (1975). In general, interest centers on identifying and estimating models in the form, described in . . . subsection [2.3] . . . ,

$$\phi_p(L)\Gamma_P(L^s)\Delta_s^D\Delta^d z_t = \Theta_q(L)\Omega_Q(L^s)a_t,$$

written as an ARIMA model of order (p, d, q) $(P, D, Q)_s$. It is assumed that a_t is white noise and that the roots of $\phi(\xi) = 0$ and $\Theta(\xi) = 0$, the monthly polynomials and $\Gamma(\xi) = 0$ and $\Omega(\xi) = 0$, the seasonal polynomials lie outside the unity circle so that $w_t = \Delta^d\Delta_s^D z_t$ is stationary and invertible.

It is important to note that, for a stationary series, the autocorrelations approach zero as the lag increases, so that persistently high values for the estimated autocorrelations at increasing lags might suggest the need for differencing. In addition, and a point that is often overlooked, is that sample autocorrelations need not have large values for a non-stationary series. All that is required is that the series generate a sample autocorrelation function that remains relatively flat. Similarly, a persistence of high or stable values at lags 12, 24, . . . , etc. (with monthly data) would suggest the need for annual or seasonal differencing.

However, in many instances, the question of the appropriateness of differencing or the question of stationarity is not readily resolved. Unfortunately, tests and test statistics that rely on the asymptotic distribution of the observation vector are questionable, since the distribution of these statistics, when the series exhibits homogeneous non-stationarity, is generally not known.[16] In light of this, it would seem inappropriate to use standard testing procedures to test for stationarity. One alternative to consider is to proceed with differencing and test for a root of one in the resulting model's MA polynomial. If there is a root of 1, the process becomes non-invertible, and there is an indication of over-differencing.[17] However, two caveats must be mentioned here. First, Nelson (1974) has shown that, for the first order moving average process, the standard error of the parameter estimate, based on an asymptotic normal distribution, is under-stated in sample sizes as large as 100. In addition, it is not clear what the distributional properties of the standard tests are under the null hypothesis, i.e. when the moving average parameter equals 1. Consequently, the approach followed in this work has been to utilize the standard techniques for the identification of the ARIMA models while, at the same time, being aware of the problems that might arise in finite samples when the stationarity of the series is in question. Recall that, based on the theory in the section on the analysis of an economic model, this problem may arise with both the model for prices and the model for interest rates.

It is useful, at this point, to make a few comments concerning the data being used in this analysis. As with all other econometric work, there is the recurring problem of finding data that adequately measure the quantities which are of theoretical interest. In this case, even the theoretical quantities are, in some instances, not universally agreed upon, such as the appropriate definition of the money stock, the appropriate measure of income, and the use of short versus long-run interest rates.

The actual data used in this study are detailed in appendix B (p. 380). The series are made up of monthly observations from January 1953 through July 1971. The net source base, as calculated by the St. Louis Federal Reserve Bank, is used as the unadjusted monetary base. The money stock is represented by M1, currency plus demand deposits. The interest rate is the yield on one-month Treasury bills, as compiled by Fama (1975). These data should constitute reliable measures of the theoretical quantities. The remaining two series are somewhat less reliable measures for the variables of interest. The price level is represented by the Consumer Price Index (CPI) and real income (output) is measured by the Index of Industrial Production (IIP). Both of these measures are apt to contain measurement error by the mere fact that they are indexes. Sampling properties of these indexes might also cause problems, because the individual components of each index are not measured every month.

[16] See White (1958) and Anderson (1959).
[17] See Plosser and Schwert (1976).

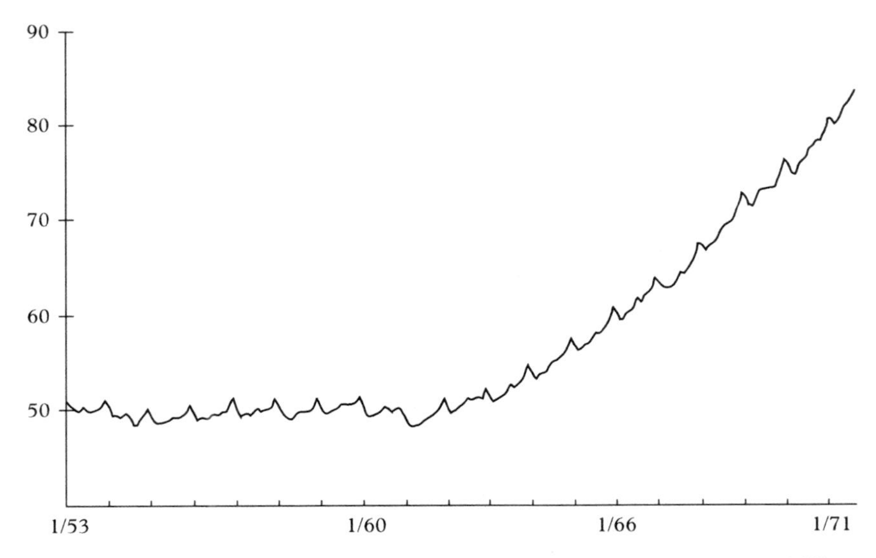

Figure 9.3 Monetary base, seasonally adjusted, 1953–1971, billion dollars

Plots of the raw data are presented in figures 9.3–9.7. Upon inspection of these charts, it becomes apparent why the issue of the appropriate level of differencing becomes difficult. In particular, the growth rates of the monetary base, M1, and the CPI seem to be increasing steadily throughout the time period. However, this does not appear as strikingly in the IIP. The interest rate appears nonstationary or highly autoregressive, which was noted in the theoretical discussions as something that might be observed.

Table 9.3 summarizes the results of a univariate time series analysis of the different series.[18] The first two series, $\ell n(B)$ and $\ell n(Y)$, represent the exogenous variables in the system, and the models shown in table 9.3 describe the processes governing them. These findings indicate that there are no autoregressive polynomials associated with the exogenous variables and that the moving average polynomials are of low order. In terms of the notation of the model, $\phi_B = \Gamma_B = \phi_Y = \Gamma_Y = \Theta_B = 1$, $\Omega_B = (1 - 0.897\ L^{12})$, $\Omega_Y = (1 - 0.915\ L^{12})$, and $\Theta_Y = (1 + 0.247\ L + 0.157\ L^2)$.

[18] These calculations, as well as many others in this chapter, were performed using a set of time series programs developed by C. R. Nelson, S. Beveridge, and G. W. Schwert, Graduate School of Business, University of Chicago. The reader is referred to Plosser (1976) for a more complete documentation of the development of these results.

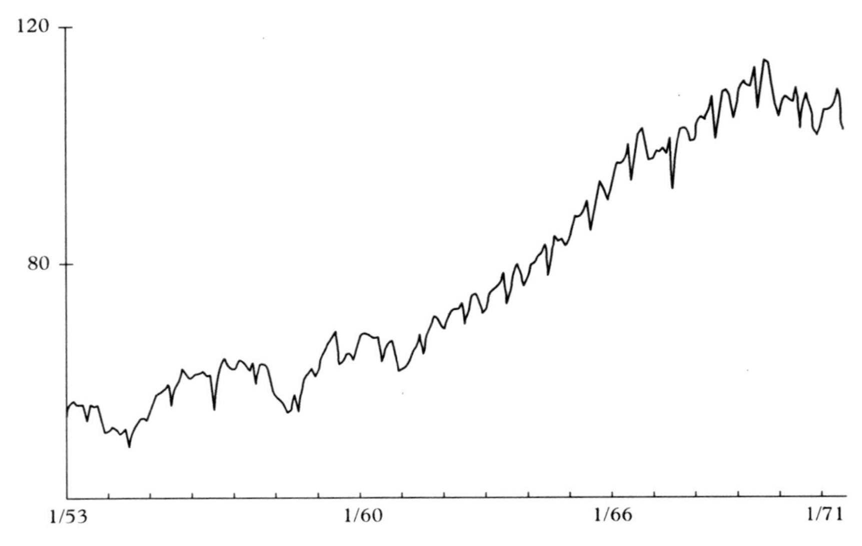

Figure 9.4 Index of industrial production, seasonally adjusted, 1953–1971 (base: 1967 = 100)

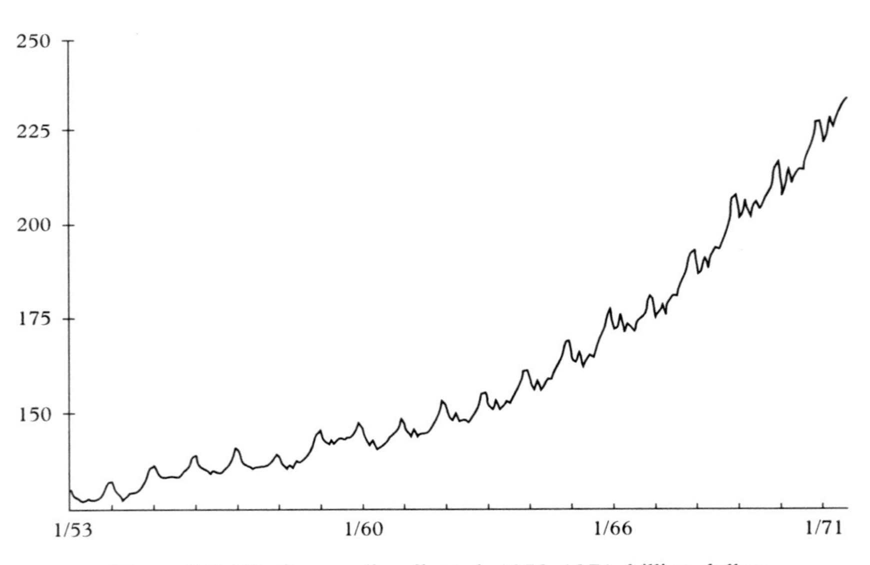

Figure 9.5 M1, Seasonally adjusted, 1953–1971, billion dollars

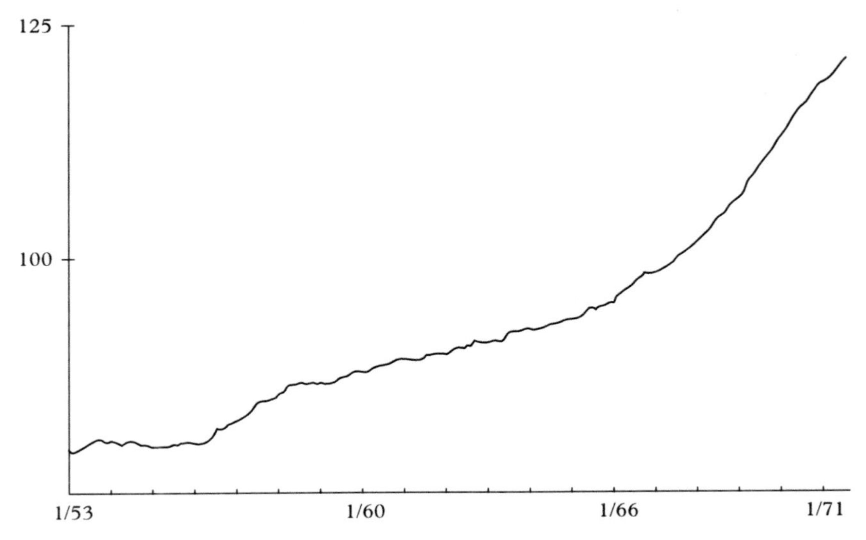

Figure 9.6 Consumer price index, seasonally adjusted, 1953–1971 (base: 1967 = 100)

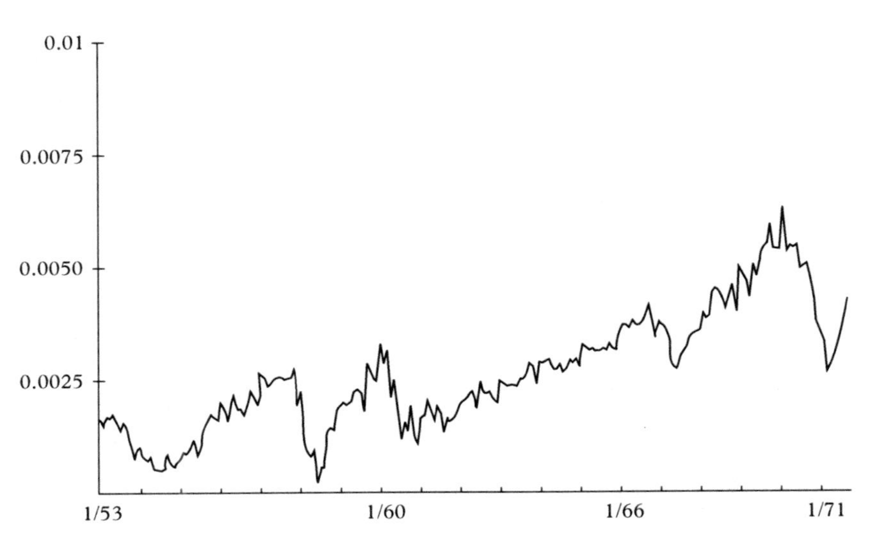

Figure 9.7 Yield on one-month Treasury bills, 1953–1971 (rate of return per month)

Table 9.3 *Estimated univariate time series models*

Variable	Model $(p,d,q)\ (P,D,Q)_s$	Period	D.F.	$\hat{\sigma}_u^{2a}$	AR1	MA1	MA2	MA3	SMA1	c	p value[b]
$\ell n\ (B)$	$(0,1,0)\ (0,1,1)_{12}$	1/53–7/71	208	0.2116×10^{-4}	(X) (X)	(X) (X)	(X) (X)	(X) (X)	0.897 (0.020)	0.429×10^{-3} (0.646×10^{-4})	0.44
$\ell n\ (Y)$	$(0,1,2)\ (0,1,1)_{12}$	1/53–7/71	206	0.1757×10^{-3}	(X) (X)	−0.247 (0.068)	−0.157 (0.069)	(X) (X)	0.915 (0.018)	0.294×10^{-4} (0.294×10^{-3})	0.69
$\ell n\ (M)$	$(0,1,3)\ (0,1,1)_{12}$	1/53–7/71	205	0.1598×10^{-4}	(X) (X)	−0.005 (0.067)	−0.123 (0.067)	−0.282 (0.066)	0.489 (0.062)	0.257×10^{-3} (0.217×10^{-3})	0.06
$\ell n\ (P)$	$(1,1,1)\ (0,1,1)_{12}$	1/53–7/71	206	0.3580×10^{-5}	0.938 (0.045)	0.800 (0.076)	(X) (X)	(X) (X)	0.917 (0.017)	0.118×10^{-4} (0.959×10^{-5})	0.06
	$(0,2,1)\ (0,1,1)_{12}$	1/53–7/71	206	0.3657×10^{-5}	(X) (X)	0.855 (0.036)	(X) (X)	(X) (X)	0.921 (0.012)	-0.340×10^{-6} (0.412×10^{-5})	0.08
i	$(1,0,2)\ (0,1,1)_{12}$	1/53–7/71	206	0.8767×10^{-7}	0.917 (0.032)	0.229 (0.074)	−0.218 (0.073)	(X) (X)	0.897 (0.060)	0.158×10^{-4} (0.731×10^{-5})	0.06
	$(0,1,2)\ (0,1,1)_{12}$	1/53–7/71	206	0.9115×10^{-7}	(X) (X)	0.266 (0.069)	−0.200 (0.069)	(X) (X)	0.897 (0.020)	0.445×10^{-6} (0.394×10^{-5})	0.06

Notes:

X Not applicable.

[a] Backforecasted residuals included.

[b] This is the p value for the Box–Pierce statistic (the Q-statistic) for the first 36 residual autocorrelations. (See Box and Pierce 1970.)

An analysis of the $\ell n(M)$ reveals that an (0, 1, 3) $(0, 1, 1)_{12}$ appears as an adequate representation of the data. (Note that this is the same model used in generating the forecasts in the first section.) Since this variable is an endogenous variable in the system, the next step is to interpret these results in light of the theoretical FEs implied by the model, (3.23). Since the SMA1 parameter for this model is much less than 1, it appears that if the theory is correct, the model can not be factored exactly, so the multiplicative model is at best an approximation. Also, since u_{4t} in (3.23) is non-seasonal by construct, the fact that the seasonality is slightly different in the process for r_{M_t} and r_{B_t} indicates the possibility of seasonal influences from u_{2t}. As noted and discussed earlier, this could be due to seasonal fluctuations in the multiplier.

Another point is that the monthly MA polynomial is of order 3 in this case, and, under the assumption of constant coefficients in the structure, the theory suggested that the order should be the same as the order of Θ_B, which is zero. This might suggest that something else is entering the FE. A likely possibility is a term that involves u_{5t}, the error from the process generating real output. This could occur, as was suggested in . . . section [3] . . ., if the base and real output were not independent. Another possibility is that u_{2t} is autocorrelated at low lags as well as at seasonal lags.

The second endogenous variable to be analyzed is the price level for which the CPI is used as representative. Recall, from the discussion in . . . section [3] . . ., that the analysis indicated that seasonal differences of the rate of inflation may very well appear as an AR process or even non-stationary in finite samples. Although an examination of the sample autocorrelation structure does not suggest non-stationarity, the results of fitting an ARIMA model to $\Delta_{12} r_{P_t}$ do point to that possibility. The model developed for this combination of differencing is a (1, 1, 1) $(0, 1, 1)_{12}$. The estimated values are presented in table 9.3. Note that the AR parameter is very close to one, suggesting non-stationarity. Unfortunately, as was indicated previously, the standard statistical tests cannot be performed here with satisfactory results.[19] If the data are differenced, the preferred model appears to be (0, 2, 1) $(0, 1, 1)_{12}$. Since it is difficult to compare these models, the question of which one is preferred is left unanswered. However, the mere fact that this situation occurs lends support to the theory which suggested that such a phenomenon might exist.

[19] It is important to note that the AR and MA parameter appear close to each other, and, therefore, the possibility of redundancy must be considered. If the (1, 1, 1) $(0, 1, 1)_{12}$ model is redundant, then the parameters are not identified and the usual test procedures can not be utilized. However, a study of the standard errors of the parameter estimates and the correlation coefficient between them does not suggest redundancy or indicate that they are not identified.

The last of the endogenous variables is the nominal interest rate. The data used are the yields on one-month Treasury bills. The same problem is experienced here that was experienced with the univariate model for prices. The theory suggests that either an AR model of the seasonal differences or even apparent non-stationarity of the seasonal differences might be observed in finite samples. It was seen that this was a result of the expectation mechanism at work. Note that, for the $(1, 0, 2)\ (0, 1, 1)_{12}$ model presented in table 9.3, the autoregressive coefficient is close to 1. Once again, the standard testing procedures can not be utilized here when the null hypothesis is that the process is non-stationary. However, the model appears reasonably well behaved, showing no signs of redundancy or instability over time. Alternatively, the $(0, 1, 2)\ (0, 1, 1)_{12}$ also appears adequate, given the data.

The univariate time series models analyzed here display a remarkable degree of consistency, not only in the monthly process but in the seasonal process as well. The actual models chosen are summarized in the following manner:

$$\begin{aligned} \ell n(B_t) &: (0, 1, 0)(0, 1, 1)_{12} \\ \ell n(Y_t) &: (0, 1, 2)(0, 1, 1)_{12} \\ \ell n(M_t) &: (0, 1, 3)(0, 1, 1)_{12} \\ \ell n(P_t) &: (0, 2, 1)(0, 1, 1)_{12} \text{ or } (1, 1, 1)(0, 1, 1)_{12} \\ i_t &: (0, 1, 2)(0, 1, 1)_{12} \text{ or } (1, 0, 2)(0, 1, 1)_{12} \end{aligned}$$

Though these parsimonious models should be interpreted as FEs with caution, the analyses indicate a reasonable amount of compatibility with the theory, as proposed in . . . section [3] . . . If one were to rely solely on these results, several hypotheses about the structural model could not be rejected. For instance, since the AR portions of the exogenous variables are of degree zero, the AR portions of the endogenous variables should be of degree zero or greater. In fact, for prices and interest rates, the cases where the AR polynomials are greater than zero were exactly those that were suggested by the expectation mechanism. This speaks well for the hypothesis regarding rational expectations. Secondly, all of the models displayed a very similar seasonal moving average polynomial. The theory suggested that if $\Omega_B = \Omega_Y = \Delta_{12}$, then the seasonal polynomials for the endogenous variables could be factored out, and the seasonal multiplicative model would be the appropriate representation with a seasonal moving average polynomial close to Δ_{12}. With the exception of the process for the money supply that has an SMA1 coefficient of 0.469, the seasonal moving average parameter for the remaining endogenous variables was indeed close to 1 (0.92 for prices and 0.90 for the interest rate).

The results described in the previous paragraph should not be regarded as conclusive but suggestive of further analysis. The results of the transfer function analysis in the following subsection will provide further checks on the adequacy of the model as well as suggestions for possible modification.

4.2 *Analysis of the transfer functions*

The next step in the analysis of the model entails the estimation of the TFs. These distributed lag models, which express the current endogenous variables in the system in terms of current and past exogenous variables, are initially developed under the assumption that the model has been properly specified concerning which variables are endogenous and which are exogenous. However, this is not an innocuous assumption and should be considered an important part of the specification of any econometric model that should be checked against the data.

In this chapter, use will be made of the cross-correlations between the residuals from the TF estimation and the various prewhitened exogenous variables. Under the null hypothesis that the model is correctly specified these cross-correlations are distributed independently with zero mean and large sample variance of $1/T$, where T is the sample size. If these cross-correlations between the current residual and future prewhitened inputs are non-zero, then the suspicion must be that feedback is likely to be occurring. This simple test should provide a check on the specification of the exogenous variables in the model.[20]

Table 9.4 reports the estimated TFs.[21] These results provide additional checks on the specification of the model presented in . . . section [3] . . . Based on the model, the TF for the nominal money stock is given by (3.17). Consequently, if the model is correctly specified, the structural coefficient α_2 can be directly estimated. Also of interest is the implication on the error structure of (3.17). As was pointed out earlier, the error structure of the estimated model should display a seasonal moving average polynomial of the form $(1 - L^{12})$. However, this polynomial may be affected by seasonal autocorrelation in u_{2t}.

The model presented in table 9.4 suggests an estimate of α_2 of 0.172 with an estimated standard error of 0.053. One would expect this parameter to be in the neighborhood of 1.0, i.e. the elasticity of M1 with respect

[20] See Haugh (1972) and Plosser (1976) for further discussion of this point.

[21] The observant reader may note that $\hat{\sigma}_a^2$ appears only slightly smaller and sometimes larger in the estimated TFs than $\hat{\sigma}_u^2$ obtained from the univariate models. This is due to the estimation techniques used. The univariate models are estimated using backforecasting, while the TFs are not. See Box and Jenkins (1970) for a discussion.

Table 9.4 *Estimated TFs*

Endogenous variable	RSS	D.F.	$\hat{\sigma}_a^{2\,b}$	Estimated IRF[b] $\Delta_{12} r_{B_t}$	Estimated IRF[b] $\Delta_{12} r_{Y_t}$	Noise model	c
$\Delta_{12} r_{M_t}$	0.3134×10^{-2}	204	0.1536×10^{-4}	$\underset{(0.053)}{0.172}$	$\underset{(X)}{(X)}$	$(1 - \underset{(0.068)}{0.464} L^{12})(1 + \underset{(0.069)}{0.009} L + \underset{(0.069)}{0.125} L^2 + \underset{(0.069)}{0.280} L^3)$	$\underset{(0.214 \times 10^3)}{0.187 \times 10^{-3}}$
$\Delta_{12} r_{P_t}$	0.8613×10^{-3}	204	0.4243×10^{-5}	$\dfrac{\underset{(0.011)}{0.020}}{1 - \underset{(0.015)}{0.966} L}$	– $\dfrac{\underset{(0.008)}{0.015}}{1 - \underset{(0.299)}{0.607} L}$	$(1 - \underset{(0.053)\,(0.053)}{0.691} L^{12})(1 + \underset{(0.070)\,(0.070)}{0.073} L + \underset{(0.070)\,(0.070)}{0.200} L^2)$	(X) –
$\Delta_{12} i_t$	0.1881×10^{-4}	201	0.9360×10^{-7}	$\dfrac{\underset{(0.004)}{0.001}}{1 - \underset{(0.543)}{0.997} L}$	$\dfrac{\underset{(0.001)}{0.004}}{1 - \underset{(0.011)}{0.989} L}$	$\dfrac{(1 - \underset{(0.050)}{759} L^{12})(1 - \underset{(0.076)}{0.282} L + \underset{(0.076)}{0.196} L^2)}{1 - \underset{(0.033)}{0.929} L}$	(X) –

Notes:

– Entry represents zero.

X Not applicable.

[a] Not estimated using backforecasting.

[b] Estimated impulse response function (IRF).

to the base should be close to 1. Furthermore, the results indicate a significant amount of autocorrelation in u_{2t}. In particular, as was concluded from the univariate model, seasonal autocorrelation seems to be present, as evidenced by the seasonal moving average parameter being substantially different from 1. These results indicate that the base should not be considered the sole source of seasonality in M1.

Finally, the diagnostic checks applied to this model indicate evidence of non-zero cross-correlations between the current residuals and future values of the base. This suggests model misspecification in the sense that either there is feedback from M1 to the base or that there is a dynamic relationship between the multiplier and the base. In either case, the results provide evidence that the base is not the sole source of seasonality in M1 and that a more complicated money supply relationship needs to be specified.

Obviously, if the TF for M1 is misspecified, the results should indicate that the other TFs are also inadequate. In both the TF for prices and interest rates, the seasonal moving average parameter is less than 1. Furthermore, cross-correlation checks suggest that neither the base nor real income can be considered exogenous relative to prices. The source of this problem for the base is the appearance of u_{2t} in the error structure of the TF for prices. The source of the problem for real income is more likely a short-run Phillips-curve relationship that is missing from the model. As for the interest rate equation, the cross-correlations indicate that the base is virtually independent of the nominal interest rate and that real income, though significantly related, shows evidence of feedback.

4.3 *Summary of empirical findings*

The results of the empirical analysis are not supportive of the economic model detailed in . . . section [3] . . .The weaknesses of the model seem to stem from the seasonality in the unobserved money multiplier and its relationship with the monetary base and from the feedback effects indicating that real income is not an exogenous variable. These problems were clearly pointed out in the TF analysis but were suggested even in the analysis of the FEs. The FEs did not present enough evidence by themselves to reject the model, but the seasonal moving average parameter in the univariate model for M1, being much less than anticipated, was indicative of the weakness of the model's specification.

Certainly, at this point, the analyst attempting to construct a satisfactory seasonal model of the economy could proceed by modifying the model presented here in such a way as to eliminate the defects that are suggested by the analysis. For example, the inclusion of a real sector or a short-run Phillips curve would be a logical extension and would make output endogenous. In addition, further consideration must be given to the specification of the money supply relationship.

Despite the weakness of the model analyzed, it proved useful in demonstrating the methodological issues and the relevance of the form of analysis presented in . . . section [2] . . .Clearly, the techniques demonstrated here can be very useful in analyzing even very simple supply and demand models for any market.

5 Discussion

In this chapter, an effort has been made to investigate seasonality in economic data from a slightly different perspective than has been common in the literature. Interest is focused on demonstrating a methodology whereby seasonality can be directly incorporated into an econometric model. Utilizing this approach, one can determine what is implied about the seasonal properties of the endogenous variables in the model and use the implications to check the specification of the model against information in the data.

There are several different perspectives from which one can view the results of this approach. First, for the econometrician it has been shown how one might build seasonality into a model and check the specification. Although, in the example analyzed here, the seasonality entered through the exogenous economic variables, there are possibilities of including seasonal structural parameters or, in other types of markets, one may consider including stochastic seasonal variables, such as rainfall, to act as the driving seasonal force. In proceeding in this manner, the model builder must carefully consider the possible sources of seasonal variation in [her] output variables. However, the analyst avoids using data that have been smoothed, using methods which may not be appropriate for [her] purposes.

There may still be reasons, however, for obtaining a seasonally adjusted series. In these cases, it would, of course, be ideal to make use of a model that would specify the source and type of seasonality. That is, through the use of an economic model and some economic analysis, understanding and insights can be gained concerning seasonal variation in economic time series, e.g. what might cause seasonality to change through time. Unfortunately, it may not be feasible to construct econometric models for every series that may need to be adjusted. Under these circumstances, obtaining a seasonally adjusted series based only on the past history of the series may prove to be necessary. However, it was demonstrated that, in general, the stochastic properties of economic variables in the system are not independent of the economic structure of which they are a part. Therefore, it would seem appropriate for those who wish to obtain adjusted series to study the stochastic behavior of the unadjusted data, investigating its form and properties, prior to adjustment. In making such an analysis, it is useful to realize that the standard linear econometric models do not imply that the multiplicative seasonal models of Box and Jenkins (1970), in general, hold. Even in the case of the exogenous variables following the multiplicative specification, the endogenous variables would be a mixture of an additive and multiplicative process.

Many issues, of course, remain unresolved, but it would do no harm to suggest that econometricians or any consumer of economic data consider carefully the objectives of seasonal adjustment and why it is that adjusted data are desired. It may prove very easy to fail to see when one should stop making corrections and alterations to data once the process has started. Furthermore, the more the basic data are changed, the more cautiously any estimated relationship must be regarded, and, certainly, its results become more difficult to interpret. If some initial model does not provide an adequate explanation of the phenomenon under study, one should try to improve the model by explicit introduction of the factors that may have been omitted.

Seasonality, for instance, can be incorporated into an economic model in various ways. For example, it seems that the concern for abstracting from seasonality in economic data has stemmed partly from the belief that somehow economic agents respond to seasonal fluctuations differently from non-seasonal fluctuations, because these seasonal movements are so highly predictable. Therefore, using data which have not been adjusted may lead to misleading inferences about the true relationship, because the estimation would be contaminated by the relationship among the seasonal components. However, one can view this aspect of the seasonality problem as being part of a broader class of issues that has received a great deal of attention in the economics literature in the last few years. This literature is concerned with making the distinction between anticipated and unaticipated effects. It would seem a natural extension to merely consider seasonality as then belonging to this larger class of anticipated phenomena. Similarly, from the standpoint of forecasting, one may be concerned only with whether a particular observation deviates from what might have been anticipated and whether or not the anticipated portion was seasonal or non-seasonal may be of less importance.

APPENDIX A DERIVATIONS OF FEs AND TFs

This appendix details the derivations of the final equations (FEs) and the transfer functions (TFs) for the monetary model described in . . . subsection [3.1] . . . . The analyzes and interpretations of these results however, are primarily conducted in the text and not undertaken here.

The model is written as

$$r^D_{M_t} = \alpha_1 r_{Y_t} + \beta_1 \Delta i_t + \gamma_1 r_{P_t} + u_{1t} \tag{A.1}$$

$$r^S_{M_t} = \alpha_2 r_{B_t} + u_{2t} \tag{A.2}$$

$$i_t = r^*_{P_{t+1}} + u_{3t} \tag{A.3}$$

$$r^*_{P_{t+1}} = \underset{t}{E}(r_{P_{t+1}} \mid \cdot) \tag{A.4}$$

$$r_{M_t} = r^D_{M_t} = r^S_{M_t}. \tag{A.5}$$

Substituting (A.5) and (A.3) into (A.1) results in

$$r_{M_t} = \alpha_1 r_{Y_t} + \beta_1 \Delta(r^*_{P_{t+1}} + u_{3t}) + \gamma_1 r_{P_t} + u_{1t}. \tag{A.6}$$

Combining this then with (A.2) and solving for r_{P_t} yields

$$\gamma_1 r_{P_t} = \alpha_2 r_{B_t} - \alpha_1 r_{Y_t} - \beta_1 (r^*_{P_{t+1}} - r^*_{P_t}) - \beta_1 \Delta u_{3t} + u_{2t} - u_{1t}. \tag{A.7}$$

Now, for convenience, define

$$X_t \equiv \alpha_2 r_{B_t} - \alpha_1 r_{Y_t} - \beta_1 \Delta u_{3t} + u_{2t} - u_{1t}, \tag{A.8}$$

and rewrite (A.7) as

$$\gamma_1 r_{P_t} = X_t - \beta_1 (r^*_{P_{t+1}} - r^*_{P_t}). \tag{A.9}$$

Under the assumption of rational expectations, the conditional expectation of $r_{P_{t+1}}$ can be calculated from (A.9)

$$\underset{t}{E}(r_{P_{t+1}} \mid \cdot) = \gamma_1^{-1} [\underset{t}{E} X_{t+1} - \beta_1 r^*_{P_{t+2}} + \beta_1 r^*_{P_{t+1}}], \tag{A.10}$$

or

$$(\gamma_1 - \beta_1) r^*_{P_{t+1}} = \underset{t}{E} X_{t+1} - \beta_1 r^*_{P_{t+2}}. \tag{A.11}$$

The same can be done for $r_{P_{t+2}}$, yielding

$$(\gamma_1 - \beta_1) r^*_{P_{t+2}} = \underset{t}{E} X_{t+2} - \beta_1 r^*_{P_{t+3}}, \tag{A.12}$$

which substituted into (A.11) gives

$$(\gamma_1 - \beta_1) r^*_{P_{t+1}} = \underset{t}{E} X_{t+1} - \frac{\beta_1}{\gamma_1 - \beta_1} \underset{t}{E} X_{t+2} + \beta_1 r^*_{P_{t+3}}. \tag{A.13}$$

Solving this recursively then yields

$$r^*_{P_{t+1}} = (\gamma_1 - \beta_1)^{-1} \sum_{j=0}^{\infty} \left(\frac{\beta_1}{\gamma_1 - \beta_1} \right)^j \underset{t}{E} X_{t+j+1}. \tag{A.14}$$

Note that the anticipated rate of inflation is a weighted average of all expected future values of the exogenous variables. The weighting scheme itself is determined by the model or structure. The importance of these future values will, in large part, be determined by the term $\frac{-\beta_1}{\gamma_1 - \beta_1}$.

In order to generate expectations of future values of the exogenous variables, the assumption that they are ARIMA processes proves convenient. In fact, based on the analysis in . . . section [2] . . ., one can interpret these processes as implying that these variables are generated by some unspecified structure and that the processes, used as inputs into this system, are merely the final equations from the model actually generating the monetary base and real income. These exogenous variables are then written as multiplicative seasonal time series models of the type described in . . . section [2] . . .

$$\phi_B(L) \Gamma_B(L^{12}) \Delta_{12} r_{B_t} = \Theta_B(L) \Omega_B(L^{12}) u_{4t} \tag{A.15}$$

$$\phi_Y(L) \Gamma_Y(L^{12}) \Delta_{12} r_{Y_t} = \Theta_Y(L) \Omega_Y(L^{12}) u_{5t}. \tag{A.16}$$

Note that Γ_B, Γ_Y, Ω_B, and Ω_Y represent the seasonal polynomials that have an explicit representation maintained to allow them to be traced through the model. Taking conditional expectations and dropping the Ls for convenience produces

$$\underset{t}{E} r_{B_{t+1}} = \Delta_{12}^{-1} \left[\left(1 - \frac{\phi_B \Gamma_B}{\Theta_B \Omega_B} \right) \Delta_{12} r_{B_{t+1}} \right] \tag{A.17}$$

$$\underset{t}{E} r_{Y_{t+1}} = \Delta_{12}^{-1} \left[\left(1 - \frac{\phi_Y \Gamma_Y}{\Theta_Y \Omega_Y} \right) \Delta_{12} r_{Y_{t+1}} \right]. \tag{A.18}$$

Equations (A.17) or (A.18) simply indicate that the one-step-ahead forecast can be written in terms of an infinite distributed lag of all past values where the weights are determined by the AR and MA polynomials. Through recursive calculations, one could obtain an expression for the expected value of any future observation, conditional on the information contained in the series at time t. That is,

$$\underset{t}{E} r_{B_{t+j}} = \Delta_{12}^{-1} \pi_j^{(B)}(L) \Delta_{12} r_{B_t} \tag{A.19}$$

$$\underset{t}{E} r_{Y_{t+j}} = \Delta_{12}^{-1} \pi_j^{(Y)}(L) \Delta_{12} r_{Y_t}, \tag{A.20}$$

where $\pi_j^{(\cdot)}(L)$ represents the polynomial in L that gives weights applied to all previous observations of the exogenous variable to produce the minimum mean-square error forecast for j periods into the future. Note that the weights are a function of j, the forecast horizon.

Using (A.3), (A.14) can now be rewritten as

$$(\gamma_1 - \beta_1)i_t = \sum_{j=0}^{\infty} \left(\frac{-\beta_1}{\gamma_1 - \beta_1} \right)^j \underset{t}{E} X_{t+j+1}, \tag{A.21}$$

or as

$$\Phi \Delta_{12} i_t = \Psi_1 \Delta_{12} r_{B_t} + \Psi_2 \Delta_{12} r_{Y_t} + \Delta_{12} v_t, \tag{A.22}$$

where

$$\begin{aligned}
\Phi &= (\gamma_1 - \beta_1) \\
\Psi_1 &= \alpha_2 \sum_{j=0}^{\infty} \left(\frac{-\beta_1}{\gamma_1 - \beta_1} \right)^j \pi_{j+1}^{(B)}(L) \\
\Psi_2 &= -\alpha_1 \sum_{j=0}^{\infty} \left(\frac{-\beta_1}{\gamma_1 - \beta_1} \right)^j \pi_{j+1}^{(Y)}(L) \\
\Delta_{12} v_t &= \Delta_{12} \sum_{j=0}^{\infty} \left(\frac{-\beta_1}{\gamma_1 - \beta_1} \right)^j \\
&\quad \times (\underset{t}{E} u_{2t+j+1} - \beta_1 \underset{t}{E} u_{3t+j+1} - \underset{t}{E} u_{1t+j+1}) \\
&\quad + (\gamma_1 - \beta_1) \Delta_{12} u_{3t}.
\end{aligned} \tag{A.23}$$

In matrix form, the system can now be written as

$$\begin{bmatrix} 1 & -\gamma_1 & -\beta_1 \Delta \\ 1 & 0 & 0 \\ 0 & 0 & \Phi \Delta_{12} \end{bmatrix} \begin{bmatrix} r_{M_t} \\ r_{P_t} \\ i_t \end{bmatrix} + \begin{bmatrix} 0 & -\alpha_1 \\ -\alpha_2 & 0 \\ -\Psi_1 \Delta_{12} & -\Psi_2 \Delta_{12} \end{bmatrix} \begin{bmatrix} r_{B_t} \\ r_{Y_t} \end{bmatrix} = \begin{bmatrix} u_{1t} \\ u_{2t} \\ \Delta_{12} v_t \end{bmatrix}. \tag{A.24}$$

Referring to the first matrix on the l.h.s. as H_{11}, its determinant can be written as

$$|H_{11}| = \gamma_1 \Phi \Delta_{12}, \tag{A.25}$$

while its adjoint is

$$H_{11}^* = \begin{bmatrix} 0 & \gamma_1 \Phi \Delta_{12} & 0 \\ -\Phi \Delta_{12} & \Phi \Delta_{12} & -\beta_1 \Delta \\ 0 & 0 & \gamma_1 \end{bmatrix}. \tag{A.26}$$

By premultiplying both sides of (A.16) by H_{11}^{-1} and then multiplying through by $|\mathrm{H}_{11}|$, the following can be obtained:

$$|H_{11}| \begin{bmatrix} r_{M_t} \\ r_{P_t} \\ i_t \end{bmatrix} = H_{11}^* \begin{bmatrix} 0 & \alpha_1 \\ \alpha_2 & 0 \\ \Psi_1 \Delta_{12} & \Psi_2 \Delta_{12} \end{bmatrix} \begin{bmatrix} r_{B_t} \\ r_{Y_t} \end{bmatrix} + H_{11}^* \begin{bmatrix} u_{1t} \\ u_{2t} \\ \Delta_{12} v_t \end{bmatrix}. \tag{A.27}$$

More explicitly,

$$\gamma_1 \Phi \Delta_{12} \begin{bmatrix} r_{M_t} \\ r_{P_t} \\ i_t \end{bmatrix} = \begin{bmatrix} \gamma_1 \Phi \alpha_2 & 0 \\ \Phi \alpha_2 - \beta_1 \Delta \Psi_1 & -\Phi \alpha_1 - \beta_1 \Delta \Psi_2 \\ \gamma_1 \Psi_1 & \gamma_1 \Psi_2 \end{bmatrix} \begin{bmatrix} \Delta_{12} r_{B_t} \\ \Delta_{12} r_{Y_t} \end{bmatrix} + \begin{bmatrix} 0 & \gamma_1 \Phi \Delta_{12} & 0 \\ -\Phi \Delta_{12} & \Phi \Delta_{12} & -\beta_1 \Delta \\ 0 & 0 & \gamma_1 \end{bmatrix} \begin{bmatrix} u_{1t} \\ u_{2t} \\ \Delta_{12} v_t \end{bmatrix}. \tag{A.28}$$

The set of equations in (A.27) or (A.28) represent the set of TFs for the system.

The set of FE can be obtained by substituting

$$\begin{aligned} \begin{bmatrix} \Delta_{12} r_{B_t} \\ \Delta_{12} r_{Y_t} \end{bmatrix} &= \begin{bmatrix} \phi_B \Gamma_B & 0 \\ 0 & \phi_Y \Gamma_Y \end{bmatrix}^{-1} \begin{bmatrix} \Theta_B \Omega_B & 0 \\ 0 & \Theta_Y \Omega_Y \end{bmatrix} \begin{bmatrix} u_{4t} \\ u_{5t} \end{bmatrix} \\ &= H_{22}^{-1} F_{22} \begin{bmatrix} u_{4t} \\ u_{5t} \end{bmatrix} \end{aligned} \tag{A.29}$$

into (A-28), resulting in

$$|H_{11}||H_{22}| \begin{bmatrix} r_{M_t} \\ r_{P_t} \\ i_t \end{bmatrix} = -H_{11}^* H_{12} H_{22}^* F_{22} \begin{bmatrix} u_{4t} \\ u_{5t} \end{bmatrix} + |H_{22}| H_{11}^* \begin{bmatrix} u_{1t} \\ u_{2t} \\ \Delta_{12} v_t \end{bmatrix}, \tag{A.30}$$

or, more explicitly,

$$
\begin{aligned}
&\gamma_1 \Phi \phi_B \phi_Y \Gamma_B \Gamma_Y \begin{bmatrix} \Delta_{12} r_{M_t} \\ \Delta_{12} r_{P_t} \\ \Delta_{12} i_t \end{bmatrix} \\
&= \begin{bmatrix} \gamma_1 \Phi \alpha_2 & 0 \\ \Phi \alpha_2 - \beta_1 \Delta \Psi_1 - \Phi \alpha_1 - \beta_1 \Delta \Psi_1 & \\ \gamma_1 \Psi_1 & \gamma_1 \Psi_2 \end{bmatrix} \begin{bmatrix} \phi_Y \Gamma_Y \Theta_B \Omega_B u_{4t} \\ \phi_B \Gamma_B \Theta_Y \Omega_Y u_{5t} \end{bmatrix} \\
&\quad + \phi_B \phi_Y \Gamma_B \Gamma_Y \begin{bmatrix} 0 & \gamma_1 \Phi \Delta_{12} & 0 \\ -\Phi \Delta_{12} & \Phi \Delta_{12} & -\beta_1 \Delta \\ 0 & 0 & \gamma_1 \end{bmatrix} \begin{bmatrix} u_{1t} \\ u_{2t} \\ \Delta_{12} v_t \end{bmatrix}
\end{aligned} \tag{A.31}
$$

The derivations presented here are summarized and analyzed in . . . section [3] . . .

APPENDIX B SOURCES OF DATA

1. Monetary Base (B) data, seasonally unadjusted, were provided by the St. Louis Federal Reserve Bank. They are averages of daily figures and have been subject to no adjustment for changes in reserve margins or the like.
2. Index of Industrial Production (IIP) data were taken from the seasonally unadjusted. Federal Reserve Board Production Index, as specified by the *Board of Governors of the Federal Reserve System Statistical Release G.12.3*, "Business Indexes."
3. Money Stock (M1) data, not seasonally adjusted, are averages of daily figures for currency plus demand deposits, as specified by the *Board of Governors of the Federal Reserve System Statistical Release H.6*, and the *Federal Reserve Bulletin*.
4. Consumer Price Index (CPI) data, not seasonally adjusted, were taken from the US Department of Labor, Bureau of Labor Statistics.
5. Interest Rate (i) data was compiled by Fama (1975) from Salomon Brothers quote sheets and represent the yields on one-month US Treasury bills.

All the above data, except for the monetary base and the yields on the Treasury bills, were taken from the data base as collected by Data Resources, Inc., as provided to the University of Chicago, Graduate School of Business, through the H. G. B. Alexander Research Foundation.

BIBLIOGRAPHY

Anderson, T. W. (1959), "On the asymptotic distributions of estimates of parameters of stochastic difference equations," *Annals of Mathematical Statistics* 30, 677–87

(1971), *The Statistical Analysis of Time Series* (New York: John Wiley)

Box, G. E. P. and G. M. Jenkins (1970), *Time Series Analysis: Forecasting and Control* (San Francisco, Holden-Day)

Box, G. E. P. and D. A. Pierce (1970), "Distribution of residual autocorrelations in autoregressive-integrated-moving average time series models." *Journal of the American Statistical Association* December (1972), 1509–26

Cleveland, W. P. (1972), "Analysis and forecasting of seasonal time series," PhD dissertation, University of Wisconsin

Dhrymes, P. J. (1970), *Distributed Lags* (San Francisco, Holden-Day)

Fama, E. F. (1975), "Short-term interest rates as predictors of inflation," *American Economic Review* 65, 269–82

Haugh, L. D. (1972), "The identification of time series interrelationships with special reference to dynamic regression models," PhD dissertation, University of Wisconsin

Jorgenson, D. W. (1966), "Rational distributed lag functions," *Econometrica* 34, 135–49

Kmenta, J. (1971), *Elements of Econometrics* (New York, Macmillan)

Kuznets, S. (1933), *Seasonal Variation in Industry and Trade* (New York, National Bureau of Economic Research)

Laffer, A. and D. Ranson (1971), "A formal model of the economy," *Journal of Business* 44, 247–70

Lucas, R. E. (1973), "Some international evidence on output–inflation tradeoffs," *American Economic Review* 63, 326–34

Muth, J. F. (1961), "Rational expectations and the theory of price movements," *Econometrica* 29, 315–35

Nelson, C. R. (1974), "The first order moving average process: identification, estimation, and prediction," *Journal of Econometrics* 2, 121–42

Pierce, D. A. and J. M. Mason (1971), "On estimating the fundamental dynamic equations of structural econometric models," Paper presented at the Econometric Society, New Orleans

Plosser, C. I. (1976), "Time series analysis and seasonality in econometric models with an application to a monetary model," PhD dissertation, University of Chicago, Graduate School of Business

Plosser, C. I. and G. W. Schwert (1976), "Regression relationships among nonstationary variables: estimation and inference in the noninvertible moving average process," University of Chicago, manuscripts

Quenouille, M. H. (1968), *The Analysis of Multiple Time Series*, 2nd edn. (New York, Hafner Press)

Sargent, T. J. (1973), "'Rational' expectations, the 'real rate of interest' and the 'Natural Rate of Unemployment,'" in A. M. Okun and G. Perry (eds.), *Brookings Papers on Economic Activity* 2 (Washington, DC, Brookings Institute)

US Department of Commerce, Bureau of the Census (1967), *The* X–11 *Variant of the Census Method II Seasonal Adjustment Program*, by J. Shiskin, A. H. Young, and J. C. Musgrave, Technical Paper 15 (Washington, DC, Goverment Printing Office)
US Department of the Interior (1962), "Economic aspects of the Pacific halibut fishery," by J. Crutchfield and A. Zellner, *Fishery Industrial Research*, April
Wallis, K. F. (1974), "Seasonal adjustment and the relations between variables," *Journal of the American Statistical Association* 69, 18–31
White, J. S. (1958), "The limiting distribution of the serial correlation coefficient in the explosive case," *Annals of Mathematical Statistics* 29, 1188–97
Zellner, A. (1975), "Time series analysis and econometric model construction," in R. P. Gupta (ed.), *Applied Statistics* (Amsterdam, North-Holland), pp. 373–98
Zellner, A. and F. C. Palm (1974), "Time series analysis and simultaneous equation econometric models," *Journal of Econometrics* 2, 17–54; chapter 1 in this volume
(1975), "Time series and structural analysis of monetary models of the US economy," *Sankhyā: The Indian Journal of Statistics*, Series C 37, 12–56; chapter 6 in this volume

Comment (1978)

Gregory C. Chow

The conference participants can probably be divided into two groups: Those who believe in using a specific model because they think they have a good one, and those who prefer not to choose a model but rather to devise robust methods. Time does not allow me to discuss these opposing positions. However, I agree with the late Chairman Mao that we should let a hundred flowers bloom.

Among those who believe in using a model, some would perform multivariate, and others would perform univariate time series analyzes. The papers in this session [of the conference] are devoted to multivariate time series analyzes. In particular, the basic proposition of Charles Plosser's [chapter 9] is that the analysis of seasonal fluctuations in economic time series, and the construction of econometric models, can be made an integrated process. My comments will be divided into three parts. First, I will summarize the main features of Plosser's particular approach to combine a seasonal model with a traditional simultaneous econometric model. Second, I will comment on the specific illustrative example used in his chapter. Third, I will suggest an alternative approach to integrate seasonal fluctuations with a simultaneous equations model.

In Plosser's approach, it is first assumed that the economic model for the endogenous variables y_t and the exogenous variables x_t can be written as

$$H_{11}(L)y_t + H_{12}(L)x_t = F_{11}(L)e_{1t} \tag{C.1a}$$
$$H_{22}(L)x_t = F_{22}(L)e_{2t}, \tag{C.1b}$$

where H_{ij} and F_{ij} are polynomials in the lag operator L with matrix coefficients, and e_{1t} and e_{2t} are serially independent random variables. Second, there exist seasonalities in x_t that can be described by the multiplicative seasonal time series models of Box and Jenkins [1970], i.e. $H_{22}(L)$ and $F_{22}(L)$ take the special form, such that the ith element x_t^i of x_t is determined by

$$\Gamma(L^s)\phi(L)(1-L^s)^D(1-L)^d x_t^i = \Omega(L^s)\Theta(L)e_{2t}^i, \tag{C.2}$$

where $s = 12$ if seasonal fluctuations in monthly data are being studied, the operators $(1-L^s)^D$ and $(1-L)^d$ will serve to difference the original series seasonally and consecutively, $\phi(L)$ and $\Theta(L)$ are the usual autoregressive and moving average operators for the process governing x_t^i, and, finally, $\Gamma(L^s)$ and $\Omega(L^s)$ are seasonal AR and MA polynomials (or polynomials in L^{12}) that help characterize the seasonalities in the process. Strictly speaking, all operators in (C.2) should be superscripted by the index i for the ith exogenous variable, but this superscript has been omitted. The basic approach is to trace the implications of the particular specifications of $\Gamma^i(L^s)$ and $\Omega^i(L^s)$, which are parts of the specifications of $H_{22}(L)$ and $F_{22}(L)$, respectively, on the transfer functions and the final equations of the model (C.1a) and (C.1b), thus imposing restrictions on the latter equations for statistical analysis and testing.

The transfer functions [TFs], often called the final form of an econometric model (to be distinguished from the final equations [FEs], which will be presented) are obtained by using (C.1a) to express y_t as a distributed lag function of x_t and e_{1t}, i.e.

$$y_t = -H_{11}^{-1}H_{12}x_t + H_{11}^{-1}F_{11}e_{1t}. \tag{C.3}$$

The final equations are obtained by using the identity $H_{11}^{-1} \equiv |H_{11}|^{-1}H_{11}^*$, H_{11}^* being the adjoint matrix of H_{11} and $|H_{11}|$ being its determinant, to isolate a (common) scalar autoregressive polynomial in L, $|H_{11}(L)|$, for each of the elements of y_t, rather than the original matrix autoregressive polynomial $H_{11}(L)$ for the vector y_t

$$|H_{11}|y_t = -H_{11}^*H_{12}x_t + H_{11}^*F_{11}e_{1t}. \tag{C.4}$$

Using $x_t = H_{22}^{-1} F_{22} e_{2t} = |H_{22}|^{-1} H_{22}^* F_{22} e_{2t}$ to substitute for x_t in (C.4), we obtain

$$|H_{11}||H_{22}|y_t = -H_{11}^* H_{12} H_{22}^* F_{22} e_{2t} + |H_{22}| H_{11}^* F_{11} e_{1t}. \tag{C.5}$$

Insofar as the specifications of the seasonal AR and MA polynomials $\Gamma^i(L^s)$ and $\Omega^i(L^s)$ affect H_{22} and F_{22}, respectively, they also impose restrictions, through $|H_{22}|$, H_{22}^*, and F_{22} on the final equations (C.5), and these restrictions can be confronted with the time series data on the endogenous variables.

Having briefly described the general approach, let me list its major characteristics as follows:

1. The linear simultaneous equations model (C.1a) explains the observed time series y_t by the observed x_t, rather than the seasonally adjusted series, as might be supposed.
2. Seasonality in y_t is explained solely by the seasonality in x_t and not by seasonality in the parameters or other mechanisms.
3. Seasonality in x_t is described by the multiplicative seasonal model of Box and Jenkins for each exogenous variable separately and not by a simultaneous multiplicative seasonal model for the vector x_t, nor by some additive model.
4. The algebraic relationships between the specifications of the seasonal variations in the exogenous variables, such as imbedded in $H_{22}(L)$ and $F_{22}(L)$ through (C.2) and the final equations (C.5), appear to be very complicated. Even for medium-size models, it appears difficult to trace explicitly the algebraic implications of the seasonal equations (C.2) on the final equations (C.5). Thus, the approach of this chapter might be applicable only to very small and very specialized models.
5. In general, the transfer functions (C.3) and even the partially final equations (C.4), where x_t has not been eliminated, do not involve $H_{22}(L)$ and $F_{22}(L)$ and are, therefore, not affected by the specifications of the seasonal pattern for x_t. It is only in the special example, treated by Plosser, that the seasonalities in x_t impose restrictions on the transfer functions. The reason is the rational expectations hypothesis adopted in the illustrative example. By this hypothesis, some endogenous variables will depend on the conditional expectations of x_t which, in turn, are distributed lag functions of past x_{t-k} as implied by the seasonal model (C.2). Hence, the relations between y_t and the lagged exogenous variables, as given by the transfer functions, are also restricted by the specification of the seasonal pattern given by (C.2).
6. The approach does not yield a decomposition of an economic time series into seasonal, trend-cycle, and irregular, components. Purely

for the purpose of measuring the changes in such important economic variables such as industrial production, the consumer price index [CP], and the rate of unemployment net of seasonal effects, the approach fails to provide an answer.

Concerning the illustrative example, the author is aware of many of its limitations and shortcomings. I will, however, emphasize several problems with this example, since they have bearing on the general applicability of the proposed approach. The simple model explains three endogenous variables – money stock, a general price index, and the rate of interest – by two exogenous variables – the monetary base and the aggregate real output – to be measured by the Index of Industrial Production [IIP]. The first problem concerns the use of the selected exogenous variables alone to account for the seasonalities in the econometric model. The first exogenous variable, the monetary base, is a policy instrument. Why should the monetary authorities determine the monetary base following a certain seasonal pattern that is independent of all the endogenous variables in the economy? There is a general problem of attributing seasonalities in the policy variables that are independent of the internal workings of the economy. Are not the increases in the demand for money during certain seasons, such as the Christmas–New Year holidays, due to the seasonal pattern of demand itself? Similarly, are not seasonal fluctuations in the Index of Industrial Production due, at least partly, also to the seasonal pattern of demand? It seems difficult to attribute all the seasonalities in any reasonable econometric model to the exogenous variables, as usually defined, without including at least some seasonal dummy variables that are not used by the author.

The second problem is that the mechanism transmitting the seasonalities in the exogenous variables to the endogenous variables, such as the first transfer function relating the changes in the money stock to the changes in the monetary base in the general formulation of table 9.1, is far too simple to be useful. It is likely that such simple formulations are used in order to keep track of the algebraic relations between the transfer functions and the final equations, on the one hand, and the seasonal specifications of the exogenous variables on the other. This example raises doubt on whether more complicated formulations can be explicitly analyzed by the approach of this chapter. Third, even in this over-simplified example, analysis and interpretation of the implications of the seasonalities in x_t on the dynamic characteristics of y_t have encountered difficulties, as exemplified by the final equation (3.22) for the rate of change in the price level. Fourth, related to the third problem is the difficulty of obtaining conclusive results from statistical analysis of the final equations. The final equations are derived from the many characteristics of the econometric model besides the seasonal characteristics of the exogenous variables. To attribute the dynamic characteristics of the final equations to the seasonal specifications alone would, therefore, be extremely difficult. Could not the same dynamic implications for the final equations have been derived from a different seasonal model combined with different specifications for the remaining parts of the econometric model? This problem would surely be more serious for larger models. Fifth, one may question whether the particular method of trend elimination by way of various differencing operations is adequate. The requirement appears to exist that, after the differencing operations, the resulting model should have autoregressive polynomials having roots which will insure that the time series are stationary. How much does one sacrifice by restricting the method of trend elimination to differencing operations and by insisting that only stationary models be studied? Sixth, few significant economic conclusions seem to have resulted from the example.

As the above comments may apply not only to the specific economic example, one would question the general applicability of the method proposed. There is no question that this chapter has suggested interesting ideas and methods for analyzing seasonalities in economic time series. However, unless these problems could be resolved and a serious and relevant economic application could be provided to demonstrate its usefulness, I am afraid that the approach would not be widely accepted by analysts of seasonalities in economic time series.

By way of providing an alternative approach to combining seasonal analysis with the construction of an econometric model for cyclical fluctuations, I would like to suggest the following specifications and methods of analysis . . . Adrian Pagan (1975) has pointed out the possibility of applying the filtering and estimation methods for state-space models to the estimation of seasonal and cyclical components in economic time series. The following suggestion is essentially a combination of an econometric model for the cyclical components with the filtering and estimation of the seasonal components formulated in a state-space form. Assume, first, that the vector y_t of endogenous variables is the sum of cyclical, seasonal, and irregular components, as given by

$$y_t = y_t^c + y_t^s + v_t, \tag{C.6}$$

and, second, that the cyclical component y_t^c is governed by the following model:

$$y_t^c = Ay_{t-1}^c + Cx_t + b + u_t, \tag{C.7}$$

where x_t is a vector of exogenous variables and u_t is a vector of random disturbances. The exogenous variables might or might not be seasonally adjusted, but this issue does not affect our analysis, since the vector x_t, seasonally adjusted or not, is treated as predetermined. Third, an autoregressive seasonal model is assumed for the seasonal component, as illustrated by, but not confined to, the simple scheme

$$y_t^s = By_{t-12}^s + w_t, \tag{C.8}$$

where w_t consists of random residuals. Combining equations (C.7) and (C.8), we can write the vector z_t of unobserved components in the form

$$z_t = Mz_{t-1} + Nx_t + \epsilon_t, \tag{C.9}$$

where z_t includes both y_t^c and y_t^s as its first two subvectors as well as the necessary lagged y_{t-k}^c and y_{t-k}^s to transform the original model (C.7) of possibly higher order and equation (C.8) of order 12 into first order, the matrix M will depend on the matrices A and B, the matrix N will depend on C and b, the vector x_t will include dummy variables to absorb the intercept b of equation (C.7), and ϵ_t will depend on u_t and w_t. Equation (C.6) can be rewritten as

$$y_t = [I \ I \ O]z_t + v_t \tag{C.10}$$

Thus, (C.9) and (C.10) are in the standard state-space form, the first explaining the unobserved state variables z_t and the second relating the observed y_t to z_t. Given observations on y_t and x_t, the conditional expectations of the unobserved components of z_t can be estimated by the well-known techniques of Kalman filtering and smoothing, provided that the parameters A, C, b, and thus M and N, are known.

In practice, the parameters A, C, and b of the econometric model (C.7) are unknown. One can employ seasonally adjusted data for y_t^c, obtained by a standard seasonal adjustment procedure, and the standard statistical estimation techniques to obtain estimates of A, C, and b. Using these estimates, one can then compute estimates of the seasonal and cyclical components in z_t by Kalman filtering and smoothing. The new estimates of y_t^c will serve as new data for the reestimation of the econometric model (C.7). New estimates of the seasonal components y_t^s will result from this process. I believe that this approach, as well as the approach suggested by Plosser to combine econometric modeling with the specifications of seasonalities in economic variables, should be further studied and pursued. In closing, I would like to thank and congratulate Charles Plosser for having provided us with an interesting, original, and thought-provoking paper.

BIBLIOGRAPHY

Box, G. E. P., and G. M. Jenkins (1970), *Time Series Analysis, Forecasting, and Control* (San Francisco, Holden-Day)

Pagan, A. (1975), "A note on the extraction of components from time series," *Econometrica* 43, 163–8

Comment and implications for policy-makers and model builders (1978)

Raymond E. Lombra

. . . Charles Plosser has [in chapter 9] presented an interesting alternative procedure for dealing with seasonality in econometric models. The fundamental premise he argues is that "economic concepts may be useful in arriving at a better understanding of seasonality. Within the context of an economic structure . . . the seasonal variation in one set of variables . . . should have implications for the seasonal variation in closely related variables." In general, this view, articulated by Nerlove (1964, p. 263) and others some time ago, leads one to look beyond the mechanical approaches for dealing with seasonality, such as the Census X–11 procedure, and instead seek a structural approach. As is well known, the major problems with the mechanical approaches revolve around defining and obtaining an optimal decomposition of the unobserved seasonal component from the observed series. On the other hand, the major difficulty associated with a structural approach concerns the identification of the correct structure. I suspect many of us would agree that the structural approach is preferable. However, the difficulties associated with making such an approach operational have led most producers and consumers of adjusted data to adopt the mechanical approaches as a kind of second-best solution.[1]

No doubt, nearly all of our empirical work suffers from problems generated by using imperfectly adjusted data. However, the major issues revolve

The author expresses his gratitude to Herbert Kaufman and Dennis Farley for helpful comments during the preparation of this comment.

[1] It should be pointed out that producers and consumers of data probably have different objective functions and face different constraints. A producer like the Census Bureau must turn out a huge number of series on a timely basis. This being the case, a structural approach may only be useful as a diagnostic tool, employed from time to time, to evaluate the output from a mechanical approach. Researchers, on the other hand, desire to minimize the distortions that seasonal fluctuations can generate in trying to identify longer-run relationships. Plosser's chapter is primarily directed at producing a technique applicable to the latter set of problems.

Table C.1 *Range of M1 growth rates yielded by alternative adjustment procedures, 1975 (growth rates are expressed as seasonally adjusted annual percentage rates)*

1975	High	Low	Range
January	−2.1	−10.9	8.8
February	4.3	−6.4	10.7
March	11.5	7.2	4.3
April	8.0	1.3	6.7
May	11.3	5.5	5.8
June	19.2	11.7	7.5
July	6.2	2.9	3.3
August	7.0	1.2	5.8
September	6.1	0.4	5.7
October	0.4	−3.3	3.7
November	13.1	3.2	9.9
December	2.8	−6.9	9.7
Average	(X)	(X)	*6.8*

Note:
X Not applicable.

around the seriousness of such problems and whether or not an alternative method for dealing with seasonality, such as Plosser's, can provide us with a better understanding of the processes generating seasonality. If the latter can be accomplished, it may assist the producers of seasonally adjusted data in improving their procedures and, thereby, aid the users of such data.

Since the technique developed in Plosser's approach is applied to a simple monetary model of the US economy, it might be useful to illustrate the type of data problems faced by the Federal Reserve System, the key user of money stock data in the United States . . . Fry (US Federal Reserve System 1976b) applied a variety of seasonal adjustment techniques to monthly money stock data. In general, he found that "a variety of plausible seasonal methods produce roughly similar turning points in the M1 series, but seasonally adjusted growth rates differ substantially in the short run" (1976b, pp. 1–2) . . . Table C.1 is part of a larger table in Fry's paper (1976b, p. 14). It shows the range of M1 growth rates for 1975, produced by applying eleven different seasonal adjustment procedures to the unadjusted data.[2]

[2] The eleven procedures included various X–11 options (multiplicative, additive moving seasonal, and constant seasonal), multiplicative and additive versions of a regression technique, developed by Stephenson and Farr (1972), and a new daily method, developed by Pierce, VanPeski, and Fry (US Federal Reserve System 1976b, 1976c).

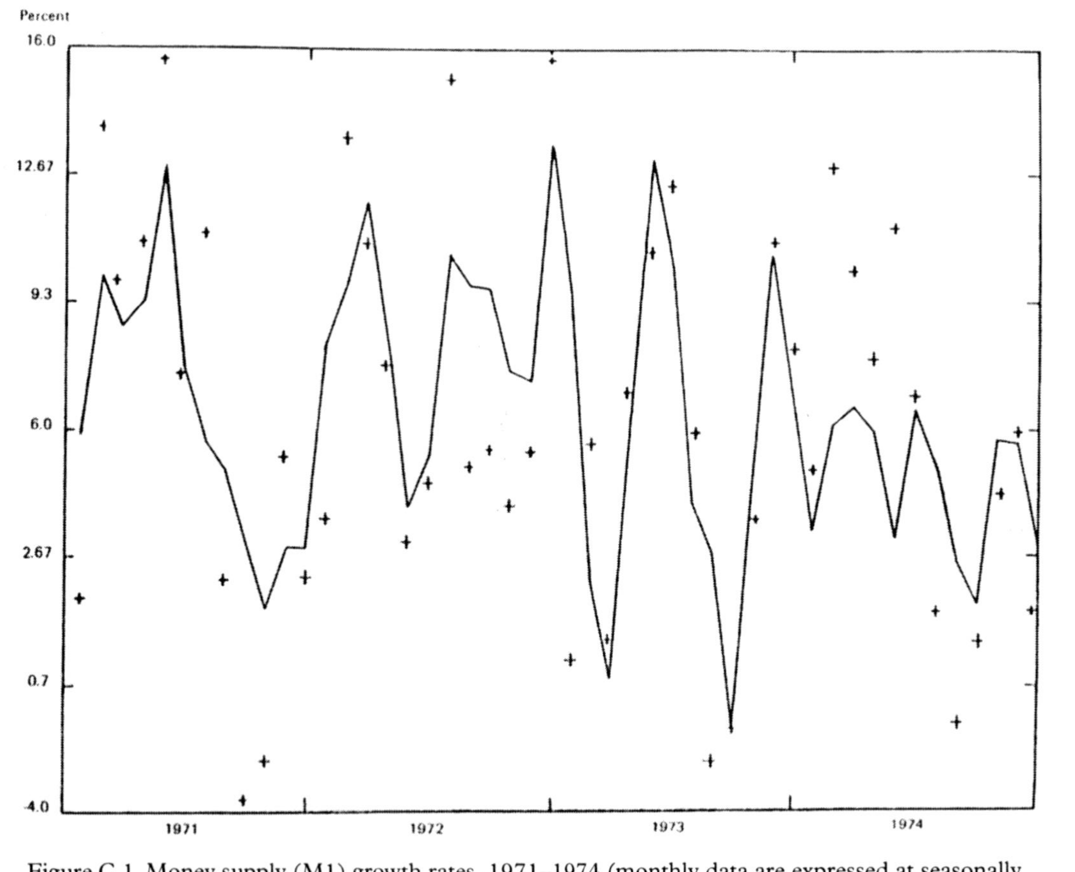

Figure C.1 Money supply (M1) growth rates, 1971–1974 (monthly data are expressed at seasonally adjusted annual rates, percent)

As can be seen, the average range of monthly growth rates, produced by the eleven procedures, was nearly 7 percentage points (or $1.7 billion). Perhaps I am overly sensitive to these results, but, in view of the fact that the short-run target ranges for M1 specified by the Federal Open Market Committee are typically 4 percentage points wide, it is a bit unsettling to learn that the implied confidence interval for the adjusted data is so wide.[3]

Another serious problem for the Federal Reserve System concerns the *ex post facto* revisions in the seasonal factors that are initially adopted. As is well known, the factors derived from an X–11-type procedure used to adjust current data (which, in effect, are forecasts of seasonal factors) will be subject to revision in following years as the extrapolations of the terminal years in the ratios and moving averages are replaced with actual data. This procedure along, with the way outliers are handled, often results in significant *ex post facto* revisions in the date relative to the data policymakers initially had available to guide their actions . . . Figure C.1 vividly illustrates this problem. The first published data for the money stock are often very different from the revised data, and the revised data tend to show considerably less variance. It seems clear that the variety of issues underlying these adjustment problems firmly establishes the need for new approaches to seasonal analysis.

The primary focus of Plosser's chapter is to build an econometric model that contains an explicit specification of the causal sequence generating seasonality in the endogenous variables. More specifically, the central hypothesis to be tested is that seasonality enters the system through the processes generating the exogenous variables. The presumption is that the structure then transmits the exogenous seasonal impulses to the endogenous variables. An alternative hypothesis, as Plosser recognizes, is that various parameters in the structure could fluctuate seasonally. The result, of course, would be observed seasonality in the endogenous variables without any seasonality in the exogenous variables. The latter hypothesis would imply a different model for each season and would be considerably more difficult to specify and estimate.

The simple macromodel constructed to test the hypothesis treats the monetary base and real income as exogenous, and, therefore, it is assumed that seasonality enters the system only through systematic movements in these variables. In addition, the real rate of interest is, in effect, also treated exogenously, since a constant expected value is assumed.

[3] See Lombra and Torto (1975) for a detailed discussion of the role of money stock target ranges in the strategy of monetary policy.

As is usually true in applied econometrics, there are a variety of concessions that a researcher must make to translate a theoretical construct into a model that can be estimated. This being the case, it is often easy to critique the compromises made in doing empirical work. Of course, there are a variety of such compromises that Plosser has made. Rather than trace the problems with the model in detail, it is sufficient to point out that, when Plosser checked his empirical findings against the restrictions implied by the model and accompanying assumptions, he found the model and assumptions deficient. More specifically, there appeared to be evidence of seasonality in the structure (particularly the money multiplier), and the assumed exogeniety of real income and the monetary base appeared to be inappropriate. Although the results are, in some sense, negative, they do reveal the major strength of Plosser's approach: By constructing a model with an explicit specification of the processes generating seasonality, various restrictions on the model were imposed and could be checked. This diagnostic checking, in turn, will lead to improved model specification.

Unfortunately, it would appear that model builders rarely check their results for the effects of seasonality and for sensitivity to alternative seasonal adjustment procedures for the input data. This void in hypothesis testing has become potentially more serious with the development of monthly and weekly models. Assuming policy-maker performance, in the short run, is dependent ultimately on the reliability of short-run data and the robustness of such models, the potential costs of poor seasonal analysis are obvious.

The results in Plosser's chapter probably come as no great surprise to many of us. For example, if the description of monetary policy in (2.2) is reasonably accurate, it seems fairly clear that, in the short run, movements in the monetary base are not exogenous, but rather, are a function in part of contemporaneous movements in income and interest rates.[4] This being the case, correctly explaining the seasonality in the base will require the specification of a reaction function for the Federal Reserve System that captures seasonal objectives.[5]

More generally, the systematic movement (in contrast to the strictly seasonal movements) in the money stock over time are the product of

[4] To illustrate, the correlation coefficient between the monthly seasonal factors estimated by X–11 for total reserves (a critical part of the monetary base) and the 90-day Treasury bill rate for 1960–75 is about 0.6.

[5] The seasonal forces the Federal Reserve System may be concerned with might include, i.e., regular Treasury financings (such as quarterly refundings) and the increase in money demand over the last half of the year as production, consumption, and borrowing rise in anticipation of Christmas.

natural seasonals on the demand side (such as seasonal movements in the currency: demand deposit ratio) and the supply side (i.e. seasonal aspects of bank behavior), and the systematic movements in policy as reflected in the variance of the monetary base. However, the systematic movements in the base may reflect both cyclical and seasonal phenomena.[6]

To illustrate, assume that there is a cyclical upswing in economic activity which lasts eighteen months (January of year 1 through June of year 2) and that the Federal Reserve system allows the base and, therefore, the money stock to expand as much as is necessary to hold short-term interest rates constant throughout the period. Subsequently, we will observe systematic movement in the money stock in the January–June period of both years, and I would guess that few would want to characterize such systematic movement as seasonal. Against this background, I would expect some systematic movement to remain in the M1 series, even after perfect seasonal filtering.

In summary, I would want to reserve judgment on the ultimate payoff of Plosser's time series approach in this specific area until it can be shown that a considerably more complex model can, in fact, be successfully estimated. However, on the other hand, Plosser has clearly demonstrated the need for model builders to consider carefully the souces of seasonal variation in the endogenous variables, and he has developed a general method to identify the sources that appears feasible. As he states, "the stochastic properties of economic variables in the system are not independent of the economic structure of which they are a part. Therefore, it would seem appropriate for those who wish to obtain adjusted series to study the stochastic behavior of the unadjusted data, investigating its form and properties prior to adjustment."[7]

Finally, Plosser's work along with several other . . . papers (Sims 1974; Wallis 1974; Kaufman and Lombra 1977), amply demonstrate that various estimated relationships can be quite sensitive to the way seasonality is handled. My own guess is that such problems will ultimately be solved only when talented teams of researchers (econometricians, theoreticians, and institutionalists), like those that combined to build the large macromodels, can be brought together to extend the work of Plosser and others on the structure of seasonality.[8]

[6] This issue is discussed in detail in Kaufman and Lombra (1977) and in the . . . report of the Advisory Committee on Monetary Statistics (US Federal Reserve System 1976a).

[7] A good example of the failure to consider structural relationships is the inattention accorded balance sheet constraints by most producers or users of seasonally adjusted data.

[8] The papers by Engle, Granger, Pierce, and Wallis, . . . [cited in their conference paper], represent important building blocks in this process.

BIBLIOGRAPHY

Kaufman, H. and R. Lombra (1977), "Short-run variations in the money stock," *Southern Economic Journal* April, 1515–27

Lombra, R. and R. Torto (1975), "The strategy of monetary policy," *Economic Review* September–October, 3–14

Nerlove, M. (1964), "Spectral analysis of seasonal adjustment procedures," *Econometrica* 32, 241–86

Sims, C. A. (1974), "Seasonality in regression," *Journal of the American Statistical Association* 69, 618–26

Stephenson, J. and H. Farr (1972), "Seasonal adjustment of economic data by application of the general linear statistical model," *Journal of the American Statistical Association* 67, 37–45

US Federal Reserve System, Board of Governors of the Federal Reserve System (1976a), *Improving the Monetary Aggregates*, Report by the Advisory Committee on Monetary Statistics (Washington, DC, Government Printing Office)

(1976b), *Seasonal adjustment of M1: Currently Published and Alternative Methods*, by Edward R. Fry, Staff Economic Studies 87 (Washington, DC, Government Printing Office)

(1976c), "Seasonal adjustment of the monetary aggregates," in D. A. Pierce and N. VanPeski *Improving the Monetary Aggregates: Staff Papers* (Washington, DC, Government Printing Office)

Wallis, K. F. (1974), "Seasonal adjustment and relations between variables," *Journal of the American Statistical Association* 69, 18–31

Response to discussants (1978)

Charles I. Plosser

In his comment, Gregory Chow lists six characteristics of the general approach for the analysis of seasonal economic models that I have proposed. I would like to take this opportunity to briefly comment on these characteristics and clarify some of the issues involved.

Chow's first point seems to suggest that econometric models should be built to explain the seasonally adjusted data rather than the observed series. The notion that seasonally adjusted data are the only data of interest or relevance to the economist for the purpose of testing economic theories seems to stem, in part, I think, from the over-used notion of breaking an economic time series into trend cycle, seasonal, and random components and assuming that the economic model exists solely between the trend-cycle components. As pointed out in . . . subsection [2.3] . . . of [chapter 9], this approach could be used and would lead to very explicit restrictions on the properties of the FEs and TFs. Alternatively, as in the approach suggested by Chow in his remarks, one can pose the restriction that the series can be decomposed to the sum of three unobserved components and obtain seasonally adjusted series first, before the model is estimated. This seems to me to be a distinct disadvantage. It would be preferable to set up hypotheses to be tested regarding the decomposition rather than assume it takes a certain form and then never bother to investigate the validity of the assumption.

There is another more fundamental reason for not proceeding in this manner where it is assumed that the appropriate economic model exists exclusively between the so-called cyclical components. As discussed briefly at the end of my chapter, the role of anticipations has played an increasingly important function in economic theory since the permanent income hypothesis was advanced by Milton Friedman. In models in which expectations play an important role, the appropriate distinction to be made is between anticipated and unanticipated phenomenon and not to some arbitrarily chosen decomposition dealing with trend, cycles, and seasonal components. In making the distinction between anticipated and unanticipated effects, it would seem appropriate to merely consider seasonal fluctuations as a contribution, primarily, to the anticipated component. If such a model is appropriate, the economist or econometrician should be considering a different type of adjustment procedure (e.g. a prewhitening filter that reduces a series to white noise), rather than one that focuses only on the seasonal component.

The second and third points listed by Chow refer more to the specific example I considered rather than the general approach. The model builder certainly has the option to allow structural parameters (elements of H_{11} or H_{12}) to vary seasonally as well as to model the exogenous variables in a different manner. Both of these approaches would have implications for the forms of the TFs and FEs that could be checked against the data.

The fourth point raised by Chow is that the algebra is very complicated, and, therefore, the approach "might be applicable only to very small and very specialized models." Clearly, more experience is needed in applying the approach suggested here, but it is not necessarily true that only small models can be considered. Many of the characteristics that cause the algebraic manipulations to be simplified are actually found in many of the larger econometric models, including either a fully recursive or a block-recursive structure. Consequently, larger models would not necessarily be more complicated to analyze.

The fifth characteristic noted by Chow concerns the role of expectations that was previously discussed. The sixth point asserts that the procedures do not yield a decomposition into the usual set of unobserved components. In light of this discussion, it is not clear whether such a decomposition is desirable or not. However, analysis of the sort suggested can be helpful in understanding the types of adjustment procedures that might be appropriate. The FEs, for example, are in the form of univariate time series models, and there have been various methods suggested for decomposing these models if such an adjustment is of interest.

I hope that this discussion clarifies some of the issues in question. Although our techniques may differ, it is clear that Chow and I both favor incorporating seasonality into econometric models and feel that there may be much to learn from doing so.

10 The behavior of speculative prices and the consistency of economic models (1985)

Robert I. Webb

1 Introduction

Modern financial economic theory suggests that changes in speculative prices should follow simple time series processes in an informationally efficient capital market. Moreover, this theoretical implication enjoys substantial support in the empirical financial economic literature (see Fama 1970). Yet, the implications of the observed time series behavior of speculative price changes for the structure of equilibrium models of asset pricing or information theory do not appear to be fully appreciated. Simply stated, financial economists have not attempted to integrate time series analysis with econometric model building along the lines suggested by Zellner (1979b) and Zellner and Palm (1974).

2 Changes in speculative prices in an efficient capital market: theory and evidence

Fama (1970) has defined an efficient capital market as one in which speculative prices fully (and correctly) reflect available information. In such a market, changes in speculative prices occur only in response to new information or to reassessments of existing information. The pioneering empirical work of Working (1934), Kendall (1953), Roberts (1959), and Bachelier (1964) suggested that changes in various speculative price series appeared to follow simple time series processes or "fair game" models (in particular, random or quasi-random walks). Excepting Bachelier, the first rigorous economic theory consistent with the above empirical evidence was developed by Samuelson (1965).

Opinions expressed herein are those of the author and do not necessarily represent the opinions of the Chicago Mercantile Exchange. Earlier versions of this chapter have benefited from comments by Victor A. Canto, Richard Leonard, Terry Marsh, and especially Arnold Zellner.

Originally published in the *Journal of Econometrics* 27 (1985), 123–30. 0304-4076/85/ $3.30 © 1985, Elsevier Science Publishers BV (North-Holland).

Samuelson demonstrates that changes in futures prices will follow martingale or submartingale processes under very general assumptions concerning the stochastic behavior of spot commodity prices. Although Samuelson's theory is illustrated with respect to futures prices it applies to fluctuations in speculative prices in general. The critical assumption is that speculative prices are determined in an informationally efficient capital market.[1] Essentially, Samuelson develops a 'general stochastic model of price changes' *given* the assumption of an informationally efficient capital market and without specifying an equilibrium model of asset pricing. Alternatively stated, Samuelson examines how speculative prices will fluctuate in informationally efficient capital markets for *any* equilibrium model of asset pricing.[2]

3 The Zellner–Palm consistency constraints

Zellner and Palm have (1974, 1975) proposed a method of integrating time series analysis and econometric model building which Zellner (1979b) has called the SEMTSA (Structural Econometric Modeling Time Series Analysis) approach. Essentially, Zellner and Palm recognize that any system of simultaneous equations for a dynamic linear model may be equivalently represented as a multiple time series process (assuming of course, that the input variables are generated by autoregressive integrated moving average (ARIMA) processes as well). Further, they contend that one may describe individual final equations [FEs] to a structural econometric model by univariate ARIMA models for the output variables. Similarly, the time series model of an input variable may be regarded as the final equation of the (presumably unknown) model generating it. The structural econometric model will imply certain restrictions on the final equations or, equivalently, on the order of the ARIMA models for the output variables. These restrictions may be tested to assess whether the

[1] The assumption of informationally efficient capital markets is not a sufficient condition for security returns to follow *simple* martingale processes – an additional assumption, such as constancy of expected returns, is required. Violation of this additional assumption, of course, does not necessarily imply that security returns will follow complicated martingale processes. The actual complexity of the time series processes for security returns or interest rates is an empirical issue.

[2] As Fama (1970) has pointed out, Samuelson's conclusions concerning the fair game properties of changes in speculative prices may be derived more easily. Essentially, one need only specify that "the conditions of market equilibrium can be stated in terms of expected returns and that equilibrium expected returns are formed on the basis of (and thus fully reflect) the information set." Simply stated, a sequence of expected returns or changes in speculative prices is assumed to follow a fair game process with respect to a temporal sequence of information sets. This suggests that the time series behavior of speculative prices has implications for both the manner in which new information is incorporated into prices as well as the time series process by which new information enters the marketplace.

structural econometric model is internally consistent.[3] Alternatively, the order of the empirically *identified* ARIMA process of an output variable implies certain restrictions on the transfer function [TF] models and on the time series processes of the input variables.

4 Applications of the Zellner–Palm consistency constraints

4.1 *Application to equilibrium models of asset pricing*

Ross (1976) has proposed the Arbitrage Pricing Theory (APT) as an alternative to the simple one-period Capital Asset Pricing Model (CAPM) of Sharpe (1964). Lintner (1965), Mossin (1966), and Black (1972). For our purposes, the APT may be conveniently represented by a linear transfer function model of m asset returns and k risk factors of the following form:

$$r_{j,t} = v_j(B)Z_{j,t} + v_i(B)X_{i,t} + \cdots + v_k(B)X_{k,t} + n_t, \qquad (4.1)$$
$$j = 1, \ldots, m, \quad i = 1, \ldots, k, \quad m > k,$$

where $r_{j,t}$ is the uncertain return on asset j at time t, $v_j(B)$ represents the transfer function of factor Z_j, and n_t represents the noise process (which may be represented by an ARIMA process). In this model, factor $Z_{j,t}$ represents the expected return on asset j, while factors $X_1 \ldots X_k$ represent other risk factors. All risk factors are assumed to follow time series processes which may be described by multiplicative seasonal ARIMA models.[4]

The primary point of contention between the two alternative models of asset pricing centers on the number and nature of the risk indices (although the APT also requires fewer restrictive assumptions than the CAPM). Roll and Ross (1980) present empirical analysis which suggests that there are at least three or four unique risk factors rather than only one as the CAPM posits.

The Zellner–Palm consistency constraints can be used to delimit the class of possible risk factors in the APT asset pricing model by indicating the time series properties that valid risk factors must possess. Moreover, the consistency constraints can be used to derive restrictions on the order of the transfer functions for the APT.

[3] Implicit in these constraints are assumptions that the input and output variables may be suitably described by ARIMA processes. The assumption that input variables follow an ARIMA process also imposes some restrictions on the form of control of input variables subject to control.

[4] Plosser (1976) argues that seasonality in certain economic time series is not well described by multiplicative seasonal ARIMA models, as Box and Jenkins (1976) suggest.

When $i = 2$, the Zellner–Palm consistency constraints are

$$\begin{aligned} p_0 &\leq (p_n + r_1 + p_1 + p_1^s + r_2 + p_2 + p_2^s), \\ q_0 &\leq \max\left[(p_n + b + r_2 + p_2 + p_2^s + q_1 + q_1^s + s_1),\right. \\ &\quad (p_n + b + r_1 + p_1^s + q_2 + q_2^s + s_2), \\ &\quad \left.(b + r_2 + p_2 + p_2^s + r_1 + p_1 + p_1^s + q_n)\right], \end{aligned} \tag{4.2}$$

where the univariate time series X_1 and X_2 and noise model n are represented by ARMA(p_1, q_1), ARMA(p_2, q_2), and ARMA(p_n, q_n) processes, respectively; the numerator and denominator polynomials in the transfer function model are of order s and r, respectively, and b is the 'delay operator' or lag in response of the output variable to changes in the input variables. (It should be noted that s need not equal r.) The subscripts denote the output variable o, input variables X_1 and X_2, or noise term n_t. The superscript s indicates a seasonal parameter. It should also be noted that when other possible models are considered the degrees of polynomials hitting the output variable would be larger than indicated in inequalities (4.2).

The time series behavior of the output variables for equilibrium models of asset pricing have already been estimated elsewhere. The empirical analyses of Working, Kendall and Roberts mentioned earlier as well as studies by Fama (1965, 1975) and Roll (1966) are consistent with a quasi-random walk for such variables as stock returns, short-term interest rates, and changes in grain futures prices. This research suggests that changes in speculative prices or changes in expected returns may be characterized by IMA(1) or MA(1) processes. The restrictions on discrete linear combined transfer function models, such as (4.1) above, and the time series processes for the input variables (when suitably described by ARIMA models) are derived by substituting in the order for the ARIMA process the output variable follows into the appropriate set of consistency constraints. Obviously, as the number of input variables increases the consistency constraints become more complex.

If changes in futures prices, stock returns, interest rates, and other output variables from financial economic models follow simple "fair game" processes such as an IMA(1), then the Zellner–Palm consistency constraints impose some fairly strong restrictions on financial structural econometric models (SEMs). Consider, the least restrictive case for the APT. Let the delay operator equal zero and assume that the impact of a change in any risk factor on returns is completely contemporaneous. Under such circumstances the Zellner–Palm consistency constraints require either that the time series process for *each* of the three or four risk factors (input variables) follows MA(1) or multiplicative seasonal MA(1) processes or a linear combination of the risk factors follows an IMA(1) process.

4.2 *Application to information theoretic models*

Goldman and Sosin (1979) posit a model of financial market disequilibrium due to uncertainty surrounding the speed of information dissemination (Type 1) and the sequence or order (Type 2) in which traders are informed. When Type 2 uncertainty is held constant but Type 1 uncertainty exists, then Goldman and Sosin contend that speculators will always "under-shoot" the price (i.e. not fully impound all information). The amount of under-shooting decreases but still remains positive as the trading interval decreases toward zero (i.e. as continuous trading is approached). In other words, serial correlation in prices results in a world where information dissemination is uncertain and information production is costly.

For convenience of exposition, speculative prices are assumed to be determined according to (4.1). The interpretation of (4.1) differs from before, in that it is no longer an equilibrium model and Z_j no longer appears in the model. As before, however, the risk factors change in response to new information which, in turn, induces price changes. Moreover, prices may respond over time to a piece of new information rather than immediately change. If changes in speculative prices follow simple time series processes then the restrictions on the transfer functions of (4.1) are the same as before regardless of whether capital markets are informationally efficient. Moreover, to the extent that new (unanticipated) information also follows a white-noise process then this would imply that information theoretic models which posit a lagged adjustment to speculative prices are inconsistent with the Zellner–Palm consistency constraints.[5] Simply stated, the simplicity of the time series processes for observed changes in speculative prices severely restricts the nature of possible informational flows and financial market *disequilibria*.

4.3 *Additional considerations*

The formulation of time series processes as ARIMA models rather than as high order AR processes serves to economize on the number of parameters. The use of Box–Jenkins (1976) techniques or *single*-equation

[5] This may not appear to hold in certain empirical applications where data collection methods induce an apparent lag. That is, data may be reported in intervals which are more frequent than the market can actually adjust to.

likelihood ratio tests in the analysis of appropriate time series models for jointly dependent variables, however, ignores information contained in the cross-equation covariances of the disturbances or lost in the transformation necessary to induce variance–covariance stationarity. This problem may be circumvented through the use of joint tests in a multiple time series framework. Univariate time series models for output variables may be identified individually or multivariate identification procedures may be used.[6]

A related issue concerns the impact of sampling variability on checking the consistency constraints. Finite sample confidence intervals are difficult to construct analytically currently, but may be available in the future. Application of the Zellner–Palm consistency constraints represents an attempt to exploit available information and is not contingent upon the identification of a "true" ARIMA model (if, indeed, one even exists) for the output variable. It should be emphasized that one need not measure the time series processes of the output variables very sharply to arrive at conclusions which essentially reject a broad class of models.[7]

5 Conclusions

The simple time series processes which changes in speculative prices apparently follow impose strong restrictions on both (dynamic linear) equilibrium models of asset pricing and models of informational flows and disequilibria in financial markets. Macroeconomic and monetary models which posit interest and inflation rates as output variables are similarly restricted.[8,9] Indeed, the absence of relationship between money and

[6] . . . Palm (1977) has shown that one "may represent any variable out of a multivariate normal ARIMA process as a univariate normal ARIMA process. Thus, the analysis of the likelihood function associated with a single final equation is legitimate." In addition, Wallis (1977) has proposed a method of joint estimation. Tiao and Tsay (1983) discuss multivariate identification methods.

[7] There is also a technical problem associated with the use of Box–Jenkins (1976) identification and estimation techniques which deserves explicit mention. Namely, economic variable may have ARIMA models of higher order than those identified. This problem arises whenever economic variables may be characterized by high degree polynomials where the higher order roots have small but non-zero coefficients. In these cases, the identification techniques used may ignore the small non-zero coefficients which results in the identification of a lower order ARIMA model for the variable than actually exists. Essentially, this results in excluding the low frequency components of the time series. Naturally, this difficulty may be largely avoided if there are sufficient data to permit a more thorough investigation.

[8] Zellner (1979b) makes a similar observation concerning capital market efficiency. Other common output variables may also follow simple time series processes. For example . . . , Hall (1978) argues that economic theory would predict that consumption should follow a random walk with drift.

[9] See Zellner (1979a). For example, in many monetary models a relationship between the nominal interest rate and the growth rates of money and real income is specified. In an

interest rates reported by Pierce (1977) and Feige and Pearce (1976) may not be an indication of tests with low power as Hsiao (1977) contends, but rather a reflection of the inadequacy of certain economic models. Obviously, these restrictions may change if more advanced time series analytic techniques lead to the identification of a "finer structure" for the time series processes which changes in speculative prices follow.

BIBLIOGRAPHY

Bachelier, L. (1964), "Theory of speculation," in P. Cootner (ed.), *The Random Character of Stock Market Prices* (Cambridge, Mass., MIT Press)

Black, F. (1972), "Capital market equilibrium with restricted borrowing," *Journal of Business* 45, 343–53

Box, G. E. P. and G. Jenkins (1976), *Time Series Analysis: Forecasting and Control* (San Francisco, Holden-Day)

Fama, E. F. (1963), "Mandlebrot and the stable Paretian hypothesis," *Journal of Business* 36, 420–9

(1965), "The behavior of stock market prices," *Journal of Business* 38, 34–105

(1970), "Efficient capital markets: a review of theory and empirical work," *Journal of Finance* 25, 383–417

(1975), "Short-term interest rates as predictors of inflation," *American Economic Review* 65, 269–82

Feige, E. L. and D. K. Pearce (1976), "Economically rational expectations: are innovations in the rate of inflation independent of innovations of measures of monetary and fiscal policy?," *Journal of Political Economy* 84, 499–522

Goldman, B. and H. Sosin (1979), "Information dissemination and market efficiency," *Journal of Financial Economics* 7, 29–61

Granger, C. W. J. and R. Joyeux (1980), "An introduction to long-memory time series models and fractional differencing," *Journal of Time Series Analysis* 1, 15–30

Hall, R. E. (1978), "Stochastic implications of the life-cycle permanent-income hypothesis: theory and evidence," *Journal of Political Economy* 86, 971–88

Haugh, L. D. and D. A. Pierce (1976), *The Assessment and Detection of Causality in Temporal Systems* (Burlington, Vt., University of Vermont)

Hsiao, C. (1977), *Money and Income Causality Detection* (*Berkeley*, Cal., University of California)

efficient capital market the appropriate specification would be to associate *changes* in the output variable and the *innovations* in the input variables. That is, in an efficient capital market, one is interested in assessing the impact of *new* (i.e. unanticipated) information on the *change* in security prices or yields – in this case, the nominal interest rate. This suggests a formulation akin to the cross-correlation of residuals technique proposed by Haugh and Pierce (1976) (provided, of course, that the time series models employed in the prewhitening technique are superior models of separating anticipated and unanticipated variables). By definition, the input variable is white noise and, hence, in this *model* the prewhitening technique does not take too much out of the data. Also, it should be noted (as Webb 1979 points out) that although this technique is formally equivalent to the cross-correlation of residuals method proposed by Haugh and Pierce, the interpretation is very different.

Kendall. M. G. (1953), "The analysis of economic time series, part I: prices," *Journal of the Royal Statistical Society*, 96, 11–35
Lintner, J. (1965), "Security prices, risk and maximal gains from diversification," *Journal of Finance* 20, 587–615
Mossin, J. (1966), "Equilibrium in a capital asset market," *Econometrica* 34, 768–83
Nelson, C. R. and G. W. Schwert (1977), "On testing the hypothesis that the real rate of interest is constant," *American Economic Review* 67, 478–86
Palm, F. C. (1977), "On univariate time series methods and simultaneous equation econometric models," *Journal of Econometrics* 5, 379–88
Pierce, D. A. (1977), "Relationships – and the lack thereof – between economic time series with special reference to money and interest rates," *Journal of the American Statistical Association* 72, 11–22
Plosser, C. (1976), "Time series analysis and seasonality in econometric models with an application to a monetary model," PhD dissertation, University of Chicago
Roberts, H. V. (1959), "Stock market patterns and financial analysis: methodological suggestions," *Journal of Finance* 14, 1–10
Roll, R. (1966), "Interest-rate risk and the term structure of interest rates: comment," *Journal of Political Economy* 74, 629–31
Roll, R. and S. A. Ross (1980), "An empirical investigation of the arbitrage pricing theory," *Journal of Finance* 35, 1073–1104
Ross, S. A. (1976), "The arbitrage theory of capital asset pricing," *Journal of Economic Theory* 13, 341–60
Samuelson, P. A. (1965), "Proof that properly anticipated prices fluctuate randomly," *Industrial Management Review* Spring, 198–206
Sharpe, W. F. (1964), "Capital asset prices: a theory of market equilibrium under conditions of risk," *Journal of Finance* 19, 425–42
Tiao, G. C. and R. S. Tsay (1983), "Multiple time series modeling and extended sample cross-correlations," *Journal of Business and Economic Statistics* 1, 43–56
Wallis, K. F. (1977), "Multiple time series analysis and the final form of econometric models," *Econometrica* 45, 1482–97
Webb, R. I. (1979), "The impact of open market operations on Treasury bill yields," PhD dissertation, University of Chicago
Working, H. (1934), "A random difference series for use in the analysis of time series," *Journal of the American Statistical Association* 29, 11–24
Zellner, A. (1979a), "Causality and econometrics," in K. Brunner and A. Meltzer (eds), *Three Aspects of Policymaking, Carnegie–Rochester Conference Series*, 10 (Amsterdam, North-Holland)
1979b, "Statistical analysis of econometric models," *Journal of the American Statistical Association* 74, 628–51; chapter 2 in this volume
Zellner, A. and F. C. Palm (1974), "Time series analysis and simultaneous equation econometric models," *Journal of Econometrics*, 2, 17–54; chapter 1 in this volume
(1975), "Time series and structural analysis of monetary models of the US economy," *Sankhyā: The Indian Journal of Statistics*; Series C 37, 12–56; chapter 6 in this volume

11 A comparison of the stochastic processes of structural and time series exchange rate models (1987)

Francis W. Ahking and Stephen M. Miller

Zellner and Palm (1974) show that comparing the actual with the implied stochastic processes generating the endogenous variables in a system of dynamic structural equations provides important information about the system's correct specification. We apply their methodology to structural exchange rate models. We find that the log of the bilateral exchange rate is generally well approximated by a random-walk model. This implies that the stochastic processes generating the exogenous variables should also be random-walk models. Our empirical results, however, show that this is not, in general, the case. We conclude by suggesting a reconciliation of our results based on a technique developed by Beveridge and Nelson (1981).

1 Introduction

. . . [T]heories of exchange rate determination have emphasized the asset approach to foreign exchange markets. As an asset price, the exchange rate is seen as adjusting rapidly and freely to maintain stock equilibrium. For example, in the pure monetary approach, the exchange rate is determined when the total stocks of outstanding foreign and domestic moneys are held willingly by economic agents (see, e.g., Dornbusch 1976; Bilson 1978; Frenkel 1978; Frankel 1983; Hoffman and Schlagenhauf 1983; Huang 1984). Moreover, expectations of future exchange rate movements play a dominant role in determining the current spot rate. This linkage of expectations to the current spot rate is usually accomplished by assuming that uncovered interest rate parity holds, that the foreign exchange market is efficient, and that expectations about the future spot rate are formed rationally.

We acknowledge the comments of P. R. Allen and two anonymous referees on previous drafts and the support of the University of Connecticut Computer Center. The usual caveat applies. A longer version of this chapter with more detail (Ahking and Miller 1986) is available from the authors.

Originally published in the *Review of Economics and Statistics* 69 (1987), 496–502.

Empirical research on exchange rate determination has two dimensions. First, the structural approach estimates the various asset models of exchange rate determination (see, e.g., Bilson 1978; Frenkel 1978; Frankel 1979; Dornbusch 1980; Driskill 1981; Edwards 1983). Some studies support the asset approach while others find little explanatory power. The . . . rational expectations models of exchange rate determination test the systematic response of expectations to the stochastic processes generating the fundamental determinants of the exchange rate. Hence, cross-equation parameter restrictions between the equations generating the fundamental determinants and the exchange rate are employed in the final system estimation (see, e.g., Hoffman and Schlagenhauf 1983; Huang 1984).

Second, the time series approach examines the foreign exchange market using the various tests of market efficiency developed to analyze equity markets. At the simplest level, researchers examine whether the stochastic behavior of the exchange rate is well approximated by a random walk (see, e.g., Poole 1967; Burt, Kaen, and Booth 1977; Logue and Sweeney 1977).[1]

Although different, the two empirical approaches are related. The time series approach tests for foreign exchange market efficiency by examining the stochastic evolution of the time series of the exchange rate and by testing for possible correlations between the exchange rate and other lagged variables, including the lagged exchange rate. It provides useful information on the stochastic behavior of the exchange rate, but sheds little light on the fundamental determinants of the exchange rate. The structural approach provides a theory of exchange rate determination; but, it is not concerned generally with testing for foreign exchange market efficiency.

Several researchers (e.g. Hodrick 1979 and Levich 1985, and especially Mussa 1983) observe that foreign exchange market efficiency and rational expectations impose testable restrictions on the stochastic behavior of the exchange rate and its fundamental determinants. That is, a systematic relationship between the stochastic processes generating both the fundamental determinants of the exchange rate and the exchange rate itself should exist. Few of the above-mentioned empirical studies, however, attempt to use this information.

We provide a systematic analysis of the relationship between the stochastic behavior of the exchange rate and its fundamental determinants, employing Zellner and Palm's (1974) and Zellner's (1984, chs. 2.1, 2.2) methodology. Zellner and Palm (1974) demonstrate that

[1] See Levich (1985) for a thorough and critical review of the tests of foreign exchange market efficiency.

the stochastic process of an endogenous variable from a dynamic simultaneous equations system is implied by the stochastic processes generating the exogenous variables. Thus, [comparing] the conformity of the actual stochastic process generating the endogenous variable with that implied by the exogenous variables constitutes an important check in formulating and estimating dynamic economic models. If the restrictions implied by the stochastic processes generating the exogenous variables are rejected when compared with the stochastic behavior of the endogenous variable, then the structural economic model is probably misspecified. In this case, new structural models must be developed that are consistent with the stochastic information in the data. Our objective here is not to test for rational expectations or for foreign exchange market efficiency. We employ Zellner and Palm's (1974) methodology because of its generality; it does not require the assumptions that expectations are rational or that exchange rates are determined in efficient markets.

We proceed as follows. Section 2 describes the various exchange rate models and examines the data base and our empirical results. Section 3 interprets the implications of our empirical analyses of exchange rate determination. Finally, a summary and conclusion are in section 4.

2 A stylized exchange rate model and empirical analysis

We specify a general exchange rate model that subsumes other models as special cases. Our stylized reduced form equation is

$$\begin{aligned} s_t = {} & b_0 s_{t-1} + b_1(m_t - m_t^*) + b_2(y_t - y_t^*) \\ & + b_3(i_t - i_t^*) + b_4\left(\pi_t^e - \pi_t^{*e}\right) \\ & + b_5(p_{t-1} - p_{t-1}^*) + b_6(y_{t-1} - y_{t-1}^*) \\ & + b_7 B_t - b_7^* B_t^* + e_t, \end{aligned} \tag{2.1}$$

where

s_t = log of the spot exchange rate, defined as the domestic price of one unit of foreign currency

m_t (m_t^*) = log of the domestic (foreign) nominal money stock

y_t (y_t^*) = log of the domestic (foreign) real income level

i_t (i_t^*) = log of one plus the domestic (foreign) nominal interest rate

π_t^e (π_t^{*e}) = log of one plus the expected domestic (foreign) inflation rate

p_t (p_t^*) = log of the domestic (foreign) price level
B_t (B_t^*) = cumulative domestic (foreign) trade balance

By appropriate parameter restrictions, specific exchange rate models are derived from (2.1): (i) The flexible price monetary model (e.g. Bilson 1978 and Frenkel 1978) results if $b_0 = b_4 = b_5 = b_6 = b_7 = b_7^* = 0$; (ii) The short-run flexible price monetary model (e.g. Bilson 1979 and Edwards 1983) emerges from the flexible price monetary model once the parameters b_2 and b_3 are interpreted as distributed-lag operators (i.e. $b_2 = b_2(L)$ and $b_3 = b_3(L)$, where $b_2(L)$ and $b_3(L)$ are polynomial functions in the lag operator L); (iii) The sticky-price monetary model (e.g. Dornbusch 1976 and Frankel 1979) results if $b_0 = b_5 = b_6 = b_7 = b_7^* = 0$; (iv) The dynamic stock-flow model of Driskill (1981) imposes $b_3 = b_4 = b_7 = b_7^* = 0$; (v) The stock-flow model of Hooper and Morton (1982) constrains $b_0 = b_5 = b_6 = 0$; and (vi) The rational expectations monetary model (e.g. Hoffman and Schlagenhauf 1983 and Huang 1984) is the same as the flexible price monetary model but for the imposing of uncovered interest rate parity and solving for the rational expectations solution after determining the stochastic processes generating the exogenous variables.

To apply the Zellner–Palm methodology, the stochastic processes generating the exogenous variables must be explicitly defined. We assume that the stochastic processes generating the exogenous variables can be well approximated by univariate ARIMA models, i.e.

$$\gamma_i(L)z_{it} = \delta_i(L)\eta_{it}, \tag{2.2}$$

where

$$z_{it} = (m_t, m_t^*, y_t, y_t^*, i_t, i_t^*, \pi_t^e, \pi_t^{*e}, p_t, p_t^*, B_t, B_t^*),$$
η_{it} = zero-mean independently distributed white-noise stochastic processes.

Given the suggestion by several researchers that the exchange rate is a random walk, (2.1) is a random-walk model if the following restrictions are true:

$$b_0 = b_5 = b_6 = 0$$
$$\gamma_i(L) = (1 - L),$$

and

$$\delta_i(L) = 1, \qquad \text{for all } i.$$

The data employed are monthly observations from 1973:4 to 1984:6 for Canada, France, Germany, Italy, Japan, Switzerland, the United Kingdom, and the United States. Generally, data are drawn from OECD, *Main Economic Indicators*, and are not seasonally adjusted.[2] The exchange rates are bilateral spot rates with the United States as the domestic economy. The price level is the consumer price index, and real income is the industrial production index; both are indexed to 1975 = 100. We use the narrow (M1) definition for the money stock and a short-term interest rate (i_t). Following Frankel (1979), we use a long-term interest rate (r_t) as a proxy for expected inflation (π_t^e). The cumulative trade balance is obtained by summing the trade balance for each country.

The Zellner–Palm technique suggests determining the univariate time series models of the exogenous variables and drawing inferences about the time series properties of the endogenous variables using the structural model. Then, a comparison of the implied with the actual time series properties of the endogenous variables suggests whether the structural economic model is misspecified.

Our empirical analysis begins with fitting univariate ARIMA models to the individual variables rather than variables in relative form (i.e. the log of the domestic variable minus the log of the foreign variable) using the Box–Jenkins (1976) time-series modeling method. Table 11.1 summarizes the estimated univariate time-series models.[3] In general, all the estimated parameters are significantly different from zero at the 5 percent level, and all models have white-noise residuals.[4]

Table 11.1, column (1) presents the final models for the log of the spot exchange rate. With one exception, the log of the spot exchange rate appears to be a white-noise process after first differencing. Thus, a random-walk model appears to describe the stochastic processes of the spot exchange rates adequately.

The exception is Canada. In addition to first differencing, the log of the United States to Canadian dollar spot exchange rate has a significant autoregressive seasonal at lag 12. The Ljung–Box *Q*-statistic, although

[2] Our preference is to use only seasonally unadjusted data. This is not possible, however, for all our variables. This does not present difficulty in our econometric work since we are not including seasonally adjusted and unadjusted data in the same regression. Rather, our econometric work involves modeling a univariate ARIMA model for each variable. A more detailed list of variables and data sources is contained in an appendix to Ahking and Miller (1986).

[3] A more complete description of the final models with separate tables for each variable is contained in Ahking and Miller (1986).

[4] The exception is the nominal money stock for Switzerland, where we can reject the hypothesis of white-noise residuals at the 5 percent level.

Table 11.1 *ARIMA models*

Country	s_t (1)	m_t (2)	i_t (3)	r_t (4)	y_t (5)	B_t (6)	p_t (7)
United States	–	(3,1,12)[a]	(12,1,0)	(7,1,0)	(13,1,0)	(12,2,1)	(2,1,0)
Canada	(12,1,0)	(15,1,0)	(8,1,0)	(7,1,0)	(0,1,15)[a]	(12,2,5)	(13,1,0)
Japan	(0,1,0)	(14,1,0)	(1,1,0)	(0,1,0)	(0,1,15)[a]	(0,2,13)[a]	(13,1,0)
France	(0,1,0)	(14,1,0)	(1,1,0)	(3,1,0)	(2,0,12)[a]	(15,1,0)	(15,1,0)
Germany	(0,1,0)	(12,1,0)	(0,1,1)	(1,1,0)	(0,1,1)	(12,2,5)	(13,1,0)
Italy	(0,1,0)	(0,1,12)[a]	(0,1,0)	(1,1,0)	(2,0,12)[a]	(15,1,0)	(2,1,0)
Switzerland	(0,1,0)	(13,1,0)	(0,1,0)	(1,1,0)	–	(16,2,0)	(13,1,0)
United Kingdom	(0,1,0)	(21,1,0)	(0,1,0)	(0,1,1)	(13,0,0)	(15,1,0)	(15,1,0)

Notes:
This table provides summary information only. The convention used is (p, d, q), where p is the highest implied order of the AR polynomials, d is the degree of ordinary differencing, and q is the highest implied order of the MA polynomials. For example, the money stock in the United Kingdom is estimated as $(1-\theta L^9)(1-\phi L^{12})(1-L)m_t = \mu + e_t$ and is identified in the table as (21, 1, 0). Details of the estimated models are in Ahking and Miller (1986). The models are estimated using the ARIMA procedure in SAS, version 82.2.
– Means that the data are not available.
[a] Denotes that the model is also seasonally differenced.

implying a white-noise process at the 5 percent level, has the largest value of all the exchange rates.

These results impose strong *a priori* restrictions on the time series properties of the exogenous variables in any structural exchange rate model. In particular, excluding Canada, a sufficient condition for the spot exchange rate to have a random-walk representation is that all the exogenous variables are also random walks.[5] For Canada, its spot exchange rate with the United States has a significant autoregressive seasonal only at lag 12; a sufficient condition requires that its exogenous variables have only significant seasonal autoregressive parameters, but no other AR or MA parameters.

Columns (2)–(7) of table 11.1 present the final models for the exogenous variables – the logs of the money stock, one plus the short-term interest rate, one plus the long-term interest rate, industrial production, the cumulative trade balance (unlogged), and the price level, respectively. The price variables appear only in Driskill's (1981) model as lagged endogenous variables and are presented for completeness.

[5] The reason why this is only a sufficient and not a necessary condition is that there is a possibility of cancellation of common AR and MA factors of the exogenous variables. See Zellner and Palm (1974, n. 2) and Palm (1977).

The results are as follows. First, all variables except the indexes of industrial production for France, Italy, and the United Kingdom require first differencing, and in some cases, seasonal differencing, to induce stationarity. Second, only the short-term interest rates of Italy, Switzerland, and the United States, and Japan's long-term interest rate, are adequately modeled as random walks. All the other variables have non-zero AR and MA parameters. Moreover, with the exception of the short-term and long-term interest rates, nearly all other exogenous variables have seasonality as indicated by significant AR or MA parameters at lag 12.

The results suggest that the *estimated stochastic processes of the spot exchange rates are inconsistent with their implied stochastic processes*. In particular, the implied stochastic processes for the exchange rate involve non-zero AR, MA, and seasonal parameters, whereas the estimated stochastic processes do not. More formal tests of alternative models using the log-likelihood ratio test against the exchange rate models reported in table 11.1 yield similar conclusions (see Ahking and Miller 1986).

The apparent inconsistency between the stochastic processes of the exchange rates and their fundamental determinants may be due to cancellation of common AR and MA factors of the exogenous variables (see Zellner and Palm 1974 and Palm 1977). For the exchange rates to be random walks, however, requires a complete cancellation of the AR and the MA polynomials of the exogenous variables, including seasonal parameters. Our results suggest that this is a remote possibility.

Structural exchange rate models are frequently estimated with the exogenous variables in relative form. This estimation procedure imposes a common stochastic structure on the domestic and foreign exogenous variables. Our results suggest that this condition is unlikely to be true. Nonetheless, we also obtained the univariate time series models of the exogenous variables in relative form except for the cumulative trade balance. These time series models are, in general, different from those reported in table 11.1. But they also include non-zero AR and MA parameters, as well as seasonal parameters. These results are not reported.

Finally, it is also possible that the apparent inconsistency between the stochastic processes of the exchange rates and their fundamental determinants is due to the fact that the exchange rate equation should be treated as one equation in a system of simultaneous equations, including real incomes, the nominal interest rates, and the cumulative trade balances as endogenous variables and nominal money stocks and possibly other explanatory variables as exogenous variables. In this system of equations, the *minimum* order of the AR and MA polynomials for the exchange rates should be the same as those implied by the order of the AR and MA polynomials of the nominal money stocks. Our results suggest that this condition is unlikely to be met.

Based on the empirical results, it appears that none of the empirical exchange rate models is consistent with the information contained in the data.

3 Interpretations and implications

Our results suggest a failure of existing exchange rate models, a strong conclusion. Of the seven bilateral spot exchange rates examined, six are adequately modeled as random walks while the United States: Canadian dollar exchange rate has a significant autoregressive seasonal lag. These results are consistent with the view that the foreign exchange markets are efficient and that exchange rates are determined like asset prices (e.g. stock prices).[6]

When seen as asset prices determined in efficient markets, observed exchange rates can also be viewed as equilibrium prices adjusting rapidly to clear foreign exchange markets. The flexible price monetary model with its purchasing power parity (PPP) condition views the exchange rate as an equilibrium relationship between two relative prices where the two prices are determined by equilibrium conditions in the domestic and foreign money markets. In sum, the simple monetary model postulates an equilibrium relationship between exchange rates, domestic and foreign nominal money supplies, real incomes, and interest rates. But, observed domestic and foreign nominal money supplies, real incomes, and interest rates are not necessarily at their equilibrium values. This may explain why the stochastic properties of the exchange rates are inconsistent with the stochastic properties implied by their fundamental determinants.

The short-run monetary models are attempts to allow for lagged adjustment to equilibrium in the money markets. The adjustment scheme, however, is ad hoc and the long distributed lags of domestic and foreign nominal money stocks, real incomes, and interest rates imply serial correlations in the exchange rate. The implications for exchange rates, moreover, are not borne out by an examination of their stochastic properties.

The sticky-price monetary models are based on the assumption that PPP does not hold in the short run because of "sticky" goods prices. Frankel (1979) specifies an equilibrium model with PPP imposed only

[6] Our results only suggest weak form efficiency in the foreign exchange market (i.e. changes in spot exchange rates are uncorrelated with past changes). Meese and Rogoff (1983a), however, find that various exchange rate models fail to perform better than a random-walk model of the exchange rate in out-of-sample forecasts, i.e. semi-strong form efficiency in the foreign exchange market.

in the long run, but the estimated model involves no lagged variables. Driskill's (1981) model includes lagged prices as explanatory variables. From table 11.1, we note that the observed price indexes are all serially correlated, appearing to justify the assumption that PPP holds only in the long run. Serial correlation in the price indexes, however, should also imply serial correlation in the exchange rates. Put differently, sticky-price models suggest that the exchange rate should immediately overshoot its long-run equilibrium value following a monetary disturbance. When the exchange rate begins adjusting toward long-run equilibrium (i.e. PPP), we should observe serial correlations in the exchange rate where the structure of the serial correlations should be the same as the structure of the serial correlations in the domestic and foreign price levels (e.g. see Dornbusch 1976, p. 1165 and Frankel 1979, p. 620). This is not borne out by the empirical results.

Also, the failure of the stock-flow model to explain the stochastic behavior of the exchange rate may be attributed to the failure to measure properly the equilibrium values of the exogenous variables.

The rational expectations models present a paradox. These models explicitly test the restrictions implied by the assumption of rational expectations. Hoffman and Schlagenhauf (1983) and Huang (1984) are unable to reject their rational expectations restrictions. Their results, therefore, can be interpreted as being consistent with foreign exchange market efficiency (see Abel and Mishkin 1983). Their reduced form exchange rate equations all include distributed lags of the exogenous variables. Serial correlations in the exogenous variables are not inconsistent with foreign exchange market efficiency. But, since the stochastic behavior of the exogenous variables implies the stochastic behavior of the endogenous variables, their exchange rate series should be serially correlated. The paradox is that if we interpret their empirical results as being consistent with foreign exchange market efficiency, then the time series of the exchange rates must be serially correlated, contrary to our results.[7] We, unfortunately, do not have a good explanation for this seeming paradox. One possibility is that the tests performed by Hoffman and Schlagenhauf and Huang are not sufficiently powerful to reject their hypotheses.

Our discussion suggests that the existing empirical models of exchange rate determination are misspecified. The primary source of misspecification, in our view, is the use of observed values of the exogenous variables rather than equilibrium values. Since the equilibrium values of the exogenous variables are typically unobservable, a satisfactory empirical

[7] Hoffman and Schlagenhauf [1983], however, report in n. 6 of their paper that all their exchange rate series can be approximated by a random walk.

model of the exchange rate may be difficult to implement. A useful methodology has been proposed . . . by Beveridge and Nelson (1981), however. They develop a time series technique for decomposing an observed time series into permanent and cyclical components. The permanent component of a non-stationary time series is always a random walk while the cyclical component is a stationary ARMA process. If the permanent component can be interpreted as the equilibrium value, then the various exchange rate theories can be tested. More specifically, the exogenous variables in a reduced form equation of the exchange rate can first be decomposed into permanent and cyclical components. The exchange rate series, since it is already a random walk, has no cyclical component. If we interpret the permanent components of the exogenous variables as equilibrium values, then they can be used as the explanatory variables in the reduced form exchange rate equations. Since the permanent components of the exogenous variables are random walks, their stochastic properties are consistent with the stochastic behavior of the exchange rates. Thus, tests of the various theories of exchange rate determination can be conducted with these permanent values.

4 Summary and conclusion

Zellner and Palm (1974) show that if a structural model is correctly specified, then the actual time series behavior of the endogenous variables is the same as that implied by the time series behavior of the exogenous variables. A comparison of the actual and implied time series properties of the endogenous variables then provides important information on the specification of any structural econometric model.

Using the Zellner–Palm methodology that does not depend upon the assumptions of rational expectations or efficient foreign exchange markets, we find that six monthly bilateral exchange rates are adequately modeled as random walks. The seventh, the United States: Canadian dollar exchange rate, has a significant autoregressive seasonal lag. These results are consistent with the view of exchange rates as asset prices determined in efficient markets. For consistency, the time series behavior of the fundamental determinants of the exchange rate must also be random walks, with the possible exception of Canada where autoregressive seasonal parameters are permissible. The time series properties of the nominal money stocks, real incomes, one plus the short-term and one plus the long-term interest rates, the cumulative trade balances, and the price levels, all of which are considered to be important fundamental determinants of exchange rates, are, with few exceptions, not random walks. Thus, existing exchange rate models appear to be misspecified.

Our results are related to the work of Meese and Rogoff (1983a, 1983b) who find that various structural exchange rate models fail to perform better than a random-walk model in out-of-sample predictions. Nevertheless, several important differences in approach and interpretation need to be identified. They suggest several possible explanations for their results, including sampling errors, simultaneous equations bias, and instability of the money demand equations. But, even if all these possible causes of the failure of the structural exchange rate models can be remedied, the structural exchange rate models would probably, in our opinion, still be misspecified in the Zellner–Palm sense. We suggest that since most exchange rate theories are formulated as equilibrium relationships between the exchange rate and its fundamental determinants, the use of the observed values of the exogenous variables instead of their equilibrium values in a reduced form equation of the exchange rate is inappropriate and leads to misspecification of the empirical models.

Our study also differs from Meese and Rogoff's on an important methodological point. The Zellner–Palm technique demonstrates that a univariate time series model can be derived from a structural model. Meese and Rogoff, in our opinion, view time series models as alternatives to, and as competing with, the structural models. When time series and structural models are seen as complementary rather than competing, it provides important insights into Meese and Rogoff's results. For example, the "correct" univariate time series model should perform no worse in out-of-sample forecasting than the "correct" structural model. Thus, if the random-walk model is the correct univariate time series model, then structural models will not perform better than the random-walk model.

A promising line of research that may resolve the conundrum raised by our work is to decompose the observed time series of the exogenous variables in an exchange rate equation into permanent and transitory (cyclical) components as suggested by Beveridge and Nelson (1981). If the permanent components can be interpreted as equilibrium values, then they can be used as explanatory variables in exchange rate equations, and empirical tests to distinguish various theories of exchange rate determination can be carried out.

BIBLIOGRAPHY

Abel, A. and F. S. Mishkin (1983), "An integrated view of tests of rationality, market efficiency and the short-run neutrality of monetary policy," *Journal of Monetary Economics* 11, 3–24

Ahking, F. W. and S. M. Miller (1986), "Structural exchange-rate models and efficient foreign exchange markets: an examination of seemingly divergent views," University of Connecticut, manuscript

Beveridge, S. and C. R. Nelson (1981), "A new approach to decomposition of economic time series into permanent and transitory components with particular attention to measurement of the 'Business Cycle'," *Journal of Monetary Economics* 7, 151–74

Bilson, J. F. O. (1978), "Rational expectations and the exchange rate," in J. A. Frenkel and H. G. Johnson (eds.), *The Economics of Exchange Rates* (Reading, Mass. Addison-Wesley), 75–96

(1979), "The Deutsche Mark/dollar rate: a monetary analysis," in K. Brunner and A. Meltzer (eds.), *Policies for Employment, Prices, and Exchange Rates*, Carnegie–Rochester Series on Public Policy (Amsterdam: North-Holland), 59–101

Box, G. E. P. and G. M. Jenkins (1976), *Time Series Analysis: Forecasting and Control*, rev. edn. (San Francisco: Holden-Day)

Burt, J., F. R. Kaen, and G. G. Booth (1977), "Foreign exchange market efficiency under flexible exchange rates," *Journal of Finance* 32, 1325–30

Dornbusch, R. (1976), "Expectations and exchange rate dynamics," *Journal of Political Economy* 84, 1161–76

(1980), "Exchange-rate economics: where do we stand?," in G. L. Perry (ed.), *Brookings Papers on Economic Activity* (Washington, DC, Brookings Institution), 143–85

Driskill, R. A. (1981), "Exchange-rate dynamics: an empirical investigation," *Journal of Political Economy* 89, 357–71

Edwards, S. (1983), "Floating exchange rates in less-developed countries: a monetary analysis of the Peruvian experience, 1950–54," *Journal of Money, Credit, and Banking* 15, 73–81

Frankel, J. (1979), "On the Mark: a theory of floating exchange rates based on real interest differentials," *American Economic Review* 69, 610–22

(1983), "Monetary and portfolio-balance models of exchange rate determination," in J. S. Bhandari and B. H. Putnam (eds.), *Economic Interdependence and Flexible Exchange Rates* (Cambridge, Mass., MIT Press), 84–115

Frenkel, J. A. (1978), "A monetary approach to the exchange rate: doctrinal aspects and empirical evidence," in J. A. Frenkel and H. G. Johnson (eds.), *The Economics of Exchange Rate* (Reading, Mass., Addison-Wesley), 1–25

Hodrick, R. J. (1979), "On the monetary analysis of exchange rates: a comment," in K. Brunner and A. Meltzer (eds.), *Policies for Employment, Prices, and Exchange Rates*, Carnegie–Rochester Series on Public Policy (Amsterdam: North-Holland), 103–21

Hoffman, D. L. and D. E. Schlagenhauf (1983), "Rational expectations and monetary models of exchange rate determination: an empirical examination," *Journal of Monetary Economics* 11, 247–60

Hooper, P. and Morton, J. (1982), "Fluctuations in the dollar: a model of nominal and real exchange rate determination," *Journal of International Money and Finance* 1, 35–56

Huang, R. D. (1984), "Exchange rate and relative monetary expansions: the case of simultaneous hyperinflation and rational expectations," *European Economic Review* 24, 189–95

Levich, R. M. (1985), "Empirical studies of exchange rates: price behavior, rate determination and market efficiency," in R. W. Jones and P. B. Kenen (eds.), *Handbook of International Economics*, 2 (Amsterdam: Elsevier Science Publishers), 979–1040

Logue, D. E. and R. J. Sweeney (1977), "'White-noise' in imperfect markets: the case of the Franc/Dollar exchange rate," *Journal of Finance* 32, 761–8

Meese, R. A. and K. Rogoff (1983a), "Empirical exchange rate models of the seventies: do they fit out of sample?," *Journal of International Economics* 14, 3–24

(1983b), "The out-of sample failure of empirical exchange rate models: sampling error or misspecification?," in J. A. Frenkel (ed.), *Exchange Rates and International Macroeconomics* (Chicago, University of Chicago Press), 67–112

Mussa, M. (1983), "Empirical regularities in the behavior of exchange rates and theories of the foreign exchange market," in K. Brunner and A. H. Meltzer (eds.), *Theories, Policy, Institutions: Papers from the Carnegie–Rochester Conference on Public Policy* (Amsterdam, North-Holland), 165–213

Palm, F. C. (1977), "On univariate time series methods and simultaneous equation econometric models," *Journal of Econometrics* 5, 379–88

Poole, W. (1967), "Speculative prices as random walks: an analysis of ten time series of flexible exchange rates," *Southern Economic Journal* 33, 468–78

Zellner, A. (1984), *Basic Issues in Econometrics* (Chicago, University of Chicago Press)

Zellner, A. and F. C. Palm (1974), "Time series analysis and simultaneous equation econometric models," *Journal of Econometrics* 2, 17–54; chapter 1 in this volume

12 Encompassing univariate models in multivariate time series: a case study (1994)

Augustín Maravall and Alexandre Mathis

Through the encompassing principle, univariate ARIMA analysis could provide an important tool for diagnosis of VAR models. The univariate ARIMA models implied by the VAR should explain the results from univariate analysis. This comparison is seldom performed, possibly due to the paradox that, while the implied ARIMA models typically contain a very large number of parameters, univariate analysis yields highly parsimonious models. Using a VAR application to six French macroeconomic variables, it is seen how the encompassing check is straightforward to perform, and surprisingly accurate.

1 Introduction

After the crisis of traditional structural econometric models, a particular multivariate time series specification, the Vector Autoregression or VAR model has become a standard tool used in testing macroeconomic hypotheses. Zellner and Palm (1974, 1975) showed that the reduced form of a dynamic structural econometric model has a multivariate time series model expression, and that this relationship could be exploited empirically as a diagnostic tool in assessing the appropriateness of a structural model. As Hendry and Mizon (1992) state, a well-specified structural model should encompass the results obtained with a VAR model; similar analyses are also found in Monfort and Rabemananjara (1990), Clements and Mizon (1991), and Palm (1986).

It is also well known that a multivariate time series model implies a set of univariate models for each of the series. Thus, as argued by Palm (1986), univariate results can, in turn, provide a benchmark for multivariate models, and should be explained by them. When done, the comparison usually takes the form of comparing the forecasting performances

Thanks are due to Stephania Fabrizio, Grayham Mizon, Franz Palm, the Associate Editor, and two referees for their valuable comments.

Originally published in the *Journal of Econometrics* 61 (1994), 197–233.

of the multivariate model versus the set of univariate models, identified with Box–Jenkins (1970) techniques, see Palm 1983. More generally, however, since the multivariate model implies a set of univariate models, these should be derived from the fitted multivariate one, and then compared to the models obtained through univariate analysis. If the two sets of univariate models are clearly different, then there is reason to suspect specification error in some of the models. Given that, in general, direct identification of the univariate model is simpler than identification of the multivariate one, lack of agreement between the two sets of univariate models may well indicate misspecification of the multivariate model and invalidate, as a consequence, its use in testing economic hypotheses.

Therefore, the use of univariate models as a diagnostic tool should include the comparison between the univariate models derived from the multivariate one and those obtained with univariate analysis (we shall refer to them as "implied" and "estimated" univariate models). This comparison, however, is seldom done. Univariate analysis is used (often wrongly) in identification of multivariate models (see, for example, Jenkins 1979 and Maravall 1981); it is hardly ever used (as it rightly should) in the diagnostics stage. Perhaps this is due to what Rose (1986) has termed "the autoregressivity paradox", which can be described as follows.

It is a well-known fact that the immense majority of ARIMA models fitted to economic series are parsimonious, including few parameters. Yet even relatively small multivariate models imply univariate models with a very large number of parameters. Therefore, if the world is multivariate (as it is), ARIMA models should be highly unparsimonious, and hence of little practical use. Yet we know that this is not the case. How can the two facts be reconciled? Rose (1985, 1986) suggests an explanation: macroeconomic variables are basically contemporaneously correlated and there are few dynamic relationships among them. The explanation is a bit drastic, and it seems sensible to seek for some alternative . . .. As pointed out by Wallis (1977), two possibilities come to mind: First, it may happen that the autoregressive (AR) and moving average (MA) polynomials of the implied ARIMA model have roots in common. Cancelling them out, the order of the model would be reduced. Second, some of those two polynomials may contain a large number of small coefficients that would be undetectable for the sample size used. The first possibility will be denoted the "root effect", and the second, the "coefficient effect".

Although both effects are certainly possible, the question remains as to whether they can be measured with enough accuracy in actual applications. For example, the autoregressive coefficient estimates in VAR models are, on occasion, unstable, and the roots of the polynomials are sensitive to small variations in those coefficients. That factor might have an effect on the detection of common roots. Furthermore, it is an empirical fact that often the factorization of the determinant of the AR matrix in VAR models yields roots with a relatively large modulus. This might affect the presence of small coefficients in the implied univariate representation.

Yet the issue of whether the root and coefficient effect can actually be detected, so as to simplify an ARMA model with perhaps forty or fifty parameters to an ARMA model with (at most) two or three parameters, is ultimately an empirical issue. Therefore, we shall look at an example consisting of a standard VAR model, for six quarterly macroeconomic variables. We shall see whether, in practice, despite the Autoregressivity Paradox, univariate analysis (a relatively familiar tool) can be of practical help in checking the adequacy of a multivariate model. Finally, we consider what the comparison says in terms of an economic application: the measurement of the persistence of macroeconomic shocks.

2 Univariate models implied by a vector autoregressive model

Let $z_t = (z_{1t}, \ldots, z_{kt})'$ be a stationary stochastic vector process which follows the VAR model

$$\Phi(L)z_t = a_t + \mu, \tag{2.1}$$

where L is the usual lag operator, $\Phi(L)$ is a $(k \times k)$ matrix with finite polynomials in L as elements, a_t is a k-dimensional white-noise variable with zero mean vector and contemporaneous covariance matrix Ω, and μ is a vector of constants. If the (i, j)th element of $\Phi(L)$ is a polynomial with coefficients Φ_{ijk}, $k = 0, 1, \ldots$, we adopt the standardization $\Phi_{ii0} = 1$, and $\Phi_{ij0} = 0$ for $i \neq j$. Finally, the stationarity of z_t implies that the roots of the equation $|\Phi(L)| = 0$ (where $|\bullet|$ denotes the determinant of a matrix) lie outside the unit circle. Following Zellner and Palm (1974), to obtain the univariate representation of z_{it} $(i = 1, \ldots, k)$, we simply need to express (2.1) as

$$z_t = [\Phi(L)]^{-1}(a_t + \mu) = |\Phi(L)|^{-1}\Phi^*(L)(a_t + \mu), \tag{2.2}$$

where $\Phi^*(L)$ is the adjoint matrix of $\Phi(L)$. For the ith element of z_t, (2.2) becomes

$$|\Phi(L)|z_{it} = \sum_{j=1}^{k} \Phi^*_{ij}(L)a_{jt} + c_i, \tag{2.3}$$

where c_i is a constant, equal to the ith element of $\Phi^*(1)\mu$. Since (ignoring the constant) the r.h.s. of (3) is the sum of k finite moving averages, it can also be represented as a finite moving average $\theta_i(L)u_{it}$, where u_{it} is a white-noise variable, such that

$$\theta_i(L)u_{it} = \sum_{j=1}^{k} \Phi_{ij}^*(L)a_{jt} \tag{2.4}$$

(see Anderson 1971). Considering (2.3) and (2.4), the univariate models implied by (2.1) are given by

$$\phi(L)z_{it} = \theta_i(L)u_{it} + c_i, \quad i = 1, \ldots, k, \tag{2.5}$$

where the autoregressive (AR) polynomial is equal to

$$\phi(L) = |\Phi(L)|, \tag{2.6}$$

and each moving average polynomial, together with the variance of the univariate innovation u_{it}, can be obtained through (2.4) as detailed in appendix A. It is worth noting:

(a) A VAR process implies univariate ARMA (not simply AR) models, and that all univariate models share the same AR polynomial (2.6)
(b) As shown in appendix A, the univariate MA polynomials are always invertible, with the orders (q_i) depending on the elements $\Phi_{ij}(L)$
(c) When $\mu \neq 0$, the univariate model for z_{it} will always contain a constant.

3 A case study: the series and univariate analysis

We consider six quarterly macroeconomic series of the French economy, taken from Deniau, Fiori, and Mathis (1989). Each series has eighty-four observations and starts in the first quarter of 1963. The series are the following:

d_t = public debt
y_t = gross domestic product (GDP)
p_t = GDP deflator
r_t = interest rate (on first-class bonds)
n_t = balance of the current account (exports–imports)
m_t = monetary aggregate (M1).

The sources of the series, as well as some (minor) modifications performed on them, are described in the above reference. Figure 12.1 plots the six series; all, except n_t (which can take negative values), have been log transformed. They seem to exhibit, in all cases, a non-constant mean and, as seen in figure 12.2, Autocorrelation Functions (ACF) that converge

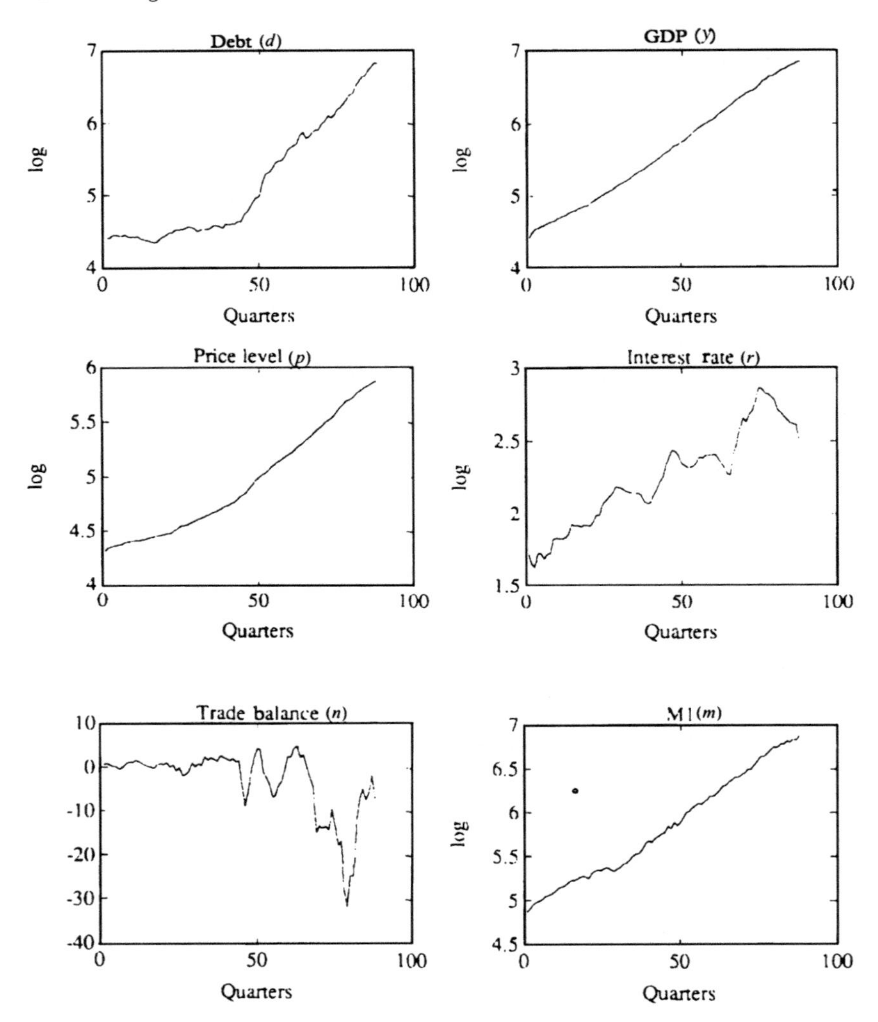

Figure 12.1 Macroeconomic series

very slowly. The Augmented Dickey–Fuller (ADF) tests, allowing for a constant (according to result (c) above), are presented in the first row of table 12.1. At the 5 percent size, the critical value – taken from MacKinnon (1991) – is 2.90, and hence in no case is the unit root hypothesis rejected. (All regressions were run with eight lags, enough to whiten all series.)

Table 12.1 *Tests on the univariate series*

	d_t	y_t	p_t	r_t	n_t	m_t
ADF-t	0.62	−0.89	1.12	−1.61	−1.90	0.67
Q_{27}	19.9	20.8	24.3	22.8	22.9	24.0

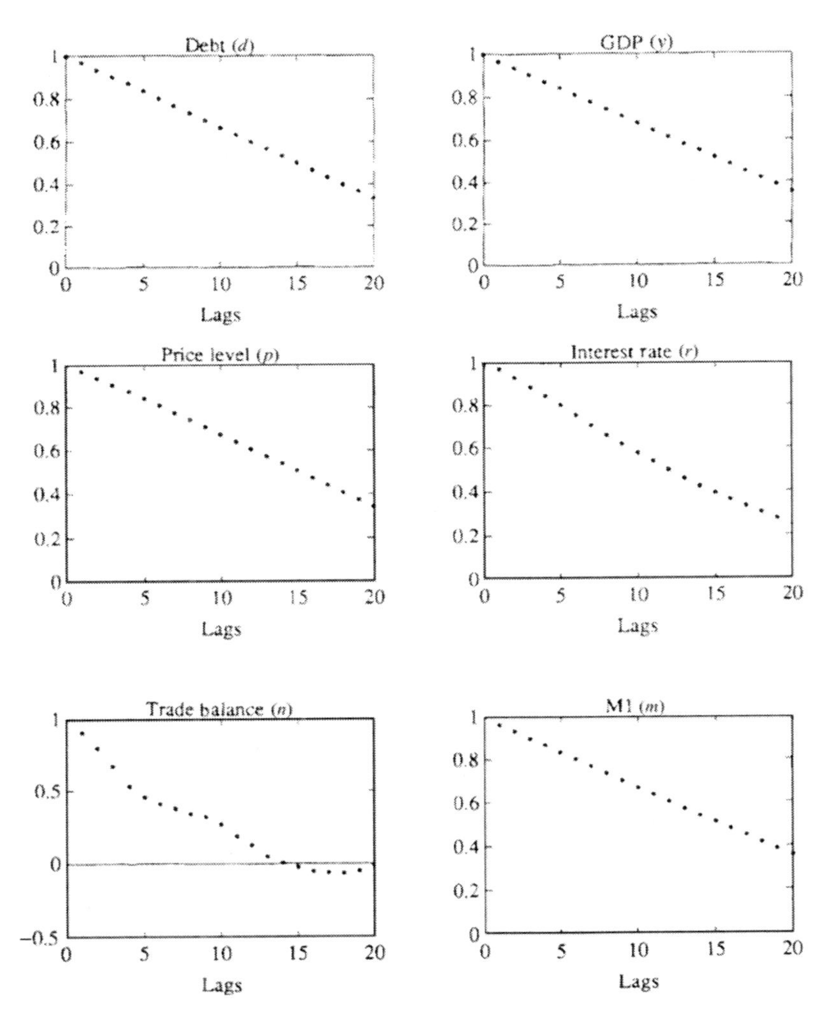

Figure 12.2 ACF: series

Table 12.2 *Residual standard deviation*

	Estimated ARIMA	Implied ARIMA	VAR
d_t	0.0359	0.0359	0.0319
y_t	0.0112	0.0092	0.0086
p_t	0.0079	0.0073	0.0070
r_t	0.0394	0.0374	0.0361
n_t	0.0288	0.0311	0.0283
m_t	0.0199	0.0205	0.0186

First differences of all series were thus taken, and the following ARIMA models were identified and estimated (all differenced series were centered around the mean):

$$(1 - \underset{(0.11)}{0.74}\, L^4)\nabla \log d_t = (1 - \underset{(0.16)}{0.32}\, L^4)u_{1t}, \tag{3.1a}$$

$$\nabla \log y_t = u_{2t}, \tag{3.1b}$$

$$\nabla^2 \log p_t = (1 - \underset{(0.07)}{0.74}))u_{3t},$$

$$(1 - \underset{(0.10)}{0.35}\, L)\nabla \log r_t = u_{4t}, \tag{3.1d}$$

$$\nabla n_t = (1 - \underset{(0.10)}{0.47}\, L^4)u_{5t}, \tag{3.1e}$$

$$(1 + \underset{(0.09)}{0.19}\, L - \underset{(0.10)}{0.50}\, L^4)\nabla \log m_t = u_{6t}. \tag{3.1f}$$

(The numbers in parentheses below the parameter estimates are the associated standard errors.) The ACF of the residuals are displayed in figure 12.3, and in all cases they are seen to be close to the ACF of white noise. The Box–Ljung–Pierce Q statistics for the first twenty-seven autocorrelations are displayed in the second row of table 12.1, and for the six series they are smaller than the corresponding χ^2 (5 percent) critical value. The residual standard deviations are displayed in the first column of table 12.2. Three comments are in order:

(i) Since our aim is to confront the parsimony of these estimated univariate models with the lack of parsimony of univariate ARIMAs derived from a VAR model, an important model selection criterion was to minimize the number of parameters. Although alternative specifications are certainly possible, the models in (3.1) passed all diagnostics and, besides the innovation variance, no model contains more than two parameters.

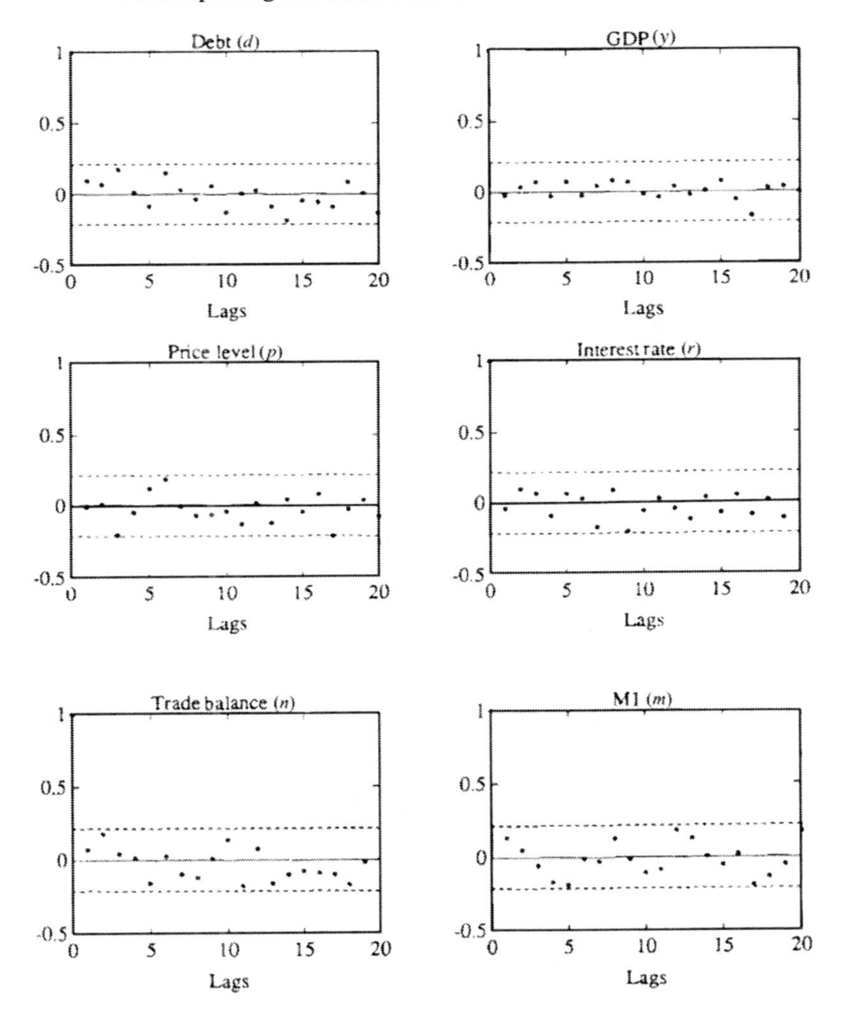

Figure 12.3 ACF: residuals

(ii) All variables are integrated of order 1 [or I(1)], except for the GNP deflator p_t. However, estimation of an ARIMA(1,1,1) model without imposing the second unit root yields

$$(1 - \underset{(0.04)}{0.94}\, L)\nabla \log p_t = (1 - \underset{(0.09)}{0.70}\, L)u_{3t}, \tag{3.1c}$$

with slightly smaller values of Q_{27} and of the residual variance. Since there are no compelling reasons to impose the second unit root,

Table 12.3 *Engle–Granger cointegration test*

	d_t	y_t	p_t	r_t	n_t	m_t
Q_{27}	34.4	28.2	26.4	23.9	36.5	30.1
ADF-t	−2.93	−1.79	−2.67	−3.38	−4.42	−3.16
Critical value (5%)	−5.22	−5.22	−5.22	−5.22	−5.22	−5.23

to preserve the order of integration, we shall use as the estimated univariate model for p_t that given by (3.1c).

(iii) Finally, concerning the model for m_t, factorization of the AR polynomial produced the following roots:

Modulus:	0.80	0.84	0.89
Frequency:	0	$\pi/2$	π

The first root $(1 - 0.8L)$ is associated with the trend, and the last two roots with the once- and twice-a-year seasonal frequencies; all roots display a relatively large modulus.

4 Testing for cointegration

Before proceeding to estimation of a multivariate model for the six variables, we need to test for the presence of cointegration relationships among them. Let $x_t = (x_{1t}, \ldots, x_{6t})$ denote the vector of the six undifferenced variables; two procedures will be applied. First, following Engle and Granger (1987), we compute the six regressions

$$x_{jt} = \alpha_0 + \alpha_j t + \sum_{t=1;i\neq j}^{6} \alpha_t x_{it} + e_{jt}, \qquad (4.1a)$$

$(j = 1, \ldots, 6)$. Then, ADF tests are run on the series of estimated residuals $\hat{e}_{jt}$. If in no case the null hypothesis of a unit root is rejected, the series are not cointegrated.

For the six regressions of the type (4.1a), the first row of table 12.3 shows the Q (27) statistics associated with the autocorrelation function of the residuals obtained in the Dickey–Fuller regression on $\hat{e}_{jt}$ (using up to four lagged values). The second and third rows present the Dickey–Fuller t-statistics to test for the hypothesis that there is a unit root, and its corresponding 5 percent critical value; these last values have been computed using the response surface regression of MacKinnon (1991). It is seen that in no case is the unit root hypothesis rejected. If the term $\alpha_j t$ is removed from (4.1a), the results remain basically unchanged, except

Table 12.4 *Johansen cointegration test*

	Number of cointegrating vectors					
	$r \leq 5$	$r \leq 4$	$r \leq 3$	$r \leq 2$	$r \leq 1$	$r = 0$
Lambda-max test	3.45	7.99	10.14	13.33	22.29	35.59
Critical value (5%)	10.25	14.17	22.30	26.58	37.76	42.04
Trace test	3.45	11.44	21.59	34.92	57.21	92.80
Critical value (5%)	10.25	16.94	32.59	47.11	72.48	98.20

when the variable n_t is the regressand, in which case the t-statistic becomes marginally significant.

The second type of cointegration test performed is that proposed by Johansen (1988), based on the rank of the matrix Π in the multivariate regression

$$\Delta x_t = \Gamma_1 \Delta x_{t-1} + \cdots + \Gamma_{p-1} \Delta x_{t-p+1} + \Pi x_{t-p} + \mu + \varepsilon_t. \quad (4.1b)$$

Since, at this stage, the model is unrestricted, each matrix of parameters in (4.1b) is of order 6×6, and hence, considering our sample size, a small value of p is required. Setting $p = 2$, the autorcorrelations of the estimated residuals were reasonably low (values of $p > 2$ implied estimation of over 100 parameters). Letting r denote the number of cointegrating vectors, table 12.4 presents the lambda-max and trace tests for the sequential testing of H_0: $r \leq j$ ($j = 5, \ldots, 1, 0$), where the 5 percent critical values have been taken from Gardeazabal and Regulez (1990). Both tests indicate that the six series can be safely assumed to be noncointegrated.

5 The vector autoregressive model (VAR)

Since we can assume that there are no cointegration relationships, the VAR model can be specified in first differences of the variables. Such a VAR model, for the six variables we consider, was estimated by Deniau, Fiori, and Mathis (1989) in order to analyze the effect of the public debt on several macroeconomic variables of the French economy. The model was identified in a manner similar to that proposed by Hsiao (1981) and Caines, Keng, and Sethi (1981). In a first step, the VAR structure is determined, equation by equation, according to the results of "causality tests" between variables; the maximum lags are found with an MFPE information criterion. The model thus specified is estimated as a SURE model. We re-estimated the same VAR with the rates of growth replaced by the differences in logs. Also, a few parameters that were not significant were removed. (The effect of these modifications was minor.) It should be stressed that the aim of the chapter is not to improve upon the VAR specification, but to look at a model already available in the literature, in order to see whether it explains (or encompasses) the results obtained in univariate analysis.

The estimated model has a total of twenty autoregressive coefficients, and hence, for a six-variate VAR, it is considerably parsimonious (an average of 3.3 parameters per equation). In terms of (2.1), letting z_t denote the vector:

$$z_t = (\nabla \log d_t, \nabla \log y_t, \nabla \log p_t, \nabla \log r_t, \nabla n_t, \nabla \log m_t)',$$

the estimated $\Phi(L)$ matrix is given by

$$\Phi(L) = \begin{bmatrix} \phi_{11} & 0 & \phi_{13} & 0 & 0 & \phi_{16} \\ \phi_{21} & \phi_{22} & 0 & \phi_{24} & 0 & 0 \\ 0 & 0 & \phi_{33} & \phi_{34} & 0 & 0 \\ 0 & 0 & 0 & \phi_{44} & \phi_{45} & 0 \\ 0 & \phi_{52} & 0 & \phi_{54} & \phi_{55} & 0 \\ \phi_{61} & \phi_{62} & 0 & 0 & 0 & \phi_{66} \end{bmatrix},$$

where 0 denotes the null polynomial, and the non-zero elements are the following polynomials in L:

$$\phi_{11} = 1 - \underset{(2.07)}{0.202}\, L^4, \quad \phi_{13} = -\underset{(3.06)}{1.08}\, L - \underset{(2.78)}{0.987}\, L^4, \quad \phi_{16} = \underset{(-3.46)}{0.548}\, L^2,$$

$$\phi_{21} = -\underset{(3.13)}{0.055}\, L^8, \quad \phi_{22} = 1, \quad \phi_{24} = -\underset{(3.62)}{0.065}\, L^7,$$

$$\phi_{33} = 1 - \underset{(1.98)}{0.137}\, L - \underset{(4.3)}{0.299}\, L^2 - \underset{(4.3)}{0.303}\, L^5, \quad \phi_{34} = -\underset{(3.76)}{0.058}\, L^2,$$

$$\phi_{44} = 1 - \underset{(3.48)}{0.34}\, L, \quad \phi_{45} = \underset{(-2.36)}{0.3}\, L,$$

$$\phi_{52} = \underset{(-2.9)}{0.964}\, L, \quad \phi_{54} = -\underset{(-2.58)}{0.192}\, L, \quad \phi_{55} = 1 + \underset{(-4.1)}{0.4}\, L^4,$$

$$\phi_{61} = -\underset{(1.94)}{0.097}\, L, + \underset{(-2.96)}{0.15}\, L^2, \quad \phi_{62} = -\underset{(6.62)}{0.73}\, L,$$

$$\phi_{66} = 1 + \underset{(-2.1)}{0.18}\, L - \underset{(3.65)}{0.35}\, L^4,$$

where the t-statistics of the parameter estimates are given in parentheses. In terms of its economic interpretation, the model implies that a positive shock in the public debt increases, in the short to medium term, aggregate demand (with a limited crowding-out effect), which in turn increases imports. As a consequence, the balance of the current account deteriorates, and there is an increase in interest rates associated with foreign capital inflows. Economic interpretation, however, is not our present concern, and we refer to the Deniau, Fiori, and Mathis (1989) paper.

An important element in the diagnosis of a VAR model is the behavior of the vector of estimated residuals $\hat{a}_t$. Table 12.5 summarizes the correlation functions among the components of $\hat{a}_t$. The "$\oplus$", "$\ominus$", and "$\bullet$" signs indicate, respectively, a positive significant correlation, a negative significant correlation, and a correlation that can be assumed to be zero (see Tiao and Box 1981). The distribution of the significant correlations appears to be random; the largest positive and negative values are 0.31 and -0.30, and the number of significant ones is seventeen, or approximately 5 percent of the total number of computed correlations. The residuals obtained behave, thus, as a white-noise vector.

The residual variances are given in the second column of table 12.2. Compared to the ones obtained in the univariate ARIMA fit, it is seen that the innovations in the multivariate model have smaller variances. The percentage reduction varies between 2 percent (variable n_t) and 23 percent (variable y_t), with an average reduction of approximately 11 percent.

To further validate the models, an out-of-sample forecasting exercise was performed. Some of the series were modified after 1985, and the last observation available on our complete set of series is for 1985:4. In order to increase the number of out-of-sample forecasts, the ARIMA and VAR models were estimated with data up to 1983:4. Then, one-period-ahead forecasts were computed for the four quarters of 1984 and of 1985. Table 12.6 presents the root mean-squared error of the out-of-sample forecasts for the six series. The results for y_t, p_t, r_t, and m_t are clearly close to the in-sample values given in table 12.2, and for d_t the out-of-sample forecast is better. For n_t, the out-of-sample forecast deteriorates and an F-test for the equality of variances in the case of the VAR model yielded the value 3.4, and hence equality could be marginally rejected. (For the other series, the corresponding F values were not significant.) As for the relative performance of ARIMA and VAR models in out-of-sample forecasting, for four variables the VAR yields better forecasts, while in two cases the ARIMA models perform better. In no case, however, is the difference between the two forecasts large. Considering the improvement in in-sample fit and the overall better performance in out-of-sample forecasting of the multivariate model, the univariate ARIMA models do not seem to "parsimoniously encompass" the VAR one (see Hendry and Mizon 1992).

In summary, both the set of estimated ARIMA models and the VAR model behave reasonably. The multivariate structure does not bring spectacular improvement, but it does bring some. Altogether, considering the simplicity of the models identified, the results represent sensible applications of univariate ARIMA and VAR modeling.

Table 12.5 *VAR estimation: auto and cross-correlations of residuals*[a]

	d_t	y_t	p_t	r_t	n_t	m_t
d_t	••••••••••••	••••••••••••	••••••••••••	••⊖•••••••••	•••⊕••••••••	••••••••••••
y_t	••••••••••••	••••••••••••	•••••⊕••••••	•⊕••⊖•••••••	•••••••••⊕••	••••••••••••
p_t	••••••••••••	••••••••••••	••••••••••••	••••⊖•••••••	••••••••••••	••••••••••••
r_t	••••••••••••	••••••••••••	••••••••••••	••••••••••••	•••••••••⊕••	••••••••••••
n_t	•⊕••••••••••	••••••••••••	••⊖•••••••••	•••••⊖••••••	•••••••⊖••••	⊖•••••••⊕•••
m_t	•••••••••••⊕	••••••••⊕•••	••••••••••••	••••••••••••	•••⊕••••••••	••••••••••••

Note: [a] ⊕ = significant positive correlation, ⊖ significant negative correlation, • = insignificant correlation.

Table 12.6 *Out-of-sample forecast: RMSE*

	Estimated ARIMA	VAR
d_t	0.0148	0.0155
y_t	0.0133	0.0119
p_t	0.0081	0.0078
r_t	0.0496	0.0435
n_t	0.0502	0.0529
m_t	0.0175	0.0154

6 Implied univariate models in the VAR and comparison with the estimated ARIMA models: ad hoc comparison

Following the derivation of section 2 and appendix A, the univariate ARIMA models implicit in the VAR model have been obtained. The third column of table 12.2 contains the innovation variances of the implied univariate models. They are similar to those obtained with the estimated ARIMA models, and slightly closer to the innovation variances of the VAR model.

Concerning the autoregressive and moving average coefficients of the implied ARIMAs, the common AR polynomial, $\phi(L)$ of (5), is of order 22. The order q_i of the six moving average polynomials are those in the first row of table 12.8, and hence, despite the parsimony of our VAR model, the example provides a good illustration of the autoregressivity paradox referred to earlier: While the univariate models implied by the VAR contain an average of forty-two AR and MA parameters, the univariate models estimated in section 3 have an average number of 1.3 parameters. We mentioned before two simple reasons that might explain the apparent paradox; let us see how they operate in practice.

It is well known that, when the matrix $\Phi(L)$ in (2.1) has a block-triangular or block-diagonal structure, exact cancellation of roots between the AR and MA polynomials in some of the implied univariate models will occur (see Goldberger 1959, Wallis 1977, and Palm 1986). The matrix $\Phi(L)$, in our case, does not have that type of structure, and hence no such root cancellations can be done. For each of the six series, the twenty-two roots of $\phi(L)$ have to be compared with the roots of the corresponding polynomial $\theta_i(L)$. Computation of the 144 roots shows that the VAR model is indeed stationary, although a high proportion of the roots are relatively large and, for example, only seven of them are smaller than 0.5 in modulus. The MA polynomials are invertible, and they also display roots that are relatively large in modulus.

Table 12.7 *Series d_t, AR and MA roots of the univariate model implied by the VAR*

Root	Modulus	Frequency
Roots of the autoregressive polynomial		
-0.92	0.92	3.14
0.91	0.91	0
$0.03 \pm 0.85_i$	0.85	1.54
$0.61 \pm 0.54_i$	0.81	0.73
$-0.67 \pm 0.41_i$	0.78	2.59
$-0.56 \pm 0.55_i$	0.78	2.36
$0.77 \pm 0.10_i$	0.77	0.13
$0.29 \pm 0.68_i$	0.73	1.17
$-0.15 \pm 0.67_i$	0.69	1.79
$-0.55 \pm 0.25_i$	0.61	2.71
$0.28 \pm 0.53_i$	0.60	1.08
0.50	0.50	0
-0.26	0.26	3.14
Roots of the moving average polynomial		
0.85	0.85	0
$0.60 \pm 0.54_i$	0.81	0.73
$-0.67 \pm 0.41_i$	0.78	2.59
$-0.04 \pm 0.78_i$	0.78	1.62
$-0.55 \pm 0.55_i$	0.77	2.36
-0.76	0.76	3.14
$0.30 \pm 0.67_i$	0.73	1.15
$0.66 \pm 0.07_i$	0.66	0.10
$0.23 \pm 0.50_i$	0.56	1.14
$-0.37 \pm 0.33_i$	0.49	2.41
-0.19	0.19	3.14

When comparing, for the six series, the AR and MA roots, in order to decide which of them cancel out, a criterion is needed. Let ω and h denote the frequency (in radians) and modulus, respectively, of a complex root, and consider, for example, the implied ARMA model for the variable d_t. Table 12.7 displays the roots of the AR and MA polynomials (to facilitate interpretation, the roots displayed are those of L^{-1}). It is easy to accept that the MA root with $\omega = 2.59$ and $h = 0.78$ will cancel out with the root with the same frequency and modulus in $\phi(L)$. But, what about the pair of roots ($\omega = 1.62$, $h = 0.78$) and ($\omega = 1.54$, $h = 0.85$)? Since, on the one hand, the roots of the polynomials in the implied ARIMA models are complicated functions of forty-one parameters (those in $\Phi(L)$ and in Ω) and, on the other hand, the comparison involves 144 roots, computed from eighty-four observations on the vector of variables, we

Table 12.8 *Order of ARMA (p, q) model*

	d_t		y_t		p_t		r_t		n_t		m_t	
	p	q	p	q	p	q	p	q	p	q	p	q
Implied ARMA	22	19	22	24	22	17	22	20	22	21	22	21
Implied ARMA after ad hoc root cancellation	13	10	10	12	5	0	6	4	10	9	13	12
Estimated ARMA	1	1	0	0	1	1	1	0	0	1	2	0

had *a priori* doubts as to whether formal testing could be of help, and hence proceeded in two ways. First, a simple ad hoc criterion is used, which is fairly restrictive and biased towards under-cancellation. Second, a formal test, adapted from Gourieroux, Monfort, and Renault (1989) is performed.

The ad hoc criterion, discussed in appendix B, is as follows: The root $(\hat{h}_1, \hat{\omega}_1)$ will cancel with the root $(\hat{h}_2, \hat{\omega}_2)$ if
(a) $|\hat{h}_1 - \hat{h}_2| \leq 0.05$,
(b) $|\hat{\omega}_1 - \hat{\omega}_2| \leq 0.05$,
(c) $|\hat{h}_1 - \hat{h}_2| + |\hat{\omega}_1 - \hat{\omega}_2| \leq 0.07$.

Applying this criterion to our example, to get an insight into the proximity of the cancelled roots, we consider the two that are most distant, in terms of the sum of the two absolute deviations. These are the root $(\omega = 0.73, h = 0.81)$ in the AR polynomial and the root $(\omega = 0.78, h = 0.79)$ in the MA polynomial, of the univariate model for r_t. The two roots generate the AR and MA polynomials $(1 - 1.21\,L + 0.66\,L^2)$ and $(1 - 1.12\,L + 0.63\,L^2)$, respectively. Assume the AR polynomial is the result of estimating an AR(2) model with $T = 84$ observations (our sample size), and that we perform the test $\phi_1 = 1.12$ and $\phi_2 = -0.63$ (i.e. the AR polynomial is equal to the MA one). Then, denoting by M the asymptotic covariance matrix of the autoregressive parameter estimators,

$$S = (\hat{\Phi} - \Phi)' M^{-1} (\hat{\Phi} - \Phi) \dot{\sim} \chi^2_2,$$

where $\hat{\Phi}' = (1.21, -0.66)$ and $\Phi' = (1.12, -0.63)$. Using the expression for M in Box and Jenkins (1970, p. 244), it is found that $S = 1.61$, certainly below the 95 percent critical value of 5.99. Since this result holds for the pair of cancelled roots that are most distant, it is clear that the criterion favors undercancellation.

Using the above criterion, roots were cancelled in the implied ARMA model; remultiplying the remaining ones, new models are obtained for the six variables. Their orders are indicated in the second row of table 12.8:

Table 12.9 *Implied ARMA models after cancellation of roots: ad hoc criterion*

	d_t	y_t	p_t	r_t	n_t	m_t
AR coefficients						
Lag 1	0.05	0.92	0.15	−0.18	−0.21	0.02
Lag 2	0.04	−0.08	0.30	0.25	0.25	0.04
Lag 3	−0.03	0.09	−0.01	−0.14	−0.08	−0.03
Lag 4	0.65	0.02	0.01	−0.08	−0.48	0.65
Lag 5	−0.06	−0.08	0.30	0.02	−0.06	−0.06
Lag 6	−0.00	−0.00	–	−0.01	0.08	−0.00
Lag 7	−0.02	0.03	–	–	−0.04	−0.02
Lag 8	−0.07	−0.04	–	–	−0.03	−0.07
Lag 9	−0.01	0.01	–	–	−0.02	−0.01
Lag 10	0.00	0.00	–	–	−0.01	0.00
Lag 11	−0.02	–	–	–	–	−0.02
Lag 12	0.01	–	–	–	–	0.01
Lag 13	0.00	–	–	–	–	0.00
MA coefficients						
Lag 1	0.03	0.98	–	−0.50	−0.21	0.29
Lag 2	−0.04	−0.18	–	0.06	0.20	−0.09
Lag 3	0.02	0.12	–	−0.10	−0.07	0.05
Lag 4	0.31	0.03	–	−0.11	−0.03	0.18
Lag 5	−0.01	−0.17	–	–	0.00	−0.02
Lag 6	0.00	0.05	–	–	−0.01	0.00
Lag 7	0.00	−0.09	–	–	−0.01	0.00
Lag 8	0.00	0.10	–	–	−0.00	−0.02
Lag 9	−0.01	−0.00	–	–	0.00	0.01
Lag 10	−0.00	−0.00	–	–	–	−0.01
Lag 11	–	0.01	–	–	–	−0.00
Lag 12	–	0.00	–	–	–	0.00

They have been considerably reduced (the average number of parameters per model drops from forty-two to seventeen), but the models are still far from the parsimony of the ARMA models from univariate analysis.

The implied ARMA models obtained after removing common roots are displayed in table 12.9. Since (as shown in appendix B) the standard deviation of $\hat{\omega}$ and of $\hat{h}$ are larger for roots with smaller modulus, cancellation will be likely to affect the roots with relatively large modulus. Thus the remaining ARMA models will mostly contain the smaller roots, which are estimated with less precision. In considering whether the ARMA models of table 12.9 can be made more parsimonious by removing small coefficients (undetectable in estimation), again a criterion is needed. Considering that most of the standard errors of the parameters in the estimated ARMA models of expressions (3.1a)–(3.1f) are in the order of 0.09 or larger, a reasonable criterion is to remove coefficients that are below 0.18 in absolute value. Proceeding in this way, the following models are obtained.

(a) Series d_t:

$$(1 - 0.65\, L^4)\nabla \log d_t = (1 - 0.31\, L^4)u_{1t}, \tag{6.1a}$$

which is quite close to (3.1a). In this case, the VAR model certainly explains the model obtained in univariate analysis.

(b) Series y_t:

$$(1 - 0.92\, L)\nabla \log y_t = (1 - 0.98\, L)u_{2t}, \tag{6.1b}$$

or, approximately, the random walk model of (3.1b). Again, the VAR model explains well the estimated univariate ARIMA model. Considering the relatively large decrease in the residual variance of y_t in the VAR model, the variable GDP seems particularly suited for multivariate analysis.

(c) Series p_t:

$$(1 - 0.15\, L - 0.30\, L^2 - 0.30\, L^5)\nabla \log p_t = u_{3t}. \tag{6.1c}$$

Model (6.1c) appears to be quite distant from (3.1c), yet if in the latter the MA polynomial is inverted and approximated up to the fourth power, the product $(1 - 0.94\, L)(1 - 0.70)^{-1}$, after deleting small coefficients, yields

$$(1 - 0.24\, L - 0.17\, L^2 - 0.23\, L^5)\nabla \log p_t = u_{3t},$$

more in line with (6.1c). However, in so far as a fourth order approximation to $(1 - 0.70\, L)^{-1}$ is a poor approximation, the series p_t illustrates how, when the univariate model contains a relatively large MA root, VAR models will have trouble capturing that behavior.

(d) Series r_t:

$$(1 + 0.18L - 0.25L^2)\nabla \log r_t = (1 + 0.50L)u_{4t},$$

somewhat different from the model (3.1d). However, expressing the model in pure autoregressive form, it is obtained as

$$(1 - 0.32\, L)\nabla \log r_t = u_{4t}, \tag{6.1d}$$

with all other parameters smaller than 0.10 and converging fast towards zero. Models (3.1d) and (6.1d) are obviously very close.

(e) Series n_t:

$$(1 + 0.21\, L - 0.25L^2 + 0.48L^4)\nabla n_t = (1 + 0.21\, L - 0.20L^2)u_{5t}$$

Expressing this time the model in pure MA form, it is obtained that

$$\nabla n_t = (1 - 0.47\, L^4)u_{5t}, \tag{6.1e}$$

with all other MA parameters smaller than 0.15. Model (6.1e) is the same as (3.1e).

(f) Series m_t:

$$(1 - 0.65L^4)\nabla \log m_t = (1 - 0.29L - 0.18L^4)u_{6t}.$$

The autoregressive expression is found to be

$$(1 + 0.29\, L - 0.46\, L^4)\nabla\log m_t = u_{6t}, \tag{6.1f}$$

with all other parameters smaller than 0.10. Again, model (6.1f) is reasonably close to model (3.1f).

In conclusion, when the (conservative) ad hoc criterion is used, careful analysis of the multivariate VAR model explains well the models obtained with univariate analysis. We turn next to the results of formal testing.

7 Comparison of the implied and estimated univariate ARIMA models: a test procedure

Despite the large number of roots we wish to compare, it is straightforward to adapt to our case an ingenious testing procedure developed by Gourieroux, Monfort, and Renault (1989). Let the AR and MA polynomials of one of the implied ARIMA models be, respectively,

$$\phi(L) = 1 + \phi_1 B + \cdots + \phi_p B^p, \quad \theta(L) = 1 + \theta_1 B + \cdots + \theta_q B^q,$$

and assume there are r common roots shared by the two polynomials. Then, there exists a polynomial $\lambda(L)$, of order r, formed by the product of all the common roots, such that

$$\phi(L) = \lambda(L)\alpha(L), \quad \theta(L) = \lambda(L)\beta(L).$$

Let $\alpha(L) = 1 + \alpha_1 L + \cdots + \alpha_a \beta^a$, $a = p - r$, and $\beta(L) = 1 + \beta_1 L + \cdots + \beta_b \mathrm{B}^b$, $b = q - r$. Removing $\lambda(L)$ from the previous two equations yields

$$\phi(L)\beta(L) = \theta(L)\alpha(L), \tag{7.1}$$

an identity between two polynomials of order $k = p + q - r$. Denote by ψ the vector of coefficients of the implied ARMA, and by δ the vector of coefficients after the common polynomial $\lambda(L)$ has been removed, i.e.

$$\psi = [\phi_1, \ldots, \phi_p, \theta_1, \ldots, \theta_q]', \tag{7.2}$$

$$\delta = [\alpha_1, \ldots, \alpha_a, \beta_1, \ldots, \beta_b]'. \tag{7.3}$$

Table 12.10 *Test for common roots*

	Order of implied ARMA	Number of common roots	Order of simplified ARMA
d_t	(22, 19)	18	(4, 1)
y_t	(22, 24)	22	(0, 2)
p_t	(22, 17)	17	(5, 0)
r_t	(22, 20)	20	(2, 0)
n_t	(22, 21)	21	(1, 0)
m_t	(22, 21)	19	(3, 2)

Equating the coefficients of L^j $(j = 1, \ldots, k)$ in (7.1) yields a system of k equations. Conditional on ψ, the system is linear in δ and can be written as

$$h = H\delta, \tag{7.4}$$

where h is a k-dimensional vector and H a $k \times (k - r)$ matrix. Conditional on δ, the system is linear in ψ and can be expressed as

$$e = E\psi,$$

where e is a k-dimensional vector and E a $k \times (p + q)$ matrix. The test consists of the following procedure:

(1) Run OLS on (7.4) to obtain $\hat{\delta}$, and with this estimator construct the matrix $\hat{E}$. Compute, then, the $(k \times k)$ matrix $\xi = \hat{E}\,\Sigma\,\hat{E}'$, where Σ denotes the covariance matrix of the estimators of the parameters in ψ.

(2) Run GLS on (7.3), using ξ as the covariance matrix of the error term, and denote by SSR the sum of squares of the residuals in this regression. For the test consisting of

$$H_0 : \phi(L) \text{ and } \theta(L) \text{ have exactly } r \text{ common roots},$$
$$H_A : \phi(L) \text{ and } \theta(L) \text{ have at most } r \text{ common roots},$$

the statistics $(T \times SSR)$ is distributed as a χ^2 variable with r degrees of freedom. In order to proceed sequentially, we start with $r = \min(p, q)$, i.e. with r equal to its maximum possible value. If H_0 is rejected, we then set $r' = r - 1$ and redo the test, until H_0 is not rejected.

Implementation of the test requires computation of the matrices H, h, E, e, and Σ. A simple procedure is described in appendix C. Table 12.10 presents the results from the test (for a 5 percent size), and it is seen that the orders of the ARMA models obtained after removal of the common

Table 12.11 *Implied ARMA models after cancellation of roots, results from the test*

	d_t	y_t	p_t	r_t	n_t	m_t
AR coefficients						
Lag 1	0.03	–	0.16	0.31	−0.12	0.91
Lag 2	0.14	–	0.28	0.09	–	−0.33
Lag 3	−0.01	–	−0.02	–	–	0.45
Lag 4	0.40	–	−0.01	–	–	–
Lag 5	–	–	0.30	–	–	–
MA coefficients						
Lag 1	0.02	−0.03	–	–	–	−0.79
Lag 2	–	−0.00	–	–	–	−0.17

roots are much smaller than the ones obtained with the ad hoc (restrictive) criterion of section 6. The coefficients of the ARMA model after removal of the common roots are the elements of δ, consistently estimated when running the test; they are displayed in table 12.11. Ignoring small parameters, table 12.11 yields the following models:

$$(1 - 0.40L^4)\nabla \log d_t = u_{1t}, \tag{7.5a}$$

$$\nabla \log y_t = u_{2t}, \tag{7.5b}$$

$$(1 - 0.16L - 0.28L^2 - 0.30L^5)\nabla \log p_t = u_{3t}, \tag{7.5c}$$

$$(1 - 0.31L)\nabla \log r_t = u_{4t}, \tag{7.5d}$$

$$\nabla n_t = u_{5t}, \tag{7.5e}$$

$$(1 - 0.91L + 0.33L^2 - 0.45L^3)\nabla \log m_t = (1 + 0.79L + 0.17L^2)u_{6t}. \tag{7.5f}$$

The first model is similar to model (3.1a), since the AR representation of the latter is, approximately, $(1 - 0.42\ L^4 - 0.13\ L^8)\nabla \log\ d_t = u_{1t}$. Model (7.5b) is the same as model (3.1b), and models (7.5d) and (3.1d) are practically identical. As for the series p_t, model (7.5c) is very close to the implied ARIMA model obtained with the ad hoc criterion (i.e. model (6.1c)), which was seen to be a rough approximation to (3.1c). For the first four series, thus, the test gives results that are in close agreement with the results of direct univariate analysis and with the implied univariate models obtained with the ad hoc criterion.

For the last two series, however, models (7.5e) and (7.5f) are markedly different from models (3.1e) and (3.1f). In both cases it happens that significant coefficients at seasonal lags are missing. This is due to the fact that the test yields a value of r which is too large, so that the AR and

MA polynomials in the simplified ARMA models are not long enough to reach the seasonal lags. Setting $r = 18$ for n_t and $r = 17$ for m_t, so as to allow for seasonal coefficients, the model for n_t can be expressed (once small coefficients have been removed) as

$$\nabla n_t = (1 - 0.40L^4)u_{5t},$$

and that for m_t as

$$(1 + 0.15L - 0.53L^4)\nabla \log m_t = u_{6t}.$$

These two models are now very similar to model (3.1e) and (3.1f). Therefore, our example shows that, when using the Gourieroux–Monfort–Renault test for cancelling common roots, care should be taken with seasonal models. Blind application of the test may over-estimate the value of r, with the consequence that seasonal coefficients may be left out from the derived model. Once this fact is taken into account, the test is seen to perform surprisingly well. In summary, it seems safe to conclude that the ARMA models obtained from univariate analysis are quite in agreement with the univariate models derived from the VAR. This result is true whether the comparison is made with an ad hoc criterion or with a testing procedure.

8 An economic application

The comparison of the VAR model with the ARIMA models estimated with univariate techniques has shown how the results obtained in the latter can be reasonably explained by the VAR. Be that as it may, since the comparison implies cancelling many roots and removing many small coefficients, it is of interest to see how, when those models are used in economic applications, inferences may be affected by the type of model used.

The application we chose is related to the effort by macroeconomists at explaining the permanent changes in aggregate output, as well as the fluctuations around this "permanent component." From an early period when the permanent component (or trend) of the series was assumed deterministic, economists have moved towards modeling trends as stochastic components. When a variable contains a stochastic trend, a shock in the series will not only affect the so-called cyclical component, but will also have an impact on the permanent one. The measurement of this long-term effect (or "persistence") of shocks has been the subject of attention by macroeconomists. In a univariate world, for I(1) series with Wold representation,

$$\nabla x_t = \psi(B)u_t, \tag{8.1}$$

the impact of a shock u_t on x_{t+k} is given by $(1 + \psi_1 + \cdots + \psi_k)u_t$. Following Campbell and Mankiw (1987), the persistence of a standardized shock $u_t = 1$ can be defined as its very long-run impact on the series or, more formally, as

$$m = \lim_{k \to \infty} \sum_{t=0}^{k} \psi_k = \psi(1).$$

There has been considerable interest in estimating persistence, in particular for the case of aggregate output, where different values of $\psi(1)$ have been assigned to different theories of the business cycle. If $\psi(1) > 1$, "real factors," typically associated with supply (such as changes in productivity), would account for both economic growth and most of the business cycle. On the contrary, if $\psi(1) < 1$, the business cycle would be more likely to be associated with transitory (typically demand) shocks; see, for example, the discussion in Lippi and Reichlin (1991).

Of the several approaches to the estimation of persistence, we shall select three that are relevant to our example. First, following Campbell and Mankiw (1987), $\psi(1)$ can be obtained from the univariate ARIMA estimation of (8.1) using Box–Jenkins methods. Second, since additional variables may provide information in explaining deviations of a variable with respect to its trend level, Evans (1989) computes the measure using the parameters from a VAR estimation. Specifically, he proposes to use $\psi(1)$ in the univariate ARIMA model implied by the VAR one. These two measures are based, in theory, on the same set of univariate innovations. Moreover, since the ARIMA models implied by the VAR should be in agreement with the ACF of the series, and this function is the basic identification tool in univariate analysis, the two measures of persistence should not be too distant. Discrepancies between them would be likely to indicate misspecification in some of the models.

Evans finds, however, that persistence of GNP, measured with the ARIMA model implied by his VAR model, is considerably different from the measures obtained by Campbell and Mankiw with univariate analysis. In order to see whether this discrepancy flags some problem with the model specification, we re-estimated the bivariate VAR model of Evans (who kindly supplied us with the data). The equation for GNP is given by

$$\begin{aligned} y_t = & \underset{(0.38)}{-0.62} + \underset{(0.11)}{0.13}\, y_{t-1} + \underset{(0.11)}{0.18}\, y_{t-2} + \underset{(0.10)}{0.02}\, y_{t-3} - \underset{(0.33)}{0.48}\, x_{t-1} \\ & + \underset{(0.50)}{1.32}\, x_{t-2} - \underset{(0.33)}{0.59}\, x_{t-3} - \underset{(0.26)}{0.85}\, d_t + a_t, \end{aligned} \tag{8.2}$$

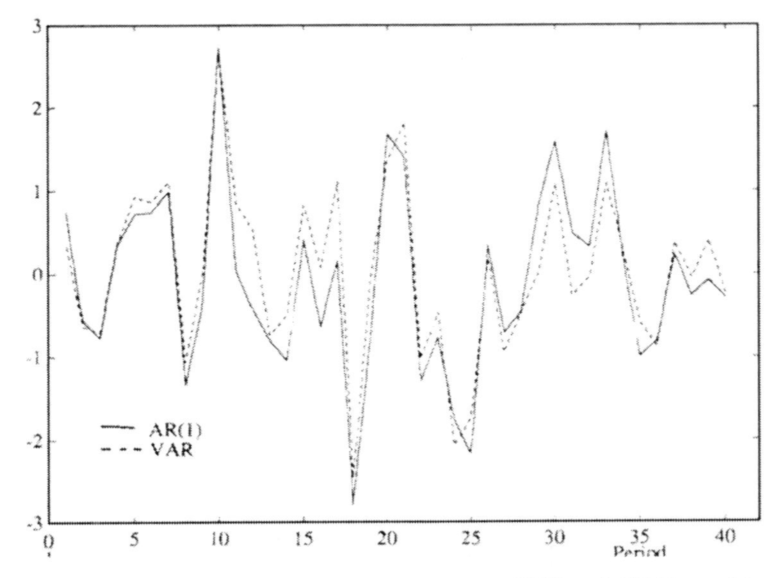

Figure 12.4 One-period-ahead forecast errors: VAR and AR(1) model

where $y_t = \nabla \log GNP$, x_t is the unemployment rate, and d_t a step dummy variable capturing a structural break; the numbers in parentheses denote standard errors.

To judge the validity of the equation it is not possible to perform a proper out-of-sample forecast exercise because the series y_t has been subsequently revised (partly because of revisions in seasonal factors). We split the sample period used by Evans into two subperiods, one with the first 100 observations and the other with the last forty observations. His VAR model was re-estimated for the first subperiod, and one-period-ahead forecasts were computed for the second subperiod. Figure 12.4 compares the associated one-period-ahead forecast errors with those obtained with a simple "AR(1) + constant" structure (with no structural break), estimated also for the first 100 observations. The two series of errors are very close, and the large number of parameters in (8.2) does not improve upon the naive AR(1) specification. The equation is overly parameterized, and this is reflected in the large standard errors of the parameter estimates. The difference between the VAR and univariate measures of persistence thus does not seem the result of a more efficient multivariate estimation; on the contrary, the VAR model obtained seems an unreliable tool for inference.

Table 12.12 *Measures of persistence*[a]

	Estimated ARIMA	Implied ARIMA	VAR
d_t	2.61 (0.80)	2.57	2.95 (0.77)
y_t	1 (0.00)	1.10	1.69 (0.28)
p_t	5.01 (1.89)	3.80	4.56 (1.35)
r_t	1.54 (0.16)	1.61	1.75 (0.19)
n_t	0.53 (0.10)	0.78	1.01 (0.15)
m_t	1.45 (0.31)	1.18	1.24 (0.19)

Note:
[a] The standard errors of the estimators (computed using linear approximations) are provided in parentheses.

Finally, Pesaran, Pierse, and Lee (1993) suggest a multivariate measure of persistence, with the innovations defined with respect to the multivariate information set. In the univariate case, if $g(\omega)$ denotes the spectrum of ∇x_t in (8.2), using a well-known result, $g(0) = \psi(1)^2\sigma_u^2$. The multivariate extension of this result, for the case of the VAR model given by (2.1), is

$$g(0) = [\Phi(1)^{-1}]\Omega[\Phi(1)^{-1}]'.$$

The measure of persistence proposed by Pesaran, Pierse, and Lee is given by the squared root of the elements of the main diagonal of this matrix, standardized by the variance of the appropriate multivariate innovation. (For a vector with only one variable, the multivariate measure becomes the univariate one.)

Table 12.12 presents the three measures of persistence for the six variables we consider. The first two measures refer to the response to the univariate innovation, which is a function of all the innovations of the multivariate model, as shown in expression (2.4). Therefore, the two measures are not strictly comparable to that obtained with the VAR model, which reflects the response to the innovation defined in a multivariate information set. It is seen, however, that, given the precision of the measurements, for four variables d_t, p_t, r_t, and m_t, the three measures are reasonably close. A unit innovation in public debt has a large permanent effect on the level of debt, and a similar result is obtained for the price level variable. In this later case, the discrepancy between the two ARIMA measures may reflect the limitations of the VAR model in capturing a series with a relatively large MA root, as mentioned in section 7. For the interest rate and the monetary aggregate series the persistence is slightly larger than 1, although for m_t it could be easily accepted as equal to 1.

For the series y_t and n_t the univariate and multivariate results are more distant. For the GDP series the univariate measure of persistence is 1, while the multivariate measure is 1.7 and, considering the standard errors, they cannot be accepted as equal. According to the interpretation mentioned above, this could be seen as evidence that, when the innovation is cleaned of the effects due to other correlated shocks (i.e. when the information set is enlarged), the real business cycle theory gains support. For the balance of trade series, the univariate measure is below 1, while the multivariate measure is 1. An economic interpretation of the persistence measures is beyond the scope of this chapter. Relevant to our discussion are the following two results:

(a) The proximity of the measures of persistence between the estimated and implied ARIMA models shows that inferences drawn from the VAR (concerning persistence) explain well the ones obtained from univariate analysis. Altogether, it is somewhat striking that the measurement is not more affected by the numerous cancellations of roots and removal of coefficients.

(b) The univariate measure of persistence may be a reasonable approximation to the persistence measured in a wider information set. But there are cases when this is clearly not true.

9 Summary and conclusions

It is well known that a linear dynamic structural econometric model has a reduced form with a multivariate linear time series model expression, which in turn implies univariate ARIMA models for each of the series. An important way to evaluate a structural econometric model, thus, is by checking for whether it encompasses the appropriate VAR model. Since the univariate models implied by VAR models have ARIMA expressions, in a similar manner, an important way to evaluate a VAR model is to see if the results obtained with univariate analysis can be explained by the VAR, i.e. if the ARIMA models implied by the VAR are close to the ones found in univariate analysis. Since identification of univariate models is easier than identification of (not too small) VAR models, if an implied ARIMA model is substantially different from the ARIMA model that fits the unvariate series, the difference may well reflect misspecification of the multivariate model (an example is provided in section 8).

Although the idea is simple, it is rarely put into practice. This may be partly due to the fear that the comparison may be worthless because of the so-called autoregressivity paradox: while the ARIMA models from univariate analysis typically have very few parameters, the implied ARIMA models, even for relatively small VAR models, have a very large number of parameters. Can we reasonably expect to bring, for example, a forty-five-parameter ARIMA model down to a one- or two-parameter one? More generally, can we expect univariate models to be useful as diagnostic tools for VAR models?

The question is general, but the answer is ultimately empirical. Thus we consider a particular application: A VAR model for six quarterly macroeconomic variables. First, using univariate analysis, ARIMA models are fitted to each one of the series. Not counting the innovation variances, all models have at most two parameters. Then, after testing for cointegration, a parsimonious VAR model is estimated; the model is a slight modification of the one used by some French economists to analyze the effect of public debt on several macroeconomic variables. It is seen that both the set of univariate ARIMA models and the VAR model provide good fits and perform reasonably well in out-of-sample forecasting.

Next, the univariate models implied by the VAR are derived (following a procedure described in appendix A). All have an AR polynomial of order 22, and the orders of the MA polynomials vary between 18 and 24. The application considered provides thus a good example for the autoregressivity paradox ignoring the innovation variances, the average number of parameters is forty-two for the implied ARIMA models and 1.3 for the ones estimated in univariate analysis.

In order to compare the two types of models, the roots of the common AR and of the six MA polynomials of the implied models are computed (a total of 144 roots). To determine which ones should cancel out, two approaches are followed. First, we use a simple ad hoc criterion (discussed in appendix C), biased towards under-cancellation. Once the common roots are removed, careful analysis of the simplified models shows that the ARIMA models from univariate estimation are remarkably close to the ones implied by the VAR. The comparison also evidences the gain from multivariate modelling for some variables (in particular, *GDP*) and, in the case of the price variable, the difficulties of the VAR specifications in handling series with a large moving average root. Second, a formal test is applied to determine the roots that could be cancelled. The test is seen to be biased towards over-cancellation, in particular when the series contains seasonality. Careful application of the test, however, yields finally implied ARIMA models that are in agreement with those obtained with the ad hoc criterion and with univariate analysis.

In summary, the VAR model explains reasonably well the results from univariate analysis and passes, thus the encompassing test. All considered, it seems safe to conclude that the improvement obtained with the multivariate model is not very large, but that it can be properly attributed to having captured some relationships among the macroeconomic variables.

Although the differences between the implied and estimated ARIMA models are relatively small, it is still of interest to see what effect they may have when the models are used for economic inference. As an example, we consider the problem of measuring the so-called persistence, or long-term effect, of shocks on macroeconomic variables. Persistence has been estimated in different ways, three of which are relevant to our discussion: First, it has been measured using ARIMA models from univariate analysis. Second, it has been measured using implied ARIMA models (derived from VAR ones). Third, we consider a multivariate measure based directly on the VAR model.

The three measures of persistence are computed for the six series. The first two measures are close, and hence the VAR model again explains well the inference obtained in univariate analysis. The comparison also shows how, although for some variables the inference based on univariate analysis may approximate the one based on multivariate models, on occasion, it can be misleading. This is clearly the case for the GDP variable.

APPENDIX A UNIVARIATE ARIMA MODELS IMPLIED BY A VECTOR AUTOREGRESSIVE MODEL

As seen in section 2, the univariate model for the *i*th series implied by the multivariate VAR is given by (2.5), where $\phi(L)$ is straightforward to obtain through (2.6) and the moving average part $\theta_i(L)u_{it}$ satisfies (2.4). We proceed to summarize a procedure (easy to implement in most available softwares) to obtain $\theta_i(L)$ and the variance of u_{it}, σ_i^2.

The adjoint matrix $\Phi^*(L)$ is directly obtained from $\Phi(L)$, and hence the elements $\Phi_{ij}^*(L)$ and the matrix Ω (with the contemporaneous covariances of the vector a_t) are assumed known. Let q denote the order of the polynomial $\theta_t(L)$. The autocovariances of the r.h.s. of (4), say γ_0, $\gamma_1, \ldots, \gamma_q$, can be obtained through the Autocovariance Generating Function(ACGF),

$$\gamma_i(L) = f_t(L)\Omega f_t(L^{-1})',$$

where $f_t(L)$ is the *i*th row of $\Phi^*(L)$. The ACGF $\gamma_i(L)$ is an expression of the type

$$\gamma_i(L) = \gamma_0 + \sum_{t=1}^{q} \gamma_i(L^i + L^{-i}), \tag{A.1}$$

and our aim is to find the moving average process $\theta_i(L)u_{it}$ that generates this set of autocovariances. We proceed as follows (for notational simplicity the subscript i is dropped). Write (A.1) as

$$\gamma(L) = L^{-q}(\gamma_q + \cdots + \gamma_0 L^q + \cdots + \gamma_q L^{2q}) = L^{-q}\Gamma(L).$$

Since the polynomial $\Gamma(L)$ is symmetric around L^q, the $2q$ roots of the equation $\Gamma(z) = 0$ can be expressed as the sets $(r_1, \ldots, r_q)$ and $r_1^{-1}, \ldots, r_q^{-1})$, with $|r_i| \geq 1 \geq |r_i^{-1}|$, $i = 1, \ldots, q$. In practice, however, there is no need to compute the $2q$ roots of $\Gamma(z)$. Using the transformation $y = z + 1/z$, the polynomial $\Gamma(z)$ is transformed into a polynomial in y of order q, say,

$$A(y) = a_0 + a_1 y + \cdots + a_q y^q, \tag{A.2}$$

where the vector of coefficients $a = (a_0, a_1, \ldots a_q)'$ is obtained as follows. Let

$$\begin{aligned} b_0 &= 2, \\ b_1 &= (0, 1), \\ b_j &= (0, b_{j-1}) - (b_{j-2}, 0, 0), \quad h = 2, \ldots, q, \end{aligned}$$

and build the $(q + 1) \times (q + 1)$ matrix $S = [s_1, \ldots, s_{q+1}]$, with the columns given by

$$\begin{aligned} s_1 &= (1, O_q)', \\ s_j &= (b_{j-1}, O_{q-j+1})', \quad j = 2, \ldots, q, \\ s_{q+1} &= b_q', \end{aligned}$$

where O_k denotes a k-dimensional row vector of zeros. Then, $a = S\gamma$, where $\gamma = (\gamma_0, \gamma_1, \ldots, \gamma_q)'$. Let $y_1, \ldots, y_q$ denote the q roots of (A.2). In each of the equations

$$z^2 - y_j z + 1 = 0, \quad j = 1, \ldots, q, \tag{A.3}$$

selecting the root z_j such that $|z_j| \geq 1$, the polynomial $\theta(L)$ is found through

$$\theta(L) = (1 - z_1 L) \ldots (1 - z_q L), \tag{A.4}$$

and σ_i^2 can be obtained from $\sigma_i^2 = \gamma_0(1 + \sum_{i=1}^q \theta_i^2)^{-1}$.

Notice that the coefficient y_j in (A.3) can be complex. In this case, the solution is found in the following way: Let $y_j = a + bi$, and define $k = a^2 - b^2 - 4$ $m = 2ab$, and $h^2 = [(|k| + (k^2 + m^2)^{1/2})/2]$. Then, if $z_j = z_j^r + z_j^i i$ is a solution of (A.3), its real and imaginary parts are given by

$$z_j^r = (-a \pm c)/2, \quad z_j^i = (-b \pm d)/2,$$

where, when $k \geq 0$, $c = h$, $d = m/2h$, and when $k < 0$, $d = [\text{sign}(m)]\ h$, $c = m/2d$.

The derivation of $\theta_i(B)u_{it}$ is valid for invertible as well as non-invertible moving averages. (In the latter case, the unit root would appear twice in $\Gamma(L)^1$.) But, as we proceed to show, the moving average part of the implied ARIMA will always be invertible, thus $|z_i| > 1$ for $i = 1, \ldots, q$.

A univariate finite order autoregressive model, by construction, is invertible. But, as seen in section 2, the univariate models implied by multivariate VAR ones are not finite autoregressive models, but full ARMAs, where the moving average part can be long and complex. There is thus the question of whether, for some values of the ϕ-parameters in the AR matrix, the MA part of an implied univariate model may include a unit root.

Consider the VAR model given by (2.1). We have seen in (A.1) that, in the factorization of $\gamma(L)$, we can always choose $\theta_i(L)$ so as to have all roots on or outside the unit circle. Thus we have only to prove that no root of $\theta_i(L)$ will be on the unit circle. If $\theta_i(L)$ has a unit root, this implies a 0 in the spectrum for an associated frequency. If the spectrum of the l.h.s. of (2.4) has a 0, all components in the r.h.s. of (2.4) have a spectrum with a 0 for that particular frequency (see Teräsvirta 1977), and hence the polynomials $\Phi^*_{ij}(L)$, $j = 1, \ldots, k$, will share the same unit root. Considering the expansion of the determinant of $\Phi(L)$ by the elements of the ith row:

$$|\Phi(L)| = \sum_{j=1}^{k} \Phi_{ij}(L)\Phi^*_{ij}(L),$$

and factorizing the unit root common to $\Phi^*_{i1}(L), \ldots, \Phi^*_{ik}(L)$, the same unit root will have to appear in $|\Phi(L)|$. The root would thus be present in the AR and MA polynomials of the implied ARIMA model, and hence it would cancel out. It follows that the univariate models implied by the VAR model are always invertible.

APPENDIX B A COMMENT ON THE PRECISION OF THE FREQUENCY AND MODULUS OF THE ROOTS IN AN ESTIMATED AUTOREGRESSIVE MODEL

When using models with AR expressions, it is often of interest to look at the roots of the AR polynomials, where the roots are expressed in terms of the frequency ω and the modulus h. Since ω and h are computed as functions of the AR parameters, it is important to know how errors in the estimators of the latter induce imprecision in the measurements of

ω and h. In our case the interest is due to the need to select a criterion to determine when two roots can be safely assumed to be close enough for cancellation. Since our comparison involves 144 roots, where the modulus and frequency of each one are non-linear functions of the forty-one parameters in the matrices $\Phi(L)$ and Ω, we seek a simple ad hoc criterion, such that only roots that are clearly close in a probabilistic sense will be cancelled. In order to do that, we consider the case of an AR(2) model for series of the same length as ours ($T = 84$). One could expect perhaps a precision somewhat similar to that of our VAR model, with 3.3 parameters per equation: the slight gain from the multivariate fit could compensate for the small increase in the number of parameters.

Let the AR(2) model be given by

$$z_t - \phi_1 z_{t-1} - \phi_2 z_{t-2} = a_t. \tag{B.1}$$

Expressing the roots of $x^2 - \phi_1 x - \phi_2 = 0$ in terms of frequency and modulus, for $0 < \omega < \pi$ (i.e. when the roots are not real), it is obtained that

$$h = \sqrt{-\phi_2}, \quad \omega = \text{a} \cos \frac{\phi_1}{2\sqrt{-\phi_2}}. \tag{B.2}$$

It is then possible to approximate the functions that relate the estimation errors in h and ω (to be denoted δ_h and δ_ω) to the estimation errors in ϕ_1 and ϕ_2 (denoted ε_1 and ε_2, respectively). Since, for the relevant range $0 > \phi_2 > -1$ and $0 < \omega < \Pi$, the functions given by (B.2) are continuous in ϕ_1 and ϕ_2, it is straightforward to obtain the linear approximation that relates $\delta = (\delta_h, \delta_\omega)$ to $\varepsilon = (\varepsilon_1, \varepsilon_2)$. The estimators of h and ω are consistent (becoming superconsistent when $h = 1$), and the asymptotic covariance matrix of δ, V_δ, can be linearly approximated by

$$V_\delta = DV_\varepsilon D', \tag{B.3}$$

where V_ε is the asymptotic covariance matrix of ε, equal to

$$V_\varepsilon = \frac{1-h^4}{T}\begin{bmatrix} 1 & \rho \\ \rho & 1 \end{bmatrix}, \quad \rho = -\frac{2h\cos\omega}{1+h^2},$$

and D is the matrix of derivatives

$$D = \left(\frac{\partial \delta}{\partial \varepsilon}\right) = \begin{bmatrix} 0 & -(2h)^{-1} \\ -(2h\sin\omega)^{-1} & -(2h^2\tan\omega)^{-1} \end{bmatrix}.$$

In our example, the vast majority of the roots have modulus in the range 0.6 to 0.9. For $h = 0.6, 0.75, 0.9$, table 12A.1 presents the standard deviations of the estimation errors for h and ϕ_2, obtained with the asymptotic approximation (they do not depend on ω). The larger the modulus, the

Table 12A.1 *Modulus estimator for a complex root in an AR(2) model, asymptotic results ($T = 84$)*

Modulus h	S.E. of $\hat{h}$	AR coeff. ϕ_2	S.E. of $\hat{\phi}_2$
0.600	0.085	0.360	0.102
0.750	0.060	0.562	0.090
0.900	0.036	0.810	0.064

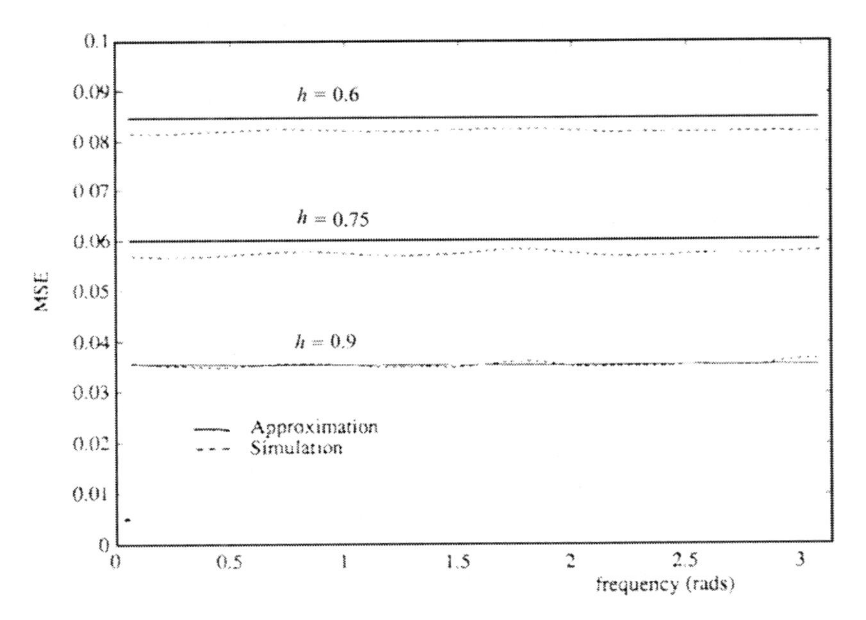

Figure 12A.1 MSE: modulus estimator

smaller the estimation error becomes for both parameters. For the values in table 12A.1, despite its larger numerical value (in absolute terms), the modulus is estimated with more precision than ϕ_2.

In order to assess the accuracy of the linear approximation, 10,000 simulations of eighty-four observations each from model (B.1) were made, for the three values of h in table 12A.1 and fifty partitions of the interval $\omega \in (0, \Pi)$. In each case, the AR coefficients were estimated, and the estimators of h and ω were obtained through (B.2). For the pairs (h, ω), figures 12A.1 and 12A.2 compare the standard errors of the modulus and frequency estimators, respectively, obtained with the simulation and

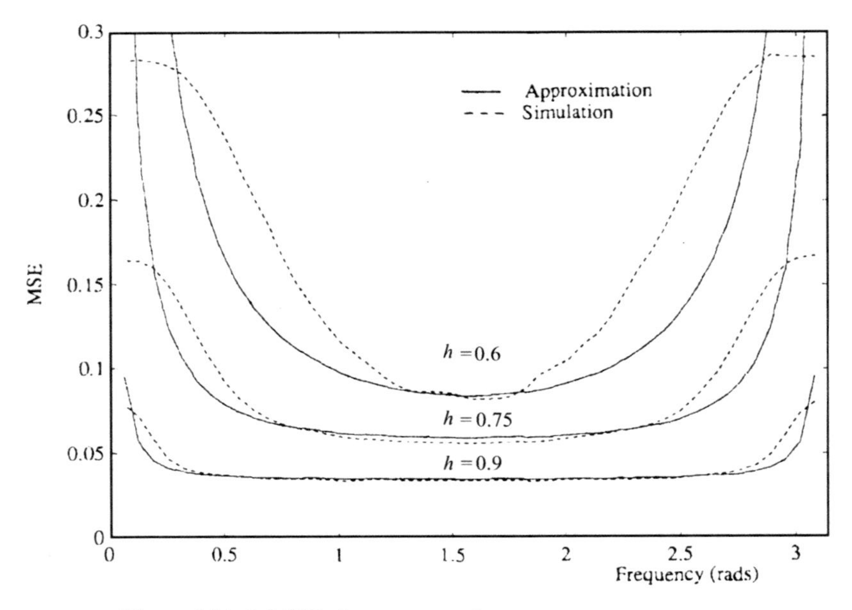

Figure 12A.2 MSE: frequency estimator

with the linear approximation. For δ_m, the approximation works reasonably well; for δ_ω, except for relatively large modulus, the approximation is less reliable. From . . . figures [12A.1 and 12A.2] it is seen that, for complex roots with values of h between 0.6 and 0.9 (our range of concern), the standard error of δ_h varies between 0.03 and 0.09, while that of δ_ω varies between 0.04 and 0.3. Considering the positive correlations between the two errors for low values of h and ω (figure 12A.3), we adopted the following simple criterion: for the two roots (h_1, ω_1) and (h_2, ω_2) to cancel, we require that the differences $h_1 - h_2$ and $\omega_1 - \omega_2$ be smaller than 0.05 (in absolute value). We require further that the sum of the two absolute differences be smaller than 0.07.

The criterion seems safe in the following sense: Consider a pair of cancelled roots in one of the implied ARIMA models, and let δ and d denote the errors in the estimators of (h, ω) in the AR and MA roots, respectively. Assume δ is distributed normally, with zero mean vector and covariance matrix (B.3), and that we wish to test $\delta = d$. For all roots actually cancelled, the p-value of the test would be smaller than 0.5. In this way, the criterion will tend towards under-cancellation, and will avoid cancelling roots measured with imprecision.

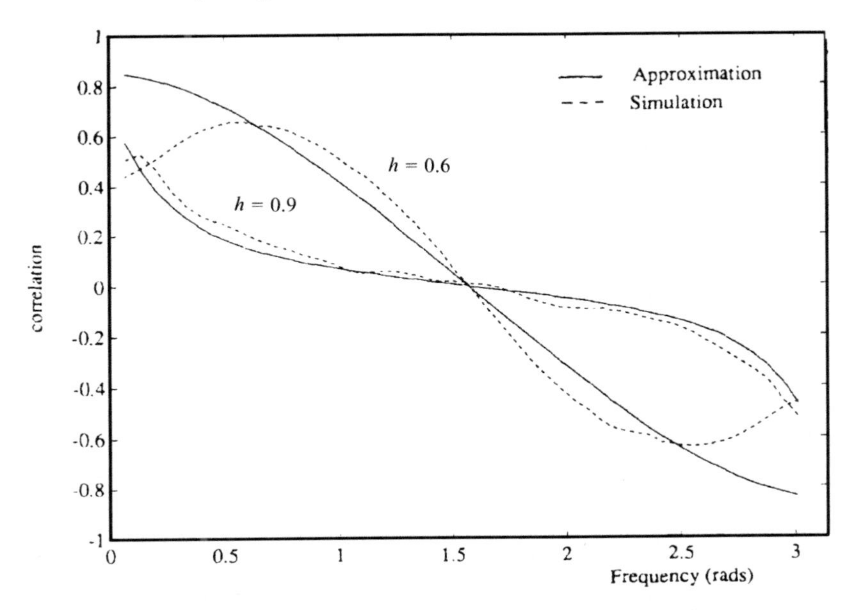

Figure 12A.3 Correlation: Modulus and frequency estimators

APPENDIX C COMMON ROOTS TEST: COMPUTATION OF THE MATRICES

To carry out the test described in section 7, the matrices H, h, E, e, and Σ need to be computed. For the first four matrices, this can easily be done in the following way: Let O_j denote a column vector of j zeros, and define the vector $c = (c_1, \ldots, c_d)'$, the $(m \times n)$ matrix

$$A(c) = \begin{bmatrix} 1 & 0 & \ldots & 0 \\ c_1 & 1 & \ldots & 0 \\ \vdots & \vdots & & \vdots \\ \vdots & \vdots & & 1 \\ c_d & \vdots & & c_1 \\ 0 & c_d & & \vdots \\ \vdots & \vdots & & \vdots \\ 0 & 0 & \ldots & c_d \end{bmatrix},$$

with $n < d < m$, and the m-dimensional vector $b(c) = (1, c', O'_{m-d-1})'$. Then,

$$h = b(\phi) - b(\theta), \qquad e = b(\alpha) - b(\beta),$$

and the matrices H and E can be obtained through

$$H = [H_1 \vdots - H_2], \quad E = [-E_1 \vdots E_2],$$

where H_1, H_2, E_1, and E_2 are given by

$$\begin{aligned} H_1 &= A(\theta), \quad n = p - r, \\ H_2 &= A(\phi), \quad n = q - r, \\ E_1 &= A(\beta), \quad n = p, \\ E_2 &= A(\alpha), \quad n = q, \end{aligned}$$

and $m = p + q - r$ in all cases.

Finally, we need an estimator of $\Sigma = \text{cov}(\psi)$, where ψ contains the parameters of the implied univariate model. These parameters are functions of the VAR model parameter estimates, as indicated by (2.4) and (2.6). The VAR model parameters are the AR coefficients $\Phi(L)$ and the elements of Ω, the residual error covariance matrix. Let Φ denote the vector of AR coefficient estimators, and σ the vector containing the estimators of the elements in Ω. (In order to simplify notation, we delete the symbol "$\wedge$" to denote an estimator.) Then, a linear approximation to Σ yields

$$\Sigma \doteq \mathcal{J} M \mathcal{J}',$$

where

$$\mathcal{J} = \left[\frac{\partial \psi}{\partial \Phi} \vdots \frac{\partial \psi}{\partial \sigma}\right], \quad M = \text{cov}(\Phi, \sigma)' = \begin{bmatrix} M_\Phi & M_{\Phi\sigma} \\ M_{\sigma\Phi} & M_\sigma \end{bmatrix}.$$

The derivatives in $\mathcal{J}$ have been computed numerically. As for the matrix M, the first submatrix $M_\Phi = \text{cov}(\Phi)$ is available from the VAR estimation results; also, asymptotically, $M_{\Phi\sigma} = M_{\sigma\Phi} = 0$. In order to obtain $M_\sigma = \text{cov}(\sigma)$, its elements are expressions of the form $\text{cov}(\sigma_{ij}\sigma_{kh})$, where $\Omega = (\sigma_{ij})$, and

$$\sigma_{ij} = T^{-1} \sum_t a_{it} a_{jt}.$$

Since the vector $a_t \sim N_6(0, \Omega)$, from its moment-generating function, it is straightforward to find that, for all values of i, j, k, h,

$$\text{cov}(\sigma_{ij}\sigma_{kh}) \doteq (\sigma_{ik}\sigma_{jh} + \sigma_{jk}\sigma_{ih})/T,$$

and hence M_σ can be easily computed.

BIBLIOGRAPHY

Anderson, T. W. (1971), *The Statistical Analysis of Time Series* (New York, John Wiley)

Box, G. E. P. and G. M. Jenkins (1970), *Time Series Analysis: Forecasting and Control* (San Francisco, Holden-Day)

Caines, P. E., C. W. Keng, and S. P. Sethi (1981), "Causality analysis and multivariate autoregressive modelling with an application to supermarket sales analysis," *Journal of Economic Dynamics and Control* 3, 267–98

Campbell, J. Y. and N. G. Mankiw (1987), "Permanent and transitory components in macroeconomic fluctuations," *American Economic Review Proceedings*, 111–17

Clements, M. P. and G. E. Mizon (1991), "Empirical analysis of macroeconomic time series: VAR and structural models," *European Economic Review* 35, 887–932

Deniau, C., G. Fiori, and A. Mathis (1989), "Impact de la dette publique sur quelques variables macroeconomiques françaises," *Economie et Prevision* 90, 87–95

Engle, R. F. and C. W. J. Granger (1987), "Cointegration and error correction: representation, estimation and testing," *Econometrica* 55, 251–76

Evans, G. W. (1989), "Output and unemployment dynamics in the United States: 1950–1985," *Journal of Applied Econometrics* 4, 213–37

Gardeazabal, J. and M. Regulez (1990), "Long run exchange rate determination in the light of cointegration theory," University of Pennsylvania, September, mimeo

Goldberger, A. S. (1959), *Impact Multipliers and Dynamic Properties of the Klein–Goldberger Model* (Amsterdam, North-Holland)

Gourieroux, C., A. Monfort, and E. Renault (1989), "Testing for common roots," *Econometrica* 57, 171–85

Hendry, D. F. and G. E. Mizon (1992), "Evaluating dynamic econometric models by encompassing the VAR," in P. C. B. Phillips and V. B. Hall (eds.), *Models, Methods and Applications of Econometrics: Essays in Honor of Rex Bergstrom* (Oxford, Basil Blackwell)

Hsiao, C. (1981), "Autoregressive modelling and money–income causality detection," *Journal of Monetary Economics* 7, 85–106

Jenkins, G. M. (1979), "Practical experiences with modelling and forecasting time series," G. J. P. publ.; also in O. D. Anderson (ed.), *Forecasting* (Amsterdam, North-Holland)

Johansen, S. (1988), "Statistical analysis of cointegration vectors," *Journal of Economic Dynamics and Control* 12, 231–54

Lippi, M. and L. Reichlin (1991), "Permanent and temporary fluctuations in macroeconomics," in N. Thygesen, K. Velupillai, and S. Zambelli (eds.), *Recent Developments in Business Cycle Theory, Proceedings of an IEA Conference* (New York, Macmillan)

MacKinnon, J. G. (1991), "Critical values for cointegration tests," in R. F. Engle and C. W. J. Granger (eds.), *Long-Run Economic Relationships: Readings in Cointegration* (Oxford, Oxford University Press)

Maravall, A. (1981), "A note on identification of multivariate time series models," *Journal of Econometrics* 16, 237–47

Monfort, A. and R. Rabemananjara (1990), "From a VAR model to a structural model, with an application to the wage–price spiral," *Journal of Applied Econometrics* 5, 203–27

Palm, F. C. (1983), "Structural econometric modeling and time series analysis: an integrated approach," in A. Zellner (ed.), *Applied Time Series Analysis of Economic Data* (Washington, DC, US Department of Commerce, Bureau of the Census), 199–233; chapter 3 in this volume

(1986), "Structural econometric modeling and time series analysis," *Applied Mathematics and Computation* 20, 349–64

Pesaran, M. H., R. G. Pierse, and K. C. Lee (1993), "Persistence, cointegration and aggregation: a disaggregated analysis of output fluctuations in the US economy," *Journal of Econometrics* 56, 57–88

Rose, A. K. (1985), "The autoregressivity paradox in macroeconomics," PhD dissertation, Department of Economics, MIT, Cambridge, Mass.

(1986), "Four paradoxes in GNP," *Economics Letters* 22, 137–41

Teräsvirta, T. (1977), "The invertibility of sums of discrete MA and ARMA processes," *Scandinavian Journal of Statistics* 4, 165–70

Tiao, G. C. and G. E. P. Box (1981), "Modeling multiple time series with applications," *Journal of the American Statistical Association* 71, 802–16

Wallis, K. F. (1977), "Multiple time series analysis and the final form of econometric models," *Econometrica* 45, 1481–97

Zellner, A. and F. C. Palm (1974), "Time series analysis and simultaneous equation econmetric models," *Journal of Econometrics* 2, 17–54; chapter 1 in this volume

Zellner, A. and F. C. Palm (1975), "Time series and structural analysis of monetary models of the US economy," *Sankhyā: The Indian Journal of Statistics*, Series C 37, 12–56; chapter 6 in this volume

Part III

Macroeconomic forecasting and modeling

13 Macroeconomic forecasting using pooled international data (1987)

Antonio Garcia-Ferrer, Richard A. Highfield, Franz C. Palm, and Arnold Zellner

1 Introduction

It has long been recognized that national economies are economically interdependent (see, e.g., Burns and Mitchell 1946 for evidence of comovements of business activity in several countries and Zarnowitz 1985 for a summary of recent evidence). Recognition of such interdependence raises the question: Can such interdependence be exploited econometrically to produce improved forecasts of countries' macroeconomic variables such as rates of growth of output, and so forth? This is the problem that we address in this chapter, using annual and quarterly data for a sample of European Economic Community (EEC) countries and the United States.

We recognize that there are several alternative approaches to the problem of obtaining improved international macroeconomic forecasts. First, there is the approach of Project Link that attempts to link together elaborate structural models of national economies in an effort to produce a world structural econometric model. A recent report on this ambitious effort was given by Klein (1985). We refer to this approach as a "top-down" approach, since it uses highly elaborate country models to approach the international forecasting problem. In our work, we report results based on a "bottom-up" approach that involves examining the properties of particular macroeconomic time series variables, building simple forecasting models for them, and appraising the quality of forecasts yielded by them. We regard this as a first step in the process of constructing more elaborate models in the structural econometric modeling time series analysis (SEMTSA) approach described by Palm (1983), Zellner (1979), and Zellner and Palm (1974). Analysis of simple models

This research was financed in part by the National Science Foundation and by income from the H. G. B. Alexander Endowment Fund, Graduate School of Business, University of Chicago. C. Hong provided valuable computational assistance.

Originally published in the *Journal of Business and Economic Statistics* 5(1) (1987), 53–67.

for the rates of growth of output for various countries provides benchmark forecasting performance against which other models' performance can be judged, much in the spirit of Nelson's (1972) work with US data. What is learned in this process can be very helpful in constructing more elaborate models. As Kashyap and Rao (1976, p. 221) remark,

> Since the number of possible classes [or models] in the multivariate case is several orders larger than the corresponding number of univariate classes, it is of utmost importance that we develop a systematic method of determining the possible classes. This is best done by considering the equations for the individual variables $y_1, \ldots, y_m$ separately.

The plan of our chapter, which reports our progress to date, is as follows. In Section 2, we analyze data on annual growth rates of real output for nine countries. The forecasting performance of several naive models is compared with that of more sophisticated models that incorporate leading indicators, common influences, and similarities in models' parameter values across countries in a Bayesian framework. (For some other studies comparing the properties of alternative forecasting procedures, see Makridakis et al. 1982; Harvey and Todd 1983; Meese and Geweke 1984; Zellner 1985; McNees 1986.) In addition, the quality of forecasts so obtained is compared with that of other available forecasts. Section 3 deals with an analysis of quarterly data for six countries. Again the forecasting performance of several naive models is compared with that of several slightly more complex models. Finally, in Section 4 a summary of results is presented and some concluding remarks regarding future research are presented.

2 Analyzes of annual data for nine countries

Annual data, 1951–81, for nine countries' output growth rates, measured as $g_t = \ln(O_t/O_{t-1})$, where O_t is real output (real GNP or real GDP), have been assembled in the main from the IMF's International Financial Statistics data base. (An appendix giving the data is available on request.) The data relate to the following countries: Belgium, Denmark, France, Germany, Ireland, Italy, the Netherlands, the United Kingdom, and the United States. Plots of the basic data are shown in figure 13.1, including data for Spain that have not as yet been analyzed.

Our procedure in analyzing our data was as follows:

1. We used the data, 1954–73, twenty observations to fit each of our models, with the 1951–3 data used for initial lagged values of variables.
2. Then the fitted models were employed to generate eight one-step-ahead forecasts for the years in our forecast period, 1974–81. In

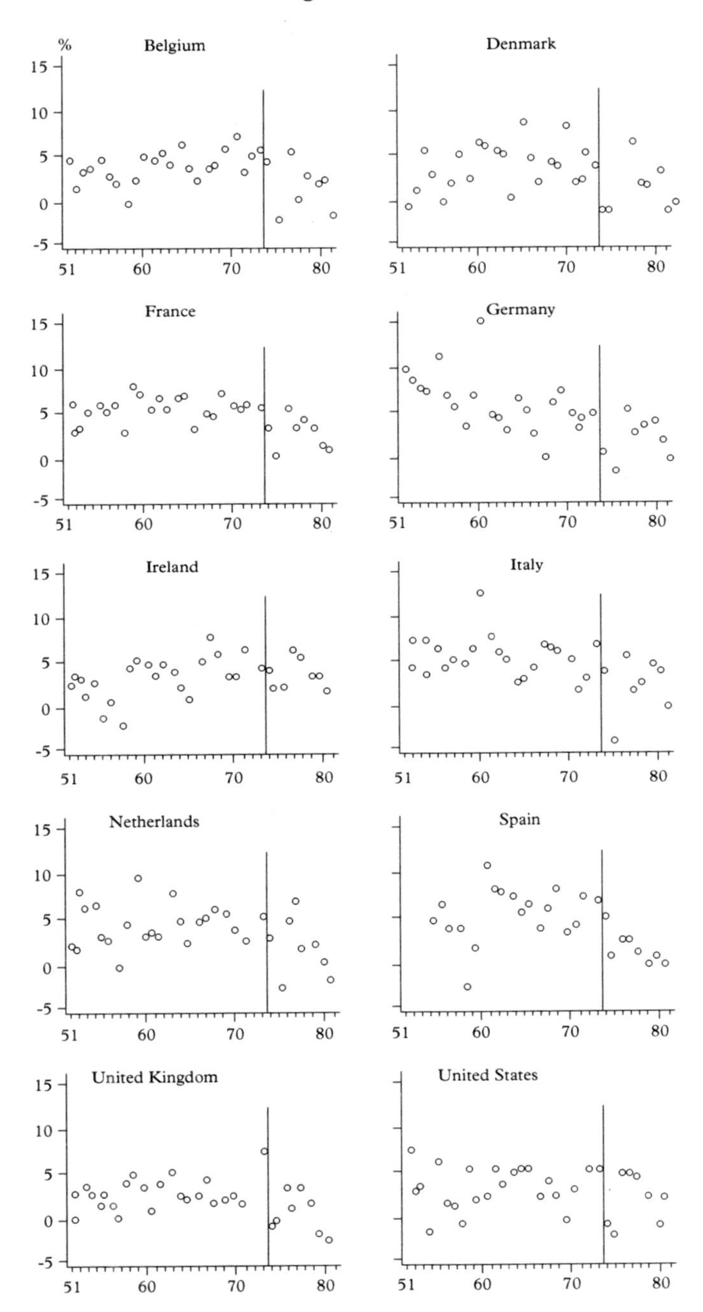

Figure 13.1 Annual growth rates of real output, by country, 1951–1981

making one-step-ahead forecasts, models were re-estimated using all past data prior to each forecast period.

3. Forecast errors were computed for each forecast period and country. The root mean-squared errors (RMSEs) by country and overall measures of forecasting precision have been computed to appraise the forecasting performance of different models.

2.1 *Individual country models*

Years ago, Milton Friedman suggested that the forecasting performance of naive models (NMs) be determined to serve as a benchmark in evaluating the forecasting performance of more complicated models and procedures, a suggestion pursued by Christ (1951) in his evaluation of the forecasting performance of a structural econometric models of the US economy and many others. Here we use the following two NMs to forecast the growth rate g_t:

$$\text{NMI: } \hat{g}_t = 0, \tag{2.1}$$

$$\text{NMII: } \hat{g}_t = g_{t-1}. \tag{2.2}$$

The forecast of NMI, $\hat{g}_t = 0$, is optimal in a mean-squared error (MSE) sense if the logarithm of output, O_t, follows a random walk – that is, $g_t \equiv \ln O_t - \ln O_{t-1} = \varepsilon_{1t}$, where ε_{1t} is a white-noise error term with zero mean. Then $\hat{g}_t = 0$ is the minimal MSE forecast. The NMII forecast is a minimal MSE forecast if g_t follows a random walk, $g_t = g_{t-1} + \varepsilon_{2t}$, where ε_{2t} is a white-noise error term with zero mean. For this latter process, $E(g_t|I_t) = g_{t-1}$ (where I_t denotes values of g_t prior to period t) is a minimal MSE forecast. For each country, the two NM forecasts in (2.1)–(2.2) were calculated for each of eight years in the forecast period 1974–81 with results shown in lines A and B of table 13.1.

From line A of table 13.1, it is seen that NMI's RMSEs in percentage points range from 4.38 for Ireland to 2.21 for the United Kingdom. The median RMSE across countries is 3.09.

In line B of table 13.1, the RMSEs for NMII are shown. These range from 2.06 percentage points for Ireland to 4.88 for Italy, with a median of 3.73. Thus the range of the RMSEs and the median RMSE for NMII's forecasts are both larger than those for NMI. Both NMs' forecast errors are rather large, however, particularly in the vicinity of turning points in the rate of growth of output that occurred in the neighborhoods of 1974–5 and 1979–80 for most countries. Thus possible improvement in forecasting performance relative to that of the NMs' performance might be attained through improved forecasting of turning points.

Table 13.1 *Root mean-squared errors (RMSEs) of one-step-ahead forecasts of annual real output growth rates for nine countries*

Model	Belgium	Denmark	France	Germany	Ireland	Italy	Netherlands	United Kingdom	United States
				Percentage points					
A NMI ($\hat{g}_t = 0$)	3.09	2.83	2.96	2.95	4.38	3.72	3.77	2.21	3.48
B NMII ($\hat{g}_t = g_{t-1}$)	4.25	3.73	2.43	3.26	2.06	4.88	4.04	3.91	3.60
C AR(3)[a]	3.66	3.46	2.89	3.39	1.69	4.75	3.52	3.50	2.48
D AR(3) with leading indicators[b]	2.70	3.29	3.21	2.14	1.82	3.95	2.29	2.89	2.31
E AR(3) model as in D plus world return[c]	1.54	3.04	2.39	2.31	1.84	3.06	2.36	2.91	2.43
F AR(3) model as in E plus money growth rate[d]	1.56	2.92	2.43	1.47	1.83	2.57	2.63	2.23	1.82

Notes: The forecast period is 1974–81; the initial fitting period is 1954–73.

[a] The AR(3) model is $y_{it} = \alpha_{0i} + \alpha_{1i} y_{it-1} + \alpha_{2i} y_{it-2} + \alpha_{3i} y_{it-3} + \varepsilon_{it}$; see (2.3).

[b] This model is that shown in note *a* with the addition of two terms $\beta_{1i} SR_{it-1} + \beta_{2i} SR_{it-2}$, where SR_{it} denotes the real stock return for the *i*th country in period *t*.

[c] This model is that described in note *b* plus the addition of a term $\gamma_i WR_{t-1}$, where WR_{t-1} is the median of the countries' real stock returns in period $t-1$.

[d] This model is that described in note *c* plus the addition of a term $\delta_i GM_{it-1}$, where GM_{it-1} is the lagged growth rate of the *i*th country's real money supply.

As a first step in the direction of attempting to improve on the forecasting performance of NMs I and II, we fitted autoregressions of order 3 (AR(3)s) to each country's data. That is,

$$g_{it} = \alpha_{oi} + \alpha_{1i}g_{it-1} + \alpha_{2i}g_{it-2} + \alpha_{3i}g_{it-3} + \varepsilon_{it},$$
$$t = 1, 2, \ldots, 20, \quad i = 1, 2, \ldots, 9, \qquad (2.3)$$

was considered for each country. We chose an AR(3) model to allow for the possibility of having two complex roots, associated with a cyclical solution, and one real root associated with a trend in the growth rate.

The AR(3) model in (2.3) was estimated for each country separately using twenty observations, 1954–73. Using a diffuse prior distribution for the parameters and given initial starting values, it is well known that the posterior means of the autoregressive parameters are identical to least squares estimates. In addition, the mean of the one-step-ahead predictive distribution is identical to the least squares point prediction. Thus these diffuse prior estimates and diffuse prior predictions are the same as those provided by application of least squares (see, e.g., Zellner 1971, ch. 7). Later, these diffuse prior predictions will be compared with those based on more informative prior distributions.

For many countries, the autoregressive parameters were not estimated very precisely, and if one performed mechanical "t-tests" of hypotheses that coefficients of lagged terms are equal to 0 at an approximate 5 percent level of significance, many coefficients appeared equal to 0 with the exception of those for Ireland. Moreover, the residuals from the fitted relations appeared to be non-autocorrelated. In cases in which the autoregressive coefficients are "truly" equal to 0 or very small in value, there should not be much if any improvement in forecasting precision relative to that of the NMs. It is recognized, however, that the preceding approximate t-tests or F-tests are not very powerful; and thus it was decided to compute AR(3) forecasts and their RMSEs that are shown in row C of table 13.1. It is seen that the largest RMSE is 4.75 for Italy, whereas the smallest is 1.69 for Ireland. The median RMSE is 3.46, that for Denmark; it is considerably higher than the median RMSE for NMI, namely 3.09. The AR(3) model resulted in a lower RMSE relative to NMI just for Ireland and the United States, where the improvement is relatively large, 1.69 versus 4.38 and 2.48 versus 3.48, and for France and the Netherlands, where the improvement is slight, 2.89 versus 2.96 and 3.52 versus 3.77.

Thus, using AR(3) models did not produce substantially improved overall performance, perhaps because unneeded parameters were added and/or because the effects of lagged terms were masked due to omitted relevant variables, a possibility explored hereafter. It was also noted that AR(3) models had large forecasting errors in the vicinity of turning points.

Since the NMs and the AR(3) models generally performed rather poorly in the vicinity of turning points, it was decided to add a leading indicator variable, lagged real stock returns for each country (see Fischer and Merton 1984, who found that lagged stock market variables were useful in forecasting US real GNP). This indicator variable is close to being white-noise and thus could be buried in the error terms of our AR(3)s that also appeared to be non-autocorrelated. By taking a measurable white-noise component out of our approximately white noise error terms, we are effectively attempting to "forecast white-noise" along the lines suggested by Granger (1983). In these experiments, we used the AR(3) model in (2.3) with each country's equation containing that country's real stock returns lagged one and two years. Thus two terms, real stock returns lagged one and two years, were added to (2.3) and the relations were fitted for each country by least squares. The results of forecasting from these "leading indicator" models are shown in row D of table 13.1. The lowest RMSE is 1.82 for Ireland and the largest is 3.95 for Italy, with a median equal to 2.70, that for Belgium. This median RMSE of 2.70 is quite a bit lower than that for the AR(3) without leading indicator terms, namely 3.46. Thus use of the leading indicator terms in the AR(3) model has led to about a 22 percent reduction in median RMSE. On comparing rows C and D of table 13.1, it is also seen that the RMSEs in row D are with two exceptions, France and Ireland, lower (many considerably lower) than the corresponding RMSEs in row C for the AR(3) model. It appears that use of each country's real stock returns has produced a noticeable and important improvement in forecasting performance relative to those of the AR(3) model and the NMs. Note that the median RMSE for NMII is 3.73, whereas that for the leading indicator model in row D is 2.70, approximately a 38 percent reduction.

The leading indicator AR(3) models' RMSEs in row D of table 13.1 do not reflect any allowance for intercountry effects, except insofar as these are reflected in each country's lagged variables. There are many ways to model interdependencies among countries. One simple way of doing this is to view the median real stock return of all countries as representing a "world effect" that exerts an influence on individual countries as an indicator of future world conditions. With this possibility in mind, we expanded the AR(3) model in (2.3) to allow for two lagged own-country real stock return terms and a one-period lagged effect of the world real stock return as measured by the median of the nine countries' one-period lagged real stock returns. With the addition of this lagged world return variable to the model used in row D of table 13.1, the model was fitted by least squares for each country. The residuals for each country's model were found not to be very highly correlated with those of other countries, suggesting that the world return variable was indeed picking up a common

Table 13.2 *Summary measures of forecasting performance for single-equation forecasts*

Model	Largest-country RMSE	Smallest-country RMSE	Median of nine countries' RMSEs
		Percentage points	
A NMI ($\hat{g}_t = 0$)	4.38	2.21	3.09
B NMII ($\hat{g}_t = g_{t-1}$)	4.88	2.06	3.73
C AR(3)	4.75	1.69	3.46
D AR(3) with two lagged own-real stock returns[a]	3.95	1.82	2.70
E AR(3) as in D plus one lag of real world return[a]	3.06	1.54	2.39
F AR(3) as in E plus one lag of real money growth rate[a]	2.92	1.47	2.23

Notes: Based on information in table 13.1.
[a] See the notes in table 13.1.

influence affecting all countries. RMSEs of forecasts for these "leading indicator, world return, AR(3)" models are shown in row E of table 13.1. The largest RMSE is 3.06 for Italy, and the smallest is 1.54 for Belgium. The median RMSE is 2.39, about a 12 percent reduction relative to the median RMSE in line D, 2.70.

Another leading indicator that has received considerable attention is the rate of growth of a country's real money supply. Thus the one-period lagged real money supply growth rate was added as a variable in each country's AR(3) model along with the two lagged stock return variables and the lagged world return variable. The forecasting performance of this model, fitted by least squares for each country, is shown in row F of table 13.1. The median RMSE is 2.23 percentage points, lower than that for all other models. The lowest country RMSE is 1.47 for Germany, and the highest is 2.92 for Denmark. Relative to the RMSEs in row E of table 13.1, those in row F show large reductions for Germany, from 2.31 to 1.47; Italy, from 3.06 to 2.57; the United Kingdom, from 2.91 to 2.23; and the United States, from 2.43 to 1.82. In the cases of Belgium, France, and the Netherlands, there are slight increases in the RMSEs. Thus it appears that introducing the lagged money growth rates has provided an improvement in forecasting performance.

For convenience some of the summary measures mentioned previously are collected in table 13.2. We see from table 13.2 that the world return, own-stock return, money growth rate leading indicator AR(3) model in row F of tables 13.1 and 13.2 has produced the lowest median RMSE of forecast, the lowest country RMSE, and the lowest, largest country RMSE. For the model in row F,

$$g_{it} = \alpha_{oi} + \alpha_{1i}g_{it-1} + \alpha_{2i}g_{it-2} + \alpha_{3i}g_{it-3} + \beta_{1i}SR_{it-1} + \beta_{2i}SR_{it-2} + \gamma_i WR_{t-1} + \delta_i GM_{it-1} + u_{it}, \quad (2.4)$$

the RMSEs of forecast ranged from 2.92 percentage points for Denmark to 1.47 for Germany, with a median of 2.23 for the nine countries. Except for Denmark and the United Kingdom, RMSEs are much smaller than those for NMI (see table 13.1). The RMSEs for Denmark and the United Kingdom are about the same as those for NMI. Relative to NMII, the RMSEs for forecasts from (2.4) are all substantially lower with the exception of France, for which they are equal. Thus, in summary, use of the model in (2.4) has produced improvement in terms of median RMSE relative to all other models. Use of (2.4) has also produced improvements for most countries.

It has been recognized in the literature that parameters may not be constant through time because of aggregation effects, policy changes, and so forth; thus time-varying parameter (TVP) versions of our models were formulated, estimated, and used in forecasting. (For some other works using Bayesian TVP models, see Harrison and Stevens 1976, Doan, Litterman, and Sims 1983, Highfield 1984, Los 1985, and West, Harrison, and Migon 1985). The TVP model that we employed for each country is in the following form:

$$g_{it} = \mathbf{x}'_{it}\boldsymbol{\beta}_{it} + u_{it} \quad (2.5a)$$

$$\boldsymbol{\beta}_{it} = \boldsymbol{\beta}_{it-1} + \mathbf{v}_{it}, \quad (2.5b)$$

where $\mathbf{x}'_{it}$ is a vector of input variables including a unit element for the intercept term, three lagged values of g_{it}, and lagged leading indicator variables. The time-varying coefficient vector $\boldsymbol{\beta}_{it}$ is assumed to follow a vector random walk, as shown in (2.5b). Further, we assume that u_{it}s have been independently drawn from a normal distribution with zero mean and variance σ_i^2 and that the $\mathbf{v}_{it}$ vectors have been independently drawn from a multivariate normal distribution with zero mean vector and covariance matrix $\phi_i\sigma_i^2 I$ – that is,

$$u_{it}\text{s: NID}(0, \sigma_i^2) \quad \text{and} \quad \mathbf{v}_{it}\text{s: NID}(\mathbf{0}, \phi_i\sigma_i^2 I). \quad (2.6)$$

If $\phi_i = 0$, this model reduces to a fixed parameter model.

The model in (2.5) was estimated using a Bayesian recursive state–space algorithm with various values of ϕ ranging from 0 to 0.50. The mean of the one-step-ahead predictive distribution was used as a forecast, since the mean is an optimal forecast relative to a squared error loss function with results shown in table 13.3. In part A of table 13.3, results for the AR(3) model with two lagged own-real stock return variables, SR_{it-1} and SR_{it-2}, included as leading indicator variables are presented. For $\phi_i = 0$, the RMSEs are identical with those in row D of table 13.1. For positive values of ϕ_i, including some not shown in the table, there were declines in the country RMSEs, some large, but the improvement was not encountered in all cases. For example, on comparing row A.1 ($\phi_i = 0$) with row A.3 ($\phi_i = 0.50$), it is seen that the RMSEs for Belgium, Denmark, France, and Italy show declines, some large, on allowing parameters to vary. In the cases of Germany, Ireland, the Netherlands, and the United States, however, the RMSEs increased slightly, except in the case of Ireland, where the increase was large. For $\phi_i = 0.50$, the median RMSE is 2.52, whereas for $\phi_i = 0$, the fixed parameter case, the median RMSE is 2.70. Thus there is a slight overall reduction in median RMSE in going from $\phi_i = 0$ to $\phi_i = 0.50$.

In part B of table 13.3, a time-varying parameter AR(3) model including own-lagged stock return variables, SR_{it-1} and SR_{it-2}, the lagged world return, WR_{t-1}, and lagged money growth rate, MG_{it-1}, was used to produce one-step-ahead forecasts for the years 1974–81. Its RMSEs of forecast are shown in rows B.1–3 of table 13.3. For $\phi_i = 0$, the model reduces to the fixed parameter model in row F of table 13.1. The median RMSE for the country forecasts with $\phi_i = 0$, the fixed parameter case is 2.23, and for the time-varying parameter cases, $\phi_i = 0.25$ and $\phi_i = 0.50$, the median RMSEs are 1.92 and 1.82, respectively, the latter about 18 percent lower than that for $\phi_i = 0$. For $\phi_i = 0.50$ relative to $\phi_i = 0$, there are large reductions in country RMSEs for France, Germany, Italy, and the United Kingdom and small reductions for Denmark and the Netherlands. For Belgium and the United States, the RMSEs are essentially the same for $\phi_i = 0.50$ as for $\phi_i = 0$, whereas for Ireland the RMSE is increased from 1.83 to 2.59 in going to time-varying parameters with $\phi_i = 0.50$. Thus although the overall reduction in median RMSE is large (approximately 18 percent in using time-varying parameter models and encountered for six of the nine countries), for three countries there was no improvement in RMSE. Current research is focused on use of time-varying parameter models along with the pooling techniques described in the next section.

2.2 *Forecasts based on pooled international data*

In this section we consider forecasts derived from models implemented with data from all countries – that is, pooled international data. There are many different pooling models. Here we report the forecasting

Table 13.3 *Root mean-squared errors (RMSEs) of one-step-ahead forecasts, 1974–1981, yielded by time-varying parameter models (2.5)–(2.6)*

Model	Belgium	Denmark	France	Germany	Ireland	Italy	Netherlands	United Kingdom	United States	Median RMSE
					Percentage points					
A AR(3) with two lagged own-real stock returns										
1. $\phi_i = 0^a$	2.70	3.29	3.21	2.14	1.82	3.95	2.29	2.89	2.31	2.70
2. $\phi_i = 0.25$	2.55	2.85	2.59	2.19	2.39	3.55	2.32	2.77	2.39	2.55
3. $\phi_i = 0.50$	2.52	2.80	2.49	2.26	2.56	3.39	2.36	2.88	2.49	2.52
B AR(3) as in A, plus one lag of world return and real money growth rate										
1. $\phi_i = 0^b$	1.56	2.92	2.43	1.47	1.83	2.57	2.63	2.23	1.82	2.23
2. $\phi_i = 0.25$	1.63	2.82	2.17	1.01	2.29	1.92	2.62	1.82	1.80	1.92
3. $\phi_i = 0.50$	1.56	2.89	2.08	0.97	2.59	1.68	2.59	1.82	1.81	1.82

Notes:

[a] $\phi_i = 0$ yields the fixed parameter model in row D of table 13.1.

[b] $\phi_i = 0$ yields the fixed parameter model in row F of table 13.1.

performance of a subset of these models. Of particular interest is the extent to which use of various pooling models results in improved forecasting performance relative to that of naive models and of single-equation models implemented just with individual countries' data. As previously, we use the annual data on countries' growth rates of output and other variables for the period 1954–73 to fit our models. Then these fitted models are employed to compute one-year-ahead forecasts for the eight years 1974–81 using updated estimates based on all data prior to each forecast period.

The first model that we consider is the AR(3) model including each country's real stock returns lagged one and two years. We have such a model for each of our nine countries. The *seemingly unrelated regression* (SUR) approach was used, with an estimated disturbance covariance matrix, to obtain pooled estimates of coefficient vectors for each country. One-year-ahead forecasts based on these pooled estimates, updated each year, were obtained for the years 1974–81. The RMSEs of these forecasts are presented in row E 2 of table 13.4. For comparison, the RMSEs of unpooled forecasts – that is, individual country least squares forecasts – are presented in row E 1 of table 13.4. In this case, the pooled forecasts' RMSEs are smaller in three cases than the corresponding unpooled forecast RMSEs. The lack of substantial improvement through pooling in this case may be due to the large number of elements in the disturbance covariance matrix, forty-five elements, that must be estimated and the fact that the sample size is not large. As shown in table 13.5, rows E 1 and E 2, the median RMSEs for the unpooled and pooled forecasts are 2.70 and 3.08, respectively. Thus overall there is not much difference in the performance of the unpooled and pooled forecasts in this instance. Both median RMSEs, however, are substantially below those for the NMs and for an unpooled AR(3) model (see rows A–D in table 13.5).

In a second pooling experiment, an AR(3) model with each country's real stock returns lagged one and two years and a world return variable lagged one year was considered. Because, as mentioned previously, introduction of the lagged world return variable in each country's equation produced error terms' contemporaneous correlations that were very low, we used the following simple pooling technique. Let $\mathbf{g}_i$ be the observation vector of the ith country's growth rates and X_i the matrix of observations of the ith country's input variables with a typical row $(1, g_{it-1}, g_{it-2}, g_{it-3}, SR_{it-1}, SR_{it-2}, WR_{t-1})$. Then the pooled coefficient vector estimate, denoted by $\bar{\beta}$, was computed as follows:

$$\bar{\beta} = (X_1'X_1 + X_2'X_2 + \cdots + X_9'X_9)^{-1} \times (X_1'X_1\hat{\beta}_1 + X_2'X_2\hat{\beta}_2 + \cdots + X_9'X_9\hat{\beta}_9), \quad (2.7)$$

Table 13.4 *Root mean-squared errors (RMSEs) for one-step-ahead forecasts from models using unpooled and pooled data, 1974–1981*

Model	Belgium	Denmark	France	Germany	Ireland	Italy	Netherlands	United Kingdom	United States
Percentage points									
A NMI ($\hat{g}_t = 0$)	3.09	2.83	2.96	2.95	4.38	3.72	3.77	2.21	3.48
B NMII ($\hat{g}_t = g_{t-1}$)	4.25	3.73	2.43	3.26	2.06	4.88	4.04	3.91	3.60
C NMIII ($\hat{g}_t$ = past average)	3.23	3.48	3.06	3.87	1.88	3.90	3.74	2.95	2.81
D AR(3): Unpooled[a]	3.66	3.46	2.89	3.39	1.69	4.75	3.52	3.50	2.48
E AR(3) plus two lagged real stock returns									
1. Unpooled[a]	2.70	3.29	3.21	2.14	1.82	3.95	2.29	2.89	2.31
2. Pooled[b]	2.47	3.49	3.03	3.08	1.86	3.88	3.61	3.10	2.68
F As in row E plus lagged world return									
1. Unpooled[a]	1.54	3.04	2.39	2.31	1.84	3.06	2.36	2.91	2.43
2. Pooled[c]	2.21	2.72	1.70	2.28	1.77	2.62	2.91	2.75	3.08
G As in row F plus lagged real money growth rate									
1. Unpooled[a]	1.56	2.92	2.43	1.47	1.83	2.57	2.63	2.23	1.82
2. Pooled(1)[c]	1.69	2.37	1.35	2.03	1.77	2.22	2.87	2.26	2.75
3. Pooled(2)[d] ($\eta = 0.5$)	1.68	2.21	1.61	1.25	1.52	2.01	2.52	2.46	1.78

Notes:

[a] Least squares forecasts for individual countries.

[b] Seemingly unrelated regression forecasts based on an estimated disturbance covariance matrix.

[c] Forecasts computed using the pooled coefficient vector estimate in (2.7).

[d] Forecasts computed using the pooling technique in (2.11) with $\eta = 0.5$.

Table 13.5 *Summary statistics for root mean-squared errors (RMSEs) in table 13.4*

Model	Median RMSE	Lowest-country RMSE	Highest-country RMSE
	Percentage points		
A NMI	3.09	2.21	4.38
B NMII	3.73	2.06	4.88
C NMIII	3.23	1.88	3.90
D AR(3): Unpooled[a]	3.46	1.69	4.75
E AR(3) plus two lagged real stock returns			
1. Unpooled[a]	2.70	1.82	3.95
2. Pooled[b]	3.08	1.86	3.88
F As in row E plus lagged world return			
1. Unpooled[a]	2.39	1.54	3.06
2. Pooled[c]	2.62	1.70	3.08
G As in row F plus lagged real money growth rate			
1. Unpooled[a]	2.23	1.47	2.92
2. Pooled(1)[c]	2.22	1.35	2.87
3. Pooled(2)[d] ($\eta = 0.5$)	1.78	1.25	2.52

[a] See note *a*, table 13.4.
[b] See note *b*, table 13.4.
[c] See note *c*, table 13.4.
[d] See note *d*, table 13.4.

a matrix-weighted average of the single-equation least squares estimates, $\hat{\beta}_i = (X_i'X_i)^{-1}X_i'\mathbf{g}_i$ $(i = 1, 2, \ldots, 9)$. The joint estimate $\bar{\beta}$ in (2.7) can be rationalized in at least three ways. First, if we assume that countries' parameter vectors and their disturbance variances are not very different, we can consider the following model for the observations:

$$\begin{bmatrix} \mathbf{g}_1 \\ \mathbf{g}_2 \\ \vdots \\ \mathbf{g}_9 \end{bmatrix} = \begin{bmatrix} X_1 \\ X_2 \\ \vdots \\ X_9 \end{bmatrix} \beta + \begin{bmatrix} \mathbf{u}_1 \\ \mathbf{u}_2 \\ \vdots \\ \mathbf{u}_9 \end{bmatrix}. \tag{2.8a}$$

Given that we use least squares to estimate the common coefficient vector β, the result is the estimate shown in (2.7).

Second, if for each country we have $\mathbf{g}_i = X_i\beta_i + \mathbf{u}_i$ and we assume that the β_is are random satisfying $\beta_i = \boldsymbol{\theta} + \mathbf{v}_i$ $(i = 1, 2, \ldots, 9)$, where $\boldsymbol{\theta}$ is a common mean vector and the $\mathbf{v}_i$ vectors are uncorrelated with the variables in X_i, then $\mathbf{g}_i = X_i\boldsymbol{\theta} + \eta_i$, where $\eta_i = X_i\mathbf{v}_i + \mathbf{u}_i$. Then, under relatively weak conditions, the estimator $\bar{\beta}$ in (2.7) is a consistent estimator for $\boldsymbol{\theta}$. If more were assumed about the properties of the $\mathbf{u}_i$s and

$\mathbf{v}_i$s it is possible to define asymptotically efficient estimators for $\boldsymbol{\theta}$ and predictors. This involves introducing possibly questionable assumptions and additional parameters, however. Since our sample size is small and we do not have much data to assess the quality of the needed additional assumptions, we decided to use $\bar{\beta}$ in (2.7) for each country to generate pooled forecasts.

A third way of rationalizing $\bar{\beta}$ in (2.7) is to consider a version of the Lindley and Smith (1972) pooling model. Here we have $\mathbf{g}_i = X_i\beta_i + \mathbf{u}_i$ $(i = 1, 2, \ldots, 9)$ or

$$\begin{bmatrix} \mathbf{g}_1 \\ \mathbf{g}_2 \\ \vdots \\ \mathbf{g}_9 \end{bmatrix} = \begin{bmatrix} X_1 & 0 & \cdots & 0 \\ 0 & X_2 & \cdots & 0 \\ \vdots & & \ddots & \vdots \\ 0 & 0 & \cdots & X_9 \end{bmatrix} \begin{bmatrix} \beta_1 \\ \beta_2 \\ \vdots \\ \beta_9 \end{bmatrix} + \begin{bmatrix} \mathbf{u}_1 \\ \mathbf{u}_2 \\ \vdots \\ \mathbf{u}_9 \end{bmatrix}, \tag{2.8b}$$

or

$$\mathbf{g} = Z\beta_a + \mathbf{u}, \tag{2.8c}$$

where $\mathbf{g}' = (\mathbf{g}_1'\mathbf{g}_2' \cdots \mathbf{g}_9')$, Z is the block-diagonal matrix in (2.8b), $\beta_a' = (\beta_1'\beta_2' \cdots \beta_9')$, and $\mathbf{u}' = (\mathbf{u}_1'\ \mathbf{u}_2' \cdots \mathbf{u}_9')$. Further, it is assumed that

$$\beta_i = \boldsymbol{\theta} + \mathbf{v}_i, \quad i = 1, 2, \ldots, 9, \tag{2.9}$$

where $\boldsymbol{\theta}$ is a mean vector for the β_is. If we assume that the $\mathbf{u}_i$s are independent, each with an $N(\mathbf{0}, \sigma^2 I)$ distribution, and that the $\mathbf{v}_i$s are independent, each with an $N(0, \sigma_v^2 I)$ distribution, the probability density function for β_a given the data, σ^2, $\lambda = \sigma^2/\sigma_v^2$, and $\boldsymbol{\theta} = \bar{\beta}$ in (2.7) has a posterior mean, $\bar{\beta}_a$, given by

$$\begin{aligned} \bar{\beta}_a &= (Z'Z + \lambda I)^{-1}(Z'\mathbf{g} + \lambda \mathcal{J}\bar{\beta}) \\ &= (Z'Z + \lambda I)^{-1}(Z'Z\hat{\beta} + \lambda \mathcal{J}\bar{\beta}), \end{aligned} \tag{2.10}$$

where $\hat{\beta} = (Z'Z)^{-1}\ Z'\mathbf{g}$, with typical subvector $\hat{\beta}_i = (X_i'X_i)^{-1}X_i'\mathbf{g}_i$, and $\mathcal{J}' = (I\ I\ \cdots\ I)$. $\bar{\beta}_a$ in (2.10) is a matrix-weighted average of $\hat{\beta}$, the vector of the equation-by-equation least squares estimates, $\hat{\beta}_i$, and $\bar{\beta}$ the pooled estimate given in (2.7). The weights involve $\lambda = \sigma^2/\sigma_v^2$. If $\sigma_v^2 \to 0$, $\lambda \to \infty$ and $\beta_i \to \boldsymbol{\theta}$, from (2.9), and $\bar{\beta}_a \to \mathcal{J}\bar{\beta}$ or each subvector of $\bar{\beta}_a$ approaches $\bar{\beta}$, the pooled estimate in (2.7). On the other hand, if $\lambda = \sigma^2/\sigma_v^2 \to 0$, then $\bar{\beta}_a \to \hat{\beta}$, the equation-by-equation least squares estimates. Using (2.10) to generate forecasts for various values of λ and for different models, it was generally found that use of a very large value for λ – that is, the pooled estimate in (2.7) – produced the most satisfactory forecasts in a RMSE sense.

Thus, with the preceding considerations and results in mind, we present the RMSEs of forecasts generated using $\bar{\beta}$ as the coefficient vector estimate for each country. The RMSEs of these pooled forecasts are presented in row F 2 of table 13.4 with the RMSEs of unpooled forecasts in row F 1. The pooled forecasts' RMSEs are smaller in six of nine cases. Also from rows F 1 and F 2 of table 13.5, however, the median RMSE for the unpooled estimates is 2.39, whereas that for the pooled estimates is 2.62, a 9 percent increase. Thus use of pooled forecasts in this case led to an improvement for six of nine countries, but an increase in the overall median RMSE.

Using the same pooling technique – that is, the common coefficient vector estimate $\bar{\beta}$ in (2.7) – for all countries and an AR(3) model with own-country real stock returns lagged one and two years, the lagged world return, and each country's lagged money growth rate, pooled forecasts were generated. The RMSEs associated with these pooled forecasts are shown in row G 2 and the unpooled forecasts' RMSEs are shown in row G 1 of table 13.4. For four of nine countries, the pooled RMSEs are smaller. From table 13.5, rows G 1 and G 2, the median RMSEs for the unpooled and pooled forecasts are 2.23 and 2.22, respectively. It is seen that pooling in this instance led to just a very small reduction in median RMSE.

Last, another method of pooling was tried for the model considered in the last paragraph and in row G of tables 13.4 and 13.5. For the *i*th country, $\hat{g}_{it}$ is its output growth rate least squares forecast for period t. Let $\bar{g}_t$ be the mean of the individual countries' forecasts for period t – that is, $\bar{g}_t = \sum_{i=1}^{9} \hat{g}_{it}/9$. Then we define a pooled forecast, $\bar{g}_{it}$, as follows:

$$\begin{aligned} \hat{g}_{it} &= (1-\eta)\hat{g}_{it} + \eta\bar{g}_t \\ &= \bar{g}_t + (1-\eta)(\hat{g}_{it} - \bar{g}_t), \end{aligned} \tag{2.11}$$

where η is a weighting factor. From the second line of (2.11), it is seen that we are "shrinking" the individual forecasts, $\hat{g}_{it}$, toward the mean forecast, $\bar{g}_t$, in a Steinlike manner.

RMSEs for country forecasts, 1974–81, based on (2.11) for selected values of η are shown in table 13.6. With $\eta = 0$, the RMSEs are identical to those shown in row G 1 in table 13.4 – that is, those for least squares forecasts for individual countries with no pooling. It is seen that for η values 0.25, 0.50, and 0.75 there are large reductions in country RMSEs relative to those for $\eta = 0$ for Denmark, France, Germany, Ireland, and Italy, small reductions for the Netherlands and the United States, and slight increases for Belgium and the United Kingdom. For $\eta = 0.5$, the median RMSE is 1.78, with the smallest RMSE being 1.25 and the

Table 13.6 *Country root mean-squared errors (RMSEs) of the forecast for 1974–1981, associated with the use of the pooling model (2.11) for various values of* η

η	Belgium	Denmark	France	Germany	Ireland	Italy	Netherlands	United Kingdom	United States	Median
					Percentage points					
0	1.56	2.92	2.43	1.47	1.83	2.57	2.63	2.23	1.82	2.23
0.25	1.59	2.48	2.01	1.30	1.59	2.24	2.55	2.32	1.73	2.01
0.50	1.68	2.21	1.61	1.25	1.52	2.01	2.52	2.46	1.78	1.78
0.75	1.80	2.18	1.26	1.34	1.64	1.89	2.53	2.63	1.97	1.89
1	1.96	2.39	1.01	1.54	1.91	1.92	2.57	2.82	2.25	1.96

Note:
Individual country forecasts, $\hat{g}_{it}$, were generated using the model in (2.4) and pooled using the relation in (2.11).

largest 2.52. This median RMSE of 1.78 is about 20 percent lower than the median RMSE of 2.23 for the unpooled least squares forecasts (see rows G 1 and G 3 of table 13.5). Thus the pooling procedure given in (2.11) has produced a substantial improvement in forecasting precision as measured by median RMSE. Plots of pooled forecasts, calculated from (2.11) and actual annual growth rates are presented in figure 13.2.

2.3 *Comparison with OECD forecasts*

In a very interesting article, Smyth (1983) presented an analysis of the annual forecasts made by the Organization for Economic Cooperation and Development (OECD) for seven countries' rates of growth of real GNP for the years 1968–79. Although this period is different from our period, 1974–81, in that the difficult (in a forecasting sense) years 1980 and 1981 are not included, it is thought that a comparison of RMSEs of forecast is of interest.

Smyth (1983, p. 37) explained that "It is probable that more policy attention in the various countries is attached to the annual forecasts than to the half-yearly ones and that is why we analyze them here." He went on (1983, p. 38) to state, "While account is taken of both official and unofficial national forecasts, the OECD forecasts are entirely the responsibility of the OECD Department of Economics and Statistics." He described the OECD forecasting procedures as follows (1983, p. 37):

The OECD's forecasting cycle is semi-annual. The forecasting "round" begins with a simulation of the interlink model to provide an initial update of the previous

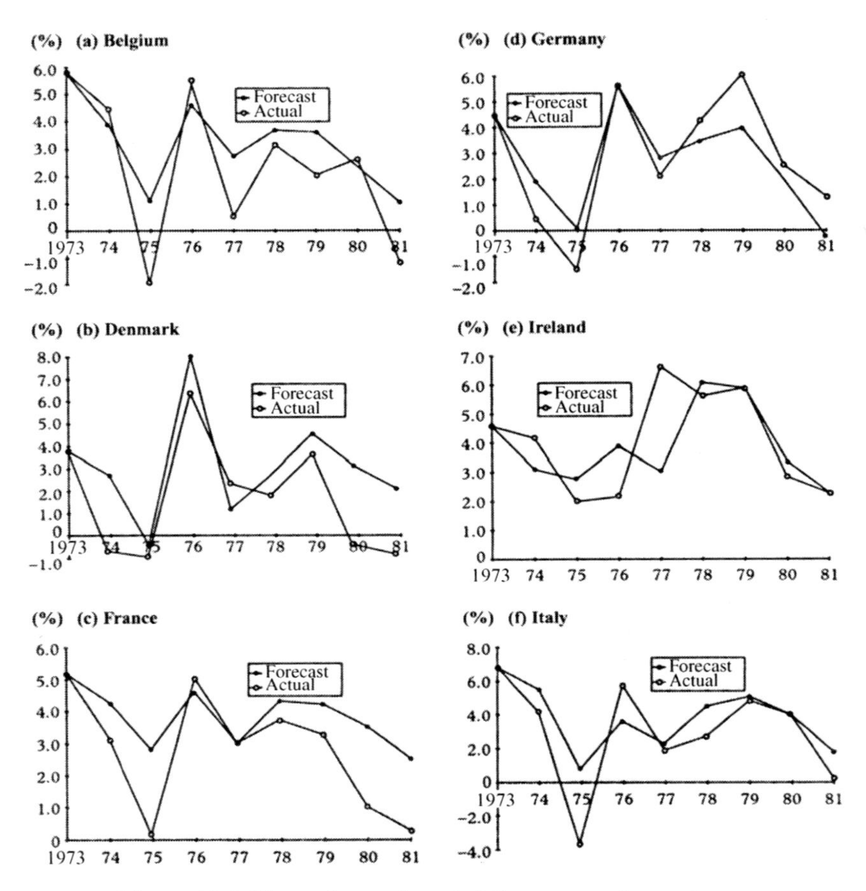

Figure 13.2 Plots of annual pooled forecasts and actual output growth rates, 1974–1981; the pooled forecasts are from row G3 of table 13.4 ($\eta = 0.5$)

set of forecasts in the light of changes in exogenous factors. This, together with an assessment of special factors influencing each economy, provides a basis for preliminary assessments of the level of demand for the individual economies, which permits initial estimates of import and export demand. Exchange rates and the real price of oil are assumed to remain unchanged over the forecast period. Fiscal and monetary policy assumptions are made on the basis of existing stated policies.

Budgetary statements are widely used to estimate public consumption and investment. Private investment components are forecast separately . . . There is quite extensive reliance on investment surveys. Private consumption depends primarily on personal disposable income. The stockbuilding forecast is often based upon the behaviour of stock–output ratios.

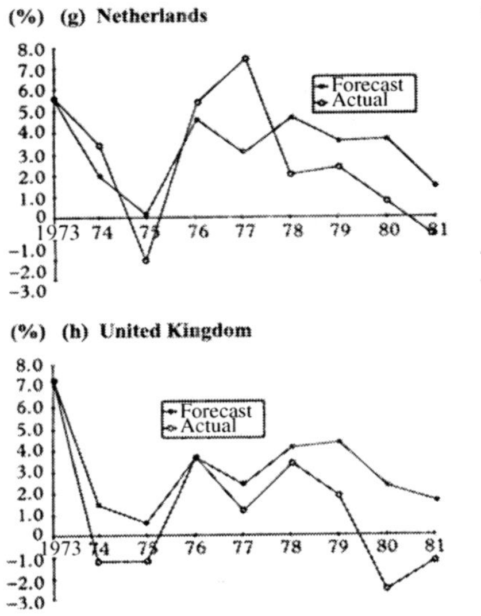

Figure 13.2 (*cont.*)

Further, he noted (1983, p. 37) that "judgements are imposed on the forecasting round by individuals not associated with the modelling process – namely, individuals from the OECD's various country desks."

It is clear that the OECD forecasting procedures are much more complicated than those presented here. Whereas we have employed various stock market return variables to reflect "outside influences and information," the OECD approach involves use of detailed data and the judgment of individuals in an attempt to capture outside influences and information. These and other differences in our and the OECD approaches are apparent to the reader from what has been presented previously.

As Smyth (1983, p. 45) showed, the OECD forecasts are better than those of naive random-walk models for all countries. The RMSEs of the OECD annual GNP growth rate forecasts, 1968–79, are given in table 13.7, part A, along with those reported in table 13.4, rows G1–G3, for countries appearing both in our and the OECD samples. The major results shown in part A of table 13.7 are the following:

1. Our unpooled least squares forecasts, 1974–81, based on (2.4), have lower RMSEs for three of the five countries – Germany, Italy, and the United Kingdom. For France and the United States, the OECD RMSEs are considerably smaller. The median RMSE of our forecasts, 2.23, is slightly larger than that for the OECD forecasts, 2.12.

Table 13.7 *Comparison of forecasting performance*

Forecasts	France	Germany	Italy	United Kingdom	United States	Median RMSE
		Percentage points				
A. RMSEs of GNP growth rate forecasts, by country						
OECD (1968–9)[a]	1.45	2.12	2.86	2.26	1.38	2.12
Table 4 (1974–81)						
G1 unpooled[b]	2.43	1.47	2.57	2.23	1.82	2.23
G2 pooled[c]	1.35	2.03	2.22	2.26	2.75	2.22
G3 pooled[d]	1.61	1.25	2.01	2.46	1.78	1.78
B. Mean absolute errors of GNP growth rate forecasts, by country						
OECD (1968–79)[a]	1.10	1.63	2.55	1.72	0.85	1.63
Table 4 (1974–81)						
G1 unpooled[b]	1.94	1.08	2.12	1.84	1.42	1.84
G3 pooled[d]	1.32	1.10	1.50	2.02	1.64	1.50
Table 4 (1974–9)						
G1 unpooled[b]	1.57	1.25	2.39	1.22	1.12	1.25
G3 pooled[d]	0.98	1.12	1.74	1.41	1.33	1.33

Notes:
[a] Taken from Smyth (1983, table 3, p. 45).
[b] Least squares forecasts for individual countries using the model in (2.4).
[c] Forecasts computed using the pooled coefficient estimate in (2.7) and the variables in (2.4).
[d] Forecasts computed using the pooling technique in (2.11) with $\eta = 0.5$ and the model in (2.4).

2. Our 1974–81 pooled forecasts, row G 2 of table 13.7, computed using the coefficient estimate in (2.7) and the variables in (2.4), have RMSEs smaller than the OECD's for three countries – France, Germany, and Italy; the same RMSE for the United Kingdom; and a much larger RMSE for the United States. The median RMSEs, 2.12 for the OECD forecasts and 2.22 for our forecasts, are very similar.
3. Our 1974–81 pooled forecasts, row G 3 of table 13.7, computed using the shrinkage formula in (2.11) with $\eta = 0.5$ and the model in (2.4), have RMSEs that are smaller for two countries, Germany and Italy, and larger for the remaining three countries. The median RMSE for our forecasts is 1.78, somewhat lower than that for the OECD forecasts, namely 2.12.
4. On comparing the OECD forecast RMSEs in table 13.7 with those for our Bayesian TVP models' forecast RMSEs for $\phi = 0.50$ in table 13.3, it is seen that the latter are much smaller for Germany, 0.97 versus 2.12; Italy, 1.68 versus 2.86, and the United Kingdom, 1.82

versus 2.26, and somewhat larger for France, 2.08 versus 1.45, and the United States, 1.81 versus 1.38. The median RMSE for the TVP models' forecasts is 1.82, lower than the OECD median RMSE, 2.12.

5. In addition to the OECD RMSEs reported in table 13.7, Smyth (1983, table 13.3, p. 47) reported the following RMSEs for countries not in our sample: Canada, 1.71, and Japan, 4.40.

Overall, it is concluded that our pooled and TVP forecasts compare favorably with the OECD forecasts except in the case of the United States, for which the OECD forecasts have a smaller RMSE. This conclusion should be qualified because the periods covered are not exactly the same. As mentioned previously, our period contains the "difficult" forecasting years 1980 and 1981. On the other hand, the OECD forecasts are "on-line" forecasts made in December of year $t - 1$ for year t and may thus be subject to difficulties associated with use of preliminary estimates of GNP and related variables, whereas we have used data that are currently available for past years. At present, it is difficult to determine how these two considerations affect the comparisons reported in table 13.7.

As a final comparison of the OECD forecasts and ours, we present mean absolute errors (MAEs) for the OECD and our forecasts, 1974–81, in part B of table 13.7. Also shown are the MAEs for the years 1974–9, a period that may be more comparable to the OECD period. Our unpooled forecasts for the 1974–81 period have lower MAEs for two countries, Germany and Italy, and larger MAEs for the remaining three countries. The OECD median MAE is 1.63, and for our unpooled forecasts it is 1.84. When the 1974–9 period is used for our unpooled forecasts, our MAEs are lower for three countries and the median MAE is 1.25, much lower than that for the OECD forecasts, namely 1.63.

As regards the MAEs for our pooled forecasts for the 1974–81 period, the pooled forecasts have lower MAEs for two of the countries and a median MAE of 1.50, almost the same as that for the OECD forecasts. When the 1974–9 period is used, our pooled forecasts have lower MAEs for four of the five countries and a larger MAE for the United States. The median MAE is 1.33 for our pooled forecasts and 1.63 for the OECD forecasts. As mentioned previously, forecasts for the six-year period 1974–9 are seen to be more accurate for most countries than those for the eight-year period 1974–81.

In conclusion, it appears that our relatively mechanical forecasts are competitive with the OECD forecasts except in the case of the United States, in which the OECD forecasts are better. Although several hypotheses to explain this finding could be considered, we shall not pursue this matter further now.

3 Forecasting quarterly output growth rates

Quarterly data, 1967:II–1977:IV, for six countries' output growth rates calculated from seasonally adjusted data and other variables have been employed to fit models. Then the fitted models have been used to generate one-quarter-ahead forecasts for each of sixteen quarters, 1978:I–1981:IV. Each model has been re-estimated quarter by quarter during the forecast period using data available prior to the forecast quarter. As with our analyzes of annual data, we are interested in determining the extent to which pooling data across countries improves forecasting accuracy. For comparative purposes, we provide forecasting results for three naive models and autoregressive integrated moving average (ARIMA) models. The latter were identified using standard Box–Jenkins (1976) procedures and the 1967:II–1977:IV data. Then they were used to produce one-quarter-ahead forecasts with parameter estimates updated quarter by quarter.

In rows A–C of table 13.8, the RMSEs of one-quarter-ahead forecasts for the sixteen quarters in the period 1978:I–1981:IV produced by three NMs are reported for each of the six countries. The median RMSEs are 1.16, 1.50, and 1.20 for NMs I, II, and III, respectively. These overall measures suggest that NMs I and III outperform NMII. For NMI, the country RMSEs range from a low of 0.71 for Spain to a high of 1.65 for the United Kingdom. The United Kingdom's RMSEs are the largest for each NM.

Shown in row D of table 13.8 are the ARIMA one-quarter-ahead forecast RMSEs country by country. They range from a low of 0.94 for Germany to a high of 1.70 for the United Kingdom. The median RMSE is 1.18, about 2 percent higher than that for NMI. Thus the ARIMA forecasts are not appreciably better overall than those obtained from use of NMI. The RMSEs for NMI are lower than those for the ARIMA models for France, Spain, and the United Kingdom and higher for Germany, Italy, and the United States.

In row E of table 13.8, unpooled and pooled forecasting results are presented. Each country's output growth rate was related linearly to a constant, C, its own-output growth (OG) rate lagged one quarter, OG(1), its own-real stock return lagged one quarter $S(1)$, the median of the country real stock returns lagged one quarter, $W(1)$, and each country's money growth rate lagged one quarter, $M(1)$. The unpooled forecasts are least squares forecasts derived from the individual country models with coefficient estimates updated quarter by quarter. These unpooled forecasts' median RMSE is 1.24, higher than that for two of the NMs and that for the ARIMA models. Thus these unpooled forecasts do not

Table 13.8 *Root mean-squared errors (RMSEs) for one-quarter-ahead forecasts of quarterly output growth rates for six countries, 1978I–1981IV*

Model	France	Germany	Italy	Spain	United Kingdom	United States	Median RMSE
			Percentage points				
A NMI ($\hat{g}_{it} = 0$)	0.97	1.06	1.53	0.71	1.65	1.25	1.16
B NMII ($\hat{g}_{it} = \hat{g}_{it-1}$)	1.09	1.45	1.70	1.13	2.63	1.56	1.50
C NMIII ($\hat{g}_{it}$ = past mean growth rate)	0.91	1.02	1.34	1.18	1.73	1.22	1.20
D ARIMA[a]	1.20	0.94	1.44	0.77	1.70	1.16	1.18
E C, OG(1), S(1), W(1), M(1)[b]							
1. Unpooled[c]	1.29	1.18	1.44	1.17	1.78	1.20	1.24
2. Pooled[d]	1.01	1.03	1.33	0.95	1.92	1.24	1.14
3. Pooled[e] (η = 0.5)	1.12	1.15	1.34	1.09	1.87	1.23	1.19
F C, OG (1, 2, 3, 4, 8), S(1), W(1), M(1)[f]							
1. Unpooled[c]	1.31	1.08	1.58	0.95	1.92	1.20	1.26
2. Pooled[d]	0.98	0.94	1.38	0.89	1.88	1.23	1.10
3. Pooled[e] (η = 0.5)	1.14	1.09	1.44	0.95	1.93	1.21	1.18

Notes: Models were fitted using the data for 1967II–1977IV and re-estimated using the data available before each forecast quarter.

[a] ARIMA models, identified and estimated using the 1967II–1977IV data, were used to generate one-quarter-ahead forecasts with parameter estimates updated quarter by quarter.

[b] C = constant, OG(1) = output growth rate lagged one quarter, S(1) = stock returns lagged one quarter, W(1) = world stock return (median of country returns) lagged one quarter, and M(1) = money growth rate lagged one quarter.

[c] Country-by-country least squares forecasts.

[d] Pooled forecasts using the procedure in (2.7).

[e] Forecasts computed using the pooling technique in (2.11) with η = 0.5.

[f] Variables as defined in note *b* but with output growth rates lagged one, two, three, four, and eight quarters for each country.

compare very favorably with those of NMs I and III and of the ARIMA models.

Pooled forecasts were obtained from the model in row E of table 13.8 by a procedure exactly the same as that described in (2.7). The Lindley–Smith procedure (see (2.10)) was also applied, and it was found that using very large values of the variance ratio parameter produced the lowest RMSEs of forecast; that is, the procedure reduced to the use of the pooling formula in (2.7). The pooled forecasts' RMSEs in row E 2 of table 13.8 are lower than those for the unpooled forecasts for four of the six countries. The median RMSE is also 8 percent lower than the corresponding median RMSE for the unpooled forecasts. The pooling techniques in (2.11) with η = 0.5 produced the results in row E 3 of table 13.8. It resulted in lower RMSEs for four of the six countries and a 4 percent decrease in the median RMSE. Thus pooling has generally increased forecast precision. Further, the pooled forecasts' median RMSEs compare favorably with those of the ARIMA and NM forecasts.

Shown in row F [of table 13.8] are forecasting results for a model similar to that considered in row E except that the output growth rate has been lagged one, two, three, four, and eight quarters rather than just one quarter. The lags of four and eight quarters were introduced to allow for possible inadequacies in the output seasonal adjustment procedures. The unpooled country least squares forecasts have RMSEs ranging from 0.95 for Spain to 1.92 for the United Kingdom, with a median RMSE equal to 1.26, about the same as that for the unpooled forecasts in row E 1 [of table 13.8]. Further, the median RMSE for the unpooled forecasts in row F 1 is slightly larger than those for two of the NMs and the ARIMA models. On the other hand, the median RMSEs for the pooled forecasts in rows F 2 and F 3 of table 13.8 are 1.10 and 1.18, lower than that for the unpooled forecasts, 1.26, and those for the NMs. Again, pooling has produced an overall gain in forecasting accuracy. Note that on comparing the country RMSEs in row F 1 with those in row F 2 the former unpooled, forecast RMSEs are all larger except for the case of the United States.

In a last set of experiments with the quarterly data, data on short-term interest rates were obtained for three countries – France, Germany, and the United States. Interest rates lagged one quarter and one through four quarters were introduced in the models of rows E and F of table 13.9. The RMSEs of one-quarter-ahead forecasts from these models with lagged interest rates included are shown in table 13.9 along with RMSEs for the NMs and the ARIMA models. Adding the lagged interest rate terms generally led to reductions in RMSEs in almost all cases. Further, the pooled country forecast RMSEs are all lower than the corresponding unpooled country forecast RMSEs. The pooled country forecast RMSEs in row F 2 of table 13.9 are all smaller than the corresponding RMSEs for the NMs and the ARIMA models. Plots of these forecasts and actual quarterly output growth rates are presented in figure 13.3. In future work, lagged interest rate terms will be introduced in other countries' models.

4 Summary of results and concluding remarks

For annual data on the growth rate of real output for eight EEC countries and the United States, one-year-ahead forecasts were generated for the years 1974–81 using different models and forecasting techniques, with the following results in terms of forecasting RMSEs:

Table 13.9 *Root mean-squared errors (RMSEs) for one-quarter-ahead forecasts of quarterly output growth rates for three countries, 1978I–1981IV*

Model	France	Germany	United States
	Percentage points		
A NMI ($\hat{g}_{it} = 0$)	0.97	1.06	1.25
B NMII ($\hat{g}_{it} = g_{it-1}$)	1.09	1.45	1.56
C NMIII ($\hat{g}_{it}$ = past mean growth rate)	0.91	1.02	1.22
D ARIMA[a]	1.20	0.94	1.16
E C, OG(1, 2, 3, 4, 8), $S(1)$, $W(1)$, $M(1)$, IN(1)[b]			
1. Unpooled[c]	1.29	1.07	1.17
2. Pooled[d]	0.92	0.91	1.13
F C, OG(1, 2, 3, 4, 8), $S(1)$, $W(1)$, $M(1)$ IN(1, 2, 3, 4)[e]			
1. Unpooled[c]	1.50	1.06	1.28
2. Pooled[d]	0.89	0.93	1.03

Notes: See the note to table 13.8.

[a] See note *a* in table 13.8.

[b] Here the model for each country is the same as that described in note *f* of table 13.8 with the addition of a country short-term interest rate lagged one-quarter, IN(1).

[c] Country-by-country least squares forecasts.

[d] Pooled forecasts using the procedure in (2.7).

[e] This model is the same as that on row E of this table except that each country's short-term interest rate enters lagged one, two, three, and four quarters, IN(1, 2, 3, 4).

1. Our AR(3)-leading indicator models outperformed several naive models and purely autoregressive models. These results underline the importance of using leading indicator variables in forecasting.
2. Our time-varying parameter models that incorporated country lagged stock returns, the lagged median return, and lagged money growth produced improved forecasts for six of the nine countries and improved overall forecast performance as measured by the median of the RMSEs of forecast for individual countries.
3. Relatively simple techniques for pooling individual countries' data or forecasts, applied to our models, led to an improvement in forecasting precision for many countries and overall.
4. The precision of our forecasts, produced by our relatively simple models and techniques, compares favorably with that of annual OECD forecasts, produced by use of much more complicated models and methods and incorporating judgmental information.

As regards our forecasting experiments with quarterly real output growth rates for several countries, one-quarter-ahead forecasts for the period 1978:I–1981:IV were calculated and compared, with the following results:

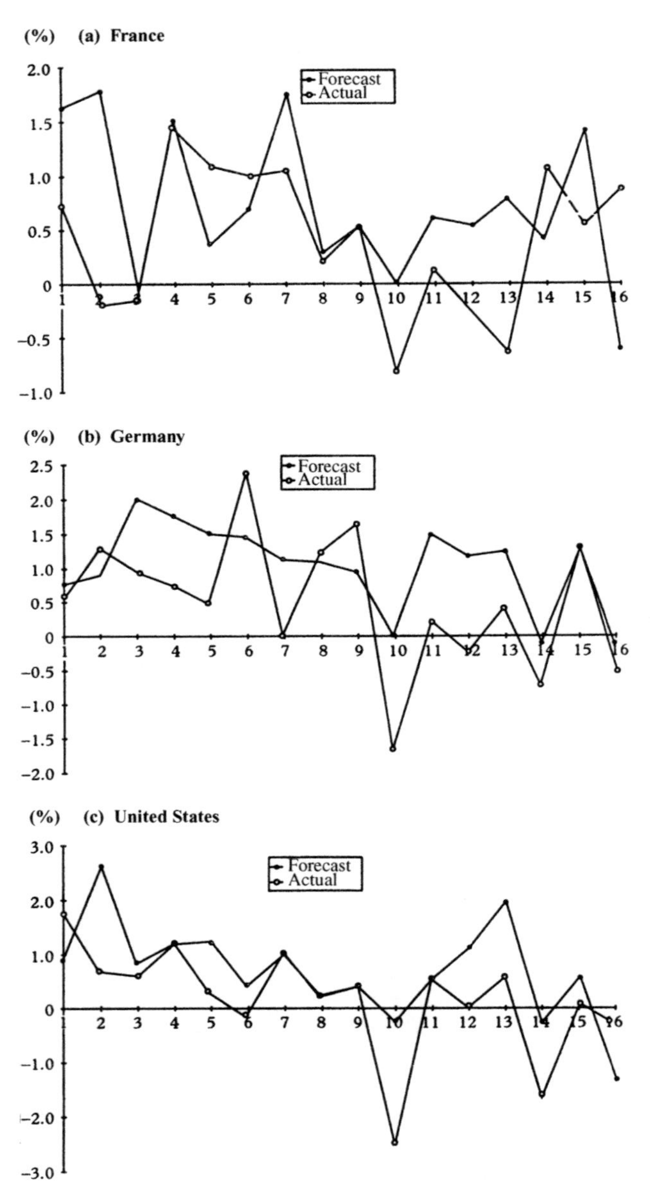

Figure 13.3 Plots of quarterly pooled forecasts and actual output growth rates, 1978:I–1981:IV; the pooled forecasts are from the model in row F of table 13.9

1. Forecasts derived from our pooled autoregressive–leading indicator models compared favorably with those produced by three naive models and ARIMA models as measured by median RMSEs.
2. As with the annual data, use of relatively simple pooling techniques led to improved forecasting precision in terms of the countries' median RMSEs and in terms of many countries' RMSEs.

The results described are encouraging and indicate that use of relatively simple models and pooling techniques leads to improved forecasting results. In future work, we shall use additional variables – for example, lagged exchange rates that Litterman (1986) has found useful – and more recent data to check further the forecasting performance of our models. We also plan to use combinations of time-varying parameter models and various pooling techniques in an attempt to obtain further improvements in forecasting precision. Finally, an effort will be made to rationalize the forms of our models using relevant macroeconomic theory and to extend them to incorporate additional variables to be forecasted. Such interaction between subject matter theory and statistical analysis is needed to develop "causal" models that explain past data and forecast reasonably well.

BIBLIOGRAPHY

Box, G. E. P. and G. M. Jenkins (1976), *Time Series Analysis, Forecasting and Control* (San Francisco, Holden-Day)

Burns, A. F. and W. C. Mitchell (1946), *Measuring Business Cycles* (New York, National Bureau of Economic Research)

Christ, C. F. (1951), "A test of an econometric model for the United States, 1921–1947," in *Conference on Business Cycles* (New York, National Bureau of Economic Research), 35–107

Doan, T., R. B. Litterman, and C. A. Sims (1983), "Forecasting and conditional projection using realistic priors," Working Paper 1202 (Cambridge, Mass., National Bureau of Economic Research)

Fischer, S. and R. C. Merton (1984), "Macroeconomics and finance: the role of the stock market," Working Paper 1291 (Cambridge, Mass., National Bureau of Economic Research)

Granger, C. W. J. (1983), "Forecasting white noise," in A. Zellner (ed.), *Applied Time Series Analysis of Economic Data* (Washington, DC, Bureau of the Census, US Department of Commerce), 308–14

Harrison, P. J. and C. F. Stevens (1976), "Bayesian forecasting," *Journal of the Royal Statistical Society*, Series B 38, 205–47

Harvey, A. C. and P. H. J. Todd (1983), "Forecasting economic time series with structural and Box–Jenkins models: a case study," *Journal of Business and Economic Statistics* 1, 299–307

Highfield, R. A. (1984), "Forecasting with Bayesian state space models," H. G. B. Alexander Research Foundation Technical Report, University of Chicago, Graduate School of Business

Kashyap, R. L. and A. R. Rao (1976), *Dynamic Stochastic Models From Empirical Data* (New York, Academic Press)
Klein, L. R. (1985), "New developments in Project LINK," *American Economic Review* 75, 223–7
Lindley, D. V. and A. F. M. Smith (1972), "Bayes estimates for the linear model," *Journal of the Royal Statistical Society,* Series B 34, 1–41
Litterman, R. B. (1986), "Forecasting with vector autoregressions – five years of experience," *Journal of Business and Economic Statistics* 4, 25–38
Los, C. A. (1985), "Regression analysis and diagnosis by an evolutionary parameter estimator," unpublished paper presented at the American Statistical Association Meetings, August 5–8, Las Vegas (available from the Federal Reserve Bank of New York)
Makridakis, S., A. Andersen, R. Carbone, R. Fildes, M. Hibon, R. Lewandowski, J. Newton, E. Parzen, and R. Winkler (1982), "The accuracy of extrapolation (time series) methods: results of a forecasting competition," *Journal of Forecasting* 1, 111–53
McNees, S. K. (1986), "Forecasting accuracy of alternative techniques: a comparison of US macroeconomic forecasts" (with discussion), *Journal of Business and Economic Statistics* 4, 5–23
Meese, R. and J. Geweke (1984), "A comparison of autoregressive univariate forecasting procedures for macroeconomic time series," *Journal of Business and Economic Statistics* 2, 191–200
Nelson, C. R. (1972), "The prediction performance of the FRB–MIT–Penn model of the US economy," *American Economic Review* 62, 902–17
Palm, F. (1983), "Structural econometric modelling and time series analysis: an integrated approach," in A. Zellner (ed.), *Applied Time Series Analysis of Economic Data* (Washington, DC, US Department of Commerce, Bureau of the Census), 199–233; chapter 3 in this volume
Smyth, D. J. (1983), "Short-run macroeconomic forecasting: the OECD performance," *Journal of Forecasting* 2, 37–49
West, M., P. J. Harrison and H. S. Migon (1985), "Dynamic generalized linear models and Bayesian forecasting," *Journal of the American Statistical Association* 80, 73–83
Zarnowitz, V. (1985), "Recent work on business cycles in historical perspective," *Journal of Economic Literature* 23, 523–80
Zellner, A. (1971), *An Introduction to Bayesian Inference in Econometrics* (New York, John Wiley)
(1979), "Statistical analysis of econometric models," *Journal of the American Statistical Association* 74, 628–51; chapter 2 in this volume
(1985), "A tale of forecasting 1001 series: the Bayesian knight strikes again," H. G. B. Alexander Research Foundation Technical Report, University of Chicago, Graduate School of Business; published in *Journal of Forecasting* 92, 491–4 (1986)
Zellner, A. and F. C. Palm (1974), "Time series analysis and simultaneous equation econometric models," *Journal of Econometrics* 2, 17–54; chapter 1 in this volume

14 Forecasting international growth rates using Bayesian shrinkage and other procedures (1989)

Arnold Zellner and Chansik Hong

1 Introduction

In our past work, Garcia-Ferrer *et al.* (1987), we employed several methods to forecast growth rates of real output (GNP or GDP) for eight European Economic Community (EEC) countries and the United States year by year for the period 1974–81. It was found that diffuse prior or least squares forecasts based on an autoregressive model of order 3 including leading indicator variables, denoted by AR(3)LI, were reasonably good in terms of forecast root mean-squared error (RMSE) relative to those of three naive models and of AR(3) models without leading indicator variables. Also, it was found that certain shrinkage forecasting techniques produced improved forecasting results for many countries and that our simple mechanical forecasts compared favorably with [Organization for Economic Cooperation and Development] (OECD) annual forecasts which were constructed using elaborate models and judgmental adjustments.

In the present chapter our main objectives are to extend our earlier work by (1) providing further analysis of shrinkage forecasting techniques, (2) providing forecasting results for an extended time period, 1974–84, for our past sample of nine countries, (3) applying our forecasting techniques to data relating to nine additional countries, and (4) reporting results of forecasting experiments using a simple modification of our AR(3)LI model.

The importance of checking the forecasting performance of our techniques using new data is reflected in objectives (2) and (3) above. The modification of our AR(3)LI model, mentioned in (4), was motivated by macroeconomic considerations embedded in structural models currently

Research financed by the National Science Foundation and by income from the H. G. B. Alexander Endowment Fund, Graduate School of Business, University of Chicago. Luis Mañas-Antón, Carla Inclan, and Michael Zellner provided valuable research assistance.

Originally published in the *Journal of Econometrics, Annals* 40 (1989) 183–202. 0304-4076/89/$3.50.

being formulated, which yield reduced form equations similar in form to our AR(3)LI forecasting model and reduced form equations for other variables (e.g. the rate of inflation, employment growth, etc.) that will be subjected to forecasting tests in the future.

The plan of the chapter is as follows. In section 2 our AR(3)LI model is explained and analysis yielding several shrinkage forecasts is presented. Also, an extended version of our AR(3)LI model is specified and it is shown how it can be employed to yield forecasts. Section 3 is devoted to a presentation of our data, and in section 4 previous forecasting results are compared with those relating to our broader data set and with those yielded by our extended AR(3)LI model. Finally, we provide a summary of results and some concluding remarks in section 5.

2 Model description and forecasting procedures

In this section, we shall describe the autoregressive-leading indicator (ARLI) model employed in our past work as well as some possible extensions of it. Then we shall consider various forecasting procedures of our ARLI models.

2.1 *Model description*

In Garcia-Ferrer *et al.* (1987), the following AR(3)LI model was employed to generate one-year-ahead forecasts of the growth rate of real output, y_{it}, for eight years, 1974–81, for nine countries:

$$\begin{aligned} y_{it} = {} & \beta_{0i} + \beta_{1i} y_{it-1} + \beta_{2i} y_{it-2} + \beta_{3i} y_{it-3} + \beta_{4i} SR_{it-1} \\ & + \beta_{5i} SR_{it-2} + \beta_{6i} GM_{it-1} + \beta_{7i} WR_{t-1} + u_{it}, \\ & i = 1, 2, \ldots, 9, \quad t = 1, 2, \ldots, T, \end{aligned} \tag{2.1a}$$

or

$$\boldsymbol{y}_i = X_i \beta_i + \boldsymbol{u}_i, \quad i = 1, 2, \ldots, 9, \tag{2.1b}$$

where, with L being the lag operator and the subscript i, t denoting the value of a variable for the ith country in the tth time period,

$$\begin{aligned} y_{it} &= \text{rate of growth of output} = (1 - L) \log O_{it} \text{ with } O_{it} \text{ real output} \\ SR_{it} &= \text{real stock return} = (1 - L) \log(SP_{it}/P_{it}) \text{ with } SP_{it} \text{ a stock price index and } P_{it} \text{ a general price index} \\ GM_{it} &= \text{growth rate of real money supply} \\ &= (1 - L) \log(M_{it}/P_{it}) \text{ with } M_{it} = \text{nominal money supply} \end{aligned}$$

$$
\begin{aligned}
WR_t &= \text{world return} \\
&= \text{median of countries' real stock returns}, SR_{it} \text{ in period } t \\
\beta_{ji} &= \text{parameters for } i\text{th country, } j = 0, 1, \ldots, 7 \\
u_{it} &= \text{disturbance term.}
\end{aligned}
$$

In (2.1b) the model for each country is expressed in matrix notation with y_{it} a typical element of $\boldsymbol{y}_i$, $(1, y_{it-1}, y_{it-2}, y_{it-3}, SR_{it-1}, SR_{it-2}, GM_{it-1}, WR_{t-1})$ a typical row of X_i, and u_{it} a typical element of $\boldsymbol{u}_i$. Some comments regarding the model in (2.1) follow:

(1) An autoregression of order 3 was chosen to permit the possibility of having two complex roots associated with a cycle and a real root associated with a trend. Past calculations indicated that estimated roots had these properties for eight of the nine countries. Also, use of just an AR(3) process without leading indicator variables did not perform well in actual forecasting. Use of leading indicator variables led to improved forecasts in most cases as measured by RMSEs of forecast.

(2) The disturbance terms in (1) were found to be practically serially uncorrelated for most countries and not highly correlated across countries, results based on least squares analyses of (1) using initial annual data of 1951–73 for estimation. The introduction of the "common effect" variable, WR_t, reduced contemporaneous disturbance terms' correlations considerably.

(3) The leading indicator stock return variables and money growth rate variable apparently caught the effects of oil price shocks, policy changes, etc. in the period of fit 1951–73 and in our previous forecast period 1974–81. Here we are employing market variables to take rough account of expectational and other effects influencing countries' output growth rates.

(4) Macroeconomic considerations suggest that a measure of world output growth and changes in countries' real exchange rates affect countries' exports and these should be included in our ARLI model. Since these variables are close to being white noise, they may be buried in the disturbance terms of (2.1). Below, we shall report some results using an ARLI model including a measure of world output growth.

(5) In our past work, forecasts from (2.1), using least squares and some shrinkage forecast procedures, were reported. Also, some forecasts yielded by a time-varying-parameter version of (2.1) were reported. Some of these results will be presented below and compared with more recently obtained results.

In our forecasting experiments, we employ annual data, usually 1954–73, twenty observations with data from 1951–53 used for initial lagged values to fit our models.[1] Then the fitted models are employed to forecast outcomes for 1974 and subsequent years with the models re-estimated year by year. Multi-year-ahead forecasts have not as yet been calculated. For the forecast period 1974–81 (eight years), least squares forecasts using (2.1) have yielded forecast RMSEs ranging from 1.47 to 2.92 with a median of 2.23 percentage points for eight EEC countries and the United States (see table 14.2, line F of Garcia-Ferrer *et al.* 1987). Our "η-shrinkage" forecasts described below, yielded forecast RMSEs ranging from 1.25 to 2.52 percentage points with a median of 1.78 percentage points (see table 14.4, line G3 of Garcia-Ferrer *et al.* 1987). Similar results for an extended time period and for nine additional countries are presented below.

2.2 *Derivation and description of shrinkage forecasts*

In this subsection, we provide derivations of several shrinkage forecasts, including the "η-forecast" and the "γ-forecast." The performance of these forecasts will be compared with those of naive models and diffuse prior forecasts or least squares forecasts derived from the ARLI model, or variants of it, shown in (2.1).

The η-forecast involves averaging a forecast from (2.1), say a diffuse prior or least squares forecast for a particular country, $\hat{y}_{if}$, with the mean of all the N countries' forecasts, $\bar{\hat{y}}_f = \sum_{i=1}^{N} \hat{y}_{if}/N$, as follows:

$$\begin{aligned} \tilde{y}^*_{if} &= \eta\bar{\hat{y}}_f + (1-\eta)\hat{y}_{if} \\ &= \bar{\hat{y}}_f + (1-\eta)(\hat{y}_{if} - \bar{\hat{y}}_f). \end{aligned} \tag{2.2}$$

From the second line of (2.2), it is seen that for $0 < \eta < 1$, a country's forecast, $\hat{y}_{if}$, is shrunk toward the average forecast $\bar{\hat{y}}_f$ for all countries.

One way to obtain an optimal forecast in the form of (2.2) is to employ the following predictive loss function:

$$L = (y_{if} - \tilde{y}_{if})^2 + c\left(\sum_{i=1}^{N} y_{if}/N - \tilde{y}_{if}\right)^2, \tag{2.3}$$

where $c > 0$ is a given constant, the y_{if}s are the future unknown values, $i = 1, 2, \ldots, N$, and $\tilde{y}_{if}$ is some forecast. Note that the loss function in (2.3) incorporates an element of loss associated with being away from

[1] The 1954–73 period was used for all countries except Australia 1960–73, Canada 1959–73, Japan 1956–73, and Spain 1958–73.

the mean outcome in its second term. Under the assumption that the y_{if}s are independent (common influences have been represented by input variables in (2.1)) and have predictive probability density functions (pdfs) with mean m_i and variance v_i, $i = 1, 2, \ldots, N$. The predictive expectation of the loss function in (2.3) is

$$\mathrm{E}L = v_i + (m_i - \tilde{y}_{if})^2 + c[\mathrm{E}(\bar{y}_f - \mathrm{E}\bar{y}_f)^2 + (\mathrm{E}\bar{y}_f - \tilde{y}_{if})^2], \quad (2.4)$$

where $\bar{y}_f = \sum_{i=1}^{N} y_{if}/N$. On minimizing (2.4) with respect to $\tilde{y}_{if}$, the result is

$$\tilde{y}^*_{if} = \eta \sum_{i=1}^{N} m_i/N + (1 - \eta)m_i, \quad (2.5)$$

where $\eta = c/(1 + c)$. If diffuse priors for the β_is in (2.1) are employed, the means of the predictive pdfs are $m_i = \mathbf{x}'_{if}\hat{\beta}_i$, $i = 1, 2, \ldots, N$, where $\mathbf{x}'_{if}$ is a vector of observed inputs for the first future period and $\hat{\beta}_i = (X_i'X_i)^{-1}X_i'\mathbf{y}_i$, the least squares estimate for country i. Under these conditions (2.5) takes the form of (2.2) with $\hat{y}_{if} = \mathbf{x}'_{if}\hat{\beta}_i$. This is the "diffuse prior η-forecast."

Another approach for obtaining relatively simple shrinkage forecasts is a slightly modified form of the Lindley–Smith (1972) procedure in which the coefficient vectors are assumed generated by

$$\beta_i = \theta + \delta_i, \quad i = 1, 2, \ldots, N, \quad (2.6)$$

with the δ_is assumed independently distributed, each having a $\mathrm{N}(\mathbf{0}, \phi^{-1}\sigma_u^2 I_k)$ distribution where $0 < \phi < \infty$, σ_u^2 is a common variance of u_{it} for all i and t, and θ is a $k \times 1$ mean vector. If the u_{it}s are assumed normally and independently distributed, each with zero mean and common variance σ_u^2, then a conditional point estimate for $\beta' = (\beta_1', \beta_2', \ldots, \beta_N')$, denoted by $\tilde{\beta}_a$, an $Nk \times 1$ vector, is given by

$$\tilde{\beta}_a = (Z'Z + \phi I_{Nk})^{-1}(Z'Z\hat{\beta} + \phi \mathcal{J}\tilde{\beta}), \quad (2.7)$$

where Z is a block-diagonal matrix with $X_1, X_2, \ldots, X_N$ on the main diagonal, $\hat{\beta} = (Z'Z)^{-1}Z'\mathbf{y}$, where $\mathbf{y}' = (\mathbf{y}_1', \mathbf{y}_2', \ldots, \mathbf{y}_N')$, $\hat{\beta}' = (\hat{\beta}_1', \hat{\beta}_2', \ldots, \hat{\beta}_N')$ with $\hat{\beta}_i = (X_i'X_i)^{-1}X_i'\mathbf{y}_i$, $\mathcal{J}' = (I_k, I_k, \ldots, I_k)$, and

$$\tilde{\beta} = \left(\sum_{i=1}^{N} X_i'X_i'\right)^{-1} \sum_{i=1}^{N} X_i'X_i\hat{\beta}_i, \quad (2.8)$$

a matrix-weighted average of the least squares estimates, the $\hat{\beta}_i$s, which replaces θ in (2.7). Also, $\tilde{\beta}$ in (2.8) can be obtained by regressing $\mathbf{y}' = (\mathbf{y}_1', \mathbf{y}_2', \ldots, \mathbf{y}_N')$ on X, where $X' = (X_1', X_2', \ldots, X_N')$, that is, from one big regression in which it is assumed that the β_is are equal. Point forecasts can be obtained using the coefficient estimate $\tilde{\beta}_a$ for various selected values of ϕ. When ϕ is very large, (2.7) reduces approximately to (2.8). Forecasts based on the estimate in (2.8) have been reported . . . in Garcia-Ferrer *et al.* (1987). It should be recognized that, while the β_is are probably not all the same, the bias introduced by assuming them to be may be more than offset in a MSE error sense by a reduction in variance.

As an alternative to the assumptions used in connection with (2.6), following the "*g*-prior" approach of Zellner (1983, 1986), we assume that the $\boldsymbol{\delta}_i$s in (2.6) are independently distributed with normal distributions $\mathrm{N}[\mathbf{0}, (X_i'X_i)^{-1}\sigma_\delta^2]$. With this assumption and the earlier assumption made about the u_{it}s, the joint pdf for $\boldsymbol{y}' = (\boldsymbol{y}_1', \boldsymbol{y}_2', \ldots, \boldsymbol{y}_N')$ and $\boldsymbol{\beta}' = (\boldsymbol{\beta}_1', \boldsymbol{\beta}_2', \ldots, \boldsymbol{\beta}_N')$ is proportional to

$$\exp\left\{-[(\boldsymbol{y} - Z\boldsymbol{\beta})'(\boldsymbol{y} - Z\boldsymbol{\beta}) + \gamma(\boldsymbol{\beta} - \mathcal{J}\boldsymbol{\theta})'Z'Z(\boldsymbol{\beta} - \mathcal{J}\boldsymbol{\theta})]/2\sigma_u^2\right\}, \tag{2.9}$$

where $\gamma = \sigma_u^2/\sigma_\delta^2$ and other quantities have been defined in connection with (2.6)–(2.7). On completing the square on $\boldsymbol{\beta}$ in the exponential terms of (2.9), the mean of $\boldsymbol{\beta}$ given $\boldsymbol{y}$, γ and $\boldsymbol{\theta}$ is

$$\overline{\boldsymbol{\beta}} = [(Z'Z)^{-1}Z'\boldsymbol{y} + \gamma\mathcal{J}\boldsymbol{\theta}]/(1+\gamma), \tag{2.10a}$$

with the *i*th subvector of $\overline{\boldsymbol{\beta}}$ given by

$$\overline{\boldsymbol{\beta}}_i = (\hat{\boldsymbol{\beta}}_i + \gamma\boldsymbol{\theta})/(1+\gamma), \tag{2.10b}$$

where $\hat{\boldsymbol{\beta}}_i = (X_i'X_i)^{-1}X_i'\boldsymbol{y}_i$, the least squares estimate for the *i*th country's data. Thus (2.10b) is a simple average of $\hat{\boldsymbol{\beta}}_i$ and $\boldsymbol{\theta}$ with $\gamma = \sigma_u^2/\sigma_\delta^2$ involved in the weights. When a diffuse prior pdf for $\boldsymbol{\theta}$, $p(\boldsymbol{\theta}) \propto$ const., is employed, the posterior pdf for $\boldsymbol{\theta}$ can be derived from (2.9) and employed to average the expression in (2.10) to obtain the marginal mean of $\boldsymbol{\beta}_i$ given γ and the data, namely,

$$\overline{\boldsymbol{\beta}}_i^m = (\hat{\boldsymbol{\beta}}_i + \gamma\tilde{\boldsymbol{\beta}})/(1+\gamma), \tag{2.11}$$

with $\tilde{\boldsymbol{\beta}}$ given in (2.8), the estimate resulting from a big regression in which the $\boldsymbol{\beta}_i$s are assumed equal. As $\gamma = \sigma_u^2/\sigma_\delta^2$ grows in value, $\overline{\boldsymbol{\beta}}_i^m \to \tilde{\boldsymbol{\beta}}$ while as $\gamma = \sigma_u^2/\sigma_\sigma^2 \to 0$, $\overline{\boldsymbol{\beta}}_i^m \to \hat{\boldsymbol{\beta}}_i$, the *i*th country's least squares estimate.

If instead of assuming that $\boldsymbol{\delta}_i$ has a $\mathrm{N}(\mathbf{0}, (X_i'X_i)^{-1}\sigma_\delta^2)$ distribution, we assume that the $\boldsymbol{\delta}_i$s are independently distributed, with a $\mathrm{N}[\mathbf{0}, (X_i'X_i)^{-1}\sigma_i^2]$ distribution, $i = 1, 2, \ldots, N$, then analysis similar to that presented in connection with (2.9) yields as the conditional mean of $\boldsymbol{\beta}_i$,

$$\overline{\boldsymbol{\beta}}_i^c = (\hat{\boldsymbol{\beta}}_i + \gamma_i\boldsymbol{\theta})/(1+\gamma_i), \quad i = 1, 2, \ldots, N, \tag{2.12}$$

where $\gamma_i = \sigma_u^2/\sigma_i^2$. If we further condition on $\boldsymbol{\theta} = \tilde{\boldsymbol{\beta}}$, with $\tilde{\boldsymbol{\beta}}$ given in (2.8), then

$$\overline{\boldsymbol{\beta}}_i^c = (\hat{\boldsymbol{\beta}}_i + \gamma_i \tilde{\boldsymbol{\beta}})/(1 + \gamma_i), \quad i = 1, 2, \ldots, N, \tag{2.13}$$

which is similar to (2.11) except that the γ_is are not all equal as is the case in (2.11).

Upon introducing prior pdfs for σ_u and σ_δ, or σ_u and the σ_is, it is possible to compute the marginal distributions of the $\boldsymbol{\beta}_i$s, a possibility to be explored in future work (see Miller and Fortney 1984 for interesting computations on a closely related problem). At present, we shall evaluate (2.13) for various values of γ_i and determine the quality of resulting forecasts. That is, the γ_i-forecast for country i is

$$\tilde{y}_{if} = \boldsymbol{x}_{if}' \overline{\boldsymbol{\beta}}_i^c, \tag{2.14}$$

with $\overline{\boldsymbol{\beta}}_i^c$ given in (2.13) and $\boldsymbol{x}_{if}'$ a given input vector.

In summary, we shall use the η-forecast in (2.2), the γ-forecast based on (2.11) and the γ_i-forecast in (2.14) in our forecasting experiments. Also, note that the η-forecasting approach can be applied to the γ-forecasts.

2.3 *Elaboration of the AR(3)LI model*

As mentioned previously, we think that it is advisable to add a variable reflecting world real income growth, denoted by w_t, to our AR(3)LI model in (2.1). Then our equation becomes

$$\boldsymbol{y}_{it} = w_t \alpha_i + \boldsymbol{x}_{it}' \boldsymbol{\beta}_i + u_{it}, \quad i = 1, 2, \ldots, N, \quad t = 1, 2, \ldots T, \tag{2.15}$$

where $\boldsymbol{x}_{it}' \boldsymbol{\beta}_i$ represents the constant and other lagged variables in (2.1) and α_i is the ith country's coefficient of the world income growth rate variable, w_i. To forecast one period ahead using (2.15), it is clear that w_t must be forecasted. To do this we introduce the following equation for w_t which will be estimated and used to forecast w_t one period in the future:

$$\begin{aligned} w_t = {} & \pi_0 + \pi_1 w_{t-1} + \pi_2 w_{t-2} + \pi_3 w_{t-3} + \pi_4 MSR_{t-1} \\ & + \pi_5 MGM_{t-1} + v_t, \quad t = 1, 2, \ldots, T, \end{aligned} \tag{2.16}$$

where MSR_t is the median of all countries' real stock returns, MGM_t the median of all countries' real money growth rates, v_t a disturbance term, and the π_is are parameters. Thus (2.16) indicates that we are employing an AR(3)LI model for w_t, the rate of growth of world real income. As a proxy for w_t, we employ the median of all countries' real output growth rates.

Viewing (2.15)–(2.16), it is seen that we have a "triangular" system. For the future period $f = T + 1$, we have

$$\mathrm{E}\boldsymbol{y}_{if} = \mathrm{E}w_f \mathrm{E}\alpha_i + \boldsymbol{x}_{if}' \mathrm{E}\boldsymbol{\beta}_i, \quad i = 1, 2, \ldots, N, \tag{2.15a}$$

$$\mathrm{E}w_f = \boldsymbol{z}_f' \mathrm{E}\pi, \tag{2.15b}$$

where $\mathrm{E}y_{if}$ and $\mathrm{E}w_f$ are means of the predictive pdfs for y_{if} and w_f, respectively, and $\mathrm{E}\alpha_i$, $\mathrm{E}\boldsymbol{\beta}_i$ and $\mathrm{E}\pi$ are posterior means of the parameters α_i, $\boldsymbol{\beta}_i$ and $\pi' = (\pi_0, \pi_1, \ldots, \pi_6)$, respectively. If the system in (2.15)–(2.16) is *fully recursive* and diffuse prior pdfs for all parameters are employed, $\mathrm{E}\pi = \hat{\pi}$, $\mathrm{E}\alpha_i = \hat{\alpha}_i$ and $\mathrm{E}\boldsymbol{\beta}_i = \hat{\boldsymbol{\beta}}_i$, where $\hat{\pi}$, $\hat{\alpha}_i$ and $\hat{\boldsymbol{\beta}}_i$ are least squares estimates (see Zellner 1971, ch. 8 and Bowman and Laporte 1975). Also, $\mathrm{E}w_f = \boldsymbol{z}_f'\hat{\pi}$ and $\mathrm{E}\boldsymbol{y}_{if} = \boldsymbol{z}_f'\hat{\pi}\hat{\alpha}_i + \boldsymbol{x}_{if}'\hat{\boldsymbol{\beta}}_i$, $i = 1, 2, \ldots, N$. Thus forecasts under these assumptions are easily computed. If the system in (2.15)–(2.16) is not fully recursive, that is, the u_{it}s and v_t are correlated, then the expectations in (2.15) have to reflect the non-recursive nature of the system. In the present work, we shall use a "conditional" forecasting procedure which is equivalent to a 2SLS point forecast. That is, the parameters α_i and $\boldsymbol{\beta}_i$ in (2.15) are estimated by 2SLS, a conditional Bayesian estimate (see Zellner 1971, p. 266), and these estimates along with a forecast of w_t from (2.16) are employed to obtain a forecast of y_{it} from (2.15). Such forecasts will be compared with those that assume that w_ts value in a forecast period is perfectly known, a "perfect foresight" assumption. In current work, an unconditional Bayesian approach for analyzing (2.15)–(2.16) when the u_{it}s and v_t are correlated is being developed.

3 Data

Annual data for eighteen countries employed in our work have been assembled in the main from the IMF's International Financial Statistics data base . . . The output data include annual rates of growth of real output (GNP or GDP), of real stock prices and of real money for each country. In computing rates of growth of real stock prices, an index of nominal stock prices was deflated by an index of the price level for each country. Nominal money, M_1, was deflated by a general price index for each country to obtain a measure of real money.

Boxplots of output growth rates, real stock price growth rates, and real money growth rates are shown in figure 14.1. It is seen that the median growth rates exhibit a cyclical pattern with that for real stock prices having a considerably greater amplitude than those for output and real money growth rates. Also, as might be expected, the interquartile ranges of the real stock price growth rates are much larger than those of output and

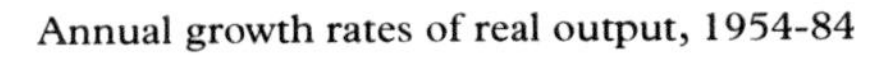

Annual growth rates of real output, 1954-84

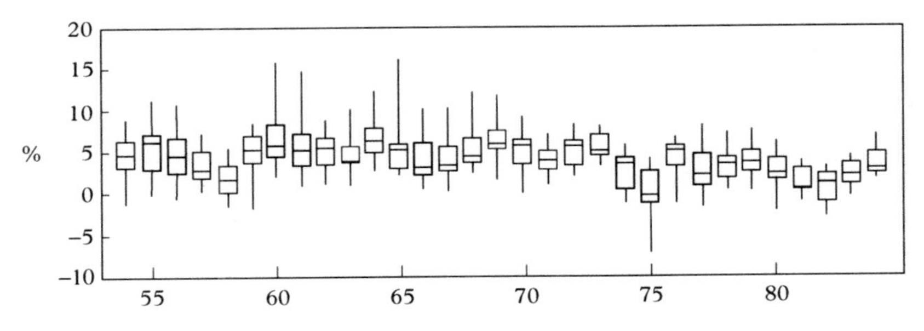

Annual growth rates of real money, 1954-84

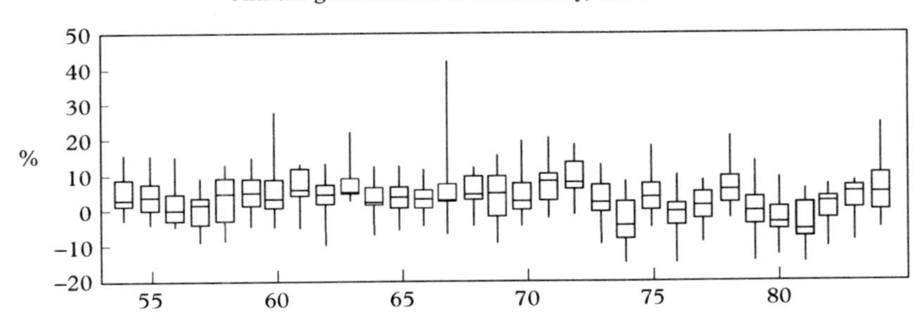

Annual growth rates of real stock prices, 1954-84

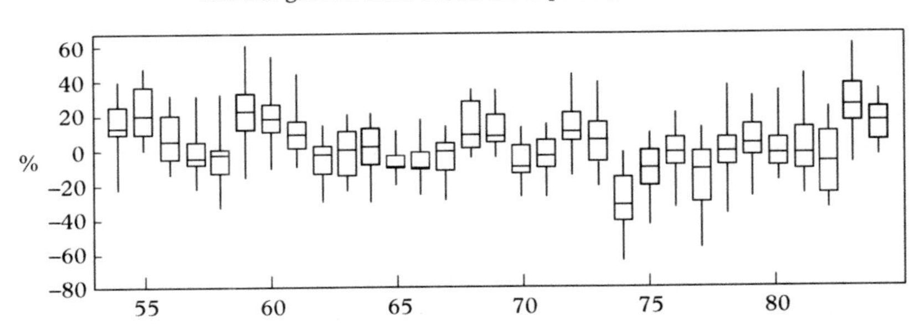

Figure 14.1 Boxplots of data for eighteen countries, 1954–1984

real money growth rates. Further, the interquartile ranges for growth rates of real stock prices appear to be slightly smaller in the vicinity of troughs than of peaks for the first half of the sample and the ranges for all three variables tend to be slightly larger in the vicinity of peaks than of troughs in many cases. Last, the plots of the real money and real stock price growth rates give some evidence of a slight lead relative to those of the output growth rates.

It will be noted from the plots in figure 14.1 that there are some apparently outlying points in the data with a number of them present in the data from the nine additional countries' data. These outlying data points are being subjected to close scrutiny and procedures for accommodating outlying data points are being considered for use in future work. In the present study, *all data were employed*, including outlying data without any special treatment given to them.

4 Forecasting results

In this section, we first compare RMSEs of one-year-ahead forecasts, 1974–81 (eight years), with those for the period 1974–84 for our original nine countries using various models and methods. Then for nine additional countries, forecasting results for the period 1974–84 are presented and compared with earlier results. Finally, the effects of introducing a world growth rate variable in our AR(3)LI model in (1a) on forecasting performance will be described.

4.1 Forecasting results for an expanded data set

Shown in table 14.1 are the RMSEs of forecast for nine countries for the periods 1974–81 and for 1974–84. Here and elsewhere, all models were re-estimated using data up to the forecast year. In the top panel of table 14.1, results for the eight one-year-ahead forecasts, 1974–81, are shown. It is seen that the median RMSE for the AR(3)LI model, 2.23 percentage points, is quite a bit lower than the median RMSEs for the naive models and for the AR(3) model. In addition, shrinkage or pooling techniques applied to the AR(3)LI model led to median RMSEs of 2.22 and 1.78, a very slight reduction in the former case and a somewhat larger reduction in the latter case, from 2.23 to 1.78 percentage points. Use of the η-shrinkage procedure, with $\eta = 0.5$, led to reduction of RMSEs for seven of the nine countries.

As regards the results for 1974–84, in the lower panel of table 14.1, the AR(3)LI model's median RMSE is 2.41 percentage points, a good deal lower than those associated with the naive models and the AR(3) model. Also, the two shrinkage or pooling procedures produced modest decreases in median RMSEs, from 2.41 to 2.26 and 2.31 and in six of nine cases for the η-shrinkage procedure. In these respects, the forecasting results parallel those obtained for the shorter period, 1974–81. However, note that there is an increase in the median RMSEs for the AR(3)LI model in

Table 14.1 *Nine countries' root mean-squared errors (RMSEs) of one-year-ahead forecasts, 1974–1981 and 1974–1984*

Model	Belgium	Denmark	France	Germany	Ireland	Italy	Netherlands	United Kingdom	United States	Median
					(Percentage points)					
(A) 1974–81										
A NMI ($\hat{y}_t = 0$)	3.09	2.83	2.96	2.95	4.38	3.72	3.77	2.21	3.48	3.09
B NMII ($\hat{y}_t = y_{t-1}$)	4.25	3.73	2.43	3.26	2.06	4.88	4.04	3.91	3.60	3.73
C NMIII ($\hat{y}_t$ = past average)	3.23	3.48	3.05	3.87	1.88	3.90	3.74	2.95	2.81	3.23
D AR(3)[a]	3.66	3.46	2.89	3.39	1.69	4.75	3.52	3.50	2.48	3.46
E AR(3) LI[b]	1.56	2.92	2.43	1.47	1.83	2.57	2.63	2.23	1.82	2.23
1. Shrinkage(1)[c]	1.69	2.37	1.35	2.03	1.77	2.22	2.87	2.26	2.75	2.22
2. Shrinkage(2)[d]	1.68	2.21	1.61	1.25	1.52	2.01	2.52	2.46	1.78	1.78
(B) 1974–84										
A NMI ($\hat{y}_t = 0$)	2.59	2.78	2.65	2.69	4.02	3.27	3.32	2.30	3.79	2.78
B NMII ($\hat{y}_t = y_{t-1}$)	3.53	3.56	2.20	2.91	2.85	4.26	3.57	3.69	3.89	3.56
C NMIII ($\hat{y}_t$ = past average)	3.01	3.05	3.08	3.85	2.35	3.98	3.87	2.62	3.09	3.08
D AR(3)[a]	2.98	2.96	2.47	3.10	2.29	4.34	3.35	3.21	3.01	3.01
E AR(3)LI[b]	1.73	2.73	2.52	2.28	2.80	3.40	2.41	2.32	2.14	2.41
1. Shrinkage(1)[c]	1.96	2.26	1.66	2.00	2.14	2.45	2.53	2.39	2.79	2.26
2. Shrinkage(2)[d]	1.81	2.37	2.07	1.94	2.31	2.73	2.50	2.63	2.03	2.31

Notes:

[a] Least squares forecasts from an AR(3) model for each country.

[b] Least squares forecasts from an AR(3) model with leading indicator variables shown in (2.1a).

[c] Least squares forecasts with use of coefficient estimate in (2.8).

[d] Use of shrinkage equation in (2.2) with $\eta = 0.5$.

going from the period 1974–81 to the longer period 1974–84 which also shows up in seven of the nine countries' RMSEs.

In table 14.2 RMSEs of forecast are shown for nine additional countries. Here the median RMSE for the AR(3)LI model, 3.33, is somewhat larger than that for the naive models and for the AR(3) model. However, the shrinkage or pooled forecasts based on the AR(3)LI model show substantial reductions in median RMSEs, from 3.33 to 2.01 and 2.32, which are similar in magnitude to those reported for the original nine countries for the periods 1974–81 and 1974–84 in table 14.1. In table 14.2, on comparing the AR(3)LI country RMSEs with the corresponding shrinkage RMSEs, it is seen that there is a reduction of RMSEs in all but one case. Thus for the nine additional countries, shrinkage results in a somewhat greater improvement in forecasting results than in the case of the original nine countries.

In table 14.3 summary forecasting results for different models, applied to all eighteen countries' data to forecast year by year for the period 1974–84, are reported. It is seen that the AR(3)LI model's median RMSE is about 13 percent or more below those of the AR(3) and naive model's median RMSEs, 2.62 versus 3.00 or greater. Also, the shrinkage or pooled forecasts median RMSEs, 2.14 and 2.32, are 18 and 11 percent lower, respectively, than the AR(3) model's median RMSE. Further, the ranges of the AR(3)LI and shrinkage forecast RMSEs for the eighteen countries are much smaller than those for the AR(3) and naive models.

In summary, the shrinkage or pooling techniques used earlier in Garcia-Ferrer *et al.* (1987) for nine countries for the period 1974–81 are effective in reducing overall median RMSEs for an extended forecast period, 1974–84, and for nine additional countries.

The η-shrinkage forecast RMSEs, reported above, are based on the same value of $\eta = 0.5$ for all countries. It is of interest to see how sensitive forecasting results for individual countries are to variation in the value of η. In table 14.4 one-year-ahead forecast RMSEs are reported for selected values of η in (2) for each of the eighteen countries in our sample. When $\eta = 0$, the forecasts are least squares forecasts for each country based on our AR(3)LI model in (2.1a). When $\eta = 1.0$, the forecast for each country is the mean of the eighteen countries' least squares forecasts from the AR(3)LI model in (2.1a). On viewing the median RMSEs at the bottom of table 14.4, it is seen that the median RMSEs are 2.43 for $\eta = 0.25$, 2.32 for $\eta = 0.5$, and 2.21 for $\eta = 0.75$, all below the median RMSE for $\eta = 0$, 2.62, that for individual countries' unshrunk least squares forecasts. For individual countries, the RMSEs change as the value of η changes. For example in the case of Belgium, the RMSEs vary from 1.73 for $\eta = 0$ to 2.35 for $\eta = 1.0$, with a minimum of 1.70

Table 14.2 *Nine additional countries' root mean-squared errors (RMSEs) of one-year-ahead forecasts, 1974–1984*

Model	Australia	Austria	Canada	Finland	Japan	Norway	Spain	Sweden[e]	Switzerland	Median[e]
					(Percentage points)					
A NMI ($\hat{y}_t = 0$)	3.10	2.91	3.51	3.39	4.10	4.15	2.51	2.29	3.03	3.10
B NMII ($\hat{y}_t = y_{t-1}$)	2.90	2.82	4.13	2.34	3.12	1.98	1.86	1.87	3.91	2.82
C NMIII ($\hat{y}_t$ = past average)	2.75	3.11	3.31	2.88	4.98	1.76	3.56	2.43	4.48	3.11
D AR(3)[a]	2.87	3.16	3.55	2.58	3.26	1.75	2.50	2.22	4.24	2.87
E AR(3)LI[b]	3.34	2.71	3.68	3.37	3.33	1.62	2.06	2.32	3.45	3.33
1. Shrinkage(1)[c]	2.01	1.77	2.39	2.01	2.40	1.68	1.65	2.01	2.71	2.01
2. Shrinkage(2)[d]	2.32	2.12	2.92	2.42	2.51	1.52	1.81	2.29	3.25	2.32

Notes:

[a] Least squares forecasts from an AR(3) model for each country.

[b] Least squares forecasts from an AR(3) model with leading indicator variables shown in (2.1a).

[c] Least squares forecasts with use of coefficient estimate in (2.8).

[d] Use of shrinkage equation in (2.2) with $\eta = 0.5$.

[e] Based on ten forecasts, 1974–83.

Table 14.3 *Summary statistics on forecasting results for eighteen countries, 1974–1984*

Model	Median RMSE	Smallest RMSE	Largest RMSE
		(Percentage points)	
A NMI ($\hat{y}_t = 0$)	3.07	2.29	4.15
B NMII ($\hat{y}_t = y_{t-1}$)	3.02	1.87	4.26
C NMIII ($\hat{y}_t$ = past average)	3.09	1.76	4.98
D AR(3)[a]	3.00	1.75	4.34
E AR(3)LI[b]	2.62	1.62	3.68
1. Shrinkage(1)[c]	2.14	1.65	2.79
2. Shrinkage(2)[d]	2.32	1.52	3.25

Notes:
[a] Least squares forecasts from an AR(3) model for each country.
[b] Least squares forecasts from an AR(3) model with leading indicator variables shown in (2.1a).
[c] Least squares forecasts with use of coefficient estimate in (2.8).
[d] Use of shrinkage equation in (2.2) with $\eta = 0.5$.

for $\eta = 0.25$. Two countries' minimal RMSEs occur for $\eta = 0$ and $\eta = 0.25$, six for $\eta = 0.50$, one for $\eta = 0.75$, and seven for $\eta = 1.0$. The median RMSE of these minimal values is 1.96, somewhat below that associated with the use of $\eta = 0.5$ for all countries, namely 2.32, or of $\eta = 0.75$, 2.21.

Just as it is of interest to determine the sensitivity of countries' η-forecasts to the values of η employed, it is of interest to determine how sensitive γ-forecasts for countries are to the value of γ employed. The coefficient estimate in (2.13) permits γ to be different for different countries. In table 14.5, RMSEs of forecast are reported for each country for selected values of γ. A zero value for γ yields AR(3)LI least squares forecasts while a very large value for γ results in an AR(3)LI forecast based on the pooled coefficient estimate in (2.8). From the median RMSEs reported at the bottom of table 14.5, it is seen that a common value of $\gamma = 5.0$ yields a median RMSE equal to 2.13, not far different from that associated with $\gamma = 10^6$, namely 2.14. However for individual countries, RMSEs show more substantial variation as γ assumes different values. For example in the case of Germany, the RMSE is 1.80 when $\gamma = 2.0$, quite a bit lower than the RMSE of 2.00 associated with $\gamma = 10^6$. Ten of the eighteen countries show minimal RMSEs for $\gamma = 10^6$, while the remaining eight countries have minima in the vicinity of $\gamma = 0.5$ to $\gamma = 1.0$ in five cases and of $\gamma = 2$ to $\gamma = 5$ for the remaining three. The median RMSE for these minimal values is 2.01, slightly lower than

Table 14.4 *Sensitivity of η-forecast root mean-squared errors (RMSEs) to value of η for AR(3)LI model in (2.1a), 1974–1984*[a]

Country	$\eta = 0$	$\eta = 0.25$	$\eta = 0.50$	$\eta = 0.75$	$\eta = 1.0$
			(Percentage points)		
Belgium	1.73	1.70	1.81	2.04	2.35
Denmark	2.73	2.49	2.37[b]	2.41	2.59
France	2.52	2.29	2.07	1.90	1.77[b]
Germany	2.28	2.04	1.94[b]	2.01	2.24
Ireland	2.80	2.51	2.31	2.24[b]	2.29
Italy	3.40	3.04	2.73	2.46	2.27[b]
Netherlands	2.41[b]	2.42	2.50	2.65	2.87
United Kingdom	2.32[b]	2.44	2.63	2.89	3.20
United States	2.14	1.97[b]	2.03	2.31	2.75
Australia	3.34	2.80	2.32	1.92	1.67[b]
Austria	2.71	2.40	2.12	1.87	1.67[b]
Canada	3.68	3.27	2.92	2.64	2.46[b]
Finland	3.37	2.86	2.42	2.10	1.95[b]
Japan	3.33	2.90	2.51	2.18	1.96[b]
Norway	1.62	1.53	1.52[b]	1.57	1.70
Spain	2.06	1.86	1.81[b]	1.91	2.14
Sweden[c]	2.32	2.28	2.29[b]	2.34	2.44
Switzerland	3.45	3.26	3.25[b]	3.42	3.75
Median RMSE	2.62	2.43	2.32	2.21	2.28
Range	1.62–3.68	1.53–3.27	1.52–3.25	1.57–3.42	1.67–3.75

Notes:
[a] The η-forecast is $\tilde{y}_{if} = \eta\hat{y}_f + (1-\eta)\hat{y}_{if}$, where $\hat{y}_f$ is the mean of the eighteen countries' least squares forecasts and $\hat{y}_{if}$ is the ith country's least squares forecast.
[b] Minimum of entries in each row. The median of these RMSEs is 1.96.
[c] Based on ten forecasts, 1974–83.

the median RMSE of 2.14 when $\gamma = 10^6$ is used for all countries. In future work it may be worthwhile to estimate γ for each country which will probably produce lower RMSEs for selected countries, e.g., Belgium, the Netherlands, the United States, and Norway. Also combinations of γ-forecast and η-forecast procedures are under consideration as well as other shrinkage techniques.

4.2 *Forecasting using a world output growth rate variable*

As mentioned previously, countries' exports are influenced by world income or output. Thus a world output growth rate variable, denoted by w_t, was added to the AR(3)LI model as shown in (2.15). The variable

Table 14.5 *Sensitivity of γ-forecast root mean-squared errors (RMSEs) to values of γ for AR(3)LI model in (2.1a), 1974–1984*[a]

Country	$\gamma = 0$	$\gamma = 0.50$	$\gamma = 1.0$	$\gamma = 2.0$	$\gamma = 3.0$	$\gamma = 5.0$	$\gamma = 10^6$
				(Percentage points)			
Belgium	1.73	1.60[b]	1.62	1.69	1.74	1.81	1.96
Denmark	2.73	2.34	2.22	2.16[b]	2.16[b]	2.18	2.26
France	2.52	2.17	2.01	1.87	1.81	1.75	1.66[b]
Germany	2.28	1.91	1.81	1.80[b]	1.82	1.87	2.00
Ireland	2.80	2.47	2.34	2.24	2.20	2.17	2.14[b]
Italy	3.40	3.00	2.82	2.67	2.61	2.55	2.45[b]
Netherlands	2.41	2.34[b]	2.35	2.38	2.41	2.44	2.53
United Kingdom	2.32	2.19	2.18[b]	2.21	2.24	2.28	2.39
United States	2.14	1.98[b]	2.06	2.24	2.35	2.48	2.79
Australia	3.34	2.72	2.46	2.25	2.16	2.09	2.01[b]
Austria	2.71	2.35	2.19	2.03	1.96	1.89	1.77[b]
Canada	3.68	3.08	2.83	2.63	2.54	2.47	2.39[b]
Finland	3.37	2.70	2.42	2.20	2.12	2.06	2.01[b]
Japan	3.33	2.89	2.71	2.57	2.51	2.46	2.40[b]
Norway	1.62	1.54[b]	1.54[b]	1.56	1.58	1.61	1.68
Spain	2.06	1.85	1.76	1.70	1.68	1.66	1.65[b]
Sweden[c]	2.32	2.18	2.12	2.07	2.05	2.03	2.01[b]
Switzerland	3.45	2.98	2.81	2.71	2.68[b]	2.68[b]	2.71
Median RMSE	2.62	2.34	2.22	2.21	2.16	2.13[b]	2.14
Range	1.62–3.68	1.54–3.08	1.54–2.83	1.56–2.71	1.58–2.68	1.61–2.68	1.65–2.79

Notes:

[a] The coefficient estimate in (2.13) was employed to compute forecasts. When $\gamma = 0$, the forecasts are least squares forecasts, and when $\gamma = 10^6$, they are produced using the coefficient estimate in (2.8).

[b] Minimum of entries in each row.

[c] Based on ten forecasts, 1974–83.

w_t is taken to be the median of the eighteen countries' output growth rates for the year t – (see figure 14.1 for a plot of w_t for the years in our sample).

To use (2.15) in forecasting, it is necessary to forecast w_t. The model for w_t in (2.16) was fitted by least squares, using data from 1954–73, and used to forecast the 1974 value and subsequent values with coefficient estimates updated year by year for the years 1974–84. The RMSEs of these one-year-ahead forecasts and of those yielded by an AR(3) model of w_t are reported, along with MAEs, in table 14.6. It is seen that the AR(3)LI model for w_t produced a RMSE of 1.48 and a MAE of 1.24, values much smaller than those associated with forecasts from an AR(3) model, 2.74 and 2.24, respectively. The forecasts from these two models for w_t, denoted by $\hat{w}_t(1)$ for the AR(3) forecasts and $\hat{w}_t(2)$ for the AR(3)LI model forecasts, were used to generate one-year-ahead forecasts

Table 14.6 *Root mean-squared errors (RMSEs) and mean absolute errors (MAEs) of one-year-ahead forecasts of the medians of eighteen countries' output growth rates, 1974–84*[a]

Model	RMSE	MAE
	(Percentage points)	
AR(3)	2.74	2.24
AR(3)LI[b]	1.48	1.24

Notes:
[a] The initial estimation period is 1954–73 (twenty years). Estimates are updated year by year in the forecast period.
[b] With w_t being median output growth rate in year t, the AR(3)LI model is $w_t = \pi_0 + \pi_1 w_{t-1} + \pi_2 w_{t-2} + \pi_3 w_{t-3} + \pi_4 x_{t-1} + \pi_5 z_{t-1} + \varepsilon_t$, where, for year t, x_t is the median of countries' growth rates of real stock prices and z_t the median of countries' growth rates of real money. This equation was employed to generate one-year-ahead least squares forecasts for each year, 1974–84.

from (2.15) for individual countries' output growth rates for 1974–84, using the γ-forecast with a very large value of γ. Also, for comparative purposes, a "perfect foresight" model, one in which it is assumed that w_t is known exactly, was employed to generate forecasts with results given in . . . column [(1)] of table 14.7. With the w_t value assumed known in each forecast period, the median RMSE of the annual forecasts for the eighteen countries, 1974–84, is 1.82, a value much lower than those reported in table 14.3, and for the AR(3)LI model without the w_t variable it is 2.14, shown in . . . column [(4)] of table 14.7. In column (2) of table 14.7 are shown forecast RMSEs when w_t was forecasted using an AR(3) model. Since the AR(3) forecasts of w_t are not very good (see table 14.6) the forecasts of country output growth rates based on them are in general not as good as those based on known values of the w_t variable. When the AR(3)LI model in (2.16) was used to produce forecasts of w_t, denoted by $\hat{w}_t(2)$, and these were used to forecast individual countries' output growth rates, the results, as shown in column (3) of table 14.7, were much better. The median RMSE associated with these forecasts is 1.90, not far different from the "perfect foresight" median RMSE of 1.82. Also when $\hat{w}_t(2)$ was employed, sixteen of eighteen countries' RMSEs were reduced relative to the RMSEs for the AR(3)LI model without the world growth variable, shown in column (4) of table 14.7. Further, the median RMSE of 1.90, associated with the AR(3)LI world growth rate model, is smaller than all of those shown in table 14.3.

Table 14.7 *Root mean-squared errors (RMSEs) of one-year-ahead forecasts of annual real output growth rates employing an AR(3)LI model including the world growth rate, 1974–1984*[a]

Country	AR(3)LI with w_t[b] (1)	AR(3)LI with $\hat{w}_t(1)$[c] (2)	AR(3)LI with $\hat{w}_t(2)$[d] (3)	AR(3)LI without w_t[e] (4)
		(Percentage points)		
Belgium	1.54	2.80	1.79	1.96
Denmark	1.77	2.74	2.04	2.26
France	1.03	1.90	1.36	1.66
Germany	0.90	2.39	1.35	2.00
Ireland	2.82	1.57	2.54	2.14
Italy	1.57	3.08	1.82	2.45
Netherlands	1.86	2.52	2.27	2.53
United Kingdom	1.87	2.67	2.21	2.39
United States	2.56	2.95	2.36	2.79
Australia	2.24	2.39	1.96	2.01
Austria	1.23	2.06	1.57	1.77
Canada	2.27	2.81	2.15	2.39
Finland	2.14	2.13	1.83	2.01
Japan	2.85	2.30	2.58	2.40
Norway	1.66	1.79	1.45	1.68
Spain	1.12	2.03	1.22	1.65
Sweden[f]	1.62	2.58	1.70	2.01
Switzerland	2.20	3.17	2.42	2.71
Median RMSE	1.82	2.46	1.90	2.14
Range	0.90–2.85	1.57–3.17	1.22–2.58	1.65–2.71

Notes:
[a] The model employed is shown in (2.15) and estimated under the assumption that coefficients are the same for all countries.
[b] The value of w_t, the median output growth rate or world growth rate, is assumed known in the forecast period, a "perfect foresight" assumption.
[c] $\hat{w}_t(1)$ is a forecast of w_t from an AR(3) model for w_t (see table 14.6).
[d] $\hat{w}_t(2)$ is a forecast of w_t from the AR(3)LI model described in note *b* of table 14.6.
[e] Least squares forecasts using AR(3)LI model in (2.1), with the use of coefficient estimate in (2.8).
[f] Based on ten forecasts, 1974–83.

4.3 *Comparisons with OECD forecast RMSEs*

Smyth (1983) has presented a description of the forecasting procedures employed by the OECD to produce annual forecasts of seven countries' annual rates of growth of output, 1968–79. The OECD forecasts are

Table 14.8 *Comparison with OECD forecast root mean-squared errors (RMSEs)*

Forecasts	Canada	France	Germany	Italy	Japan	United Kingdom	United States	Median	Range
				(Percentage points)					
1 OECD, 1968–79	1.71	1.45	2.12	2.86	4.40	2.26	1.38	2.12	1.38–4.40
2 AR(3)LI, 1974–84									
a Least squares	3.68	2.52	2.28	3.40	3.33	2.32	2.14	2.52	2.14–3.68
b γ-forecasts[a]	2.39	1.66	2.00	2.45	2.40	2.39	2.79	2.39	1.66–2.79
c η-forecasts[b]	2.92	2.07	1.94	2.73	2.51	2.63	2.03	2.51	1.94–2.92
3 AR(3)LI with forecasted w_t, 1974–84[c]	2.15	1.36	1.35	1.82	2.58	2.21	2.36	2.15	1.35–2.58

Notes:

[a] $\gamma = 10^6$ in model for eighteen countries.

[b] $\eta = 0.5$ in model for eighteen countries.

[c] w_t, the world output growth rate was forecasted from (2.16) and used in (2.15) to produce γ-forecasts for eighteen countries with $\gamma = 10^6$.

derived from elaborate country econometric models and are subjected to judgmental adjustments by individuals not associated with the modeling process – that is, by individuals from the OECD's various country desks (see Smyth 1983, p. 37). In Garcia-Ferrer *et al.* (1987, pp. 61ff.) comparisons of RMSEs of OECD forecasts for five countries with those provided by AR(3)LI models were presented and discussed. In table 14.8, OECD RMSEs of forecast for seven countries, 1968–79, are presented along with forecast RMSEs for the same countries, 1974–84, computed in the present study. While the forecast periods 1968–79 and 1974–84 are somewhat different and different methodologies were employed, it is still of interest to consider the relative forecasting performance of OECD and our forecasts.

From table 14.8, it is seen that the OECD forecast RMSEs have a median of 2.12 percentage points with a range of 1.45–4.40. The OECD's RMSE for Japan, 4.40 is quite large. The median RMSEs, in our study range from 2.52 for least squares forecasts from our AR(3)LI model in (2.1) to 2.15 for the forecasts obtained from our "AR(3)LI world income" model in (2.15) using a forecasted value of w_t from (2.16) and a γ-forecast with $\gamma = 10^6$. The range of the forecast RMSEs in this latter case is 1.35–2.58. The RMSEs in line 3 of table 14.8 are smaller than the corresponding OECD RMSEs in five of seven cases. Large reductions were encountered for Germany, Italy, and Japan, while smaller reductions appeared for France and the United Kingdom. In the cases of the United States and Canada, the OECD RMSEs were smaller, much smaller for the United States. These comparisons, however, must be qualified since the forecast period and data used by OECD are different from those employed in this study.

5 Summary and concluding remarks

We have presented the results of forecasting experiments for nine countries' annual output growth rates for the periods 1974–81 and 1974–84 and similar results for an additional sample of nine countries for the period 1974–84. In general, the forecasting experiments revealed that methods and models employed . . . in Garcia-Ferrer *et al.* (1987) worked reasonably well when applied to an extended sample of data and countries. The shrinkage forecasting procedures produced larger reductions in RMSEs of forecast for the nine additional countries than for the original sample of nine countries. The AR(3)LI model, incorporating a world income growth rate variable, forecasted the best and provided forecast RMSEs that compared favorably with the RMSEs of OECD forecasts for a subsample of seven countries that were produced using complex country models and judgmental adjustments.

In future work, we shall extend our forecasting experiments to include forecasts derived from time-varying parameter Bayesian state–space models – see Garcia-Ferrer *et al.* (1987) and Highfield (1986) for macroeconomic forecasting results obtained with such models. In addition, as stated in section 1, macroeconomic structural equation systems are under consideration that have reduced form equations for the rate of growth of output that are similar in form to the forecasting equations used in our present and past work. These structural macroeconomic equation systems also yield reduced form equations for several additional variables that will be appraised in future forecasting experiments. With several forecasting equations per country, there will be an opportunity to experiment with a broader range of shrinkage forecasting techniques. Hopefully, this work will yield a set of forecasting equations for each country that yield reasonably good forecasts for a broad sample of countries and for a temporally expanded data set. Further, disaggregation will be studied by modeling components of GNP as fractions or log-fractions of relevant aggregates. In this way, we hope to "iterate in" to satisfactory structural models for countries in the structural econometric modeling, time series analysis (SEMTSA) approach put forward in our past work (see Zellner and Palm 1974, 1975; Palm 1983; and Zellner 1979, 1984, 1987).

BIBLIOGRAPHY

Bowman, H. W. and A. M. Laporte (1975), "Stochastic optimization in recursive equation systems with random parameters with an application to control of the money supply," in S. E. Fienberg and A. Zellner (eds.), *Studies in Bayesian Econometrics and Statistics in Honor of Leonard J. Savage* (Amsterdam, North-Holland), 441–62

Garcia-Ferrer, A., R. A. Highfield, F. C. Palm, and A. Zellner (1987), "Macroeconomic forecasting using pooled international data," *Journal of Business and Economic Statistics* 5(1), 53–67; chapter 13 in this volume

Highfield, R. A. (1986), "Forecasting with Bayesian state – space models," Unpublished doctoral dissertation, Graduate School of Business, University of Chicago

Lindley, D. V. and A. F. M. Smith (1972), "Bayes' estimates for the linear model," *Journal of the Royal Statistical Society*, B 34, 1–41

Miller, R. B. and W. G. Fortney (1984), "Industry-wide expense standards using random coefficient regression," *Insurance: Mathematics and Economics* 3, 19–33

Palm, F. C. (1983), "Structural econometric modeling and time series analysis: an integrated approach," in A. Zellner (ed.), *Applied Time Series Analysis of Economic Data, Proceedings of the Conference on Applied Time Series Analysis of Economic Data*, October 13–15, 1981 (Washington, DC, US Government Printing Office), 199–233; see chapter 3 in this volume

Smyth, D. J. (1983), "Short-run macroeconomic forecasting: the OECD performance," *Journal of Forecasting* 2, 37–49

Zellner, A. (1971), *An Introduction to Bayesian Inference in Econometrics* (New York, John Wiley); reprinted (Melbourne, Fla., Krieger, 1987)

(1979), "Statistical analysis of econometric models," *Journal of the American Statistical Association* 74, 628–51; chapter 2 in this volume

(1983), "Applications of Bayesian analysis in econometrics," *The Statistician* 32, 23–34

(1984), *Basic Issues in Econometrics* (Chicago, University of Chicago Press)

(1986), "On assessing prior distributions and Bayesian regression analysis with g-prior distributions," in P. K. Goel and A. Zellner (eds.), *Bayesian Inference and Decision Techniques: Essays in Honor of Bruno de Finetti* (Amsterdam, North-Holland), 233–43

(1987), "Macroeconomics, econometrics and time series analysis," *Revista Espanola de Economia* 4, 3–9

Zellner, A. and F. C. Palm (1974), "Time series analysis and simultaneous equation econometric models," *Journal of Econometrics* 2, 17–54; chapter 1 in this volume

(1975), "Time series and structural analysis of monetary models of the US economy," *Sankhyā: The Journal of the Indian Statistical Association*, Series C 37, 12–56; chapter 6 in this volume

15 Turning points in economic time series, loss structures, and Bayesian forecasting (1990)

Arnold Zellner, Chansik Hong, and Gaurang M. Gulati

1 Introduction

In a letter commenting on a draft of Zellner (1987), Barnard (1987) wrote, "I very much liked your emphasis on the need for sophisticated, simple model building and testing in social science." Apparently, Barnard and many other scientists are disturbed by the complexity of many models put forward in econometrics and other social sciences. And indeed we think that they should be disturbed since not a single complicated model has worked very well in explaining past data and in predicting as yet unobserved data. In view of this fact, in Garcia-Ferrer *et al.* (1987) and Zellner and Hong (1989), a relatively simple, one-equation model for forecasting countries' annual output growth rates was formulated, applied, and found to produce good forecasts year by year, 1974–84 for eighteen countries. This experience supports Barnard's and many others' preference for the use of sophisticatedly simple models and methods. See Zellner (1988) for further discussion of this issue.

In the present chapter, we extend our previous work to consider the problem of forecasting future values and turning points of economic time series given explicit loss structures. Kling (1987, pp. 201–4) has provided a good summary of past work on forecasting turning points by Moore (1961, 1983), Zarnowitz (1967), Wecker (1979), Moore and Zarnowitz (1982), Neftci (1982), and others. In this work there is an emphasis on the importance and difficulty of forecasting turning points. Also, in our opinion, not enough attention has been given to the role of loss structures in forecasting just turning points and in forecasting turning points *and* the future values of economic variables.

Research financed in part by the National Science Foundation and by income from the H. G. B. Alexander Endowment Fund, Graduate School of Business, University of Chicago.

Originally published as part VI in S. Geisser, J. S. Hodges, S. J. Press, and A. Zellner (eds.), *Bayesian and Likelihood Methods in Statistics and Econometrics: Essays in Honor of George A. Barnard* (Amsterdam, North-Holland, 1990), 371–93.

The plan of our chapter is as follows. In section 2, we introduce loss structures and explain how optimal forecasts of the occurrence of turning points and future values of economic time series can be computed. Applications of our procedures are presented in section 3 based on annual output growth rate data for eighteen countries and an autoregressive model with leading indicator variables employed in our previous work (see Garcia-Ferrer *et al.* 1987 and Zellner and Hong 1989). In section 4, a summary of results and some concluding remarks are presented.

2 Turning points, loss structures, and forecasting

As Wecker (1979) and Kling (1987) recognize, given a model for past observations and a definition of a turning point, probabilities relating to the occurrence of future turning points can be evaluated. In our work, we, along with Kling (1987), evaluate these probabilities using predictive probability density functions for as yet unobserved, future values of variables which take account of uncertainty regarding the values of model parameters as well as the values of future error terms. We also indicate how to take account of model uncertainty in evaluating probabilities relating to future events such as the occurrence or non-occurrence of a turning point at a future time.

2.1 Forecasting turning points

As regards the definition of turning points in an economic time series, we recognize that there are many possible definitions and that no one definition has achieved universal acceptance. Thus in what follows, we present analysis that can be employed for whatever definition of a turning point is adopted. We shall employ one, particularly simple definition of a turning point to illustrate our analysis and then go on to use a broader definition in our applied work.

Let the given past measurements of a variable, say the growth rate of real GNP, be denoted by $y' = (y_1, y_2, \ldots, y_{n-1}, y_n)$ and $z \equiv y_{n+1}$ be the first future value of the series. Particular definitions of a downturn (DT) and of an upturn (UT) and their negations, based on just y_{n-1}, y_n and z are given by,

$$y_{n-1} < y_n \text{ and } \begin{cases} z < y_n \equiv \text{Downturn } (DT_1) \\ z \geq y_n \equiv \text{No Downturn } (NDT_1), \end{cases} \tag{2.1a}$$

and

$$y_{n-1} > y_n \text{ and } \begin{cases} z > y_n \equiv \text{Upturn } (UT_1) \\ z \leq y_n \equiv \text{No Upturn } (NUT_1), \end{cases} \quad (2.1\text{b})$$

where the subscript 1 on DT, UT, etc., denotes that one past observation, y_{n-1} and one future observation, $z \equiv y_{n+1}$ have been employed in defining downturns and upturns. Below, we shall introduce broader definitions of DTs and UTs.

Given a model assumed to generate the observations, $\underline{y}$, let $p(\underline{y} \mid \underline{\theta})$ be the likelihood function for the past data, where $\underline{\theta}$ is a vector of parameters, $\pi(\underline{\theta} \mid I)$ a prior distribution for $\underline{\theta}$, and $p(\underline{\theta} \mid D) \propto \pi(\theta \mid I)\, p(y \mid \underline{\theta})$ the posterior distribution for $\underline{\theta}$, where $D = (\underline{y}, I)$ represents the past sample and prior information. Then with $p(z \mid \underline{\theta}, D)$ representing the probability density function (pdf) for $z \equiv y_{n+1}$, given $\underline{\theta}$ and the past data, the predictive pdf for z is given by

$$p(z \mid D) = \int_{\Theta} p(z \mid \underline{\theta}, D)\, p(\underline{\theta} \mid D) d\underline{\theta} \quad -\infty < z < \infty, \quad (2.2)$$

where $\underline{\theta} \subset \Theta$, the parameter space. As is well known, $p(z \mid D)$ can be viewed as an average of the conditional pdf, $p(z \mid \underline{\theta}, D)$ with the posterior pdf, $p(\underline{\theta} \mid D)$ serving as the weighting function.

The predictive pdf in (2.2) can be employed to obtain probabilities associated with the events in (2.1). For example, if $y_{n-1} < y_n$, the probability of a downturn, P_{DT_1} is given by

$$P_{DT_1} = \int_{-\infty}^{y_n} p(z \mid D) dz, \quad (2.3)$$

while the probability of no downturn is $P_{NDT_1} = 1 - P_{DT_1}$. Note that with $y_{N-1} < y_N$, the probability of an upturn, given the definitions in (2.1), is zero.

With $y_{n-1} < y_n$ and the probability of a downturn P_{DT_1} as given by (2.3), we now wish to make a decision as to whether to forecast a downturn or no downturn. To solve this problem, consider the loss structure shown in table 15.1. The two possible outcomes are DT_1 and NDT_1. If the act forecast a DT_1 is chosen, loss is scaled to be 0 if the forecast is correct and to be $c_1 > 0$ if it is incorrect. If the act forecast a NDT_1 is chosen, loss is 0 if the outcome is NDT_1 and $c_2 > 0$ if it is DT_1. In many circumstances, $c_1 \neq c_2$. From (2.3), probabilities associated with the outcomes DT_1 and NDT_1 are available and can be used to compute expected losses associated with the two acts shown in table 15.1 as follows:

$$EL \mid \text{Forecast } DT_1 = 0 \cdot P_{DT_1} + c_1(1 - P_{DT_1}) = c_1(1 - P_{DT_1}), \quad (2.4\text{a})$$

Table 15.1 *Forecasting loss structure given* $y_{n-1} < y_n^a$

Acts	Possible outcomes	
	DT_1	$N\,DT_1$
Forecast DT_1	0	c_1
Forecast $N\,DT_1$	c_2	0

Note:
[a] c_1 and c_2 are given positive quantities.

and

$$EL|\ \text{Forecast } NDT_1 = c_2 P_{DT_1} + 0(1 - P_{DT_1}) = c_2 P_{DT_1}. \tag{2.4b}$$

If (2.4a) is less than (2.4b), that is $c_1(1 - P_{DT_1}) < c_2 P_{DT_1}$ or, equivalently

$$1 < \frac{c_2}{c_1}\left(\frac{P_{DT_1}}{1 - P_{DT_1}}\right), \tag{2.5a}$$

or

$$P_{DT_1} > \frac{c_1}{c_1 + c_2}, \tag{2.5b}$$

then choosing the act forecast a DT_1 will lead to lower loss than choosing the act forecast $N\,DT_1$. Note that if $c_2 = c_1$, this rule leads to a forecast of a DT_1 if $P_{DT_1} > 1/2$. On the other hand, if c_2/c_1 is much larger than 1, say $c_2/c_1 = 2$, then the condition in (2.5) would be satisfied for $P_{DT_1}/(1 - P_{DT_1}) > 1/2$ or $P_{DT_1} > 1/3$. This example indicates that the decision to forecast a downturn is very sensitive to the value of the ratio c_2/c_1 as well as to the value of P_{DT_1}. Thus, considering just the value of P_{DT_1} in forecasting turning points is not usually satisfactory except in the special case of symmetric loss, $c_1 = c_2$.[1]

2.2 *Point forecasts using different loss functions*

Above, we have considered the problem of forecasting turning points. Now we consider the problem of deriving point forecasts using two different loss functions, one appropriate for a downturn situation and the other for a no-downturn situation. That loss functions can be different in

[1] Since analysis of forecasting an upturn, UT_1, given that $y_{n-1} > y_n$ is similar to that for forecasting a downturn, it will not be presented.

these two situations is an important consideration which should be taken into account in producing forecasts.

Consider forecasting the value of $z \equiv y_{n+1}$ given $y_{n-1} < y_n$ and the probability of a DT_1 from (2.3) employing squared error loss functions. Let

$$L_{DT_1} = k_1(\hat{z} - z)^2 \quad k_1 > 0 \tag{2.6a}$$

be the loss incurred if a downturn occurs and $\hat{z}$ is used as a point forecast of z and

$$L_{NDT_1} = k_2(\hat{z} - z)^2 \quad k_2 > 0 \tag{2.6b}$$

be the loss incurred if no downturn occurs and $\hat{z}$ is used as a point forecast of z. Expected loss is

$$EL = P_{DT_1} k_1 E(\hat{z} - z)^2 | DT_1 + (1 - P_{DT_1}) k_2 E(\hat{z} - z)^2 | NDT_1. \tag{2.7}$$

On minimizing (2.7) with respect to the choice of $\hat{z}$, the minimizing value for $\hat{z}$, denoted by $\hat{z}^*$, is[2]

$$\hat{z}^* = \frac{P_{DT_1} k_1 \bar{z}_{DT_1} + (1 - P_{DT_1}) k_2 \bar{z}_{NDT_1}}{P_{DT_1} k_1 + (1 - P_{DT_1}) k_2}, \tag{2.8}$$

a weighted average of the conditional mean of z given $z < y_n$, $\bar{z}_{DT_1}$, and the conditional mean of z given $z \geq y_n$, $\bar{z}_{NDT_1}$, with weights $P_{DT_1} k_1$ and $(1 - P_{DT_1}) \times k_2$. Note that if $k_1 = k_2$, $\hat{z}^* = \bar{z}$, the mean of the predictive pdf $p(z \mid D)$ in (2). However, if $k_1 \neq k_2$, $\hat{z}^*$ in (2.8) will not be equal to $\bar{z}$. For example, if $P_{DT_1} = 1/2$, (2.8) reduces to $\hat{z}^* = (k_1 \bar{z}_{DT_1} + k_2 \bar{z}_{NDT_1})/(k_1 + k_2)$ which differs from $\bar{z}$. Thus while $\bar{z}$ is optimal relative to an overall squared error loss function ($k_1 = k_2$), it is *not* optimal in the case that different loss functions ($k_1 \neq k_2$) are appropriate for downturn and no-downturn situations.

In (2.6a–b), we have allowed for the possibility that loss functions may be different for downturn and no-downturn cases. However, the use of *symmetric*, squared error loss functions may not be appropriate in all circumstances. If $L_{DT_1}(z, \hat{z})$ and $L_{NDT_1}(z, \hat{z})$ are general convex loss functions for DT_1 and NDT_1, respectively, then expected loss is given by

$$EL = P_{DT_1} EL_{DT_1}(z, \hat{z}) | DT_1 + (1 - P_{DT_1}) EL_{NDT_1}(z, \hat{z}) | NDT_1 \tag{2.9}$$

[2] To derive $\hat{z}^*$ in (2.8) express (2.7) as $EL = P_{DT_1} k_1 [\mathrm{E}(z - \bar{z}_{DT_1})^2 \mid DT_1 + (\hat{z} - z_{NDT_1})^2] + (1 - P_{DT_1}) k_2 [E(z - \bar{z}_{NDT_1})^2 \mid NDT_1 + (\hat{z} - \bar{z}_{NDT_1})^2]$ and minimize with respect to $\hat{z}^*$.

where the expectations on the r.h.s. of (2.9) are computed using the conditional predictive pdfs $p(z \mid z < y_n, D)$ and $p(z \mid z \geq y_n, D)$. Then EL in (2.9) can be minimized, analytically or by computer methods to obtain the minimizing value of $\hat{z}$.

To illustrate the approach described in the previous paragraph, it may be that given a DT, over-forecasting is much more serious than under-forecasting by an equal amount. A loss function, the LINEX loss function, employed in Varian (1975) and Zellner (1986), is a convenient loss function which captures such asymmetric effects. It is given by:

$$L_{DT_1} = b_1[e^{a_1(\hat{z}-z)} - a_1(\hat{z}-z) - 1] \quad \begin{array}{l} b_1 > 0 \\ a_1 > 0. \end{array} \tag{2.10}$$

When $\hat{z} = z$, loss is zero and when $\hat{z} - z > 0$, a case of over-forecasting, loss rises almost exponentially with $a_1 > 0$. When $\hat{z} - z < 0$, loss rises almost linearly. Choice of the value of a_1 governs the degree of asymmetry. For example when a_1 has a small value, the loss function in (2.10) is close to a symmetric squared error loss function as can be seen by noting that $e^{a_1(\hat{z}-z)} \doteq 1 + a_1(\hat{z}-z) + a_1^2(\hat{z}-z)^2/2$ and substituting this expression in (2.10).

With NDT_1, it may be that under-forecasting is a more serious error than over-forecasting by an equal amount. The following LINEX loss function provides such asymmetric properties.

$$L_{NDT_1} = b_2[e^{a_2(\hat{z}-z)} - a_2(\hat{z}-z) - 1] \quad \begin{array}{l} b_2 > 0 \\ a_2 < 0. \end{array} \tag{2.11}$$

The loss functions in (2.10) and (2.11) can be inserted in (2.9) and an optimal point forecast can be computed. Note that the necessary condition for a minimum of (2.9) is

$$P_{DT_1}\frac{dEL_{DT_1} \mid DT_1}{d\hat{z}} + (1 - P_{DT_1})\frac{dEL_{NDT_1} \mid NDT_1}{d\hat{z}} = 0. \tag{2.12}$$

If the derivatives in this last expression are approximated by expanding them around $\hat{z}_1$ and $\hat{z}_2$, values which set them equal to zero, respectively, the approximate value of $\hat{z}$, $\hat{z}^*$, which satisfies (2.12) is given by

$$\hat{z}^* = w\hat{z}_1 + (1-w)\hat{z}_2, \tag{2.13}$$

where $w = P_{DT_1}b_1a_1^2/[P_{DT_1}b_1a_1^2 + (1 - P_{DT_1})b_2a_2^2]$. Thus it is seen that when $[P_{DT_1}/(1 - P_{DT_1})](b_1a_1^2/b_2a_2^2]$ is large, $\hat{z}^*$ is close to $\hat{z}_1$ and when it is small, $\hat{z}^*$ is close to $\hat{z}_2$, with $\hat{z}_1 = -\ell n(Ee^{-a_1z} \mid DT_1)/a_1$ and $\hat{z}_2 = -\ell n(Ee^{-a_2z} \mid NDT_1)/a_2$. Again, the optimal point forecast in (2.12) is sensitive not only to the values of P_{DT_1} and $1 - P_{DT_1}$, but also to the values of the loss functions' parameters, a_1, a_2, b_1, b_2.

Table 15.2 *Extended forecasting loss structure*[a]

Acts	Possible outcomes DT_1	$N\,DT_1$
Forecast DT_1	$k_1(\hat{z}-z)^2$	$k_2(\hat{z}-z)^2+c_1$
Forecast $N\,DT_1$	$k_1(\hat{z}-z)^2+c_2$	$k_2(\hat{z}-z)^2$

Note:
[a] It is assumed that $y_{n-1} < y_n$ and DT_1 and $N\,DT_1$ are defined as in (2.1a) of the text. The random variable $z \equiv y_{n+1}$ denotes the as yet unobserved value of y_{n+1} and $\hat{z}$ is some point forecast of z. k_1, k_2, c_1, and c_2 are given constants.

2.3 *Turning point and point forecasting combined*

Above, we have considered forecasting turning points and point forecasting *separately*. In this section, we provide a loss structure within the context of which it is possible to obtain a minimum expected loss solution to the joint problem of forecasting turning points and obtaining point forecasts.

The loss structure which we shall employ is shown in table 15.2. It is seen that we allow for different quadratic losses, $k_1(\hat{z}-z)^2$ and $k_2(\hat{z}-z)^2$ for DT_1 and $N\,DT_1$ cases as in (2.6). Also, we incorporate additive elements of loss, c_1 and c_2 to reflect errors in turning point forecasts as in table 15.1.

To obtain the minimum expected loss solution, we consider choice of the DT_1 forecast, derive the optimal point prediction and evaluate expected loss using it. We do the same for a $N\,DT_1$ forecast and then choose the act, forecast DT_1 or forecast $N\,DT_1$ which has the lower evaluated expected loss. As will be seen, the optimal point forecast of z has a value consistent with the selected optimal turning point forecast.

Given a choice of the DT_1 forecast, expected loss is given by:

$$\begin{aligned} EL_{DT_1} &= P_{DT_1} k_1 E[(\hat{z}-z)^2 \mid DT_1] \\ &\quad + (1-P_{DT_1})\{k_2 E[(\hat{z}-z)^2 \mid N\,DT_1] + c_1\}. \end{aligned} \tag{2.14}$$

The value of $\hat{z}$ which minimizes (2.14) subject to the side condition $\hat{z} < y_n$ is denoted by $\hat{z}_{DT_1}$. Then (2.14) is evaluated at $\hat{z} = \hat{z}_{DT_1}$. Similarly, if the $N\,DT_1$ forecast is chosen, expected loss is given by:

$$\begin{aligned} EL_{N\,DT_1} &= P_{DT_1}\{k_1 E(\hat{z}-z)^2 \mid DT_1 + c_2\} \\ &\quad + (1-P_{DT_1}) k_2 E(\hat{z}-z)^2 \mid N\,DT_1. \end{aligned} \tag{2.15}$$

The value of $\hat{z}$ which minimizes (2.15) subject to $\hat{z} \geq y_n$ is denoted by $\hat{z}_{NDT_1}$ and it can be used to evaluate (2.15). Then, if $EL_{DT_1}(\hat{z} = \hat{z}_{DT_1}) < EL_{NDT_1}(\hat{z} = \hat{z}_{NDT_1})$, we forecast a DT_1 and use $\hat{z}_{DT_1}$ as our point forecast. Alternatively, if $EL_{DT_1}(\hat{z} = \hat{z}_{DT_1}) > EL_{NDT_1}(\hat{z} = \hat{z}_{NDT_1})$, we forecast a NDT_1 and use $\hat{z}_{NDT_1}$ as our point forecast.

In a special case $k_1 = k_2$ and $c_1 = c_2 = 0$, the above procedure yields the mean of the predictive pdf for z, denoted by $\bar{z}$, as the optimal point prediction and a DT_1 forecast if $\bar{z} < y_n$ or a NDT_1 forecast if $\bar{z} > y_n$.

As another special case, consider $k_1 = k_2 = 1$ and $c_1, c_2 > 0$. Then minimization of (2.14) with respect to $\hat{z}$ subject to $\hat{z} < y_n$, leads to

$$\hat{z}_{DT_1} = \begin{cases} \bar{z} & \text{if } \bar{z} < y_n \\ y_n - \Delta & \text{if } \bar{z} \geq y_n, \end{cases} \tag{2.16}$$

where $\Delta > 0$ is arbitrarily small and

$$EL_{DT_1}(\hat{z} = \hat{z}_{DT_1}) = \begin{cases} (1 - P_{DT_1})c_1 + E(z - \bar{z})^2 & \text{if } \bar{z} < y_n \\ (1 - P_{DT_1})c_1 + E(z - \bar{z})^2 + (y_n - \bar{z})^2 & \text{if } \bar{z} \geq y_n. \end{cases}$$

Similarly, minimization of (2.15) with respect to $\hat{z}$ subject to $\hat{z} \geq y_n$ leads to:

$$\hat{z}_{NDT_1} = \begin{cases} y_n & \text{if } \bar{z} < y_n \\ \bar{z} & \text{if } \bar{z} \geq y_n, \end{cases} \tag{2.17}$$

and

$$EL_{NDT_1}(\hat{z} = \hat{z}_{NDT_1}) = \begin{cases} P_{DT_1}c_2 + E(z - \bar{z})^2 + (y_n - \bar{z})^2 & \text{if } \bar{z} < y_n \\ P_{DT_1}c_2 + E(z - \bar{z})^2 & \text{if } \bar{z} \geq y_n \end{cases}$$

Thus if $\bar{z} < y_n$ the optimal turning point forecast is DT_1 if

$$(1 - P_{DT_1})c_1 + E(z - \bar{z})^2 < P_{DT_1}c_2 + E(z - \bar{z})^2 + (y_n - \bar{z})^2,$$

or

$$1 < \frac{P_{DT_1}c_2 + (y_n - \bar{z})^2}{(1 - P_{DT_1})c_1}, \tag{2.18a}$$

or

$$P_{DT_1} > \frac{c_1 - (y_n - \bar{z})^2}{c_1 + c_2}. \tag{2.18b}$$

Also, with $\bar{z} \geq y_n$ and (2.18) holding, $\bar{z}$ is the optimal point forecast for z. By similar considerations, if $\bar{z} < y_n$ and $P_{DT_1} < [c_1 - (y_n - \bar{z})^2]/(c_1 + c_2)$ then the optimal turning point forecast is NDT_1 and the optimal point forecast is $\hat{z}_{NDT_1} = y_n$.

With $\bar{z} \geq y_n$, then by the same line of reasoning, if

$$P_{DT_1} > \frac{c_1 + (y_n - \bar{z})^2}{c_1 + c_2}, \tag{2.19}$$

the optimal forecasts are DT_1 and $\hat{z}_{DT_1} = y_n - \Delta$. On the other hand, if the inequality in (2.19) is reversed, the optimal forecasts are $N\,DT_1$ and $\hat{z}_{N\,DT_1} = \bar{z}$.

The analysis leading to (2.18) and (2.19) was based on the assumption that $k_1 = k_2 = 1$. The above analysis, while tedious, can be extended to cover the case $k_1 \neq k_2$ with $k_1, k_2 > 0$ and $c_1, c_2 > 0$.

2.4 *Summary and additional considerations*

In summary, it has been shown how the loss structures in tables 15.1 and 15.2 can be used to obtain forecasts of turning points and how the probability of a DT and various loss functions can be employed to obtain optimal point forecasts. The main conclusion which emerges is that forecasts of turning points and of a future value are quite sensitive to assumptions regarding loss structures.

Above we have concentrated attention on the first future observation, $z = y_{n+1}$ given $y_{n-1} < y_n$ or given $y_{n-1} > y_n$ (see (2.1)). Given a predictive pdf for the next q future observations, $\underline{z}' = (z_1, z_2, \ldots, z_q) \equiv (y_{n+1}, y_{n+2}, \ldots, y_{n+q})$, namely $p(\underline{z} \mid D)$, $-\infty < z_i < \infty$, $i = 1, 2, \ldots, q$, it is possible to calculate the probabilities associated with various inequalities involving the z_is, using the predictive pdf, $p(\underline{z} \mid D)$. For example, the probability that $z_1, z_2 < z_3 > z_4, z_5$ can be computed.

The definitions in (2.1) can be broadened to include more past and future observations, as Wecker (1979) and Kling (1987) indicate. As an example consider, with $z_1 \equiv y_{n+1}$ and $z_2 \equiv y_{n+2}$,

$$y_{n-2} < y_{n-1} < y_n \text{ and } \begin{cases} z_1 < y_n \text{ and } z_2 < z_1 \equiv DT_2 \\ \text{otherwise } \equiv N\,DT_2, \end{cases} \tag{2.20}$$

where $DT_2 \equiv$ downturn based on two previous observations and two future observations relative to the given value y_n and $N\,DT_2 \equiv$ no such downturn. Then given $y_{n-2} < y_{n-1} < y_n$, the probability of a DT_2, P_{DT_2}, is given by

$$P_{DT_2} = \int_{-\infty}^{y_n} \int_{-\infty}^{z_1} p(z_1, z_2 \mid D)\, dz_2\, dz_1, \tag{2.21}$$

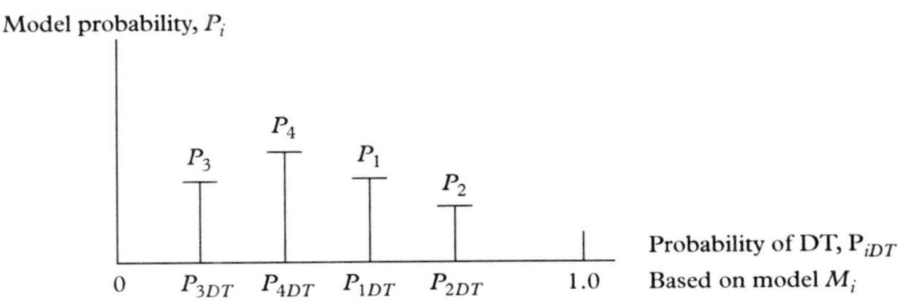

Figure 15.1 Probability mass function for probabilities of a DT

where it has been assumed that $\infty < z_i < \infty$, $i = 1, 2$.[3] This probability can be employed along with the loss structure in table 15.1 to choose between the acts forecast DT_2 and forecast $N\ DT_2$ so as to minimize expected loss. Also, this analysis can be combined with the problem of obtaining optimal point predictions of z_1 and z_2. Finally, using m past values and m future values, a DT_m can be defined and its probability calculated and used in forecasting.

Above, we have considered just one model for the observations, $\underline{y}' = (y_1, y_2, \ldots, y_{n-1}, y_n)$. Often forecasters utilize several alternative models, $M_1, M_2, \ldots, M_r$, for example autoregressive models of differing orders, autoregressive models with various leading indicator variables, etc. Let $p_i(\underline{y} \mid M_i, D_i)$ be the marginal pdf for $\underline{y}$, based on model M_i and sample and prior information D_i.[4] Then the posterior probability associated with the ith model M_i, denoted by P_i is given by

$$P_i = \Pi_i p_i(\underline{y} \mid M_i, D_i) \Big/ \sum_{i=1}^{r} \Pi_i p_i(\underline{y} \mid M_i, D_i) \tag{2.22}$$

where Π_i, $i = 1, 2, \ldots, r$ is the prior probability associated with M_i.[5] Also, for each model the predictive pdf for $z = y_{n+1}$ can be derived and is denoted by $p_i(z \mid M_i, D_i)$. It can be used, for example, to compute the probability of a DT, P_{iDT}. The P_{iDT}s and the P_is in (2.22) allow us to form a probability mass function as shown in figure 15.1 for the case of $r = 4$. Figure 15.1 reveals the effects of model uncertainty on the

[3] Note that $y_{n-2}, y_{n-1} < y_n > z_1, z_2$ is an alternative definition of a DT_2 and its probability of occurring is $\int_{-\infty}^{y_n} \int_{-\infty}^{y_n} p(z_1, z_2 \mid D) dz_1 dz_2$, somewhat different from (2.21).

[4] As is well known, $p_i(\underline{y} \mid M_i, D_i) = \int_{\Theta_i} p_i(\underline{y} \mid M_i, \underline{\theta}_i) \pi(\underline{\theta}_i \mid I_i) d\underline{\theta}_i$ where $p_i(\underline{y} \mid M_i, \underline{\theta}_i)$ is the likelihood function given model i, $\underline{\theta}_i$ is a vector of parameters with prior pdf $\pi(\underline{\theta}_i \mid I_i)$, where I_i denotes the prior parameters, $D_i = (\underline{y}, I_i)$ and $\underline{\theta}_i \subset \underline{\Theta}_i$, the parameter space.

[5] Here we assume that the r models are mutually exclusive and exhaustive. It is possible to relax the assumption that the collection of models is exhaustive.

probability of a DT. That is, instead of having a single probability of a DT, when there is model uncertainty, we have several probabilities of a DT, $P_{1DT}, P_{2DT}, \ldots, P_{rDT}$ and their respective probabilities, $P_1, P_2, \ldots, P_r$. The usual practice of selecting one model and viewing it as "absolutely true" in deriving a single probability of a DT obviously abstracts from model uncertainty and is inappropriate when model uncertainty is present.

Formally, we have the marginal predictive pdf, $p(z \mid D)$, given by

$$p(z \mid D) = \sum_{i=1}^{r} P_i p_i(z \mid M_i, D_i), \tag{2.23}$$

where P_i is given in (2.22) and D is the union of the D_i. Then, for example, the probability of a DT_1, P_{DT_1} is given by

$$\begin{aligned} P_{DT_1} &= \int_{-\infty}^{y_n} p(z \mid D)dz \\ &= \sum_{i=1}^{r} P_i P_{iDT_1}, \end{aligned} \tag{2.24}$$

where $P_{iDT_1} = \int_{-\infty}^{y_n} p_i(z \mid M_i, D_i)dz$ is the probability of DT_1 based on model M_i. It is seen from the second line of (2.24) that P_{DT_1} is an average of the P_{iDT_1} with the posterior model probabilities, the P_is serving as weights. Further, various measures can be computed to characterize the dispersion and other features of the P_{iDT_1}s. For example their variance is given by

$$\text{var}(P_{iDT_1}) = \sum_{i=1}^{r} P_i (P_{iDT_1} - P_{DT_1})^2. \tag{2.25}$$

Given a loss structure such as that in table 15.1, P_{DT_1} in (2.24) can be employed to make an optimal choice between forecast DT_1 or forecast $N\ DT_1$. Similar analysis yields results for forecasting an upturn or no upturn. For broader definitions of turning points, predictive pdfs for several future observations would replace $p_i(z \mid M_i, D_i)$ in (2.23) and the integral in (2.24) would have to be modified along the lines shown in (2.22).

Finally, the problem of forecasting turning points in two or more time series is of interest. Given a definition of a turning point, and a model for two or more time series, probabilities of downturns or upturns can be computed from the joint predictive pdf of future values of the several time series. Then, given a loss structure, optimal turning point forecasts can be derived. In the case of two time series, for example the rates of growth of output and of inflation, the set of possible forecasts will involve forecasting downturns for both variables, a downturn for one variable and no downturn for the other, or no downturn for both variables. Given m such possible forecasts and m possible outcomes, an $m \times m$ loss structure can be defined, an expanded version of the 2×2 case shown in table 15.1. Using probabilities associated with possible outcomes, the forecast that minimizes expected loss can be determined along the lines shown in (2.4) for the 2×2 case. Also, the multiple models, multiple time series case can be addressed using a generalization of the multiple models, one time series case analyzed above.

3 Data and applications

In this section some of the techniques described above will be applied in the analysis of data relating to *annual* output growth rates for eighteen countries used in our previous work, Garcia-Ferrer *et al.* (1987), and Zellner and Hong (1987) . . .

Shown in figure 15.2 is a boxplot of the annual rates of output growth for eighteen countries over the period, 1951–85. It is seen that the annual median growth rates, given by the horizontal line in each box, appear to follow a cyclical path with peaks in 1955, 1960, 1964, 1969, 1973, 1976, 1979, and 1984 and troughs in 1952, 1958, 1963, 1966, 1971, 1975, and 1981. The average time between peaks is 4.1 years and between troughs is 4.8 years. To provide more detail, figures. 15.3a and 15.3b provide the number of countries experiencing peaks and troughs in each year. It is seen that many countries experienced peaks in years close to or at 1955, 1960, 1964, 1969, 1973, 1976, 1979, and 1984. As regards troughs, they were encountered for many countries in years close to or at 1954, 1958, 1962, 1966–7, 1971, 1975, 1977, and 1982. These descriptive measures reflect the well-known fact that economies' output growth rates, while not perfectly synchronized, tend to move up and down together, as noted by Burns and Mitchell (1946), Zarnowitz (1985) and others.

We now turn to the problem of forecasting turning points for the eighteen countries' growth rates of annual output. *Here a downturn is defined to be a sequence of observations satisfying* $y_{n-1}, y_{n-2} < y_n > y_{n+1}$ *and an upturn a sequence of observations satisfying* $y_{n-2}, y_{n-1} > y_n < y_{n+1}$. While these are not the only possible definitions, they will be used since if there were a deterministic four-year cycle in the data, these definitions would be exactly appropriate for identifying peaks and troughs.

Our forecasting model, an autoregression of order 3 with leading indicator variables, denoted by AR(3)LI model, used in our previous work,

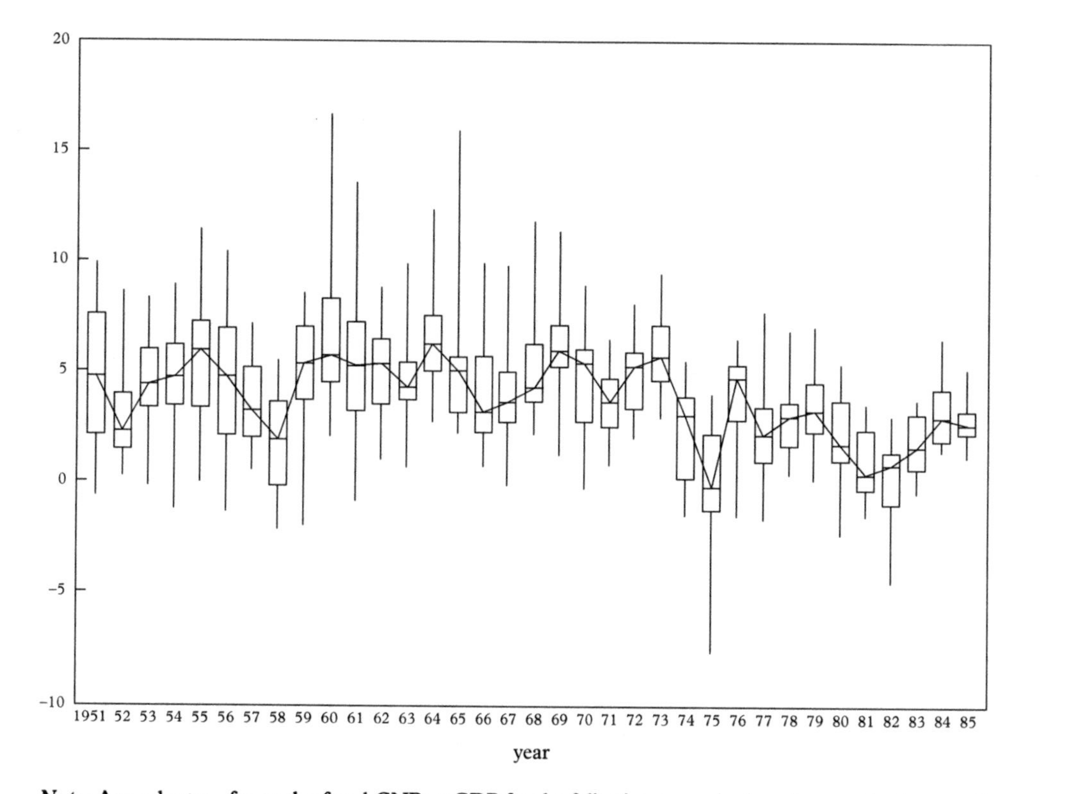

***Note*: Annual rates of growth of real GNP or GDP for the following countries have been utilized: Australia, Austria, Belgium, Canada, Denmark, Finland, France, Germany, Ireland, Italy, Japan, the Netherlands, Norway, Spain, Sweden, Switzerland, the United Kingdom, and the United States.**

Figure 15.2 Boxplot of annual real output growth rates for eighteen countries, 1951–1985
Source: Data from University of Chicago, Graduate School of Business IMF data base.

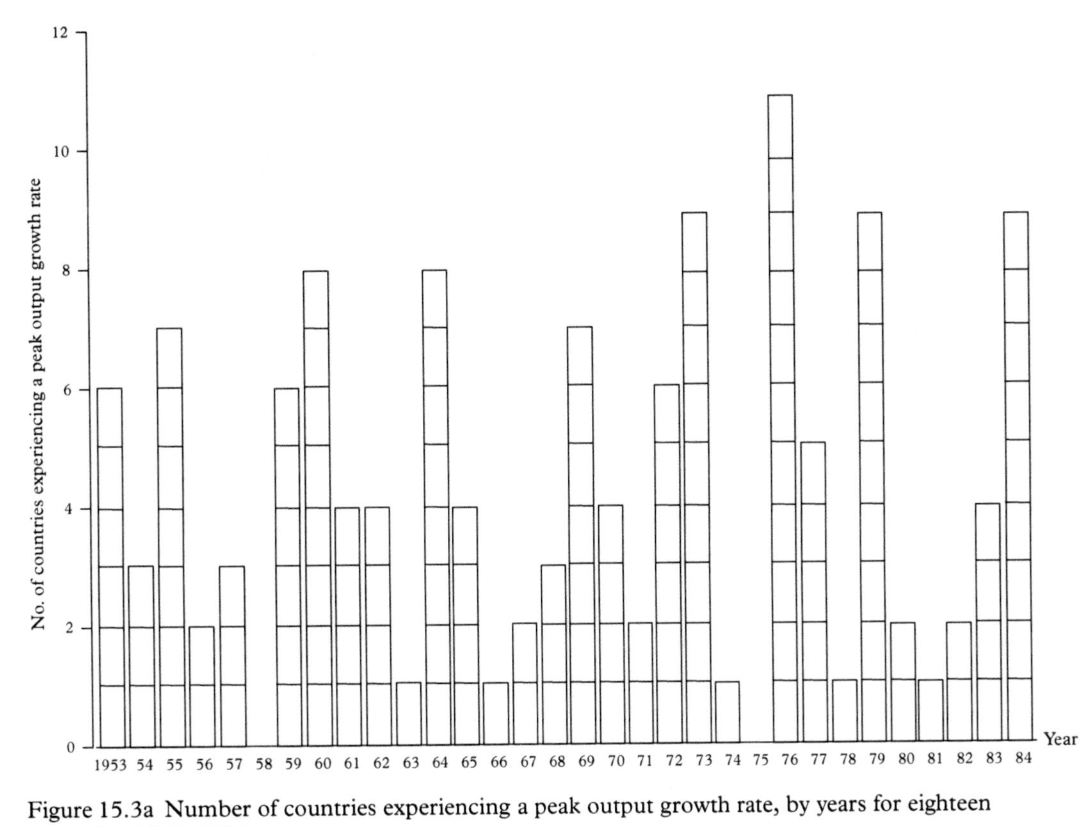

Figure 15.3a Number of countries experiencing a peak output growth rate, by years for eighteen countries, 1953–1984

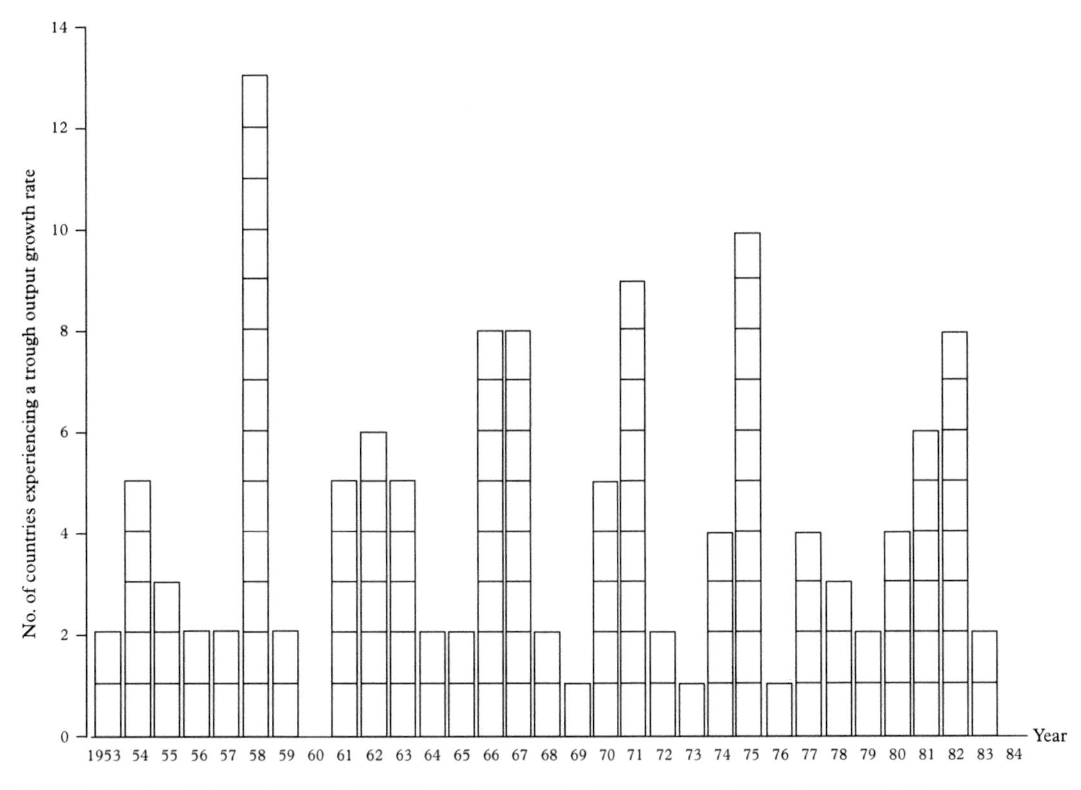

Figure 15.4b Number of countries experiencing a trough output growth rate, by years for eighteen countries, 1953–1984

Garcia-Ferrer *et al.* (1987), and Zellner and Hong (1987) is[6]

$$\begin{aligned} y_{it} = {} & \alpha_{i0} + \alpha_{1i} y_{it-1} + \alpha_{2i} y_{it-2} + \alpha_{3i} y_{it-3} && i = 1, 2, \ldots, 18 \\ & + \beta_{1i} SR_{it-1} + \beta_{2i} SR_{it-2} + \beta_{3i} GM_{it-1} && \\ & + \beta_{4i} WR_{t-1} + \epsilon_{it}, && t = 1, 2, \ldots, T \end{aligned} \tag{3.1}$$

where the subscripts i and t denote the ith country and tth year, respectively and

$$\begin{aligned} y_{it} &= \text{growth rate of real output} \\ SR_{it} &= \text{growth rate of real stock prices} \\ GM_{it} &= \text{growth rate of real money} \\ WR_t &= \text{“world return,” the median of the } SR_{it}\text{s} \\ \epsilon_{it} &= \text{error term} \end{aligned}$$

The ϵ_{it}s are assumed independently drawn from a normal distribution with zero mean and variance σ_i^2 for all i and t. Using annual data, 1951–73, a diffuse prior pdf for the parameters, the countries' Student-t predictive pdfs for 1974 were computed, updated year by year, and used to compute probabilities of downturns and upturns. Then using the loss structure in table 15.2 with $c_1 = c_2$ forecasts of downturns (DTs) and upturns (UTs) were made. In this case, a DT is forecasted when $P_{DT} > 1/2$ and NDT is forecasted when $P_{DT} < 1/2$. Also, a UT is forecasted when $P_{UT} > 1/2$ and NUT when $P_{UT} < 1/2$. For all eighteen countries over the period 1974–85, under the definition two preceding observations below (or above) a current observation and the following observation below (or above) the current observation for a DT (UT), our forecasting procedure yielded the results shown in table 15.3.

We see from panel A of table 15.3 that forty-five out of sixty-eight or 66 percent of the DT/NDT forecasts are correct. Of the forty DT forecasts thirty-five or 88 percent are correct. However, only ten of twenty-eight or 36 percent of the NDT forecasts are correct. As regards forecasts of UTs or NUTs, sixty of eighty-two forecasts, 73 percent are correct. Of the sixty-three UT forecasts, forty-three or 68 percent are correct while for the NUT forecasts, seventeen of nineteen, or 89 percent are correct. Thus except for the NDT forecasts, the turning point forecasts are quite good.

[6] See appendix tables [15A.1 and 15A.2] for a description of the statistical fits of (2.26) to our data and Garcia-Ferrer *et al.* (1987) and Zellner and Hong (1987) for the quality of point forecasts yielded by (2.26).

Table 15.3 *Forecasts of turning points for eighteen countries' output growth rates, 1974–1985*[a]

Forecast	Correct	Incorrect	Total
A Downturn (*DT*) and No Downturn (*N DT*)			
DT	35	5	40
N DT	10	18	28
Total	45	23	68
Percent	(66)	(34)	(100)
B Upturn (*UT*) and No Upturn (*NUT*)			
UT	43	20	63
NUT	17	2	19
Total	60	22	82
Percent	(75)	(25)	(100)

Note:
[a] With y_t denoting a growth rate in year t, the following definitions have been employed: $y_{n-2}, y_{n-1} < \dot{y}_n \leq y_{n+1} \equiv DT$; $y_{n-2}, y_{n-1} < y_n \leq y_{n+1} \equiv N\,DT$; $y_{n-2}, y_{n-1} > y_n < y_{n+1} \equiv UT$; and $y_{n-2}, y_{n-1} > y_n \geq y_{n+1} \equiv NUT$.

As regards the eighteen incorrect *N DT* forecasts, four occurred for calculated P_{DT}s between 0.40 and 0.49, eight for calculated P_{DT}s between 0.30 and 0.39 and four for calculated P_{DT}s between 0.20 and 0.29. Thus sixteen of the eighteen incorrect *N DT* forecasts had substantial probabilities of *DT*s associated with them. If a more conservative rule for forecasting *DT*s had been employed – say, forecast a downturn if $P_{DT} > 0.2$ – then only two of the *N DT* forecasts would be incorrect. The results of using a cut-off probability of 0.2, that is forecast a *DT* if $P_{DT} > 0.2$, which would be optimal given a cost parameter ratio $c_2/c_1 = 4$ in table 15.1, are: fifty-one or 85 percent of sixty *DT* forecasts correct and six or 75 percent of eight *N DT* forecasts correct. These figures indicate very well the sensitivity of turning point forecasting to asymmetry of loss structures.

4 Summary and concluding remarks

The problems of forecasting turning points and future values of economic time series were considered with the major result being that such forecasts are very sensitive to properties of loss structures. An operational procedure for forecasting turning points was formulated and applied to forecast turning points in output growth rates for eighteen countries, 1974–85, utilizing an AR(3)LI model fitted with pre-1974 data and updated year by year in the forecast period. In general, with the exception of no-downturn forecasts, the results were encouraging, namely 66 percent of the *DT* and *N DT* forecasts correct and 75 percent of the *UT* and *NUT* forecasts correct. These results indicate that our AR(3)LI model and our forecasting techniques may be of practical value to applied economic forecasters not only in providing forecasts regarding turning points but also in computing probabilities associated with future, as yet unobserved values of economic variables.

BIBLIOGRAPHY

Barnard, G. A. (1987), Personal communication

Burns, A. F. and W. C. Mitchell (1946), *Measuring Business Cycles* (New York, National Bureau of Economic Research)

Garcia-Ferrer, A., R. A. Highfield, F. C. Palm, and A. Zellner (1987), "Macroeconomic forecasting using pooled international data," *Journal of Business and Economic Statistics* 5 (1), 53–67; chapter 13 in this volume

Kling, J. L. (1987), "Predicting the turning points of business and economic time series," *Journal of Business* 60, 201–38

Moore, G. H. (1961), *Business Cycle Indicators* 1 (Princeton, Princeton University Press)

(1983), *Business Cycles, Inflation and Forecasting*, 2nd edn. (Cambridge, Mass., Ballinger)

Moore, G. H. and V. Zarnowitz (1982), "Sequential signals of recession and recovery," *Journal of Business* 55, 57–85

Neftci, S. H. (1982), "Optimal prediction of cyclical downturns," *Journal of Economic Dynamics and Control* 4, 225–41

Press, S. J. and A. Zellner (1978), "Posterior distribution for the multiple correlation coefficient with fixed regressors," *Journal of Econometrics* 8, 307–21

Varian, H. (1975), "A Bayesian approach to real estate assessment," in S. E. Fienberg and A. Zellner (eds.), *Studies in Bayesian Econometrics and Statistics in Honor of Leonard J. Savage* (Amsterdam, North-Holland), 195–208

Wecker, W. E. (1979), "Predicting the turning points of a time series," *Journal of Business* 52, 35–50

Zarnowitz, V. (1967), *An Appraisal of Short-Term Economic Forecasts* (New York, National Bureau of Economic Research)

(1985), "Recent work on business cycles in historical perspective," *Journal of Economic Literature* 23, 523–80

Zellner, A. (1986), "Bayesian estimation and prediction using asymmetric loss functions," *Journal of the American Statistical Association* 81, 446–51

(1987), "Science, economics and public policy," *American Economist* 31, 3–7

(1988), "Causality and causal laws in economics," in D. Aigner and A. Zellner (eds.), *Causality*, special issue of the *Journal of Econometrics* 39, 7–21

Zellner, A. and C. Hong (1989), "Forecasting international growth rates using Bayesian shrinkage and other procedures," *Journal of Econometrics, Annals* 40, 183–202; chapter 14 in this volume

APPENDIX

Table 15A.1 *Bayesian diffuse prior regression output for the AR(3)LI model in (3.1), eighteen countries, 1954–1973*[a]

	Posterior means and standard deviations[b]										
Country	Const.	y_{it-1}	y_{it-2}	y_{it-3}	$S R_{it-1}$	$S R_{it-2}$	$M G_{it-1}$	$W R_{t-1}$	R^{2c}	$E(\sigma \mid D)^d$	$r_1{}^e$
Belgium	−0.11 (1.86)	0.26 (0.25)	0.54 (0.35)	0.18 (0.26)	−0.09 (0.08)	−0.09 (0.04)	−0.06 (0.11)	0.22 (0.11)	0.55	1.68	0.08
Denmark	4.26 (1.92)	−0.31 (0.31)	−0.13 (0.33)	0.10 (0.24)	−0.03 (0.08)	0.02 (0.06)	0.36 (0.15)	0.03 (0.08)	0.40	2.25	−0.06
France	5.26 (1.68)	−0.10 (0.29)	0.02 (0.25)	0.13 (0.21)	−0.04 (0.03)	0.01 (0.02)	−0.03 (0.05)	0.10 (0.03)	0.56	0.98	−0.29
Germany	4.17 (2.36)	−0.23 (0.27)	−0.22 (0.25)	0.08 (0.20)	0.03 (0.05)	0.00 (0.04)	0.44 (0.21)	0.18 (0.11)	0.72	2.48	−0.11
Ireland	2.00 (1.38)	0.41 (0.41)	0.15 (0.40)	−0.13 (0.34)	0.05 (0.06)	−0.04 (0.05)	0.05 (0.14)	−0.10 (0.08)	0.43	2.57	−0.06
Italy	5.68 2.39	0.10 0.37	0.21 0.35	−0.44 0.27	0.04 0.04	−0.03 0.04	0.05 0.11	0.06 0.07	0.54	1.89	0.06
Netherlands	1.41 (2.05)	0.35 (0.19)	−0.02 (0.20)	0.10 (0.20)	−0.11 (0.07)	−0.08 (0.03)	0.39 (0.12)	0.33 (0.10)	0.78	1.67	0.22
United Kingdom	4.90 (2.01)	−0.53 (0.43)	−0.10 (0.55)	−0.11 (0.41)	0.01 (0.06)	0.00 (0.07)	0.20 (0.12)	0.03 (0.10)	0.42	1.71	−0.11
United States	4.98 (1.20)	−0.23 (0.19)	−0.04 (0.21)	−0.33 (0.17)	0.67 (0.05)	−0.03 (0.04)	0.67 (0.21)	−0.09 (0.05)	0.76	1.52	0.18
Australia	7.66 (3.17)	−0.43 (0.29)	−0.54 (0.23)	0.28 (0.29)	0.06 (0.06)	0.11 (0.06)	0.38 (0.15)	−0.13 (0.07)	0.81	1.66	−0.36
Austria	6.34 (2.34)	−0.03 (0.29)	−0.16 (0.30)	−0.25 (0.26)	−0.02 (0.06)	0.03 (0.04)	0.17 (0.14)	0.09 (0.10)	0.45	2.11	0.00

Canada	6.57 (2.85)	−0.29 (0.39)	0.08 (0.30)	−0.12 (0.32)	0.12 (0.06)	0.01 (0.06)	0.00 (0.05)	−0.12 (0.08)	0.61	1.61	−0.14
Finland	4.94 (1.91)	0.08 (0.25)	−0.27 (0.20)	0.11 (0.21)	−0.01 (0.07)	−0.09 (0.05)	0.15 (0.13)	0.17 (0.11)	0.70	2.71	0.00
Japan	6.89 (5.63)	0.29 (0.38)	−0.13 (0.42)	0.14 (0.40)	−0.08 (0.08)	−0.02 (0.06)	0.02 (0.12)	0.13 (0.16)	0.28	3.61	−0.04
Norway	3.46 (1.96)	0.08 (0.28)	0.12 (0.30)	−0.12 (0.29)	0.02 (0.06)	0.01 (0.05)	0.13 (0.15)	0.00 (0.06)	0.15	1.91	0.10
Spain	6.23 (5.00)	0.18 (0.52)	−0.36 (0.47)	0.12 (0.43)	0.09 (0.11)	−0.01 (0.10)	0.05 (0.31)	−0.03 (0.20)	0.33	3.89	0.05
Sweden	2.41 (2.03)	0.50 (0.31)	−0.10 (0.32)	0.00 (0.27)	−0.03 (0.03)	−0.06 (0.03)	−0.18 (0.09)	0.10 (0.06)	0.57	1.39	0.06
Switzerland	5.47 (1.35)	−0.25 (0.17)	−0.13 (0.17)	−0.28 (0.14)	−0.08 (0.03)	0.02 (0.02)	0.34 (0.07)	0.21 (0.06)	0.84	1.18	0.11

Notes:

[a] The estimation period is 1954–73, twenty observations for all countries except for Australia, 1960–73, Canada, 1959–73, Japan, 1956–73, and Spain, 1958–73.

[b] Figures not in parentheses are posterior means which are equal to least squares estimates. Figures in parentheses, standard deviations of marginal posterior pdfs, are $\sqrt{\nu(\nu-2)}$ times the usual asymptotic least squares standard errors where $\nu = n - k$, where n = number of observations and k = number of parameters estimated, that is ν = usual degrees of freedom.

[c] R^2, the sample squared multiple correlation coefficient, is approximately the mean of the posterior distribution of the squared population multiple correlation coefficient (see Press and Zellner 1978).

[d] $E(\sigma \mid D)$ is the posterior mean of the error term standard deviation, σ, namely $E(\sigma \mid D) = s(\nu/2)^{1/2}.\ \Gamma[(\nu-1)/2]/\Gamma(\nu/2)$, where $\nu = n - k$, the degrees of freedom and s^2 = sum of squared residuals divided by ν. For $\nu = 20 - 8 = 12$, $E(\sigma \mid D)$ is equal to 1.068 times the non-Bayesian standard error of estimate, σ.

[e] r_1, the sample first order residual serial correlation coefficient is an approximate posterior mean of ρ_1, the population first order serial correlation.

Table 15A.2 *Bayesian diffuse prior regression output for the AR(3)LI model in (3.1), eighteen countries, 1954–1985*[a]

	Posterior means and standard deviations[b]										
Country	Const.	y_{it-1}	y_{it-2}	y_{it-3}	$S\,R_{it-1}$	$S\,R_{it-2}$	$M\,G_{it-1}$	$W\,R_{t-1}$	R^{2c}	$E(\sigma \mid D)^d$	$r_1{}^e$
Belgium	−0.43 (0.80)	0.20 (0.14)	0.66 (0.15)	0.15 (0.16)	−0.03 (0.03)	−0.10 (0.03)	0.00 (0.08)	0.19 (0.05)	0.71	1.49	0.01
Denmark	1.71 (1.08)	−0.01 (0.19)	−0.15 (0.20)	0.17 (0.19)	0.01 (0.03)	−0.03 (0.02)	0.16 (0.07)	0.01 (0.05)	0.37	2.28	0.03
France	0.46 (0.86)	0.17 (0.21)	0.34 (0.21)	0.34 (0.19)	−0.03 (0.02)	−0.01 (0.02)	0.01 (0.06)	0.10 (0.04)	0.57	1.59	0.23
Germany	1.61 (0.89)	0.01 (0.18)	−0.05 (0.17)	0.14 (0.14)	0.05 (0.03)	−0.01 (0.03)	0.40 (0.10)	0.08 (0.05)	0.72	2.19	−0.21
Ireland	2.02 (1.16)	0.27 (0.27)	−0.05 (0.27)	0.14 (0.25)	0.04 (0.05)	−0.02 (0.03)	0.05 (0.10)	−0.04 (0.06)	0.28	2.43	−0.08
Italy	1.94 (1.09)	−0.05 (0.20)	0.23 (0.18)	0.04 (0.15)	0.01 (0.03)	−0.02 (0.02)	0.19 (0.08)	0.10 (0.05)	0.60	2.10	0.34
Netherlands	−0.09 (0.73)	0.42 (0.12)	0.03 (0.14)	0.28 (0.12)	−0.01 (0.05)	−0.05 (0.02)	0.33 (0.07)	0.14 (0.06)	0.76	1.71	−0.13
United Kingdom	1.76 (0.84)	−0.10 (0.19)	0.07 (0.20)	0.24 (0.19)	0.05 (0.03)	0.00 (0.03)	0.18 (0.07)	−0.02 (0.05)	0.44	1.79	−0.08
United States	3.89 (0.73)	−0.16 (0.14)	0.12 (0.15)	−0.27 (0.13)	0.09 (0.04)	−0.06 (0.03)	0.52 (0.13)	−0.08 (0.04)	0.72	1.59	−0.01

Australia	1.83 (1.43)	0.11 (0.22)	−0.04 (0.20)	0.45 (0.21)	0.06 (0.04)	−0.02 (0.03)	0.05 (0.12)	−0.02 (0.05)	0.43	2.22	0.09
Austria	3.01 (1.26)	0.20 (0.21)	−0.24 (0.21)	0.19 (0.18)	0.03 (0.05)	0.01 (0.03)	0.17 (0.07)	0.04 (0.05)	0.46	2.13	−0.04
Canada	2.17 (1.19)	0.10 (0.23)	0.37 (0.23)	−0.06 (0.21)	0.04 (0.04)	−0.05 (0.03)	0.09 (0.05)	0.01 (0.05)	0.44	2.24	−0.14
Finland	3.15 (1.28)	0.29 (0.20)	−0.28 (0.18)	0.19 (0.17)	0.00 (0.05)	−0.07 (0.03)	0.12 (0.08)	0.10 (0.07)	0.57	2.60	0.01
Japan	2.24 (1.59)	0.43 (0.26)	−0.03 (0.27)	0.27 (0.21)	−0.05 (0.06)	−0.04 (0.04)	0.08 (0.08)	0.09 (0.08)	0.57	3.02	−0.04
Norway	3.02 (1.21)	0.23 (0.19)	−0.01 (0.20)	−0.09 (0.20)	−0.01 (0.02)	−0.03 (0.02)	0.14 (0.07)	−0.01 (0.04)	0.26	1.60	0.05
Spain	2.99 (1.65)	0.32 (0.28)	−0.23 (0.25)	0.22 (0.22)	0.01 (0.05)	0.02 (0.04)	0.15 (0.15)	0.08 (0.06)	0.55	2.67	−0.01
Sweden	1.95 (0.85)	0.81 (0.27)	−0.44 (0.26)	0.16 (0.20)	−0.04 (0.02)	−0.01 (0.02)	−0.18 (0.09)	0.05 (0.03)	0.48	1.66	−0.05
Switzerland	1.75 (0.66)	0.21 (0.16)	−0.02 (0.16)	0.05 (0.13)	−0.02 (0.04)	0.00 (0.03)	0.21 (0.06)	0.15 (0.05)	0.72	1.93	−0.21

Notes:

[a] The estimation period is 1954–85, thirty-two observations for all countries except for Australia, 1960–85, Canada, 1959–85, Japan, 1956–85, Spain, 1958–84, and Sweden, 1954–83.

[b] For notes *b*, *c*, *d*, and *e*, see table 15A.1.

16 Forecasting turning points in international output growth rates using Bayesian exponentially weighted autoregression, time-varying parameter, and pooling techniques (1991)

Arnold Zellner, Chansik Hong, and Chung-ki Min

1 Introduction

In previous work (Zellner, Hong, and Gulati 1990 and Zellner and Hong 1989), the problem of forecasting turning points in economic time series was formulated and solved in a Bayesian decision theoretic framework. The methodology was applied using a fixed parameter autoregressive, leading indicator (ARLI) model and unpooled data for eighteen countries to forecast turning points over the period 1974–85. In the present chapter, we investigate the extent to which use of exponential weighting, time-varying parameter ARLI models, and pooling techniques leads to improved results in forecasting turning points for the same eighteen countries over a slightly extended period, 1974–86.

The methodology employed in this work has benefited from earlier work of Wecker (1979), Moore and Zarnowitz (1982), Moore (1983), Zarnowitz (1985), and Kling (1987). Just as Wecker and Kling have done, we employ a model for the observations and an explicit definition of a turning point, for example a downturn (*DT*) or an upturn (*UT*). Along with Kling, we allow for parameter uncertainty by adopting a Bayesian approach and computing probabilities of a *DT* or *UT* given past data from a model's predictive probability density function (pdf) for future observations. Having computed such probabilities from the data, we use them in a decision theoretic framework with given loss structures to obtain optimal turning point forecasts which can readily be computed.

The plan of our chapter is as follows. In section 2, we explain our models and methods. Section 3 is devoted to a description of our data.

Research financed in part by the National Science Foundation and by income from the H. G. B. Alexander Endowment Fund, Graduate School of Business, University of Chicago.

Originally published in the *Journal of Econometrics* 49 (1991), 275–304. 0304-4076/91/$03.50.

In section 4, the results of the present chapter are compared to those computed using earlier methods and to turning point forecasting results yielded by naive models. Finally, section 5 presents a summary of results and some concluding remarks.

2 Models and methods

In this chapter we use two alternative models for each country. The first is a fixed parameter, autoregressive model of order 3 with lagged leading indicator variables, denoted by FP/ARLI and given by

$$y_{it} = \boldsymbol{x}_{it}'\beta_i + u_{it}, \quad t = 1, 2, \ldots, T, \quad i = 1, 2, \ldots, N, \tag{2.1}$$

where the subscripts i and t denote the value of a variable for the ith country in the tth year. Further,

$$\begin{aligned} y_{it} &= \text{rate of growth of output, real GNP or GDP} \\ \beta_i &= \text{a } k \times 1 \text{ parameter vector} \\ u_{it} &= \text{disturbance term} \\ x_{it}' &= (y_{it-1}, y_{it-2}, y_{it-3}, 1, SP_{it-1}, SP_{it-2}, GM_{it-1}, WSP_{t-1}) \end{aligned}$$

with

$$\begin{aligned} SP_{it} &= \text{rate of growth of real stock prices} \\ GM_{it} &= \text{rate of growth of real money, } M_1, \text{ divided by a} \\ &\quad\ \text{general price index} \\ WSP_t &= \text{median of the } SP_{it}\text{s for the } t\text{th year} \end{aligned}$$

Equation (2.1) incorporates an autoregression of order 3 to allow for the possibility of having two complex roots and a real root for the process. After using this assumption in Garcia-Ferrer *et al.* (1987) and in Zellner and Hong (1989), calculations reported in Hong (1989) show that with high posterior probabilities, there are two complex roots and one real root in the FP/ARLI model for each of the eighteen countries in our sample. Further, lagged stock price change variables are introduced to capture expectational, policy change and other chaotic effects. The lagged growth rate of real money is employed to represent real balance effects and WSP_{t-1} to allow for common shocks hitting all countries. (See Zellner, Hong, and Gulati 1990 for estimates of the parameters of (2.1) for each of the eighteen countries in our sample and for evidence that the u_{it}s are not very highly autocorrelated.) In our present work, we shall assume that the u_{it}s are independently distributed, each with a normal distribution with zero mean and variance σ_i^2.

We consider two variants of model (2.1), namely a fixed parameter variant as shown in (2.1), and a time-varying parameter (TVP) version, denoted by TVP/ARLI, as shown in (2.2),

$$y_{it} = \boldsymbol{x}'_{it}\boldsymbol{\beta}_{it} + u'_{it}. \tag{2.2}$$

Aggregation effects, policy changes, etc. may cause parameters to vary through time and hence the use of a time-varying parameter vector in (2.2). As in past work (Garcia-Ferrer *et al.* 1987), we assume that the $\boldsymbol{\beta}_{it}$s are generated by a vector random-walk process,

$$\boldsymbol{\beta}_{it} = \boldsymbol{\beta}_{it-1} + \boldsymbol{v}_{it}, \quad i = 1, 2, \ldots, N, \quad t = 1, 2, \ldots, T, \tag{2.3}$$

where the $\boldsymbol{v}_{it}$s are assumed independently and identically normally distributed with zero mean vector and covariance matrix $\phi\sigma_i^2 I_k$, with $0 < \phi < \infty$. When ϕ has a very small value, (2.2) behaves like a fixed parameter model.

The second model which we shall employ is that given in (2.1) with the addition of a world real income (WI) growth rate variable, w_t, namely,

$$y_{it} = w_t\alpha_i + \boldsymbol{x}'_{it}\boldsymbol{\beta}_i + \eta_{it}, \quad i = 1, 2, \ldots, N, \quad t = 1, 2, \ldots, T, \tag{2.4}$$

where w_t is a measure of world income growth, which we represent by the median of the individual country growth rates, the y_{it}s in the tth year and α_i is a scalar parameter. To use (2.4) to forecast, there is need for an equation to forecast the w_ts. It is assumed that the η_{it}s are zero mean, independent normal error terms with variance $\sigma_{\eta i}^2$. In previous work we have found that an AR(3)LI model for w_t works well in forecasting (see Zellner and Hong 1989). This model is given by

$$w_t = \boldsymbol{z}'_t\boldsymbol{\pi} + \varepsilon_t, \quad t = 1, 2, \ldots, T, \tag{2.5}$$

where

π = coefficient vector

ε_t = white noise, zero mean normal disturbance term with variance σ_ε^2

$\boldsymbol{z}'_t = (w_{t-1}, w_{t-2}, w_{t-3}, 1, WSP_{t-1}, WGM_{t-1})$,

with

WSP_t = median of countries' SP_{it}s for year t, and

WGP_t = median of countries' GM_{it} for year t

Thus the world output growth rate as represented by w_t is assumed generated by an AR(3)LI model.

As written in (2.4) and (2.5), the model is a fixed parameter model, denoted by FP/ARLI/WI. Just as above, it is possible to allow the coefficient vectors in (2.4) and (2.5) to be time-varying and generated by independent, vector random-walk processes. This yields our time-varying parameter, autoregressive, leading indicator, world income model, denoted by TVP/ARLI/WI (see appendix, p. 554 for further details).

As regards pooling techniques (see Garcia-Ferrer *et al.* 1987 and Zellner and Hong 1989 for effects of pooling on the RMSEs of one-year-ahead point forecasts), we shall employ an extreme form of pooling in the present chapter. That is we assume that the coefficient vectors in (2.1) and (2.2) are the same for all countries. Under this assumption, (2.1) and (2.2) become

$$y_{it} = \boldsymbol{x}'_{it}\boldsymbol{\beta} + u_{it}, \tag{2.1a}$$

and

$$y_{it} = \boldsymbol{x}_{it}\boldsymbol{\beta}_t + u'_{it}. \tag{2.2a}$$

Similarly, with respect to (2.4), we assume

$$y_{it} = w_t\alpha + \boldsymbol{x}'_{it}\boldsymbol{\beta} + \eta_{it}, \tag{2.4a}$$

and in the TVP version,

$$y_{it} = w_t\alpha_t + \boldsymbol{x}'_{it}\boldsymbol{\beta}_t + \eta'_{it}, \tag{2.4b}$$

$$w_t = \boldsymbol{z}'_t\boldsymbol{\pi}_t + \varepsilon'_t. \tag{2.5a}$$

While the assumptions embedded in (2.1a), (2.2a), (2.4a), and (2.4b) may seem extreme, it is the case that they have led to good point forecasts in previous work and, as will be seen, to good turning point forecasts in our present work.

Finally, we note that exponential weighting or "discounting" (see, e.g., West, Harrison, and Migon 1985 and Highfield 1986) can be employed when parameters are thought to be changing in value through time. By putting heavier weight on recent observations and lighter weight on observations distant in time from the present, use of exponential weighting produces parameter estimates which can adapt to local changes in values of parameters. Thus exponential weighting is a procedure that permits adaptation to deterministic and/or random changes in parameters' values (see appendix for recursions for computing TVP estimates and exponentially weighted parameter estimates).

2.1 *Optimal turning point forecasts*

As mentioned above, if we have a model for the observations, a definition of a turning point, a predictive pdf, and a loss structure, we can compute an optimal turning point forecast. Above we have indicated several models which we shall use in our analyzes. As regards definitions of turning points, we shall use the following definitions employed in our previous work:

Definition of a downturn (DT) in year $T+1$: If the annual growth rate observations for country i, y_{iT-2}, y_{iT-1}, y_{iT}, and y_{iT+1} satisfy

$$y_{iT-2}, y_{iT-1} < y_{iT} > y_{iT+1}, \tag{2.6a}$$

then a *DT* has occurred in period $T+1$ for country i, while if the growth rates satisfy

$$y_{iT-2}, y_{iT-1} < y_{iT} \leq y_{iT+1}, \tag{2.6b}$$

no *DT* (*NDT*) has occurred for country i in period $T+1$. Note that we condition the definition of a DT on two previous observations being below a third observation. Similarly, we define an upturn as follows:

Definition of a upturn (UT) in year $T+1$: If the growth rate observations satisfy

$$y_{i,T-2}, y_{iT-1} > y_{iT} < y_{iT+1}, \tag{2.7a}$$

an UT has occurred in period $T+1$, while if

$$y_{iT-2}, y_{iT-1} > y_{iT} \geq y_{iT+1}, \tag{2.7b}$$

no UT (*NUT*) has occurred for country i in period $T+1$.

While the definitions in (2.6) and (2.7) are not the only possible definitions, we shall use them because they are explicit and seem appropriate for data which may display a cycle with about a four-year period. Note that if there were a deterministic cycle with an exact four-year period, the above definitions would be completely appropriate.

Having defined a *DT* in (2.6a), the problem of forecasting arises when we have observed y_{iT-2}, y_{iT-1}, and y_{iT}, and y_{iT+1} is a future, as yet unobserved, value. If a predictive pdf for y_{iT+1} given all past data, a

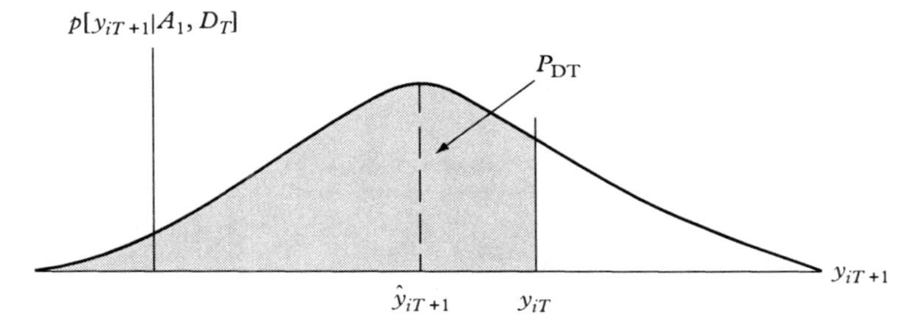

Figure 16.1 Calculation of probability of a DT in period $T+1$.

model, and prior information through period T are available, then we can compute the probability of a *DT*, P_{DT}, as shown below.

As is well known, in general a predictive pdf for a future observation, e.g. y_{iT+1}, $-\infty < y_{iT+1} < \infty$, is given by

$$p(y_{iT+1} \mid D_T) = \int_\theta f(y_{iT+1} \mid \boldsymbol{\theta}, D_T)\pi(\boldsymbol{\theta} \mid D_T)\,\mathrm{d}\boldsymbol{\theta}, \tag{2.8}$$

where $f(y_{iT+1} \mid \boldsymbol{\theta}, D_T)$ is the pdf for y_{iT+1} given the parameter vector $\boldsymbol{\theta} \subset \Theta$, the parameter space, and D_T, the past sample and prior information as of time T, and $\pi(\boldsymbol{\theta} \mid D_T)$ is the posterior pdf for $\boldsymbol{\theta}$ yielded by Bayes' Theorem.

The predictive pdf in (2.8) is calculated for each of our models, as shown in the appendix, and used to evaluate probabilities of *DT*s and *UT*s. For example, using the definition in (2.6a) and the predictive pdf in (2.8), the probability of a *DT* for country i in the future period $T+1$, denoted by P_{DT}, is

$$P_{DT} = \Pr(y_{iT+1} < y_{iT} \mid A_1, D_T) = \int_{-\infty}^{y_{iT}} p(y_{iT+1} \mid A_1, D_T)\,\mathrm{d}y_{iT+1}, \tag{2.9}$$

where A_1 denotes the condition: $y_{iT-2}, y_{iT-1} < y_{iT}$. In figure 16.1, the calculation of (2.9) is illustrated. The shaded area in figure 16.1 is just the probability of a *DT*. Note that it will be larger than $\frac{1}{2}$ if the given value $y_{iT} > \hat{y}_{iT+1}$, the modal value of the predictive pdf.

Once P_{DT} has been computed, we use its value in conjunction with the loss structure in table 16.1 to choose an optimal turning point forecast. In table 16.1 losses are scaled to be zero if forecasts are correct, while losses $c_1, c_2 > 0$ are incurred for incorrect forecasts. For example, if the forecast *DT* is selected and the outcome is *NDT*, then loss c_1 is incurred. Thus c_1 and c_2 represent costs associated with two possible errors, namely

Table 16.1 *Loss structure for forecasting downturn (DT) or no downturn (NDT)*

	Outcomes		
	DT	*NDT*	Expected losses
Forecasts			
DT	0	c_1	$c_1(1 - P_{\text{DT}})$
NDT	c_2	0	$c_2 P_{\text{DT}}$
Probabilities	P_{DT}	$1 - P_{\text{DT}}$	

(1) forecast *DT* and the outcome is *NDT* and (2) forecast *NDT* and the outcome is *DT*. (See Zellner, Hong, and Gulati 1990 for analysis of expanded loss structures which take account of not only cost c_1 and c_2 but also costs associated with errors of point forecasts.)

The expected losses, shown in table 16.1, associated with the choice of a *DT* or a *NDT* forecast, can be employed to choose the forecast which minimizes expected loss. For example, if $c_1(1 - P_{\text{DT}}) < c_2 P_{\text{DT}}$, or equivalently,

$$1 < (c_2/c_1)P_{\text{DT}}/(1 - P_{\text{DT}}), \tag{2.10}$$

the optimal forecast is *DT*. Note that the decision rule in (2.10) involves a cost parameter ratio, c_2/c_1, and the odds on a *DT* occurring, $P_{\text{DT}}/(1 - P_{\text{DT}})$. If $c_2/c_1 = 1$, the case of a symmetric loss structure, (2.10) will be satisfied if $P_{\text{DT}} > 0.5$. Thus for a symmetric loss structure, forecast *DT* if $P_{\text{DT}} > \frac{1}{2}$ and *NDT* if P_{DT} does not satisfy this condition. On the other hand if $c_2/c_1 = 4$, (2.10) will be satisfied for $P_{\text{DT}} > 0.2$. Thus, if the computed value of P_{DT} is larger than 0.2 and $c_2/c_1 = 4$, the optimal forecast is *DT*. In this case, a less stringent condition on P_{DT} is employed to help avoid the relatively large cost c_2 associated with an incorrect *NDT* forecast. See Zellner and Hong (1988) for the results of forecasting turning points using various cost parameter ratios.

Since the methodology of forecasting *UT*s using the definition in (2.7), a predictive pdf for y_{iT+1}, and a loss structure similar to that in table 16.1 with positive costs c_3 and c_4, associated with incorrect *UT* and *NUT* forecasts, respectively, is similar to that explained above for forecasting *DT*s, it will not be presented. To minimize expected loss in choosing between *UT* and *NUT* forecasts, one chooses an *UT* forecast if $c_3(1 - P_{UT}) < c_4 P_{UT}$ or, equivalently,

$$1 < (c_4/c_3)P_{\text{UT}}/(1 - P_{\text{UT}}), \tag{2.11}$$

where P_{UT} is the probability of an upturn in period $T+1$ given by

$$P_{UT} = \Pr(y_{iT+1} > y_{iT} \mid A_2, D_T) = \int_{y_{iT}}^{\infty} p(y_{i,T+1} \mid A_2, D_T)\, dy_{iT+1}, \tag{2.12}$$

where A_2 is the condition $y_{iT-2}, y_{iT-1} > y_{iT}$. Thus when A_2 is satisfied, the probability of an UT, P_{UT}, can be computed from (2.12) and used to select between *UT* and *NUT* forecasts so as to minimize expected loss.

In our empirical work, from the annual output growth rate data for each country, we identify sequences of data satisfying condition $A_1: y_{it-2}, y_{it-1} < y_{it}$. Then we use the data through year t to compute the predictive pdf for y_{it+1} and use it to choose optimally between *DT* and *NDT* forecasts, as explained above. A similar procedure is employed with respect to making *UT* and *NUT* forecasts. These procedures were carried out for each of our two models using fixed parameters, time-varying parameters and exponential weighting, and the pooling techniques mentioned above. Similar calculations have been performed without employing pooling. The results will be presented after a brief discussion of our data.

3 Description of data

The data employed in our calculations have been taken from the IMF International Financial Statistics data base and are available on a diskette on request for a small fee to cover costs. The annual data 1954–73 were used to estimate our models with data 1951–3 serving as initial values. The models were used to compute probabilities of *DT*s and *UT*s for each country whenever a sequence of points satisfying the conditions for turning points, defined above, was encountered, using all data prior to the year in which a turning point could occur.

The output data for each country, either annual real GNP or real GDP, were logged and first-differenced to yield annual growth rates. Similarly, nominal money, M_1, at the end of each year was deflated by a general price index, logged and first-differenced to yield the rate of change of real money. A general price index was used to deflate an annual stock price index, which was logged and first-differenced to yield a rate of growth of real stock prices for each country and year.

Shown in figure 16.2 are boxplots for the three variables' data used in this study, rates of growth of real output, real money, and real stock prices. In each box, the horizontal line is the median of the eighteen countries' values of a variable in a particular year, the height of each box is the interquartile range of the eighteen countries' values of a variable

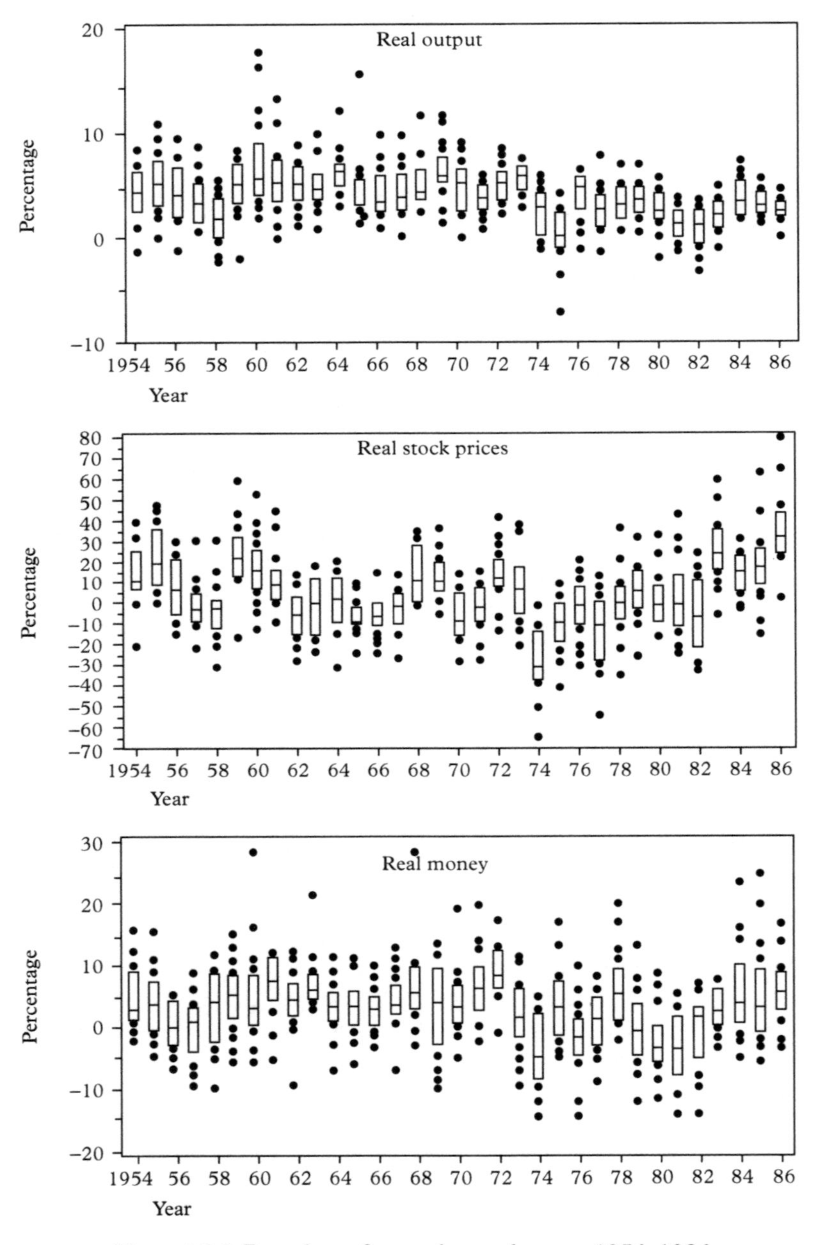

Figure 16.2 Box plots of annual growth rates, 1954–1986

in a given year while the points outside each box are observations in the tails of the distribution of values of a variable in a particular year. From figure 16.2, it is seen that the countries' output, money, and real stock price growth rates tend to move up and down together with the variation of the growth rate of real stock prices considerably larger than that of the other two variables. It will be noted that there are several outlying points in the data. While we plan to deal with these outlying points in future work, in the present study *all data were employed with no adjustments for outliers or anything else.*

In figure 16.2, the horizontal lines in the boxes are the median values of variables in each year used in our model for w_t in (2.5). It is indeed interesting to note that the annual median growth rates of output, real money, and real stock prices follow rather smooth paths through time which appear to exhibit fluctuations with about a four-year period as measured from peak to peak or trough to trough. (See Zellner and Hong 1989 for evidence on the forecasting performance of the ARLI model for w_t in (2.5).)

4 Results of forecasting turning points, eighteen countries, 1974–1986

For each country's data, 1974–86, we identified sequences of observations which satisfied our definitions of turning points given in (2.6) and (2.7). Then we used data from 1954 through year T, the year prior to the year in which a turning point did or did not occur, to compute the probabilities that a turning point would occur in year $T + 1$ from the predictive pdfs of six alternative models. These probabilities were employed to make optimal turning point forecasts as explained above. Initially, we report results based on a symmetric loss structure for which a turning point is forecasted when its probability of occurrence is greater than 0.5. After reporting these results, others relating to asymmetric loss structures will be presented and discussed.

The six models used in making turning point forecasts are the following third-order ARLI models:

1. TVP/ARLI, a time-varying parameter ARLI model
2. TVP/ARLI/WI, a time-varying parameter ARLI-world income model
3. EW/ARLI, an exponentially weighted ARLI model
4. EW/ARLI/WI, an exponentially weighted ARLI-world income model
5. FP/ARLI, a fixed parameter ARLI model
6. FP/ARLI/WI, a fixed parameter ARLI-world income model.

Each model is implemented with no pooling of data across countries and with pooling using the methods discussed in section 2. Our calculations provide information relating to the following questions:

(a) Do fixed parameter models perform as well as TVP and EW ARLI models?
(b) Do the models containing the world income variable perform better than those that do not?
(c) Does pooling of data across countries improve forecast performance?
(d) Are DT and NDT forecasts better than UT and NUT forecasts?
(e) Is the percentage of correct DT forecasts greater than the percentage of correct NDT forecasts?
(f) Is the percentage of correct UT forecasts greater than the percentage of correct NUT forecasts?

The forecasting results presented below for 158 turning point forecasts will help to provide answers to these and other questions.

In table 16.2 and figures 16.3 and 16.4 are shown results of using our six models to forecast 158 turning points for eighteen countries, 1974–86. Using symmetric loss structures, we forecast a *DT* when $P_{\mathrm{DT}} > \frac{1}{2}$ and an *UT* when $P_{\mathrm{UT}} > \frac{1}{2}$. In the upper panel of table 16.2, the forecasting results using no data pooling are presented. The model that appears to have a slight edge over the others is the TVP/ARLI/WI model. Use of it led to 82 percent of 158 turning point forecasts being correct, that is 129 of 158 forecasts are correct. Of the 76 *DT* and *NDT* forecasts, use of this model produces 83 percent correct forecasts, while for 82 *UT* and *NUT* forecasts, 80 percent are correct. For the other models, of 158 turning point forecasts, the percentages correct vary from 72 to 77, somewhat lower than for the TVP/ARLI/WI model. Also for the 76 *DT* and *NDT* forecasts, use of these other models leads to percentages correct that range from 68 to 76, while for the 82 *UT* and *NUT* forecasts, the percentages correct vary from 74 to 77. Thus use of the TVP/ARLI/WI model produces approximately 80 percent correct forecasts overall and for *DT*/*NDT* and *UT*/*NUT* forecasts separately.

Note further, from panel A of table 16.2, that use of TVP models produces better results than fixed parameter models or models with exponential weighting. However, in most cases the differences in performance as measured by percentage of correct forecasts are not large. Further, from lines 1–4 of table 16.2, it is seen that addition of the world income variable leads to small improvements in forecasting performance relative to that of models not including it.

From panel B of table 16.2, which provides results based on data pooled across countries, use of the TVP/ARLI/WI and EW/ARLI/WI models results in 80 and 81 percent, respectively, of 158 turning point forecasts

Table 16.2 *Forecasting turning points in annual output growth rates, eighteen countries, 1974–1986, using various models*

Model	Percentage of correct forecasts			Number of incorrect forecasts		
	158 turning point forecasts	76 *DT* and *NDT* forecasts	82 *UT* and *NUT* forecasts	158 turning point forecasts	76 *DT* and *NDT* forecasts	82 *UT* and *NUT* forecasts
(A) *No pooling*						
1 TVP/ARLI	77	76	77	37	18	19
2 TVP/ARLI/WI	82	83	80	29	13	16
3 EW/ARLI	72	68	74	45	24	21
4 EW/ARLI/WI	76	75	77	38	19	19
5 FP/ARLI	74	71	77	41	22	19
6 FP/ARLI/WI	72	70	74	44	23	21
(B) *With pooling*						
7 TVP/ARLI	74	80	68	41	15	26
8 TVP/ARLI/WI	80	84	76	32	12	20
9 EW/ARLI	76	86	67	38	11	27
10 EW/ARLI/WI	81	88	74	30	9	21
11 FP/ARLI	75	82	68	40	14	26
12 FP/ARLI/WI	79	84	74	33	12	21

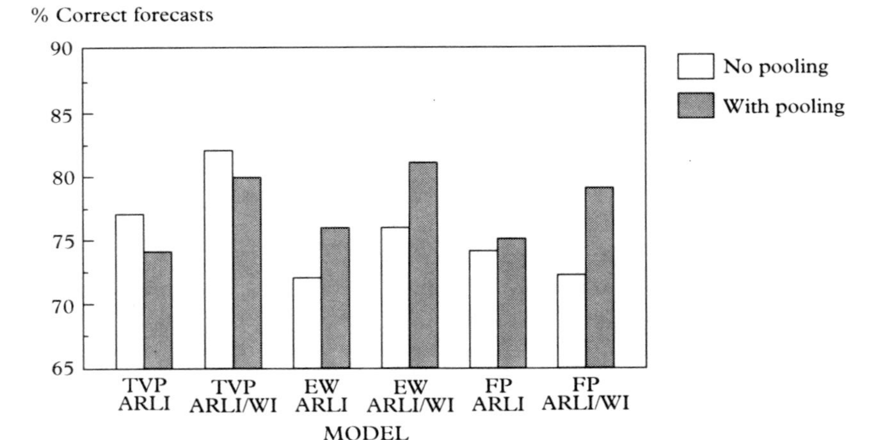

Figure 16.3 Percentages of correct forecasts, 158 turning point forecasts, 1974–1986

being correct. The percentages correct for the other models range from 74 to 79, with models incorporating the WI variable performing slightly better than those not including it. Also, the TVP and EW models tend to perform slightly better in forecasting turning points relative to the pooled fixed parameter models.[1] Similar conclusions emerge with respect to results for the 76 *DT* and *NDT* and 82 *UT* and *NUT* forecasts. In addition, with the use of pooling, the performance of the models for *DT*/*NDT* forecasting, 80–88 percent correct, is somewhat better than that for the *UT*/*NUT* forecasts, 67–76 percent correct.

As regards the comparison of the results in the upper and lower panels of table 16.2, it appears that pooling has led to improved *DT* and *NDT* forecasts but no improvement with respect to *UT* and *NUT* forecasts. It is the case that pooling leads to much better performance of the NDT forecasts, particularly for the FP models.

To summarize the results in table 16.2 and figures 16.3 and 16.4, use of our methods and models leads to approximately 70–80 percent correct turning point forecasts based on 158 cases. While the differences in performance are not large in many instances, the results tend to favor the TVP/ARLI/WI model with or without pooling which produces 80 and 82 percent correct turning point forecasts, respectively.

[1] The fixed parameter model denoted by FP/ARLI was employed in our [earlier] work without pooling (see Zellner, Hong, and Gulati 1990).

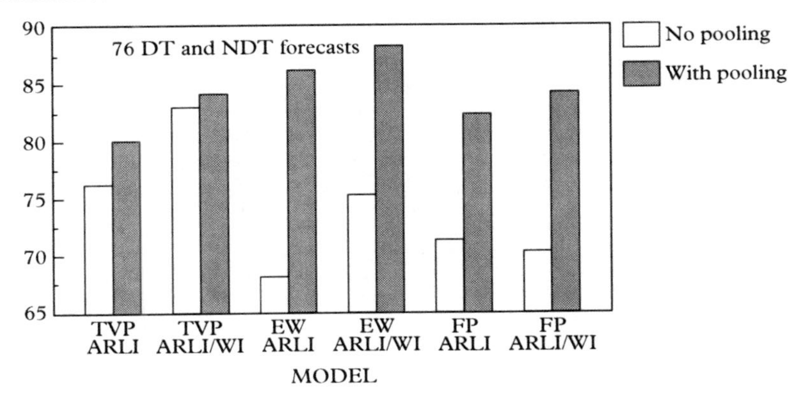

Figure 16.4 Percentages of correct forecasts, 1974–1986

To appraise the forecasting performance of various models and methods further, we have shown the numbers of incorrect forecasts in table 16.2. For example, the TVP/ARLI/WI model in line 2 produces 29 incorrect turning point forecasts, 13 associated with *DT* and *NDT* forecasts and 16 associated with *UT* and *NUT* forecasts. If the loss structure in table 16.1 is symmetric with $c_1 = c_2 = 1$ and if a similar one for *UT*/*NUT* forecasts is also symmetric with costs of incorrect forecasts equal to 1, then 29 is the realized loss associated with the use of the unpooled TVP/ARLI/WI model in forecasting turning points. From table 16.2, it is seen that realized losses range from 29 to 45 for panel A of the table and from 30 to 41 for panel B. These results can be compared to the results that are

Table 16.3 *Performance of various procedures for forecasting turning points in annual output growth rates for eighteen countries, 1974–1986*

	Number and percentage of incorrect forecasts[a]		
Forecasting procedures	76 *DT* and *NDT* forecasts	82 *UT* and *NUT* forecasts	158 turning point forecasts
1 Eternal optimist[b]	59 (78)	34 (41)	93 (59)
2 Eternal pessimist[c]	17 (22)	48 (59)	65 (41)
3 Deterministic four-year cycle[d]	17 (22)	34 (41)	51 (32)
4 TVP/ARLI/WI unpooled	13 (17)	16 (20)	29 (18)
TVP/ARLI/WI pooled	12 (16)	20 (24)	32 (20)

Notes:

[a] Figures in parentheses are percentages of incorrect forecasts.

[b] An eternal optimist forecasts either *NDT* or *UT*.

[c] An eternal pessimist forecasts either *DT* or *NUT*.

[d] Forecasts either *DT* or *UT*.

associated with an eternally pessimistic forecaster who always chooses *DT* and *NUT* forecasts (see table 16.3, line 2). Since there were 59 actual *DT*s out of 76 cases, the eternal pessimist has 17 of his *DT* forecasts wrong. Also, since there were 34 *NUT*s in 82 cases, 48 of his *NUT* forecasts are incorrect. Thus his total number of incorrect forecasts is $17 + 48 = 65$. As is evident, this eternally pessimistic forecaster experiences a realized loss under symmetric loss of 65 which is much higher than those associated with the models and methods in table 16.2. Thus, our turning point forecasts are much better than those of an "eternal pessimist."

Also shown in table 16.3 are the forecasting results of an "eternal optimist" who always forecasts *NDT* and *UT*. The eternal optimist has 59 of his 76 *NDT* forecasts incorrect and 34 of his 82 *UT* forecasts incorrect. Thus his realized loss is $59 + 34 = 93$, much higher than that associated with the unpooled or pooled TVP/ARLI/WI model's realized losses, 29 and 32, respectively.

The forecasting performance of a person who believes in a deterministic four-year cycle is shown in line 3 of table 16.3. Such a forecaster always forecasts *DT* and *UT* with the result that 17 of his 76 *DT* forecasts and 34 of his 82 *UT* forecasts are incorrect. Thus his total realized loss is 51, again higher than those produced using the unpooled or pooled TVP/ARLI/WI model, namely 29 and 32, respectively. Further, as can be seen from the

Table 16.4 *Expected number of turning points, DTs, and UTs*

Model	Unpooled	Pooled
1 TVP/ARLI	104	98
2 TVP/ARLI/WI	101	98
3 EW/ARLI	101	102
4 EW/ARLI/WI	98	99
5 FP/ARLI	100	98
6 FP/ARLI/WI	96	94
Actual number of turning points	107	107
Total number of cases	158	158

figures in table 16.2, the realized losses for all of our models are much lower than those of the eternal optimist, the eternal pessimist, and the deterministic four-year cycle forecaster.

In summary, in terms of overall realized losses, the results in tables 16.2 and 16.3 indicate that all of our models performed better than the eternal optimist, the eternal pessimist, and the deterministic four-year cycle forecasters. The addition of the WI variable has led to improved forecasting as measured by overall realized loss.

We can also ask if our turning point forecasts are better than those yielding by coin-flipping, that is use of a probability of a $DT = \frac{1}{2}$ on all occasions or probability of an $UT = \frac{1}{2}$ on all occasions. Under these assumptions, the expected number of turning points $= \frac{1}{2} \times 158 = 79$. However, we observe 59 *DT*s and 48 *UT*s in our data using the definitions in (2.6) and (2.7). Thus there are 107 actual turning points observed. The number of [turning points] observed, 107 of 158 cases, is much greater than the number predicted by a coin-flipping forecaster, 79. On the other hand, if we sum our computed probabilities of turning points for the 158 possible turning points, we obtain the results shown in table 16.4.

It is seen that the forecasting results of table 16.4 indicate that our methods and models yield an expected number of turning points much closer to the observed outcome, 107, than those of the pessimistic forecaster, the coin-flipping forecaster, and a deterministic four-year cycle forecaster who would forecast a *DT* or an *UT* on every occasion and thus would expect 158 turning points.

In tables 16.5 and 16.6 we display results for *DT* and *NDT* and *UT* and *NUT* forecasts separately for each model, with and without pooling. For the results without pooling, shown in panel A, we see that for all models the percentages of correct *DT* forecasts are much higher than

Table 16.5 *Forecasting downturns (DTs) and no downturns (NDTs) in annual output growth rates, eighteen countries, 1974–1986*

Decision rule: forecast *DT* if $P_{DT} > 0.5$						
		Number				
Model	Forecast	Correct	Incorrect	Total	% correct	Realized loss
(A) *No pooling*						
TVP/ARLI	*DT*	47	6	53	89	18
	NDT	11	12	23	48	
TVP/ARLI/WI	*DT*	49	3	52	94	13
	NDT	14	10	24	58	
EW/ARLI	*DT*	40	5	45	89	24
	NDT	12	19	31	39	
EW/ARLI/WI	*DT*	45	5	50	90	19
	NDT	12	14	26	46	
FP/ARLI	*DT*	40	3	43	93	22
	NDT	14	19	33	42	
FP/ARLI/WI	*DT*	38	2	40	95	23
	NDT	15	21	36	42	
(B) *With pooling*						
TVP/ARLI	*DT*	50	6	56	89	15
	NDT	11	9	20	55	
TVP/ARLI/WI	*DT*	52	5	57	91	12
	NDT	12	7	19	63	
EW/ARLI	*DT*	52	4	56	93	11
	NDT	13	7	20	65	
EW/ARLI/WI	*DT*	53	3	56	95	9
	NDT	14	6	20	70	
FP/ARLI	*DT*	47	2	49	96	14
	NDT	15	12	27	56	
FP/ARLI/WI	*DT*	49	2	51	96	12
	NDT	15	10	25	60	

those for *NDT* forecasts, with the former ranging from 89 to 95 percent and the latter from 39 to 58 percent. The poor performance of *NDT* forecasts was also encountered in Zellner, Hong, and Gulati (1990) using the unpooled FP/ARLI model. The percentages of the TVP/ARLI/WI model's *DT* and *NDT* forecasts correct, 94 and 58, respectively, are higher than the corresponding percentages for all other models except for the *DT* percentage correct for the FP/ARLI/WI model, 95. As regards the effects of pooling on forecasting performance, panel B of table 16.5 indicates that pooling has resulted in an increase in the percentage of correct *NDT*

Table 16.6 *Forecasting upturns (UTs) and no upturns (NUTs) in annual output growth rates, eighteen countries, 1974–1986*

Decision rule: forecast *UT* if $P_{UT} > 0.5$.

Model	Forecast	Number Correct	Number Incorrect	Total	% correct	Realized loss
(A) *No pooling*						
TVP/ARLI	*UT*	44	15	59	75	19
	NUT	19	4	23	83	
TVP/ARLI/ WI	*UT*	42	10	52	81	16
	NUT	24	6	30	80	
EW/ARLI	*UT*	45	18	63	71	21
	NUT	16	3	19	84	
EW/ARLI/WI	*UT*	42	13	55	76	19
	NUT	21	6	27	78	
FP/ARLI	*UT*	45	16	61	74	19
	NUT	18	3	21	86	
FP/ARLI/WI	*UT*	43	16	59	73	21
	NUT	18	5	23	78	
(B) *With pooling*						
TVP/ARLI	*UT*	39	17	56	70	26
	NUT	17	9	26	65	
TVP/ARLI/WI	*UT*	40	12	52	77	20
	NUT	22	8	30	73	
EW/ARLI	*UT*	40	19	59	68	27
	NUT	15	8	23	65	
EW/ARLI/WI	*UT*	41	14	55	74	21
	NUT	20	7	27	74	
FP/ARLI	*UT*	41	19	60	68	26
	NUT	15	7	22	68	
FP/ARLI/WI	*UT*	39	12	51	76	21
	NUT	22	9	31	71	

forecasts for all models and not much change in the percentages of correct *DT* forecasts.

Just as above, we can ask how many *DT*s a coin-flipper, an optimist, a pessimist, or a believer in a deterministic four-year cycle would predict. Since there are 76 cases meeting the conditions for forecasting either *DT* or *NDT*, a coin-flipper using $P_{DT} = \frac{1}{2}$ in each case would expect 38 *DT*s. The optimist would expect zero *DT*s, while the other two forecasters would expect 76 *DT*s. The number of downturns actually observed is 59, far from zero, 38, or 76. If we sum the probabilities of *DT*s associated with the 76 cases in which a *DT* could occur according to our definition, the sums for our different models range from 42 to 49 for our unpooled cases and from 44 to 51 for our pooled cases. Thus the expected number of *DT*s computed from our models are much closer to the actual number of observed *DT*s, 59, than are those of a coin-flipper, an optimist, a pessimist, and a believer in a deterministic four-year cycle.

In table 16.6, information is provided on the performance of our models and methods in forecasting *UT*s and *NUT*s. Without pooling, 71–81 percent of the *UT* forecasts are correct and 78–86 percent of the *NUT* forecasts are correct. The numbers of incorrect forecasts range from 16 for the TVP/ARLI/WI model to 21 for the EW/ARLI and FP/ARLI/WI models. Again the TVP/ARLI/WI model performs slightly better than the other models. The same is true when pooling is used as shown in panel B of table 16.6, although the differences are rather small.

Further, there are 82 cases in which either an *UT* or a *NUT* forecast can be selected. A coin-flipper, $P_{\mathrm{UT}} = \frac{1}{2}$, would expect 41 *UT*s. A pessimist who forecasts *NUT* in each case, equivalent to $P_{\mathrm{UT}} = 0$, would expect zero *UT*s. On the other hand, an optimist and a believer in a deterministic four-year cycle would expect 82 *UT*s. Actually, there were 48 *UT*s observed in our data, much greater than zero, much lower than 82, and somewhat larger than the coin-flipper's expected number, 41. On the basis of the sum of our computed probabilities, we found that our expected number of *UT*s ranged from 53 to 58 for our unpooled models and from 48 to 54 for our pooled models. The pooled TVP/ARLI/WI model's summed probabilities equals 47.6, very close indeed to the number of *UT*s observed, namely 48.

From the above turning point forecasting performance, based on symmetric loss structures or equivalently on the use of the decision rules, forecast a *DT* if $P_{\mathrm{DT}} > 0.5$, otherwise forecast a *NDT*, and forecast an *UT* if $P_{\mathrm{UT}} > 0.5$, otherwise forecast *NUT*, we conclude that the TVP/ARLI/WI model, with or without pooling, produces relatively good results. The margin of superiority of this model is generally small relative to other models which included the WI variable. These models perform quite a bit better than do four naive turning point forecasting procedures.

Shown in table 16.7 are probabilities of *DT*s by year and country computed from the pooled TVP/ARLI/WI model using the definition of a *DT* given in (2.6a). For example, the probability of a *DT* in 1974 for Belgium is 0.82 based on data through 1973. The asterisk affixed to this probability indicates that the Belgian output growth rate did indeed turn down in 1974. It is seen that eight countries have high *DT* probabilities in 1974 and that all experienced a *DT* in this year. In 1977, thirteen countries have high probabilities of a *DT* and twelve experienced a

Table 16.7 *Computed probabilities of downturns (DTs), by year and country from the pooled TVP/ARLI/WI model, 1974–1986*[a]

	Year												
Country	1974	1975	1976	1977	1978	1979	1980	1981	1982	1983	1984	1985	1986
Australia				0.68*			0.83*					0.94*	
Austria				0.77*			0.95*				0.27*		0.36*
Belgium	0.82*			0.90*				0.89*				0.54*	
Canada	0.94*			0.91*							0.37–	0.88*	
Denmark				0.95*			0.80*			0.65*		0.33–	0.36*
Finland						0.13–	0.93*						0.69*
France				0.88*						0.44*			0.51–
Germany	0.87*			0.92*			0.85*				0.27–	0.64*	
Ireland					0.74*				0.92*			0.60*	
Italy	0.81*			0.87*			0.81*					0.66*	
Japan				0.83–	0.55*							0.67*	
Netherlands	0.88*			0.84*			0.70*				0.23–	0.58*	
Norway		0.98*		0.95*			0.78*				0.58–	0.66*	
Spain					0.51*						0.44–	0.31–	0.37–
Sweden	0.68*					0.12–	0.80*				0.37–	0.80*	
Switzerland					0.47*		0.74–	0.94*				0.57–	0.77*
United Kingdom	0.99*			0.85*		0.31*				0.32–	0.54*		0.44*
United States	0.86*			0.85*		0.52*					0.42–	0.96*	

Note:
[a] An asterisk (*) indicates that a downturn occurred while a dash (–) indicates that a downturn did not occur. See (2.6a) in the text for the definition of a downturn.

downturn, Japan, with a probability of a *DT* equal to 0.83 did not experience a *DT*, as indicated by the dash affixed to its probability. However, one year later in 1978, the probability of a *DT* for Japan is 0.55, based on data through 1977, and it did experience a *DT* in 1978. In the majority of cases for which the probabilities of *DT*s are high, a *DT* was experienced. In addition to the exception noted for Japan in 1977, other exceptions are encountered for Switzerland in 1980, Norway in 1984, and Switzerland in 1985. In each of these cases, the probability of a *DT* in the next year is high and a *DT* occurred in that year. Thus for these cases, there is an early false *DT* forecast followed by a correct *DT* forecast in the next year. There are also several cases in which the probability of a *DT* is low and yet a *DT* occurred, for example for the United Kingdom in 1979 and for Austria, Denmark, and Switzerland in 1986. In general, however, when the probabilities of a *DT* are high, countries experienced *DT*s, while when they are low they did not experience *DT*s.

In table 16.8, the computed probabilities of *UT*s by country and year based on the pooled TVP/ARLI model are displayed. In general, when the computed probabilities are high, *UT*s occurred, while when they are low, *UT*s did not occur. For example in 1975, the probabilities of an *UT* are smaller than 0.50 for ten countries and eight of them did not experience an *UT*. In this year four countries had probabilities of an *UT* equal to or greater than 0.50 and two experienced *UT*s. Germany, with a probability of an *UT* of 0.70, did not experience an *UT* in 1975; however, for the next year, 1976, its probability of an *UT* is 0.98 and it did experience an upturn. Similarly for the United States, its 1975 probability of an *UT* is 0.52 and it did not experience an *UT*, but its 1976 probability is 0.90 and its output growth rate did turn up in 1976. In several other cases, for example Denmark, Germany, and Sweden in 1981 and 1982, a similar pattern is encountered, namely high probabilities of *UT*s in 1981 and 1982 with *UT*s occurring in the latter but not in the former year. Also, in some cases, the probabilities of an *UT* are low and yet upturns occurred, for example for Australia and Denmark in 1975 and for Ireland in 1980. Overall, however, when the computed probabilities are high, *UT*s usually occurred, while when they are low, *UT*s did not occur.

We now turn to a presentation of some forecasting results for asymmetric loss structures. In table 16.9 are shown the results of *DT*/*NDT* forecasts for various values of the cost parameter ratio, c_2/c_1, where c_1 and c_2 are the cost parameters in the loss structure given in table 16.1. When $c_2/c_1 = 4$, it is optimal to forecast a *DT* when the probability of a *DT* is greater than 0.2. For this case, 74 *DT* forecasts were made with 59 or 80 percent correct. Only two *NDT* forecasts were made and both were correct. With $c_2 = 4c_1$, the cost of an erroneous *NDT* forecast is much

Table 16.8 *Computed probabilities of upturns (UTs), by year and country from the pooled TVP/ARLI model, 1974–1986*[a]

	Year												
Country	1974	1975	1976	1977	1978	1979	1980	1981	1982	1983	1984	1985	1986
Australia		0.11*			0.45*			0.77*		0.31*			
Austria	0.42–	0.12–	0.90*			0.92*			0.57*				
Belgium		0.04–	0.96*						0.74*				
Canada		0.05–	0.77*					0.56*		0.95*			
Denmark		0.25*				0.86*		0.80–	0.95*				
Finland		0.20–	0.89–	0.51*					0.71*				
France	0.39–	0.19–	0.89*					0.41–	0.55*		0.90*		
Germany		0.70–	0.98*					0.61–	0.64–	0.80*			
Ireland		0.04–	0.67–	0.73*			0.34*			0.15–	0.98*		
Italy			0.96*						0.53–	0.42*			
Japan		0.50*						0.12–	0.48–	0.31*			
Netherlands			0.97*			0.74*		0.63–	0.70–	0.89*			
Norway	0.82*				0.38*			0.12–	0.63–	0.71*			
Spain		0.09–	0.70*			0.79–	0.81*		0.81*				
Sweden			0.54–	0.60–	0.82*			0.69–	0.82*				0.57–
Switzerland	0.30–	0.16–	0.99*						0.28–	0.89*			
United Kingdom		0.55*						0.80*					
United States		0.52–	0.90*				0.57–	0.76*		0.91*			0.80*

Note:
[a] An asterisk (*) indicates that an upturn occurred while a dash (–) indicates that an upturn did not occur. See (2.7a) in the text for the definition of an upturn.

Table 16.9 *Forecasting downturns (DTs) and no downturns (NDTs) using asymmetric and symmetric loss structures and pooling*

Cost parameter			TVP/ARLI/WI pooled		EW/ARLI/WI pooled		FP/ARLI/WI pooled	
ratio, c_2/c_1[a]	Decision rule	Forecasts	Number	% correct	Number	% correct	Number	% correct
4	$P_{DT} > 0.2$	*DT*	74	80	74	80	74	80
		NDT	2	100	2	100	2	100
7/3	$P_{DT} > 0.3$	*DT*	71	82	68	84	68	84
		NDT	5	80	8	75	8	75
3/2	$P_{DT} > 0.4$	*DT*	62	89	61	88	61	88
		NDT	14	71	15	67	15	67
1	$P_{DT} > 0.5$	*DT*	57	91	56	95	56	95
		NDT	19	63	20	70	20	70
2/3	$P_{DT} > 0.6$	*DT*	47	96	45	96	45	96
		NDT	29	52	31	48	31	48
3/7	$P_{DT} > 0.7$	*DT*	38	95	31	97	31	97
		NDT	38	40	45	36	45	36
1/4	$P_{DT} > 0.8$	*DT*	30	97	24	96	24	96
		NDT	46	35	52	31	52	31

Note:
[a] c_1 and c_2 are cost parameters in the loss structure in table 16.1.

Table 16.10 *Brier scores for turning point forecasts, eighteen countries, 1974–1986*[a]

Model	All turning points	Downturns	Upturns
(A) *No pooling*			
1 TVP/ARLI	0.32	0.30	0.35
2 TVP/ARLI/WI	0.31	0.30	0.33
3 EW/ARLI	0.37	0.35	0.38
4 EW/ARLI/WI	0.34	0.33	0.35
5 FP/ARLI	0.37	0.36	0.37
6 FP/ARLI/WI	0.35	0.36	0.35
(B) *With pooling*			
7 TVP/ARLI	0.31	0.26	0.35
8 TVP/ARLI/WI	0.28	0.24	0.31
9 EW/ARLI	0.32	0.28	0.37
10 EW/ARLI/WI	0.29	0.26	0.32
11 FP/ARLI	0.34	0.30	0.37
12 FP/ARLI/WI	0.29	0.27	0.31

Note:
[a] The Brier score is defined in (4.1) in the text.

greater than that of an erroneous *DT* forecast and thus fewer *NDT* forecasts are made. On the other hand, when $c_2/c_1 = \frac{1}{4}$, the optimal decision rule is to forecast a *DT* when the probability of a *DT* is greater than 0.8, a rule which leads to fewer *DT* forecasts being made. In this case, only 30 *DT* forecasts were made with 29 or 97 percent being correct. Of 46 *NDT* forecasts made in this case, 16 or 35 percent are correct. These results indicate that the turning point forecasting procedure adapts to variations in the cost parameter ratio, c_2/c_1. Similar results have been obtained for forecasting *UT*s and *NUT*s when the cost parameter ratio, or equivalently the "cut-off" probability of an *UT*, is varied.

In table 16.10 the Brier scores for our turning point forecasts are presented. The Brier score, or quadratic probability score, denoted by *B* is defined as follows:

$$B = \sum_{i=1}^{N} 2(P_i - R_i)^2/N, \qquad (4.1)$$

where N = number of forecasts, P_i = computed probability associated with the *i*th forecast, and R_i = realized outcome, equal to one if the forecasted event occurred and equal to zero if it did not. *B* ranges from 0 to 2, with $B = 0$ denoting perfect accuracy. B in (4.1) was computed separately for all our turning point forecasts and for *DT/NDT* and *UT/NUT* forecasts separately.

From the results in table 16.10, it is seen that the Brier scores for all turning points range from 0.31 to 0.37 for the six models when there is no pooling, and from 0.28 to 0.34 with pooling. With no pooling the TVP/ARLI/WI model has the lowest Brier score, namely 0.31. With pooling, this same model has the lowest score, 0.28. Thus according to the Brier score criterion, the TVP/ARLI/WI model performed the best with pooling resulting in about a 10 percent reduction in Brier score relative to the unpooled case. Further, the other models' performance also shows improvement with pooling in every case. The pooled EW/ARLI/WI model's Brier score, 0.29, is only slightly larger than that of the TVP/ARLI/WI model. It is also the case that inclusion of the WI variable improved performance in every case relative to that of corresponding models without this variable. Last, the FP/ARLI model performed the worst both in the case of no pooling and with pooling.

In the case of computed probabilities for *DT*s, the Brier scores range from 0.30 to 0.36 for models with no pooling and from 0.24 to 0.30 with pooling. In each case, the score is lower with pooling, indicating that pooling improves performance. The lowest and best score, 0.24, is encountered for the TVP/ARLI/WI model with pooling. A score of 0.26 is obtained for the pooled TVP/ARLI and EW/ARLI/WI models. That these three models, which allow for non-constant parameters, exhibit the best performance is noteworthy.

With respect to computed probabilities for *UT*s, the models with no pooling have Brier scores of 0.33 to 0.38, while those with pooling range from 0.31 to 0.37. In this case, pooling did not produce a uniform improvement in the scores. For the TVP/ARLI/WI model, its scores are 0.33 with no pooling and 0.31 with pooling, the latter being the lowest and best score. Also, the results in table 16.10 indicate that the Brier scores are somewhat better for probabilities associated with *DT*s than for those associated with *UT*s for all models except the FP/ARLI/WI model with no pooling where they were equal. Thus, using the Brier score measure, forecasting performance with respect to *UT*s is slightly poorer than that for *DT*s.

Finally, in table 16.11 are shown the Brier scores for the naive turning point forecasters, the eternal optimist, the eternal pessimist, the coin-flipper, and the believer in a deterministic four-year cycle. As can be seen, the Brier scores for all these naive forecasters are larger than all those in table 16.10 indicating that naive forecasters' performance is poorer than that associated with the models and methods reported in table 16.10.

Table 16.11 *Brier scores[a] for eternal optimist, eternal pessimist, coin-flipper, and deterministic four-year cycle forecasters*

Forecaster	76 DT and NDT forecasts	82 UT and NUT forecasts	158 turning point forecasts
Eternal optimist[b]	1.55	0.83	1.18
Eternal pessimist[c]	0.45	1.17	0.82
Coin-flipper[d]	0.50	0.50	0.50
Deterministic four-year cycle[e]	0.45	0.83	0.65

Notes:
[a] The Brier score is defined in (4.1) in the text.
[b] The eternal optimist assumes that the probability of a DT = 0 and the probability of an *UT* = 1.0 on all occasions.
[c] The eternal pessimist assumes that the probability of a DT = 1.0 and the probability of an UT = 0 on all occasions.
[d] A coin-flipper assumes that the probability of a DT or of an UT = $\frac{1}{2}$ on all occasions.
[e] A deterministic four-year cycle forecaster assumes that the probability of an *UT* or a DT = 1.0 on all occasions.

5 Summary and concluding remarks

In this chapter we have found that several ARLI models have produced rather good turning point forecasting performance. Using a Bayesian decision theoretic approach to generate optimal turning point forecasts, it was found that about 80 percent of 158 turning point forecasts for eighteen countries, 1974–86, are correct using the TVP/ARLI/WI and EW/ARLI/WI models with pooling (see table 16.2). For other models, with and without pooling, the percentage of correct turning point forecasts is 72 or higher. Similar good results are obtained for *DT*/*NDT* and *UT*/*NUT* forecasts separately. Also these forecasts are superior to those of several naive turning point forecasting procedures.

The weakest performance was encountered in the case of *NDT* forecasts where, with pooling, the percentage of correct forecasts ranges from 55 to 70. In the case of the pooled TVP/ARLI/WI model *NDT* forecasts, 7 of 19 *NDT* forecasts were incorrect. For these 7 cases, the computed probabilities of *DT*s have values from 0.27 to 0.47. Thus in each of these cases there is a substantial probability of a *DT*. If a more conservative rule for forecasting *DT*s were employed, say forecast a *DT* if $P_{\mathrm{DT}} > 0.2$, then, as shown in table 16.9, all *NDT* forecasts are correct.

Finally, the good performance of our models and methods in forecasting turning points in annual growth rates of real output for eighteen countries, 1974–86, is surprising in view of the fact that this period is a difficult one from the point of view of forecasting. That is, the period is one in which two oil crises occurred, countries moved from fixed to flexible exchange rates, monetary policy in the United States changed dramatically in 1979, inflation and unemployment rates were very high in the late 1970s, etc. The fact that our models and methods performed so well without any ad hoc adjustments for the effects of these major events suggests that they may be useful to private and public decision-makers who are interested in computing probabilities of *DT*s and *UT*s and forecasting turning points.

APPENDIX CALCULATION OF POSTERIOR AND PREDICTIVE DISTRIBUTIONS

A.1 Fixed parameter models

Dropping the subscript i, the model in (2.1) of the text is $y_t = \boldsymbol{x}_t'\boldsymbol{\beta} + u_t$, $t = 1, 2, \ldots, n$. The u_ts are assumed independently drawn from a zero-mean normal distribution with variance σ^2. A spread out normal-inverted gamma natural conjugate prior distribution for $\boldsymbol{\beta}$ and σ was employed, namely $p(\boldsymbol{\beta} \mid \sigma) \sim N(\boldsymbol{0}, \sigma^2 I \times 10^6)$ and $p(\sigma) \sim IG(\nu_0 s_0)$ with ν_0 and s_0 assigned very small values. Then at $t = 0$, with given initial values, the predictive probability density function (pdf) for y_1, denoted by $p(y_1 \mid D_0)$, is in the univariate Student-t form. That is, $t_{\nu_0} = (y_1 - \boldsymbol{x}_i'\hat{\boldsymbol{\beta}}_0)/s_0 a_0$ has a univariate Student-t pdf with ν_0 degrees of freedom where $a_0^2 = 1 + \boldsymbol{x}_i'\boldsymbol{x}_i \cdot 10^6$ and $\hat{\boldsymbol{\beta}}_0 = \boldsymbol{0}$. By updating the posterior pdfs period by period by use of Bayes' Theorem, they can be employed to obtain predictive pdfs period by period. For example, the predictive pdf for y_{t+1} is in the univariate Student-t form and such that

$$t_{\nu_t} = (y_{t+1} - \boldsymbol{x}_{t+1}'\hat{\boldsymbol{\beta}}_t)/s_t a_t \tag{A.1}$$

has a univariate Student-t pdf where

$$\hat{\boldsymbol{\beta}}_t = (Z_t'Z_t)^{-1}Z_t'y_t, \tag{A.2}$$

with $\boldsymbol{y}_t' = (y_t, y_{t-1}, y_{t-2}, \ldots, y_1)$, $Z_t' = (\boldsymbol{x}_t, \boldsymbol{x}_{t-1}, \ldots, \boldsymbol{x}_1)$, and $a_t^2 = 1 + \boldsymbol{x}_{t+1}'(Z_t'Z_{t-1})^{-1}\boldsymbol{x}_{t+1}$. The following up-dating formulas were employed:

$$\begin{aligned}
\hat{\boldsymbol{\beta}}_t &= \hat{\boldsymbol{\beta}}_{t-1} + (Z_{t-1}'Z_{t-1})^{-1}\boldsymbol{x}_t(y_t - \boldsymbol{x}_t'\hat{\boldsymbol{\beta}}_{t-1})/[1 + \boldsymbol{x}_t'(Z_t'Z_t)^{-1}\boldsymbol{x}_t], \\
\nu_t &= \nu_{t-1} + 1, \\
\nu_t s_t^2 &= v_{t-1}s_{t-1} + (y_t - \boldsymbol{x}_t'\hat{\boldsymbol{\beta}}_t)^2 + (\hat{\boldsymbol{\beta}}_t - \hat{\boldsymbol{\beta}}_{t-1})'Z_{t-1}'Z_{t-1}(\hat{\boldsymbol{\beta}}_t - \hat{\boldsymbol{\beta}}_{t-1}), \\
(Z_t'Z_t)^{-1} &= (Z_{t-1}'Z_{t-1})^{-1} - (Z_{t-1}'Z_{t-1})^{-1}\boldsymbol{x}_t\boldsymbol{x}_t'(Z_{t-1}'Z_{t-1})^{-1}/ \\
&\qquad [1 + \boldsymbol{x}_t'(Z_{t-1}'Z_{t-1})^{-1}\boldsymbol{x}_t].
\end{aligned}$$

To compute, for example, the probability that $y_{t+1} < y_t$, given information through t, $\Pr(y_{t+1} < y_t \mid D_t)$, use is made of (A.1) from which $y_{t+1} = s_t a_t t_{\nu_t} + \boldsymbol{x}'_{t+1}\hat{\boldsymbol{\beta}}_t$. Then $\Pr(y_{t+1} < y_t \mid D_t) = \Pr[t_{\nu_t} < (y_t - \boldsymbol{x}'_{t+1}\hat{\boldsymbol{\beta}}_t)/s_t a_t \mid D_t]$, which was evaluated using the Student-t distribution with ν_t degrees of freedom.

For the pooled ARLI model, all countries' parameters were assumed to be the same, that is, $\boldsymbol{\beta}_1 = \boldsymbol{\beta}_2 = \cdots = \boldsymbol{\beta}_{18}$ and $\sigma_1 = \sigma_2 = \cdots = \sigma_{18}$, and a spread-out natural conjugate prior pdf and the above methods were used to compute posterior and predictive pdfs period by period. The predictive pdfs were employed to compute probabilities of downturns and upturns, as explained above.

Similar procedures were employed to update posterior and predictive pdfs for the FP/ARLI/WI model in (2.4) and (2.5) of the text under the assumption that (2.4) and (2.5) constitute a fully recursive econometric model with constant variance, zero-mean independent normal disturbance terms. Using spread-out natural conjugate prior pdfs for the parameters, posterior and one-step-ahead predictive pdfs were computed using regression-like results in the literature (see Zellner 1971, ch. 8 and Bowman and Laporte 1975). The one-step-ahead predictive pdf for y_{t+1} from (2.4) was approximated by a univariate Student-t pdf conditioned on $w_{t+1} = \hat{w}_{t+1}$ the mean of the predictive pdf for w_{t+1}. Such predictive pdfs were employed to compute probabilities of turning points. For the pooled FP/ARLI/WI model, countries' parameters were all assumed to be the same and the procedures described above were employed.

A.2 Time-varying parameter (TVP) models

For the TVP model (2.2)–(2.3), on dropping the subscript i for convenience, we have for $t = 1, 2, \ldots, n$

$$y_t = \boldsymbol{x}'_t\boldsymbol{\beta}_t + u_t, \quad u_t\text{'s NID}(0, \sigma^2), \tag{A.3a}$$

$$\boldsymbol{\beta}_t = \boldsymbol{\beta}_{t-1} + v_t, \quad v_t\text{'s NID}(0, \phi\sigma^2 I_k). \tag{A.3b}$$

The distribution on the initial coefficient vector, $\boldsymbol{\beta}_0$, was taken to be a spread-out normal distribution for $\boldsymbol{\beta}_0$ given σ with a zero-mean vector and an inverted gamma pdf for σ with small values for its parameters, ν_0 and s_0, $p(\sigma \mid \nu_0, s_0^2) \propto \sigma^{-(\nu_0+1)} \exp\{-\nu s_0^2/2\sigma^2\}$. Then by direct calculations, at time t, we have

$$\begin{aligned}
\hat{\boldsymbol{\beta}}_t &= \hat{\boldsymbol{\beta}}_{t-1} + V_{t-1}\boldsymbol{x}_t(y_t - \boldsymbol{x}'_t\hat{\boldsymbol{\beta}}_{t-1})/(1 + \boldsymbol{x}'_t V_{t-1}\boldsymbol{x}_t), \\
\nu_t s_t^2 &= \nu_{t-1}s_{t-1}^2 + (y_t - \boldsymbol{x}_t'\hat{\boldsymbol{\beta}}_t)^2 + (\hat{\boldsymbol{\beta}}_t - \hat{\boldsymbol{\beta}}_{t-1})' V_{t-1}^{-1}(\boldsymbol{\beta}_t - \hat{\boldsymbol{\beta}}_{t-1}), \\
\nu_t &= \nu_{t-1} + 1, \\
V_t &= V_{t-1} - V_{t-1}\boldsymbol{x}_t\boldsymbol{x}'_t V_{t-1}/(1 + \boldsymbol{x}_t' V_{t-1}\boldsymbol{x}_t) + \phi I_k,
\end{aligned}$$

and, for the predictive pdf,

$$p(y_{t+1} \mid D_t) \rightarrow t_{\nu_t} = (y_{t+1} - \boldsymbol{x}'_{t+1}\hat{\beta}_t)/s_t a_t,$$

where

$$a_t^2 = 1 + \boldsymbol{x}'_{t+1} V_t \boldsymbol{x}_{t+1}.$$

The parameter ϕ was assigned a value equal to 0.30 in the above calculations.

For the pooled TVP/ARLI model, the equations in (A.3a)–(A.3b) are interpreted as relating to all countries, and recursions similar to those presented above, with $\phi = 0.01$, were employed to compute predictive pdfs and associated quantities. For the TVP/ARLI/WI model, it was assumed to be fully recursive, and recursions similar to those presented above were employed for each of its two equations country by country for the unpooled case and for the entire set of countries in the pooled case. The ϕ parameter was assigned a value of 0.20 for the unpooled case and 0.10 for the pooled case. The values of ϕ were selected to provide a minimum for the median of the eighteen countries' RMSEs of one-year-ahead forecasts, 1974–86, for each model.

In the case of models employing exponential weighting, we have for the ith country (with the subscript i omitted for convenience):

$$W_t^{1/2}\boldsymbol{y}_t = W_t^{1/2}X_t\beta + \varepsilon_t, \tag{A.4a}$$

or

$$\boldsymbol{y}_t^* = X_t^*\beta + \varepsilon_t, \tag{A.4b}$$

where

$$W_t = \begin{bmatrix} 1 & 0 & 0 & \dots & 0 \\ 0 & \mu & 0 & \dots & 0 \\ 0 & 0 & \mu^2 & \dots & 0 \\ \vdots & \vdots & \vdots & \ddots & \vdots \\ 0 & 0 & 0 & \dots & \mu^{t-1} \end{bmatrix},$$

$$\boldsymbol{y}_t = \begin{pmatrix} y_t \\ y_{t-1} \\ \vdots \\ y_1 \end{pmatrix} \quad \text{and} \quad X_t = \begin{pmatrix} \boldsymbol{x}'_t \\ \boldsymbol{x}'_{t-1} \\ \vdots \\ \boldsymbol{x}'_1 \end{pmatrix}.$$

We assume that $\varepsilon_t \sim N(\mathbf{0}, \sigma^2 I_t)$. As is evident from (A.4a) or (A.4b), this can be considered to be a fixed parameter model with the current observations given heavier weights than those more removed from the present. With a diffuse prior for the parameters, the posterior mean of β with data through t is

$$\hat{\beta}_t = (X_t' W_t X_t)^{-1} X_t' W_t \boldsymbol{y}_t,$$

and

$$s_t^2 = (\boldsymbol{y}_t - X_t\hat{\beta}_t)'(\boldsymbol{y}_t - X_t\hat{\beta}_t)/(t-k).$$

From the predictive pdf for y_{t+1},

$$t_{\nu_t} = (\boldsymbol{y}_{t+1} - \boldsymbol{x}_{t+1}'\hat{\beta}_t)/s_t a_t$$

has a univariate Student-t distribution with $\nu_t = t - k$ degrees of freedom, where $a_t^2 = 1 + \boldsymbol{x}_{t+1}'(X_t' W_t X_t)^{-1}\boldsymbol{x}_{t+1}$.

In the calculations presented in the text, the parameter μ was assigned a value equal to 0.95 for the unpooled ARLI and ARLI/WI models. In the case of pooling, wherein the value of β was assumed the same for all countries, μ was taken equal to 0.85 for the ARLI model and 0.80 for the ARLI/WI model. These μ values provided a minimum value of the median RMSEs of one-year-ahead forecasts, 1974–86, for the eighteen countries.

BIBLIOGRAPHY

Bowman, H. W. and A. M. Laporte (1975), "Stochastic optimization in recursive equation systems with random parameters with an application to control of the money supply," in S.E. Fienberg and A. Zellner (eds.), *Studies in Bayesian Econometrics and Statistics* (Amsterdam, North-Holland), 441–62

Garcia-Ferrer, A., R. A. Highfield, F. C. Palm, and A. Zellner (1987), "Macroeconomic forecasting using pooled international data," *Journal of Business and Economic Statistics* 5, 53–67; chapter 13 in this volume

Highfield, R. A. (1986), "Forecasting with Bayesian state space models," PhD dissertation, University of Chicago, Graduate School of Business

Hong, C. (1989), "Forecasting real output growth rates and cyclical properties of models: a Bayesian approach," PhD dissertation, University of Chicago, Department of Economics

Kling, J. L. (1987), "Predicting the turning points of business and economic time series," *Journal of Business* 60, 201–38

Moore, G. H. (1983), *Business Cycles, Inflation and Forecasting*, 2nd edn. (Cambridge, Mass., Ballinger)

Moore, G. H. and V. Zarnowitz (1982), "Sequential signals of recession and recovery," *Journal of Business* 55, 57–85

Wecker, W. E. (1979), "Predicting the turning points of a time series," *Journal of Business* 52, 35–50

West, M., P. J. Harrison, and H. S. Migon (1985), "Dynamic generalized linear models and Bayesian forecasting," *Journal of the American Statistical Association* 80, 73–83

Zarnowitz, V. (1985), "Recent work on business cycles in historical perspective," *Journal of Economic Literature* 23, 523–80

Zellner, A. (1971), *An Introduction to Bayesian Inference in Econometrics* (New York, John Wiley)

Zellner, A. and C. Hong (1988), "Bayesian methods for forecasting turning points in economic time series: sensitivity of forecasts to asymmetry of loss structures," in K. Lahiri and G. Moore (eds.), *Leading Economic Indicators: New Approaches and Forecasting Records* (Cambridge, Cambridge University Press)

(1989), "Forecasting international growth rates using Bayesian shrinkage and other procedures," *Journal of Econometrics, Annals* 40, 183–202; chapter 14 in this volume

Zellner, A., C. Hong, and G. M. Gulati (1990), "Turning points in economic time series, loss structures, and Bayesian forecasting," in S. Geisser, J. S. Hodges, S. J. Press, and A. Zellner (eds.), *Bayesian and Likelihood Methods in Statistics and Econometrics: Essays in Honor of George A. Barnard* (Amsterdam, North-Holland), 371–89; chapter 15 in this volume

17 Bayesian and non-Bayesian methods for combining models and forecasts with applications to forecasting international growth rates (1993)

Chung-ki Min and Arnold Zellner

1 Introduction

In past work, Garcia-Ferrer *et al.* (1987) and Zellner and Hong (1989), variants of a relatively simple autoregressive model of order three containing lagged leading indicator variables, called an ARLI model, provided good one-year-ahead forecasts of annual output growth rates for eighteen industrial countries, 1974–84. In Zellner, Hong, and Gulati (1990) and Zellner, Hong, and Min (1991), this ARLI model and variants of it produced good turning point forecasts, about 70–80 percent of 158 turning points correctly forecasted. In Hong (1989), the ARLI model's cyclical properties were analyzed and its forecasting performance was shown to be slightly superior to that of a version of Barro's "money surprise" model. LeSage (1989) and LeSage and Magura (1990) have used ARLI models to forecast employment growth rates and turning points in them for eight metropolitan labor markets with satisfactory results. Blattberg and George (1991) used similar techniques in successfully forecasting sales of different brands of a product.

Some of our past work has involved use of fixed parameter models (FPMs) and time-varying parameter models (TVPMs). In the present chapter, we derive and compute posterior odds relating to our FPMs and TVPMs using data for eighteen countries, 1973–87. While there are many reasons – Lucas effects, aggregation effects, wars, etc. – for believing that parameters may be time-varying, economic theorists' models are generally fixed parameter models. Our calculated posterior odds will

Research financed by the National Science Foundation and by income from the H. G. B. Alexander Endowment Fund. Graduate School of Business, University of Chicago. Cheng Hsiao and two referees provided comments which were helpful in revising an earlier version of this chapter.

Originally published in the *Journal of Econometrics, Annals* 56 (1993), 89–118. 0304/4076/93-$05.00.

shed some light on the parameter constancy issue and are used to choose between FPMs' and TVPMs' forecasts year by year. As an alternative approach, we consider procedures for combining models and/or their forecasts (see Clemen 1989, an extensive review paper discussing theoretical and empirical work on various methods of combining forecasts). While a good portion of the literature suggests that combining forecasts will produce better forecasting performance, we show that this argument is not true in general. Clearly, combining a good forecast with a bad forecast will not usually produce a combined forecast that performs better than the good forecast, a point also made in Diebold (1990a). To cope with this problem, we develop a Bayesian forecast selection rule based on a predictive loss criterion which indicates which of several forecasts, including a combined forecast, is optimal. This rule is applied in forecasting experiments and results are compared to those of other procedures. Thus, our chapter provides analytical and empirical results on the questions: (1) Fixed or time-varying parameters? (2) To combine or not to combine? (3) If so, how? and (4) How do different procedures perform in actual forecasting?

The plan of the chapter is as follows. Section 2 deals with the issue of whether or not to combine forecasts. In section 3, it is indicated how posterior odds can be used to choose between or among models in such a way as to minimize expected loss. Then, building on the work of Geisel (1975) and Zellner (1989), posterior odds are employed to combine forecasts of two or more exhaustive or non-exhaustive models. Further, a Bayesian forecast selection rule is derived using a predictive loss structure which can be employed to decide when to use a combined forecast and when to use some other forecast. Section 4 contains a brief review of our data and models and presentation of the results of our forecasting experiments. A summary of results and some concluding remarks are presented in section 5.

2 To combine or not to combine forecasts?

In previous analyses of whether to combine or average different forecasts, it has been customary to assume that all forecasts being combined are unbiased (see, e.g., Reid 1968, 1969 and Bates and Granger 1969 which, according to Clemen 1989, "are considered by most forecasters to be the seminal works in the area of combining forecasts"). While some works (e.g. Clemen 1989, Diebold and Pauly 1987, and Palm and Zellner 1990) discuss the possibility of forecasts being biased, the earlier analytical results, based on the assumption of no bias (see also Winkler 1981 for a Bayesian analysis involving the assumption of no bias), are still influential. Therefore we review briefly the analysis leading to the conclusion that combining forecasts leads to improved performance in terms of a mean-squared error reduction. Let f_1 and f_2 be two unbiased forecasts, that is, $f_1 = y + e_1$ and $f_2 = y + e_2$, where y is the actual, random outcome and e_1 and e_2 are zero mean errors with variances σ_1^2 and σ_2^2, respectively, and covariance σ_{12}. Bates and Granger (1969) showed that with $f_c = wf_1 + (1-w)f_2$, a combined forecast, then $\mathrm{E}(f_c - y)^2$ will be minimized if $w = w^* = (\sigma_2^2 - \sigma_{12})/(\sigma_1^2 + \sigma_2^2 - 2\sigma_{12})$ and $\mathrm{E}(f_c - y)^2 < \mathrm{E}(f_i - y)^2$, $i = 1, 2$. Thus using a value of w different from 0 or 1 is considered to be optimal. However, as is almost obvious, this conclusion depends critically on the unbiasedness assumption, as shown below. Also, it should be noted that only linear combinations of the two forecasts are being considered. It may be that limiting the analysis to linear combinations is unduly restrictive. Further, it has been noted that the optimal weight, w^*, depends on parameters with unknown values which must be estimated. How estimation of the weights affects the optimality conclusion requires analysis. If past observations are available on $y - f_1$ and $y - f_2$ and it is assumed that these pairs are iid bivariate normal with a zero mean vector, then the predictive density can be derived and used to evaluate $\mathrm{E}(y - f_c)^2$ and to obtain an optimal value of w. Again it will be optimal always to combine forecasts. In this case the weights are functions of sample quantities and prior parameters. However, the analysis is based on the assumption that the *e*s have zero means and constant covariance matrix for all sample points, very strong assumptions which clearly will not hold in all situations.

We now consider what happens when the condition of zero bias is relaxed. Let f_1 and f_2 be two forecasts and y the random variable being forecasted. Assume $f_1 - y = e_1$ and $f_2 - y = \theta + e_2$, where e_1 and e_2 have zero means, variance σ^2, and covariance σ_{12}, and θ is the unknown bias associated with f_2. Then we have $MSE_1 = \mathrm{E}(y - f_1)^2 = \sigma^2$ and $MSE_2 = \mathrm{E}(y - f_2)^2 = \theta^2 + \sigma^2$. If we consider a combined forecast, $f_c = wf_1 + (1-w)f_2$, where w is a weight, $0 < w < 1$, then $f_c - y = w(f_1 - y) + (1-w)(f_2 - y)$ and $MSE_c = \mathrm{E}(f_c - y)^2 = w^2\sigma^2 + (1-w)^2(\sigma^2 + \theta^2) + 2w(1-w)\sigma_{12}$. It follows that

$$MSE_c - MSE_1 = (1-w)\sigma^2[(1-w)(\theta/\sigma)^2 - 2w(1-\rho_{12})], \tag{2.1}$$

where $\rho_{12} = \sigma_{12}/\sigma^2$. From (1), $MSE_c - MSE_1 > 0$ if $(1 - w)(\theta/\sigma)^2 - 2w(1-\rho_{12}) > 0$, or

$$(\theta/\sigma)^2 > (2w/(1-w))(1-\rho_{12}). \tag{2.2}$$

As $\rho_{12} \to 1$, this condition approaches $(\theta/\sigma)^2 > 0$ which will be satisfied for $|\theta| > 0$. As $\rho_{12} \to -1$, the condition approaches $(\theta/\sigma)^2 > 4w/(1-w)$.

For example, if $w = \frac{1}{2}$, the value of w^* above when $\sigma_1^2 = \sigma_2^2 = \sigma^2$, (2.2) yields $(\theta/\sigma)^2 > 2(1 - \rho_{12})$ or $|\theta|/\sigma > 1.414(1 - \rho_{12})^{1/2}$ or $|\theta|/\sigma > 1.414$ if $\rho_{12} = 0$. If $w = \frac{1}{2}$, and $\rho_{12} = 0$, $MSE_c - MSE_1 = (\sigma^2/2)\ [(\theta/\sigma)^2/2 - 1]$ which can be large if $|\theta|/\sigma$ is large. Thus use of weights appropriate for unbiased forecasts can be quite suboptimal when forecasts are biased.

In summary, it is seen that the presence of bias in forecasts can result in combined forecasts being worse in terms of *MSE* than individual forecasts. See also Diebold (1990a, 1990b) for additional analysis of these issues. Forecasts may be biased for many reasons including use of shrinkage techniques or asymmetric loss functions and presence of specification, measurement, and other errors. Correcting for such time-varying biases and defining "optimal weights" is difficult. Thus it is important to appraise models and their forecasts carefully before combining them. In the next section, we show how posterior odds can be utilized to compare models and choose between or among them and their forecasts.

3 Posterior odds and choosing models and their forecasts

3.1 *Posterior odds and choosing models*

Many, including Jeffreys (1967), Zellner (1971, 1984), Geisel (1975), Leamer (1978), Schwarz (1978), and Rossi (1983, 1985), have indicated how posterior odds can be calculated and used to discriminate between or among different hypotheses and models. In the present context, we compute posterior odds for a fixed parameter model (FPM) versus a time-varying parameter model (TVPM), denoted by $K = P_1/P_2$, where P_1 and P_2 are the posterior probabilities associated with the FPM and the TVPM, respectively (see appendix A, p. 582, for a derivation of the explicit expression for K used in our calculations). It has been recognized in many works that parameters of aggregate time series models may not be constant because of (1) aggregation effects, (2) effects of policy changes, that is Lucas effects, (3) effects of major events such as strikes, wars, oil crises, etc., (4) adaptive optimization on the part of economic agents, (5) changes in factor prices that induce changes in choice of production techniques, and (6) changes in tastes and preferences, perhaps due to generational, advertising, and educational effects. Given so many reasons for anticipating parameters to be time-varying, it is indeed of interest to compute odds on FPMs versus TVPMs country by country to gain information relating to this important issue.

Further, if we have a standard two-state, two-action loss structure, shown in table 17.1, we can make the choice of model which yields lower

Table 17.1 *A loss structure*[a] *for model selection*

	State of world		
	FPM[b]	TVPM[c]	Expected loss
Choices FPM	0	c_1	$c_1 P_2$
TVPM	c_2	0	$c_2 P_1$
Posterior probability	P_1	P_2	

Notes:
[a] c_1 and c_2 are positive losses associated with incorrect choices.
[b] FPM denotes a fixed parameter model.
[c] TVPM denotes a time-varying parameter model.

expected loss. If the loss structure is symmetric, $c_1 = c_2$, then we choose FPM if $K > 1$ and TVPM if $K < 1$. As has been explained in the literature, posterior odds reflect goodness of fit, extent to which prior information is in accord with sample information, relative precision of sample, and prior information and sample size (see the references cited above for discussions of these points).

For each year in our forecast period, we compute the posterior odds $K = P_1/P_2$ based on data up to the forecast year. If $K > 1$, we choose the FPM and use the mean of its predictive density for the forecast year's output growth rate as our point forecast. If $K < 1$, we choose the TVPM and use its one-year-ahead predictive mean as our point forecast. These calculations are performed year by year, with K being updated, beginning with forecast year 1974 and ending with forecast year 1987.

Since it is possible that neither the FPM nor the TVPM may perform well relative to a benchmark model (BMM), we expand the 2×2 loss structure in table 17.1 to a 3×3 loss structure relating to choices of the FPM, the TVPM, and the BMM using posterior odds computed for pairs of these models year by year. Using these posterior odds and assuming that the 3×3 loss structure is symmetric, we choose among the three models so as to minimize expected loss and use the mean of the predictive density of the model so chosen as our point forecast. The BMM used in these calculations is an autoregressive model of order 3.

3.2 *Combining and choosing among individual and combined forecasts*

Given that posterior odds are available relating to two mutually exclusive models, the issue arises as to how they can be used to combine the models and/or their forecasts (see Diebold 1990b for further discussion). If the models are considered to be exhaustive, then their probabilities add to one and the expectation of a future observation, y, is just

$$\begin{aligned} E(y) &= P_1 E(y \mid \text{FPM}) + (1 - P_1) E(y \mid \text{TVPM}) \\ &= (K/(1 + K)) E(y \mid \text{FPM}) + (1/(1 + K)) E(y \mid \text{TVPM}). \end{aligned} \tag{3.1}$$

This last expression is the predictive mean of a future observation given that we average over the two models. Clearly if K is large, favoring the FPM, its mean will get a large weight and that of the TVPM will receive a small weight. Since the large value of K reflects good *relative* performance of the FPM in the past, it seems appropriate to give its forecast heavier weight in the combined forecast. The predictive mean in (3.1) will be used as a point forecast in our forecasting experiments.

It should be appreciated that the posterior odds is just a relative measure which does not assure that the absolute performance of either model is satisfactory. For example, if $K = 19$, $P_1 = 0.95$, a high probability associated with the FPM. However, this high probability is attained under the assumption that the two models are exhaustive. If they are not, then it is clear that with $K = 19$, P_1 might have any value and the mean in (3.1) cannot be computed. Since there is another rationalization of the combining formula in (3.1) other than the exhaustive assumption (see below), we shall use the predictive mean in (3.1), the Bayesian combined forecast (BCF) in our forecasting experiments.

When the models considered, say a FPM and a TVPM, do not constitute an exhaustive set, as is usually the case since there are many additional models which can be considered, the posterior odds $K = P_1/P_2$ can be computed, but it is not possible to evaluate P_1 and P_2 separately and thus the first line of (3.1) cannot be implemented. For this problem, Zellner (1989) introduced the concept of a combining predictive density which is closest in a distance metric sense to the predictive densities of the non-exhaustive models and suggested use of the mean of the combined predictive density as a point forecast. With a weighted quadratic distance measure, it was shown that for two non-exhaustive models (e.g. FPM and TVPM) the mean of the combined predictive density is given by the second line of (3.1). Note that to compute the second line of (3.1) only the value of $K = P_1/P_2$ and the means of the two models' predictive densities are needed. The values of P_1 and P_2 individually are not needed.

In (3.1) it has been assumed that the available forecasts will be averaged. If we consider the predictive squared error loss structure in table 17.2, wherein possible forecast choices and different states of the world

Table 17.2 *Predictive squared error loss structure*[a]

Model Choice set	State: M_1 (FPM)	M_2 (TVPM)	M_c	Expected loss $(P_1 v_1 + P_2 v_2 + P_c \bar{v}) +$
M_1 (FPM)	v_1	$v_2 + \Delta_{12}$	$\bar{v} + w_2 \Delta_{12}$	$\Delta_{12} w_2 (P_1 + P_2 + P_c)$
M_2 (TVPM)	$v_1 + \Delta_{12}$	v_2	$\bar{v} + w_1 \Delta_{12}$	$\Delta_{12} w_1 (P_1 + P_2 + P_c)$
M_c	$v_1 + w_2^2 \Delta_{12}$	$v_2 + w_1^2 \Delta_{12}$	$\bar{v} + w_1 w_2 \Delta_{12}$	$\Delta_{12} w_1 w_2 (P_1 + P_2 + P_c)$
Posterior probability	P_1	P_2	P_c	

Note:
[a] In the table, $v_i = E(y - \bar{y}_i)^2$, the variance of M_i's predictive density, $\bar{v} = w_1 v_1 + w_2 v_2$, with $w_i = P_i/(P_1 + P_2)$ and $\Delta_{12} = (\bar{y}_1 - \bar{y}_2)^2$, where $\bar{y}_1$ and $\bar{y}_2$ are predictive means of M_1's and M_2's predictive densities, respectively. Each non-diagonal entry in the table is given by $E_j(y - \bar{y}_i)^2$, where E_j indicates that M_j's predictive density is employed.

are shown, it is optimal to choose the combined forecast when only three alternatives are available, namely choose the FPM's forecast, the TVPM's forecast, or the combined forecast, shown in (3.1). This can be seen from consideration of table 17.2. The entries in table 17.2 are expected squared error losses given that model i's forecast, $\bar{y}_i$, the predictive mean, is used and model j is appropriate. In this situation, expected loss, computed using M_j's predictive density, is $\mathrm{E}_j(y - \bar{y}_i)^2 = \mathrm{E}_j[y - \bar{y}_j - (\bar{y}_i - \bar{y}_j)]^2 = \mathrm{E}_j(y - \bar{y}_j)^2 + (\bar{y}_i - \bar{y}_j)^2 \equiv +v_j + \Delta_{ij}$, where $\bar{y}_j$ is the mean of M_j's predictive density, v_j is the variance of M_j's predictive density, and $\Delta_{ij} \equiv (\bar{y}_i - \bar{y}_j)^2$. On comparing expected losses shown in table 17.2, which can be done without assuming the models (M_1, M_2, and M_c) are exhaustive, we have $\mathrm{E}(L \mid M_c)/\mathrm{E}(L \mid M_1) = P_1/(P_1 + P_2) < 1$ and $\mathrm{E}(L \mid M_c)/\mathrm{E}(L \mid M_2) = P_2/(P_1 + P_2) < 1$. Thus, choice of the combined forecast (3.1) leads to minimal expected loss for the predictive squared error loss structure in table 17.2.

To allow for other models not explicitly considered in table 17.2, we consider the 3×4 loss structure in table 17.3 which includes M_4, e.g. a benchmark model. On comparing expected losses given in the last column of table 17.3, it is seen that the combined forecast is not always preferred because of additional terms involving $\Delta_{14} = (\bar{y}_1 - \bar{y}_4)^2$, $\Delta_{24} = (\bar{y}_2 - \bar{y}_4)^2$, and $\Delta_{c4} = (\bar{y}_c - \bar{y}_4)^2$, where $\bar{y}_4$ is the predictive mean of M_4 and $\bar{y}_c$ is the predictive mean of M_c.

Given that M_4 is specified, say an AR (3) benchmark model, posterior odds can be calculated and used in connection with table 17.3 to choose the forecast which leads to minimal expected loss.

Table 17.3 *Simplified extended predictive squared error loss structure*[a]

Model Choice set	State				Expected loss[b]
	M_1 (FPM)	M_2 (TVPM)	M_c	M_4 (BMM)	
M_1 (FPM)	0	Δ_{12}	$w_2\Delta_{12}$	Δ_{14}	$\Delta_{12}w_2(P_1 + P_2 + P_c) + P_4\Delta_{14}$
M_2 (TVPM)	Δ_{12}	0	$w_1\Delta_{12}$	Δ_{24}	$\Delta_{12}w_1(P_1 + P_2 + P_c) + P_4\Delta_{24}$
M_c	$w_2^2\Delta_{12}$	$w_1^2\Delta_{12}$	$w_1w_2\Delta_{12}$	Δ_{c4}	$\Delta_{12}w_1w_2(P_1 + P_2 + P_c) + P_4\Delta_{c4}$
Posterior probability	P_1	P_2	P_c	P_4	

Notes:

[a] See note [a] of table 17.2. Also $\Delta_{i4} = (\bar{y}_i - \bar{y}_4)^2$, where $\bar{y}_i$ and $\bar{y}_4$ are predictive means of M_i and M_4, respectively, $\bar{y}_c = w_1\bar{y}_1 + w_2\bar{y}_2$, where $w_1 = P_1/(P_1 + P_2) = K/(1 + K)$ and $w_2 = P_2/(P_1 + P_2) = 1/(1 + K)$ where $K = P_1/P_2$.

[b] Common elements in the columns of the loss structure (e.g. v_1, v_2, $\bar{v}$, etc., see table 17.2, have been omitted).

3.3 *Regression combining techniques*

Having discussed general Bayesian procedures for combining models and their forecasts, we now briefly take up regression procedures which have been used extensively in the literature (see, e.g., Nelson 1972, Granger and Ramanathan 1984, Diebold and Pauly 1987, and the discussion and references in Clemen 1989). In this approach past actual values of a variable are related to forecasts of them provided by two or more forecasting methods; that is, the following regression relation is considered:

$$A_t = \alpha + \beta_1 f_{1t} + \beta_2 f_{2t} + u_t, \quad t = 1, \ldots, n. \tag{3.2}$$

In regressing A_t on f_{1t} and f_{2t}, two forecasts, some impose the constraint that the intercept is zero and the regression coefficients add to one. Others allow for a non-zero intercept to allow for possible bias and may or may not impose the condition that the regression coefficients sum to one. Further, others have used ridge regression techniques, time-varying parameter techniques, etc. in analyzing the relation between A_t and f_{1t} and f_{2t}. In our opinion, the basic problem is one of identifying or determining the form of this relationship. If it is not determined very well, it is not at all clear that the regression approach will yield satisfactory results. Note that in comparison to the Bayesian combining relationship in (3.1), there the weights or coefficients on the individual forecasts, the predictive means, are not least squares weights. Further, if f_{1t} and f_{2t} are derived from imperfect models, say models with important variables omitted, then they will have biases that usually vary through time in a complicated way. Modeling such time-varying biases seems difficult. Be that as it may, below we implement variants of the regression approach to produce combined forecasts and compare measures of their forecasting performance to those of other approaches described above and in section 4.

4 Models, methods, and their performance in forecasting

In this section, we first describe the models which we shall employ to forecast annual output growth rates for eighteen countries year by year, 1974–87, using data 1951–73 to "start up" the forecasting experiments. Second, we enumerate the forecasting techniques which will be employed. Third, we briefly describe the data employed in our calculations and then present the empirical forecasting results for uncombined, combined, unpooled, and pooled forecasts derived from use of fixed and time-varying parameter models.

4.1 Models and forecasts

The two models that we employ, used in our previous work, are (i) a univariate autoregression of order 3 containing lagged leading indicator variables, denoted by ARLI, and (ii) the ARLI model expanded to include a measure of current world income growth, denoted by ARLI/WI. As explained in our past work, an autoregression of order 3 is employed to allow for the possibility of there being two complex conjugate roots, giving rise to an oscillatory component, and one real root associated with a local trend. In Hong (1989) it is shown that ARLI models for eighteen countries have two complex roots and one real root with high posterior probability in each case. Further, Hong established that the oscillatory components are damped in each case and have periods in the vicinity of four–six years. Also, each estimated real root has amplitude less than one. As regards the leading indicator variables, lagged growth rates of real stock prices and of real money, they are introduced to take account of informational, expectational, and policy effects as well as a real balance effect exerted by money supply changes (see Zellner, Huang, and Chau 1965 for discussion and estimation of a real balance effect). The world income variable, introduced in Zellner and Hong (1989), is a variable affecting countries' exports. Further discussion of the economic rationale of the ARLI models is provided by Hong (1989) who shows that they can be produced as the reduced form equations of particular IS-LM models. Also, Zellner and Manas-Anton (1986) link up the ARLI models to reduced form equations obtained from particular aggregate supply and demand models and Min (1990) shows their relation to generalized real business cycle models.

In what follows, we utilize fixed parameter (FP) and time-varying parameter (TVP) versions of the ARLI and ARLI/WI models. The FP versions of these models, denoted by FP/ARLI and FP/ARLI/WI, are utilized with and without pooling. The unpooled FP/ARLI model is

$$y_{it} = \boldsymbol{x}_{it}'\boldsymbol{\beta}_i + u_{it}, \quad i = 1, 2, \ldots, N, \quad t = 1, 2, \ldots, T, \tag{4.1}$$

where y_{it} = annual output growth rate for ith country in the tth year, $\boldsymbol{\beta}_i = k \times 1$ parameter vector, $\boldsymbol{x}_{it}' = (1, y_{it-1}, y_{it-2}, y_{it-3}, SR_{it-1}, SR_{it-2}, GM_{it-1}, MSR_{t-1})$, where SR_{it} = rate of growth of real stock prices, GM_{it} = rate of growth of real money, M_1, MSR_t = median of SR_{it}s in year t, and u_{it} is an error term. We assume that the u_{it}s are NID(0, σ_i^2). Using a diffuse prior density for $\boldsymbol{\beta}_i$ and σ_i, it is well known that given initial values 1951–3 and data 1954–73, the predictive density for 1974's output growth rate is in the univariate Student-t form with mean equal to the least squares forecast of the 1974 growth rate (see, e.g., Zellner

1971). This is our 1974 forecast for the FP/ARLI model without pooling. To forecast 1975's output growth rate, we update our predictive density to incorporate the 1974 data and use its mean, the least squares forecast of 1975's growth rate, as our forecast. Proceeding in this way year by year, we obtain a sequence of unpooled FP/ARLI forecasts, recursively calculated, for the years 1974–87.

The FP/ARLI/WI model, introduced and used in Zellner and Hong (1989), involves adding a scalar world income growth rate variable, w_t, to (4.1) as follows:

$$y_{it} = w_t \alpha_i + \boldsymbol{x}'_{it} \boldsymbol{\beta}_i + u_{it}, \tag{4.2}$$

where α_i is a scalar parameter and w_t is the median of the y_{it}s in the tth year. The term $\boldsymbol{x}'_{it}\boldsymbol{\beta}_i$ is exactly the same as that in (4.1). To use (4.2) to forecast, it is necessary to forecast w_t. To accomplish this, we use the following ARLI model for w_t, an autoregression of order 3 including lagged leading indicator variables:

$$w_t = \boldsymbol{z}'_t \boldsymbol{\pi} + \varepsilon_t, \tag{4.3}$$

where $\boldsymbol{z}'_t = (1, w_{t-1}, w_{t-2}, w_{t-3}, MSR_{t-1}, MGM_{t-1})$, with MSR_t = median growth rate of real stock prices for year t and MGM_t = median growth rate of real money for year t with the median taken relative to the eighteen countries for year t. The ε_ts in (4.3) are assumed independently drawn from a zero mean normal distribution with variance σ^2_w. For country i, (4.2) and (4.3) constitute a triangular system which we assume to be fully recursive, that is u_{it} and ε_t are assumed to be independently distributed for all t. Using the recursive assumption and a diffuse prior density for the parameters, we implement the model using data 1951–3 for initial values and 1954–73, twenty years, for estimation and obtain an approximate predictive density for the output growth rate in 1974. The mean of this predictive density is our forecast for 1974. On updating the posterior and predictive densities year by year, a sequence of forecasts for the remaining years, 1975–87, is obtained. These are our unpooled FP/ARLI/WI forecasts.

Our unpooled, time-varying parameter models are simple elaborations of the above fixed parameter models. Our unpooled TVP/ARLI model is (4.1) with $\boldsymbol{\beta}_i$ made time-varying, that is,

$$y_{it} = \boldsymbol{x}'_{it} \boldsymbol{\beta}_{it} + u_{it}, \tag{4.4}$$

and

$$\boldsymbol{\beta}_{it} = \boldsymbol{\beta}_{it-1} + \boldsymbol{v}_{it}. \tag{4.5}$$

Thus in our unpooled TVP/ARLI model, used in Garcia-Ferrer *et al.* (1987), the random coefficient vector is assumed to be generated by a vector random-walk process. Given start-up data, 1951–73, and an initial state distribution, recursive calculations provide a predictive density for the 1974 output growth rate, the mean of which is our point forecast. On recursively updating the predictive densities year by year, their means are our unpooled TVP/ARLI model forecasts, 1974–87.

Our unpooled TVP/ARLI/WI model is that shown in (4.2) and (4.3) with the parameters made random as shown below:

$$y_{it} = w_t \alpha_{it} + \boldsymbol{x}'_{it} \boldsymbol{\beta}_{it} + u_{it}, \tag{4.6}$$

$$w_t = \boldsymbol{z}'_t \boldsymbol{\pi}_t + \varepsilon_t, \tag{4.7}$$

$$\begin{pmatrix} \alpha_{it} \\ \boldsymbol{\beta}_{it} \\ \boldsymbol{\pi}_t \end{pmatrix} = \begin{pmatrix} \alpha_{it-1} \\ \boldsymbol{\beta}_{it-1} \\ \boldsymbol{\pi}_{t-1} \end{pmatrix} + \begin{pmatrix} \eta_{it} \\ \boldsymbol{v}_{it} \\ \boldsymbol{\delta}_t \end{pmatrix}. \tag{4.8}$$

Recursive algorithms and needed assumptions for obtaining and updating predictive densities for this system are given in appendix B (p. 583). The means of these successively updated predictive densities, 1974–87, are our unpooled TVP/ARLI/WI forecasts.

We have also computed separate forecasts and combined forecasts for alternative models under various pooling assumptions. In pooling the FP/ARLI model in (4.1), we assumed that the coefficient vectors, the $\boldsymbol{\beta}_i$s, satisfy $\boldsymbol{\beta}_{it} = \boldsymbol{\theta} + \varepsilon_{it}$ with $\varepsilon_{it} \sim \mathrm{N}(\boldsymbol{0}, \sigma^2, \Sigma)$, an assumption implying that the $\boldsymbol{\beta}_{it}$s have a common mean vector $\boldsymbol{\theta}$, a model similar to those used by, among others, Swamy (1971) and Lindley and Smith (1972). In connection with the FP/ARLI/WI model in (4.2) and (4.3), we make a similar assumption, namely that $\alpha_{it} = \alpha + v_{it}$ and $\beta_{it} = \boldsymbol{\theta} + \varepsilon_{it}$, where $\varepsilon^*_{it} = (\boldsymbol{v}_{it} \varepsilon'_{it})'$ is assumed $\mathrm{N}(\boldsymbol{0}, \sigma^2\ \Sigma^*)$. Thus we allow the βs to be time-varying with a fixed hyperparameter vector $\boldsymbol{\theta}$ and call these models fixed hyperparameter ARLI models, denoted by FHP–ARLI. Using these pooling assumptions, individual and combined forecasts were computed year by year, 1974–87 (see appendix C, p. 587, for details).

The pooled TVHP/ARLI model is given by

$$\boldsymbol{y}_t = X_t \boldsymbol{\beta}_t + \boldsymbol{u}_t, \quad \boldsymbol{u}_t \sim \mathrm{N}(\boldsymbol{0}, \sigma^2 \mathrm{I}_m),$$

with $\boldsymbol{y}'_t = (y_{1t}, y_{2t}, \ldots, y_{mt})$, $\boldsymbol{\beta}'_t = (\boldsymbol{\beta}'_{1t}, \boldsymbol{\beta}'_{2t}, \ldots, \boldsymbol{\beta}'_{mt})$, $\boldsymbol{u}'_t = (u_{1t}, u_{2t}, \ldots, u_{mt})$, and

$$X_t = \begin{pmatrix} \boldsymbol{x}'_{1t} & \boldsymbol{0}' & \cdots & \boldsymbol{0}' \\ \boldsymbol{0}' & \boldsymbol{x}'_{2t} & \cdots & \boldsymbol{0}' \\ \vdots & \vdots & \ddots & \vdots \\ \boldsymbol{0}' & \boldsymbol{0}' & \cdots & \boldsymbol{x}'_{mt} \end{pmatrix}.$$

Further it is assumed that $\beta_t = B\theta_t + \varepsilon_t$ and $\theta_t = \theta_{t-1} + \eta_t$, where ε_t is N($\boldsymbol{0}$, σ^2 Ω) and is independent of $\boldsymbol{\eta}_t$ which is N($\boldsymbol{0}$, $\phi\sigma^2$ I_k) with Ω = $I_n \otimes \Sigma$ and $B' = [I_k I_k \ldots I_k]$.

It is seen that with these pooling assumptions the β_{it} given $\boldsymbol{\theta}_t$ have a common mean vector, $\boldsymbol{\theta}_t$, which is assumed generated by a random walk. See appendix B for further description and analysis of this pooling model and a similar one for the TVP/ARLI/WI model.

4.2 *Description of data*

The data used in our study have been taken from the University of Chicago IMF *International Financial Statistics* data base . . . Annual data 1954–73 were used to estimate our models with data 1951–3 serving as initial values. Then the one-year-ahead forecasts described in the preceding sections were computed year by year, 1974–87, with models being updated each year.

The output data for each country, either annual real GNP or real GDP, were logged and first-differenced to yield annual growth rates. For each country nominal money, M_1, at the end of each year was deflated by a general price index, logged and first-differenced to yield the rate of change of real money. Similarly, for each country a general price index was used to deflate an annual stock price index, which was logged and first-differenced to yield a rate of growth of real stock prices. Symbolically the variables are given by: $y_{it} = (1 - L)\ln(O_{it}/P_{it})$, $SR_{it} = (1 - L)\ln(SP_{it}/P_{it})$, and $GM_{it} = (1 - L)\ln(M_{it}/P_{it})$, where L is the lag operator, $L^n x_t = x_{t-n}$, and for the ith country in the tth year, O_{it}=output level, P_{it} = general price index, SP_{it} = stock price index, and M_{it} = nominal M_1. Data plots are given in Garcia-Ferrer *et al.* (1987) and Zellner and Hong (1989).

4.3 *Empirical results*

In tables 17.4a and 17.4b, we present posterior odds for FP versus TVP models, based on data 1973–87. For the unpooled ARLI and ARLI/WI models, twelve and eight countries, respectively, have odds of less than 0.50, that is greater than 2:1 in favor of the TVP models. For the ARLI

Table 17.4a *Posterior odds for fixed versus time-varying parameter models computed from annual data, 1973–1987*[a]

	Models			
	Unpooled		Pooled	
Country	ARLI[b]	ARLI/WI[c]	ARLI[d]	ARLI/WI[e]
Australia	0.16	0.34	1.01	0.94
Austria	0.01	0.07	0.21	1.02
Belgium	1.64	0.45	0.40	1.57
Canada	0.41	2.21	0.57	1.72
Denmark	0.01	0.39	0.05	0.99
Finland	0.78	1.68	0.36	0.76
France	0.01	1.30	0.47	1.42
Germany	0.00	0.01	0.16	1.43
Ireland	6.18	3.04	1.87	1.50
Italy	0.01	0.06	0.32	1.21
Japan	0.45	0.48	6.81	0.91
Netherlands	0.41	2.32	0.10	1.92
Norway	6.60	4.95	1.36	1.09
Spain	0.27	0.75	0.57	1.35
Sweden	0.25	0.34	0.04	0.58
Switzerland	0.71	4.25	0.21	0.91
United Kingdom	0.22	8.67	0.13	0.82
United States	9.18	17.31	0.24	0.81

Notes:

[a] Posterior distributions for models' parameters, computed using annual data 1954–72, were used to form Bayes factors for the period 1973–87. Prior odds were set 1:1 in all cases. The entries in the table are odds in favor of fixed parameter (FP) models.

[b] Unpooled FP/ARLI versus unpooled TVP/ARLI.

[c] Unpooled FP/ARLI/WI versus unpooled TVP/ARLI/WI.

[d] Pooled FHP/ARLI versus pooled TVHP/ARLI.

[e] Pooled FHP/ARLI/WI versus pooled TVHP/ARLI/WI.

model, only three countries have odds greater than or equal to 2:1 in favor of the FP model. Thus without the WI variable, there is a general preference for TVP for twelve countries. However, with the WI variable included, the odds favor TVPs for eight countries and FPs for seven. For the pooled ARLI model, TVHPs are favored for twelve countries. However, for the pooled ARLI/WI model, all countries' odds are between 0.50 and 1.99, somewhat inconclusive with respect to whether the hyperparameter vector is fixed or time-varying.

Table 17.4b *Number of countries by values of posterior odds, fixed versus time-varying parameter models, annual data, 1973–1987*[a]

	Models			
	Unpooled		Pooled	
Posterior odds, FP versus TVP[b]	ARLI	ARLI/WI	ARLI	ARLI/WI
0–0.49	12	8	12	0
0.50–0.99	2	1	2	8
1.00–1.99	1	2	3	10
$\geq$ 2.00	3	7	1	0

Notes:
[a] Tabulation of data from table 17.4a. See notes of table 17.4a and appendix A for procedures employed in computing odds.
[b] Posterior odds for fixed parameter (FP) versus time-varying parameter (TVP) models.

Tables 17.5, 17.6a, and 17.6b provide RMSEs of forecasts for unpooled and pooled FP and TVP models as well as those for two decision theoretic model selection procedures. For the unpooled ARLI model, it is seen from table 17.6a that the lowest median RMSEs are those for the TVP and LS_2 cases, where LS_2 denotes a 2×2 symmetric loss structure, table 17.1 with $c_1 = c_2$, relating to the choice of FP and TVP models' forecasts. For the pooled ARLI model, the lowest median RMSE is that for the TVHP version, 1.95, with 1.97 being the RMSE for the LS_2 forecasts. Note that pooling has resulted in about a 15 percent reduction in median RMSEs. Further, in the case of the pooled ARLI forecasts, the minimal and maximal RMSEs are lower in every case. For the four methods reported in table 17.6a, it appears that for both the pooled cases the TVP and LS_2 forecasts are overall the best and are not very different. For each of the forecasting methods, pooling models performed better than corresponding unpooled models according to the median RMSE criterion and generated improved RMSEs in many individual cases.

Table 17.6b shows that pooled ARLI/WI models' forecasts and model selection procedures perform better than their unpooled counterparts. For example, the median RMSEs for the pooled models and procedures range from 1.74 to 1.86, while for the unpooled models and procedures they range from 2.24 to 2.60. For the unpooled case, the LS_2 forecast had a median RMSE of 2.24, slightly below those of the forecasts of the FP, TVP, and LS_3 procedures. For the pooled ARLI/WI model all procedures worked about equally well in terms of median RMSE.

Table 17.5 *Root mean-squared errors (RMSEs) of the one-year-ahead forecasts of annual real output growth rates for 1974–1987 for several variants of fixed parameter autoregressive leading indicator (FP/ARLI), world income (FP/ARLI/WI) models, and an AR(3) benchmark model*[a]

	Models				
Country	AR(3)	Unpooled ARLI	Unpooled ARLI/WI	Pooled ARLI[b]	Pooled ARLI/WI[b]
			(percentage points)		
Australia	2.72	3.21	3.45	2.10	2.27
Austria	2.58	2.75	2.84	1.97	1.80
Belgium	2.80	1.71	1.54	1.87	1.94
Canada	2.79	2.70	2.52	2.08	1.74
Denmark	2.89	3.40	3.21	2.77	2.39
Finland	2.45	3.22	3.32	1.85	1.69
France	2.16	2.51	2.27	1.58	1.62
Germany	2.90	2.82	2.82	1.97	1.56
Ireland	2.38	2.56	2.68	2.19	2.48
Italy	3.88	3.19	3.68	2.35	2.14
Japan	2.93	3.02	3.46	2.13	2.23
Netherlands	2.86	1.92	1.67	1.92	1.60
Norway	1.76	1.72	1.84	1.71	1.44
Spain	2.35	2.17	2.28	1.68	1.21
Sweden	1.98	2.11	2.15	2.30	1.76
Switzerland	3.77	3.24	3.15	2.56	2.43
United Kingdom	2.94	2.13	1.50	2.20	1.92
United States	2.69	2.31	1.99	2.53	2.08
Median	2.76	2.63	2.60	2.09	1.86
Minimum	1.76	1.71	1.50	1.58	1.21
Maximum	3.88	3.40	3.68	2.77	2.48

Notes:
[a] See text for discussion of models.
[b] In these cases the hyperparameters are fixed.

To show more clearly the effects of using pooled and unpooled models, table 17.6c presents RMSEs for unpooled and pooled ARLI/WI forecasts by RMSE interval and by country. On comparing panels A and B, it is seen that the pooled TVHPM forecast RMSEs are much more highly concentrated in the vicinity of 1.5 to 2.5 than is the case of RMSEs for forecasts of the unpooled TVPM. The same conclusion holds for the comparison of the distribution of countries by RMSE for the pooled and unpooled FPMs. Finally, on comparing panels A and C, it is seen that the results for the pooled TVHPMs and the pooled FHPMs are similar.

Table 17.6a *Root mean-squared errors (RMSEs) of various one-year-ahead forecasts of annual real output growth rates, 1974–1987, using an autoregressive leading indicator (ARLI) model*

	Unpooled ARLI				Pooled ARLI			
Country	FP	TVP	LS_2[a]	LS_3[b]	FHP	TVHP	LS_2[a]	LS_3[b]
				(percentage points)				
Australia	3.21	3.01	3.06	2.72[c]	2.10[c]	2.16	2.36	2.46
Austria	2.75	1.99	1.92[c]	2.33	1.97	1.67[c]	1.82	2.32
Belgium	1.71	1.70[c]	1.71	1.71	1.87	1.70[c]	1.93	1.93
Canada	2.70[c]	2.96	2.96	3.28	2.08	2.05[c]	2.07	2.07
Denmark	3.40	2.67[c]	2.67[c]	2.67[c]	2.77	2.48	2.45	2.38[c]
Finland	3.22	2.97	3.22	2.45[c]	1.85	1.53[c]	1.54	1.69
France	2.51	2.27[c]	2.27[c]	2.61	1.58	1.23[c]	1.37	1.49
Germany	2.82	1.91[c]	1.91[c]	2.81	1.97	1.54[c]	1.55	1.55
Ireland	2.56	2.90	2.56	2.36[c]	2.19[c]	2.26	2.28	2.33
Italy	3.19	2.32[c]	2.32[c]	3.39	2.35	2.18[c]	2.22	2.22
Japan	3.02[c]	3.16	3.16	3.16	2.13[c]	2.46	2.13[c]	2.13[c]
Netherlands	1.92[c]	2.05	2.04	2.29	1.92	1.42[c]	1.59	1.59
Norway	1.72[c]	2.31	1.96	1.76	1.71[c]	1.95	1.94	1.76
Spain	2.17	1.87[c]	1.93	2.08	1.68	1.51[c]	1.52	1.52
Sweden	2.11	1.93[c]	1.97	2.03	2.30	1.90[c]	1.99	2.03
Switzerland	3.24	3.18[c]	3.25	3.25	2.56	2.11[c]	2.35	2.52
United Kingdom	2.13	1.90[c]	1.98	1.98	2.20	1.96[c]	1.96[c]	1.96[c]
United States	2.31[c]	2.46	2.43	2.37	2.53	2.46[c]	2.49	3.09
Median	2.63	2.31	2.29[c]	2.41	2.09	1.95[c]	1.97	2.05
Minimum	1.71	1.70	1.71	1.71	1.58	1.23	1.37	1.49
Maximum	3.40	3.18	3.25	3.39	2.77	2.48	2.49	3.09

Notes:

[a] Year-by-year model selection based on the 2×2 symmetric loss structure (LS_2) in table 17.1. The forecast of the model favored by posterior odds is employed.

[b] Year-by-year model selection based on a 3×3 symmetric loss structure (LS_3), involving FP, TVP, and an AR(3) benchmark model. The forecast of the model chosen to minimize expected loss was used.

[c] Minimal value in row.

In table 17.7, we form 2×2 tables relating models preferred by the posterior odds criterion and relative size of RMSEs of FPMs and TVPMs. In general there appears to be a positive association present. For example, for the unpooled ARLI model, when the posterior odds favor the FPM, three of four RMSEs were smaller for the FPM model, while when they favor the TVPM, eleven of fourteen favored TVPMs had lower RMSEs. In the case of the pooled ARLI models, the association is perfect, while for the pooled ARLI/WI model the association is somewhat weaker.

Table 17.6b *Root mean-squared errors (RMSEs) of various one-year-ahead forecasts of annual real output growth rates, 1974–1987, using an autoregressive leading indicator – world income (ARLI/WI) model*[a]

	Unpooled ARLI/WI				Pooled ARLI/WI			
Country	FP	TVP	LS_2[a]	LS_3[b]	FHP	TVHP	LS_2[a]	LS_3[b]
				(percentage points)				
Australia	3.45	3.23	3.23	3.10[c]	2.27[c]	2.27[c]	2.29	2.29
Austria	2.84	2.13[c]	2.14	2.42	1.80	1.70[c]	1.70[c]	1.70[c]
Belgium	1.54[c]	1.78	1.77	1.77	1.94	1.84[c]	1.94	1.94
Canada	2.52[c]	2.78	2.64	3.01	1.74[c]	1.78	1.74[c]	1.74[c]
Denmark	3.21	2.68[c]	2.95	2.95	2.39	2.14[c]	2.26	2.26
Finland	3.32	3.16	3.32	2.91[c]	1.69	1.58[c]	1.58[c]	1.58[c]
France	2.27	1.74[c]	2.29	2.33	1.62	1.43[c]	1.60	1.60
Germany	2.82	1.98[c]	1.98[c]	2.84	1.56	1.23[c]	1.42	1.42
Ireland	2.68	3.15	3.35	2.38[c]	2.48[c]	2.53	2.49	2.49
Italy	3.68	2.51[c]	2.58	3.52	2.14	1.98[c]	2.18	2.18
Japan	3.46	3.32[c]	3.32[c]	3.32[c]	2.23[c]	2.30	2.28	2.28
Netherlands	1.67[c]	1.74	1.86	2.28	1.60	1.39[c]	1.64	1.64
Norway	1.84	2.56	2.19	1.76[c]	1.44[c]	1.55	1.56	1.56
Spain	2.28	1.85[c]	2.15	2.37	1.21	1.17[c]	1.21	1.21
Sweden	2.15	1.90[c]	1.97	2.09	1.76	1.50[c]	1.51	1.51
Switzerland	3.15	3.04[c]	3.17	3.17	2.43	2.23[c]	2.38	2.38
United Kingdom	1.50	1.39[c]	1.50	1.50	1.92	1.69[c]	1.87	1.87
United States	1.99[c]	2.22	2.19	2.19	2.08	1.94[c]	2.02	2.02
Median	2.60	2.37	2.24[c]	2.40	1.86	1.74[c]	1.80	1.80
Minimum	1.50	1.39	1.50	1.50	1.21	1.17	1.21	1.21
Maximum	3.68	3.32	3.35	3.52	2.48	2.53	2.49	2.49

Notes:
[a, b, c] See notes of table 17.6a.

In tables 17.8a and 17.8b, we present forecasting results relating to two combining methods. One, $COMB_1$, is an average of the predictive means of FP and TVP models with posterior probabilities or posterior odds involved in the weights (see (3.1) above). The other, $COMB_2$, is the forecast provided by the Bayesian predictive loss model selection rule described in connection with the predictive loss structures in table 17.3 . . . We see from tables 17.8a and 17.8b that, while pooling resulted in improved performance of the combining procedures, the combining procedures did not in general produce large or even moderately large decreases in median RMSEs. Further there was not much difference in the performance of the two combining procedures. The pooled TVHP forecasts tended to perform best and also better in terms of median RMSE

Table 17.6c *Root mean-squared errors (RMSEs) for pooled and unpooled ARLI/WI models' forecasts, by country, 1974–1987*

RMSE (%)	Countries									Freq.	Prop.
(A) Pooled TVHPM											
1.00–1.49	FRN	GER	NET	SPN						4	0.22
1.50–1.99	AUR	BEL	CAN	FIN	ITY	NOR	SWD	UKM	USA	9	0.50
2.00–2.49	AUL	DEN	JAP	SWZ						4	0.22
2.50–2.99	IRE									1	0.06
3.00–3.49	–									0	0.00
		Median = 1.74			Minimum = 1.17			Maximum = 2.53		18	1.00
(B) Unpooled TVPM											
1.00–1.49	UKM									1	0.06
1.50–1.99	BEL	FRN	GER	NET	SPN	SWD				6	0.33
2.00–2.49	AUR	USA								2	0.11
2.50–2.99	CAN	DEN	ITY	NOR						4	0.22
3.00–3.49	AUL	FIN	IRE	JAP	SWZ					5	0.28
		Median = 2.37			Minimum = 1.39			Maximum = 3.32		18	1.00
(C) Pooled FHPM											
1.00–1.49	NOR	SPN								2	0.11
1.50–1.99	AUR	BEL	CAN	FIN	FRN	GER	NET	SWD	UKM	9	0.50
2.00–2.49	AUL	DEN	IRE	ITY	JAP	SWZ	USA			7	0.39
2.50–2.99										0	0.00
3.00–3.49										0	0.00
		Median = 1.86			Minimum = 1.21			Maximum = 2.48		18	1.00
(D) Unpooled FPM											
1.00–1.49										0	0.00
1.50–1.99	BEL		NET	NOR	UKM	USA				5	0.28
2.00–2.49	FRN		SPN	SWD						3	0.17
2.50–2.99	AUR		CAN	GER	IRE					4	0.22
3.00–3.49	AUL		DEN	FIN	JAP	SWZ				5	0.28
3.50–3.99	ITY									1	0.06
		Median = 2.60		Minimum = 1.50				Maximum = 3.68		18	1.01

Table 17.7 *Number of countries by posterior odds and root mean-squared error (RMSE) of forecast by type of model, annual output growth rates, 1974–1987*

		ARLI Lower RMSE for:			ARLI/WI Lower RMSE for:		
		FPM	TVPM	Total	FPM	TVPM	Total
(A) Unpooled models							
Posterior	FPM	3	1	4	5	4	9
odds favor:	TVPM	3	11	14	1	8	9
	Total	6	12	18	6	12	18
		ARLI Lower RMSE for:			ARLI/WI Lower RMSE for:		
		FHPM	TVHPM	Total	FHPM	TVHPM	Total
(B) Pooled models							
Posterior	FHPM	4	0	4	3	7	10
odds favor:	TVHPM	0	14	14	2	6	8
	Total	4	14	18	5	13	18

than equally weighted averages of FP and TVP forecasts (results available on request).

Tables 17.9a and 17.9b present the results of employing three regression combining techniques. It is seen that the regression combining technique A, involving a zero intercept and free regression coefficients, produces the lowest median RMSEs in all but one case, with values of 2.24 and 2.21 for the unpooled cases (see table 17.9a) and 2.25 and 1.89 for the pooled cases (see table 17.9b). These values are not far different from corresponding RMSEs for the Bayesian $COMB_1$ method, namely 2.35, 2.22, 1.99, and 1.77. However, the regression combining methods A, B, and C produced rather different results. Further none of the regression combining techniques produced a median RMSE lower than that associated with the pooled FHP and TVHP ARLI/WI models, namely 1.86 and 1.74.

5 Summary and concluding remarks

From our empirical results, we have reached the following conclusions:

(1) While there is some evidence in terms of posterior odds values for time-varying parameter models in the case of unpooled models and

Table 17.8a *Root mean-squared errors (RMSEs) of one-year-ahead forecasts of annual real output growth rates, 1974–1987, using model combining approaches based on posterior odds and a predictive model selection rule*

	Unpooled ARLI				Pooled ARLI			
Country	FP	TVP	COMB$_1$[a]	COMB$_2$[b]	FHP	TVHP	COMB$_1$[a]	COMB$_2$[b]
Australia	3.21	3.01	3.07	2.93[c]	2.10	2.16	2.11	2.06[c]
Austria	2.75	1.99[c]	2.05	2.11	1.97	1.67[c]	1.77	1.81
Belgium	1.71	1.70[c]	1.72	1.72	1.87	1.70[c]	1.75	1.80
Canada	2.70	2.96	2.97	2.65[c]	2.08	2.05[c]	2.05[c]	2.06
Denmark	3.40	2.67[c]	2.79	2.69	2.77	2.48[c]	2.51	2.48[c]
Finland	3.22	2.97[c]	3.08	3.03	1.85	1.53[c]	1.63	1.62
France	2.51	2.27[c]	2.32	2.43	1.58	1.23[c]	1.34	1.41
Germany	2.82	1.91[c]	1.96	1.96	1.97	1.54[c]	1.66	1.67
Ireland	2.56[c]	2.90	2.69	2.66	2.19	2.26	2.20	2.08[c]
Italy	3.19	2.32[c]	2.38	2.76	2.35	2.18[c]	2.23	2.33
Japan	3.02[c]	3.16	3.07	3.06	2.13[c]	2.46	2.23	2.24
Netherlands	1.92[c]	2.05	1.98	1.92[c]	1.92	1.42[c]	1.55	1.64
Norway	1.72[c]	2.31	1.99	1.74	1.71	1.95	1.79	1.58[c]
Spain	2.17	1.87[c]	1.93	1.87[c]	1.68	1.51[c]	1.56	1.60
Sweden	2.11	1.93[c]	1.96	1.95	2.30	1.90[c]	1.97	1.96
Switzerland	3.24	3.18[c]	3.25	3.21	2.56	2.11[c]	2.29	2.40
United Kingdom	2.13	1.90[c]	1.97	1.97	2.20	1.96[c]	2.00	2.00
United States	2.31[c]	2.46	2.40	2.35	2.53	2.46[c]	2.48	2.49
Median	2.63	2.31[c]	2.35	2.39	2.09	1.95[c]	1.99	1.98
Minimum	1.71	1.70	1.72	1.72	1.58	1.23	1.34	1.41
Maximum	3.40	3.18	3.25	3.21	2.77	2.48	2.51	2.49

Notes:
[a] A Bayesian model combining procedure in which two models, such as a FP and a TVP model, are combined with weights based on their posterior odds ratio (see (3.1) in text).
[b] A Bayesian predictive model selection rule in which the model with the minimum expected loss is used to forecast at each time period. Choice is made among the forecasts of three models, a FPM, a TVPM, and a combined model, using the predictive loss structure in table 17.3.
[c] Minimal value in row.

pooled ARLI models, for the pooled ARLI/WI models the evidence is not as decisive with respect to models with time-varying hyperparameters. However, use of TVP models led to reductions in RMSEs of forecast in many cases.

(2) The use of various Bayesian and non-Bayesian forecast combining methods did not produce much, if any, reduction in RMSEs *vis-à-vis* those for uncombined forecasts. Different Bayesian combining methods did not produce very different results, whereas different regression combining techniques did and were not as good as certain Bayesian procedures.

(3) Several new procedures for selecting forecasts and for combining forecasts were developed. Further investigation of their theoretical

Table 17.8b *Root mean-squared errors (RMSEs) of one-year-ahead forecasts of annual real output growth rates, 1974–1987, using model combining approaches based on posterior odds and a predictive model selection rule*

	Unpooled ARLI/WI				Pooled ARLI/WI			
Country	FP	TVP	COMB$_1$[a]	COMB$_2$[b]	FHP	TVHP	COMB$_1$[a]	COMB$_2$[b]
Australia	3.45	3.23	3.29	3.13[c]	2.27	2.27	2.26[c]	2.27
Austria	2.84	2.13[c]	2.22	2.27	1.80	1.70[c]	1.73	1.75
Belgium	1.54[c]	1.78	1.74	1.74	1.94	1.84[c]	1.89	1.91
Canada	2.52[c]	2.78	2.66	2.61	1.74	1.78	1.74	1.73[c]
Denmark	3.21	2.68[c]	2.86	2.89	2.39	2.14[c]	2.25	2.22
Finland	3.32	3.16[c]	3.22	3.25	1.69	1.58[c]	1.62	1.58[c]
France	2.27	1.74[c]	2.00	2.03	1.62	1.43[c]	1.52	1.55
Germany	2.82	1.98[c]	2.01	1.99	1.56	1.23[c]	1.39	1.43
Ireland	2.68	3.15	3.03	2.55[c]	2.48	2.53	2.50	2.45[c]
Italy	3.68	2.51[c]	2.75	3.03	2.14	1.98[c]	2.05	2.12
Japan	3.46	3.32[c]	3.40	3.36	2.23[c]	2.30	2.26	2.30
Netherlands	1.67	1.74	1.71	1.66[c]	1.60	1.39[c]	1.51	1.53
Norway	1.84[c]	2.56	2.22	1.87	1.44	1.55	1.48	1.40[c]
Spain	2.28	1.85[c]	1.99	1.98	1.21	1.17	1.16[c]	1.19
Sweden	2.15	1.90[c]	1.92	2.00	1.76	1.50[c]	1.59	1.61
Switzerland	3.15	3.04[c]	3.15	3.17	2.43	2.23[c]	2.32	2.40
United Kingdom	1.50	1.39[c]	1.47	1.48	1.92	1.69[c]	1.79	1.79
United States	1.99[c]	2.22	2.10	2.12	2.08	1.94[c]	2.00	2.01
Median	2.60	2.37	2.22	2.20[c]	1.86	1.74[c]	1.77	1.77
Minimum	1.50	1.39	1.47	1.48	1.21	1.17	1.16	1.19
Maximum	3.68	3.32	3.40	3.36	2.48	2.53	2.50	2.45

Notes:
[a, b, c] See notes of table 17.8a.

properties and more experience in using them would be desirable. Also, posterior odds for fixed versus time-varying parameter models have been derived, computed, and shown to be useful in evaluating alternative models.

(4) As shown graphically in table 17.6c, the pooling techniques that we utilized in connection with fixed and time-varying parameter models led to substantial reductions in forecast RMSEs (see also tables 17.5, 17.6a, and 17.6b).

Finally, our forecasting methods and models have been applied using data for a period, 1951–87, which includes the Korean War, the Vietnamese War, price and credit controls, strikes, oil crises, exchange rate policy changes, major changes in US monetary policy, etc. That our models and methods worked to produce good forecasting results without any special adjustments for these and other major events is indeed satisfying. It appears that our leading indicator stock market and money variables reflect information about major events before output

Table 17.9a *Root mean-squared errors (RMSEs) of combined annual output growth rate forecasts of FP and TVP models using regression methods, 1974–1987*[a]

	Combined unpooled ARLI forecasts[b]			Combined unpooled ARLI/WI forecasts[b]		
Country	A	B	C	A	B	C
Australia	2.40[c]	3.43	2.49	2.61	3.26	2.55[c]
Austria	1.86[c]	2.12	2.31	1.70[c]	2.03	2.17
Belgium	1.71[c]	1.79	3.08	1.85[c]	1.85[c]	2.48
Canada	2.54[c]	3.46	2.73	3.22	3.22	2.90[c]
Denmark	2.02[c]	2.70	2.44	2.35[c]	2.70	2.67
Finland	2.43	3.07	2.39[c]	2.46	3.46	2.40[c]
France	3.97[c]	4.19	4.25	1.95[c]	1.97	2.65
Germany	1.59[c]	1.95	1.62	1.59[c]	2.04	1.60
Ireland	3.03	2.83	2.80[c]	3.45	3.09	3.07[c]
Italy	2.23[c]	3.20	2.75	2.84[c]	3.51	2.86
Japan	3.51[c]	3.55	4.06	4.20	4.55	3.88[c]
Netherlands	1.54[c]	2.06	1.94	1.76[c]	2.51	2.15
Norway	2.40	2.92	1.96[c]	2.59	3.02	1.94[c]
Spain	1.41[c]	2.42	2.65	1.48[c]	2.99	2.87
Sweden	1.75[c]	1.95	1.77	1.57[c]	1.96	1.85
Switzerland	2.96[c]	3.90	3.49	2.95[c]	3.66	3.37
United Kingdom	2.10[c]	2.25	2.38	1.41[c]	1.51	2.11
United States	2.24[c]	2.50	2.34	2.08[c]	2.30	2.18
Median	2.24[c]	2.77	2.46	2.21[c]	2.84	2.51
Minimum	1.41	1.79	1.62	1.41	1.51	1.60
Maximum	3.97	4.19	4.25	4.20	4.55	3.88

Notes:

[a] Regression weights for the 1974 regression combined forecast were obtained by regressing actual outcomes on FP and TVP models' forecasts for the years 1965–73. Weights for subsequent years were obtained by updating the 1974 weights year by year.

[b] Methods A, B, and C involve regression of actual outcomes on FP and TVP models' forecasts under the following conditions. For regression A, the intercept is zero and the regression coefficients are unrestricted. For regression B. the intercept is zero and the regression coefficients are constrained to add to one. For regression C, the intercept and the coefficients are unrestricted. See Granger and Ramanthan (1984) for further discussion of these methods.

[c] Minimal value in row.

adjustments take place. Thus the output growth rate responds with a lag to changes in lagged indicator variables and such variables reflect anticipated systematic and non-systematic shocks, which produce oscillatory behavior of output growth rates, a topic to be explored further in future research.

Table 17.9b *Root mean-squared errors (RMSEs) of combined annual output growth rate forecasts of FHP and TVHP models using regression methods, 1974–1987*[a]

	Combined pooled ARLI forecasts[b]			Combined pooled ARLI/WI forecasts[b]		
Country	A	B	C	A	B	C
Australia	2.69	2.54	2.49[c]	2.64	2.61	2.40[c]
Austria	1.86[c]	1.96	2.00	1.68[c]	1.77	1.91
Belgium	2.25	2.06[c]	2.62	1.90[c]	1.92	2.61
Canada	2.36	2.30	2.21[c]	1.97[c]	2.16	2.25
Denmark	2.26[c]	2.98	2.51	2.15[c]	2.39	2.18
Finland	1.82	1.70[c]	1.79	1.73	1.74	1.72[c]
France	2.62	2.25	1.93[c]	1.46[c]	1.48	2.15
Germany	1.69	1.88	1.63[c]	1.08[c]	1.10	1.16
Ireland	2.66	2.36[c]	2.78	2.67	2.66	2.37[c]
Italy	2.55[c]	2.67	2.66	2.04	2.03[c]	2.31
Japan	2.86	2.56[c]	4.19	3.47[c]	3.48	3.82
Netherlands	1.50	1.47[c]	1.59	1.89	1.84[c]	2.21
Norway	1.90	1.88[c]	2.01	1.60	1.57[c]	1.95
Spain	1.69[c]	1.69[c]	2.54	1.19	1.18[c]	1.63
Sweden	1.94	1.83[c]	2.00	1.63	1.56[c]	1.78
Switzerland	2.46[c]	2.58	2.56	2.47	2.44[c]	2.66
United Kingdom	1.90[c]	2.08	2.11	1.84[c]	2.08	1.95
United States	2.70	2.65[c]	2.83	2.25[c]	2.27	2.41
Median	2.25	2.17[c]	2.35	1.89[c]	1.98	2.20
Minimum	1.50	1.47	1.59	1.08	1.10	1.16
Maximum	2.86	2.98	4.19	3.47	3.48	3.82

Notes:
[a, b, c] See notes of table 17.9a.

APPENDIX A POSTERIOR ODDS FOR FIXED VERSUS TIME-VARYING PARAMETER MODELS

Denote the hypothesis of fixed parameters by H_0 and that of time-varying parameters by H_1. Then, as is well known, the posterior odds, denoted by K, is given by

$$K = \Pr(H_0 \mid y, Z, I_0)/\Pr(H_1 \mid y, Z, I_1)$$

$$= \frac{\Pr(H_0 \mid I_0)}{\Pr(H_1 \mid I_1)} \frac{f(y \mid H_0, Z, I_0)}{f(y \mid H_1, Z, I_1)}, \tag{A.1}$$

where $Pr(H_0 \mid I_0)/Pr(H_1 \mid I_1)$ is the prior odds, taken equal to one in our calculations, and the second factor on the r.h.s. of (A.1) is the Bayes factor. Here y = data, Z = data on the exogenous or input variables, and I_0 and I_1 denote prior information under hypotheses H_0 and H_1, respectively.

In our calculations, we employed the data 1951–72 to obtain posterior densities for the parameters (see appendix B) which were employed as prior densities in computing the posterior odds incorporating data for 1973. The Bayes factors were computed readily by exploiting the fact that the marginal density of the observation vector $\boldsymbol{y}' = (y_{t+1}, y_{t+2}, \ldots, y_T)$ can be expressed as follows:

$$\begin{aligned} f(\boldsymbol{y} \mid H_i, Z, I_i) &= p(y_{t+1}, y_{t+2}, \ldots, y_T \mid Z, H_i, I_i) \\ &= p(y_{t+1} \mid Z_{t+1}, H_i, I_i)\, p(y_{t+2} \mid y_{t+1} Z_{t+1}, Z_{t+2}, H_i, I_i) \\ &\quad \times \ldots \times p(y_T \mid y_{t+1}, y_{t+2}, \ldots, y_{T-1} Z_{t+1}, \ldots, Z_T, H_i, I_i), \end{aligned} \tag{A.2}$$

for $i = 0, 1$, with each factor on the r.h.s. of (A.2) in the univariate Student-t form. Thus the computation of the Bayes factor in (A.1), period by period, is relatively straightforward. Explicitly.

$$\begin{aligned} &P(y_\tau \mid y_{t+1}, \ldots, y_{\tau-1}, Z_{t+1}, \ldots, Z_\tau, H_i, I_i) \\ &\quad = a_{i\tau}\{1 + h_{i\tau}(y_\tau - \hat{y}_{i\tau \mid \tau-1})^2 / y_{\tau-1}\}^{-(v_{\tau-1}+1)/2}. \end{aligned}$$

Then,

$$f(\boldsymbol{y} \mid H_i, Z, I_i) = \prod_{\tau=t+1}^{T} a_{i\tau}\{1 + h_{i\tau}(y_\tau - \hat{y}_{i\tau \mid \tau-1})^2 / y_{\tau-1}\}^{-(v_{\tau-1}+1)/2}.$$

APPENDIX B UPDATING PROCEDURES FOR PREDICTIVE DENSITY OF TIME-VARYING PARAMETER MODELS

B.1 Unpooled case

The model for the ith country, where for convenience we drop the subscript i, is

$$\begin{aligned} y_t &= \boldsymbol{x}_t' \boldsymbol{\beta}_t + u_t, \quad u_t \text{s} \, \text{NID}(\boldsymbol{0}, \sigma^2), \\ \boldsymbol{\beta}_t &= \boldsymbol{\beta}_{t-1} + \varepsilon_t, \quad \varepsilon_t \text{s} \, \text{NID}(\boldsymbol{0}, \phi\sigma^2 I), \end{aligned}$$

for $t = 1, 2, \ldots$. Our distributions on the initial state, with I_0 denoting initial information, are given by

$$p(\boldsymbol{\beta}_0 \mid \sigma, I_0) \sim \mathrm{N}(\boldsymbol{0}, 10^6 \times \sigma^2 I),$$
$$p(\sigma \mid I_0) \sim \mathrm{IG}(v_0, s_0^2),$$

where v_0 and s_0^2 have been assigned small values in the inverted gamma density for σ.

Then the recursive algorithm for updating parameters is Posterior mean:

$$\hat{\boldsymbol{\beta}}_t = \hat{\boldsymbol{\beta}}_{t-1} + V_{t-1}\boldsymbol{x}_t(y_t - \boldsymbol{x}_t'\hat{\boldsymbol{\beta}}_{t-1})/(1 + \boldsymbol{x}_t' V_{t-1}\boldsymbol{x}_t),$$

Residual SS:

$$v_t s_t^2 = v_{t-1}s_{t-1}^2 + (y_t - \boldsymbol{x}_t'\hat{\boldsymbol{\beta}}_t)^2 + (\hat{\boldsymbol{\beta}}_t - \hat{\boldsymbol{\beta}}_{t-1})' V_{t-1}^{-1}(\hat{\boldsymbol{\beta}}_t - \hat{\boldsymbol{\beta}}_{t-1}),$$

Degrees of freedom:

$$v_t = v_{t-1} + 1,$$

where

$$V_t \equiv \sigma^{-2}\mathrm{cov}(\boldsymbol{\beta}_{t+1} \mid \sigma, D_t) = C_t + \phi \boldsymbol{I},$$
$$C_t \equiv \sigma^{-2}\mathrm{cov}(\boldsymbol{\beta}_t \mid \sigma, D_t) = V_{t-1} - V_{t-1}\boldsymbol{x}_t\boldsymbol{x}_t' V_{t-1}/(1 + \boldsymbol{x}_t' V_{t-1}\boldsymbol{x}_t),$$
$$D_t = \text{information available at } t.$$

At time t, the predictive density for y_{t+1} is

$$p(y_{t+1} \mid \boldsymbol{x}_{t+1}, D_t, I_0) = \int\int p(y_{t+1} \mid \boldsymbol{x}_{t+1}, \boldsymbol{\beta}_{t+1}, \sigma) \times p(\boldsymbol{\beta}_{t+1}, \sigma \mid D_t, I_0)\mathrm{d}\boldsymbol{\beta}_{t+1}\mathrm{d}\sigma,$$
$$\propto \left\{ v_t + \left(\frac{y_{t+1} - \boldsymbol{x}_{t+1}'\hat{\boldsymbol{\beta}}_t}{s_t a_t} \right)^2 \right\}^{-(v_t+1)/2},$$

and thus

$$t_{v_t} = (y_{t+1} - \boldsymbol{x}_{t+1}'\hat{\boldsymbol{\beta}}_t)/s_t a_t,$$

has a univariate Student-t density with v_t degrees of freedom where $a_t^2 = 1 + \boldsymbol{x}_{t+1}' V_t \boldsymbol{x}_{t+1}$.

B.2 A pooled case

Using the notation introduced in section 4, the model for the observation vector $\boldsymbol{y}_t' = (y_{1t}, y_{2t}, \ldots, y_{mt})$ is

$$\boldsymbol{y}_t = X_t\boldsymbol{\beta}_t + \boldsymbol{u}_t, \quad \boldsymbol{u}_t \sim \text{NID}(\boldsymbol{0}, \sigma^2 I_m), \tag{B.1a}$$

$$\boldsymbol{\beta}_t = \boldsymbol{\beta}\boldsymbol{\theta}_t + \varepsilon_t, \quad \varepsilon_t \sim \text{NID}(\boldsymbol{0}, \sigma^2\Omega) \tag{B.1b}$$

$$\boldsymbol{\theta}_t = \boldsymbol{\theta}_{t-1} + \eta_t \quad \eta_t \sim \text{NID}(\boldsymbol{0}, \phi\sigma^2 I_k), \tag{B.1c}$$

with $\boldsymbol{u}_t$, ε_t, and $\boldsymbol{\eta}_t$ assumed independently distributed. Then the following densities are relevant: The joint density for $(\boldsymbol{y}_t, \boldsymbol{\beta}_t, \boldsymbol{\theta}_t)$ given $(\sigma, \phi, \Omega, \boldsymbol{\theta}_{t-1})$ is

$$\begin{aligned}
&p(\boldsymbol{y}_t, \boldsymbol{\beta}_t, \boldsymbol{\theta}_t \mid \boldsymbol{\theta}_{t-1}, \sigma, X_t, \phi, \Omega) \\
&\quad = p(\boldsymbol{y}_t \mid \boldsymbol{\beta}_t, \sigma, X_t)\, p(\boldsymbol{\beta}_t \mid \boldsymbol{\theta}_t, \sigma, \Omega)\, p(\boldsymbol{\theta}_t \mid \boldsymbol{\theta}_{t-1}, \sigma, \phi) \\
&\quad\quad \propto \sigma^{-m} \exp\left\{-(\boldsymbol{y}_t - X_t\boldsymbol{\beta}_t)'(\boldsymbol{y}_t - X_t\boldsymbol{\beta}_t)/2\sigma^2\right\} \\
&\quad\quad \times \sigma^{-mk}|\Omega|^{-1/2} \exp\left\{-(\boldsymbol{\beta}_t - B\boldsymbol{\theta}_t)'\Omega^{-1}(\boldsymbol{\beta}_t - B\boldsymbol{\theta}_t)/2\sigma^2\right\} \\
&\quad\quad \times (\phi\sigma^2)^{-k/2} \exp\left\{-(\boldsymbol{\theta}_t - \boldsymbol{\theta}_{t-1})'(\boldsymbol{\theta}_t - \boldsymbol{\theta}_{t-1})/2\phi\sigma^2\right\}.
\end{aligned} \tag{B.2}$$

The posterior joint density for $\boldsymbol{\beta}_t$, $\boldsymbol{\theta}_t$, σ given $\boldsymbol{y}_t$, X_t, D_{t-1}, ϕ, Ω, where $\mathrm{D}_{t-1} \equiv$ information available at time $t-1$ is

$$\begin{aligned}
p(\boldsymbol{\beta}_t, \boldsymbol{\theta}_t, \sigma \mid \boldsymbol{y}_t, X_t, D_{t-1}, \phi, \Omega) \propto\; & p(\boldsymbol{y}_t \mid \boldsymbol{\beta}_t, \boldsymbol{\theta}_t, \sigma, X_t, D_{t-1}, \phi, \Omega) \\
& \times p(\boldsymbol{\beta}_t, \boldsymbol{\theta}_t, \sigma \mid X_t, D_{t-1}, \phi, \Omega).
\end{aligned} \tag{B.3}$$

On substituting from (B.1b) in (B.1a), we have

$$\boldsymbol{y}_t = X_t(B\boldsymbol{\theta}_t + \varepsilon_t) + \boldsymbol{u}_t = X_t B\boldsymbol{\theta}_t + X_t\varepsilon_t + \boldsymbol{u}_t. \tag{B.4}$$

Then, using (B.4) and (B.1c), with a normal-IG density for $(\boldsymbol{\theta}_1, \sigma)$, namely $p(\boldsymbol{\theta}_1, \sigma \mid D_0, \phi) = p(\theta_1 \mid \sigma, D_0, \phi)\; p(\sigma \mid D_0)$ with the first factor $\mathrm{N}(\boldsymbol{\theta}_0, \sigma^2 V_0)$ and the second $\mathrm{IG}(v_0, s_0^2)$, we obtain the following joint posterior density for $(\boldsymbol{\theta}_t, \sigma)$:

$$p(\boldsymbol{\theta}_t, \sigma \mid D_t, \phi, \Omega) \sim \mathrm{N}(\hat{\boldsymbol{\theta}}_{t|t}, \sigma^2 C_t) \times \mathrm{IG}\left(v_t, s_t^2\right), \tag{B.5}$$

and

$$p(\boldsymbol{\theta}_t \mid D_t, \phi, \Omega) = c_t\, |C_t|^{-1/2}\left\{v_t s_i^2 + (\boldsymbol{\theta}_t - \hat{\boldsymbol{\theta}}_{t|t})' C_t^{-1}(\boldsymbol{\theta}_t - \hat{\boldsymbol{\theta}}_{t|t})\right\}^{-(k+v_t)/2}, \tag{B.6}$$

where

$$c_t = \Gamma\left(\frac{v_t + k}{2}\right)\left(v_t s_t^2\right)^{v_t/2} / \pi^{k/2}\,\Gamma(v_t/2),$$

$$
\begin{aligned}
\hat{\boldsymbol{\theta}}_{t|t} &= \{V_{t-1}^{-1} + B'X_t'M_tX_tB\}^{-1}\{V_{t-1}^{-1}\hat{\boldsymbol{\theta}}_{t-1|t-1} + B'X_t'M_t\boldsymbol{y}_t\},\\
M_t &\equiv \{I_m + X_t'\Omega X_t\}^{-1},\\
C_t &\equiv \sigma^{-2}\,\mathrm{var}(\theta_t \mid D_t) = \{V_{t-1}^{-1} + B'X_t'M_tX_tB\}^{-1},\\
V_t &= C_t + \phi I,\\
v_t &= v_{t-1} + m_t,\\
v_ts_t^2 &= v_{t-1}s_{t-1}^2 + (\boldsymbol{y}_t - X_tB\hat{\boldsymbol{\theta}}_{t|t})'M_t(\boldsymbol{y}_t - X_tB\hat{\boldsymbol{\theta}}_{t|t})\\
&\quad + (\hat{\boldsymbol{\theta}}_{t|t} - \hat{\boldsymbol{\theta}}_{t-1|t-1})'V_{t-1}^{-1}(\boldsymbol{\theta}_{t|t} - \hat{\boldsymbol{\theta}}_{t-1|t-1}).
\end{aligned}
$$

The marginal density for β_t given D_t, σ, and Ω is obtained from (B.5) in conjunction with (B.1b), namely,

$$
\begin{aligned}
p(\beta_t, \sigma \mid D_{t-1}, \phi, \Omega) &= \mathrm{N}[B\hat{\boldsymbol{\theta}}_{t-1|t-1}, \sigma^2(BV_{t-1}B' + \Omega)]\\
&\quad \times \mathrm{IG}(v_{t-1}, s_{t-1}^2).
\end{aligned}
\tag{B.7}
$$

On combining (B.7) with the likelihood function for (B.1a) via Bayes' Theorem and integrating out σ, the result is

$$
\begin{aligned}
p(\beta_t \mid D_t, \phi, \Omega) &= \frac{\Gamma[(v_t + mk)/2]}{\pi^{mk/2}\Gamma(v_t/2)}(v_ts_t^2)^{v_t|2}\,|R_t|^{-1/2}\\
&\quad \times\{v_ts_t^2 + (\beta_t - \hat{\beta}_{t|t})'R_t^{-1}(\beta_t - \hat{\beta}_{t|t})\}^{-(mk+v_t)/2},
\end{aligned}
\tag{B.8}
$$

where, in addition to quantities defined above.

$$
\begin{aligned}
\hat{\beta}_{t|t} &= \{X_t'X_t + A_t\}^{-1}\{X_t'\boldsymbol{y}_t + A_tB\hat{\boldsymbol{\theta}}_{t-1|t-1}\},\\
A_t &\equiv (BV_{t-1}B' + \Omega)^{-1}, \quad R_t \equiv \sigma^{-2}\,\mathrm{var}(\beta_t \mid D_t) = (X_t'X_t + A_t)^{-1}.
\end{aligned}
\tag{B.9}
$$

Finally, the predictive density for $\boldsymbol{y}_{t+1}$ is obtained by integrating the joint density for $\boldsymbol{y}_{t+1}$ and σ given X_{t+1}, D_t with respect to σ. We have

$$
p(\boldsymbol{y}_{t+1}, \sigma \mid X_{t+1}, D_t) = \int p(\boldsymbol{y}_{t+1} \mid X_{t+1}, \beta_{t+1}, \sigma)\,p(\beta_{t+1}, \sigma \mid D_t)\mathrm{d}\beta_{t+1},
\tag{B.10}
$$

and on integrating (B.10) over σ, the result is

$$
\begin{aligned}
p(\boldsymbol{y}_{t+1} \mid X_{t+1}, D_t) &= a_t\,|P_t|^{-1/2}\{v_ts_t^2 + (y_{t+1} - \hat{\boldsymbol{y}}_{t+1})'P_t^{-1}\\
&\quad \times(\boldsymbol{y}_{t+1} - \hat{\boldsymbol{y}}_{t+1})\}^{-(m+v_t)/2},
\end{aligned}
\tag{B.11}
$$

where

$$
a_t = \Gamma\left(\frac{v_t + m}{2}\right)(v_ts_t^2)^{v_t/2}/\pi^{m/2}\Gamma(v_t/2),
$$

$$\hat{\boldsymbol{y}}_{t+1} = X_{t+1} B \hat{\boldsymbol{\theta}}_{t\,|\,t},$$
$$P_t \equiv \sigma^{-2} \operatorname{var}(\boldsymbol{y}_{t+1} \mid D_t) = X_{t+1} A_{t+1}^{-1} X_{t+1}' + I_m.$$

It is seen that (B.11) is in the form of a multivariate Student-t density, and thus the marginal density for an element of $\boldsymbol{y}_{t+1}$, say $y_{i,t+1}$, has a univariate Student-t density (see, e.g., Zellner 1971, app. B.2).

B.3 Values for hyperparameters (ϕ, Σ)

The values of (ϕ, Σ), with $\Omega = I_n \otimes \Sigma$, have been selected to minimize RMSEs of one-year-ahead forecasts for the pre-forecast period 1969–73. (The values of (ϕ, Σ) presented below are for the models with variables in percentage points.) Subsequent calculations for 1974 and following years have been conditioned on these values.

(1) Unpooled TVP models:

$$\begin{cases} \phi = 0.3 & \text{for} \quad \text{ARLI}, \\ \phi = 0.2 & \text{for} \quad \text{ARLI/WI}; \end{cases}$$

(2) Pooled TVP models:

$$\begin{cases} \phi = 0.1 \quad \text{for ARLIand ARLI/W}, \\ \Sigma = \begin{bmatrix} 0.1 & & & 0 \\ & 0.002 & & \\ & & \ddots & \\ 0 & & & 0.002 \end{bmatrix}; \end{cases}$$

(3) Pooled FP models:

$$\begin{cases} \Sigma = \begin{bmatrix} 0.1 & & & 0 \\ & 0.002 & & \\ & & \ddots & \\ 0 & & & 0.002 \end{bmatrix}; \end{cases}$$

(4) The equation for world real income growth rate: $\phi = 0.5$.

APPENDIX C UPDATING PROCEDURES FOR PREDICTIVE DENSITY OF A POOLED FHP MODEL

Using the notation introduced in section 4, the model for the observation vector $\boldsymbol{y}_t' = (y_{1t}, y_{2t}, \ldots, y_{mt})$ is

$$\boldsymbol{y}_t = X_t \beta_t + \boldsymbol{u}_t, \qquad \boldsymbol{u}_t \sim \text{NID}(\boldsymbol{0}, \sigma^2 I_m),$$
$$\beta_t = B\boldsymbol{\theta} + \varepsilon_t, \qquad \varepsilon_t \sim \text{NID}(\boldsymbol{0}, \sigma^2 \Omega).$$

These will be equivalent to model (B.1a)–(B.1c) when ϕ is set to zero. Therefore the same updating procedures can be used with ϕ set to zero.

BIBLIOGRAPHY

Bates, J. M. and C. W. J. Granger (1969), "The combination of forecasts," *Operations Research Quarterly* 20, 451–68

Blattberg, R. C. and E. I. George (1991), "Shrinkage estimation of price and promotional elasticities: seemingly unrelated equations", *Journal of the American Statistical Association* 86, 304–15

Clemen, R. T. (1989), "Combining forecasts: a review and annotated bibliography," *International Journal of Forecasting* 5, 559–83

Diebold, F. X. (1990a), "Forecast combination and encompassing: reconciliation of two divergent literatures," *International Journal of Forecasting* 5, 589–92

(1990b), "A note on Bayesian forecast combination procedures," in A. Westlund and P. Haski (eds.), *Economic Structural Change: Analysis and Forecasting* (New York, Springer-Verlag), 1–8

Diebold, F. X. and P. Pauly (1987), "Structural change and the combination of forecasts," *Journal of Forecasting* 6, 21–40

Garcia-Ferrer, A., R. A. Highfield, F. C. Palm, and A. Zellner (1987), "Macroeconomic forecasting using pooled international data," *Journal of Business and Economic Statistics* 5, 53–67; chapter 13 in this volume

Geisel, M. S. (1975), "Bayesian comparisons of simple macroeconomic models," in S. E. Fienberg and A. Zellner (eds.), *Studies in Bayesian Econometrics and Statistics in Honor of Leonard J. Savage* (Amsterdam, North-Holland), 227–56

Granger, C. W. J. and R. Ramanathan (1984), "Improved methods of combining forecasts," *Journal of Forecasting* 3, 197–204

Hong, C. (1989), "Forecasting real output growth rates and cyclical properties of models: a Bayesian approach," PhD dissertation, University of Chicago

Jeffreys, H. (1967), *Theory of Probability* (London, Oxford University Press)

Leamer, E. E. (1978), *Specification Searches* (New York, John Wiley)

LeSage, J. P. (1989), "Forecasting turning points in metropolitan employment growth rates using Bayesian exponentially weighted autoregression, time-varying parameter and pooling techniques"; published as "Forecasting turning points in metropolitan employment growth rates using Bayesian techniques," *Journal of Regional Science* 30(4) (1990), 533–48, chapter 21 in this volume

LeSage, J. P. and M. Magura (1990), "Using Bayesian techniques for data pooling in regional payroll forecasting," *Journal of Business and Economic Statistics* 8 (1), 127–36; chapter 20 in this volume

Lindley, D. and A. F. M. Smith (1972), "Bayes estimates for the linear model," *Journal of the Royal Statistical Society* B 34, 1–41

Min, C. (1990), "Forecasting models with time-varying parameter, pooling and Bayesian model-combining approaches: economic theory and application," Thesis proposal paper, University of Chicago

Nelson, C. R. (1972), "The predictive performance of the FRB–MIT–Penn model of the US economy," *American Economic Review* 62, 902–17

Palm, F. C. and A. Zellner (1990), "To combine or not to combine forecasts?," H. G. B. Alexander Research Foundation, University of Chicago

Reid, D. J. (1968), "Combining three estimates of gross domestic product," *Economica* 35, 431–44

1969, "A comparative study of time series prediction techniques on economic data," PhD dissertation, University of Nottingham

Rossi, P. E. (1983), "Specification and analysis of econometric production models," PhD dissertation, University of Chicago

(1985), "Comparison of alternative functional forms in production," *Journal of Econometrics* 30, 345–61

Schwarz G. (1978), "Estimating the dimension of a model," *Annals of Statistics* 6, 441–64

Swamy, P. A. V. B. (1971), *Statistical Inference in Random Coefficient Models* (Berlin, Springer-Verlag)

Winkler, R. L. (1981), "Combining probability distributions from dependent information sources," *Management Science* 27, 479–88

Zellner, A. (1971), *An Introduction to Bayesian Inference in Econometrics* (New York, John Wiley)

(1984), *Basic issues in Econometrics* (Chicago, University of Chicago Press)

(1989), "Bayesian and non-Bayesian methods for combining models and forecasts," H. G. B. Alexander Research Foundation, University of Chicago, manuscript

Zellner, A. and C. Hong (1989), "Forecasting international growth rates using Bayesian shrinkage and other procedures," *Journal of Econometrics* (Annals) 40, 183–202; chapter 14 in this volume

Zellner, A. and L. A. Manas-Anton (1986), "Macroeconomic theory and international macroeconomic forecasting," H. G. B. Alexander Research Foundation, University of Chicago, manuscript

Zellner, A., C. Hong, and G. M. Gulati (1990), "Turning points in economic time series, loss structures, and Bayesian forecasting," in S. Geisser, J. S. Hodges, S. J. Press, and A. Zellner (eds.), *Bayesian and Likelihood Methods in Statistics and Econometrics: Essays in Honor of George A. Barnard* (Amsterdam, North-Holland), 371–89; chapter 15 in this volume

Zellner, A., C. Hong, and C. Min (1991), "Forecasting turning points in international output growth rates using Bayesian exponentially weighted autoregression, time-varying parameter, and pooling techniques," *Journal of Econometrics* 49, 275–304; chapter 16 in this volume

Zellner, A., D. S. Huang, and L. C. Chau (1965), "Further analysis of the short-run consumption function with emphasis on the role of liquid assets," *Econometrica* 33, 571–81

18 Pooling in dynamic panel data models: an application to forecasting GDP growth rates (2000)

André J. Hoogstrate, Franz C. Palm, and Gerard A. Pfann

In this chapter, we analyze issues of pooling models for a given set of N individual units observed over T periods of time. When the parameters of the models are different but exhibit some similarity, pooling may lead to a reduction of the mean squared error of the estimates and forecasts. We investigate theoretically and through simulations the conditions that lead to improved performance of forecasts based on pooled estimates. We show that the superiority of pooled forecasts in small samples can deteriorate as the sample size grows. Empirical results for postwar international real gross domestic product growth rates of 18 Organization for Economic Cooperation and Development countries using a model put forward by Garcia-Ferrer, Highfield, Palm, and Zellner and Hong, among others illustrate these findings. When allowing for contemporaneous residual correlation across countries, pooling restrictions and criteria have to be rejected when formally tested, but generalized least squares (GLS)-based pooled forecasts are found to outperform GLS-based individual and ordinary least squares-based pooled and individual forecasts.

Panel data are used more and more frequently in business and economic studies. Sometimes a given number of entities is observed over a longer period of time, whereas traditionally panel data are available for a large and variable number of entities observed for a fixed number of time periods (e.g. see Baltagi 1995 for a[n] . . . overview; Maddala 1991; Maddala, Trost, and Li 1994). In this chapter, we analyze issues of pooling models for a given set of N individual units observed over T periods of

The authors thank Arnold Zellner for kindly providing them with the data and for most helpful discussions and comments on this article. We also acknowledge the useful comments of Ruey Tsay, an associate editor, and a referee. This research was sponsored by the Economics Research Foundation, which is part of the Netherlands Organization for Scientific Research.

Originally published in the *Journal of Business and Economic Statistics* 18(3) (2000), 274–283.

time. Large T asymptotics, with N fixed, provide the benchmark against which to evaluate the methods considered. Pooling estimates in panel-data models is appropriate if parameters are the same for the individual units observed. When the parameters are different but exhibit some similarity, pooling may also lead to a reduction of the mean-squared error (MSE) of the estimates. A reduction of the MSE will be achieved when the square of the bias resulting from imposing false restrictions is outweighed by the reduction of the variance of the estimator due to restricted estimation. The existence of this trade-off has generated the literature on MSE criteria and associated tests for the superiority of restricted over unrestricted least squares estimators (e.g. see Wallace and Toro-Vizcarrondo 1969; Goodnight and Wallace 1972; McElroy 1977; Wallace 1972).

Pooling techniques have been successfully applied, for instance, to test the market efficiency hypothesis (e.g. see Bilson 1981) and to forecast multicountry output growth rates (e.g. see Mittnik 1990). In a study of real gross national product (GNP) growth rates of nine Organization for Economic Cooperation and Development (OECD) countries for the period 1951–81, Garcia-Ferrer *et al.* (1987) showed that pooled estimates of an (autoregressive) AR(3) model with leading indicator (LI) variables, denoted by AR(3)LI, provided superior forecasting results. Forecasting results for an extended time period, 1974–84, and an extended number of countries, eighteen OECD countries, provided by Zellner and Hong (1989) were in favor of the earlier findings. Leading economic indicators have come to play a dominant role in forecasting business cycle turning points on a single-country level (Stock and Watson 1989) as well as on a multicountry level (Zellner, Hong, and Min 1991). Cross-country, cross-equation restrictions have also been imposed successfully to analyze convergence of annual log real *per capita* output for fifteen OECD countries from 1900 to 1987 (Bernard and Durlauf 1995).

The objective of this chapter is threefold. First, in Section 1, we investigate theoretically whether the improvement of forecasting performance using pooling techniques instead of single-country forecasts remains valid as T grows large(r) while N remains constant. The model that we investigate consists of a set of dynamic regression equations with contemporaneously correlated disturbances. It nests the specifications put forward by Garcia-Ferrer *et al.* (1987) and used in many studies since (e.g. see Zellner 1994). Second, in section 2 we present simulation results that give some insights into the importance of the gains from pooling data when the sample size is small. The simulations also provide evidence on the statistical properties of some test procedures for pooling restrictions. Third, the theoretical results are investigated empirically. The Zellner, Hong, and Min (1991) data, slightly modified to have a consistent set of real GDP growth rates of eighteen OECD countries for an extended period of 1948 to 1990, are used to forecast international growth rates using individual and pooled estimates of the AR(3)LI model. Section 3 concludes.

1 The model and the forecasting procedures

Consider a linear regression model for N units/countries and T successive observations:

$$y_{it} = x_{it}'\beta_i + \varepsilon_{it}, \quad i = 1, 2, \ldots N, t = 1, 2, \ldots T, \tag{1.1}$$

or

$$\underset{T\times 1}{y_i} = \underset{(T\times k)}{X_i} \underset{(k\times 1)}{\beta_i} + \underset{T\times 1}{\varepsilon_i}, \quad i = 1, 2, \ldots, N, \tag{1.2}$$

where y_{it} denotes the value of the endogenous variable y for country i in period t, x_{it} is a vector of explanatory variables, β_i is a vector of k regression coefficients for country i, and ε_{it} denotes a disturbance term. Alternatively, the model (1.2) for N countries can be written as

$$\underset{n\times 1}{y} = \underset{(n\times Nk)}{X^*} \underset{(Nk\times 1)}{\beta_a} + \underset{n\times 1}{\varepsilon}, \tag{1.3}$$

with $y = (y_1', y_2', \ldots, y_N')'$, $X^* = \text{diag}(X_1, X_2, \ldots, X_N)$, $\beta_a = (\beta_1', \beta_2', \ldots, \beta_N')'$, $\varepsilon = (\varepsilon_1', \varepsilon_2', \ldots, \varepsilon_N')'$, and $n = TN$.

We allow the vector x_{it} to contain lagged values of y_{it}. The disturbances ε_{it} are assumed to be normally distributed with mean 0 and zero serial correlations, possibly contemporaneously correlated; that is $\varepsilon \sim N(0, \Omega)$ with $\Omega = \Sigma \otimes I_T$, where Σ denotes the contemporaneous covariance matrix of dimension N. The regressors x_{it} are predetermined:

$$E(x_{it}\varepsilon_{js}' | x_{lt-1}, \ldots, l = 1, 2, \ldots, N) = 0 \text{ for all } i, j \text{ and } t \leq s.$$

The system of seemingly unrelated regressions (SUR) in (1.3) has been extensively studied in the literature, from both a classical and a Bayesian point of view. For instance, Haitovsky (1990) generalized Zellner's (1971) Bayesian analysis of the SUR, using a linear hierarchical structure similar to that in the Lindley and Smith (1972) pooling model. Chib and Greenberg (1995) carried out a hierarchical analysis of SUR models with serially correlated errors and time-varying parameters. Nandram and Petruccelli (1997) considered pooling autoregressive time series panel data using a hierarchical framework for a model with a highly structured covariance matrix Σ, with off-diagonal elements that depend on one unknown parameter.

For $\Sigma = \sigma^2 I_N$, assuming a hierarchical structure, Garcia *et al.* (1987), Zellner and Hong (1989), and Min and Zellner (1993) considered shrinkage forecasts based on an estimate of β_a in (1.3), which is a matrix-weighted average of the equation-by-equation least squares estimates $\hat{\beta}_i$ and a pooled estimate (i.e. the least squares estimate of β restricting the β_is in (1.3) to be the same).

A scalar-weighted average of these estimators results, with the scalar weight depending on the ratio of σ^2 and the prior variance of the β_is, if the "*g*-prior" approach of Zellner (1986) is adopted (see Zellner and Hong 1989).

In this chapter, we consider the problem of choosing among forecasts based on various estimators of β_a in (1.3) from a sampling theory point of view using an F-statistic that has been proposed to test for pooling. Note that there is a direct link between an F-statistic for linear restrictions on the coefficients of a linear regression model and the posterior odds used to choose among the models associated with the hypothesis to be tested, although their interpretations will differ (see Zellner 1984).

The reasons for considering the problem of choosing among forecasts are fourfold.

First, under a diagonal loss structure, it is optimal to select a forecast rather than to combine forecasts (e.g. see Min and Zellner 1993).

Second, as described previously, the problem of combining forecasts has been extensively studied in the literature, using a hierarchical structure, whereas the problem of choosing among forecasts based on alternative estimators for an SUR has received less attention. In particular, SUR models for N as large as 18 have not been extensively used.

Third, adopting a fully-fledged hierarchical Bayesian procedure in a model with unrestricted covariance matrix Σ, when N is large, requires integration in high dimensions (at least $N(N+1)/2$). For applications with N as large as 18, this requirement may make it prohibitive to use such procedures if it is not appropriate to impose some structure on Σ.

Fourth, unpooled and pooled forecasts are the ingredients required for combining forecasts in a hierarchical structure. Our results can be interpreted as a benchmark against which combined forecasts can be judged.

We consider the following one-step-ahead forecasts of y_{iT+1}, $\hat{y}_{iT+1} = x_{iT+1}'\hat{\beta}_i$, with $\hat{\beta}_i$ being an estimate of β_i:

1. The individual forecast is based on the least squares estimator of β_i,

$$\hat{\beta}_i = (X_i'X_i)^{-1}X_i'y_i. \tag{1.4}$$

2. The pooled (p) forecast is based on the OLS estimator for the pooled data,

$$\hat{\beta}^p = (X'X)^{-1}X'y, \tag{1.5}$$

where $X = (X_1', X_2', \ldots, X_N')'$.

3. The forecast (g) is based on a feasible SUR estimator of β_i, $\hat{\beta}_i^g$, being the ith subvector of the generalized least squares (GLS) estimator of β_a in (1.3),

$$\hat{\beta}_a^g = (X^{*\prime}\hat{\Omega}^{-1}X^*)^{-1}X^{*\prime}\hat{\Omega}^{-1}y, \tag{1.6}$$

with $\hat{\Omega}$ being a consistent estimate of Ω.

4. The forecast (pg) is based on a pooled feasible GLS estimator of β_i,

$$\hat{\beta}^{pg} = (X'\hat{\Omega}^{-1}X)^{-1}X'\hat{\Omega}^{-1}y. \tag{1.7}$$

The predictors based on the estimators (1.4)–(1.5) were used by Garcia-Ferrer *et al.* (1987) and Zellner and Hong (1989) to obtain point forecasts of GNP growth rates and by Zellner, Hong, and Min (1991) to forecast turning points in GNP growth. We investigate the behavior of the four previously mentioned predictors when the time dimension of the sample becomes large. We compare the performance of individual forecasts with the pooled forecasts when T grows while N remains fixed and investigate the conditions under which pooling leads to improved forecast performance.

To compare the forecast performance, we use the mean-squared forecast error (MSFE) criterion

$$\text{MSFE}(\hat{y}_{iT+1}) \doteq x_{iT+1}'\text{MSE}(\tilde{\beta}_i)x_{iT+1} + \sigma_i^2, \tag{1.8}$$

where $\tilde{\beta}_i$ denotes an estimator of β_i (one of the estimators (1.4)–(1.7)) and σ_i^2 is the ith diagonal element of Σ. The cross-term $x_{iT+1}'(\beta_i - \tilde{\beta}_i)\varepsilon_{iT+1}$ has been deleted from the r.h.s. of (1.8) as it vanishes asymptotically. Comparing the MSFEs of various forecasts basically reduces to comparing the MSEs of the estimators used to compute the forecasts.

The results obtained by McElroy (1977) can be applied to compare the GLS estimator $\hat{\beta}_a^g$ (6) for β_a with the pooled GLS estimator (1.7) $\hat{\beta}_a^{pg} = (\iota \otimes \hat{\beta}^{pg})$, with ι being an $N \times 1$ unit vector. When Ω is unknown and a consistent estimate is used, the results of McElroy (1977) hold for a large sample. When $\Omega = \sigma^2 I_n$, McElroy's (1977) results specialize accordingly and can be used to compare the OLS estimators $\hat{\beta}_i$ in (1.4) with the pooled estimator $\hat{\beta}^p$ in (1.5). When Ω is known, the restricted GLS estimator $\hat{\beta}_a^{pg}$ is preferred to $\hat{\beta}_a^g$ by the strong MSE criterion defined by McElroy if

$$l'[\mathrm{MSE}(\sqrt{n}\hat{\beta}_a^g) - \mathrm{MSE}(\sqrt{n}\hat{\beta}_a^{pg})]l \geq 0 \tag{1.9}$$

holds for all $Nk \times 1$ vectors $l \neq 0$. Notice that the condition (1.9) implies superiority of the pooled estimator for each country. Imposing restrictions erroneously produces a biased estimator. When the condition (1.9) holds, the size of the bias is outweighed by the reduction of the variance of the estimator.

Using the results of McElroy, (1.9) can be shown to hold iff $\lambda_n \leq 1/2$ with $\lambda_n' = \delta_n'(RV_nR')^{-1}\delta_n/2$ and $\delta_n = \sqrt{n}(R\beta_a)$. Notice that $R\beta_a = 0$ denotes the restrictions on β_a when we pool across countries, and the $q \times Nk$ matrix R with $q = (N-1)k$ is

$$R = [\iota_{N-1} \otimes I_k, -I_{N-1} \otimes I_k], \tag{1.10}$$

with ι_{N-1} being an $(N-1)$ unit vector. $V_n = n(X^{*\prime}\Omega^{-1}X^*)^{-1}$ is the covariance matrix of $\sqrt{n}\hat{\beta}_a^g$.

The null hypothesis $R\beta_a = 0$ is true iff $\lambda_n = 0$. The restricted estimator $\hat{\beta}_a^{pg}$ is preferred to $\hat{\beta}_a^g$ by the first weak MSE criterion, which requires that the trace of the difference of the MSE matrices in (1.9) be non-negative. This holds whenever

$$\lambda_n \geq \theta_n, \tag{1.11}$$

with $\theta_n = 1/2\mu_n \mathrm{tr}[V_nR'(RV_nR')^{-1}RV_n]$ and μ_n being the smallest characteristic root of V_n^{-1}. Finally, the pooled GLS estimator is better than the GLS estimator by the second weak MSE criterion defined by McElroy (1977) as

$$E[n(\hat{\beta}_a^g - \beta_a)'V_n^{-1}(\hat{\beta}_a^g - \beta_a) - n(\hat{\beta}_a^{pg} - \beta_a)'V_n^{-1}(\hat{\beta}_a^{pg} - \beta_a)] \geq 0, \tag{1.12}$$

which holds iff $\lambda_n \leq q/2$ – that is, iff $2\lambda_n$ is smaller than the number of restrictions on regression coefficients in the system.

As shown by McElroy (1977), when the X_is are strictly exogenous (standard regression model) the test statistic

$$F^*(\Omega) = \frac{n(R\hat{\beta}_a^g)'(RV_nR')^{-1}(R\hat{\beta}_a^g)/q}{\mathrm{SSE}(\hat{\beta}_a^g)/(n-q)} \sim F(q, n-q, \lambda_T), \tag{1.13}$$

where $\mathrm{SSE}(\hat{\beta}_a^g) = (y - X^*\hat{\beta}_a^g)'\Omega^{-1}(y - X^*\hat{\beta}_a^g)$ has a non-central F-distribution. It can be used to test hypotheses about λ. Rejecting the null hypothesis when the statistic is large provides a uniformly most powerful test for $\lambda_n \leq \lambda^*$ against $\lambda_n > \lambda^*$, where $\lambda^* = 1/2$, θ_n, or $q/2$ depending on the chosen MSE criterion.

As $n \to \infty$ (e.g. for $T \to \infty$ and fixed N), qF^* in (1.13) converges to a (non-)central $\chi^2(q, \lambda)$ distribution with $\lambda = \lim_{n\to\infty} \lambda_n$, when the sequence of alternative hypotheses is chosen in such a way that λ_n converges to a finite limit λ. When Ω is replaced by a consistent estimator and/or in the presence of predetermined variables among the regressors, the same limiting distribution for qF^* results. Moreover, as $n \to \infty$ the covariance matrices for both $\sqrt{n}\hat{\beta}_a^g$ and $\sqrt{n}\hat{\beta}_a^{pg}$ converge to constant matrices. If the restrictions $R\beta_a = 0$ do not hold, δ_n, and hence the bias of $\sqrt{n}\hat{\beta}_a^{pg}$, increases without bound while the unrestricted estimator $\sqrt{n}\hat{\beta}_a^g$ remains unbiased. Therefore, for each of the three MSE criteria, there exists a sufficiently large n to make $\hat{\beta}_a^{pg}$ worse than $\hat{\beta}_a^g$ in terms of MSE. In the case in which the restrictions are false, the non-centrality parameter λ_n increases without bound. As a result, the power of a test of $H_0 : \lambda_n \leq \lambda^*$ tends to 1 and the test is consistent.

In section 2, we shall report the findings of an empirical analysis of the MSFE of forecasts based on unrestricted and pooled estimators. In particular, we shall investigate under which conditions and for which sample size it pays to use a pooled estimator rather than an unrestricted estimator.

2 Empirical analyses

In this section we investigate an AR(3) model and an AR(3)LI model used by Garcia-Ferrer *et al.* (1987) and Zellner and Hong (1989). The data set consists of the post-Second World War real GDP growth rates of eighteen OECD countries. The parameters of the individual countries are estimated using samples for which the starting date varies between 1949 and 1957 and ends in 1990 (see appendix A, p. 609). In subsection 2.1, we present the models, estimate, and test them. Subsection 2.2 is devoted to a simulation study of the finite-sample properties of McElroy's criteria for pooling. The empirical models for a subset of the eighteen OECD countries are used in the simulations. Finally, in subsection 2.3 we check the pooling restrictions for the eighteen countries using McElroy's criteria.

2.1 The models

The following AR(3)LI model was employed to generate one-year-ahead forecasts of the growth rate of real GDP, for the period 1981–90, for

Table 18.1 *The root mean-squared forecast error (RMSFE) for the individual country forecast and for the pooled forecast using the AR(3) model when up to fifteen observations are excluded from the estimation period*

	Number of observations excluded							
	0		5		10		15	
	Unpooled (1)	Pooled (2)	Unpooled (3)	Pooled (4)	Unpooled (5)	Pooled (6)	Unpooled (7)	Pooled (8)
Australia	3.66	3.36	3.66	3.41	3.60	3.35	3.26	3.30
Austria	1.53	0.92	1.30	0.89	0.92	0.74	1.03	0.75
Belgium	1.38	1.79	1.54	1.71	1.59	1.69	1.97	1.74
Canada	3.90	3.47	4.26	3.44	4.39	3.43	4.59	3.35
Denmark	1.86	1.81	2.04	1.77	2.45	1.81	2.66	1.72
Finland	1.49	1.37	1.29	1.38	1.23	1.52	1.26	1.69
France	1.44	1.59	1.67	1.52	1.69	1.47	1.70	1.46
Germany	4.27	4.50	4.40	4.30	4.47	4.39	4.37	4.34
Ireland	2.23	1.68	2.48	1.67	3.78	1.64	3.95	1.74
Italy	2.40	1.55	2.39	1.47	2.42	1.52	2.43	1.52
Japan	1.09	0.99	1.05	0.98	1.14	1.06	1.32	1.23
Netherlands	1.67	1.46	1.60	1.42	1.42	1.26	1.61	1.27
Norway	4.95	3.98	5.00	3.91	5.52	3.86	5.66	3.71
Spain	1.03	1.19	2.03	1.22	2.00	1.34	1.79	1.54
Sweden	1.41	1.19	1.47	1.16	1.72	1.20	1.69	1.22
Switzerland	1.54	1.48	1.56	1.48	1.65	1.40	2.00	1.43
United Kingdom	2.38	1.90	2.51	1.88	2.77	1.93	2.90	1.96
United States	3.15	2.74	3.59	2.64	4.20	2.76	4.13	2.75
Median	1.76	1.63	2.04	1.59	2.21	1.58	2.22	1.71

eighteen OECD countries:

$$y_{it} = \beta_{0i} + \beta_{1i} y_{it} + \beta_2 y_{it-2} + \beta_3 y_{it-3} + \beta_{4i} SR_{it-1} + \beta_{5i} SR_{it-2} + \beta_{6i} GM_{it-1} + \beta_{7i} WR_{it-1} + \varepsilon_{it}, \quad (2.1)$$

where y_{it} denotes the first difference of the logarithm of real output, SR_{it} denotes the first difference of the log of a stock price index divided by a general price index, GM_{it} denotes the first difference of the log of the nominal money supply divided by a general price index, and WR_{it} denotes world return, which equals the median of countries' real stock return in period t.

The AR(3) model arises as a special case of model (2.1) when $\beta_{4i} = \beta_{5i} = \beta_{6i} = \beta_{7i} = 0$.

To illustrate the gains from pooling when T is small, we performed the actual one-step-ahead forecasts for the period 1983–90 using an AR(3) model. The results are shown in table 18.1, in which columns (1) and (2) are based on models estimated by all data, columns (3) and (4) are from models estimated by excluding the first five observations for each country, columns (5) and (6) are from models estimated by excluding the first ten observations, and finally columns (7) and (8) are from models estimated by excluding the first fifteen observations. As expected, as T gets smaller, the forecasting performance of the pooled forecast dominates the individual predictor. This is apparent in two ways in table 18.1. First, the number of countries for which the individual forecast is better declines as T becomes smaller. Second, the difference in median-root mean squared errors (RMSEs) increases as T decreases.

Next we compare the forecasting performance for different periods of time. Table 18.2 reports RMSEs for forecasts (RMSFEs) based on country-specific OLS parameter estimates and pooled parameter estimates for the forecasting periods 1974–87, 1974–90, and 1983–90. For the three forecast periods and for both models, the pooled forecasts dominate the individual forecasts in most instances. Moreover, the median of the RMSFEs is lowest for the pooled forecasts in the six cases. This finding is in line with those of Garcia-Ferrer *et al.* (1987) for a forecast period 1974–81 and Zellner and Hong (1989) for the periods 1974–81 and 1974–84. Notice also that for the forecast period 1983–90, surprisingly the AR(3) model performs better than the AR(3)LI model in terms of RMSFE. For the forecast period 1974–87, the AR(3)LI clearly performs better than the AR(3) model in most instances. There are small differences with results reported by Min and Zellner (1993) for the same forecast period. These are because they used GNP and GDP data and an estimation period 1954–73 with data for 1951–3 serving as initial values whereas in the present study strictly GDP data are used for all countries and the estimation period is 1961–73 with data for 1958–60 serving as initial values.

The finding that, as T grows, the difference between the MSFEs of forecasts based on OLS and pooled forecasts becomes larger suggests that the restrictions of identical parameters across countries are not literally true. Before testing these restrictions, we shall examine the presence of contemporaneous correlation between the disturbances for the eighteen countries in the AR(3) and AR(3)LI models, respectively.

For the estimation period 1961–80, as expected, the residuals of the AR(3) model show more contemporaneous correlation than the residuals of the AR(3)LI model. Including leading indicators, which are approximately white noise, accounts for a major part of the contemporaneous residual correlation present in the AR(3) model.

From the estimated residual correlations, most of which are positive as expected, it also appears that the countries can be clustered in regional groups exhibiting much within-group contemporaneous residual

Table 18.2 *Eighteen countries root mean-squared forecast errors (RMSFEs) of one-year-ahead forecasts, 1974–1987, 1974–1990, and 1983–1990*

	AR(3)						AR(3)LI					
	1974–87		1974–90		1983–90		1974–87		1974–90		1983–90	
Period γ	Unpooled	Pooled	Unpooled	Pooled	Unpooled	Pooled	Unpooled	Pooled	Unpooled	Pooled	Unpooled	Pooled
Australia	4.73	3.21	4.34	2.92	3.66	3.36	4.94	2.63	4.77	2.42	3.55	1.91
Austria	3.30	2.60	3.02	2.36	1.53	0.92	3.42	2.51	3.15	2.29	3.02	1.88
Belgium	3.19	3.24	2.95	2.98	1.38	1.79	1.87	2.41	2.31	2.38	2.93	2.21
Canada	3.78	3.37	3.48	3.12	3.90	3.47	3.99	2.91	3.65	2.67	4.08	3.48
Denmark	3.50	3.40	3.36	3.30	1.86	1.81	4.00	3.92	4.02	3.94	3.73	3.06
Finland	3.63	3.34	3.34	3.12	1.49	1.37	3.74	2.92	3.61	2.73	2.66	1.53
France	3.15	2.42	2.91	2.30	1.44	1.59	3.13	2.72	2.90	2.68	1.51	2.30
Germany	5.05	4.67	4.62	4.36	4.27	4.50	4.19	3.98	3.90	3.75	3.90	3.69
Ireland	4.81	4.36	4.45	4.02	2.23	1.68	4.48	4.12	4.25	3.88	2.60	1.78
Italy	3.85	3.23	3.56	3.00	2.40	1.55	3.66	2.42	3.54	2.46	3.47	1.85
Japan	4.36	2.98	3.98	2.73	1.09	0.98	4.17	2.77	3.95	2.58	2.45	1.23
Netherlands	3.39	2.58	3.13	2.41	1.67	1.45	3.23	2.47	4.00	2.46	2.90	2.46
Norway	2.85	3.02	3.70	3.15	4.94	3.97	3.37	3.12	4.20	3.88	5.36	5.43
Spain	3.00	2.57	2.74	2.40	1.03	1.19	3.03	2.28	2.95	2.09	2.32	1.24
Sweden	2.91	2.76	2.66	2.52	1.41	1.18	2.90	2.48	2.65	2.34	2.58	1.75
Switzerland	3.73	3.77	3.46	3.43	1.54	1.48	3.89	3.46	3.71	3.19	2.51	2.13
United Kingdom	3.49	3.27	3.36	3.10	2.38	1.90	2.83	2.64	4.48	3.35	5.64	3.76
United States	4.01	3.79	3.66	3.50	3.15	2.74	3.50	3.42	3.49	3.30	3.75	2.62
Median	3.57	3.24	3.41	3.05	1.76	1.63	3.58	2.74	3.63	2.68	2.79	2.17

Table 18.3 *Testing for contemporaneous error covariances*

	Hypotheses about Σ		
H_0: H_1:	Diagonal Unrestricted	Diagonal Block-diagonal[a]	Block-diagonal[a] Unrestricted
AR(3)LI 1961–80	167.68 df = 153 $p = 0.197$	99.40 df = 17 $p < 0.001$	77.28 df = 136 $p > 0.5$
AR(3)LI 1961–90	290.88 df = 153 $p < 0.01$	149.67 df = 17 $p < 0.001$	141.21 df = 136 $p = 0.362$
AR(3) 1961–80	353.46 df = 153 $p < 0.001$	105.94 df = 17 $p < 0.001$	247.5 df = 136 $p < 0.001$
AR(3) 1961–90	474.67 df = 153 $p < 0.001$	137.79 df = 17 $p < 0.001$	336.88 df = 136 $p < 0.01$

Note:
[a] We distinguish the following seven blocks: (1) Canada, the United States; (2) Australia, Japan; (3) Denmark, Finland, Norway, Sweden; (4) Belgium, France, Germany, the Netherlands; (5) the United Kingdom, Ireland; (6) Austria, Switzerland; (7) Italy, Spain.

correlation and little between-group contemporaneous residual correlation indicating that shocks to real GDP growth are partly synchronized within blocks and uncorrelated between blocks. We distinguish the following seven regional blocks – (1) Canada, the United States; (2) Australia, Japan; (3) Denmark, Finland, Norway, Sweden; (4) Belgium, France, Germany, the Netherlands; (5) Ireland, the United Kingdom; (6) Austria, Switzerland; (7) Italy, Spain.

To formally test for contemporaneous residual correlation, we use a Lagrange multiplier (LM) statistic proposed by Breusch and Pagan (1980) for testing the null hypothesis of a diagonal Σ. Under H_0, $\lambda_{\mathrm{LM}} = T\sum_{i=2}^{N}\sum_{j=1}^{i-1} r_{ij}^2$, with r_{ij} being the sample correlation coefficients between the residuals of the OLS estimates for countries i and j, has an asymptotic $X^2[\mathrm{N}(N-1)/2]$ distribution. The results for the LM test are given in table 18.3.

The X^2 statistics reported in table 18.3 clearly indicate that the diagonality of Σ is not rejected for the AR(3)LI model for the observation period 1961–80 when tested against an unrestricted Σ matrix. When tested against a block-diagonal matrix, diagonality is rejected. For the longer period 1961–90, we reject the null hypothesis of a diagonal Σ matrix. This is possibly due to small sample size or structural changes that occurred in the 1980s. For this latter period, the block-diagonal structure is not rejected for the AR(3)LI model. For the AR(3) model the null hypothesis has to be rejected in all instances. This is not surprising because the common leading indicator WR_{t-1}, which is approximately white noise, accounts for interdependencies among the white-noise disturbances of the countries. Note that Chib and Greenberg (1995) reported that, when a time-varying parameter version of the AR(3)LI model is employed for the output growth rate of five countries (Australia, Canada, Germany, Japan, and the United States) in the period 1960–87, the matrix Σ is found to be diagonal.

2.2 *Properties of pooling restriction tests*

Before we check the appropriateness of pooling in a system of eighteen equations, we investigate the small-sample properties of the F-statistic given in (1.13) for testing the pooling restriction $H_0 : R\beta_a = 0$ against $H_1 : R\beta_a \neq 0$ and of McElroy's strong and weak pooling criteria allowing for a block-diagonal matrix Σ. The simulation results have been obtained using models for the following sets of countries – Belgium, Germany France, and the Netherlands, and Canada and the United States. The general model consists of a set of six third order autoregressions with two leading indicators:

$$\underset{(6\times 6)}{\Phi(L)}\ \underset{(6\times 1)}{y_t} = \underset{(6\times 6)}{B}\ \underset{(6\times 1)}{x_{1t}} + \underset{6\times 1}{\gamma}\ x_{2t} + \underset{6\times 1}{\varepsilon_t}, \tag{2.2}$$

where $\Phi(L)$ is a diagonal lag polynomial matrix with a third-degree polynomial on the main diagonal and where B and γ denote, respectively, a matrix and a vector of coefficients. The leading indicators x_{1t} and x_{2t} and the disturbance vector ε_t satisfy the following properties:

$$x_{1t} \sim \mathrm{IIN}(0, \Sigma_1), \quad x_{2t} \sim \mathrm{IIN}\big(0, \sigma_2^2\big), \quad \varepsilon_t \sim \mathrm{IIN}(0, \Sigma_3), \tag{2.3}$$

with Σ_1 being diagonal and Σ_3 being a block-diagonal covariance matrix. The variables x_{1t}, x_{2t}, and ε_t are mutually independent. The vector x_{1t} can be interpreted as a country-specific leading indicator. The variable x_{2t} can be interpreted as a common leading indicator (e.g., WR_{t-1}). The block-diagonal structure of Σ_3 reflects the finding that the disturbances within European and North American subgroups are correlated and that the between-subgroup correlations are 0. The model (2.2)–(2.3) implies a third order vector autoregressive (VAR) model for y_t with a diagonal VAR

Table 18.4 *Testing for pooling for a subset of six countries (Canada, the United States, Belgium, France, Germany, the Netherlands)*

Model	Σ	F-value	p-value $\lambda = 0$	p-value $\lambda = q/2$
AR(3)	$\sigma^2 I$	0.5763	0.9244	0.9944
AR(3)	Full	1.0532	0.4032	0.8839
AR(3)LI	$\sigma^2 I$	1.2580	0.2156	0.6772
AR(3)LI	Block	1.1099	0.3320	0.8453

matrix and a full-disturbance covariance matrix $\Sigma = B\Sigma_1 B' + \sigma_2^2 \gamma\gamma' + \Sigma_3$. The error-component structure of the disturbance term of the AR(3) model will be ignored in the sequel.

Both the RA(3)LI model (2.2) and the implied VAR(3) model have been simulated. The models have been simulated under parameter heterogeneity across countries and under regression parameter homogeneity. The parameter values used in the simulations for the AR(3)LI model are set equal to the OLS estimates of model (2.2), taking x_{1_t} to be the vector of observed GM_{it-1} and $x_{2t} = WR_{t-1}$. Under parameter homogeneity, OLS estimates of the equation in (2.2) for the United States are used for all six countries. The parameter values of the AR(3) model are derived from those of the AR(3)LI model under, respectively, parameter heterogeneity and homogeneity. On the basis of the data for the six countries, parameter homogeneity is not rejected for model (2.1) when testing regression parameter equality across countries using an F-test. Obviously, McElroy's criteria do not reject pooling either. The details for these tests are given in table 18.4.

The empirical distribution of the F-statistic in (1.13) and rejection frequencies were obtained by simulation. The number of runs is 1,000. Results on the empirical distributions of the F-statistic in (1.13) are not reported here. In most instances, the empirical distributions resemble an F-distribution. Rejection frequencies when the test statistic is compared with the critical value of, respectively, a central F-distribution with q and $n - q$ df and that of a non-central $F(q, n - q, \lambda_T)$ with $\lambda_T = q/2$ are reported in table 18.5. A nominal significance level of 5 percent is used.

Under parameter homogeneity, the rejection frequencies are very small when McElroy's second weak criterion is tested. With the exception of the AR(3) model with unrestricted disturbance covariance matrix, the rejection frequencies when a central F-distribution is used are also substantially smaller than 5 percent for the model under parameter homogeneity. An F-test appears to be too conservative whether the correct disturbance covariance is assumed or not.

Table 18.5 *Rejection frequencies for the F-test at a nominal significance level of 5 percent*

			Homogeneous		Heterogeneous	
T	Estimator	Σ	$\lambda=0$	$\lambda=q/2$	$\lambda=0$	$\lambda=q/2$
25	AR(3)LI	$\sigma^2 I$	0.024	0.000	0.722	0.251
50	AR(3)LI	$\sigma^2 I$	0.010	0.000	0.996	0.871
100	AR(3)LI	$\sigma^2 I$	0.023	0.000	1.000	1.000
25	AR(3)	$\sigma^2 I$	0.017	0.000	0.270	0.038
50	AR(3)	$\sigma^2 I$	0.019	0.001	0.672	0.207
100	AR(3)	$\sigma^2 I$	0.032	0.000	0.984	0.810
25	AR(3)LI	Block	0.019	0.000	0.369	0.095
50	AR(3)LI	Block	0.007	0.000	0.821	0.354
100	AR(3)LI	Block	0.005	0.000	1.000	0.950
25	AR(3)	Full	0.254	0.026	0.623	0.266
50	AR(3)	Full	0.160	0.011	0.824	0.388
100	AR(3)	Full	0.112	0.002	0.983	0.829

Under parameter heterogeneity, as expected, the power is found to increase as the sample size increases. For values of T equal to or larger than 50, the rejection frequency is found to be fairly large (larger than 80 percent). For $T=25$, McElroy's second weak criterion rejects rather infrequently the incorrect parameter restrictions. As T increases, the gain resulting from trading off some bias against a decrease in the variance of the estimates decreases as expected on the basis of asymptotic theory.

Next, information on distributions of the mean and median RMSFE using, respectively, unrestricted and pooled parameter estimates is given in table 18.6. As expected, as T increases, the distributions become more concentrated. Left-skewness of a distribution means that the median (across countries) RMSFE of the pooled forecasts is larger than that of the forecasts based on unpooled estimates. In column (4) and (9) of table 18.6, we report the [proportion] of the number of times forecasts using unpooled estimates outperformed those based on pooled estimates. For $T=25$, the simulations indicate that pooling is appropriate even under parameter heterogeneity. For $T=50$, under parameter heterogeneity, it seems to be advisable to use pooled forecasts, based on a VAR model, which leaves a major part of the contemporaneous correlation in the disturbances and therefore yields forecasts that are genuinely less accurate than those for the AR(3)LI model. These findings are in line with the theoretical results presented in section 2 and the conclusions drawn previously for the F-tests.

Table 18.6 *Mean and median (median root mean-squared forecast errors RMSFE) for the simulations*

			Homogeneous					Heterogeneous				
			Mean		Median			Mean		Median		
T	Estimator	Σ	Unpooled	Pooled	Unpooled(1)	Pooled(2)	$\Delta < 0$	Unpooled	Pooled	Unpooled(1)	Pooled(2)	$\Delta < 0$
(1)	(2)	(3)	(4)	(5)	(6)	(7)	(8)	(9)	(10)	(11)	(12)	(13)
25	AR(3)LI	$\sigma^2 I$	0.016	0.015	0.016	0.014	0.149	0.016	0.016	0.016	0.016	0.448
50	AR(3)LI	$\sigma^2 I$	0.011	0.010	0.010	0.010	0.272	0.011	0.011	0.011	0.011	0.667
100	AR(3)LI	$\sigma^2 I$	0.007	0.007	0.007	0.007	0.309	0.007	0.008	0.007	0.008	0.782
25	AR(3)	$\sigma^2 I$	0.022	0.021	0.022	0.020	0.212	0.017	0.017	0.017	0.017	0.336
50	AR(3)	$\sigma^2 I$	0.015	0.015	0.015	0.014	0.286	0.012	0.012	0.012	0.018	0.467
100	AR(3)	$\sigma^2 I$	0.010	0.010	0.010	0.010	0.348	0.008	0.008	0.008	0.008	0.547
25	AR(3)LI	Block	0.015	0.015	0.015	0.015	0.309	0.016	0.016	0.016	0.016	0.611
50	AR(3)LI	Block	0.010	0.010	0.010	0.010	0.411	0.011	0.011	0.010	0.011	0.781
100	AR(3)LI	Block	0.007	0.007	0.007	0.007	0.420	0.007	0.008	0.007	0.008	0.832
25	AR(3)	Full	0.022	0.021	0.022	0.021	0.226	0.018	0.017	0.017	0.017	0.337
50	AR(3)	Full	0.015	0.015	0.015	0.014	0.299	0.012	0.012	0.012	0.012	0.485
100	AR(3)	Full	0.010	0.010	0.010	0.010	0.354	0.008	0.008	0.008	0.008	0.560

Note: $\Delta < 0$ represents the proportion of the number of times that the forecasts based on unpooled (1) estimates outperforms those based on pooled (2) estimates.

2.3 *Analyzes of international data*

In this section the results of an empirical analysis of the pooling restrictions for model (2.1) and the associated VAR(3) model using international data for eighteen countries are presented.

Tests of the pooling restrictions $H_0 : R\beta_a = 0$ against the alternative $H_1 : R\beta_a \neq 0$ are reported in table 18.7 for the AR(3)LI model and the AR(3) model, respectively, for estimation periods varying from 1961–80 to 1961–90. We report the values of the test-statistic F^* given in (1.13) and the p-values for the asymptotically justified tests of exact linear restrictions $R\beta_a = 0$ and of the MSE criterion for these linear restrictions. The matrix Σ is assumed to be, respectively, $\Sigma = \sigma^2 I_N$, diagonal, and block-diagonal (as explained previously).

The values of the F-statistic are given in column (2) of table 18.7. Column (3) contains the value of the non-centrality parameter θ_n given in (1.13). In columns (4)–(6), the p-values are reported for the tests of $H_0 : \lambda = 0$ versus $H_1 : \lambda \neq 0$, and $H_0 : \lambda \leq \lambda^*$ against $H_1 : \lambda > \lambda^*$ for λ^* being equal to $q/2$. The two criteria correspond to the tests of McElroy's strong criterion and second weak criterion, respectively. For the AR(3)LI and the AR(3) models, under the assumption that $\Sigma = \sigma^2 I_N$ or that Σ is diagonal, an F-test usually does not lead to rejection of the pooling restrictions $R\beta_a = 0$. Consequently, the less stringent restrictions of the second weak MSE test for pooling given by McElroy (1977) are not rejected either in these two cases. A similar conclusion is reached if the first weak criterion is used to test for pooling. The p-values are never lower than 0.64 when the estimation period is varied from 1961–80 to 1961–90. McElroy's second weak pooling criterion is not rejected in general, except for the AR(3) model using an unrestricted covariance matrix Σ. Notice that a similar conclusion holds for the first weak criterion. Moreover, the pooled AR(3) model has to be rejected when compared with the pooled AR(3)LI model.

The true size of a test based on an "F-statistic" using a spherical Σ will be different from the assumed size when the true Σ is non-spherical. The evidence for the F-test in table 18.7 for $\Sigma = \sigma^2 I_N$ or Σ being diagonal supports the null hypothesis because neglecting residual correlation in estimation generally leads to rejection frequencies for the null hypothesis that are much lower than the nominal size of the test (e.g. see Palm and Sneek 1984 for results on the F-test in a regression model when neglecting serial correlation in the disturbance).

When Σ is estimated as a block-diagonal matrix in the AR(3)LI model, $H_0 : R\beta_a = 0$ usually has to be rejected at conventional significance levels. The less restrictive second weak MSE tests do not lead to rejecting

Table 18.7 *Testing for pooling in the AR(3)LI and AR(3) model*

	AR(3)LI				*AR(3)*			
Year	*F*-value	θ	*p*-value $\lambda = 0$	*p*-value $\lambda = 68$	*F*-value	θ	*p*-value $\lambda = 0$	*p*-value $\lambda = 34$
		Σ: $\sigma^2 I$				Σ: $\sigma^2 I$		
1981	1.11	4.14	0.24	0.98	0.92	5.97	0.66	0.99
1982	1.06	3.81	0.35	0.99	0.78	5.52	0.89	1.00
1983	1.10	7.89	0.26	0.98	0.75	7.59	0.92	1.00
1984	0.97	6.97	0.57	1.00	0.78	7.74	0.90	1.00
1985	0.97	7.04	0.57	1.00	0.78	7.51	0.89	1.00
1986	0.97	6.79	0.58	1.00	0.79	7.38	0.88	1.00
1987	0.94	6.53	0.66	1.00	0.82	7.23	0.84	1.00
1988	1.01	7.03	0.46	1.00	0.81	7.55	0.86	1.00
1989	1.03	8.35	0.42	1.00	0.84	7.48	0.81	1.00
1990	1.04	8.11	0.38	1.00	0.86	7.42	0.78	1.00
		Σ: *diagonal*				Σ: *diagonal*		
1981	1.39	10.35	0.01	0.70	1.21	10.27	0.14	0.87
1982	1.20	9.87	0.10	0.93	0.98	10.68	0.52	0.99
1983	1.28	9.64	0.05	0.86	0.88	10.04	0.73	1.00
1984	1.20	12.46	0.10	0.94	0.94	9.67	0.62	0.99
1985	1.15	11.85	0.16	0.97	0.94	9.33	0.61	0.99
1986	1.13	10.14	0.18	0.97	0.87	9.60	0.76	1.00
1987	1.06	9.46	0.34	0.99	0.87	9.10	0.75	1.00
1988	1.10	8.59	0.24	0.99	0.82	9.73	0.84	1.00
1989	1.11	10.04	0.23	0.98	0.86	9.49	0.78	1.00
1990	1.15	10.12	0.15	0.97	0.86	9.49	0.77	1.00
		Σ: block				Σ: full		
1981	1.63	4.89	0.00	0.29	4.55	4.38	0.00	0.00
1982	1.34	4.75	0.02	0.78	3.80	5.10	0.00	0.00
1983	1.63	5.32	0.00	0.28	2.40	4.98	0.00	0.00
1984	1.52	2.24	0.00	0.45	1.75	4.80	0.00	0.18
1985	1.35	5.22	0.02	0.77	1.61	4.67	0.00	0.32
1986	1.41	5.81	0.00	0.66	2.49	5.37	0.00	0.00
1987	1.16	5.58	0.14	0.96	2.16	5.59	0.00	0.01
1988	1.04	6.05	0.38	1.00	2.01	5.72	0.00	0.04
1989	0.95	7.44	0.64	1.00	2.12	5.72	0.00	0.02
1990	0.95	7.34	0.63	1.00	2.17	5.53	0.00	0.01

H_0 in this case. For the AR(3) model, with full disturbance covariance matrix, the *p*-values for the pooling restrictions are very small. In this case, the second weak MSE criterion also leads to rejecting pooling (see table 18.7). Note that we are using asymptotically justified procedures in relatively small samples. The procedures are asymptotically justified because Σ has to be estimated and because of the presence of lagged dependent variables among regressors. There is an earlier literature on the incorrect sizes of asymptotic tests indicating that these tests reject the null hypothesis too often in finite samples. The simulations in subsection 2.2, however, indicate that asymptotic theory provides rather good guidance in small samples. F-statistics neglecting the presence of residual correlation are expected to reject the null hypothesis less often than they should according to the nominal size. Therefore, we conclude that, on the whole, the evidence from table 18.7 supports the pooling restrictions.

This conclusion is supported by the results given in table 18.8. For the forecast period 1983–90, pooling leads to a substantial reduction in the RMSFE for the AR(3) and the AR(3)LI model when a pooled GLS estimator is used with, respectively, estimated full and block-diagonal covariance matrices. For the forecast period 1983–90, GLS-based unpooled forecasts perform slightly less [well] than OLS-based unpooled forecasts.

3 Conclusions

In this chapter, we studied the problem of whether forecasts of a set of panel data generated by models with similar but not necessarily identical parameter structures can be improved by using pooled parameter estimates. Results obtained by McElroy (1977) for a regression model with non-spherical disturbances can be generalized in a straightforward way to apply to systems of regression models used to study panel data with large T and fixed N. The gain in forecast and estimator performance measured by the reduction in MSFE or MSE results from a trade-off between the bias implied by the use of (slightly) false pooling restrictions and the reduction in the covariance matrix of the estimators due to imposing these restrictions. Moreover, as the sample size increases, the covariance matrices of restricted and unrestricted estimates converge to constant matrices but the bias of the restricted estimator (multiplied by $\sqrt{n}$) increases without bound. Therefore, beyond some given sample size, the forecasts based on unrestricted estimates will outperform the pooled forecasts.

Our simulation results show that for small and moderate values of T, reductions in MSFE can be achieved through pooling, even under parameter heterogeneity. The asymptotic properties of the pooling criteria put forward by McElroy (1977) provide a fairly accurate insight into their properties for finite T.

We applied these results to growth rates for eighteen OECD countries for the periods starting in the 1950s until 1991 using models put forward by Garcia-Ferrer *et al.* (1987) and Zellner and Hong (1989). Our empirical findings can be summarized as follows.

Table 18.8 *Root mean-squared forecast errors (RMSFEs) for the AR(3)LI and AR(3) models*

	AR(3)LI						*AR(3)*					
	Block		Diagonal		$\sigma^2 I$		Full		Diagonal		$\sigma^2 I$	
Country	Unpooled	Pooled	Unpooled	Pooled	Unpooled	Pooled	Unpooled	Pooled	Unpooled	Pooled	Unpooled	Pooled
Canada	3.58	2.58	3.57	2.26	3.57	2.26	2.64	1.94	2.81	2.16	2.81	2.17
United States	2.45	2.75	2.64	2.34	2.64	2.39	2.04	1.80	2.13	2.23	2.13	2.27
Australia	4.23	3.09	3.97	2.73	3.97	2.67	3.32	3.69	3.69	3.53	3.69	3.50
Japan	2.91	1.52	3.21	0.90	3.21	0.87	1.97	0.79	0.78	0.91	0.78	0.88
Denmark	3.65	3.24	3.46	2.91	3.46	2.95	1.89	1.69	2.13	1.65	2.13	1.72
Finland	5.17	3.07	4.27	2.50	4.27	2.50	2.28	2.31	2.19	2.26	2.19	2.26
Norway	4.36	3.46	4.01	3.51	4.01	3.54	7.14	2.19	4.36	2.31	4.36	2.34
Sweden	2.17	2.65	2.08	2.28	2.08	2.27	2.83	2.02	1.87	1.88	1.87	1.91
Belgium	2.93	2.67	3.19	2.03	3.19	2.05	2.43	2.01	1.39	1.53	1.39	1.51
France	3.29	2.77	1.75	2.22	1.75	2.28	3.13	1.65	1.21	1.27	1.21	1.27
Germany	3.43	2.68	3.21	2.61	3.21	2.69	3.62	2.96	2.64	2.71	2.64	2.70
Netherlands	2.23	2.27	2.03	1.81	2.03	1.88	4.24	1.32	1.23	0.88	1.23	0.87
Ireland	4.02	2.94	4.10	2.36	4.10	2.38	5.01	1.91	2.94	1.69	2.94	1.66
United Kingdom	6.40	4.10	5.67	3.93	5.67	3.89	2.83	2.56	2.45	2.55	2.45	2.57
Austria	2.17	2.17	2.52	1.76	2.52	1.74	1.89	1.10	0.95	0.80	0.95	0.79
Switzerland	2.14	2.50	2.23	1.95	2.23	1.95	1.69	1.41	1.47	1.09	1.47	1.10
Italy	3.98	2.24	3.49	1.95	3.49	1.99	1.29	1.36	1.46	1.11	1.46	1.11
Spain	3.63	1.70	3.43	1.26	3.43	1.23	3.00	1.18	0.92	1.04	0.92	1.02
Median	3.51	2.67	3.32	2.27	3.32	2.27	2.73	1.85	2.00	1.67	2.00	1.69

First, there is contemporaneous residual correlation in the form of a block-diagonal structure corresponding to regional groups present in the models for the eighteen countries.

Second, when formally tested using an estimated residual covariance matrix, the pooling restrictions and the MSE criteria for pooling put forward by McElroy (1977) are rejected only for the AR(3) model. They are not rejected when a diagonal or an identity residual covariance matrix is used. We should bear in mind that asymptotically justified *F*-test criteria tend to reject too often in finite samples. Moreover, the pooling restrictions and MSE criteria for eighteen countries were jointly tested even not allowing for individual fixed effects in the form of country-specific intercepts. Comparing pooled and unpooled models using posterior odds is probably a sensible alternative in relatively small samples to asymptotically justified test criteria.

Third, in actual forecasting, the median MSFE of OLS-based pooled forecasts is found to be smaller than that of OLS-based individual forecasts. A fairly large sample size is needed for the OLS-based pooled forecasts to be outperformed by a forecast based on unrestricted estimates. Using unpooled GLS with an estimated residual covariance matrix leads to slightly improved forecast performance. Pooled GLS-based forecasts have a much lower median MSFE than pooled OLS-based forecasts. Although we did not present results for shrinkage procedures, we like to note that shrinkage forecasts are convex combinations of individual and pooled forecasts. Therefore, results for shrinkage forecasts lie between the two polar cases of forecasts based on unrestricted estimates and those based on pooled estimates. Our findings parallel results obtained by Blattberg and George (1991). When modeling sales using a chain-brand model, they found that GLS added little to their data, whereas pooling and shrinkage estimation procedures provided superior estimates to OLS. Finally, the question of whether restricting the contemporaneous residual correlations to be the same within groups of countries and possibly across groups leads to further improvement of GLS-based pooled forecasts remains to be investigated.

APPENDIX DATA

The data set used is an updated set as used by Min and Zellner (1993) and consists of annual postwar data for eighteen OECD countries for the period 1948–1990. The data are obtained from the main IMF International Financial Statistics Data Base and contain the following four variables – (1) real stock prices, (2) an index of nominal stock prices as price index, (3) nominal money M1, and (4) GDP. Because of missing values and to make a fair comparison between the AR(3) model and the AR(3)LI model, we included only those years that could be used to estimate both models. The countries, with starting year given between parentheses, are Australia (1956), Austria (1949), Belgium (1953), Canada (1955), Denmark (1950), Finland (1950), France (1950), Germany (1950), Ireland (1948), Italy (1951), Japan (1953), the Netherlands (1950), Norway (1949), Spain (1954), Sweden (1950), Switzerland (1953), the United Kingdom (1957), and the United States (1955).

BIBLIOGRAPHY

Baltagi, B. H. (1995), *Econometric Analysis of Panel Data* (New York, John Wiley)

Bernard, A. B. and S. N. Durlauf (1995), "Convergence in international output," *Journal of Applied Econometrics* 10, 97–108

Bilson, J. F. O. (1981), "The 'Speculative Efficiency' Hypothesis," *Journal of Business* 54, 435–51

Blattberg, R. C. and E. I. George (1991), "Shrinkage estimation of promotional elasticities: seemingly unrelated equations," *Journal of the American Statistical Association* 86, 304–15

Breusch, T. S. and A. R. Pagan (1980), "The Lagrange multiplier test and its applications to model specifications in econometrics," *Review of Economic Studies* 47, 239–53

Chib, S. and E. Greenberg (1995), "Hierarchical analysis of SUR models with extensions to correlated serial errors and time-varying parameter models," *Journal of Econometrics* 68, 339–60

Garcia-Ferrer, A., R. A. Highfield, F. C. Palm, and A. Zellner (1987), "Macroeconomic forecasting using pooled international data," *Journal of Business and Economic Statistics* 5(1), 53–67; chapter 13 in this volume

Goodnight, J. and T. D. Wallace (1972), "Operational techniques and tables for making weak MSE tests for restrictions in regressions," *Econometrica* 40, 699–709

Haitovsky, Y. (1990), "Seemingly unrelated regression with linear hierarchical structure," in S. Geisser, J. S. Hodges, J. S. Press, and A. Zellner (eds.), *Bayesian and Likelihood Methods in Statistics and Econometrics* (Amsterdam, North-Holland, 163–77

Lindley, D. V. and A. F. M. Smith (1972), "Bayesian estimates for the linear model," *Journal of the Royal Statistical Society* Series B 34, 1–41

Maddala, G. S. (1991), "To pool or not to pool: that is the question," *Journal of Quantitative Economics* 7, 255–64

Maddala, G. S., R. P. Trost and H. Li (1994), "Estimation of short run and long run elasticities of energy demand from panel data using shrinkage estimators," Paper presented at the Econometric Society European Meeting, Maastricht, August 29–September 2

McElroy, M. B. (1977), "Weaker MSE criteria and tests for linear restrictions in regression models with nonspherical disturbances," *Journal of Econometrics* 6, 389–94

Min, C. and A. Zellner (1993), "Bayesian and non-Bayesian methods for combining models and forecasts with applications to forecasting international growth rates," *Journal of Econometrics, Annals* 56, 89–118; chapter 17 in this volume

Mittnik, S. (1990), "Macroeconomic forecasting using pooled international data," *Journal of Business and Economic Statistics* 8, 205–8

Nandram, B. and J. D. Petruccelli (1997), "A Bayesian analysis of autoregressive time series panel data," *Journal of Business and Economic Statistics* 15, 328–34

Palm, F. C. and J. M. Sneek (1984), "Significance tests and spurious correlation in regression models with autocorrelated errors," *Statistical Papers* 25, 87–105

Stock, J. H. and M. W. Watson (1989), "New indexes of coincident and leading economic indicators," in O. J. Blanchard and S. Fischer (eds.), *NBER Macroeconomics Annual 1989* (Cambridge, Mass., MIT Press), 351–94

Wallace, T. D. (1972), "Weaker criteria and tests for linear restrictions in regressions," *Econometrica* 40, 689–709

Wallace, T. D. and C. E. Toro-Vizcarrondo (1969), "Tables for the mean square error test for exact linear restrictions in regression," *Journal of the American Statistical Association* 64, 1649–63

Zellner, A. (1971), *An Introduction to Bayesian Econometrics* (New York, Wiley)

(1984), *Basic Issues in Econometrics* (Chicago: University of Chicago Press)

(1986), "On assessing prior distributions and Bayesian regression analysis with *g*-prior distributions," in P. K. Goel and A. Zellner (eds.), *Bayesian Inference and Decision Techniques: Essays in Honour of Bruno de Finetti* (Amsterdam, North-Holland), 233–43

(1994), "Time series analysis, forecasting, and econometric modelling: the structural econometric modelling, time series analysis (SEMTSA) approach," *Journal of Forecasting* 13, 215–33; chapter 4 in this volume

Zellner, A. and C. Hong (1989), "Forecasting international growth rates using Bayesian shrinkage and other procedures," *Journal of Econometrics, Annals* 40, 183–202; chapter 14 in this volume

Zellner, A., C. Hong, and C. Min (1991), "Forecasting turning points in international output growth rates using Bayesian exponentially weighted autoregression, time-varying parameter, and pooling techniques." *Journal of Econometrics* 49, 275–304; chapter 16 in this volume

19 Forecasting turning points in countries' output growth rates: a response to Milton Friedman (1999)

Arnold Zellner and Chung-ki Min

In our past work (Zellner, Hong, and Min, 1991), we used variants of a simple autoregressive-leading indicator (ARLI) model and a Bayesian decision theoretic method to obtain correct forecasts in about 70 per cent of 158 turning point forecasts for eighteen industrialized countries' annual output growth rates during the period 1974–86. IMF data for 1951–73 were employed to estimate our models that were then employed to forecast downturns and upturns in annual growth rates for the period 1974–86. When Milton Friedman learned of our positive results, in a personal communication he challenged us to check our methods with an extended data set. This is indeed an important challenge since it is possible that we were just "lucky" in getting the positive results reported above. Earlier, we recognized such problems in that we began our forecasting experiments with just nine countries' data and forecasted for the period 1974–81. Later, in Zellner and Hong (1989) and in Zellner, Hong, and Min (1991), we expanded the number of countries from nine to eighteen and extended the forecast period to 1974–86 to check that the earlier positive results held up with an expanded sample of countries and data. Fortunately, results were positive and now we report such new results for eighteen countries' revised data involving 211 turning point episodes during the forecast period 1974–90.

In table 19.1, the results of forecasting 211 possible turning points in eighteen countries' growth rates for the period 1974–90 are compared with earlier results for 158 turning point episodes in the same eighteen countries' growth rates for the period 1974–86. An upper turning point episode is defined as two successive annual growth rates below a third and the fourth either below the third, a downturn, or not below the third, no downturn. In a lower turning point episode, two successive annual growth rates are above the third and the fourth is either above the third,

Originally published in the *Journal of Econometrics* 88 (1999) 203–6. 0304-4076/99/$.

Table 19.1 *Forecasting turning points in rates of growth of real output growth rates for eighteen industrialized countries*[a]

Model	Forecasts of 211 turning points, rev. data, 1974–90[b] (percentage of correct turning point forecasts)	Forecasts of 158 turning points, 1974–86[c]
A No pooling		
1 TVP/ARLI	64	77
2 TVP/ARLI/WI	69	82
3 EW/ARLI	66	72
4 EW/ARLI/WI	67	76
5 FP/ARLI	64	74
6 FP/ARLI/WI	72	72
B With pooling		
1 TVP/ARLI	74	74
2 TVP/ARLI/WI	78	80
3 EW/ARLI	73	76
4 EW/ARLI/WI	79	81
5 FP/ARLI	74	75
6 FP/ARLI/WI	80	79

Notes:
[a] The countries are: Australia, Austria, Belgium, Canada, Denmark, Finland, France, Germany, Ireland, Italy, Japan, the Netherlands, Norway, Spain, Sweden, Switzerland, the United Kingdom and the United States. A downturn is defined to occur when two successive growth rates are below the third and the fourth in below the third while an upturn is defined to occur when two successive growth rates are above the third and the fourth is above the third.
[b] See Zellner, Hong, and Min (1991) for information about models and forecasting technique.
[c] Taken from Zellner, Hong, and Min (1991, table 2, p. 288).

an upturn, or not above the third, no upturn. In part A of . . . table [19.1], results for six different models estimated without pooling data across countries are presented. It is seen that for the current case of 211 turning point forecasts, the percentage of correct forecasts ranges from 64 to 72 while for the earlier case of 158 turning point forecasts, the percentage correct ranges from 72 to 82. Thus for the relations fitted individually without pooling, there appears to be a slight deterioration in performance. However, when the countries' relationships are fitted using pooling, the percentages of correct forecasts shown in part B of table 19.1 are very similar for cases of 211 and 158 turning point forecasts, namely 73–80 percent and 74–81 percent correct, respectively, for the earlier period and data and for the latter period and revised data. As found in previous work, use of Bayesian pooling techniques, here complete "shrinkage,"

Table 19.2 *Performance of naive forecasters and ARLI models in forecasting 211 turning points in eighteen countries' output growth rates, 1974–1990*

Forecaster[a]	116 down turn/no down turn forecasts (percentage of correct forecasts)	95 up turn/no up turn forecasts	211 turning point forecasts
1 Eternal optimist	36	64	49
2 Eternal pessimist	64	36	51
3 Deterministic four-year cycle	64	64	64
4 TVP/ARLI/WI	66	73	69
5 TVP/ARLI/WI pooled	77	79	78

Notes:
[a] See text for descriptions of the forecast procedures used by the eternal optimist, the eternal pessimist and the deterministic four-year cycle forecasters. TVP/ARLI/WI denotes a time-varying parameter autoregressive leading indicator variable model that includes a world income variable.

produces results that are better than those obtained from individually fitted relations.

The results in table 19.1 indicate that our turning point methods work well in both the old data set, 1974–86, and the revised, extended data set, 1974–90. Also, in independent calculations for the current chapter, Zellner, Tobias, and Ryu (1998), we have collected newly revised data extending to 1995 for the eighteen countries in our sample and found that our models and techniques for forecasting turning points continue to perform well.

Regarding other aspects of our turning point forecasting procedure, in our past work we compared the performance of our procedures with that of several naive turning point forecasters, namely (1) an "eternal optimist" who always forecasts "no down turn" and "up turn," (2) an "eternal pessimist" who always forecasts "down turn" and "no up turn" and (3) a "deterministic four-year cycle forecaster" who always forecasts "down turn" and "up turn". (See table 19.2 for the performance of these forecasters compared with the forecast performance of two of our ARLI models.) As is evident, the ARLI models' performance is superior to that of these naive forecasters and also to that of a coin flipper as shown in table 19.3. This use of naive models in evaluating turning point forecasting procedures parallels that of naive random walk and other such models that Christ, Friedman, Nelson, Plosser, Cooper, and others have employed to check the quality of macroeconometric and other models' point forecasts.

Table 19.3 *Coin-flipper's and models' expected number*[a] *and actual number of turning points, 1974–1990*

Model	Unpooled estimation	Pooled estimation
1 TVP/ARLI	131	134
2 TVP/ARLI/WI	125	129
3 EW/ARLI	106	136
4 EW/ARLI/WI	106	131
5 FP/ARLI	128	131
6 FP/ARLI/WI	121	122
7 Coin-flipper	105.5	105.5
Actual number of turning points	135	135
Total number of cases	211	211

Note:
[a] The expected number of turning points for each model is the sum of that model's probabilities of turning points for the 211 cases in which a turning point could occur.

We thank Milton Friedman for his constructive interest in our work and hope that the results reported herein satisfy his curiosity. Also, that the lagged rate of growth of real money is one of our important leading indicator variables is compatible with much of Friedman's well-known theoretical and empirical research in monetary economics. Use of this and other of our leading indicator variables was suggested by the fundamental empirical research of Burns and Mitchell in their classic work [1946] . . .

BIBLIOGRAPHY

Burns, A. F. and W. C. Mitchell (1946), Measuring Business Cycles using pre-World War II Data for the US, UK, French and German economies (New York, National Bureau of Economic Research)

Zellner, A. (1997), *Bayesian Analysis in Econometrics and Statistics: The Zellner View and Papers* (Cheltenham, Edward Elgar), http://www.e-elgar.co.uk and info@e-elgar.co.uk.

Zellner, A. and C. Hong (1989), "Forecasting international growth rates using Bayesian shrinkage and other procedures," *Journal of Econometrics, Annals* 40, 183–202; chapter 14 in this volume. Also published in A. Zellner, *Bayesian Analysis in Econometrics and Statistics: The Zellner View and Papers* (Cheltenham, Edward Elgar)

Zellner, A., C. Hong, and C. Min (1991), "Forecasting turning points in international output growth rates using Bayesian exponentially weighted autoregression, time-varying parameter, and pooling techniques," *Journal of Econometrics* 49, 275–304; chapter 16 in this volume. Also published in A. Zellner, *Bayesian Analysis* in *Econometrics and Statistics: The Zellner View and Papers* (Cheltenham, Edward Elgar)

Zellner, A., J. Tobias, and H. Ryu (1998), "Bayesian method of moments analysis of time series models with applications to forecasting turning points." [Published in *Estadistica Journal of the Inter-American Statistical Institute* 49–51 (152–157) (1997–1999), 3–63, with discussion by Enrique de Aelsa.] Manuscript in preparation.

Part IV

Disaggregation, forecasting, and modeling

20 Using Bayesian techniques for data pooling in regional payroll forecasting (1990)

James P. LeSage and Michael Magura

1 Introduction

This chapter adapts to the regional level a multi-country technique . . . used by Garcia-Ferrer, Highfield, Palm, and Zellner (1987) (hereafter GHPZ) and extended by Zellner and Hong (1987) (hereafter ZH) to forecast the growth rates in GNP across nine countries. We apply this forecasting methodology to a model of payroll formation in seven Ohio metropolitan areas. The technique applied to our regional setting involves using a Bayesian shrinkage scheme that imposes stochastic restrictions that shrink the parameters of the individual metropolitan-area models toward the estimates arising from a pooled model of all areas. This approach is motivated by the prior belief that all of the individual equations of the model reflect the same parameter values. Lindley and Smith (1972) labeled this an "exchangeable" prior.

There are several reasons to believe that the multi-country, exchangeable-priors forecasting methodology introduced by GHPZ will be successful in our multi-regional setting. First, it is well known that dependencies exist among regional economies. Numerous econometric modeling approaches have been proposed to exploit this information. Most multi-regional models take a structural approach, employing linkage variables such as relative cost, adjacent-state demand, and gravity variables. Ballard and Glickman (1977), Ballard, Glickman, and Gustely (1980), Milne, Glickman, and Adams (1980), and Baird (1983) presented multiregional models of this type. LeSage and Magura (1986) investigated a non-structural approach using statistical time series techniques to link regional models. Second, as evidence for the existence

We thank an associate editor and two anonymous referees for helpful comments. In addition, Arnold Zellner provided comments on an earlier draft of this chapter. This research was supported by a grant from the Ohio Board of Regents Urban Universities Research Program to the Urban Affairs Center at the University of Toledo.

Originally published in the *Journal of Business and Economic Statistics* 8(1) (1990), 127–135.

of interdependencies, GHPZ pointed to similar movements in real output growth rates induced by broad business cycle swings across their nine-country sample. We find similar comovements in payroll variation across our seven metropolitan areas. The idea of shrinkage toward the pooled estimates from all metropolitan areas is motivated by these patterns of similar movement, which result in much contemporaneous correlation between the individual time series. Finally, one problem with many of the previously proposed procedures for linking regions is that many variables that are highly correlated tend to enter these models. The collinear relations may degrade the precision of the estimates and result in poor forecasts. The shrinkage aspect of the exchangeable priors procedure should improve forecasting performance, since our model likewise exhibits collinear relations among the leading indicator variables used as explanatory variables.

Our approach, like that of GHPZ, represents a "bottom-up" attempt to build a simple model based on the time series properties of the payroll data that will forecast well. This is in contrast to the metropolitan payroll forecasting work of Liu and Stocks (1983), which we would classify as "top-down." They attempted to build elaborate models of individual metropolitan economies to approach the multi-regional forecasting problem. One focus of this study then is to determine if the same exchangeable-priors procedures employed by GHPZ and ZH (in a multi-country setting) can produce superior forecasts in a multi-regional setting.

Another focus of this study is to provide a more detailed analysis of the sources of improvement in the forecasting performance of the GHPZ approach to estimating the model. This GHPZ procedure can be viewed as modifying the traditional ordinary least squares (OLS) estimator in two ways. First, the procedure introduces multi-regional data information about the linkages that exist between the regions by using the pooled estimates as the mean of a Bayesian prior. Introducing this multi-regional information should provide one source of improved forecasting performance. Second, the shrinkage toward the pooled data prior mean produces a corresponding augmentation of the smallest eigenvalues of the data matrix during estimation. It is well known (Belsley, Kuh, and Welsch 1980) that collinear relations of the type found among the explanatory variables in these models will produce a data matrix with very small eigenvalues. Thus this augmentation may help stabilize the estimates and result in better forecasts. Our analysis attempts to separate out the forecasting value of these two aspects of the GHPZ procedure. We do this by comparing the forecasting performance of the model estimated with the GHPZ procedure to that of the model estimated using a simple ridge estimator.

The ridge estimator allows shrinkage and the corresponding augmentation of the smallest eigenvalues of the data matrix to take place without reference to the pooled data information, since this procedure pulls the least squares estimates toward a prior mean of 0. The experiments carried out here suggest that, for our sample of seven metropolitan payroll models, the forecasts produced by ridge estimates of the model are equal to or better than those produced by the GHPZ and ZH estimates.

The article proceeds as follows. Section 2 describes the model that is based on a leading indicator approach to forecasting payroll. Section 3 briefly describes the techniques proposed by GHPZ and ZH and discusses their relation to the ridge estimator. Section 4 presents the results from forecasting experiments based on five alternative approaches to estimating the model. We compare the forecasts produced by the model using least squares estimation, ridge estimation, and the two exchangeable-priors Bayesian estimation procedures proposed by GHPZ (1987) and ZH (1987). [Section 5 briefly concludes.]

2 Analysis of the data

The time series to be forecast represent quarterly total payroll of firms covered by unemployment insurance in each of seven Ohio metropolitan areas: Akron, Cincinnati, Cleveland, Columbus, Dayton, Toledo, and Youngstown. Since all seven local governments collect quarterly payroll taxes, their budget officials have a great deal of interest in payroll forecasts. In many cases, the payroll tax revenue represents more than half of the total budget of these local governments. Metropolitan-area payroll data are available from the Ohio Bureau of Employment Services (OBES), Labor Market Information (LMI) Division. One Ohio metropolitan area, Canton, was excluded from this study because the quality of the sample data for this region is suspect according to sources at the LMI Division. The individual parameter estimates for the Canton area were drastically different from those of all other areas, confirming the suspect nature of these data. The data cover the period 1978[:1] through 1987[:3], and are in nominal terms. These data were transformed to payroll growth rates by taking first differences of the logged data. Following GHPZ, the use of such transformed data in the estimation of the model results in parameters for the individual metropolitan payroll equations that are on the same order of magnitude. The requirement that the parameters be comparable in magnitude is necessitated by the procedure's attempt to shrink the parameters of individual metropolitan relations toward the pooled value of the parameter estimates for all seven metropolitan areas.

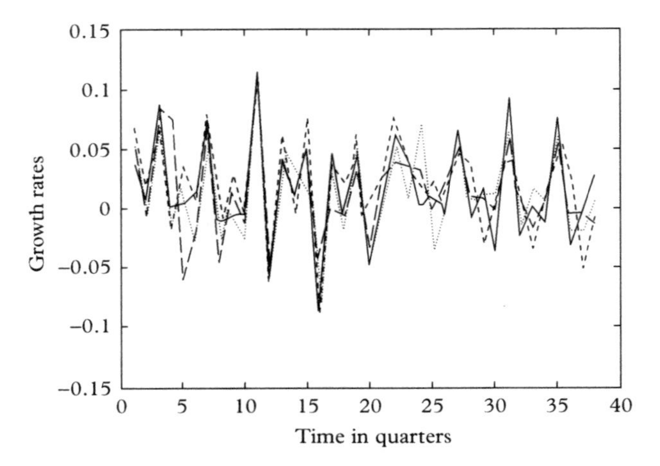

Figure 20.1 Payroll growth rates for the metropolitan areas

Figure 20.1 shows a plot of the growth rates in payroll for the seven metropolitan areas over the time period 1978:1–1987:3. It seems clear that these regional economies exhibit similar patterns or comovements. It is this information concerning common influences from seasonality, business cycles, and so forth that the procedures proposed by GHPZ attempt to exploit. The period covering 1979:1–1985:3 was used to estimate the models and the data covering 1985:4–1987:3 was used to provide eight quarters of out-of-sample forecasting experience. A vertical grid in the plot separates the forecast period from the period used to fit the models.

The specification of the payroll model used here is based on a leading indicator model for Ohio metropolitan-area employment developed by LeSage and Magura (1987), the results of which [were] . . . published on a monthly basis by the OBES beginning in 1989. The model is similar to that of GHPZ in that it includes a number of leading indicator explanatory variables.

Three different versions of the model to be discussed were used to generate one-, two-, and three-step-ahead dynamic forecasts during the out-of-sample period. The model that generated one-step-ahead forecasts was based on two lags of the dependent variable and a set of leading indicator variables that were lagged one period, the one-step-ahead version of this model is

$$\begin{aligned} P_{it} = {} & \beta_0 + \beta_1 P_{it-1} + \beta_2 P_{it-2} + \beta_3 S_{it-1} + \beta_4 C_{it-1} \\ & + \beta_5 N_{it-1} + \beta_6 H_{it-1} + \beta_7 A_{t-1} + \beta_8 L_{t-1} + \varepsilon_{it}, \end{aligned} \tag{2.1}$$

where $i = 1, \ldots, 7$ denotes the seven metropolitan areas, P_{it} is payroll in the ith city at time t, S_{it} is housing starts in the ith city at time t, C_{it} is unemployment insurance initial claims in the ith city at time t, N_{it} is total metropolitan employment in the ith city at time t, H_{it} is total metropolitan average weekly hours in the ith city at time t, A_t is national domestic auto sales at time t, L_t is the national index of twelve leading indicators at time t, and ε_{it} is a Gaussian disturbance term for city i at time t.

The appendix (p. 634) contains a description of the leading indicator variables and the sources for these data. A detailed description of the methodology used to determine the variables in the model and the specification of a second-order autoregression can be found in LeSage and Magura (1987).

GHPZ and ZH considered only one-step-ahead forecasts. It was of interest to see how their exchangeable-priors approach would perform over longer forecast horizons, such as two or three steps ahead. To generate two-period-ahead forecasts, the same leading indicator explanatory variables were used as in (2.1), but a two-period lag of these variables was used. This allowed forecasts extending two steps ahead to be produced without the need to develop separate equations to forecast the leading indicator explanatory variables. A similar approach was taken with the model that generated forecasts extending three steps ahead. These two models are

$$\begin{aligned} P_{it} = {} & \beta_0 + \beta_1 P_{it-1} + \beta_2 P_{it-2} + \beta_3 S_{it-2} + \beta_4 C_{it-2} \\ & + \beta_5 N_{it-2} + \beta_6 H_{it-2} + \beta_7 A_{t-2} + \beta_8 L_{t-2} + \varepsilon_{it}, \end{aligned} \tag{2.2}$$

and

$$\begin{aligned} P_{it} = {} & \beta_0 + \beta_1 P_{it-1} + \beta_2 P_{it-2} + \beta_3 S_{it-3} + \beta_4 C_{it-3} \\ & + \beta_5 N_{it-3} + \beta_6 H_{it-3} + \beta_7 A_{t-3} + \beta_8 L_{t-3} + \varepsilon_{it}. \end{aligned} \tag{2.3}$$

The models in (2.1)–(2.3) represent an attempt to explain variation in payroll growth rates by using "suites" of variables denoting national, local, and autoregressive influences. The index of leading indicators and automobile sales reflect national influences, and metropolitan employment, hours, unemployment claims, and housing starts reflect local influences. The lagged payroll terms capture autoregressive influences.

Following GHPZ, the models in (2.1)–(2.3), along with some simpler variants of these models, were estimated and forecast. The various versions can be summarized as follows:

1. A naive model that used no change as a forecast; that is, $\hat{P}_{it} = 0$ (labeled NM1 in table 20.1)
2. A naive model that used last period's payroll growth rate as the forecast; that is, $\hat{P}_{it} = P_{it-1}$ (labeled NM2 in table 20.1)

Table 20.1 *Average root mean-squared error forecast (RMSE) of forecast over 1985:4–1987:3*

Model	Akron	Cincinnati	Cleveland	Columbus	Dayton	Toledo	Youngstown
One-step-ahead forecasting model							
NM1	3.00	2.93	2.71	2.35	2.62	3.17	2.39
NM2	6.17	5.08	5.06	3.35	4.33	4.75	4.07
AR2	2.97	1.91	2.56	1.96	2.20	3.79	2.43
NAT	2.56	1.27*	1.90	1.17*	1.39*	2.82*	2.34*
Full-1	1.99*	1.86	1.77*	1.29	1.62	2.85	3.00
Two-step-ahead forecasting model							
NM1	2.67	2.69	2.47	2.08	2.39	3.20	2.31*
NM2	6.04	5.21	4.88	3.27	4.36	4.85	4.22
AR2	2.98	2.34	2.64	2.38	2.29	3.56	2.50
NAT	3.03	2.18	2.60	2.17	2.22	3.71	2.40
Full-2	1.18*	1.12*	1.49*	1.58*	1.93*	3.11*	3.05
Three-step-ahead forecasting model							
NM1	2.67	2.76	2.47	2.11	2.48	3.27	2.48
NM2	5.90	5.24	4.76	3.23	4.40	4.67	4.46
AR2	2.98	2.37	2.78	2.17	2.24*	3.51	2.59
NAT	3.21	2.67	3.13	1.95*	2.33	3.78	2.58
Full-3	1.58*	1.50*	2.42*	2.09	3.11	2.79*	2.06*

Note:
* The model with the smallest average RMSE for each forecast horizon.

3. An autoregressive model of order 2 – that is, consisting of a constant term and two lagged values of payroll (labeled AR2 in table 20.1)
4. An autoregressive order 2 model plus the national index of leading indicators and domestic automobile production variables (labeled NAT in table 20.1)
5. The full model from (2.1)–(2.3) (labeled Full-1, Full-2, and Full-3 in table 20.1).

The simpler models represent an attempt to analyze by decomposition the information content contained in the "suites" of variables representing national, local, and autoregressive influences.

Table 20.1 presents a comparison of the forecasting ability of the alternative models for one-, two-, and three-step-ahead forecasting horizons. The forecasts used to produce table 20.1 were generated by estimating the models with least squares using data from 1979:1 to 1985:3, calculating a forecast, then updating the estimates with an additional quarterly observation and calculating another forecast. This process of updating the estimates before each forecast was continued until the end of the sample in 1987:3. The one-step-ahead forecast average root-mean squared errors (RMSEs) were calculated by averaging eight one-quarter forecasts extending from 1985:4 through 1987:3. The two-step-ahead average RMSEs were based on the average of seven two-quarter forecasts, and the three-step-ahead average RMSEs were calculated as the average of six sets of three-quarter forecasts. The shorter forecast intervals arose because the two-step-ahead model arrives at the end of the eight-quarter forecast period after producing seven two-quarter-ahead forecasts and, similarly, the three-step-ahead model reaches the end of the eight-quarter forecast period after only six three-quarter-ahead forecasts.

In table 20.1 the model with the smallest average RMSE for each forecast horizon is denoted by an asterisk. Examining the models that produce the smallest errors, we see that the addition of national and local information to the models greatly reduces the forecast error relative to the naive and autoregressive models. For the one-step-ahead forecasting model, the NAT model produced the lowest errors in five of seven cases and the Full-1 model in the remaining two cases. For the two-step-ahead forecasting model, the Full-2 model produced the best forecasts for six of the seven cities. The three-step-ahead models also demonstrated that better forecasts were associated with the Full-3 model relative to the simpler models, producing the best forecasts in five of the seven cities.

The experiments we carry out in section 4 use the forecasting performance of the least squares estimated Full models from table 20.1 (along with the best model when this is different from the Full model) as a benchmark against which to compare the improvement in forecasting performance produced by models estimated with the ridge and the GHPZ and ZH techniques. Note that, with regard to any enumeration of which model is best, some of the average percentage RMSEs are not very different from one model to the next, making any type of counting scheme somewhat deceptive.

3 The Bayesian pooling technique

This section describes and motivates the exchangeable-priors Bayesian techniques for pooling data information introduced by GHPZ and modified by ZH. These procedures are described with reference to our metropolitan-area payroll models.

The GHPZ estimator is composed of the following formulation. Let X_i denote the matrix of explanatory variables in the ith metropolitan area and Y_i be the dependent variable vector for payroll in city i. Using this notation,

$$\hat{\beta}_i^{\text{GHPZ}} = (X_i'X_i + \lambda_i I)^{-1}(X_i'X_i\hat{\beta}_i + \lambda_i\tilde{\beta}), \tag{3.1}$$

where $\hat{\beta}_i$ is the least squares estimate for metropolitan area i and $\tilde{\beta}$ is a pooled estimate:

$$\tilde{\beta} = (X_1'X_1 + X_2'X_2 + \cdots + X_7'X_7)^{-1} \times (X_1'Y_1 + X_2'Y_2 + \cdots + X_7'Y_7). \tag{3.2}$$

The expression for $\tilde{\beta}$ is a matrix-weighted average of the individual metropolitan data information that compactly summarizes the data variation occurring at the metropolitan level. This average summary information about the rest of the region is then mixed with the data contained in the X_i data matrix and Y_i vector for the ith metropolitan area to achieve the Bayesian estimate for the ith model. Equation (3.1) shows that the mixing of this information is again a matrix-weighted average of the information for the individual areas contained in X_i and Y_i with the pooled information contained in $\tilde{\beta}$. The λ_i parameter controls the relative weighting of the two types of information, that pertaining to the individual metropolitan area and that representing all areas pooled. This parameter can be given an interpretation as the relative confidence in the two types of information. Thus, as $\lambda_i \to \infty$, the individual estimates, $\hat{\beta}_i^{\text{GHPZ}}$, approach the pooled estimate, $\tilde{\beta}$; on the other hand, for very small values of λ_i the $\hat{\beta}_i^{\text{GHPZ}}$ estimates for city i approach $\hat{\beta}_i$, the least squares estimates based solely on the individual metropolitan-area information.

The value of λ_i was set for each equation by conditioning on the value of λ_i for which the RMSE was at a minimum; that is, we generated forecasts for a range of settings of the λ_i parameter for each metropolitan area and chose the value of λ_i that produced the minimum average percentage of RMSE over the eight-quarter forecast period. Since our objective function was to produce a forecasting model, this procedure is not unreasonable. This represents a generalization of the procedure suggested by GHPZ, since they used a single λ value for all countries in their sample. The generalization was later incorporated by ZH [1987]. It seems appropriate to use this approach because it should provide better forecasts by lessening the restriction imposed by choosing a single setting for the λ parameter for all of the metropolitan areas.

One point to note here is that a proper forecasting procedure would require that the value of λ_i be determined *a priori* rather than *ex post*. This might be done using forecasted realizations over some time period near the end of the available data sample. The optimal settings for the λ_i parameters determined in this way could then be used in the forecasting of future values. Our experience indicates that these parameters are not overly sensitive to the addition of a small number of observations, suggesting that there may be no need to recalibrate the model with respect to the λ_i over short time intervals.

If we alter the estimator $\hat{\beta}_i^{\mathrm{GHPZ}}$ to shrink towards a vector of 0 instead of the β-pooled estimate, we would have the traditional ridge estimator,

$$\hat{\beta}_i^{\mathrm{ridge}} = (X_i'X_i + \lambda_i I)^{-1}(X_i'X_i\hat{\beta}_i). \tag{3.3}$$

The ridge estimator could be given the same Bayesian interpretation as that of the GHPZ estimator but with a prior mean of 0 for the coefficient estimates. It is well known that in cases of severe collinearity the ridge estimator tends to outperform least squares. It does so by arbitrarily increasing the smallest eigenvalues in the data matrix $X'X$ using the λ_i parameter. One aspect of the GHPZ estimator is that it likewise performs this augmentation with a similar λ_i parameter. This reduces the ill conditioning associated with the least squares problem and creates estimates that are more stable and of greater precision.

We believe that an important question concerning the procedures proposed by GHPZ is: To what extent is the augmentation of the small eigenvalues that occurs in their estimator responsible for the improved forecasting performance? The importance of this question is that if the augmentation introduced by ridge estimation can produce forecasts of equal or better quality than the GHPZ procedure, it would be much simpler to implement computationally. Since ridge would seem to be a simple cousin of the estimator family in which the GHPZ estimator resides, it is curious that GHPZ did not explore this issue. In addition, if ridge produces forecasts of equal or better quality, it is indicative that augmentation of small eigenvalues without reference to the pooled data prior mean can accomplish the task of improving forecasting performance.

ZH proposed some modifications to the GHPZ procedure. They followed the "g-prior" approach of Zellner (1983, 1986) and arrived at the modified estimator

$$\beta_i^{\mathrm{ZH}} = (\hat{\beta}_i + \gamma_i\tilde{\beta})/(1 + \gamma_i). \tag{3.4}$$

This modification produces an estimator that is a simple weighted average of the least squares $\hat{\beta}_i$ and the pooled $\tilde{\beta}$ estimates with a weighting factor of γ_i.

This estimator does not augment the smallest eigenvalues in the data matrix and is different from the GHPZ estimator in this regard. This is an important point, since it suggests that the ZH estimator would not benefit from the augmentation of small eigenvalues inherent in the GHPZ or ridge estimators. An examination of the relative forecasting performance arising from these separate procedures may shed some light on the question we address concerning the value of augmenting the small eigenvalues.

Note that the ZH procedure can be viewed just as the GHPZ in that the γ_i parameter moves the estimate outcomes between the pooled, $\tilde{\beta}$, and the least squares, $\hat{\beta}_i$, magnitudes, the difference being that the ZH procedure does not augment the data matrix during this movement.

4 A comparison of the forecasts

In this section, we estimate the models shown in (2.1)–(2.3) in section 2 using five different procedures: GHPZ, ZH, OLS, ridge, and pooling. The pooled estimates were derived by using (3.2); that is, the data for each metropolitan area were "stacked" to form a single Y vector and X matrix containing all observations. A single $\tilde{\beta}$ vector is estimated using least squares on this pooled data.

Before presenting the results of these forecasting experiments, we consider the collinearity problem. A collinearity diagnostic procedure suggested by Belsley Kuh, and Welsch (1980) was employed to determine whether nearlinear dependencies existed between the columns of the explanatory variables matrix in the metropolitan-area models. This technique produces a variance-decomposition-proportions table based on a singular-value decomposition of the data matrix X_i for each metropolitan area. A necessary condition for a severe collinearity problem is indicated by a maximum-condition index (which represents the ratio of the largest eigenvalue to the smallest eigenvalue of the data matrix X) in excess of 30. Since the largest condition index reflects the ratio of the largest to the smallest eigenvalue in the data matrix, an index in excess of 30 indicates that the smallest eigenvalue is 1/30th the size of the largest eigenvalue. For the $X'X$ matrix of least squares, the eigenvalues would be those of the X matrix squared, so the relative sizes associated with a condition index of 30 would indicate that the smallest eigenvalue is 1/900th the size of the largest. The maximum-condition indexes for our seven metropolitan-area data sets were in excess of 90 for every metropolitan area, indicating that the smallest eigenvalue in the $X'X$ matrix is 1/8,100th the size of the largest. For the worst-conditioned metropolitan areas, where the condition index is on the order of 175, we have a smallest eigenvalue for our $X'X$ least squares matrix that is 1/30,000th the size of the largest. Thus the use of estimators that augment the small eigenvalues in the $X'X$ matrix could be useful in stabilizing the estimates and thereby improving forecast accuracy. It seems likely that collinearity would present a potential problem for most models of the type examined here and in GHPZ.

Table 20.2 *Average root mean-squared error forecast (RMSE) of forecast over 1985:4–1987:3 using pooled and other models*

Model	Akron	Cincinnati	Cleveland	Columbus	Dayton	Toledo	Youngstown
One-step-ahead forecasting model							
Best	1.99	1.27*	1.77	1.17*	1.39*	2.82	2.34
Full	1.99*	1.86	1.77	1.29	1.62	2.85	3.00
GHPZ	2.02	1.57	1.73*	1.19*	1.59	2.87	2.71
ZH	2.00	1.32	1.73*	1.29	1.53	2.85	2.78
Pool	2.41	1.56	2.14	1.24	1.65	3.02	2.71
Ridge	1.99	1.37	1.77	1.23	1.47	2.54*	2.30*
Two-step-ahead forecasting model							
Best	1.18	1.12	1.49	1.58	1.93	3.11	2.31
Full	1.18	1.12	1.49	1.58	1.93	3.11	3.06
GHPZ	1.15*	1.10	1.21	1.09	1.82	2.92	2.46
ZH	1.18	1.09*	1.18*	1.57	1.90	2.85	2.35
Pool	1.28	1.11	1.45	1.07	1.82	2.96	2.85
Ridge	1.21	1.09*	1.19*	1.03*	1.66*	2.43*	1.77*
Three-step-ahead forecasting model							
Best	1.59	1.51	2.42	1.95	2.39	2.80	2.06
Full	1.59*	1.51	2.42	2.09	3.11	2.80	2.06
GHPZ	1.87	1.29	2.38	1.45	2.35	2.75	2.26
ZH	1.59	1.38	2.03*	1.93	2.43	2.69*	2.02*
Pool	2.15	1.29*	2.50	1.43	2.31	2.84	2.70
Ridge	1.59	1.38	2.16	1.30*	2.04*	2.74	2.03*

Note:
* The model with the smallest average RMSE for each forecast horizon.

We now turn to a discussion of the forecasting experiments. These experiments involved estimating the three models shown in (2.1)–(2.3) and producing one-step-ahead, two-step-ahead, and three-step-ahead forecasts with the various estimation procedures. In implementing the GHPZ and ZH procedures, we varied the value of the individual λ_i and γ_i parameters for each metropolitan area over a large range of values, producing forecasts over the entire eight-quarter horizon for each setting of this parameter. The average RMSE of the forecasts were used to determine an "optimal" setting for these parameters in each city. More specifically, for the two- and three-step-ahead forecasting models, we averaged over the error of the two- and three-quarter horizon forecasts to determine the best value for these parameters.

The average RMSE results of our out-of-sample forecasting experiments are reported in table 20.2. To focus on the relative performance of the proposed procedures with that of the estimators shown in table 20.1, we have replicated the best out-of-sample forecast from table 20.1 in table 20.2 and labeled it "Best." In addition, we replicate the Full model least squares forecast RMSE from table 20.1 in table 20.2 to facilitate a comparison of how much improvement the alternative procedures yield relative to this model.

We chose to define a clear-cut advantage as at least an average RMSE difference of 0.03, which is somewhat arbitrary. Given that all of the average RMSE magnitudes in table 20.2 lie between 1.0 and 3.06, the 0.03 magnitude reflects a reduction in average RMSE of more than 1 percent but less than 3 percent. Another point to note when examining table 20.2 is that the ridge, GHPZ, and ZH techniques collapse on the least squares procedure as $\lambda_i \rightarrow 0$. When our optimal setting for this parameter converges to 0, we should, of course, count the least squares estimator (labeled "Full" in table 20.2) as producing the best forecasts. Similarly, the GHPZ and ZH techniques collapse on the pooled estimate as $\lambda_i \rightarrow \infty$, in which case we should credit the pooled estimator (labeled "Pool" in table 20.2) with producing the best forecasts. Using these criteria, we can summarize the table 20.2 results in the following way: In the one-step-ahead model, ridge was a clear-cut winner in two out of seven cities, and GHPZ and ZH tied in one out of seven cities. In the two-step-ahead model, ridge was a clear-cut winner in four out of seven cities, ridge and ZH tied in two out of seven cities, and GHPZ was a clear-cut winner in one out of seven cities. In the three-step-ahead model, ridge was a clear-cut winner in two out of seven cities, ZH was a clear-cut winner in two out of seven cities, and ZH and ridge tied in one out of seven cities.

Table 20.2 and the preceding summary indicate that the ridge estimator provided forecasting performance equal to or better than the techniques of GHPZ and ZH. From these results, it seems clear that there are potential gains to be had from adopting a technique that arbitrarily augments the small eigenvalues in these types of models. It is somewhat interesting that the ZH technique, which does not augment the eigenvalues, produced forecasting results that equal the ridge technique in three of the cases summarized. This issue needs further study.

A graphical depiction of the RMSEs for various λ_i settings for the two-step-ahead version of the model are shown in figures 20.2–20.8. The graphs for these models represent only one of the three versions of the model used to produce the results in table 20.2, but the graphs for the other versions are similar. Note that the numbers on the horizontal axes do not reflect the actual values of λ_i, since the magnitudes of these differed for each of the estimation procedures. The ticks on the horizontal axes merely reflect different values of λ_i, allowing us to track the RMSE of forecasts on the vertical axes as we vary these values for the different

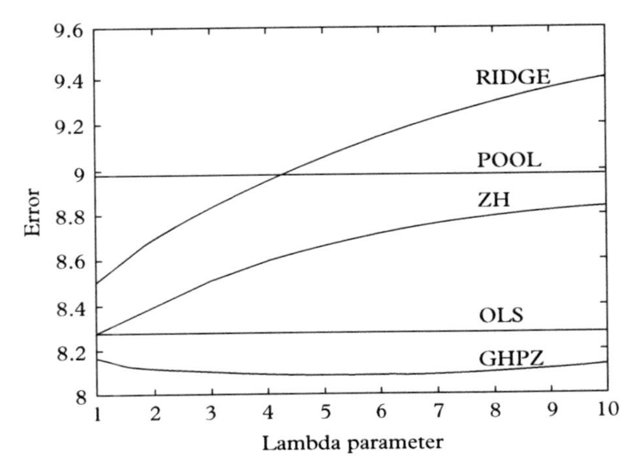

Figure 20.2 Forecast performance and the λ settings: Akron

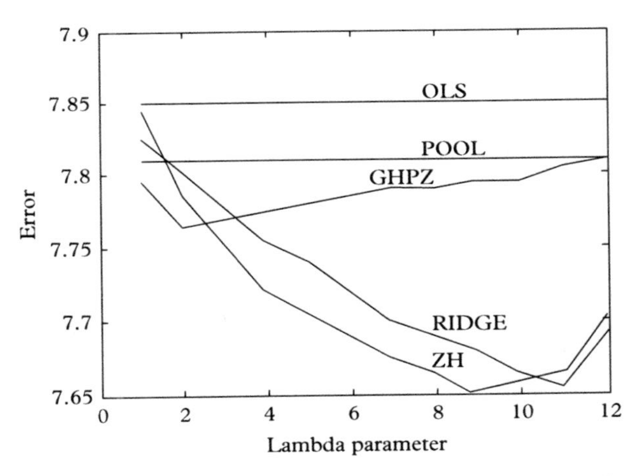

Figure 20.3 Forecast performance and the λ settings: Cincinnati

estimation procedures. The horizontal solid line in each metropolitan-area graph represents the average two-step-ahead RMSE for the least squares model, and the horizontal dashed line represents that for the pooled model. As we would expect, the GHPZ, ZH, and ridge models approach the least squares results in each metropolitan area for low settings of λ. Moreover, the GHPZ and ZH models approach the pooled results for large λ settings; the graph for the Dayton area makes this particularly clear.

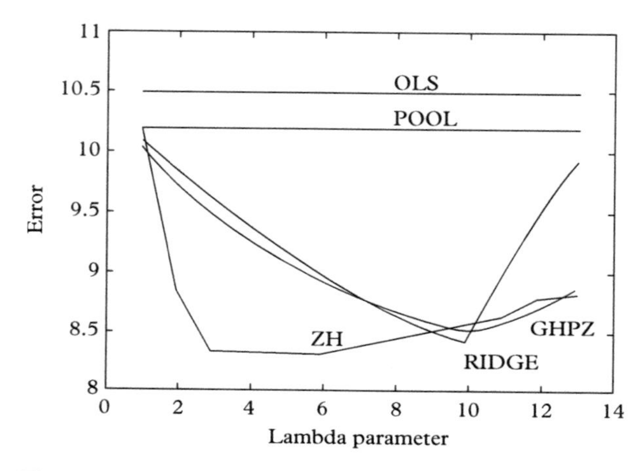

Figure 20.4 Forecast performance and the λ settings: Cleveland

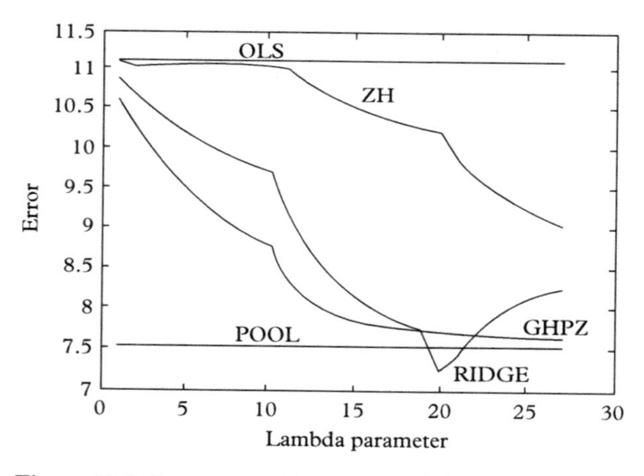

Figure 20.5 Forecast performance and the λ settings: Columbus

Given the RMSE information in table 20.2, we would expect to see three patterns emerging in the graphs. One pattern would be that in which the forecasting performance of ridge is clearly superior. We see this in three of the seven metropolitan areas, Dayton, Toledo, and Youngstown. A second pattern would be that where the GHPZ and ZH procedures produce superior forecasts. This occurs in only one case, Akron. Finally, there would be a pattern in which ridge, GHPZ, and ZH produce essentially the same forecast. We see this in three metropolitan areas, Cincinnati, Cleveland, and Columbus.

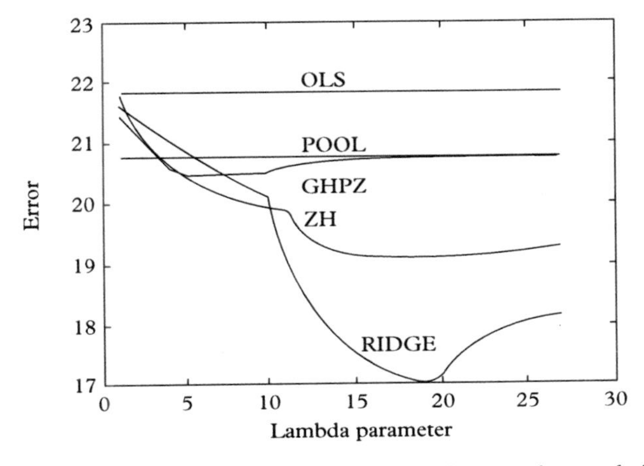

Figure 20.6 Forecast performance and the λ settings: toledo

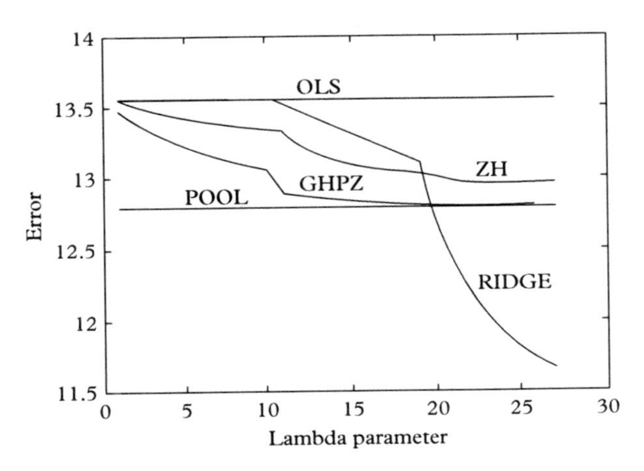

Figure 20.7 Forecast performance and the λ settings: Dayton

5 Conclusions

The forecasting performance associated with the exchangeable-priors estimation techniques proposed by GHPZ [1987] and ZH [1987] were compared with that arising from a simple ridge estimator using a model of local payroll formation. An important finding was that the ridge estimator produces forecasts equal to or better than the estimators proposed by GHPZ and ZH. We feel that GHPZ and ZH overlooked ridge estimation,

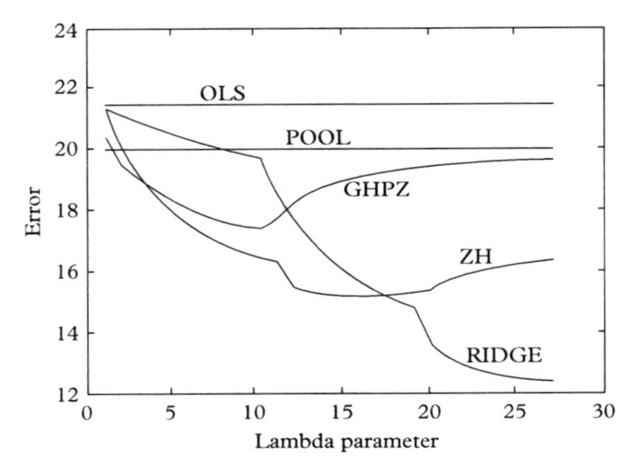

Figure 20.8 Forecast performance and the λ settings: Youngstown

which appears to be an important competing technique for these types of multi-regional or multi-country models in which severe collinearity exists.

This suggests that the value of the pooled data information may be limited in some circumstances. GHPZ may have incorrectly attributed the improved forecasts to the shrinking toward the pooled estimates, when in fact our findings indicate that, for our data set, shrinkage toward a vector of zeros (ridge) can produce forecasts equal to or better than those generated by shrinkage toward the pooled estimate (GHPZ and ZH).

Finally, we believe that the procedures introduced in the original multi-country studies of GHPZ [1987] and ZH [1987] represent valuable approaches to compactly summarizing multi-regional information among national or local regions. We do, however, feel that the findings of this study suggest an important role for the ridge estimator as a benchmark against which to judge the relative performance of the new techniques proposed by GHPZ and ZH.

APPENDIX DATA AND DATA SOURCES

Total payroll

Total payroll for the metropolitan areas of Akron, Cleveland, Cincinnati, Columbus, Dayton, Toledo, and Youngstown is the product of quarterly metropolitan data on non-agricultural employment, quarterly average hourly earnings, and quarterly average hours. The source is *Labor Market Review*, a monthly publication of OBES, LMI Division.

Average manufacturing workweek (hours)

Average manufacturing workweek measured in hours for the seven areas is the average weekly hours worked in the manufacturing sector. The source is *Labor Market Review*, a monthly publication of OBES, LMI Division.

Initial claims for unemployment insurance

Initial claims for unemployment insurance is the total initial claims under Ohio law for each of the metropolitan areas. For purposes of maintaining consistency in the data, the following claims-office reports were used to construct the claims in each metropolitan area:

1. Akron: Akron and Barberton
2. Cincinnati: Cincinnati and Cincinnati Roselawn (District 2)
3. Cleveland: Cleveland Downtown, South, and East
4. Columbus: Columbus East, North, and West
5. Dayton: Dayton
6. Toledo: Toledo and Toledo Southwyck
7. Youngstown: Youngstown

The source is "Selected Unemployment Compensation Workload Items," form RS 237.1, OBES, LMI Division.

Domestic automobile production

This is the monthly, annualized number of domestic automobiles produced in the United States. The source is Cambridge Planning Associates (data base service for RATS econometric package).

Index of 12 leading economic indicators

The national index of 12 leading indicators is the composite index constructed by the Bureau of Economic Analysis (BEA), US Department of Commerce. The source is *Business Conditions Digest*, a publication of the BEA.

Housing permits dollar valuation

The dollar value of housing permits is the total value of housing authorized by building permits in each of the metropolitan areas. The source is *Construction Reports: Housing Authorized by Building Permits and Public Contracts*, a monthly publication of the US Department of Commerce.

Note that since this publication no longer contains housing permit valuation for Akron, Canton, Dayton, Toledo, and Youngstown, Robert Benedict of the Building Permits Branch of the Department of Commerce was contacted to obtain photocopies of the data for these five areas. Line 109, "Total New Residential," was used. Further, data for the Cincinnati and Cleveland metropolitan statistical areas include only Cincinnati and Cleveland (i.e. they exclude Hamilton and Akron).

BIBLIOGRAPHY

Baird, C. A. (1983), "A multiregional econometric model of Ohio," *Journal of Regional Science* 23, 501–15

Ballard, K. P. and N. J. Glickman (1977), "A multiregional econometric forecasting system: a model for the Delaware Valley," *Journal of Regional Science* 17, 161–77

Ballard, K. P., N. J. Glickman and R. D. Gustley (1980), "A bottom-up approach to multiregional modeling: NRIES," in F. G. Adams and N. J. Glickman (eds.), *Modeling the Multiregional Economic System* (Lexington, Mass., Lexington Book, 147–60

Belsley, D. A., E. Kuh, and R. E. Welsch (1980), *Regression Diagnostics* (New York, John Wiley)

Garcia-Ferrer, A., R. A. Highfield, F. C. Palm, and A. Zellner (1987), "Macroeconomic forecasting using pooled international data," *Journal of Business and Economic Statistics* 5(1), 53–67; chapter 13 in this volume

LeSage, J. P. and M. Magura (1986), "Econometric modeling of interregional labor market linkages," *Journal of Regional Science* 26 567–77

(1987), "A leading indicator model for Ohio SMSA employment," *Growth and Change* 18, 36–48

Lindley, D. V. and A. F. M. Smith (1972), "Bayes estimates for the linear model," *Journal of the Royal Statistical Society*, Series B 34, 1–41

Liu, Y. and A. H. Stocks (1983), "A labor oriented quarterly econometric forecasting model of the Youngstown–Warren SMSA," *Regional Science and Urban Economics* 13, 317–40

Milne, W. J., N. J. Glickman, and F. G. Adams (1980), "A framework for analyzing regional decline: a multiregional econometric model of the US," *Journal of Regional Science* 20, 173–90

Zellner, A. (1983), "Application of Bayesian analysis in econometrics," *The Statistician* 32, 23–34

(1986), "On assessing prior distributions and Bayesian regression analysis with *g*-prior distributions," in P. K. Goel and A. Zellner (eds.), *Bayesian Inference and Decision Techniques: Essays in Honor of Bruno de Finetti* (Amsterdam, North-Holland), 233–43

Zellner, A. and Hong, C. (1987), "Forecasting international growth rates using Bayesian shrinkage and other procedures," Working Paper, University of Chicago, H. G. B. Alexander Research Foundation, Graduate School of Business; see also chapter 14 in this volume

21 Forecasting turning points in metropolitan employment growth rates using Bayesian techniques (1990)

James P. LeSage

1 Introduction

Zellner, Hong, and Gulati (1990) and Zellner and Hong (1989) formulated the problem of forecasting turning points in economic time series using a Bayesian decision theoretic framework. The methodology was . . . applied by Zellner, Hong, and Min (1990) (hereafter ZHM) to a host of models to forecast turning points in the international growth rates of real output for eighteen countries over the period 1974–86. They compared the performance of fixed parameter autoregressive leading indicator models (FP/ARLI), time-varying parameter autoregressive leading indicator models (TVP/ARLI), exponentially weighted autoregressive leading indicator models (EW/ARLI), and a version of each of these models that includes a world income variable – FP/ARLI/WI, TVP/ARLI/WI, EW/ARLI/WI. In addition, they implemented a pooling scheme for each of the models. A similar host of models is analysed here in order to assess whether these techniques hold promise for forecasting turning points in regional labor markets.

The innovative aspect of the ZHM study is not the models employed, but the use of the observations along with an explicit definition of a turning point, either a downturn (*DT*) or upturn (*UT*). This allows for a Bayesian computation of probabilities of a *DT* or *UT* given the past data from a model's predictive probability density function (pdf) for future observations. After computing these probabilities from the data, they can be used in a decision theoretic framework along with a loss structure in order to produce an optimal turning point forecast. The focus here is on an analysis of the turning point forecasting performance of the models given a relatively simple symmetric loss structure. Since ZHM report results for their model on the basis of this type of loss structure, it provides a benchmark against which to judge the success of these types of models in the regional forecasting setting studied here.

Originally published in the *Journal of Regional Science* 30(4) (1990), 533–48.

There are a number of reasons to believe that the multi-country forecasting methodology, first introduced by Garcia-Ferrer *et al.* (1987) and employed in the ZHM study, will be successful in our multi-regional setting. First, it is well known that dependencies exist among regional economies. Numerous econometric modeling approaches have been proposed to exploit this information. Most multi-regional models take a structural approach, employing linkage variables such as relative cost, adjacent-state demand, and gravity variables (e.g. Ballard and Glickman 1977, Ballard, Glickman, and Gustely 1980, Milne, Glickman, and Adams 1980 and Baird 1983). Second, as evidence for the existence of interdependencies, ZHM point to similar movements in real output growth rates induced by broad business cycle swings across their eighteen-country sample. Similar comovements in metropolitan employment variation exist across the eight metropolitan areas used here. Finally, LeSage and Magura (1990) show that the Bayesian shrinkage techniques employed in Garcia-Ferrer *et al.* (1987) produce good payroll forecasts (in terms of RMSE) using a set of metropolitan-area models incorporating similar variables to those employed here and metropolitan-area data for seven of the eight areas used in this study.

Our approach, like that of Garcia-Ferrer *et al.* (1987), represents a "bottom-up" attempt to build simple models based on the time series properties of the employment data that will forecast well. This is in contrast to the work of Liu and Stocks (1983), which I would classify as "top-down." They attempt to build elaborate models of the individual metropolitan-area economies in order to approach the multi-regional forecasting problem. One focus of this study, then, is to determine if the procedures employed by Garcia-Ferrer *et al.* (1987) and ZHM (in a multi-country setting) can produce accurate forecasts of turning points in a multi-regional setting.

The findings indicate that these simple models correctly forecast 70 percent of the downturns and 80 percent of the upturns. In addition, the comparison of five model specifications and five estimation methods shows that variation in turning point forecasting accuracy is much greater across the estimation methods than across the model specifications. This suggests that from an applied standpoint, the gains from exploring alternative methods of estimating regional leading indicator models are much greater than those from changing the variables used in specifying the models. Some caveats associated with this inference are mentioned in the concluding section.

In section 2 of the chapter, I explain the data and models along with a brief recap of the Bayesian turning point methods of Zellner and Hong (1989). In section 3, I present the results from the turning point forecasting analysis. In addition, I make a comparison between the turning point forecasting accuracy from the host of models studied here and a set metropolitan-area leading indicators published by the Ohio Bureau of Employment Services [OBES], Labor Market Information Division [LMI] (1989). Finally, in section 4 I provide a summary of the results and some concluding remarks.

2 Models and methods

In this chapter, I will analyse a host of alternative autoregressive leading indicator models for explaining variation in the growth rates of employment over time for each metropolitan area. The models are based on quarterly data covering the period from . . . 1976[:I] to . . . 1989[:I].[1] All variables are transformed by seasonal differencing to eliminate seasonality and converted to annual rates of growth as described in the appendix (p. 652). Figure 21.1 shows a box plot of the employment growth rates for the eight metropolitan areas in Ohio, demonstrating that a great deal of comovement in the rates exists across the regions.[2] This comovement is similar to that found by Garcia-Ferrer *et al.* (1987) for the international data sample. A vertical bar in figure 21.1 delineates the part of the data sample used for estimation (1976:I to 1983:I) and the part used for the forecasting experiments (1983:II to 1989:I).

As explanatory variables for the employment (*EMP*) growth rates, two types of leading indicator variables are used in the study: (1) metropolitan-area-specific leading indicator variables such as housing permits (*HOUSE*), average workweek in manufacturing (*HOURS*), and initial claims for unemployment insurance (*CLAIMS*); and national leading indicator variables: domestic automobile sales (*AUTO*) and the index of twelve leading indicators (*LEAD*). These variables were used in the development of leading indicator series for employment in the eight Ohio metropolitan areas published by the Ohio Bureau of Employment Services. LeSage and Magura (1987) provide a detailed discussion of the

[1] The quarterly data represent averages of monthly values which were available for all of the time series used here.

[2] A box plot shows a box formed by using the 75 percent quartile of the eight time series as the top of the box and the 25 percentile quartile as the bottom. The median of the series is used as the middle of the box. The lines extending out of the box at the top (called "whiskers") range up to the largest value of the eight time series and that extending from the bottom range down to the smallest value. The intent of this type of plot is to easily show the extent to which the time series exhibit similar movements over time. Very small boxes illustrate a high degree of comovement, whereas large boxes denote a lack of such comovement.

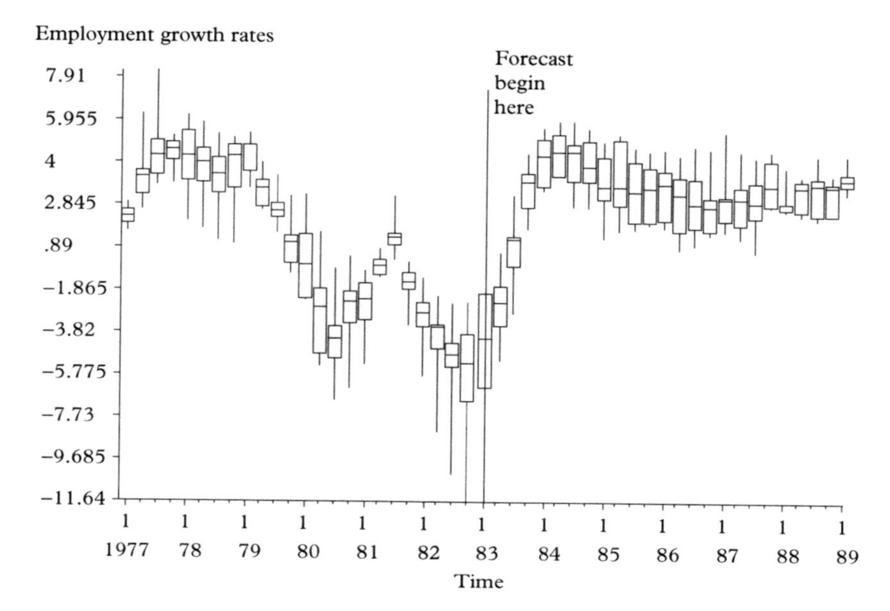

Figure 21.1 Box plot of employment growth rates for eight metropolitan areas, 1977–1989

development of these composite indices, and the appendix describes these variables and their sources in detail.[3]

A simple model which provides a benchmark against which to judge the improvements arising from the introduction of leading indicator variables is an autoregressive model of order three (AR3) in employment. Zellner and Hong (1990) provide a convincing argument for the use of an AR3 model in order to allow for the possibility of having two complex roots and a real root for the process. They perform calculations that show high posterior probabilities for the presence of two complex roots and one real

[3] It might seem plausible to use the composite leading indices themselves as explanatory variables in the model. I would advise against this for the following reason. Labeling the composite leading indicator series $\bar{y}_t$, and the state of the economy variable y_t, the relation between these two variables would be, $y_{t+1} = \bar{y}_t\gamma + \epsilon$, where the scalar parameter γ is included to indicate that the composite index $\bar{y}$ may not be of the same scale as the variable y and the error vector ϵ is included to denote errors in accurately predicting the future values of y. Given this, it would be inappropriate to include this variable in a model with lagged (autoregressive) terms for y_t unless the composite leading indicators were devised to reflect the marginal relationship with the state of the economy variable after factoring out autoregressive influences. The leading indicators used here, developed by LeSage and Magura (1987), were not developed in such a manner. Moreover, the relationship expressed in (2.1) most likely reflects the general case for composite leading indicator series.

Table 21.1 *Leading indicator variables used in the models*

Model	*EMPLAG*	*HOUSE*	*HOURS*	*CLAIMS*	*AUTO*	*LEAD*
AR3 (autoregressive order 3)	X					
LLI (local leading indicator)	X	X	X	X		
NLI (national leading indicator)	X				X	X
CLI (combined local/ national)	X			X		X
ALL (all leading indicator)	X	X	X	X	X	X

root in the fixed parameter autoregressive leading indicator (FP/ARLI) model for each of the eighteen countries in their sample.

The FP/ARLI model in which the leading indicator variables were used is shown below

$$y_{it} = x'_{it}\beta_i + u_{it} \quad (t = 1, 2, \ldots, T; i = 1, 2, \ldots,), \tag{2.1}$$

where the index i denotes which of the eight metropolitan areas is modeled. The vector x_{it} in (2.1) contains three lags of the employment growth rate (*EMPLAG*) as well as one-quarter lagged values of various combinations of the leading indicator variables: dollar value of housing permits issued (*HOUSE*), average workweek in manufacturing (*HOURS*), initial claims for unemployment insurance (*CLAIMS*), domestic auto sales (*AUTO*), and the index of twelve national leading indicators (*LEAD*). Alternative combinations of these variables were used to form five different specifications shown in table 21.1. The use of one-quarter lagged values allows the model to produce one-quarter-ahead forecasts, which will be the focus of our turning-point forecasting analysis.

The alternative specifications were chosen in order to investigate some issues that seemed of interest. First, the AR3 model will be used as a benchmark against which to judge the value of the leading indicator variables in forecasting turning points. Second, the LLI and NLI models were specified in order to investigate whether local leading indicators alone or national leading indicators alone could produce adequate forecasts. The CLI specification represents a parsimonious version of the ALL model containing a single national and local leading indicator variable. The choice of the two variables used in the CLI model, *CLAIMS* and *LEAD*, was based on some experimentation in order to find which two variables produced the best turning point forecasts.

All five specifications of the model were estimated with ordinary least squares [OLS] and forecasted in order to produce a set of five fixed parameter (FP) one-quarter-ahead forecasts, which we label FP/AR3, FP/LLI, FP/NLI, FP/CLI, and FP/ALL. The approach taken was to iteratively estimate and calculate one-step-ahead forecasts based on the updated parameter estimates using the most recently available data. This iterative scheme of estimation and forecasting was carried out over the period from . . . 1983[:I] to . . . 1989[:I], producing a total of twenty-four one-step-ahead forecasts for each of the eight metropolitan areas.[4]

Following ZHM, the five specifications were also estimated and forecast using a time-varying parameter (TVP) method taking the form shown in (2.2). This represents a random-walk scheme for the parameter movement over time in order to investigate whether the turning point forecasting performance of the models is improved by allowing the parameters to change.

$$\begin{aligned} y_{it} &= x_{it}'\beta_{it} + u_{it} \quad (t = 1, 2, \ldots, T; i = 1, 2, \ldots, N) \\ \beta_{it} &= \beta_{i-1} + v_{it}. \end{aligned} \tag{2.2}$$

In (2.2) the v_{it}s are assumed independently and identically normally distributed with zero mean and covariance matrix $\phi\sigma_i^2 I_k$, where $0 < \phi < \infty$. Small values of ϕ reflect very little variation in the parameters over time, whereas large values allow for rapid adjustments in the β_is over time.

The u_{it}s in (2.2) are assumed to be independent and normal with a zero mean and variance σ. ZHM show that, assuming the prior distribution for β_{i0} is taken to be a spread out normal distribution with a zero mean, further assuming that σ has an inverted gamma pdf with small values for its parameters, ν_0 and s_0, $p(\sigma | \nu_0, s_0^2) \propto \sigma^{-(\nu_0+1)} \exp[-\nu s_0^2/2\sigma^2]$, the updating equations shown in (2.3) can be used to estimate the parameters for the TVP model

$$\begin{aligned} \hat{\beta}_{it} &= \hat{\beta}_{it-1} + V_{it-1}x_t(y_t - x_t'\hat{\beta}_{it-1})/(1 + x_t'V_{it-1}x_t) \\ V_{it} &= V_{it-1} - V_{it-1}x_t x_t' V_{it-1}/(1 + x_t'V_{it-1}x_t) + \phi I_k. \end{aligned} \tag{2.3}$$

The TVP estimates were used to produce the one-step-ahead forecasts based on $\hat{y}_{it+1} = x_{it+1}'\hat{\beta}_{it}$. The TVP method was applied to all five specifications of the model, resulting in a set of forecasts which for presentation purposes we will denote: TVP/AR3, TVP/LLI, TVP/NLI, TVP/CLI, and TVP/ALL. The value of ϕ in (2.3) was set in order to produce a minimum for the median of the eight metropolitan area RMSEs of one-step-ahead forecasts over the 1983 to 1989 period for each of the five specifications. The selected values of ϕ were all between 0.6 and unity, reflecting a fair

[4] The first quarter of 1983 represented the last data point used in estimation with the forecasts beginning in . . . 1983[:II] and extending to . . . 1989[:I], for a total of 24 one-quarter-ahead forecasts.

degree of parameter variability over time. The values of ϕ found by ZHM were all between 0.01 and 0.30, reflecting less parameter variation.

In addition to the least squares and TVP estimation methods, an exponential-weighting estimation method was employed. This method downweights or discounts the observations in the data sample that occur in the distant past. Since the parameter estimates arising from this estimation method place more weight on the recent observations, the estimates can adapt the values to recent local changes in the data. This method was suggested by West, Harrison, and Migon (1985) in order to model situations where the parameters are thought to be changing through time. This model is shown in (2.4), with the estimation formulas in (2.5). We designate the forecasts from these procedures as: EW/AR3, EW/LLI, EW/NLI, EW/CLI, and EW/ALL

$$\sqrt{W_{it}}Y_{it} = \sqrt{W_{it}}X_{it}\beta_i + \epsilon_{it}, \tag{2.4}$$

where

$$\hat{\beta}_{it} = (X'_{it}W_tX_{it})^{-1}X'_{it}Y_{it} \quad W_{it} = \begin{bmatrix} 1 & 0 & 0 & \cdots & 0 \\ 0 & \mu & 0 & \cdots & 0 \\ 0 & 0 & \mu^2 & \cdots & 0 \\ \cdot & \cdot & \cdot & \cdot & \cdot \\ \cdot & \cdot & \cdot & \cdot & \cdot \\ \cdot & \cdot & \cdot & \cdot & \cdot \\ 0 & 0 & 0 & \cdots & \mu^{t-1} \end{bmatrix}. \tag{2.5}$$

The time subscripts on the GLS estimator in (2.5) simply denote that data through time period t were used in the vector Y_{it} and matrix X_{it} in order to estimate the parameters β_{it} used at time t to produce a one-step-ahead forecast. The lack of subscript i on the weighting matrix W_t denotes that a single value of μ was used for all eight metropolitan-area models. The parameter μ was set at a value of 0.95, since this value produced a minimum value for the median RSMEs of one-step-ahead forecasts over the 1983 to 1989 period for the eight metropolitan areas for all five model specifications.

Finally a simple pooling scheme, where the parameters from all eight metropolitan-area equations were assumed to be the same, was employed to estimate and forecast the models. The pooling scheme was applied to the fixed parameter and exponential weighting estimation methods for all five specifications, producing sets of forecasts, which I label: FP/AR3/PO, FP/LLI/PO, FP/NLI/PO, FP/CLI/PO, FP/ALL/PO, and EW/AR3/PO, EW/LLI/PO, EW/NLI/PO, EW/CLI/PO, EW/ALL/PO, respectively.

As noted above, the innovative aspect of the ZHM study is the use of the observations along with an explicit definition of a turning point. This allows a Bayesian computation of the probability of a turning point for future observations from a model's predictive . . . [pdf]. The emphasis in ZHM is on the ability of a model to forecast turning points as well as on the role that loss structures play in determining an optimal forecast in this regard. After computing the probability of a turning point, these probabilities can be used in a decision theoretic framework to produce an optimal turning point forecast and to study the impact of the various loss structures upon this optimal forecast. In this study, we abstract from the loss structure aspect of the problem by relying upon a simple symmetric loss function, so that, whenever the probability of a downturn is greater than 50 percent, we forecast a downturn and, similarly, a probability of an upturn greater than 50 percent results in an upturn forecast. The emphasis here is on a comparison of the turning point forecasting performance of the models described above with the performance of the ZHM models for the international growth rates of real output. The comparison between the international and regional models is made using results reported by ZHM for the same simple loss structure.

Designating the time-series observations on the employment growth rates by $y' = (y_1, y_2, \ldots, y_{t-1}, y_t)$ and letting $z = y_{t+1}$ be the first future value of the series, we follow ZHM in defining a downturn (DT) and upturn (UT)

$$y_{t-2}, y_{t-1} < y_t \quad \text{and} \quad \begin{cases} z < y_t \equiv \text{downturn } (DT) \\ z \geqslant y_t \equiv \text{no downturn } (NDT) \end{cases}, \tag{2.6}$$

$$y_{t-2}, y_{t-1} < y_t \quad \text{and} \quad \begin{cases} z > y_t \equiv \text{upturn } (UT) \\ z \leqslant y_t \equiv \text{no upturn } (NUT) \end{cases}. \tag{2.7}$$

Zellner and Hong (1989) show how to employ alternative definitions of an upturn or downturn in this type of analysis. In order to maintain comparability with the ZHM study, we employ the definitions in (2.6) and (2.7). Given these definitions, we turn to the calculation of the probabilities of a *DT* or *UT*. ZHM show how the predictive pdf can be used to compute the probability of a DT for each of the estimation methods presented above.[5]

As an example, the predictive pdf for y_{t+1} can be obtained period by period for the fixed parameter estimation method and takes the form of a univariate Student t-distribution shown in (2.8), where I have dropped the subscript i denoting the metropolitan areas.

$$t_{vt} = (y_{t+1} - x_{t+1}\hat{\beta}_t)/s_t a_t. \tag{2.8}$$

[5] Zellner (1971, 72–5) describes the predictive pdf for the simple regression model.

The β_t in (2.8) denotes an estimate based on all data available at time t and the values of s_t and a_t in (2.8) can be found using the recursive expressions shown in (2.9)

$$\begin{aligned} \nu_t &= \nu_{t-1} + 1 \\ \nu_t s_t^2 &= \nu_{t-1} s_{t-1} + (y_t - x_t' \hat{\beta}_t)^2 + (\hat{\beta}_t - \hat{\beta}_{t-1})' X_{t-1}' X_{t-1} (\hat{\beta}_t - \hat{\beta}_{t-1}) \\ a_t^2 &= 1 + x_{t+1}' (X_t' X_t)^{-1} x_{t+1}, \end{aligned} \tag{2.9}$$

where X_t represents the explanatory variables matrix containing data available at time t.

Similar formulas for the predictive pdfs of the other model exist and can be used to compute the probability of a DT or UT in the following way. The probability that $y_{t+1} < y_t$, given information through time t is $\Pr[y_{t+1} < y_t | D_t] = \Pr[t_{\nu_t} < (y_t - x_{t+1}' \hat{\beta}_t)/s_t a_t | D_t]$. This probability can be evaluated using a Student t-distribution. I label this probability of a downturn as P_{DT} and, similarly, let P_{UT} denote the probability of an upturn.

For the simple case of a symmetric loss structure which we rely on here, it can be shown that the probability of a DT will be larger than 0.5 whenever the given value of $y_t > \hat{y}_{t+1}$, the modal value of the predictive pdf. Of course, this is conditional on $y_{t-2}, y_{t-1} < y_t$. In these cases, I forecast a downturn which is then compared to the actual occurrence in order to assess the accuracy of the forecasted turn. Given the definitions of a turning point in (2.6) and (2.7), the probability of no downturn is simply $\Pr_{NDT} = (1 - \Pr_{DT})$ and, similarly, the probability of no upturn is $\Pr_{NUT} = (1 - \Pr_{UT})$. For the simple case considered here, an upturn is forecast whenever $\Pr_{UT} > 0.5$, which is true whenever $y_t < \hat{y}_{t+1}$, given that $y_{t-2}, y_{t-1} > y_t$.

3 The results from the forecasting experiments

Table 21.2a presents the results from the turning point forecasting experiments. Using our definitions of turning points from (2.6) and (2.7) and the employment growth rate series for all eight metropolitan areas over the period 1983:I to 1989:I, there were 85 actual DT and NDT and 70 UT and NUT points, producing a total of 155 turning points.

Table 21.2a shows both the percentage and the number of correct turning-point forecasts based on the rules

Table 21.2a *Results from the turning point forecasting experiments*

	Percentage of correct forecasts			Number of correct forecasts		
Model	Downturns ($DT + NDT$)	Upturns ($UT + NUT$)	Total turning point events	$DT + NDT$ Out of 85	$UT + NUT$ Out of 70	Turning point events out of 155
FP/AR3	64.7	58.6	61.9	55	41	96
FP/LLI	64.7	60.0	62.6	55	42	97
FP/NLI	74.1	58.6	67.7	63	41	105
FP/CLI	76.5*	65.7	71.6	65*	46	111
FP/ALL	68.2	55.7	62.6	58	39	97
Mean	69.6	59.7	65.3			
TVP/AR3	48.2	44.3	46.5	41	31	72
TVP/LLI	50.6	60.0	54.8	43	42	85
TVP/NLI	48.2	57.1	52.3	41	40	81
TVP/CLI	43.5	67.1	54.2	37	47	84
TVP/ALL	55.3	60.0	57.4	47	42	89
Mean	49.2	57.7	53.0			
EW/AR3	68.2	57.1	63.2	58	40	98
EW/LLI	68.2	65.7	67.1	58	46	104
EW/NLI	69.4	51.4	61.3	59	36	95
EW/CLI	65.9	57.1	61.9	56	40	96
EW/ALL	72.9	57.1	65.8	62	40	102
Mean	68.9	57.7	63.9			
FP/AR3/PO	70.6	84.3	76.8	60	59	119
FP/LLI/PO	74.1	80.0	76.8	63	56	119
FP/NLI/PO	65.9	80.0	72.3	56	56	112
FP/CLI/PO	51.8	65.7	58.1	44	46	90
FP/ALL/PO	65.9	82.9	73.5	56	58	114
Mean	65.6	78.6	71.5			
EW/AR3/PO	70.6	87.1*	78.1*	60	61*	121*
EW/LLI/PO	72.9	82.9	77.4	62	58	120
EW/NLI/PO	70.6	81.4	75.5	60	57	117
EW/CLI/PO	70.6	77.1	73.5	60	54	114
EW/ALL/PO	70.6	82.9	76.1	60	58	118
Mean	71.1†	82.3†	76.1†			
OBES LMI	72.9	74.2	73.5	62	52	114

Notes:
* Designates a column maximum showing the most accurate model from all specification and estimation methods.
† Designates the most accurate estimation method using the mean over all model specifications for each estimation method.

(1) when $\Pr_{UT} > 0.5$ and $y_{t-2}, y_{t-1} > y_t$, forecast an upturn
(2) when $\Pr_{DT} > 0.5$ and $y_{t-2}, y_{t-1} < y_t$, forecast a downturn
(3) when $\Pr_{NUT} > 0.5$ and $y_{t-2}, y_{t-1} > y_t$, forecast no upturn and
(4) when $\Pr_{NDT} > 0.5$ and $y_{t-2}, y_{t-1} < y_t$, forecast a no downturn.

Explicitly, in the experiments carried out here, I forecast a downturn whenever the previous two quarters exhibited values of the employment growth rate below the current value, ($y_{t-2}, y_{t-1} < y_t$) and the current value, y_t, is greater than the one-step-ahead forecast, $\hat{y}_{t+1}$. I then compare this to the actual event by determining if y_{t+1} is above y_t, in which case there is a no downturn event, or below y_t, indicating a downturn. Similarly, for the upturn events I forecast an upturn whenever the conditions $y_{t-2}, y_{t-1} > y_t$ hold, and our forecast $\hat{y}_{t+1}$ is above the current value y_t. The actual occurrence is determined by comparing the value of y_{t+1} to y_t, with $y_{t+1} > y_t$ indicating an upturn and $y_{t+1} \leq y_t$ a no upturn event.

Table 21.2a is organized according to the five estimation techniques, FP, TVP, EW, FP/PO, and EW/PO. In addition to the results from the five estimation techniques, the results from using the Ohio Bureau of Employment Services, Labor Market Information Division (OBES LMI), eight metropolitan leading indicator series are reported as the last line in Table 21.2a. Forecasts of turning points using the OBES leading indicators were made in the following way. If the change in the indicator from time $t-1$ to t is positive, forecast an upturn and when the change is negative, forecast a downturn. This is analogous to treating the indicator as a one-step-ahead forecast series, and consistent with the scheme used to convert the one-step-ahead forecasts into turning point forecasts.

The motivation for examining the results on the basis of techniques as well as specifications is that we might expect to have model uncertainty. Given uncertainty about the precise model specification, we would be interested in the question of whether a particular estimation technique outperforms other techniques regardless of the particular specification we choose. Table 21.2a presents the results for downturn and upturn forecasts by aggregating the results of *DT* and *NDT* forecasts and *UT* and *NUT* forecasts. The mean for each of the five estimation techniques of the percentage of correct turning points forecast are reported to facilitate an examination of the five estimation approaches. Using these means, we see that the pooled exponential-weighting technique produced, on average across the five specifications, the most accurate turning point forecasts for both upturns and downturns.

The worst forecasts were produced by the TVP technique, a technique that worked well in the ZHM study. It is interesting to compare our results to those of ZHM, shown in Tables 21.2b and 21.3b.[6] Table 21.2b presents the results from the ZHM international study in the same format as our table 21.2a, and table 21.3b similarly parallels our table 21.3a.

[6] It should be noted that ZHM (1987) used annual data in their study in contrast to the quarterly data used here.

Table 21.2b *ZHM results from turning point forecasting experiments*

	NT + DT Turning-point events	*NUT + UT* Turning-point events	TOTAL of All Turning-point events
Number	76	82	158
Model	% Correct	% Correct	% Correct
FP/ARLI	71	77	74
TVP/ARLI	76	77	77
EW/ARLI	68	74	72
FP/ARLI/PO	82	68	75
EW/ARLI/PO	86	67	74

The first point to note is that our percentages of total turning points correct forecasts range between 53 and 76 percent whereas table 21.2b indicates that the ZHM study found less variation over the estimation techniques, showing a range of from 72–77 percent. The overall quality of the best models in forecasting turning points is quite similar with our best technique, (EW/PO), producing on average 71.1 percent and 82.3 percent correct downturn and upturn forecasts respectively, and the best ZHM technique (TVP), producing 71 percent and 77 percent correct downturn and upturn forecasts.

Another interesting comparison with the ZHM results is found in table 21.3b where the individual *NDT*, *DT*, *NUT*, and *UT* forecasting results are presented. The ZHM study showed a relatively lower accuracy in forecasting *NDT* points than *DT* points across all estimation techniques. The ZHM accuracy on *NDT* points ranges from 39 to 65 percent and is consistently below the accuracy for the *DT* points. Our study shows this same pattern for the TVP, FP/PO, and EW/PO techniques. In contrast, however, is the FP technique, which shows a relatively equal ability on these two types of turning points, and the EW technique, which exhibits more accuracy on the *NDT* points than the *DT* points.

With respect to the *NUT* and *UT* forecasting results from ZHM, we see that the pooling reduced the accuracy of the *NUT* points. This is in contrast to the results from table 21.3a for this study, where pooling generally produced an increase in accuracy for the *NUT* turning point forecasts.

Finally, the Ohio Bureau of Employment Services, Labor Market Information Division, distributes a monthly publication containing composite leading indicators for employment in the eight Ohio metropolitan areas.[7]

[7] The publication is entitled *Leading Indicators* and is part of the *Labor Market Information* publications produced and distributed monthly by OBES. A description of the model and its development can be found in LeSage and Magura (1987).

Table 21.3a *Results categorized according to the four types of events forecast*

Model	Percentage of events correctly forecast				
	NDT events	*DT* events	*NUT* events	*UT* events	All types of events
FP/AR3	61.7	72.0	100.0*	0.0	61.9
FP/LLI	68.3	56.0	70.7	44.8	64.7
FP/NLI	76.7	68.0	56.1	62.1	74.1
FP/CLI	78.3	72.0	68.3	62.1	76.5
FP/ALL	68.3	68.0	48.8	65.5	68.2
Mean	70.7	67.2	68.8	46.9	65.3
TVP/AR3	28.3	96.0*	7.3	96.6*	46.5
TVP/LLI	40.0	76.0	53.7	69.0	54.8
TVP/NLI	28.3	96.0*	48.8	69.0	52.3
TVP/CLI	28.3	80.0	61.0	75.9	54.2
TVP/ALL	45.0	80.0	56.1	65.5	57.4
Mean	34.0	85.6	45.4	75.2	53.0
EW/AR3	76.7	48.0	95.1	3.4	63.2
EW/LLI	85.0	28.0	75.6	51.7	67.1
EW/NLI	86.7	28.0	43.9	62.1	61.3
EW/CLI	86.7	16.0	51.2	65.5	61.9
EW/ALL	91.7*	28.0	51.2	65.5	65.8
Mean	85.3	29.6	63.4	49.7	63.9
FP/AR3/PO	63.3	88.0	97.6	65.5	76.8
FP/LLI/PO	70.0	84.0	80.5	79.3	76.8
FP/NLI/PO	61.7	76.0	78.0	82.8	72.3
FP/CLI/PO	33.3	88.0	80.5	44.8	58.1
FP/ALL/PO	61.7	76.0	80.5	86.2	73.5
Mean	58.0	82.4	83.4	71.7	71.5
EW/AR3/PO	63.3	88.0	97.6	72.4	78.1
EW/LLI/PO	68.3	84.0	85.4	79.3	77.4
EW/NLI/PO	65.0	84.0	82.9	79.3	75.5
EW/CLI/PO	65.0	84.0	80.5	82.8	73.5
EW/ALL/PO	65.0	84.0	82.9	82.8	76.1
Mean	65.3	84.8	85.9	79.3	76.1†
OBES LMI	64.6	80.0	75.6	72.4	73.5

Notes:
* Designates a column maximum showing the most accurate model and estimation method for this type of event.
† Designates the most accurate estimation method over all specifications using the mean over model specifications for each estimation method.

Table 21.3b *ZHM results categorized according to the type of event forecast*

	NDT turning-point events	*DT* turning-point events	*NUT* turning-point events	*UT* turning-point events
Number	23	53	23	59
Model	% Correct	% Correct	% Correct	% Correct
FP/ARLI	42	93	86	74
TVP/ARLI	48	89	83	75
EW/ARLI	39	89	84	71
FP/ARLI/PO	56	96	68	68
EW/ARLI/PO	65	93	68	65

These monthly indices, which contain seasonal variation, were converted to quarterly series by averaging, then seasonally differenced, and converted to growth rates to make them comparable with the employment series used in this study. A set of turning point forecasts were then generated using the same rules. These forecasts used the value of the leading indicator series in place of the one-step-ahead forecast in order to calculate Pr_{DT}, Pr_{UT}, Pr_{NDT}, and Pr_{NUT} values upon which to base the forecasts. The last lines in tables 21.2a and 21.3a show the results from using these leading indicators to produce the turning point forecasts. As can be seen from the tables, these composite indices for the eight metropolitan areas compare favorably with some of the best models from the host of specifications and estimation techniques studied here.

Turning attention to the results from the perspective of model uncertainty, the table 21.2a and 21.3a results can be considered from the perspective of movement across the five different model specifications. Viewed from this perspective, tables 21.2a and 21.3a indicate that there is a great deal less variation in the accuracy of the forecasts as we vary the model specification than that encountered when we move across estimation methods. For example, considering the table 21.2a results for the five model specifications using the FP estimation method, we see a minimum of 61.9 percent correct total turning point forecasts and a maximum of 71.6 percent. Similarly, for the TVP we have a minimum and maximum of 46.5 and 57.4 percent, respectively. The EW forecasts range between 61.3 and 67.1 percent, the FP/PO between 58.1 and 76.8 percent, and the EW/PO between 73.5 and 78.1 percent. With the exception of the FP/PO, the variation here is around 10 percent or less, quite small by comparison with that encountered as we alter the estimation method. Considering the TVP estimation results reported in table 21.2a, we see around 55 percent of the total turning point forecasts correct, whereas the EW/PO produces around 75 percent correct for these same forecasts, a difference of 20 percent. The implication here is that risk associated with model uncertainty is less than that from using a non-optimal estimation technique. From the standpoint of a practitioner, the gains from the time spent exploring alternative estimation methods are likely to be greater than those from alternative model specifications.

4 Conclusions

This study introduced a set of models to forecast growth rates in metropolitan employment similar to those used by Zellner, Hong, and Min (ZHM) (1989) to forecast international growth rates in real output for a sample of eighteen countries. Five estimation techniques from the ZHM study were used to estimate and forecast the employment growth rates for the sample of eight Ohio metropolitan areas using five model specifications. The focus of the experiments here, as in ZHM, was an analysis of the accuracy in forecasting turning points. The Bayesian decision theoretic approach to this problem introduced in Zellner and Hong (1989) and implemented in the ZHM study was replicated here in order to study whether these techniques could produce metropolitan-area turning point forecasts of similar accuracy as that found in ZHM for the international sample of eighteen countries.

The findings indicate a greater amount of dispersion in the accuracy of the turning point forecasts associated with the five estimation techniques than found by ZHM. These findings are important in that the variation across the five different model specifications examined here was far less than that associated with the five estimation methods. This indicates that model uncertainty and the risk from chosing the wrong specification is small in comparison to that associated with using a suboptimal estimation technique. Some caution may be required here in drawing this conclusion since, first, it may be that the quality of the data used here was not as good as that employed in the ZHM study. A second reason to be cautious regarding this conclusion is that the variation in model specifications that would be possible in a multi-equation structural model are much greater than those examined here in the single-equation model.[8]

The findings also show that the accuracy of our turning point forecasts compare favorably with those of ZHM. This suggests that, the relatively simple leading-indicator models introduced by Garcia-Ferrer *et al.* (1987) can produce accurate forecasts of turning points in a multi-regional

[8] Both of these points were brought to my attention by an anonymous referee.

setting as well as [in] the multi-country setting explored by ZHM. Another finding is that the composite leading-indicator series for metropolitan employment published by the Ohio Bureau of Employment Services, Labor Market Information Division, works quite well in forecasting turning points.

Finally, the methods employed here are quite general and could be used to compare the turning point forecasting performance of a host of alternative regional econometric forecasting models,[9] a subject that has been ignored in the regional forecasting literature where the primary emphasis has been on the root mean-squared error [RMSE] criterion.

APPENDIX DATA AND DATA SOURCES

(1) Total employment (*EMP*)

Total employment for the SMSAs of Akron, Canton, Cleveland, Cincinnati, Columbus, Dayton, Toledo, and Youngstown is monthly, non-agricultural employment, converted to quarterly averages. These quarterly data were then seasonal differenced and converted to growth rates using $(Y_t - Y_{t-4})/Y_{t-4}$. Where Y_t represents quarterly employment. *Source*: OBES, LMI Division, *Labor Market Review*, a monthly publication.

[9] Extending the techniques of ZHM to a monthly situation represents a fairly straightforward exercise conceptually. Consider that in the context of a monthly model, we need to change the definition of a turning point to include more past and future observations because of the increased variation and frequency of the data. We would not define the necessary conditions for a downturn event based on two or three periods of monthly upward movement as was done for the quarterly series. Suppose we broaden the definition of a turning point event to include a three-horizon period so that we define the downturn event to be when $y_{t+1} < y_t$ and $y_{t+2} < y_t$ and $y_{t+3} < y_t$. This is straightforward enough, and, so is the analytical expression for the probability of a downturn event

$$\Pr_{DT} = \int_{-\infty}^{yt} \int_{-\infty}^{yt+1} \int_{-\infty}^{yt+2} \Pr[y_{t+1}, y_{t+2}, y_{t+3} \mid (y, I)],$$

where (y, I) denotes the past sample and prior information. The problem is that this involves trivariate numerical integration that must take place at every step in the iterative estimation-forecast loop during the forecasting experiments. This computational limitation may not be much of a problem in practice since, many regional econometric models are quarterly even though the data is available on a monthly basis (e.g. Liu and Stocks' 1983 Youngstown model and Shapiro and Fulton's 1985 Michigan regional models). This is because the benefits from updating and running the model monthly rather than quarterly may not outweight the costs.

(2) Average manufacturing workweek (*HOURS*)

Average manufacturing workweek measured in hours for the eight SMSAs is the average weekly hours worked in the manufacturing sector. These monthly data were converted to quarterly by averaging and then seasonal differenced and converted to growth rates as described above. *Source*: OBES, LMI Division, *Labor Market Review*, a monthly publication.

(3) Initial claims for unemployment insurance (*CLAIMS*)

Initial claims for unemployment insurance is the total initial claims under Ohio law for each of the SMSAs in Ohio. For purposes of maintaining consistency in the data, the following were included in each SMSA:

Akron: Akron and Barberton
Canton: Alliance, Canton, Massilon
Cincinnati: Cincinnati, Cincinnati Roselawn (District 2)
Cleveland: Cleveland Downtown, South, East
Columbus: Columbus East, North, West
Dayton: Dayton
Toledo: Toledo, Toledo Southwyck
Youngstown: Youngstown

Source: OBES, LMI Division, "Selected Unemployment Compensation Work-load Items," form RS 237.1. These monthly data were also converted to quarterly averages and then seasonal differenced and converted to growth rates as described above.

(4) Domestic automobile production (*AUTO*)

This is the monthly, annualized number of domestic automobiles produced in the United States, converted to quarterly averages and then seasonal differenced and converted to the growth rates as described above. *Source*: Citibase Data.

(5) Index of 12 leading economic indicators (*LEAD*)

National index of twelve leading indicator series, which is the composite index constructed by the Bureau of Economic Analysis, US Department of Commerce. These monthly data were also converted to quarterly averages and then seasonal differenced and converted to growth rates as described above. *Source*: *Business Conditions Digest*, a publication of the BEA, Department of Commerce.

(6) Housing permits dollar valuation (*HOUSE*)

The dollar value of housing permits, the total value of housing authorized by building permits in each of the SMSAs. These monthly data were also converted to quarterly averages and then seasonal differenced and converted to growth rates as described above. *Source*: US Department of Commerce, *Construction Reports: Housing Authorized by Building Permits and Public Contracts*, a monthly publication.

Note: Since this publication no longer contains housing permit valuation for Akron, Canton, Dayton, Toledo, and Youngstown, the Building Permits Branch of the Department of Commerce was contacted to obtain photocopies of the recent data for these five SMSAs.

BIBLIOGRAPHY

Baird, C. A. (1983), "A multiregional econometric model of Ohio," *Journal of Regional Science* 23, 501–15

Ballard, K. P. and N. J. Glickman (1977), "A multiregional econometric forecasting system: a model for the Delaware Valley," *Journal of Regional Science* 17, 161–77

Ballard, K. P., N. J. Glickman, and R. D. Gustely (1980), "A bottom-up approach to multiregional modeling: NRIES," in F. G. Adams and N. J. Glickman (eds.), *Modeling the Multiregional Economic System* (Lexington, Mass., Lexington Books, 147–60

Garcia-Ferrer, A., R. A. Highfield, F. C. Palm, and A. Zellner (1987), "Macroeconomic forecasting using pooled international data," *Journal of Business and Economic Statistics*, 5(1), 53–68; chapter 13 in this volume

LeSage, J. P. and M. Magura (1987), "A leading indicator model for Ohio SMSA employment," *Growth and Change* 18, 36–48

(1990), "Using Bayesian techniques for data pooling in regional payroll forecasting," *Journal of Business and Economics Statistics* 8(1), 127–36; chapter 20 in this volume

Liu, Y. and A. H. Stocks (1983), "A labor-oriented quarterly econometric forecasting model of the Youngstown–Warren SMSA," *Regional Science and Urban Economics* 13, 317–40

Milne, W. J., N. J. Glickman, and F. G. Adams (1980), "A framework for analyzing regional decline: a multiregional econometric model of the US," *Journal of Regional Science* 20, 173–90

Ohio Bureau of Employment Services (1989), *Leading Indicators*, Labor Market Information Division, Columbus, Ohio

Shapiro, H. T. and G. A. Fulton (1985), *A Regional Econometric Forecasting System* (Ann Arbor, University of Michigan Press)

West, M., P. J. Harrison, and H. S. Migon (1985), "Dynamic generalized linear models and Bayesian forecasting," *Journal of the American Statistical Association* 80, 73–83

Zellner, A. (1971), *An Introduction to Bayesian Inference in Econometrics* (New York, John Wiley)

Zellner, A. and C. Hong (1989), "Forecasting international growth rates using Bayesian Shrinkage and other procedures," *Journal of Econometrics, Annals*, 40, 183–202; chapter 14 in this volume

(1990), "Bayesian methods for forecasting turning points in economic time series: sensitivity of forecasts to asymmetricy of loss structures," in K. Lahiri and G. Moore (eds.), *Leading Economic Indicators: New Approaches and Forecasting Records* (Cambridge, Cambridge University Press)

Zellner, A., C. Hong, and G. M. Gulati (1990), "Turning points in economic time series, loss structures, and Bayesian forecasting," in S. Geisser, J. S. Hodges, S. J. Press, and A. Zellner (eds.), *Bayesian Likelihood Methods in Statistics and Econometrics: Essays in Honor of George A. Barnard* (Amsterdam, North-Holland), 371–89; chapter 15 in this volume

Zellner, A., C. Hong, and C. Min (1990), "Forecasting turning points in international output growth rates using Bayesian exponentially weighted autoregression, time-varying parameter, and pooling techniques," Working Paper, H. G. B. Alexander Research Foundation, Graduate School of Business, University of Chicago; see also chapter 16 in this volume

22 A note on aggregation, disaggregation, and forecasting performance (2000)

Arnold Zellner and Justin Tobias

1 Models

In this chapter we report the results of an experiment to determine the effects of aggregation and disaggregation in forecasting the median growth rate of eighteen industrialized countries' annual output (GDP) growth rates; see figure 22.1 for a plot of our data and table 22A.3 (p. 665) for the names of the countries in our sample. In one approach, following Zellner and Hong (1989), we model the aggregative annual median growth rate, w_t, as an autoregression of order 3 with lagged leading indicator input variables, denoted by AR(3)LI, as follows:

$$w_t = \alpha_0 + \beta_1 w_{t-1} + \beta_2 w_{t-2} + \beta_3 w_{t-3} + \beta_4 MGM_{t-1} + \beta_5 MSR_{t-1} + \varepsilon_t, \qquad (1.1)$$

where MGM_t is the median annual growth rate of real money in year t, MSR_t denotes the median annual growth rate in real stock prices in year t, and ε_t is a zero mean, non-autocorrelated, constant variance error term. Given data on eighteen industrialized countries' annual output growth rates, it is possible to compute annual median growth rates, and use them and data on the other input variables appearing in (1.1) to obtain point and turning point forecasts for future median annual growth rates of the eighteen countries. The results of such calculations will be reported below after describing alternative approaches to forecasting the median growth rate using disaggregated data and disaggregated forecasting models.

This research was financed in part by the National Science Foundation and by income from the H. G. B. Alexander Endowment Fund, Graduate School of Business, University of Chicago.

Contract grant sponsor: US National Science Foundation and H. G. B. Alexander Endowment Fund, Graduate School of Business, University of Chicago.

Originally published in the *Journal of Forecasting* 19 (2000), pp. 457–69.

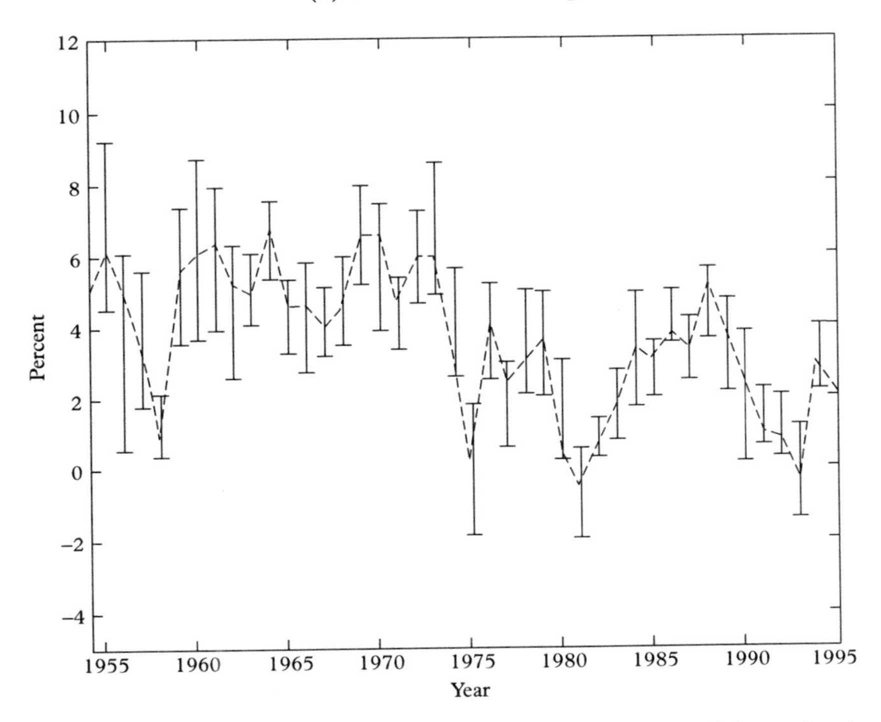

Figure 22.1a Medians and interquartile ranges for growth rates of real output for eighteen industrialized countries 1954–1995; the dashed line connects the annual median growth rates (the w_ts) and the vertical lines give the interquartile ranges. See table 22A.3 for a list of the eighteen countries included in our sample.

As an alternative to (1.1), we can employ the disaggregated ARLI relationships.

$$y_{it} = \gamma_i + \delta_{1i} y_{it-1} + \delta_{2i} y_{it-2} + \delta_{3i} y_{it-3} + \delta_{4i} GM_{it-1} + \delta_{5i} SR_{it-1} + \delta_{6i} SR_{it-2} + \delta_{7i} MSR_{t-1} + u_{it}, \quad (1.2)$$

where the subscripts i and t denote the value of a variable for the ith country in the ith year, u_{it} is an error term, and y, GM, SR, and MSR denote the annual growth rates of real GDP, real money, real stock prices and the median growth rate of real stock prices, respectively. See Garcia-Ferrer *et al.* (1987) and Zellner and Hong (1989) for discussions and uses of (1.2) in forecasting. In (1.2) we allow the regression coefficients

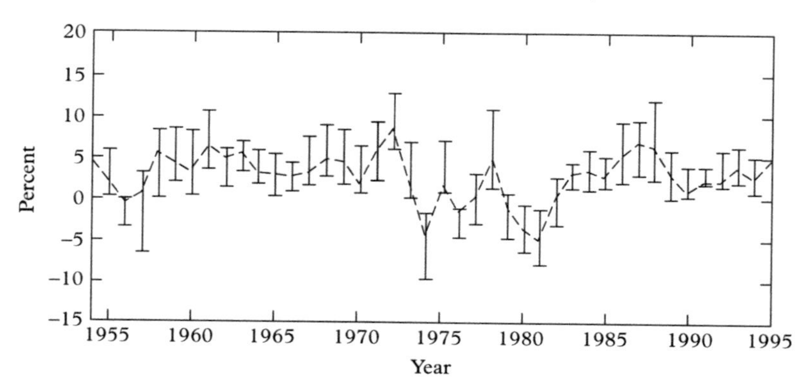

Figure 22.1b Medians and interquartile ranges for growth rates of real money for eighteen industrialized countries 1954–1995; the dashed line connects the annual median growth rates (the $w_t s$) and the vertical lines give the interquartile ranges. See table 22A.3 for a list of the eighteen countries included in our sample.

to vary in value across countries. We also consider two variants of (1.2) which involve restrictions on the regression coefficients. In the first model, the coefficients are restricted to be the same across countries; that is, $\gamma_i = \gamma, \delta_{ji} = \delta_j, j = 1, 2, \ldots 7$. In the second model, we assume that the coefficients associated with the leading indicators are the same across countries while the AR and intercept coefficients vary across countries; that is, $\delta_{ji} = \delta_j j = 4, 5, 6, 7$.[1] Then, given forecasts of eighteen countries' annual output growth rates from (1.2) and its variants, it is clearly possible to compute the medians of the eighteen forecasts year by year and use them as forecasts of the median output growth rates and compare them to the forecasts obtained by use of (1.1), as will be done below.[2]

[1] We thank a referee for suggesting these alternative models.

[2] For the model in (1.2), we also estimated a random effects specification: $y_{it} = X_{it}\theta + \alpha_i + u_{it}$, where $u_{it} = \alpha_i + \varepsilon_{it}$, and X_{it} denotes all the input variables in (1.2), including lagged ys, and θ denotes the associated regression parameters. Unlike the model in (1.2), this specification does not assume that the error terms for the same country are uncorrelated over time. We made the standard assumptions that the αs are uncorrelated across countries with common variance σ_α^2, and $E(\varepsilon_{it}^2) = \sigma_{it}^2$, $E(\varepsilon_{it}\varepsilon_i' t') = 0 \forall i \neq i', t \neq t'$. We found that predictive RMSEs and MAEs are often lower using least squares forecasts from (1.2) than results obtained using the above random effects specification. For example, updating the one-year-ahead forecasts through the hold-out period 1985–95, least squares on (1.2) gives RMSE and MAE of 1.55 and 1.37, while forecasts from the random effects model above give 1.68 and 1.44 for RMSE and MAE, respectively.

(C) Growth rates of real stock prices

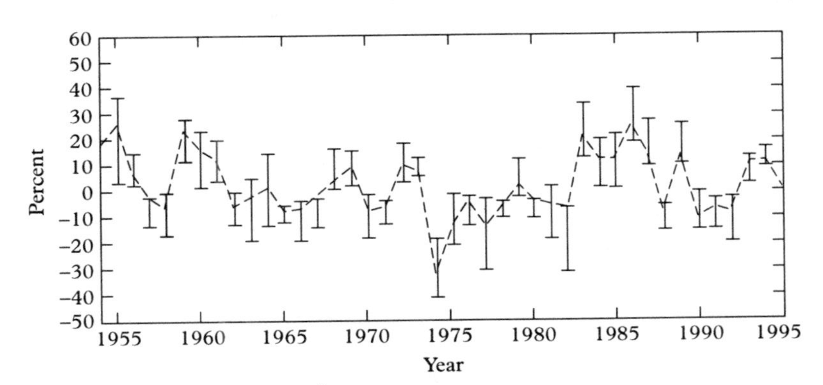

Figure 22.1c Medians and interquartile ranges for growth rates of real stock prices for eighteen industrialized countries 1954–1995; the dashed line connects the annual median growth rates (the w_ts) and the vertical lines give the interquartile ranges. See table 22A.3 for a list of the eighteen countries included in our sample.

Further, (1.2) can be expanded to include the current median output growth variable to obtain the following equation:

$$y_{it} = \phi_i w_t + \gamma_i + \delta_{1i} y_{it-1} + \delta_{2i} y_{it-2} + \delta_{3i} y_{it-3} + \delta_{4i} GM_{it-1} + \delta_{5i} SR_{it-1} + \delta_{6i} SR_{it-2} + \delta_{7i} MSR_{t-1} + u_{it}. \tag{1.3}$$

As Zellner and Hong (1989) do, we use (1.1) to obtain forecasts of the w_t's and use them and (1.3) to forecast the output growth rates of the eighteen countries year by year. As described above, we analyse two additional variants of (1.3). In the first case all coefficients are restricted to be equal across countries (i.e. $\phi_i = \phi$, $\gamma_i = \gamma$, $\delta_{ji} = \delta_j$, $j = 1, 2, \ldots 7$), and in the second case only the coefficients of the current median growth rate and the leading indicator variables are restricted to be equal across countries (i.e. $\phi_i = \phi$, $\delta_{ji} = \delta_j$, $j = 4, 5, 6, 7$). Thus, we analyse the performance of seven different models: the aggregate specification in (1.1), and models for each of the specifications in (1.2) and (1.3).

2 Experiments

In the first experiment, we use annual data for 1954–73 to fit our three models (see the appendix, p. 663 for estimation results) and then employ them to forecast the median of the annual output growth rates year by year for the period 1974–84, updating our parameter estimates as we move

Table 22.1 *Root mean-squared errors (RMSEs) and mean absolute errors (MAEs) of one-year-ahead aggregate and disaggregate forecasts of the median of eighteen countries' annual real GDP growth rates*[a]

A: 1974–84			
	Disaggregated AR(3) LI with $\hat{w}_t$ ((1.3))	Aggregate AR (3)LI for w_t ((1.1))	Disaggregated AR(3) LI ((1.2))
RMSE	1.30	1.60	1.66
MAE	1.21	1.43	1.44
B: 1980–95			
		Aggregate AR(3) LI for w_t ((1.1))	
RMSE		1.60	
MAE		1.42	
	Disaggregated AR(3) LI with $\gamma_i = \gamma$, $\delta_{ji} = \delta_j$ ((1.2))	Disaggregated AR(3) LI with $\delta_{ji} = \delta_j$, $j = 4, 5, 6, 7$ ((1.2))	Disaggregated AR(3) LI ((1.2))
RMSE	1.61	1.76	1.84
MAE	1.35	1.46	1.56
	Disaggregated AR(3) LI with $\hat{w}_t\phi_i = \phi$, $\gamma_i = \gamma$, $\delta_{ji} = \delta_j$ ((1.3))	Disaggregated AR(3) LI with $\hat{w}_i\phi_i = \phi$, $\delta_{ji} = \delta_j$, $j = 4, 5, 6, 7$((1.3))	Disaggregated AR(3) LI with $\hat{w}_t$ ((1.3))
RMSE	1.42	1.46	1.52
MAE	1.25	1.28	1.35

Note:
[a] The data are taken from the IMF computerized data base at the University of Chicago. We use data for the following eighteen countries: Australia, Austria, Belgium, Canada, Denmark, Finland, France, Germany, Ireland, Italy, Japan, the Netherlands, Norway, Spain, Sweden, Switzerland, the United Kingdom and the United States. Observations are available from 1954 to 1995 (with some countries data dating back to 1948) for most countries, but begin in 1971 for Germany. Omitting Germany from the analysis produced similar results. We define RMSE $= \sqrt{\sum_{t=1}^{T}(\hat{w}_t - w_t)^2/T}$, and similarly MAE $= (\sum_{t=1}^{T} |\hat{w}_t - w_t|)/T$.

through the forecast period. This time period was chosen to be similar to the forecasting period used in previous work by Zellner and Hong (1989). For this time period, we present results only for the aggregate model in (1.1) and the models in (1.2) and (1.3) with all coefficient vectors restricted to be the same across countries. Forecasting results are shown in table 22.1, panel A. It is seen that use of the disaggregated equations in (1.3) which include the aggregate variable w_t, perform the best with a [root mean-squared error] (RMSE) and [mean absolute error] (MAE) of prediction of 1.30 and 1.21, respectively. Second best is the performance of the aggregate relation in (1.1), with RMSE = 1.60 and

MAE = 1.43. Last in performance are the disaggregated relations in (1.2) that do not include the variable w_t with RMSE = 1.78 and MAE = 1.44. The empirical results for (1.1) using revised data are very similar to those reported in Zellner and Hong (1989) using unrevised data and which are better than those using just an AR(3) for w_t without leading indicator variables.

As a second experiment, we employed annual data for 1954–79 to fit our models and forecasted the median growth rate of the eighteen countries year by year for the period 1980–95 with results shown in table 22.1, panel B. For this experiment, we obtain results using the seven models based on equations (1.1), (1.2), and (1.3) described above. Again it is the case that use of the disaggregated equations in (1.3) with coefficients pooled over countries including the aggregate variable w_t, performed best with RMSE = 1.42 and MAE = 1.25. The results also indicate that RMSEs and MAEs increase as we add extra parameters to the models and allow some or all of the regression coefficients to vary across countries. Further, RMSEs and MAEs using the model in (1.3) that includes the forecasted current median growth rate are smaller than the RMSEs and MSEs associated with the corresponding models based on (1.2) which do not include the current median growth rate. These results also show improvement over some naive forecasting rules. For example, forecasting 0 percent as the median growth rate in each year yields RMSE = 2.69, while a 3 percent forecasting rule yields RMSE = 1.66.

Using the same data, we also performed calculations to determine which of the three models performed best in forecasting turning points in the median growth rate of the eighteen countries over the period 1980–95. As in previous work, we define a downturn (*DT*) in period $T+1$ as occurring if the following median output growth rate sequence occurs:

$$w_{T-2}, w_{T-1} < w_T > w_{T+1}.$$

Also, by definition, no downturn (*NDT*) occurs when the following sequence is observed:

$$w_{T-2}, w_{T-1} < w_T \leq w_{T+1}.$$

Similarly, an upturn (*UT*) is said to occur in period $T+1$ if the following sequence of observations occurs:

$$w_{T-2}, w_{T-1} > w_T < w_{T+1}.$$

No upturn (*NUT*) occurs in period $T+1$ if the following sequence is observed :

$$w_{T-2}, w_{T-1} > w_T \geq w_{T+1}.$$

Table 22.2 *Results of forecasting turning points in the median output growth rate of eighteen countries, 1980–1995, number of correct forecasts for alternative models*

	DT	NDT	UT	NUT
	Observed outcomes			
	3	4	2	3
Model	Number correctly forecasted			
Aggregate AR(3)LI for w_t. ((1.1))	1	4	1	1
Disaggregated AR(3)LI ((1.2)) with	1	4	2	0
$\gamma_i = \gamma, \delta_{ji} = \delta_j \forall j$	–	–	–	–
Disaggregated AR(3)LI ((1.2)) with	2	3	2	0
$\delta_{ji} = \delta_j, j = 4, 5, 6, 7$	–	–	–	–
Disaggregated AR(3)LI ((1.2))	1	3	2	0
Disaggregated AR(3)LI ((1.3)) with $\hat{w}_t$	1	4	1	1
$\phi_i = \phi, \gamma_i = \gamma, \delta_{ji} = \delta_j, \forall j$	–	–	–	–
Disaggregated AR(3)LI ((1.3)) with $\hat{w}_t$	1	4	1	0
$\phi_i = \phi, \delta_{ji} = \delta_j, j = 4, 5, 6, 7$	–	–	–	–
Disaggregated AR(3)LI ((1.3)) with $\hat{w}_t$	1	4	2	0

Given that we have a predictive density for w_{T+1}, we can easily compute the probability of *DT* and of *NDT* and use these probabilities along with a 2×2 loss structure to determine the forecast that minimizes expected loss. If the 2×2 loss structure is symmetric, a *DT* is the optimal forecast given that the probability of a *DT* is greater than 1/2.[3] If the probability of a DT is less than 1/2, then the optimal forecast is NDT. Similar considerations relate to forecasting *UT*s and *NUT*s. See LeSage (1996), Zellner, Hong, and Min (1991) and Zellner, Tobias, and Ryu (1998) for further discussion and applications of this turning point forecasting methodology.

In table 22.2 are the number of *DT*, *NDT*, *UT*, and *NUT* events that actually occurred in our sample and the number of correct forecasts using the procedure described above with each of our three models, shown in (1.1), (1.2), and (1.3). As regards the poor *NUT* forecasts, in Zellner, Tobias, and Ryu (1998), it was found that use of "trend" add factors that represent inertia effects produced improved *NUT* forecasts for individual countries' annual output growth rates. It is important to note that the restricted version of (1.3) with ω_t which had the lowest RMSE and MAE

[3] We use the forecasted median from the alternative models as an approximation to the posterior predictive mean of w_{T+1}, and assume the predictive density is symmetric about the forecasted median.

Table 22A.1 *Diffuse prior posterior means and standard deviations for coefficients of AR(3)LI model for the annual median output growth rate ((1.1)),* $R^2 = 0.68$[a]

	Constant	w_{-1}	w_{-2}	w_{-3}	MGM_{-1}	MSR_{-1}
Coefficient mean	1.10	0.278	0.081	0.011	0.012	0.409
Posterior std dev.	(0.527)	(0.145)	(0.152)	(0.142)	(0.0220)	(0.080)

Note:
[a] With use of diffuse priors and an i.i.d. normal likelihood function, coefficients' posterior means are equal to least squares estimates. Posterior standard deviations are equal to usual least squares asymptotic standard error times $\sqrt{v_i/(v_i - 2)}$ where $v_i = n_i - k_i$. Estimation results are computed using the full sample, 1954–95. R^2 is an approximate mean of the posterior density of the population R^2 parameter.

is also among the five models which correctly forecast seven of twelve turning points in the median output growth rate.

3 Conclusion

In summary, our forecasting experiments provide some evidence that improved forecasting results can be obtained by disaggregation given that an aggregate variable, w_t, appears in the disaggregated relations, as shown in (1.3). With disaggregation, there are more observations to estimate parameters and given that the disaggregated relations are reasonably specified, it is possible to obtain improved forecasts of an aggregate variable, here w_t, the median growth rate, a result that is in accord with some views expressed in the literature (see, e.g., Espasa (1994) and Palm and Zellner (1992)).

APPENDIX ESTIMATION RESULTS

The output, money, and stock price variables used in this chapter are first converted to real quantities by dividing each variable by a country-specific price index. The variables are then logged, first-differenced, and multiplied by 100 to convert to growth rates. Estimation results for the AR(3)LI model in (1.1) are presented in table 22A.1, and coefficient posterior means and standard deviations for the models in (1.2) and (1.3) with coefficients restricted to be equal across countries are presented in table 22A.2. On computing the roots of the AR(3) process for the

Table 22A.2 *Diffuse prior posterior means and standard deviations for coefficients of (1.2) and (1.3) with coefficients restricted to be the same across countries*

	Disaggregated AR(3)LI with $\hat{w}_t$ ((1.3))		Disaggregated AR(3)LI ((1.2))	
Variable	Coeff. mean	Std dev.	Coeff. mean	Std dev.
$\hat{w}_t$	0.863	(0.055)	–	–
Constant	−0.400	(0.211)	1.53	(0.200)
y_{-1}	0.228	(0.035)	0.373	(0.039)
y_{-2}	−0.103	(0.037)	−0.048	(0.042)
y_{-3}	0.080	(0.031)	0.157	(0.036)
SR_{-1}	0.033	(0.007)	0.025	(0.008)
SR_{-2}	−0.008	(0.006)	−0.021	(0.006)
GM_{-1}	0.045	(0.012)	0.077	(0.014)
MSR_{-1}	−0.043	(0.011)	0.016	(0.012)
R^2	0.48		0.29	

countries' output growth rates from (1.2),[4] we obtain one real root equal to 0.6646, and two complex conjugate roots, $0.1488 \pm 0.4676i$, associated with a damped oscillatory component with estimated amplitude = 0.491 and estimated period = 4.98. See Geweke (1986) and Hong (1989) for Bayesian procedures for making posterior inferences about the properties of roots of an AR(3) process. Using earlier data for eighteen countries included in our sample, Hong found that there is a high posterior probability that: (a) there are two complex roots and one real root, (b) the real root has amplitude less than one, and (c) the complex roots are associated with a damped oscillatory component with a period of about four–six years. We arrive at similar conclusions using updated and expanded data.

The addition of the "world return" variable, MSR_t, has been shown in past work to reduce contemporaneous correlation among the error terms. See also Zellner, Hong and Gulati (1990) and Min and Zellner (1993) for discussions of time-varying parameter models and shrinkage techniques for obtaining point and turning point forecasts, and the uses

[4] Estimation results are obtained by including the leading indicators in (1.2). The roots, amplitude, and period are then computed from the AR(3) relationship using the posterior mean as a point estimate of the regression parameters. By drawing from the multivariate Student-t distribution (the posterior distribution for the regression coefficients), we can also compute the probability that the AR(3) will have one real root associated with a cycle and two complex conjugate roots associated with a trend. Posterior distributions of the amplitude and period can also be obtained. See Hong (1989) for discussions.

Table 22A.3 *Diffuse prior posterior means and standard deviations allowing regression coefficients to vary across countries*[a]

Country	$\hat{\phi}_i$ $\hat{w}_t$	γ_i Const	δ_{1i} y_{-1}	δ_{2i} y_{-2}	δ_{3i} y_{-3}	δ_{4i} GM_{-1}	δ_{5i} SR_{-1}	δ_{6i} SR_{-2}	δ_{7i} MSR_{-1}	R^2
Australia	0.873 (0.316)	0.497 (1.20)	−0.131 (0.179)	−0.090 (0.177)	0.179 (0.170)	0.146 (0.082)	0.021 (0.040)	−0.014 (0.029)	−0.058 (0.058)	0.62
Austria	0.807 (0.194)	−0.016 (0.829)	0.199 (0.141)	−0.070 (0.145)	0.103 (0.128)	0.059 (0.069)	0.025 (0.019)	−0.003 (0.017)	−0.004 (0.034)	0.43
Belgium	0.916 (0.117)	− 1.21 (0.434)	0.123 (0.104)	0.110 (0.105)	0.053 (0.100)	0.038 (0.049)	−0.002 (0.025)	−0.009 (0.016)	0.025 (0.029)	0.86
Canada	0.844 (0.274)	−0.254 (0.914)	0.244 (0.185)	−0.003 (0.201)	−0.068 (0.171)	0.052 (0.048)	0.057 (0.044)	−0.036 (0.037)	−0.071 (0.054)	0.57
Denmark	1.34 (0.316)	−0.702 (0.886)	0.047 (0.164)	−0.461 (0.174)	−0.044 (0.145)	0.079 (0.078)	−0.014 (0.028)	0.039 (0.025)	−0.071 (0.039)	0.52
Finland	1.29 (0.319)	−1.00 (1.16)	0.242 (0.163)	−0.407 (0.145)	0.109 (0.116)	−0.002 (0.032)	0.060 (0.033)	−0.008 (0.031)	−0.022 (0.059)	0.63
France	0.868 (0.202)	−0.152 (0.655)	0.154 (0.130)	−0.172 (0.143)	0.139 (0.130)	0.080 (0.061)	−0.043 (0.028)	0.027 (0.016)	0.023 (0.046)	0.70
Germany	0.450 (0.432)	−0.255 (1.41)	0.286 (0.193)	−0.170 (0.203)	0.098 (0.201)	0.261 (0.124)	0.080 (0.060)	0.016 (0.048)	−0.053 (0.074)	0.75
Ireland	0.541 (0.269)	2.56 (1.39)	0.014 (0.189)	−0.091 (0.187)	−0.073 (0.150)	0.057 (0.094)	0.068 (0.048)	−0.024 (0.031)	−0.171 (0.058)	0.36
Italy	1.02 (0.246)	0.594 (0.949)	−0.013 (0.131)	−0.028 (0.137)	0.017 (0.130)	0.088 (0.041)	0.028 (0.021)	−0.014 (0.016)	0.012 (0.046)	0.67
Japan	0.812 (0.408)	−0.646 (1.07)	0.378 (0.173)	0.190 (0.199)	0.020 (0.166)	−0.027 (0.119)	0.116 (0.052)	−0.083 (0.039)	−0.113 (0.064)	0.70
Netherlands	1.41 (0.220)	− 1.14 (0.670)	0.108 (0.118)	−0.099 (0.122)	0.014 (0.117)	−0.093 (0.081)	0.018 (0.027)	−0.008 (0.019)	−0.037 (0.038)	0.74
Norway	0.364 (0.220)	2.48 (1.32)	0.275 (0.170)	−0.398 (0.153)	0.017 (0.168)	0.011 (0.057)	−0.002 (0.028)	−0.031 (0.020)	−0.002 (0.049)	0.45
Spain	1.10 (0.272)	−0.269 (1.03)	0.219 (0.173)	−0.053 (0.172)	−0.107 (0.154)	0.104 (0.118)	0.032 (0.034)	−0.001 (0.028)	−0.036 (0.060)	0.66
Sweden	0.988 (0.208)	−0.678 (0.690)	0.085 (0.178)	−0.179 (0.168)	−0.076 (0.175)	0.119 (0.100)	0.034 (0.022)	0.023 (0.021)	−0.082 (0.038)	0.60
Switzerland	0.967 (0.246)	− 1.22 (0.757)	0.389 (0.167)	−0.081 (0.192)	−0.059 (0.159)	−0.014 (0.064)	0.034 (0.034)	0.008 (0.025)	−0.017 (0.044)	0.75
United Kingdom	0.720 (0.242)	0.900 (0.766)	−0.133 (0.231)	−0.174 (0.202)	−0.265 (0.187)	0.060 (0.035)	0.041 (0.035)	0.018 (0.026)	−0.094 (0.049)	0.54
United States	0.798 (0.184)	1.13 (0.756)	−0.277 (0.315)	−0.050 (0.118)	−0.161 (0.095)	0.215 (0.075)	0.140 (0.038)	−0.001 (0.030)	−0.145 (0.039)	0.63

Note:
[a] The coefficient posterior means presented above are numerically equivalent to least squares estimates. The posterior standard deviations equal least squares asymptotic standard errors times $\sqrt{v_i/(v_i - 2)}$ where $v_i = n_i - k$. Estimation results are computed using the full sample, 1954–95.

of posterior odds for comparing, choosing between and/or combining alternative forecasting models.

BIBLIOGRAPHY

Espasa, A. (1994), "Comment on 'Time series analysis, forecasting and econometric modeling: the structural econometric modeling, time series analysis (SEMSTA) approach,'" *Journal of Forecasting* 13, 234–5

Garcia-Ferrer, A., R. A. Highfield, F. C. Palm, and A. Zellner (1987), "Macroeconomic forecasting using pooled international data," *Journal of Business and Economic Statistics* 5(1), 53–67; chapter 13 in this volume

Geweke, J. (1986), "Exact inference in the inequality constrained normal linear regression model," *Journal of Applied Econometrics* 1, 127–41

Hong, C. (1979), "Forecasting real output growth rates and cyclical properties of models: a Bayesian approach," PhD thesis, Department of Economics, University of Chicago

LeSage, J. P. (1990), "Forecasting turning points in metropolitan employment growth rates using Bayesian techniques," *Journal of Regional Science* 30(4), 533–48; see chapter 21 in this volume

Min, C.-K. and A. Zellner (1993), "Bayesian and non-Bayesian methods for combining models and forecasts with applications to forecasting international growth rates," *Journal of Econometrics, Annals* 56, 89–118; chapter 17 in this volume; reprinted in Zellner (1997)

Palm, F. C. and A. Zellner (1992), "To combine or not to combine? Issues of combining forecasts," *Journal of Forecasting* 11, 687–701; reprinted in Zellner (1997)

Zellner, A. (1994), "Time series analysis, forecasting, and econometric modeling: the structural econometric modeling, time series analysis (SEMSTA) approach," *Journal of Forecasting* **13**, 215–33; chapter 4 in this volume

(1997), *Bayesian Analysis in Econometrics and Statistics: The Zellner View and Papers* (Cheltenham, Edward Elgar)

Zellner, A. and C. Hong (1989), "Forecasting international growth rates using Bayesian shrinkage and other procedures," *Journal of Econometrics, Annals* 40, 183–202; chapter 14 in this volume; reprinted in Zellner (1997)

Zellner, A., C. Hong, and G. M. Gulati (1990), "Turning points in economic time series, loss structures, and Bayesian forecasting," part VI in S. Geisser, J. S. Hodges, S. J. Press, and A. Zellner (eds.), *Bayesian and Likelihood Methods in Statistics and Econometrics: Essays in Honor of George A. Barnard* (Amsterdam, North-Holland), 371–89; chapter 15 in this volume

Zellner, A., C. Hong, and C. Min (1991), "Forecasting turning points in international output growth rates using Bayesian exponentially weighted autoregression, time-varying parameter, and pooling techniques," *Journal of Econometrics* 49, 275–304; chapter 16 in this volume; reprinted in A. Zellner, *Bayesian Analysis in Econometrics and Statistics: The Zellner View and Papers* (Cheltenham, Edward Elgar)

Zellner, A., J. Tobias, and H. Ryu (1998), "Bayesian method of moments analysis of time series models with an application to forecasting turning points in output growth rates," H. G. B. Alexander Research Foundation, University of Chicago; [published in *Estadistica* 49–51 (152–157), 3–63].

23 The Marshallian macroeconomic model (2000)

Arnold Zellner

In this paper, background information on the origins and features of the Marshallian Macroeconomic Model (MMM) are presented. MMMs based on two alternative production functions are presented and compared. In addition, some empirical forecasting results for one of them are reviewed. Last, attention is focused on further development and implementation of the MMM.

1 Introduction

It is an honor and a pleasure to present my paper at this research conference honoring Professor Ryuzo Sato, a superb colleague and most productive scholar. His outstanding research analyzing production and technological change, Sato (1999a, 1999b) has been appreciated worldwide. Indeed, these topics play a central role in almost all models of industrial sectors and economies, including the models to be discussed below.

On the origins of the [Marshallian Macroeconomic Model] (MMM), in my experience it was a pleasure teaching undergraduate and graduate students the properties and uses of the Marshallian model of a competitive industry. On the other hand, teaching students macroeconomics was quite a different matter since there was no such comparable, operationally successful model available (See, e.g., Belongia and Garfinkel 1992 for an excellent review of alternative macroeconomic models, including monetarist, neo-monetarist, Keynesian, post-Keynesian, and real business cycle models and Fair 1992 and Zellner 1992, who pointed out that not enough empirical testing of alternative models had been done and

Research financed by the National Science Foundation, the CDC Management Corp. and the Alexander Fund, Graduate School of Business, University of Chicago.

Originally an invited paper presented at the research conference in honor of Professor Ryuzo Sato; published in T. Nagishi, R. V. Ramachandran, and K. Mino (eds.), *Economic Theory, Dynamics and Markets: Essays in Honor of Ryuzo Sato* (Boston, Kluwer, 2000), 19–29.

more is needed to produce macroeconomic models that explain the past, predict well, and are useful in making policy.)

To help achieve these objectives, the structural econometric time series analysis approach, described and applied in Zellner and Palm (1974, 1975), Palm (1976, 1977, 1983), Zellner (1979, 1994, 1997), Garcia-Ferrer *et al.* (1987), and Hong (1989), Min (1992), was used to generate an equation, a third order autoregression with lagged leading indicator variables (ARLI) for the annual rate of growth of real GDP that worked fairly well in point forecasting one year ahead and forecasting downturns and upturns in rates of growth with about 70 percent correct turning point forecasts in 211 turning point episodes for eighteen industrialized countries (see, e.g., Zellner and Min 1999 and references therein). While Hong (1989), Min (1992), and Zellner (1999) showed that this ARLI equation is mathematically implied by variants of a Hicksian IS-LM macro model, a generalized real business cycle model and an aggregate demand and supply model, respectively, it is not clear that these macroeconomic models are entirely adequate. In particular, they abstract from important phenomena involved in business cycles and growth, namely, industrial sectors with different cyclical properties, entry and exit of firms, sector linkages, etc. The challenge is to formulate a relatively simple model that accommodates such characteristics, works well in forecasting and is flexible enough to elaborate in various dimensions if needed.

One morning while shaving, the idea came to me to go back to my favorite Marshallian demand, supply AND entry model (see Veloce and Zellner 1985 for a formulation of such a model with an application to a Canadian industry) and use one for each sector of an economy. Note that many macroeconomic, general equilibrium and demand and supply models do not include an entry equation. Indeed, in some models, there is just a representative firm and one wonders what happens if this firm shuts down. Also, in many rational expectation models of competitive industries (see, e.g., Muth 1961) the number of firms is assumed constant. On aggregating supply functions over firms, the number of firms in operation, N, appears in the industry supply equation and thus an entry equation is needed, along with demand and supply functions, to close the model. Indeed, when the N variable is omitted from the supply function, very strange estimation results are obtained (see some examples in Veloce and Zellner 1985).

In the MMM model, we utilize industrial sectors, each with consumer demand, sector supply and entry equations, and productive units that buy factors of production, labor, capital and money services, and other inputs to production in national factor markets. With the addition of demand and supply models for factor markets and a foreign sector, we have a

MMM model. Of course such a model can be elaborated in many ways. However, consistent with the SEMTSA approach, we start simply and complicate if necessary, in contrast to the "general to specific" approach used by many. Note that there are many general models, including VAR, MVARMA, non-linear VARs, etc. and if the wrong one is chosen, users of the general to specific approach will be disappointed.

In what follows, in section 2, the Veloce–Zellner sector model, based on use of Cobb–Douglas production functions, will be reviewed and extended to include technical change and factor augmentation effects. It will be shown how the model has been used in Zellner and Chen (2000) to forecast growth rates of sectors' real sales and of total real GDP of the US economy. Then in section 3, another production function, a "generalized production function," will be utilized and the model resulting from its use will be compared with that based on the Cobb–Douglas production function with neutral technical change and factor augmentation. Last, in section 4 consideration is given to several properties of the models considered in this chapter that can be relaxed or modified. Also, as is apparent, it will be pointed out that factor markets for labor, capital, money, and bonds can be added to the sector models to complete the MMM. Last, it is noted that the MMM can accommodate the birth of new and the death of old sectors.

2 A competitive Marshallian sector model of an economy

In this section, a slightly modified version of the competitive demand, supply and entry model put forward and estimated in Veloce and Zellner (1985) is presented. We assume a competitive industry with $N(t)$ firms in operation at time t, each with a Cobb–Douglas production function, $q = AL^{\alpha}K^{\beta}$, where $A = A(t) = A_N(t)A_L(t)^{\alpha}A_K(t)^{\beta}$, the product of a neutral technological change factor and labor and capital augmentation factors that reflect changes in the quality of labor and capital inputs. Additional inputs, e.g. services of money, inventories and raw materials, can be added without much difficulty. On assuming profit maximization with respect to inputs given factor prices, $w = w(t)$, the real wage rate and $r = r(t)$, the real price of capital services and given real product price, $p = p(t)$, the sector's real sales supply function is $S = \mathrm{N}pq = NA^{*}\ p^{1/\theta}w^{-\alpha/\theta}r^{-\beta/\theta}$, where $A^{*} = (A^{\alpha}\beta^{\beta})^{1/\theta}$, and $0 < \theta = 1 - \alpha - \beta < 1$. On logging both sides of the equation for S, real sales, and differentiating with respect to time, we obtain the industry real sales supply equation:

$$\dot{S}/S = \dot{N}/N + \dot{A}^{*}/A^{*} + (1/\theta)\dot{p}/p - (\alpha/\theta)\dot{w}/w - (\beta/\theta)\dot{r}/r \quad \mathit{SUPPLY}, \tag{2.1}$$

where $\dot{x}/x = (1/x)dx/dt$. Further if we multiply both sides of the industry demand function by p, we obtain real sales $= S = pQ = Bp^{1-\eta}x_1^{\eta_1}x_2^{\eta_2}\ldots x_k^{\eta_k}$, where the x variables are demand shifters such as real income, real money balances, number of consumers, etc. On logging and differentiating this last equation with respect to time, the result is:

$$\dot{S}/S = (1-\eta)\dot{p}/p + \sum_{i=1}^{k}\eta_i\dot{x}_i/x_i \quad DEMAND, \tag{2.2}$$

Finally, the following entry equation will be utilized to complete the model for the three variables, price, p, real sales, S, and number of firms in operation, N:

$$\dot{N}/N = \gamma'(\Pi - F_e) = \gamma(S - F), \quad ENTRY \tag{2.3}$$

where profits $\Pi = \theta S$, $\gamma = \gamma'\theta$, with $0 < \gamma < 1$, and $F = F_e / \theta$, with F_e the equilibrium level of profits taking account of discounted entry costs.

In Veloce and Zellner (1985), data for the Canadian furniture industry were employed to estimate discrete versions of (2.1) and (2.3) taking factor prices and demand shift variables as exogenous variables. If it is assumed that all sectors sell in the final product market, similar analysis provides supply, demand, and entry equations for each sector. Note that the above demand equations can be elaborated to take account of substitution and complementarity effects. Also, parameters' values will usually differ over sectors.

When (2.1)–(2.3) are solved for $\dot{S}/S$ by substituting from (2.3) in (2.1) and then eliminating the $\dot{p}/p$ variable, the result is

$$\dot{S}/S = (a+g)[1 - S/(a+g)F], \tag{2.4}$$

where $a = \gamma\theta F/(\alpha + \beta - \eta)$ and $g =$ a linear function of rates of change of A^*, w, r, and the xs, the demand shifters. If in (2.4), $g = 0$ or $g =$ const., it is seen that the differential equation for S has a solution in the form of the well-known logistic function. Further, if $g = g(t)$, a given function of t, the equation is in the form of Bernoulli's differential equation (see Veloce and Zellner 1985, p. 463, for its general solution). Note that g may change through time because of changes in the rate of growth of neutral technical change, factor augmentation and/or in the rates of change of exogenous variables affecting demand and supply, say real money balances, real income, the real wage rate, etc.

In the SEMSTA approach, mentioned above, it is considered important to test the forecasting performance of equations derived from theoretical models, such as that in (2.4). In Zellner and Chen (2000) we employed, among others, the following discrete approximation to (2.4) in analyzing real GDP of eleven sectors of the US economy including agriculture, mining, construction, etc.

$$(1 - L)\log S_t = \alpha_o + \alpha_1 S_{t-1} + \alpha_2 S_{t-2} + \alpha_3 S_{t-3} + \beta_1(1 - L)\log Y_t + \beta_2(1 - L)\log M_{t-1} + \beta_3(1 - L)\log w_t + \beta_4(1 - L)\log SR_{t-1} + u_t, \quad (2.5)$$

where L is the lag operator, S sector real GDP, Y aggregate real GDP, M real money balances, w the real wage rate, SR a real stock price index, and u an error term. Lagged values of S were employed to reflect lags in the entry equation. Note that the rate of change of S is related to lagged levels of S, a "cointegration" effect that flows from the model. With w and Y assumed exogenous relative to eleven individual sectors of the US economy, equations in the form of (2.5), and variants of it, with sector-specific parameters, were fitted as a set of seemingly unrelated regressions, using US annual data, 1949–79, and then employed to produce one-year-ahead forecasts, with estimates updated year by year for the period, 1980–97. While the models for certain sectors, namely agriculture, mining, and construction did not perform very well given the great variability of these sectors' outputs, when the annual sector real GDP forecasts were added to provide a one-year-ahead forecast of total US real GDP and its growth rate, it was found that such forecasts are better than those of benchmark models implemented with aggregate annual data including an AR(3) model for the rate of change of Y, the AR(3) model with added lagged leading indicator variables, and the same model with lagged levels of Y and a time trend variable. Thus, in this case it appears that it pays to disaggregate, mainly because sector-specific variables can be employed along with many more observations than in an aggregate, one-equation approach utilizing aggregate data. Similar results were obtained using (2.5), with Y and w assumed to be endogenous variables and various estimation techniques including OLS, 2SLS, Bayesian minimum expected loss (MELO), and shrinkage techniques. Results using a real currency variable were somewhat better than using a real M1 variable in sectors' demand equations. For the one-year-ahead forecasts of the annual growth rates of real US GDP, 1980–97, the mean absolute errors (MAEs) ranged from 1.17 to 1.38 percentage points for the alternative approaches mentioned above applied to (1.5). For a benchmark AR(3) model that missed all the turning points, the MAE = 1.71, about 24–46 percent larger than the MAEs for the disaggregate forecasts. With improvement of the models for certain sectors, e.g. the highly variable agricultural, mining, and construction sectors, it may be that additional improvements in forecasting performance, etc.

can be realized. Last, joint estimation and forecasting using the three equations (1.3)–(1.5) for each sector, may produce even better results.

Thus in terms of aggregate forecasting, the disaggregate approach described above seems worthwhile. Further, it not only yields forecasts of aggregate output but, obviously, of its components sector by sector. In addition, the possibility of pooling data of various countries' sector models in estimation and prediction may result in further improvement in forecasting results. (See, e.g., Garcia-Ferrer *et al.* 1987, Zellner and Hong 1989, Zellner 1994, 1997, and Quintana, Chopra, and Putnam 1995 for empirical results showing that use of pooling or shrinkage techniques results in improved forecasts.)

3 Sector model based on a generalized production function

While the results above are useful, it is of interest to see how the sector models' form is affected by use of a "generalized production function," (GPF) rather than a Cobb–Douglas production function. In general a GPF is a monotonic function of homogeneous function, say a Cobb–Douglas function. That is, GPFs are in the class of homothetic production functions with prespecified forms for the returns to scale function (see, e.g., Zellner and Revanker 1969, Greene 1993, pp. 324–8), and Zellner and Ryu 1998. GPFs, in contrast to the Cobb–Douglas function have associated long-run average cost curves that are U- or L-shaped with a unique minimum. Further, the profit-maximizing labor's share is not a constant but varies with the level of output. A simple GPF that will be employed below is given by $\log q + \alpha q = \beta_o + \beta_1 \log L + \beta_2 \log K + \log A$, with α and the βs strictly positive parameters and where $A = A(t)$ denotes both general and factor augmenting technical change. This function has an elasticity of substitution = 1. In Zellner and Ryu (1998) many GPFs are presented and estimated, some with variable elasticity of substitution parameters, etc. Assuming competitive conditions and profit maximization on the part of firms in a sector yields the following expression for labor's share, $wL/pq = \beta_1/(1 + \alpha q)$. Thus labor's share is not constant but varies as q varies, say over the business cycle. Further, the supply function for an individual firm is given by:

$$\ln Q/N = c_o + c_1 Q/N + c_2 \ln(1 + \alpha Q/N) + c_3 \ln p + c_4 \ln w + c_5 \ln r + c_6 \ln A, \quad (3.1)$$

where $Q/N = q$, with N = number of identical firms in operation and the cs are parameters. A sector demand function is given by:

$$\ln p = \alpha_o + \alpha_1 \ln Q + \alpha_3' x, \quad (3.2)$$

where α_3 is a vector of parameters and x is a vector of demand shifters including real income, money balances, etc. Last the entry equation is given by:

$$\dot{N}/N = \delta(\pi/pq) = \delta\alpha(q - q_m)/(1 + \alpha q), \tag{3.3}$$

where δ is a positive adjustment coefficient and q_m is the output level associated with the minimum of the long-run average cost function associated with the GPF shown above. When $q = Q/N = q_m$, there is no change in the number of firms in operation, when $q > q_m$, there is entry and when $q < q_m$ there is exit of firms.

After differentiating (3.1) and (3.2) with respect to t, we can solve (3.1)–(3.3) for a differential equation for q with the exogenous variables as inputs, namely,

$$\dot{q}/q = [v_1(q - q_m) + (1 + \alpha q)g]/[(1 + \alpha q)^2 + v_2(1 + \alpha q) + v_3]. \tag{3.4}$$

In (3.4), the vs are parameters, g denotes a linear combination of the rates of change of demand and supply shift variables (e.g. real income, real money, real wage rate, etc.), and α is the parameter of the GPF given above. Note that the effect of a change in g on the growth rate is dependent on the level of output. If $g = 0$ or a constant, it is seen that the proportionate rate of growth of $q = Q/N$ is given by the ratio of a linear function of q divided by a quadratic function. It is possible to solve the differential equation in (3.4), and variants of it, and study their properties. Also solutions for the paths of p and N can be obtained, analytically or numerically.

As regards empirical work, if data are available on Q and N, then $q = Q/N$ can be formed and used to fit discrete approximations to (3.4) that can be implemented sector by sector and tested in forecasting experiments. If these results are satisfactory, further work on the structural equations (3.1)–(3.3) can be undertaken. Given data on N, Q, and p, discrete approximations to the structural equations in (3.1)–(3.3) can be estimated and checked for reasonableness and performance in prediction. On obtaining a joint predictive density for p, Q, and N, it can be employed to compute the density of $S = pQ$, real sales for each sector. Forecasts of S so obtained can be compared to those provided by the Cobb–Douglas model, described above, and actual outcomes. Simulation experiments can be performed to study dynamic properties of the two models and their responses to various policy changes, structural breaks, etc.

4 Concluding remarks

In this chapter, two sector models have been presented that are in the form of a Marshallian demand, supply, and entry model for competitive sectors or industries that sell in a final product market. Of course such models can be modified in many different ways. For example, it is possible to take account of many additional elements, namely substitution effects in demand, additional factor inputs, monopolistic competition, expectations, inventories and inventory investment, birth and death of sectors, different forms for the entry equation, etc. Indeed, much work on these topics has appeared in the literature over the years. It is thought that integration of such research results in the MMM will provide improved explanatory and predictive performance. Further, by adding labor, capital, and money service factor demands over the sectors and introducing supply functions for labor, capital, and money services along with export and government sectors, a complete MMM is obtained. As mentioned above, new sectors can be introduced, e.g. using available models for new products and services, and thus make the model sensitive to Schumpeterian-like waves of innovation associated with new industries such as the computer industry, etc.

The starting point for this modeling work is not a very general model but relatively simple sector models that perform reasonably well in explanation and prediction. These tested components can then be improved step by step to accommodate the additional features mentioned above. In this way, in accordance with the SEMTSA approach, we shall have valuable components that work well empirically and when combined will result in a sensible, satisfactory MMM that has good explanatory and predictive properties and will probably be useful to policy-makers.

BIBLIOGRAPHY

Belongia, M. and M. Garfinkel (eds.) (1992), "The business cycle: theories and evidence," Proceedings of the 16th Annual Economic Policy Conference of the Federal Reserve Bank of St. Louis (Boston, Kluwer Academic)

Fair, R. (1992), "How might the debate be resolved?," in M. Belongia and M. Garfinkel (eds.), "The business cycle: theories and evidence," Proceedings of the 16th Annual Economic Policy Conference of the Federal Reserve Bank of St. Louis (Boston, Kluwer Academic),

Garcia-Ferrer, A., R. A. Highfield, F. C. Palm, and A. Zellner (1987), "Macroeconomic forecasting using pooled international data," *Journal of Business and Economic Statistics* 5(1), 53–67; chapter 13 in this volume

Greene, W. (1993), *Econometric Analysis*, 2nd edn. (New York, Macmillan)

Hong, C. (1989), "Forecasting real output growth rates and cyclical properties of models: a Bayesian approach," PhD thesis, Department of Economics, University of Chicago

Min, C. (1992), "Economic analysis and forecasting of international growth rates using Bayesian techniques," PhD thesis, Department of Economics, University of Chicago

Muth, J. (1961), "Rational expectations and the theory of price movements," *Econometrica* 29, 315–35

Palm, F. C. (1976), "Testing the dynamic specification of an econometric model with an application to Belgian data," *European Economic Review* 8, 269–89

(1977), "On univariate time series methods and simultaneous equation econometric models," *Journal of Econometrics* 5, 379–88

(1983), "Structural econometric modeling and time series analysis: an integrated approach," in A. Zellner (ed.), *Applied Time Series Analysis of Economic Data* (Washington, DC: US Bureau of the Census, Department of Commerce), 199–233; chapter 3 in this volume

Quintana, J., V. Chopra, and B. Putnam (1995), "Global asset allocation: stretching returns by shrinking forecasts," in Proceedings of the Section on Bayesian Statistical Science, American Statistical Association, available at www.amstat.org

Sato, R. (1999a), *Theory of Technical Change and Economic Invariance* (Cheltenham, Edward Elgar); reprint, with amendments, of 1981 edn.

(1999b), *Production, Stability and Dynamic Symmetry: The Selected Essays of Ryuzo Sato,* in Economists of the Twentieth Century Series (Cheltenham, Edward Elgar)

Veloce, W. and A. Zellner (1985), "Entry and empirical demand and supply analysis for competitive industries," *Journal of Econometrics* 30, 459–71

Zellner, A. (1979), "Statistical analysis of econometric models," invited paper with discussion, *Journal of the American Statistical Association* 74, 628–51; chapter 2 in this volume

(1984), *Basic Issues in Econometrics* (Chicago, University of Chicago Press); reprinted in 1987

(1992), "Comment on Ray Fair's thoughts on 'How might the debate be resolved?'," in M. Belongia and M. Garfinkel (eds.), "The business cycle: theories and evidence," Proceedings of the 16th Annual Economic Policy Conference of the Federal Reserve Bank of St. Louis (Boston, Kluwer Academic), 148–57

(1994), "Time series analysis, forecasting, and econometric modeling: the structural econometric modeling, time series analysis (SEMTSA) approach," *Journal of Forecasting* 13, 215–33, invited paper with discussion; chapter 4 in this volume

(1997), *Bayesian Analysis in Econometrics and Statistics: The Zellner View and Papers* (Cheltenham, Edward Elgar)

(1999), "Bayesian and non-Bayesian approaches to scientific modeling in economics and econometrics," invited keynote paper presented at the Ajou University Research Conference in Honor of Professor Tong Hun Lee, South Korea, August; [published in Special Issue of the *Korean Journal of Money and Finance* 5(2) (November 2000)]

Zellner, A. and B. Chen (2000), "Bayesian modeling of economies and data requirements," invited keynote address at meeting of the International Institute of Forecasters and International Journal of Forecasting, Lisbon

June; [published in *Macroeconomic Dynamics* 5(2001), 673–700]; chapter 24 in this volume

Zellner, A. and C. Hong (1989), "Forecasting international growth rates using Bayesian shrinkage and other procedures," *Journal of Econometrics, Annals* 40, 183–202; chapter 14 in this volume

Zellner, A. and C. Min (1999), "Forecasting turning points in countries' output growth rates: a response to Milton Friedman," *Journal of Econometrics* 88, 203–6; chapter 19 in this volume

Zellner, A. and F. C. Palm (1974), "Time series analysis and simultaneous equation econometric models," *Journal of Econometrics* 2, 17–54; chapter 1 in this volume

(1975), "Time series analysis of structural monetary models of the US economy," *Sankyā: The Indian Journal of Statistics*, Series C 37, 12–56; chapter 6 in this volume

Zellner, A. and N. Revankar (1969), "Generalized production functions," *The Review of Economic Studies* 36, 241–50

Zellner, A. and H. Ryu (1998), "Alternative functional forms for production, cost and returns to scale functions," *Journal of Applied Econometrics* 13, 101–27

24 Bayesian modeling of economies and data requirements (2000)

Arnold Zellner and Bin Chen

1 Introduction

For many years, theoretical and empirical workers have tried to model national economies in order to (1) understand how they operate, (2) forecast future outcomes, and (3) evaluate alternative economic policies. While much progress has been made in the decades since Tinbergen's pioneering work [1940], it is the case that no generally accepted model has as yet appeared. On the theoretical side, there are monetary, neo-monetary, Keynesian, neo-Keynesian, real business cycle, generalized real business cycle, and other theoretical models (see, Belongia and Garfinkel 1992 for an excellent review of many of these models and Min 1992 for a description of a generalized real business cycle model). Some empirical testing of alternative models has appeared in the literature. However, in Fair (1992) and Zellner (1992) (invited contributions to a St. Louis Federal Reserve Bank conference on alternative macroeconomic models), it was concluded that there is a great need for additional empirical testing of alternative macroeconomic models and production of improved models.

Over the years many structural econometric and empirical statistical models have been constructed and used. These include large structural econometric models (e.g. the Tinbergen, Klein, Brookings–SSRC, Federal Reserve–MIT–PENN, OECD, Project Link, and other models). While progress has been made, there does not yet appear to be a structural model that performs satisfactorily in point and turning point forecasting. Indeed, the forecasting performance of some of these models is not as good as that of simple benchmark models, e.g., random-walk, autoregressive, Box–Jenkins univariate ARIMA, and autoregressive leading indicator (ARLI) models (see, e.g., Cooper 1972, Garcia-Ferrer *et al.* 1987,

Research financed by the National Science Foundation, the H. G. B. Alexander Endowment Fund, Graduate School of Business, University of Chicago, and the CDC Investment Management Corp. Originally presented as an invited keynote address at the June 2000 meeting of the International Institute of Forecasters and the International Journal of Forecasting, Lisbon; published in *Macroeconomics Dynamics*, 5 (2001), 673–700.

Hong 1989, and Nelson and Plosser 1982). Further, some have implemented vector autoregressive (VAR) and Bayesian VAR models in efforts to obtain improved forecasts (see, e.g., Litterman 1986 and McNees 1986). However these VARs have not in general been successful in point and turning point forecasting performance as noted by Zarnowitz (1986) and McNees (1986). See also the simulation experiments performed by Adelman and Adelman (1959) and Zellner and Peck (1973) that revealed some rather unusual properties of two large-scale econometric models.

Given the need for improved models, in Garcia-Ferrer *et al.* (1987) an empirical implementation of the structural econometric time series analysis (SEMTSA) approach of Zellner and Palm (1974, 1975), Palm (1976, 1977, 1983), and Zellner (1979, 1994) was reported. In line with the SEMTSA general approach, relatively simple forecasting equations, autoregressive leading indicator (ARLI) models were formulated and tested in forecasting output growth rates for nine industrialized countries with some success. In later work, the sample of countries was expanded to eighteen and the forecast period extended to include more out-of-sample growth rates of real GDP to be forecast. Building on the work of Wecker (1979) and Kling (1987), a Bayesian decision theoretic procedure for forecasting turning points was formulated and applied that yielded correct forecasts in about 70 percent of 211 turning point episodes (see Zellner and Min 1999, Zellner, Tobias, and Ryu 1999, and the references cited in these papers). Further, the ARLI models were shown to be compatible with certain aggregate supply and demand, Hicksian "IS-LM," and generalized real business cycle models in Hong (1989), Min (1992), and Zellner (1999).

In a continuing effort to improve our models, in the present chapter, we use a relatively simple, Marshallian model in section 2 that features demand, supply, and entry equations for each sector of an economy (see Veloce and Zellner 1985 for a derivation of this model and an application of it in the analysis of data for a Canadian industry). The model is solved to produce a sectoral relation that can be employed to forecast sectoral output. These sectoral output forecasts are summed to produce forecasts of total output that are compared to forecasts derived from models implemented with aggregate data. Some possible advantages of disaggregation have been discussed earlier by Orcutt *et al.* (1961), Espasa and Matea (1990), de Alba and Zellner (1991), and Espasa (1994), among others. Actual comparisons of such forecasts for US annual real GDP growth rates, 1980–97 will be reported in section 4 after statistical estimation and forecasting techniques, employed to implement the MMM, are presented in section 3. In section 4, the data used in our empirical forecasting work are described and forecasting results using the MMM and other models with no disaggregation and with disaggregation are reported. Also, MMM models' forecast performance is compared to that of various benchmark and ARLI models. In section 5, some comments on data requirements, a summary of conclusions, and remarks on future research, are presented.

2 The Marshallian macroeconomic model (MMM)

In the MMM, we have three basic rather well-known equations, described and applied in Veloce and Zellner (1985), namely the usual (1) demand for output, (2) supply of output, and (3) entry equations encountered in Marshall's famous economic analyses of the behavior of industries. While many macro models have included demand and supply equations, they have not included an entry equation. For example, in some models there is just a representative firm and one wonders what happens when the representative firm shuts down. In our MMM model, supply depends on the number of firms in operation and thus an equation governing the number of firms in operation, an entry equation, is introduced.

We shall use two variants of the MMM model, namely an aggregate, reduced form variant, and a disaggregated structural equation variant. In the aggregate variant, we shall adopt a "one-sector" view of an economy while in the disaggregated variant, we adopt a multi-sector view of an economy. As regards the multi-sectoral view, many assumed structures are possible, all the way from the multi-sectoral view of traditional Leontief input–output analysis to the simple view that we shall employ, namely an economy in which each sector sells in a final product market. Herein, we do not take up the interesting problem of classifying economies by the nature of their sectoral interrelations. However, we shall show that by adopting our sectoral view, we are able to improve forecasts of aggregate output growth rates since disaggregation provides more observations to estimate relationships and permits use of sectoral-specific variables to help improve forecasts. Of course, if the disaggregated relations are misspecified and/or the disaggregated data are faulty, then there may be no advantages, and perhaps some disadvantages, in using disagregated data, as is evident. Also, there are some circumstances even when data are good and relations are well formulated when disaggregation does not lead to improved forecasts. However the issue can not be completely settled theoretically and hence our current empirical work.

As explained in Veloce and Zellner (1985), the equations for a sector that we use are a demand equation for output, an industry supply equation for output, and a firm entry equation. While we could elaborate the system in many ways, we shall go forward to determine how well this simplest system peforms empirically, our "Model T" that can be improved in many different ways in the future. When these three equations are solved for the implied equation for the sectoral output growth rate (see Veloce and Zellner 1985 for details), the result is the following differential equation for total industry sales, denoted by $S = S(t)$:

$$(1/S)dS/dt = a(1 - S/F) + g \tag{2.1}$$

where a and F are postive parameters and g is a linear function of the growth rates of the wage rate, the price of capital, and of demand shifters such as real income, real money balances, etc. If $g = 0$ or $g = c$, a positive constant, it is the case that (2.1) is the differential equation with a logistic curve solution that is employed in many sciences, including economics. Also, note that (2.1) incorporates both the rate of change of S and the level of S, a "cointegration" effect. Also, see Veloce and Zellner (1985, p. 463), for analysis of (2.1) when $g = g(t)$, a special form of Bernouilli's differential equation and its solution.

In our empirical work we shall use the discrete approximations to (2.1) shown in table 24.1 and denoted by MMM(DA)I–IV. In these equations, the rate of growth of S, real output, is related to lagged levels of S, lagged rates of change of real stock prices, SR, and real money, m, and current rates of change of real wage rates, W, and real GDP, Y. The variables m and Y are "demand shifters" while W is the price of labor and SR is related to the price of capital. As noted in the literature and in our past work, the rates of change of m and SR are effective leading indicator variables in a forecasting context and their use has led to improved forecasts in our past work; see, references cited above for empirical evidence.

Shown under Sectoral forecast equations in table 24.1 are three benchmark models that will be used to produce sectoral one-year-ahead forecasts of the rates of change of output for each of our eleven sectors. The first is an AR(3) that has been used in many earlier studies as a benchmark model. The second is an AR(3) that incorporates lagged leading indicator variables and current values of W and Y but no lagged level variables. The third "Distributed lag" model is like the second except for the inclusion of lagged rates of change of W and of Y.

At the top of table 24.1, under Reduced form equations, are shown reduced form equations for the rate of change of Y, annual real GDP. The first is a benchmark AR(3) model. The second is an AR(3) with lagged leading indicator variables that is denoted by AR(3)LI. The third model, denoted MMM(A) is the same as the AR(3)LI model except for the inclusion of two lagged Y variables, where Y = real GDP, and a time trend, t.

Table 24.1 *Forecasting equations*

Reduced form equations:	
Real US GDP:	
AR(3)(A)	$(1-L)\log Y_t = \alpha_0 + \alpha_1(1-L)\log Y_{t-1} + \alpha_2(1-L)\log Y_{t-2} + \alpha_3(1-L)\log Y_{t-3} + u_t$
AR(3)LI(A)	$(1-L)\log Y_t = \alpha_0 + \alpha_1(1-L)\log Y_{t-1} + \alpha_2(1-L)\log Y_{t-2} + \alpha_3(1-L)\log Y_{t-3} + \beta_1(1-L)\log SR_{t-1} + \beta_2(1-L)\log m_{t-1} + u_t$
MMM(A):	$(1-L)\log Y_t = \alpha_0 + \alpha_1(1-L)\log Y_{t-1} + \alpha_2(1-L)\log Y_{t-2} + \alpha_3(1-L)\log Y_{t-3} + \alpha_4 \mathrm{Y}_{t-1} + \alpha_5 \mathrm{Y}_{t-2} + \alpha_6 t + \beta_1(1-L)\log SR_{t-1}$ $+ \beta_2(1-L)\log m_{t-1} + u_t$
Real wage:	
AR(3) (A)	$(1-L)\log W_t = \alpha_0 + \alpha_1(1-L)\log W_{t-1} + \alpha_2(1-L)\log W_{t-2} + \alpha_3(1-L)\log W_{t-3} + u_t$
AR(3)LI (A)	$(1-L)\log W_t = \alpha_0 + \alpha_1(1-L)\log W_{t-1} + \alpha_2(1-L)\log W_{t-2} + \alpha_3(1-L)\log W_{t-3}$ $+ \beta_1(1-L)\log SR_{t-1} + \beta_2(1-L)\log m_{t-1} + u_t$
MMM(A):	$(1-L)\log W_t = \alpha_0 + \alpha_1(1-L)\log W_{t-1} + \alpha_2(1-L)\log W_{t-2} + \alpha_3(1-L)\log W_{t-3} + \gamma_1 \mathrm{W}_{t-1} + \gamma_2 \mathrm{W}_{t-2} + \gamma_3 t$ $+ \beta_1(1-L)\log SR_{t-1} + \beta_2(1-L)\log m_{t-1} + u_t$
Sector forecast equations:	
AR(3)(DA)	$(1-L)\log S_t = \alpha_0 + \alpha_1(1-L)\log S_{t-1} + \alpha_2(1-L)\log S_{t-2} + \alpha_3(1-L)\log S_{t-3} + u_t$
AR(3)LI(DA)	$(1-L)\log S_t = \alpha_0 + \alpha_1(1-L)\log S_{t-1} + \alpha_2(1-L)\log S_{t-2} + \alpha_3(1-L)\log S_{t-3} + \beta_1(1-L)\log SR_{t-1} + \beta_2(1-L)\log m_{t-1}$ $+ \beta_3(1-L)\log W_t + \beta_4(1-L)\log Y_t + u_t$
Distrib. Lag (DA)	$(1-L)\log S_t = \alpha_0 + \alpha_1(1-L)\log S_{t-1} + \beta_1(1-L)\log SR_{t-1} + \beta_2(1-L)\log m_{t-1} + \beta_3(1-L)\log W_t + \beta_4(1-L)\log Y_t$ $+ \beta_5(1-L)\log W_{t-1} + \beta_6(1-L)\log Y_{t-1} + u_t$
MMM(DA)I:	$(1-L)\log S_t = \alpha_0 + \alpha_1 \mathrm{S}_{t-1} + \beta_1(1-L)\log SR_{t-1} + \beta_2(1-L)\log m_{t-1} + \beta_3(1-L)\log \mathrm{W}_t + \beta_4(1-L)\log Y_t + u_t$
MMM(DA)II:	$(1-L)\log S_t = \alpha_0 + \alpha_1 \mathrm{S}_{t-1} + \alpha_2 \mathrm{S}_{t-2} + \beta_1(1-L)\log SR_{t-1} + \beta_2(1-L)\log m_{t-1} + \beta_3(1-L)\log \mathrm{W}_t + \beta_4(1-L)\log Y_t + u_t$
MMM(DA)III:	$(1-L)\log S_t = \alpha_0 + \alpha_1 \mathrm{S}_{t-1} + \alpha_2 \mathrm{S}_{t-2} + \alpha_3 \mathrm{S}_{t-3} + \beta_1(1-L)\log SR_{t-1} + \beta_2(1-L)\log m_{t-1} + \beta_3(1-L)\log \mathrm{W}_t$ $+ \beta_4(1-L)\log Y_t + u_t$
MMM(DA)IV:	$(1-L)\log S_t = \alpha_0 + \alpha_1 \mathrm{S}_{t-1} + + \alpha_2 \mathrm{S}^2_{t-1} + \beta_1(1-L)\log SR_{t-1} + \beta_2(1-L)\log m_{t-1} + \beta_3(1-L)\log \mathrm{W}_t + \beta_4(1-L)\log Y_t + u_t$

For our aggregate analyses, we use the Reduced form equations in table 24.1 to produce one-year-ahead forecasts of the rate of change of real GDP, Y, that we refer to as "aggregate forecasts." These are means of diffuse prior Bayesian predictive densities for each model that are simple one-year-ahead least squares forecasts. As explained below, the MMM(A) reduced form equations for the rates of change of Y and of W will be employed in the estimation of the Sectoral forecast equations and in computing one-year-ahead forecasts of sectoral outputs growth rates. These sectoral growth rate forecasts are transformed into forecasts of levels, added across the sectors and converted into a forecast of the rate of change of real GDP, Y. Root mean-squared errors (RMSEs) and mean absolute errors (MAEs) are computed for each forecasting procedure and are shown in the table below.

3 Estimation and forecasting methods

3.1 Notation and equations

In what follows, we shall use the following notation. For each sector, we have:

1. Endogenous or random current exogenous variables:

$$y_{1t} = (1 - L)\log S_t; \quad y_{2t} = (1 - L)\log W_t; \quad y_{3t} = (1 - L)\log Y_t,$$

where S_t = sectoral real output, W_t = national real wage rate, and Y_t = real GDP.

2. Predetermined variables:

$$x'_{1t} = (1, S_{t-1}, S_{t-2}, S_{t-3}, (1 - L)\log SR_{t-1}, (1 - L)\log m_{t-1}),$$

where SR_t = real stock price and m_t = real money.

We use these variables to form the following structural equation for each sector:

$$y_{1t} = y_{2t}\gamma_{21} + y_{3t}\gamma_{31} + x'_{1t}\beta_1 + u_{1t} \quad t = 1, 2, \ldots, T,$$

or

$$y_1 = Y_1\gamma_1 + X_1\beta_1 + u_1, \tag{3.1}$$

where the vectors y_1 and u_1 are $Tx1$, Y_1 is $Tx2$ and X_1 is $Tx5$ and $\delta_1{}' = (\gamma_1{}', \beta_1{}')$ is a vector of structural parameters.

The MMM unrestricted reduced form equations, shown in table 24.1, are denoted by:

$$y_1 = X\pi_1 + v_1, \tag{3.2a}$$

and

$$Y_1 = X\Pi_1 + V_1, \tag{3.2b}$$

where X $= (X_1, X_0)$ with X_0 containing predetermined variables in the system that are not included in (3.1).

By substituting from (3.2b) in (3.1), we obtain the following well-known restricted reduced form equation for y_1:

$$y_1 = X\Pi_1\gamma_1 + X_1\beta_1 + v_1 \tag{3.3a}$$

$$= \overline{Z}\delta_1 + v_1, \tag{3.3b}$$

where $\overline{Z} = (X\Pi_1, X_1)$, that is assumed of full column rank.

Further, if we consider the regression of v_1 on V_1,

$$v_1 = V\eta_1 + e_1 = (Y - X\Pi_1)\eta_1 + e_1, \tag{3.4}$$

we can substitute for v_1 in (3.3) to obtain:

$$y_1 = X\Pi_1\gamma_1 + X_1\beta_1 + (Y_1 - X\Pi_1)\eta_1 + e_1. \tag{3.5}$$

In (3.5), for given Π_1, we have a regression of y_1 on $X\Pi_1$, X_1 and $Y_1 - X\Pi_1$. Given that e_1 is uncorrelated with the the elements of V_1, the system (3.2b) and (3.5) is a non-linear in the parameters SUR system with an error covariance matrix restriction. Pagan (1979) has earlier recognized a connection of the above model in (3.1) and (3.2b) to the SUR model given the "triangularity" of the system and reported an iterative computational procedure for obtaining maximum likelihood estimates of the structural coefficients. In our case, we shall use (3.2b) and (3.5) as a basis for producing a convenient algorithm for computing posterior and predictive densities.

Note further that if $\gamma_1 = \eta_1$, (3.5) becomes:

$$y_1 = Y_1\gamma_1 + X_1\beta_1 + e_1, \tag{3.6}$$

the same as (3.1) except for the error term. It is possible to view (3.6) as a regression with Y_1 containing observations on stochastic independent variables given that the elements of e_1 and V_1 are uncorrelated. The above restriction however may not hold in general. Another interpretation that permits (3.6) to be viewed as a regression with stochastic input variables is that the variables y_{2t} and y_{3t} are stochastic exogenous variables *vis-à-vis* the sectoral model. In such a situation, (3.1) can be treated as a regression equation with stochastic independent variables. However, we are not sure that this exogeneity assumption is valid and thus will use not only least squares techniques to estimate (3.1) but also special simultaneous equations techniques.

3.2 *Estimation techniques*

The sampling theory estimation techniques that we shall employ in estimating the parameters of (3.1) are the well-known "ordinary least squares" (OLS) and "two-stage least squares" (2SLS) methods. As shown in Zellner (1998), in very small samples, but not in large samples, the OLS estimate is an optimal Bayesian estimate relative to a generalized quadratic "precision of estimation" loss function when diffuse priors are employed. Also, the 2SLS estimate has been given an interpretation as a conditional Bayesian posterior mean using (3.3) conditional on $\Pi_1 = \hat{\Pi}_1 = (X'X)^{-1}X'Y$, a normal likelihood function and diffuse priors for the other parameters of (3.3). A similar conditional result is obtained without the normality assumption using the assumptions of the Bayesian method of moments (BMOM) approach (see, e.g., Zellner 1997b, 1998). Since the "plug in" assumption $\Pi_1 = \hat{\Pi}_1$, does not allow appropriately for the uncertainty regarding Π_1's value, the 2SLS estimate will not be optimal in small samples (see, e.g., Monte Carlo experiments reported by Park 1982, Tsurumi 1990, and Gao and Lahiri 1999). However, since OLS and 2SLS are widely employed methods, we shall employ them in our analyzes of the models for individual sectors.

In the Bayesian approach, we decided to use the "Extended Minimum Expected Loss" (EMELO) optimal estimate put forward in Zellner (1986, 1998) that has performed well in Monte Carlo experiments of Tsurumi (1990) and Gao and Lahiri (1999). It is the estimate that minimizes the posterior expectation of the following extended or balanced loss function:

$$\begin{aligned} L(\delta_1, \hat{\delta}_1) &= w(y_1 - \overline{Z}\hat{\delta}_1)'(y_1 - \overline{Z}\hat{\delta}_1) + (1-w)(\delta_1 - \hat{\delta}_1)'\overline{Z}'\,\overline{Z}(\delta_1 - \hat{\delta}_1) \\ &= w(y_1 - \overline{Z}\hat{\delta}_1)'(y_1 - \overline{Z}\hat{\delta}_1) + (1-w)(X\pi_1 - \overline{Z}\hat{\delta}_1)'(X\pi_1 - \overline{Z}\hat{\delta}_1), \end{aligned} \tag{3.7}$$

where w has a given value in the closed interval 0 to 1, $\hat{\delta}_1$ is some estimate of δ_1, and in going from the first line of (3.7) to the second, the identifying restrictions, multiplied on the left by X, namely $X\pi_1 = \overline{Z}\delta_1$ have been employed.

Relative to (3.3), the first term on the r.h.s. of (3.7) reflects goodness of fit while the second reflects precision of estimation or, from the second line of (3.7), the extent to which the identifying restrictions are satisfied when an estimate of δ_1 is employed. When the posterior expectation of the loss function in (3.7) is minimized with respect to $\hat{\delta}_1$ the minimizing value is:

$$\hat{\delta}_1 = (E\overline{Z}'\,\overline{Z})^{-1}[wE\overline{Z}'y_1 + (1-w)E\overline{Z}'X\pi_1]. \tag{3.8}$$

On evaluation of the moments on the r.h.s. of (3.8), we have explicit value for the optimal estimate. For example, with the assumption that for the unrestricted reduced form system in (3.2), the rows of (v_1, V_1), are iid $N(0, \Omega)$, where Ω is a pds covariance matrix, combining a standard diffuse prior for the reduced form parameters with the normal likelihood function yields a marginal matrix t-density for the reduced form coefficients. Thus the moments needed to evaluate (3.8) are readily available (see Zellner 1986 for details), and the result is surprisingly in the form of a double K-class estimate shown in (3.9):

$$\hat{\delta}_1 = \begin{bmatrix} \hat{\gamma}_1 \\ \hat{\beta}_1 \end{bmatrix} = \begin{bmatrix} Y_1'Y_1 - K_1\hat{V}_1'\hat{V}_1 & Y_1'X_1 \\ X_1'Y & X_1'X_1 \end{bmatrix}^{-1} \begin{bmatrix} (Y_1 - K_2\hat{V})'y_1 \\ X_1'y_1 \end{bmatrix}, \tag{3.9}$$

with $\hat{V}_1 = Y_1 - X\Pi_1$, $\hat{\Pi}_1 = (X'X)^{-1}X'Y_1$ and

$$K_1 = 1 - k/(T - k - m - 2) \quad \text{and} \quad K_2 = K_1 + wk/(T - k - m - 2). \tag{3.10}$$

K-class and double K-class estimates are discussed in most econometrics texts (see, e.g., Judge, *et al.* 1987) and the choice of optimal values for the Ks has been the subject of much sampling theory research. The Bayesian approach provides optimal values of these parameters quite directly on use of goodness of fit, precision of estimation, or balanced loss functions.

When the form of the likelihood function is unknown and thus a traditional Bayesian analysis is impossible, we used the Bayesian method of moments (BMOM) approach in Zellner (1998) to obtain a postdata maxent density for the elements of $\Pi = (\Pi_1, \pi 1)$ that was used to evaluate the expectation of the balanced loss function in (3.7) and derive an optimal value of $\hat{\delta}_1$ that is also in form of a double K-class estimate, shown in (3.9) but with slightly different values of the K-parameters, namely $K_1 = 1 - k/(T - k)$ and $K_2 = K_1 + wk/(T - k)$. In our calculations based on the extended MELO estimate, we used the BMOM K-values and $w = 0.75$, the value used by Tsurumi (1990) in his Monte Carlo experiments.

SUR estimates for the system were computed by assuming that the y_2 and y_3 variables in (3.1) are stochastic exogenous variables for each sector and treating the eleven sectoral equations as a set of seemingly unrelated regression equations. We estimated the parameters by "feasible" generalized least squares. The parameter estimates so obtained are means of conditional posterior densities in traditional Bayesian and BMOM approaches.

Complete shrinkage estimation utilized the assumption that all sectors' parameter vectors are the same. Under this assumption and the assumption that the y_2 and y_3 variables are stochastic exogenous variables, estimates of the restricted parameter vector were obtained by least squares that are also posterior means in Bayesian and BMOM approaches.

Exact posterior densities for the structural parameters in (3.5) can readily be calculated in the Bayesian approach by using diffuse priors for the parameters of (3.5) given Π_1, that is, a uniform prior on elements of δ_1, β_1, η_1, and $\log \sigma_e$, where σ_e is the standard deviation of each element of e. Further, the usual diffuse priors are employed for Π_1 and Ω_1, a marginal uniform prior on the elements of the reduced form matrix Π_1 in (3.2) and a diffuse prior on Ω_1, the covariance matrix for the independent, zero mean, normal rows of V_1. With use of these priors, the usual normal likelihood function for the system and Bayes' Theorem, we obtain the following joint posterior density for the parameters, where D denotes the given data (see Zellner *et al.* 1994 and Currie 1996):

$$\underset{\text{MVN}}{f(\gamma_1, \beta_1, \eta_1 \mid \sigma_e, \Pi_1, D)}\ \underset{\text{IG}}{g(\sigma_e \mid \Pi_1, D)}\ \underset{\text{MVN}}{h(\Pi_1 \mid \Omega_1, D)}\ \underset{\text{IW}}{j(\Omega_1 \mid D)} \tag{3.11}$$

where MVN denotes a multivariate normal density, IG an inverted gamma density, and IW an inverted Wishart density. A similar factorization of the joint BMOM postdata density is available (see Zellner 1997a).

Given (3.11), we can draw from the IW density and insert the drawn values in h and make a draw from it. The Π_1 value so drawn is then inserted in g and a draw from it and the drawn values of σ_e and Π_1 are inserted in f and a draw of the structural coefficients in f is made. This direct Monte Carlo procedure can be repeated many times to yield moments, fractiles, and marginal densities for all parameters appearing in (3.11). Also, a similar approach, described below, can be employed to compute predictive densities. Some of these calculations have been performed using sectoral models and data that will be reported in a future paper.

3.3 *Forecasting techniques*

For one-year-ahead forecasts of the rates of growth of real GDP using the aggregate models in table 24.1, we employed least squares forecasts that are means of Bayesian predictive densities when diffuse priors are employed and the usual normal likelihood functions are employed. Predictive means are optimal in terms of providing minimal expected loss *vis-à-vis* squared error predictive loss functions. Further, since these predictive densities are symmetric, the predictive mean is equal to the predictive median that is optimal relative to an absolute error predictive loss function.

One-year-ahead forecasts for the sectoral models in table 24.1 were made using one-year-ahead MMM(A) reduced form forecasts of the y_{2T+1} and y_{3T+1} variables on the r.h.s. of (3.1) and using the parameter estimates provided by the methods described above. That is, the one-year-ahead forecast is given by:

$$\hat{y}_{1T+1} = \hat{y}_{2T+1}\hat{\gamma}_{21} + \hat{y}_{3T+1}\hat{\gamma}_{31} + x'_{1T+1}\hat{\beta}_1. \tag{3.12}$$

The "eta" shrinkage technique, derived and utilized in Zellner and Hong (1989) involves shrinking a sector's forecast toward the mean of all eleven sectors' forecasts by averaging a sector's forecast with the mean of all sectors' forecasts as follows:

$$\hat{y}_{1t+1} = \eta\hat{y}_{1t+1} + (1-\eta)\bar{y}_{t+1},$$

where $\hat{y}_{1t+1}$ is the sector forecast, $\bar{y}_{t+1}$ is the mean of all the sectors' forecasts, and η is assigned a value in the closed interval zero to 1.

Gamma shrinkage, discussed and applied in Zellner and Hong (1989), involves assuming that the individual sector's coefficient vectors are distributed about a mean, say θ, and then using an average of an estimate of the sector's coefficient vector with an estimate of the mean θ of the parameter vectors. That is,

$$\hat{\delta}_\eta = (\hat{\delta}_1 + \gamma\hat{\theta})/(1+\gamma\hat{\theta}), \tag{3.13}$$

with $0 < \gamma < \infty$. This coefficient estimate can be employed to produce one-year-ahead forecasts using the structural equations for each sector and MMM(A) reduced form forecasts of the endogenous variables $(1-L)\log W_{t+1}$ and $(1-L)\log Y_{t+1}$. Various values of η and γ will be employed in forecasting sectoral growth rates that are used to construct an aggregate forecast of the growth rate of real GDP.

We can also compute a predictive density for a sector's one-year-ahead growth rate as follows. From (3.5), we can form the conditional density $q(y_{1T+1} \mid \Pi_1, \gamma_1, \beta_1, \eta_1, \sigma_e, y_{2T+1}, y_{3T+1}, D)$, that will be in a normal form given error term normality. Thus, each draw from (3.11) and a draw from the predictive density for (y_{2T+1}, y_{3T+1}) can be inserted in q and and a value of y_{1T+1} drawn from q. Repeating the process will produce a sample of draws from q from which the complete predictive density, its moments, etc. can be computed. Shown in figure 24.4 (p. 701) are two such predictive densities, one for the durables sector and the other for the services sector. The densities are slightly skewed to the left and rather spread out. However, the means that are optimal relative to squared error loss are not too far from the actual values being forecasted. Also, these densities are valuable in making probability statements about future outcomes, including turning point forecasts.

With this said about estimation and forecasting methods, we now turn to consider plots of the data and reports of forecasting results in the next section.

4 Discussion of data and forecasting results

In figure 24.1a are shown plots of the rates of growth of real GDP, real M1, real currency, real stock prices, and real wage rates, 1949–97. Peaks and troughs in the plots occur roughly at about four–six-year intervals. Note the sharp declines in real GDP growth rates in 1974 and 1982 and a less severe drop in 1991. The money and stock price growth rate variables tend to lead the real GDP growth rate variable, as observed in earlier work of many. While the two money growth rate variables show similar patterns before the 1990s, in the 1990s their behavior is somewhat different for some unknown (to us) reason. In our forecasting results, we find that use of the currency variable yields somewhat better results than use of the M1 variable.

Figure 24.1b presents a plot of the output growth rates for eleven sectors of the US economy. It is seen that except for the agriculture and mining sectors, the sectoral output growth rates tend to move together over the business cycle, while the agricultural and mining sectors show extreme variation. In contrast, the other sectors have much smaller interquartile ranges and fewer outlying growth rates. See also the boxplot for the sectoral growth rates in figure 24.1c.

In figure 24.2a are shown the one-year-ahead, aggregate forecasts plotted as solid lines and the actually observed rates of growth plotted as circles. In the first panel of figure 24.2a, labeled AR(3), an aggregate AR(3) model for the real GDP growth rates (see table 24.1) was employed to generate one-year-ahead forecasts year by year, 1980–97, with estimates being updated each year. The plot shows dramatically the failure of the AR(3) model to forecast turning points successfully. Very large errors occurred in 1982 and 1991. Use of the AR(3)LI model (see table 24.1) that incorporates two lagged leading indicator variables, the rates of growth of real currency and of real stock prices, produced the forecasts shown by the solid lines in the second panel of figure 24.2a. There are improvements in forecasts for 1982 and 1984 *vis-à-vis* use of the AR(3) model. However, there is still a large error in the 1991 forecast. Use of the MMM(A) model (see table 24.1) that incorporates two lagged level GDP variables and a linear time trend in the AR(3)LI model produced

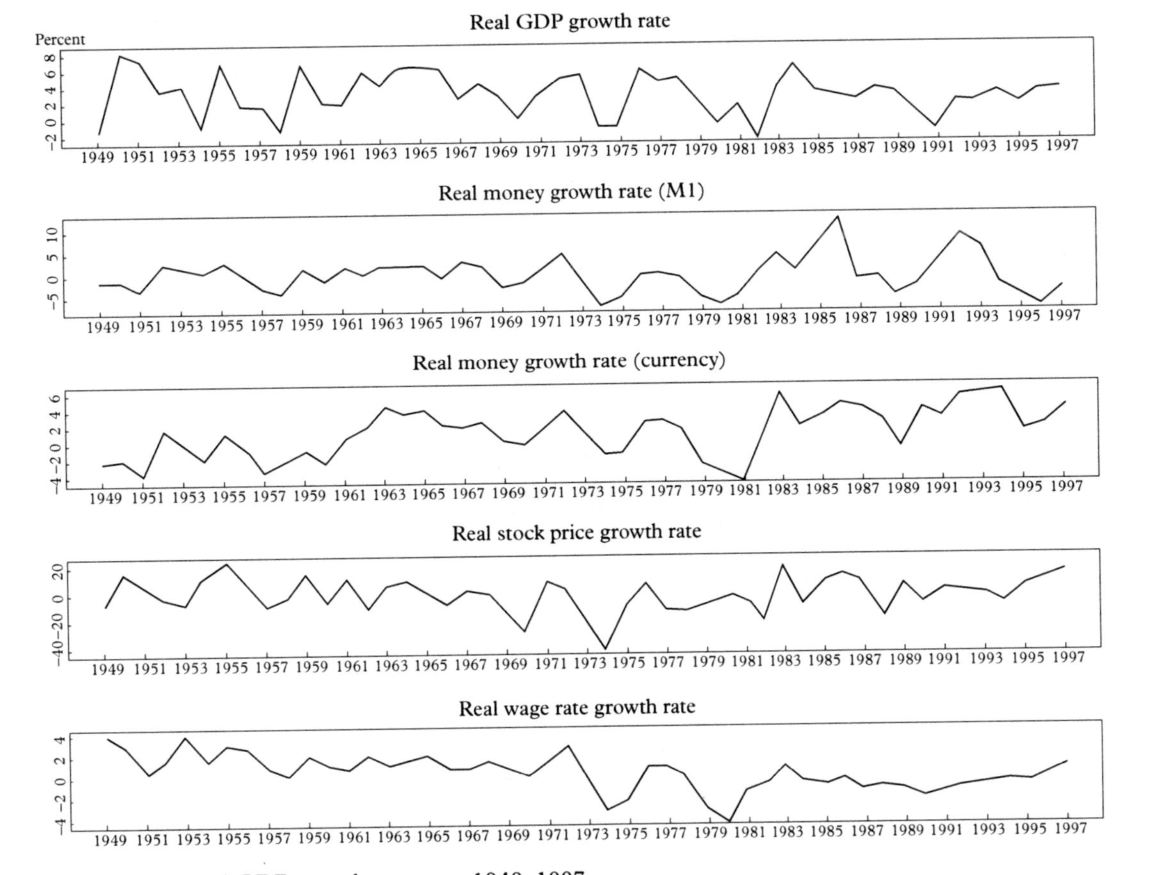

Figure 24.1a Real GDP growth rate, etc., 1949–1997

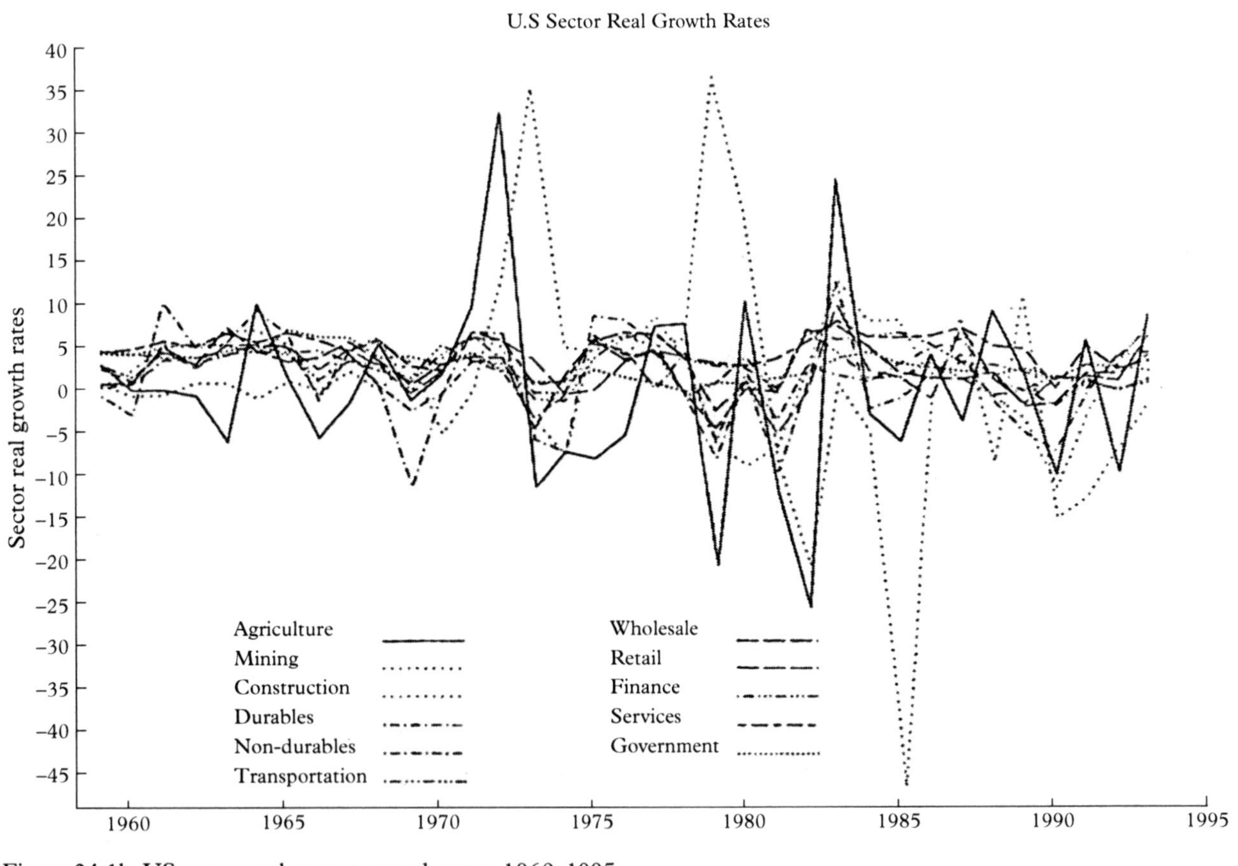

Figure 24.1b US sector real output growth rates, 1960–1995

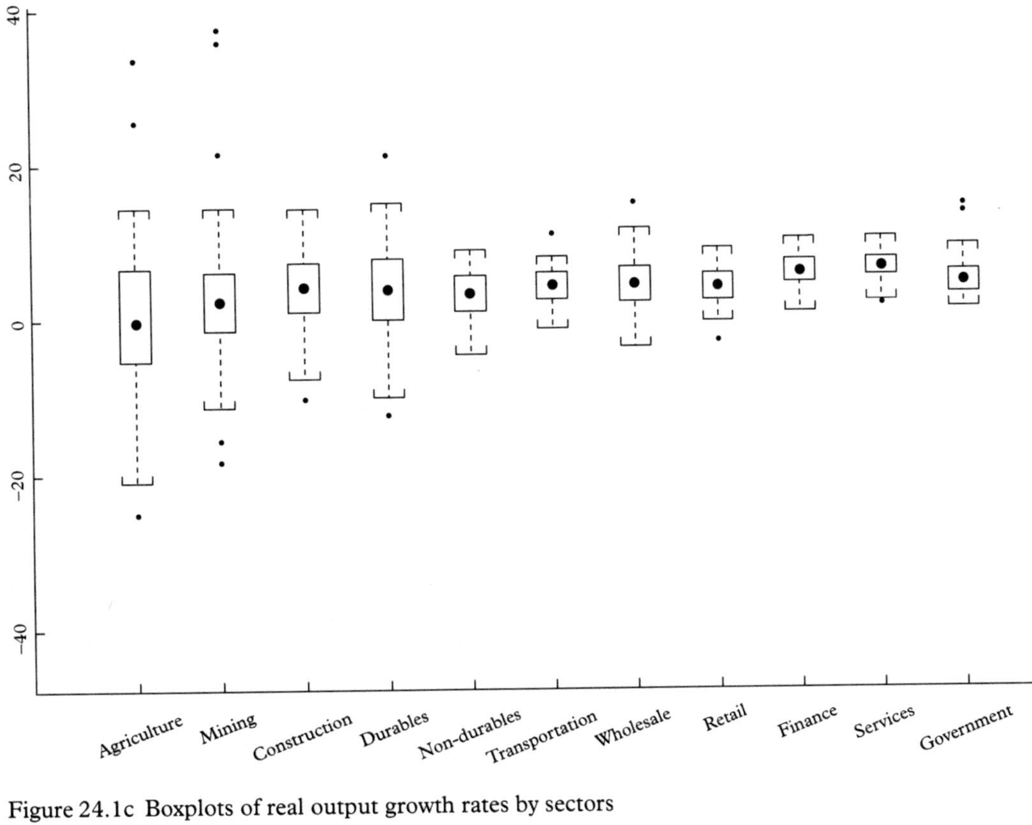

Figure 24.1c Boxplots of real output growth rates by sectors

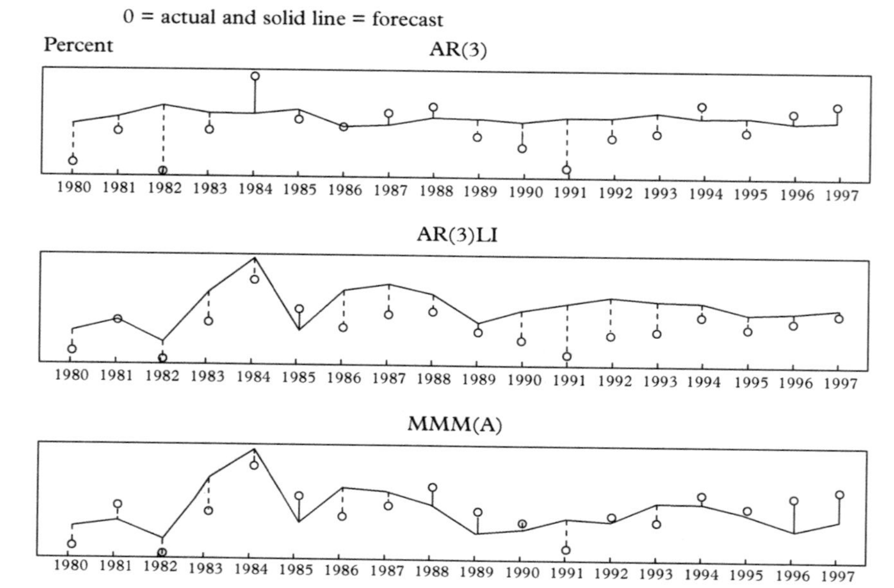

Figure 24.2a Aggregate GDP growth rate forecasts using the currency variable, 1980–1997

the forecasts shown in the third panel of figure 24.2a. Here there are improvements, as compared to the use of the AR(3) model in most years, especially 1982, 1990, and 1991. Similar use of the MMM(A) model led to improved forecasts as compared to those provided by the AR(3)LI, especially in the 1990s.

In figure 24.2b are shown the disaggregated, one-year-ahead forecasts, plotted as solid lines and the observed real GDP growth rate data, plotted as circles. Here, for each year, the eleven sectoral forecasts are employed to generate a forecast of the growth rate of aggregate real GDP using annually updated estimates of relations. Again, even though the sectoral AR(3) forecasts were employed, there is little improvement as compared with the aggregate AR(3) forecasts, shown in figure 24.2a. The AR(3)LI and Distributed lag(DL) models were used to generate forecasts for each of the eleven sectors and these were employed to calculate a forecast of the annual growth rates of real GDP with results shown in the second panel of the first column of figure 24.2b. The forecast performance of the DL model is seen to be better than that of the AR(3) model and about the same as that of the AR(3)LI model. With use of disaggregation and of the MMM(DA) models, I–IV (see table 24.1 for their definitions) that

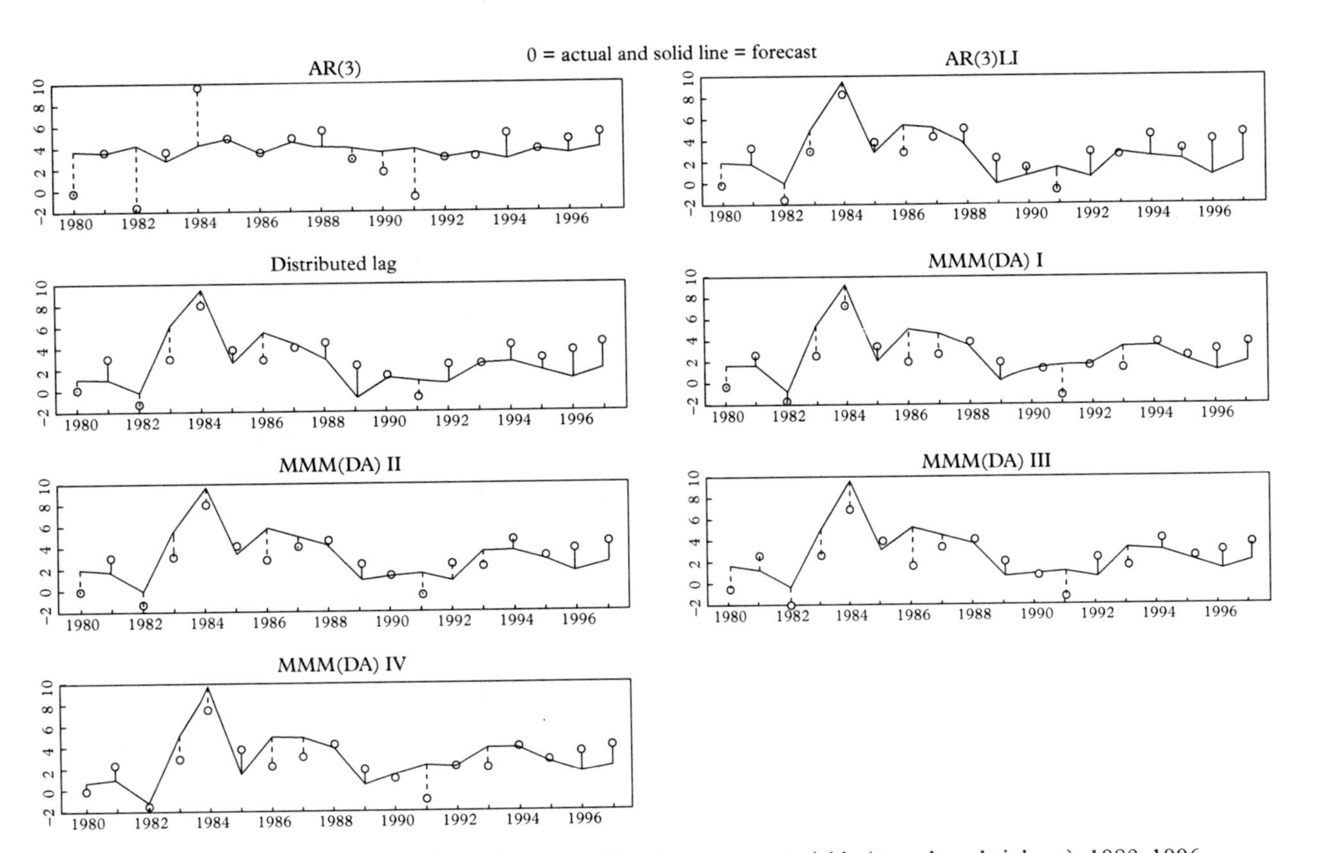

Figure 24.2b Disaggregate GDP growth rate forecasts using the currency variable (complete shrinkage), 1980–1996

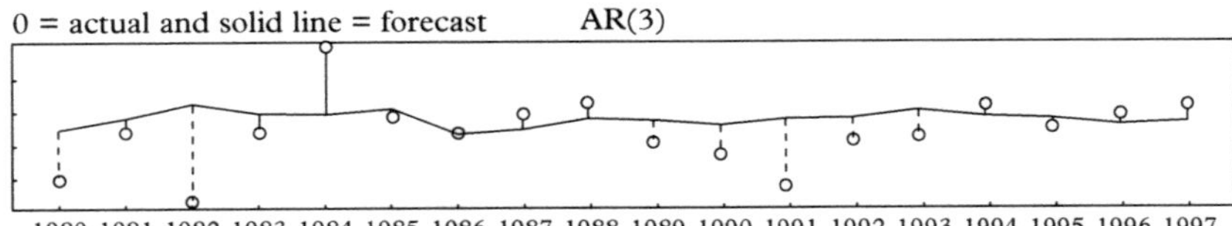

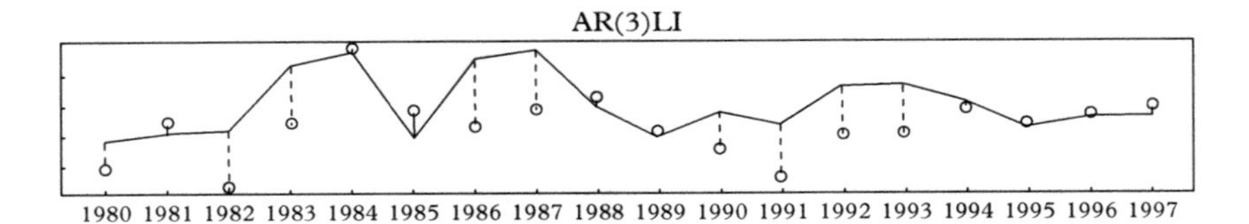

MMM(A)

1980 1981 1982 1983 1984 1985 1986 1987 1988 1989 1990 1991 1992 1993 1994 1995 1996 1997

Figure 24.3a Aggregate GDP growth rate forecasts using the M1 variable, 1980–1997

include lagged level variables, the forecasting results shown in figure 24.2b were obtained. The MMM(DA) models outperformed the AR(3) model by a wide margin and the disaggregated Distributed lag and AR(3)LI models by smaller margins. Also, from a comparison of figures 24.2a and 24.2b, the MMM(DA) models performed better than all the aggregate models.

With respect to the four MMM(DA) models, it appears that MMM(DA)III has a slight edge on the other three MMM(DA) models. It caught the 1982 downturn and subsequent upturn rather well and its performance in later years, particularly the 1990s, is slightly better than that of the alternative models considered in figures 24.2a–24.2b. However, it missed the 1991 trough growth rate.

When the lagged rate of growth of real M1 is used as a leading indicator variable, rather than the lagged rate of growth of real currency, the results in figures 24.3a–b were obtained. The results in figure 24.3a are similar to those in figure 24.2a, in that both the AR(3)LI and MMM(A) models' forecasting performance was much better than that of the AR(3) model. Use of the M1 variable rather than the currency variable led to a slight deterioration of the forecasting performance of the MMM(A).

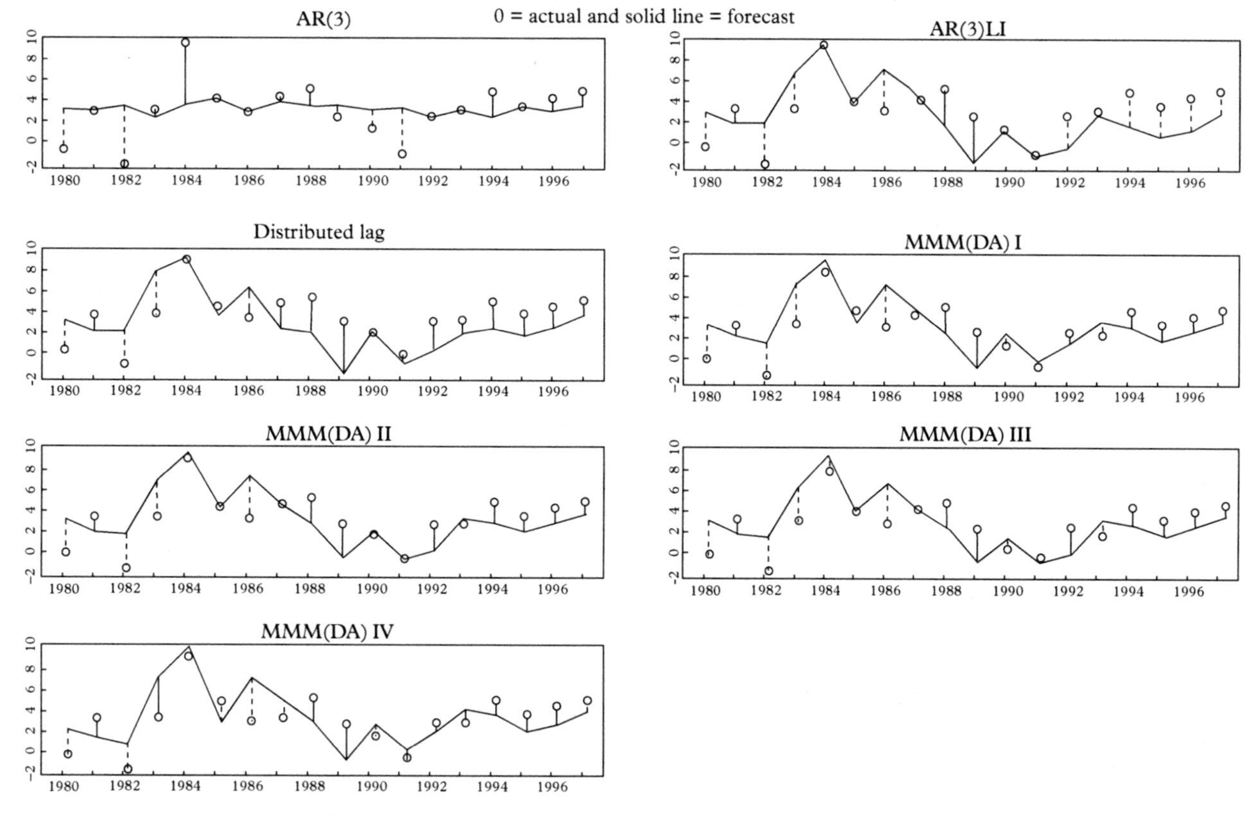

Figure 24.3b Disaggregate GDP growth rate forecasts using the M1 variable (complete shrinkage), 1980–1996

Table 24.2 *Forecast root mean-squared errors(RMSEs) and mean absolute errors(MAEs) for aggregate and disaggregated models using currency as the money variable (percentage points)*

Money variable using currency:
Aggregate forecast: 1952–79 ⇒ 1980–97

	AR(3)	AR(3)LI	MMM(A)
Income Y_t (Real GDP)			
RMSE	2.32	2.61	1.72
MAE	1.71	2.19	1.48
Real wage rate W_t			
RMSE	1.43	1.71	1.49
MAE	0.98	1.10	1.11

Disaggregated forecast: 1952–79 ⇒ 1980–97
* *Using MMM reduced form equation to forecast real income and real wage rate growth*

				MMM(DA)			
	AR(3)	AR(3)LI	Distributed lag	I	II	III	IV
OLS							
RMSE	2.26	1.62	1.61	1.61	1.52	1.47	1.80
MAE	1.65	1.32	1.35	1.31	1.28	1.25	1.47
Extended MELO							
RMSE	2.26	1.58	1.62	1.55	1.55	1.50	1.80
MAE	1.65	1.23	1.34	1.26	1.31	1.26	1.46
2SLS							
RMSE	2.26	1.60	1.63	1.63	1.49	1.48	1.78
MAE	1.65	1.31	1.38	1.33	1.25	1.24	1.45
SUR							
RMSE	2.21	1.70	1.66	1.68	1.61	1.40	1.92
MAE	1.52	1.41	1.36	1.39	1.38	1.17	1.60
Complete shrinkage							
RMSE	2.11	1.73	1.82	1.76	1.57	1.59	1.70
MAE	1.45	1.57	1.60	1.46	1.37	1.38	1.43
γ-Shrinkage							
$\gamma = 0$ (same as OLS)							
$\gamma = 0.25$							
RMSE	2.21	1.62	1.61	1.61	1.49	1.46	1.74
MAE	1.59	1.36	1.38	1.34	1.26	1.25	1.41
$\gamma = 0.5$							
RMSE	2.18	1.62	1.63	1.62	1.49	1.46	1.71
MAE	1.56	1.39	1.42	1.36	1.27	1.27	1.38
$\gamma = 1$							
RMSE	2.15	1.64	1.66	1.64	1.49	1.48	1.69
MAE	1.52	1.44	1.46	1.38	1.29	1.29	1.39

Table 24.2 (*cont.*)

	AR(3)	AR(3)LI	Distributed lag	MMM(DA) I	II	III	IV
$\gamma = 2$							
RMSE	2.13	1.66	1.70	1.67	1.51	1.50	1.68
MAE	1.49	1.48	1.51	1.41	1.32	1.32	1.40
$\gamma = 5$							
RMSE	2.11	1.69	1.75	1.71	1.53	1.54	1.68
MAE	1.47	1.52	1.56	1.44	1.34	1.35	1.41
$\gamma = 10^6$ (same as complete shrinkage)							
η-Shrinkage							
$\eta = 0$ (same as OLS)							
$\eta = 0.25$							
RMSE	2.26	1.66	1.62	1.59	1.49	1.44	1.77
MAE	1.63	1.36	1.40	1.34	1.25	1.24	1.48
$\eta = 0.5$							
RMSE	2.28	1.73	1.70	1.60	1.48	1.42	1.76
MAE	1.63	1.43	1.50	1.37	1.23	1.23	1.52
$\eta = 0.75$							
RMSE	2.33	1.82	1.82	1.62	1.48	1.42	1.79
MAE	1.68	1.55	1.64	1.41	1.21	1.23	1.57
$\eta = 1$							
RMSE	2.40	1.93	1.98	1.65	1.49	1.44	1.83
MAE	1.76	1.67	1.79	1.44	1.22	1.25	1.62

To a lesser degree, the same conclusion holds for the AR(3)LI models's performance. In figure 24.3b, the use of the M1 variable produced results similar to those reported in figure 24.2b. Note, however, that use of M1 and the models other than the AR(3) led to better forecasts of the low 1991 real GDP growth rate and slightly worse forecasts of the low 1982 growth rate.

Shown in table 24.2 are the RMSEs and MAEs associated with various models' one-year-ahead forecasts of annual real GDP growth rates, 1980–97, using data 1952–79 to estimate models which were then re-estimated year by year in the forecast period. Currency was used as the money variable. From the Aggregate forecast part of the table, it is seen that the MMM(A) model has a RMSE = 1.72 and a MAE = 1.48, lower than those associated with the AR(3) and AR(3)LI models. For the rates of change of the real wage rate, the AR(3) model's RMSE = 1.43 and

MAE = 0.98 are somewhat smaller than those of the MMM(A) and AR(3)LI models. These results indicate that the MMM(A) model for the growth rate of the real wage needs improvement, perhaps by inclusion of demographic and other variables.

As regards the disaggregated forecasts for the rate of growth of real GDP, shown in the second part of table 24.2, it is seen that all the disaggregated forecasts have smaller RMSEs and MAEs than those for the aggregate and disaggregated AR(3) model. For example, the disaggregated AR(3) model has RMSE = 2.26 and MAE = 1.65, whereas the disaggregated AR(3)LI, Distributed lag, and MMM(DA) models have RMSEs ranging from 1.40 to 1.98 and MAEs ranging from 1.17 to 1.62. As regards just the MMM(DA) models shown in table 24.2, their associated RMSEs and MAEs ranged from 1.40 to 1.92 and 1.17 to 1.62, respectively. The lowest RMSE and MAE are encountered for the MMM(DA) III model fitted using the SUR approach, namely RMSE = 1.40 and MAE = 1.17. However, quite a few other MMM(DA) models had RMSEs in the 1.4–1.5 range and MAEs in the 1.2–1.4 range.

In table 24.3 results similar to those presented in table 24.2 are shown for models incorporating a lagged rate of change of real M1 rather than the real currency variable. In general the use of the M1 variable resulted in a generally small deterioration in forecasting precision for all the models. However, again disaggregation led to improved forecasting precision for the AR(3)LI and MMM models in all cases. Use of the MMM(DA)III model generally led to slightly lower RMSEs and MAEs than other MMM(DA) models. The lowest RMSE and MAE, namely 1.84 and 1.52, respectively, are associated with the use of MMM(DA)III and η shrinkage with a value of $\eta = 0.50$ that can be compared to the aggregate MMM(A) model's RMSE = 2.23 and MAE = 1.90 and the aggregate AR(3)LI model's RMSE = 2.32 and MAE = 1.98. The RMSE and MAE for the aggregate AR(3) benchmark model are 2.32 and 1.71. Clearly use of disaggregation has led to improved forecasting performance again, about a 20 percent reduction in both RMSE and MAE.

In tables 24.2 and 24.3, use of alternative methods of estimation, OLS, Extended MELO, and 2SLS did not have much influence on the precision of forecasts. It may be that, for the present model, the rates of change of real income and of the real wage rate are stochastic exogenous or independent variables in the sector models and thus endogeneity is not a problem. However, these two variables must be forecast in order to forecast sectoral output growth rates and thus there is a need for the reduced form equations shown in table 24.1 whether these variables are stochastic exogenous or endogenous variables.

Table 24.3 *Forecast root mean-squared errors (RMSEs) and mean absolute errors (MAEs) for aggregate and disaggregated models using M1 as the money variable (percentage points)*

Money variable using M1:
Aggregated forecast: 1952–79 ⇒ 1980–97

	AR(3)	AR(3)LI	MMM(A)
Income Y_t (Real GDP)			
RMSE	2.32	2.57	2.23
MAE	1.71	1.98	1.90
Real wage rate W_t			
RMSE	1.43	1.73	1.66
MAE	0.98	1.07	1.29

Disaggregated forecast: 1952–79 ⇒ 1980–97
** Using MMM reduced form equation to forecast real income and real wage rate growth*

				MMM(DA)			
	AR(3)	AR(3)LI	Distributed lag	I	II	III	IV
OLS							
RMSE	2.26	2.03	2.01	2.04	1.96	1.89	2.17
MAE	1.65	1.77	1.74	1.78	1.76	1.67	1.88
Extended MELO							
RMSE	2.26	1.97	2.07	1.95	2.00	1.93	2.14
MAE	1.65	1.74	1.81	1.73	1.83	1.76	1.89
2SLS							
RMSE	2.26	2.06	2.06	2.07	1.91	1.89	2.12
MAE	1.65	1.78	1.76	1.77	1.73	1.69	1.85
SUR							
RMSE	2.21	2.21	2.07	2.14	2.01	2.00	2.30
MAE	1.52	1.87	1.74	1.87	1.79	1.75	1.96
Complete Shrinkage							
RMSE	2.11	2.21	2.34	2.05	1.94	1.93	1.89
MAE	1.45	1.81	2.02	1.70	1.55	1.56	1.63
γ-Shrinkage							
$\gamma = 0$ (same as OLS)							
$\gamma = 0.25$							
RMSE	2.21	2.04	2.02	2.01	1.91	1.85	2.06
MAE	1.59	1.75	1.67	1.74	1.68	1.60	1.78
$\gamma = 0.5$							
RMSE	2.18	2.06	2.04	1.99	1.89	1.84	2.01
MAE	1.56	1.73	1.65	1.71	1.62	1.55	1.71
$\gamma = 1$							
RMSE	2.15	2.08	2.09	1.99	1.88	1.84	1.95
MAE	1.52	1.74	1.71	1.69	1.56	1.52	1.66

Table 24.3 (*cont.*)

	AR(3)	AR(3)LI	Distributed lag	MMM(DA) I	MMM(DA) II	MMM(DA) III	MMM(DA) IV
$\gamma = 2$							
RMSE	2.13	2.12	2.16	2.00	1.88	1.85	1.91
MAE	1.49	1.75	1.80	1.69	1.53	1.53	1.65
$\gamma = 5$							
RMSE	2.11	2.16	2.24	2.02	1.90	1.89	1.89
MAE	1.47	1.76	1.91	1.69	1.53	1.54	1.64
$\gamma = 10^6$ (same as complete shrinkage)							
η-Shrinkage							
$\eta = 0$ (same as OLS)							
$\eta = 0.25$							
RMSE	2.26	2.04	2.03	2.00	1.93	1.85	2.13
MAE	1.63	1.73	1.70	1.72	1.70	1.60	1.82
$\eta = 0.5$							
RMSE	2.28	2.07	2.15	1.99	1.92	1.84	2.13
MAE	1.63	1.71	1.82	1.66	1.63	1.52	1.79
$\eta = 0.75$							
RMSE	2.33	2.14	2.33	2.00	1.92	1.86	2.16
MAE	1.68	1.74	2.00	1.64	1.61	1.53	1.78
$\eta = 1$							
RMSE	2.40	2.23	2.57	2.03	1.95	1.89	2.22
MAE	1.76	1.82	2.19	1.70	1.65	1.61	1.86

For the MMM(DA)III model, predictive densities for the sectoral output growth rates for the Services and durables sectors were calculated for 1980 and are shown in figure 24.4. Plotted are 1,000 draws from the BMOM predictive density made using the methods described in section 3. The Services predictive density has a mean equal to 3.14 percentage points and a standard deviation equal to 2.05 percentage points. The actual growth rate for the Services sector's output in 1980 is 3.57 percentage points. For the Durables sector, the 1980 predictive mean is 6.31 percentage points with a standard deviation of 8.06. The actual 1980 growth rate for this sector is 7.33. Both predictive densities appear to be slightly skewed to the left and rather spread out. As is well known, such densities can be employed in making probability statements regarding possible outcomes, for example a downturn in the growth rate and in implementing a decision theoretic approach for making optimal turning point forecasts. Also, these predictive densities and predictive densities for other models

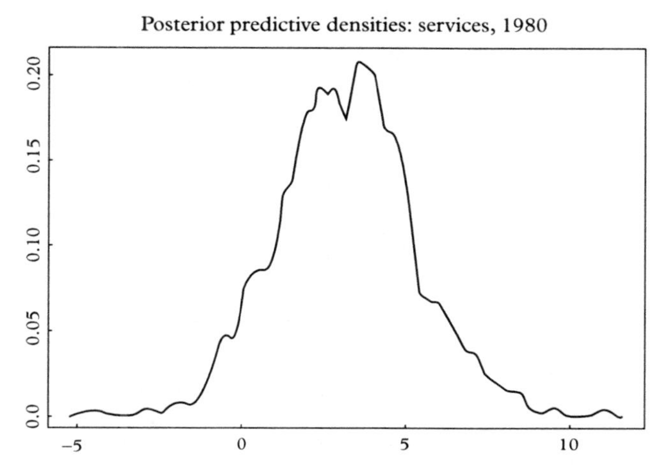

mean = 3.14, standard deviation = 2.05 actual growth rate = 3.57

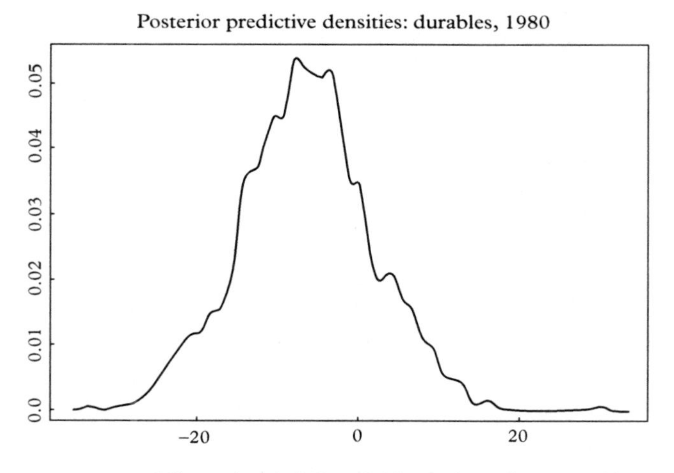

mean = -6.31, standard deviation = 8.06, actual growth rate = -7.33

Figure 24.4 Posterior predictive densities: services and durables, 1980

can be used to form Bayes' factors for model comparison and/or model combining. That these predictive densities can be computed relatively easily using the "direct" Monte Carlo approach described in section 3 is fortunate.

Last in table 24.4, we present some MAEs of forecast for various types of forecasts of one-year-ahead growth rates of real GDP for the

Table 24.4 *Mean absolute errors (MAEs) for annual forecasts of growth rates of real GNP made by various forecasters*

Period	MAEs	Average
1953–67	1.3^e, 1.0^d	1.2
1962–76	1.1^a, 1.4^d	1.2
1969–76	1.2^a, 1.0^b, 1.6^d, 0.9^c	1.2
1977–84	1.2^a, 1.0^b, 1.0^d, 1.0^c	1.0

Notes:
The forecasts are those of [a] Council of Economic Advisers, [b] ASA & NBER Surveys, [c] Wharton Newsletter, University of Pennsylvania, [d] University of Michigan and [e] An average of forecasts from the following sources: *Fortune Magazine*, Harris Bank, IBM, NICB, National Securities and Research Corp., University of Missouri, Prudential Insurance Co. and University of California at Los Angeles.
Source: Zarnowitz (1986), Table 24.1, p. 23.

United States compiled by Zarnowitz (1986). For several different periods and forecasting units, the average of their MAEs associated with annual forecasts of the growth rate of real GNP in 1972 dollars are given in table 24.4.

Many of the MAEs in table 24.4 are of magnitude comparable to those associated with the MMM(DA) annual one-year-ahead, reproducible forecasts for the period 1980–97 shown in table 24.2. Some forecasters use informal judgment along with models and data to produce forecasts. Adding "outside" information through the use of informative prior distributions may possibly improve the precision or MMM(DA) forecasts. See Zellner, Tobias, and Ryu (1999) for some examples of the use of judgmental information in forecasting turning points in output growth rates. Also, averaging forecasts from different sources may improve forecast precision, as many have pointed out. Note that the MAEs labeled b and d in table 24.4 are such averages. On the other hand, on-line forecasters have problems associated with the use of preliminary estimates of economic variables that we do not have in our forecasting experiments using revised data throughout. The results of some on-line forecasting experiments would be of great value in assessing the importance the "preliminary data" problem.

5 Summary and conclusions

In the present research, we found that several disaggregated MMM forecasting equations performed the best in our forecasting experiments.

Given the theoretical appeal of the Marshallian sector models, it is indeed satisfying that they yielded forecasts of the growth rates of aggregate real GDP, 1980–7, that were quite a bit better than those yielded by several aggregate benchmark models and competitive with other forecasting models and techniques. Shrinkage techniques and use of currency as the money variable in our models led to improved forecasts. However, the performance of various of our sector models, particularly those for the agricultural and mining sectors, has to be improved. In addition, factor market models for labor, capital, money and bonds as well as other equations are in the process of being formulated to complete our MMM.

Bayesian and certain non-Bayesian point forecasts performed about equally well for our disaggregated MMM models. However, the Bayesian approach provides exact finite sample posterior densities for parameters and predictive densities. The latter are very useful for forecasting turning points and making probabilistic statements about various future outcomes. The "direct" Monte Carlo numerical procedure for computing finite sample Bayesian posterior densities for parameters and predictive densities for simultaneous equations models appears convenient and useful.

We recognize that writing a single structural equation in restricted reduced form and allowing for error term effects in the equation, along with other unrestricted reduced form equations, yields a non-linear SUR system with an important restriction on the error term covariance matrix. This form for the system is very useful in terms of understanding it, analyzing it, and computing posterior and predictive densities.

While much can be said about the topic of data improvement, here we shall just remark that better data on (1) the numbers of firms and plants in operation within sectors, (2) sectoral stock price and wage rate indices, (3) weather variables, and (4) quality corrected output price data would be very useful and might lead to improved forecasts. Further, having monthly or quarterly data for individual sectors would be useful in dealing with temporal aggregation problems. However, seasonality must be treated carefully. Mechanical seasonal adjustment procedures may not be the best alternative. Improvement of preliminary estimates of variables is another important issue in "on-line" forecasting. Preliminary estimates of variables that are contaminated with large errors can obviously lead to poor forecasts.

Last, with better data for sectors of a number of economies and reasonably formulated MMMs, past work on use of Bayesian shrinkage forecasting and combining techniques can be extended in an effort to produce improved point and turning point forecasts for many countries.

BIBLIOGRAPHY

Adelman, I. and F. Adelman (1959), "The dynamic properties of the Klein–Goldberger model," *Econometrica* 27, 569–625

Belongia, M. and M. Garfinkel (eds.) (1992), "The business cycle: theories and evidence," Proceedings of the 16th Annual Economic Policy Conference of the Federal Reserve Bank of St. Louis (Boston, Kluwer Academic),

Cooper, R. (1972), "The predictive performance of quarterly econometric models of the United States," in B. Hickman (ed.), *Econometric Models of Cyclical Behavior* 2 (New York, Columbia University Press), 813–936

Currie, J. (1996), "The geographic extent of the market: theory and application to the US petroleum markets," PhD thesis, Department of Economics, University of Chicago

de Alba, E. and A. Zellner (1991), "Aggregation, disaggregation, predictive precision and modeling," H. G. B. Alexander Research Foundation, Graduate School of Business, University of Chicago, manuscript

Espasa, A. (1994), "Comment," *Journal of Forecasting* 13, 234–5

Espasa, A. and L. Matea (1990), "Underlying inflation in the Spanish economy: estimation and methodology," Working Paper, Bank of Spain

Fair, R. (1992), "How might the debate be resolved?," in M. Belongia and M. Garfinkel (eds.), "The business cycle: theories and evidence," Proceedings of the 16th Annual Economic Policy Conference of the Federal Reserve Bank of St. Louis (Boston, Kluwer),

Gao, C. and K. Lahiri (1999), "A comparison of some recent Bayesian and non-Bayesian procedures for limited information simultaneous equations models," Paper presented at the American Statistical Association's meeting, Baltimore, MD, August

Garcia-Ferrer, A., R. A. Highfield, F. C. Palm and A. Zellner (1987), "Macroeconomic forecasting using pooled international data," *Journal of Business and Economic Statistics* 5(1), 53–67; chapter 13 in this volume

Hong, C. (1989), "Forecasting real output growth rates and cyclical properties of models: a Bayesian approach," PhD thesis, Department of Economics, University of Chicago

Judge, G., W. Griffiths, R. Hill, H. Lütkepohl, and T. Lee (1987), *The Theory and Practice of Econometrics* (New York, John Wiley)

Kling, J. (1987), "Predicting the turning points of business and economic time series," *Journal of Business* 60, 201–38

Litterman, R. (1986), "Forecasting with Bayesian vector autoregressions: five years of experience," *Journal of Business and Economic Statistics* 4, 25–38

McNees, S. (1986), "Forecasting accuracy of alternative techniques: a comparison of US macroeconomic forecasts," *Journal of Business and Economic Statistics* 4, 5–23

Min, C. (1992), "Economic analysis and forecasting of international growth rates using Bayesian techniques," PhD thesis, Department of Economics, University of Chicago

Min, C. and A. Zellner (1993), "Bayesian and non-Bayesian methods for combining models and forecasts with applications to forecasting international growth rates," *Journal of Econometrics, Annals* 56, 89–118; chapter 17 in this volume

Nelson, C. and C. Plosser (1982), "Trends and random walks in macroeconomic time series," *Journal of Monetary Economics* 10, 139–62

Orcutt, G., M. Greenberger, J. Korbel, and A. Rivlin (1961), *Microanalysis of Socioeconomic Systems* (New York, Harper)

Pagan, A. (1979), "Some consequences of viewing LIML as an iterated Aitken estimator," Working Paper 18, Australian National University Faculty of Economics and Research School of Social Sciences

Palm, F. C. (1976), "Testing the dynamic specification of an econometric model with an application to Belgian data," *European Economic Review* 8, 269–89

(1977), "On univariate time series methods and simultaneous equation econometric models," *Journal of Econometrics* 5, 379–88

(1983), "Structural econometric modeling and time series analysis: an integrated approach," in A. Zellner (ed.), *Applied Time Series Analysis of Economic Data* (Washington, DC, US Department of Commerce, Bureau of the Census, 199–233; chapter 3 in this volume

Park, S. (1982), "Some sampling properties of minimum expected loss (MELO) estimators of structural coefficients," *Journal of Econometrics* 18, 295–311

Tsurumi, H. (1990), "Comparing Bayesian and non-Bayesian limited information estimators," in S. Geisser, J. S. Hodges, J. Press, and A. Zellner (eds.), *Bayesian and Likelihood Methods in Statistics and Econometrics: Essays in Honor of George A. Barnard* (Amsterdam, North-Holland), 179–202

Veloce, W. and A. Zellner (1985), "Entry and empirical demand and supply analysis for competitive industries," *Journal of Econometrics* 30, 459–71

Wecker, W. E. (1979), "Predicting the turning points of a time series," *Journal of Business* 52, 35–50

Zarnowitz, V. (1986), "The record and improvability of economic forecasting," *Economic Forecasts* 3, 22–31

Zellner, A. (1979), "Statistical analysis of econometric models," invited paper, with discussion, *Journal of the American Statistical Association* 74, 628–51; chapter 2 in this volume

(1986), "Further results on Bayesian minimum expected loss (MELO) estimates and posterior distributions for structural coefficients," in D. Slottje (ed.), *Advances in Econometrics* 5, 171–82

(1992), "Comment on Ray C. Fair's thoughts on 'How might the debate be resolved?'," in M. Belongia and M. Garfinkel (eds.), "The business cycle: theories and evidence," Proceedings of the 16th Annual Economic Policy Conference of the Federal Bank of St. Louis (Boston, Kluwer), 148–57

(1994), "Time series analysis, forecasting, and econometric modeling: the structural econometric modeling, time series analysis (SEMTSA) approach," *Journal of Forecasting* 13, 215–33, invited paper with discussion; chapter 4 in this volume

(1997a), "The Bayesian method of moments (BMOM): theory and applications," *Advances in Econometrics* 12, T. Fomby and R. Hill (eds.), 85–105

(1997b), *Bayesian Analysis in Econometrics and Statistics: The Zellner View and Papers* (Cheltenham, Edward Elgar)

(1998), "The finite sample properties of simultaneous equations' estimates and estimators: Bayesian and non-Bayesian approaches," *Annals Issue of Journal of Econometrics* 83, L. Klein (ed.), 185–212

(1999), "Bayesian and non-Bayesian approaches to scientific modeling and inference in economics and econometrics," invited keynote paper presented at the Ajou University Research Conference in honor of Professor Tong Hun Lee, South Korea, August; [published in Special Issue of the *Korean Journal of Money and Finance* 5(2) (November 2000)]

Zellner, A. and C. Hong (1989), "Forecasting international growth rates using Bayesian shrinkage and other procedure," *Journal of Econometrics, Annals* 40, 183–202; chapter 14 in this volume

Zellner, A. and C. Min (1999), "Forecasting turning points in countries' growth rates: a response to Milton Friedman," *Journal of Econometrics* 88, 203–6; chapter 19 in this volume

Zellner, A., C. Min, D. Dallaire, and J. Currie (1994), "Bayesian analysis of simultaneous equation, asset pricing and related models using Markov chain Monte Carlo techniques," H. G. B. Alexander Research Foundation, Graduate School of Business, University of Chicago, manuscript

Zellner, A. and F. C. Palm (1974), "Time series analysis and simultaneous equation econometric models," *Journal of Econometrics* 2, 17–54; chapter 1 in this volume

(1975), "Time series and structural analysis of monetary models of the US economy," *Sankyā: The Indian Journal of Statistics*, Series C 37, 12–56; chapter 6 in this volume

Zellner, A. and S. Peck (1973), "Simulation experiments with a quarterly macroeconometric model of the US economy," in A. Powell and R. Williams (eds.), *Econometric Studies of Macro and Monetary Relations* (Amsterdam, North-Holland), 149–68

Zellner, A., J. Tobias, and H. Ryu (1999), "Bayesian method of moments analysis of time series models with an application to forecasting turning points in output growth rates"; [published in *Estadistica* 49–51 (152–157), 3–63, with discussion]

Subject index

Author index

Lightning Source UK Ltd.
Milton Keynes UK
20 January 2011

166038UK00007B/1/P